MICROSTRIP ANTENNAS

Edited by **Nasimuddin**

Microstrip Antennas
Edited by Nasimuddin

Published by InTech
Janeza Trdine 9, 51000 Rijeka, Croatia

Publishing Process Manager Katarina Lovrecic

Technical Editor Teodora Smiljanic

Cover Designer InTech Design Team

Additional hard copies can be obtained from orders@intechopen.com

Microstrip Antennas, Edited by Nasimuddin
p. cm.
ISBN 978-953-307-247-0

For EU product safety concerns:
IN TECH d.o.o., Prolaz Marije Krucifikse Kozulić 3, 51000 Rijeka, Croatia,
info@intechopen.com or visit our website at intechopen.com.

Meet the editor

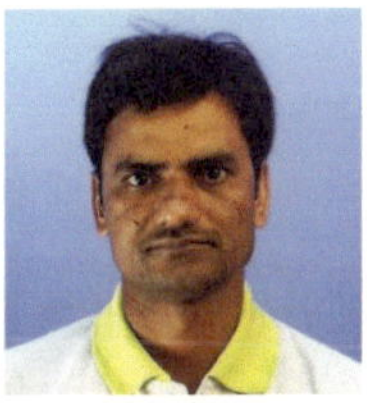

Nasimuddin was born in Bulandshahar, India. He received the B.Sc. degree from the Jamia Millia Islamia, NewDelhi, India, in 1994, the M.Tech. (Microwave Electronics) degree from the University of Delhi, India, in 1998 and the Ph.D. degree from the University of Delhi, India, in 2004, for theoretical work in the field of multilayered slow-wave microstrip structures and microstrip patch antennas. From 2004 to 2006, he was an Australian Postdoctoral Fellow at the Macquarie University, Sydney, Australia. Currently, he is working as a Research Fellow at the Department of RF and Optical, Institute for Infocomm Research, Singapore. He has published 83 journal and conference papers in area of microstrip based structures. Dr. Nasimuddin is a senior member of the IEEE and IEEE Antennas and Propagation Society. He was awarded a very prestigious Senior Research Fellowship from the CSIR, Government of India in Engineering Science (2001–2003). He was awarded a Discovery Projects Fellowship from the Australian Research Council (2004–2006) and the Young Scientist Award from the International Union.

Contents

Preface

The microstrip antennas are low-profile, low weight, ease of fabrication, conformable to planar and non-planar surfaces and mechanically robust. In the last 40 years, the microstrip antenna has been developed for many communication systems such as radars, sensors, wireless, satellite, broadcasting, ultra-wideband, radio frequency identifications (RFIDs), reader devices etc.. The progress in modern wireless communication systems has increased dramatically the demand for microstrip antennas, capable to be embedded in portable, handheld devices such RFID handheld reader, devices which provide a wireless network. Recently, demands of these devices with smaller in size and therefore antennas required smaller and light weight especially at the low microwave frequency range. The microstrip antennas can be designed in very small size with lower gain and bandwidth. For portable and handheld devices, gain and bandwidth of the antenna is not so important. However antenna meets some gain with desired bandwidth constraint. For millimeter wave applications, the antenna has to be high gain with broadband impedance bandwidth.

In this book some recent advances in the microstrip antennas are presented while highlighting the theoretical and practical design techniques for various wireless system applications. The microstrip antennas on various available substrate materials such as artificial material, uni-axial and ferrite are analyzed and designed for reconfigurable, dual and tunable applications. The small microstrip antennas can be designed using artificial materials. Various shaped radiators are also studied for compact antenna size and circular polarization radiation. The circularly polarized microstrip antennas with different feeding system and various shaped slotted microstrip patch radiators is also studied and compared for compact size and broadband applications. The microstrip antennas are also considered as a sensor for detection of materials properties. Finally, the microstrip antennas for millimeter-wave applications are also covered in this book. New emerging wireless systems that operate at millimeter wave frequencies, such as high data rate 60-GHz transceivers for wireless personal area networks (WPAN), use integrated antennas. Therefore, antennas for these systems are commonly implemented on in-package solutions. The integration of antenna-in-package is also covered by using wire bonding or flip-chip bonding interconnections. Lastly, the 3D electromagnetic software tools for microstrip antennas designing is demonstrated for helping the microstrip antenna designers. The proposed microstrip antennas book is useful for students, researchers and microstrip antenna design engineers.

The microstrip antennas book covers different types of the microstrip antennas and arrays. The book chapters are from experts/scientists in the area of the microstrip antennas and applied electromagnetics. First book chapter begins introduction of the microstrip antennas with low-cost probe-fed microstrip antenna design methods. Analysis of the rectangular microstrip antennas on uni-axial and artificial material substrates are presented in chapters 2 and 3, respectively. A particle-swarm-optimization based selective neural network ensemble and its application to modeling resonant frequency of the microstrip antenna are described in chapter 4. Chapters 5-8 present analysis of the microstrip antennas on the spherical surfaces, cylindrical wraparound, and dielectric lens. Various shapes with slotted/slit microstrip antennas are presented in chapters 9-17 for various wireless system applications such as multiband, reconfigurable antennas, compact microstrip antennas and circularly polarized microstrip antennas etc.. These chapters are also presented in comparison with slotted/slit microstrp antennas based on fixed overall antenna size. In chapters 18-19, the microstrip antennas are proposed for detection of material properties. The hybrid microstrip antennas and integration of the microstrip antennas with CMOS Chip for millimeter applications are described in chapters 20-21. The last book chapter is a practical guide to 3D electromagnetic software tools for analysis of the planar antennas and this helps reader with general guidelines for antenna design using the 3D electromagnetic software tools.

Nasimuddin
Institute for Infocomm Research
Singapore

1

Design of Low-Cost Probe-Fed Microstrip Antennas

D. C. Nascimento and J. C. da S. Lacava
Technological Institute of Aeronautics
Brazil

1. Introduction

The concept of microstrip radiators, introduced by Deschamps in 1953, remained dormant until the 1970s when low-profile antennas were required for an emerging generation of missiles (James & Hall, 1989; Garg et al., 2001; Volakis, 2007). Since then, but mainly over the last three decades, the international antenna community has devoted much effort to theoretical and experimental research on this kind of radiator (Lee & Chen, 1997). Currently, low-loss RF laminates are used in their fabrication and many of their inherent limitations have been overcome (Garg et al., 2001). On the other hand, low-cost solutions are in demand now that both market and technology are ready for mass production (Gardelli et al., 2004). Recently, the design of single-fed circularly-polarized (CP) microstrip antennas manufactured with FR4 substrate was reported (Niroojazi & Azarmanesh, 2004). Unfortunately, the use of low-cost FR4 as the substrate introduces some additional complexity on the antenna design. This is due to the inaccuracy of the FR4 relative permittivity and its high loss tangent (around 0.02). Variations in the FR4 electrical permittivity can shift the operating frequency and the high loss tangent dramatically affects the antenna axial ratio and gain, resulting in poor radiation efficiency. To increase the efficiency, microstrip antenna on moderately thick substrate must be designed. However, the technique used to compensate for the probe inductance, when the patch is fed by a coaxial probe (a known practical way to feed microstrip antennas), still relies on the designer's expertise. For instance, a series capacitor, which may be constructed in several ways, has been utilized to neutralize this inductance (Hall, 1987; Alexander, 1989; Dahele et al., 1989; Vandenbosch & Van de Capelle, 1994; Nascimento et al., 2006), or the probe geometry has been modified (Haskins & Dahele, 1998; Teng et al., 2001; Chang & Wong, 2001; Tzeng et al., 2005). Unfortunately, due to their complexity, many such techniques are not suitable when the antennas are series-produced in an assembly line.
To overcome some of the abovementioned issues, two efficient techniques for designing low-cost probe-fed microstrip antennas are proposed. Using only their intrinsic characteristics, linearly- and circularly-polarized microstrip antennas can now be designed without the need for any external matching network. Limitations of the proposed approach will also be discussed. The chapter is organized as follows: Section 2 covers the design of linearly-polarized microstrip antennas; results obtained with the new approach are compared with those using the standard design technique. Circularly-polarized antennas

are addressed in Section 3 and experimental results are shown in Section 4. Other applications using the new design approach are presented in Section 5, and finally in Section 6, conclusions are drawn from the obtained results.

2. Linearly-polarized microstrip antennas

The typical geometry of a rectangular-patch linearly-polarized (LP) microstrip antenna is shown in Fig. 1, where a denotes the patch length, b the radiating edge width, p the probe position along the x-axis, and h the substrate thickness. The patch is printed on a finite rectangular substrate of dimensions (L by W) in order to avoid the excitation of surface waves, and the antenna is directly fed by a 50-Ω SMA connector. The analysis carried out in this section is focused on this particular radiator.

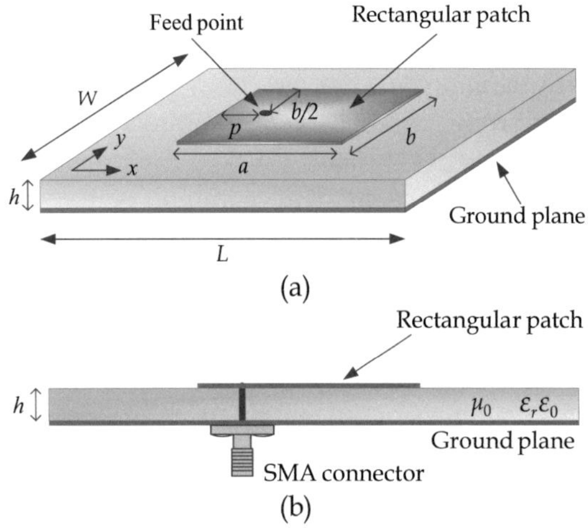

Fig. 1. Linearly-polarized probe-fed microstrip antenna: (a) top view - (b) side view.

2.1 Radiation efficiency

The radiation efficiency is defined as the ratio of the total power radiated (P_r) by an antenna to the net power accepted (P_{in}) by the antenna from the connected transmitter (IEEE Std 145, 1993). Since the antenna under consideration has its dielectric truncated, the cavity model, although originally developed for the analysis of electrically thin microstrip antennas, can be used for estimating the efficiency behavior. For the microstrip antenna shown in Fig. 1, the geometry of its equivalent cavity, neglecting the fringe effect, is given in Fig. 2. Under the condition $h << a < b$, the electric field of the TM_{mn} resonant mode excited within the cavity is expressed by

$$E_z = \frac{V_{mn}}{h}\cos\left(\frac{m\pi x}{a}\right)\cos\left(\frac{n\pi y}{b}\right), \tag{1}$$

where V_{mn}/h denotes the electric field intensity on the magnetic walls.

In case of linearly-polarized antenna excited in the fundamental TM_{10} mode, the dielectric (P_d) and metallic (P_m) losses can be calculated by means of equations (2) and (3), respectively.

$$P_d = \frac{\sigma_d}{2}\iiint_V |E_z|^2 dV = \frac{V_{10}^2\, ab\sigma_d}{4h}, \tag{2}$$

$$P_m = \frac{R_s}{2}\iint_S |\vec{J}_s|^2 dS = \frac{V_{10}^2\, R_s\, ab\varepsilon_0\varepsilon_r}{2h^2\mu_0}, \tag{3}$$

where ε_r is the substrate relative permittivity, σ_d its electric conductivity, $\vec{J}_s$ the surface electric current density on the metallic walls, R_s the surface resistance, and ε_0 and μ_0 are the electric permittivity and magnetic permeability of free space, respectively.

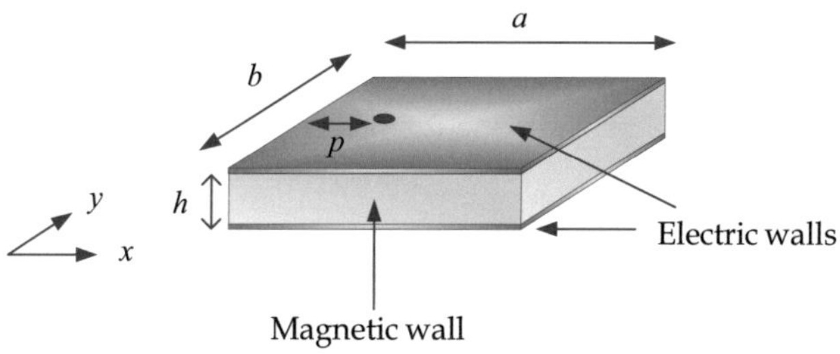

Fig. 2. Geometry of the antenna equivalent cavity.

The radiated power can be obtained by computing

$$P_r = \int_0^{2\pi}\int_0^{\pi/2} \frac{1}{2\eta_0}\left(|E_\theta|^2 + |E_\phi|^2\right) r^2 \mathrm{sen}\theta d\theta d\phi, \tag{4}$$

where η_0 denotes the free-space intrinsic impedance and E_θ and E_ϕ are the components of the far electric field radiated by the antenna, evaluated using Huygens's magnetic current source approach (Lumini et al., 1999).

Neglecting the surface wave losses, since the antenna has its dielectric truncated, the radiation efficiency can be estimated by the following expression

$$\eta = \frac{P_r}{P_d + P_m + P_r}. \tag{5}$$

Using equations (1) - (5), the radiation efficiency of LP antennas, designed to operate at 1.575 GHz in the fundamental TM_{10} mode, were calculated, and the results are shown in Fig. 3. In Fig. 3(a), the radiation efficiency curves of a 1.524-mm thick LP antenna are plotted as a function of the dielectric loss tangent, with the substrate relative permittivity as a parameter. In Fig. 3(b), graphics of radiation efficiency are presented, for the case of a rectangular patch printed on the FR4 laminate (ε_r = 4.2), as a function of the substrate thickness, with the loss tangent as a parameter. These graphics, although obtained from the cavity model, make visible the behavior of the radiation efficiency of these microstrip antennas. Thus, if low-cost materials are used in the antenna manufacture, then moderately thick substrates must be adopted for good radiation efficiency. In the case of commercial FR4 laminates (ε_r = 4.2 and tan δ = 0.02), a radiation efficiency close to 70% can be obtained if a 6.5 mm thick antenna is designed.

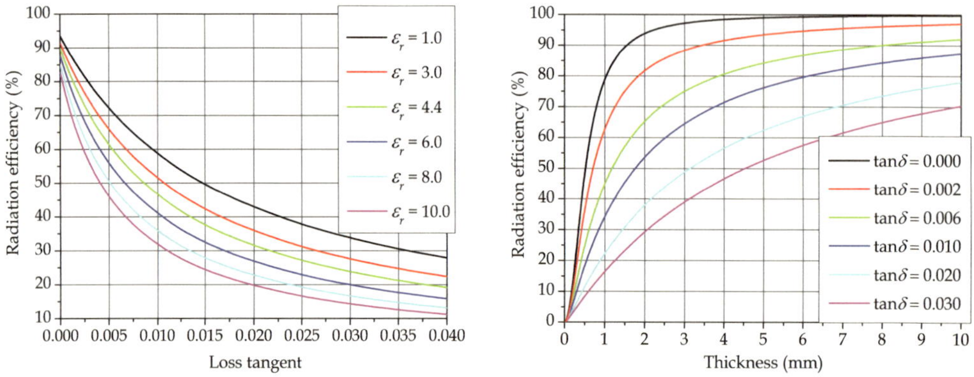

(a) (b)

Fig. 3. Radiation efficiency of LP microstrip antennas.

2.2 Rectangular patch: standard design

According to the standard procedure (James & Hall, 1989; Garg et al., 2001; Volakis, 2007) for designing a LP patch in the fundamental mode TM_{10}, the operating frequency is set up at the maximum input resistance point. Following this procedure and using the commercial software HFSS (HFSS, 2010) for optimizing the radiator dimensions, a rectangular antenna consisting of a h = 6.6 mm moderately thick (to obtain good radiation efficiency), FR4 (ε_r = 4.2 and tan δ = 0.02) substrate, fed by a 1.3-mm diameter coaxial probe, was designed to operate at 2 GHz. Utilizing a rectangular ground plane (L = 90 mm; W = 100 mm), the following optimal dimensions were obtained: a = 31.25 mm, b = 40.6 mm and p = 10.4 mm. Results for the input impedance and the reflection coefficient magnitude ($|\Gamma|$) are shown in Fig. 4(a) and (b) respectively. As expected, the radiation efficiency is 77.9% and the directivity is 7 dB at the operating frequency.

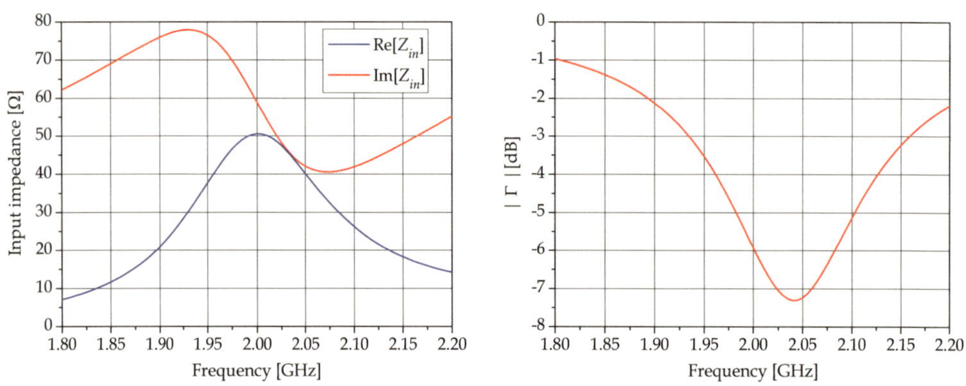

(a) (b)

Fig. 4. Standard design: (a) input impedance - (b) reflection coefficient magnitude.

It can be seen from Fig. 4(a) that the maximum input resistance occurs per design at the operating frequency (2 GHz). As a result, the antenna input impedance is highly inductive (Z_{in} = 50 + j59 Ω, at 2 GHz) and can not be perfectly matched to a 50-Ω SMA connector (i. e. $|\Gamma|$ = –6 dB, Fig. 4(b)) without an external network. Nowadays, this behavior is well known and can be properly modeled by a parallel *RLC* network with a series inductance L_p (Richards et al., 1981). As a consequence, the radiator bandwidth is asymmetrical with respect to the operating frequency (Fig. 4(b)). To overcome this limitation, a new approach for designing probe-fed moderately thick microstrip antennas is proposed next.

2.3 Rectangular patch: new design

The new procedure, differently from the standard one, consists of designing the patch to operate at the zero input reactance $X_{in} = 0$ condition. This takes two steps; first, initial values for the patch dimensions are found, using the standard approach for example. Then, its feed probe is positioned close to the radiating edge ($p \cong 0$ mm). This action is performed to check if capacitive (i.e. negative reactance) input impedances can be reached at frequencies above the operating frequency (2 GHz), as shown in Fig. 5; if so, it is clear that the antenna could be perfectly matched to the 50-Ω SMA connector for a certain intermediate feed probe position, though at a frequency greater than the operating one.

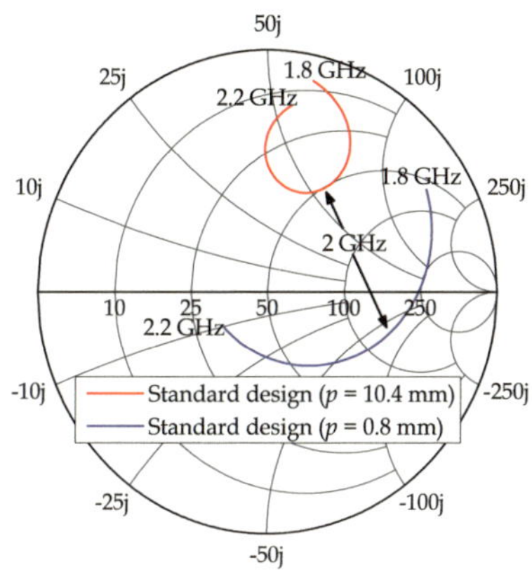

Fig. 5. Input impedances: standard design.

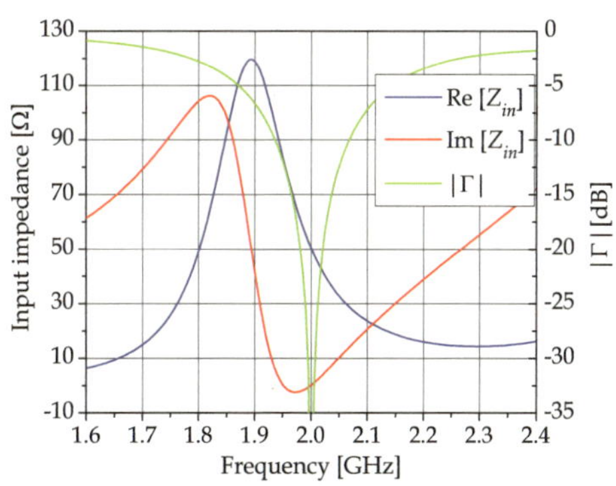

Fig. 6. Input impedance and reflection coefficient: new design.

To carry this out, starting from the initial patch dimensions, the probe position p is gradually reduced until the desired impedance is reached. The frequency where this happens can now be shifted down to the desired operating frequency through rescaling of the antenna geometry by increasing the patch dimensions (a and b). If on the other hand, upon repositioning the probe close to the radiating edge the antenna input impedance remains inductive, a perfect match to the SMA connector may not be achieved, but even so a reflection coefficient that is better than the one obtained from the standard design can still be realized. This particular situation, which depends on the antenna thickness, substrate permittivity and the operating frequency, is not treated in this chapter.

Using the new approach, the antenna is redesigned to operate at 2 GHz for the same dielectric thickness, ground plane dimensions and probe diameter used in the standard design. The new optimized dimensions are: a = 32.9 mm, b = 42.8 mm and p = 7.0 mm. As expected, the new patch dimensions are now larger, whereas the new probe position is shorter, than their corresponding initial values. Results for the input impedance and the reflection coefficient magnitude of the new antenna are shown in Fig. 6. Now, the antenna matches perfectly the 50-Ω SMA connector and presents a symmetrical bandwidth with respect to the operating frequency (2 GHz). In addition, as observed in Fig. 6, the maximum resistance is greater than 50 Ω and occurs at a frequency below 2 GHz.

It is important to point out that the new method was conceived based on the properties of the antenna input impedance equivalent circuit (parallel *RLC* network with a series inductance L_p), typical of linearly-polarized probe-fed radiators. Consequently, this procedure is independent of the patch shape so it can be equally well applied to the design of other patch geometries, like the circular and the triangular ones.

2.4 Comparison between standard and new designs

To complete the analysis of the probe-fed patch antenna designed according to this new approach, it is important to verify the behavior of its radiation patterns. For this purpose, radiation patterns of the E_{theta} and E_{phi} components for both antennas (new and standard) are plotted on the *yz* and *xz* planes, as shown in Fig. 7(a) and (b). As noted, there are no significant differences between these patterns, even in the cross-polarization case. Besides, the directivity of the standard antenna is 7 dB whereas the new one is 7.1 dB. The new antenna efficiency is close to 79.1%.

This new design approach has been successfully used recently in (Tinoco S. et al., 2008) for designing thin microstrip antennas for educational purposes.

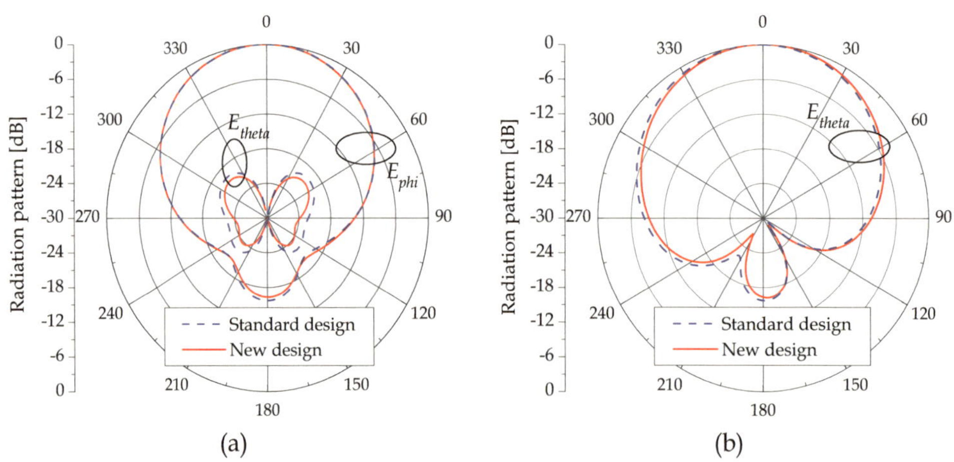

Fig. 7. Radiation patterns: (a) *yz* plane - (b) *xz* plane.

3. Circularly-polarized microstrip antennas

Singly-fed circularly-polarized microstrip antennas are largely employed in GPS receivers (Nascimento et al., 2006). Nearly-square and truncated-corner square patches (Figs. 8(a) and (b), respectively), have typically been used to obtain circular polarization (Garg et al., 2001). In this section, however, a new design approach is applied to the CP nearly-square patch. The geometry is given in Fig. 8(a), where a and b are its dimensions and p_x and p_y define the x and y coordinates of the probe position.

3.1 Nearly-square patch: standard design

Singly-fed rectangular CP microstrip antennas operate through a perturbation technique. The classical case is the rectangular thin radiator (James & Hall, 1989; Garg et al., 2001) that can be properly analyzed by means of an equivalent cavity model. According to this analysis, a CP nearly-square ($a \cong b$) radiator can be designed by feeding the patch along one of its diagonals. A left-hand CP radiation is obtained by positioning the probe along the dashed line illustrated in Fig. 8(a), were ($a > b$). For right-hand CP operation, the probe must be positioned along the other patch diagonal. A step-by-step design procedure is given in (Lumini et al., 1999).

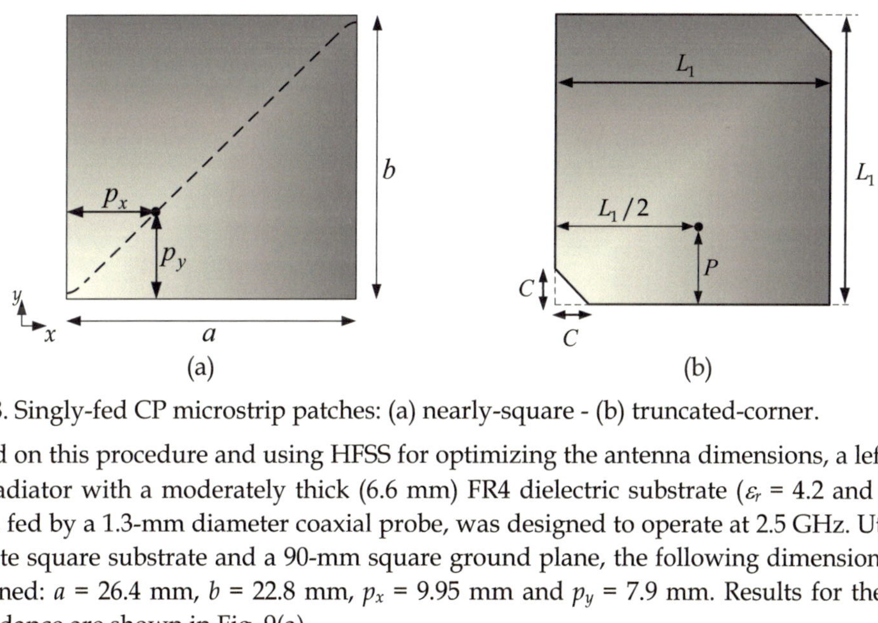

Fig. 8. Singly-fed CP microstrip patches: (a) nearly-square - (b) truncated-corner.

Based on this procedure and using HFSS for optimizing the antenna dimensions, a left-hand CP radiator with a moderately thick (6.6 mm) FR4 dielectric substrate (ε_r = 4.2 and tanδ = 0.02), fed by a 1.3-mm diameter coaxial probe, was designed to operate at 2.5 GHz. Utilizing a finite square substrate and a 90-mm square ground plane, the following dimensions were obtained: a = 26.4 mm, b = 22.8 mm, p_x = 9.95 mm and p_y = 7.9 mm. Results for the input impedance are shown in Fig. 9(a).

As expected, the operating frequency occurs between the fundamental TM_{10} and TM_{01} modes, each one corresponding to the input resistance maxima (at 2.39 GHz and 2.6 GHz, respectively). As a consequence, its input impedance is highly inductive (Z_{in} = 50 + j60 Ω) at the operating frequency (2.5 GHz) and the antenna is not properly matched to the 50-Ω SMA coaxial connector. Since the radiator under consideration exhibits an asymmetrical bandwidth, the best axial ratio (calculated in the broadside region) and the best reflection coefficient magnitude occur at different frequencies, as shown in Fig. 9(b).

The equivalent circuit shown in Fig. 10 (James & Hall, 1989) for singly-fed CP microstrip antennas can be used for better characterization of the mismatch problem. Taking this circuit into account, curves for the impedances (Z_{in1} and Z_{in2}) of each individual fundamental mode are presented in Fig. 11. At the operating frequency, however, the condition $n_1 = n_2 = 1$ must be satisfied. In this situation, the equivalent circuit is simplified and the following identities $\mathrm{Im}[Z_{in1}] = -\mathrm{Im}[Z_{in2}]$ and $\mathrm{Re}[Z_{in1}] = \mathrm{Re}[Z_{in2}]$ are verified. Consequently, at the operating frequency, the antenna input reactance is given only by the (inductive) reactance of the probe (L_p). Besides, frequencies f_1 and f_2 are tied up to the dimensions a and b, respectively, and the amplitude of the modes to the probe positions p_x and p_y.

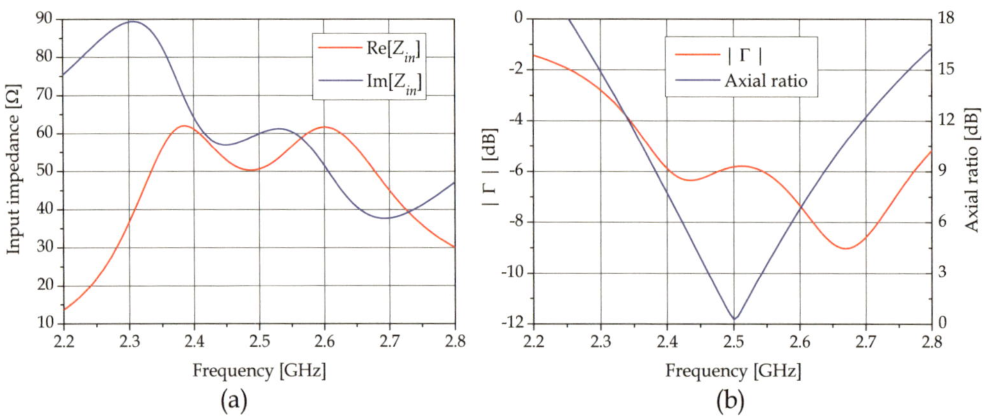

Fig. 9. Singly-fed CP microstrip patch: (a) input impedance - (b) axial ratio and reflection coefficient magnitude.

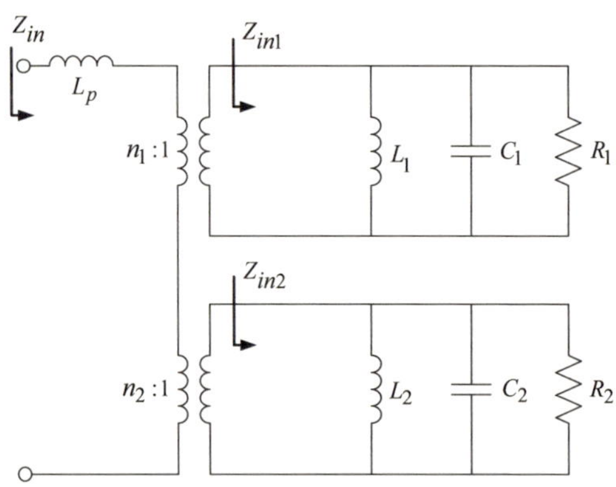

Fig. 10. Equivalent circuit for singly-fed CP microstrip antennas.

To overcome the aforementioned limitations, an innovative approach for designing singly-fed CP patch antennas is presented next. The goal is to get the best axial-ratio and $|\Gamma|$ both at the same frequency.

3.2 Rectangular patch: new design

The new strategy starts from a rectangular patch ($a \neq b$), instead of a nearly-square ($a \cong b$) one, and aims at establishing a capacitive reactance from the combination of the reactances

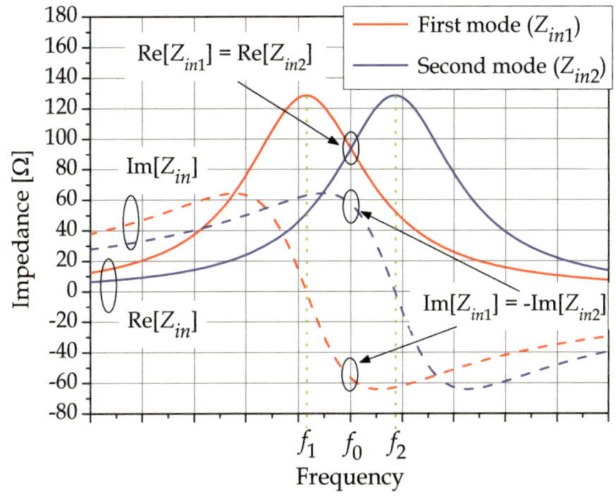

Fig. 11. Impedances of the fundamental modes: standard design.

of the two fundamental TM_{10} and TM_{01} modes at the frequency where Re[Z_{in1}] = Re[Z_{in2}]. This capacitive reactance is then used to compensate for the probe's inductive reactance, resulting in an excellent matching with the SMA coaxial connector (of 50-Ω characteristic impedance in the present case). This situation is shown in Fig. 12.

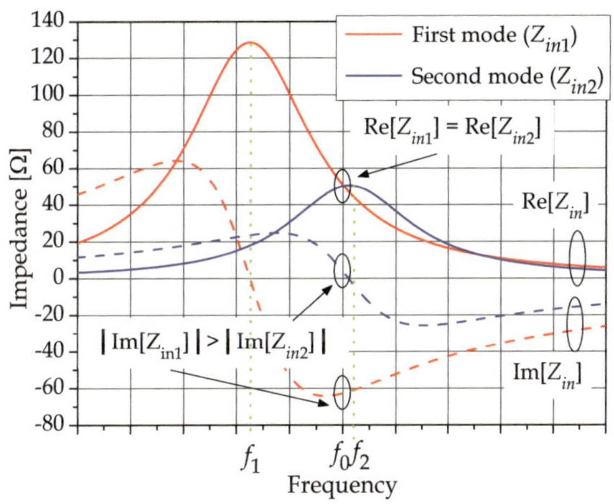

Fig. 12. Impedances of the fundamental modes: new design.

Compared with the standard approach, one notes that frequencies f_1 (of the first mode) and f_2 (of the second mode) are both shifted down. Frequency f_2 is now positioned near the operating frequency (2.5 GHz) and the amplitude of the first mode is greater than that of the second one. As a consequence, the new patch dimensions will be larger than those from the standard design, whereas the probe position p_x is reduced while p_y increases, under the conditions ($a > b$) and ($p_x < a/2$). On the other hand, if ($a > b$) and ($p_x > a/2$), then p_x increases while p_y is reduced. But before proceeding, a critical question needs to be posed at this point: can the proposed geometry establish CP radiation at the frequency where the antenna input impedance is purely real?

To answer this question, modifications in the standard design approach (Lumini et al., 1999) for nearly-square ($a \cong b$) patches were performed to encompass the present situation ($a \neq b$). As a result, a new feed locus equation, given by (6), is obtained for CP operation

$$p_y = \frac{b}{\pi}\cos^{-1}\left(\frac{2B\cos(\pi p_x / a)}{C\Delta \pm \sqrt{-4C^2B^2 + C^2\Delta^2}}\right), \tag{6}$$

where

$$C = \frac{b}{a}\mathrm{sinc}[\pi(a-b)/(2a)], \tag{7}$$

$$\Delta = \pi(b^{-1} - a^{-1}), \tag{8}$$

$$B = k_d(\delta_{ef} / 2), \tag{9}$$

$$k_d = \omega\sqrt{\mu_0\varepsilon_0\varepsilon_r}, \tag{10}$$

δ_{ef} is the antenna effective loss tangent and $\mathrm{sinc}(x) = \sin(x)/(x)$.

In the standard design δ_{ef} is calculated as an intermediate value between the effective loss tangent of the first and the second fundamental modes. Instead, as in the new approach frequency f_2 is close to the operating frequency f_0, the value of δ_{ef} can now be approximated by that calculated at f_2 only.

The positive sign before the square root in (6) results in the dashed line shown in Fig. 13. From this figure, one can see that the feed locus does not fall upon the patch diagonal as in the standard design, behaving as predicted in (Engest & Lo, 1985). Consequently, p_x must be reduced and p_y increased (if $p_x < a/2$), in comparison to the standard design. Thus, by positioning the probe along the aforementioned dashed line, such left-hand circular polarization antenna can be matched to the SMA connector.

On the other hand, if the negative sign before the square root in (6) is taken into account, the probe position p_x increases while p_y is reduced (if $p_x < a/2$). In this case, the antenna input impedance becomes more inductive and the proposed approach can not be applied.

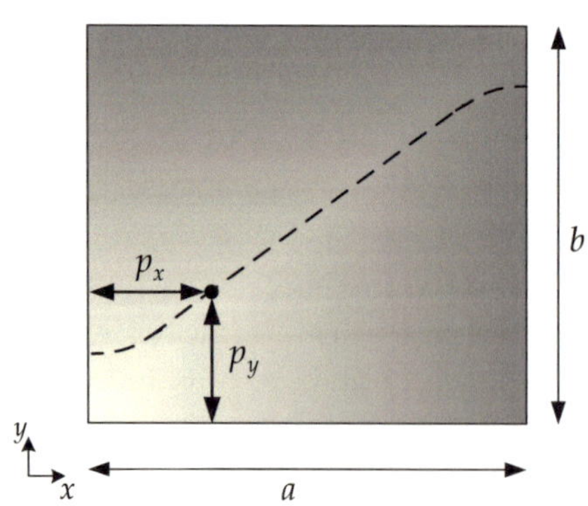

Fig. 13. Singly-fed CP rectangular patch.

To exemplify the application of the proposed strategy, a new left-hand CP antenna is designed to operate at 2.5 GHz with the same substrate characteristics, ground plane dimensions and probe diameter specified in the former design. As in the LP case, the HFSS software is used for optimizing the antenna dimensions. Consequently, it is necessary to set up the initial patch dimensions as well as the probe coordinates. In this chapter, the standard design dimensions and the coordinates given by (6) were used as the starting point.

The next step is the optimization of the probe position such that Re[Zin] = 50 Ω at the operating frequency (f_0) and the best axial ratio point occurs as close as possible to this frequency. In order to simultaneously obtain the best axial ratio and $|\Gamma|$ at f_0, the optimization procedure depends on each of the following conditions, where f_r denotes the best axial ratio frequency:

- If Im[Z_{in}] > 0 at f_0, and $f_r > f_0$, then both antenna dimensions and their ratio (a/b) must be increased.
- If Im[Z_{in}] > 0 at f_0, and $f_r < f_0$, then the antenna dimensions must be reduced whereas the ratio (a/b) must be increased.
- If Im[Z_{in}] > 0 at f_0, and $f_r = f_0$, then the ratio (a/b) must be increased, but reducing b and increasing a in the same proportion.
- If Im[Z_{in}] < 0 at f_0, and $f_r > f_0$, then the antenna dimensions must be increased whereas the ratio (a/b) must be reduced.
- If Im[Z_{in}] < 0 at f_0, and $f_r < f_0$, then both antenna dimensions and their ratio (a/b) must be reduced.
- If Im[Z_{in}] < 0 at f_0, and $f_r = f_0$, then the ratio (a/b) must be reduced, but increasing b and reducing a in the same proportion.
- If Im[Z_{in}] = 0 at f_0, and $f_r > f_0$, then both antenna dimensions and their ratio (a/b) must be increased.
- If Im[Z_{in}] = 0 at f_0, and $f_r < f_0$, then both antenna dimensions and their ratio (a/b) must be reduced.

After this procedure, the following dimensions were obtained: a = 28.25 mm, b = 23.60 mm, p_x = 6.00 mm and p_y = 8.10 mm. Results for the input impedance are presented in Fig. 14(a). The value of the antenna input impedance at the operating frequency is now purely 50 Ω. Graphics for the axial ratio and the reflection coefficient magnitude are depicted in Fig. 14(b).

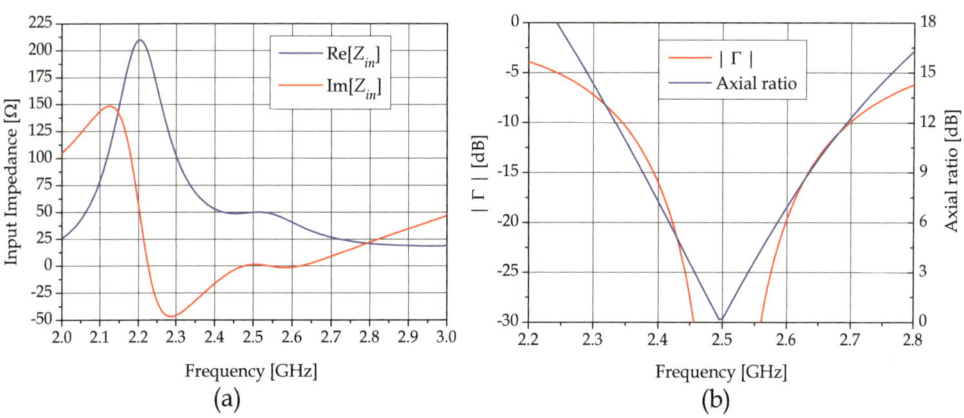

Fig. 14. Rectangular patch: (a) input impedance - (b) axial ratio and reflection coefficient magnitude.

One can see from these figures that the design goals were reached: both curves exhibit their best values right around the operating frequency. For verification purposes, Fig. 15 presents the feed locus plotted on the antenna patch, based on the data in Table 1. As expected, this confirms the behavior illustrated in Fig. 13.

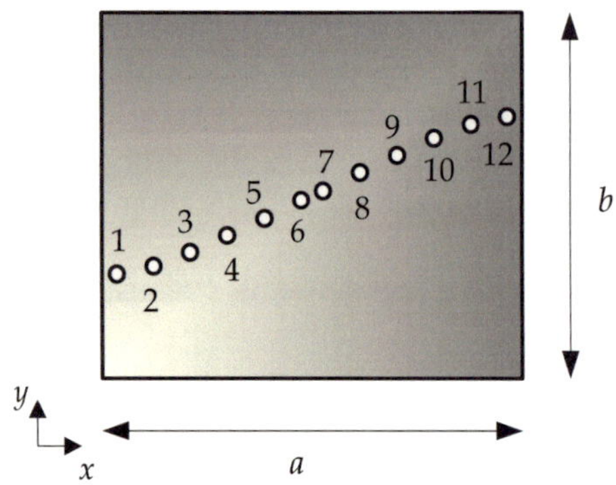

Fig. 15. Feed locus for CP antenna: new design.

	p_x (mm)	p_y (mm)
1	1.00	6.70
2	3.50	7.20
3	6.00	8.10
4	8.50	9.20
5	11.00	10.30
6	13.50	11.50
7	15.00	12.10
8	17.50	13.30
9	20.00	14.40
10	22.50	15.50
11	25.00	16.40
12	27.50	16.90

Table 1. Probe position for CP operation: new design.

The applicability of this new approach depends on the substrate thickness. In case of electrically thicker antennas, the proposed approach may fail since the feed locus deviates from the patch diagonal, tending to be parallel to the x-axis, if $a > b$. Besides, the probe position must be moved toward the antenna edge ($p_x \cong 0$ mm) for matching the antenna to a SMA connector.

3.3 Comparison between standard and new designs

To complete the analysis of the singly-fed CP antenna designed according to the new approach, its radiation patterns are now considered. Graphics of its E_{theta} and E_{phi} components on the yz and xz planes are compared to those from the antenna designed following the standard procedure in Figs. 16-17.

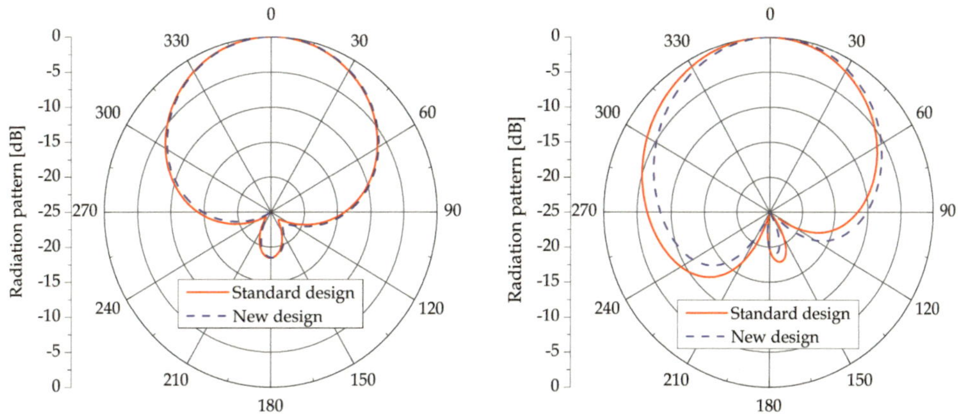

(a) (b)

Fig. 16. Radiation patterns in *xz* plane: (a) E_{phi} component - (b) E_{theta} component.

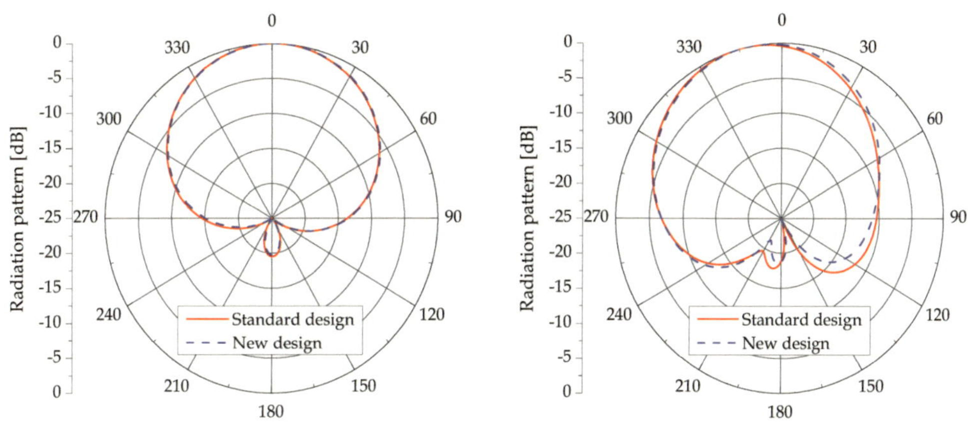

(a) (b)

Fig. 17. Radiation patterns in *yz* plane: (a) E_{phi} component - (b) E_{theta} component.

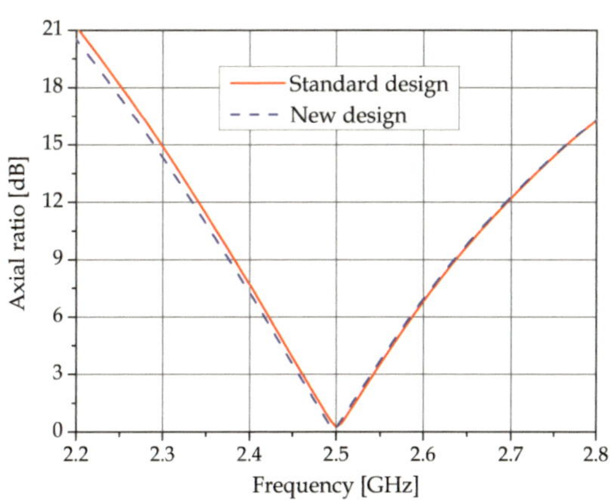

Fig. 18. Axial ratio comparison.

As seen, there are no significant differences between these patterns. Besides, the directivities are 6.6 dB for the nearly-square antenna and 6.8 dB for the new design. Although moderately thick, the antennas' efficiencies are 79.1% for nearly-square and 79.7% for the new design. The behavior of their axial ratios can be observed in Fig. 18.

4. Prototypes

To gather experimental confirmation of the effectiveness of the abovementioned new strategies, two prototypes, one linearly and the other circularly polarized, were designed, manufactured and tested using the FR4 substrate specified in this chapter.

4.1 Linearly polarized microstrip antenna

The geometry of the LP antenna designed to operate at 2 GHz is presented in Fig. 1. The following dimensions were calculated in Section 2.3 (a = 32.9 mm, b = 42.8 mm and p = 7.0 mm). A photo of the prototype is shown in Fig. 19.

Fig. 19. Photo of the LP microstrip antenna prototype.

Experimental results for the input impedance and the reflection coefficient magnitude are presented in Figs. 20(a) and (b). As seen, they are in excellent agreement with the HFSS simulation, thus validating the newly proposed design strategy for LP radiators.

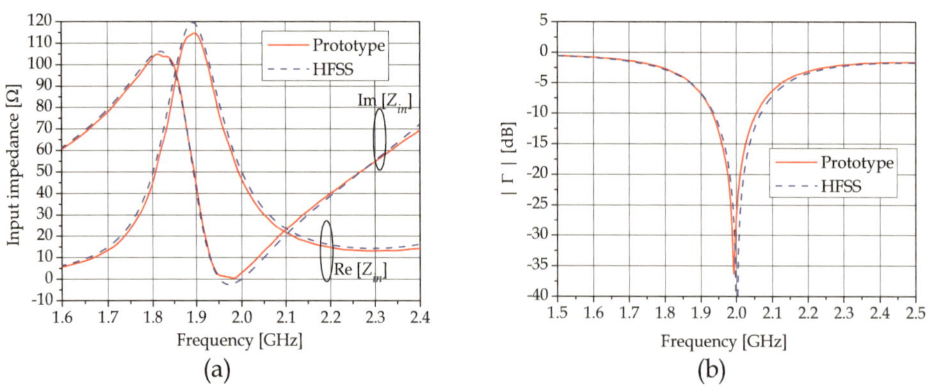

Fig. 20. LP microstrip antenna: (a) input impedance - (b) reflection coefficient magnitude.

4.2 Circularly-polarized microstrip antenna

The geometry of the CP antenna designed to operate at 2.5 GHz is presented in Fig. 13. The following dimensions were calculated in Section 3.2 (a = 28.25 mm, b = 23.60 mm, p_x = 6.00 mm and p_y = 8.10 mm). A photo of the prototype is shown in Fig. 21.

Fig. 21. Photo of the CP microstrip antenna prototype.

Experimental results for the input impedance and the reflection coefficient magnitude are presented in Figs. 22(a) and (b), and for the broadside axial ratio in Fig. 23. As seen, they are in very good agreement with HFSS simulations.

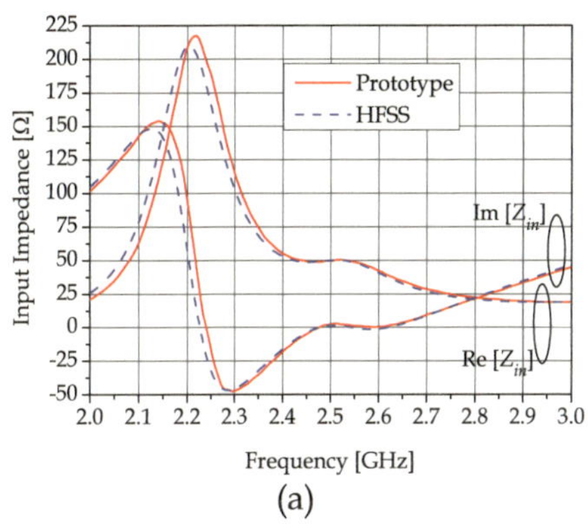

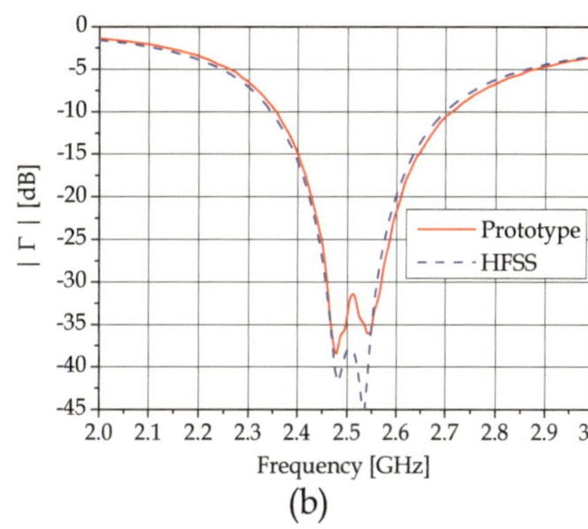

Fig. 22. CP microstrip antenna: (a) input impedance - (b) reflection coefficient magnitude.

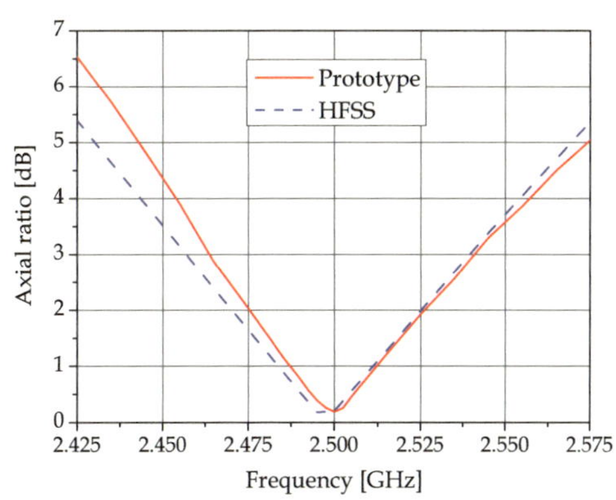

Fig. 23. Axial ratio comparison.

5. Other applications

As previously mentioned the new method was conceived based on the behavior of the input impedance of LP and CP antennas. Consequently, this procedure can be applied to the design of other patch geometries since their input impedance can be represented by the same equivalent circuit. Moreover, the method is not restricted to low-cost substrates, applying equally well to the design of LP or CP microstrip patches printed on any moderately thick commercial microwave laminates. These cases are discussed next.

5.1 Linearly polarized microstrip antennas

Different patch geometries are used nowadays to radiate LP electromagnetic waves (James & Hall, 1989; Garg et al., 2001; Volakis, 2007). Among them, the circular and PIFA radiators, since their input impedance can be represented by the same equivalent circuit used for rectangular antennas, were chosen to exemplify the application of the new proposed procedure in the case of a typical low-loss microwave laminate (ε_r = 2.55 and tan δ = 0.0022).

5.1.1 Circular patch

In the first application, a moderately thick (h = 6.35 mm) LP circular patch was designed to operate at 2.85 GHz using the HFSS software. The antenna geometry and dimensions are presented in Fig. 24(a). A photo of the prototype is shown in Fig. 24(b).

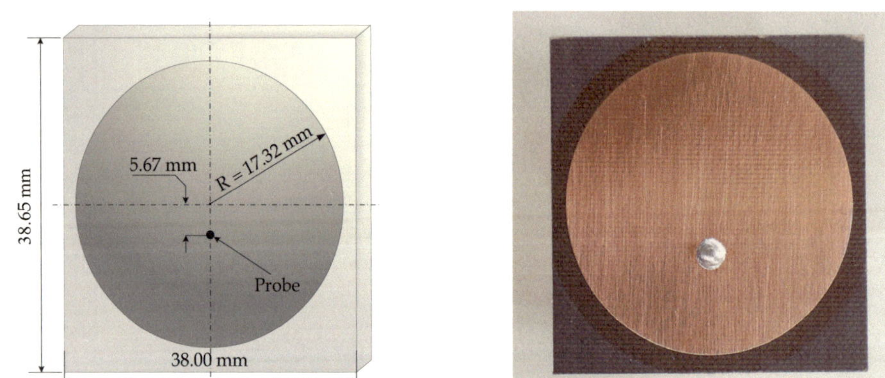

(a) (b)

Fig. 24. LP circular patch: (a) geometry - (b) photo of the prototype.

Results for the input impedance and reflection coefficient magnitude are presented in Figs. 25(a) and (b), respectively. Very good agreement is observed between the simulated and measured results.

5.1.2 PIFA antenna

In the second application, the focus is on the planar inverted F antenna (PIFA), a well know topology used for LP antennas of reduced dimensions (Chen et al., 2005). Following (Tinoco S. et al., 2008), an electrically thin antenna with h = 3.048 mm of substrate thickness was designed to operate near 2 GHz, using the HFSS software. The antenna geometry and dimensions are presented in Fig. 26(a). A photo of the prototype is shown in Fig. 26(b).

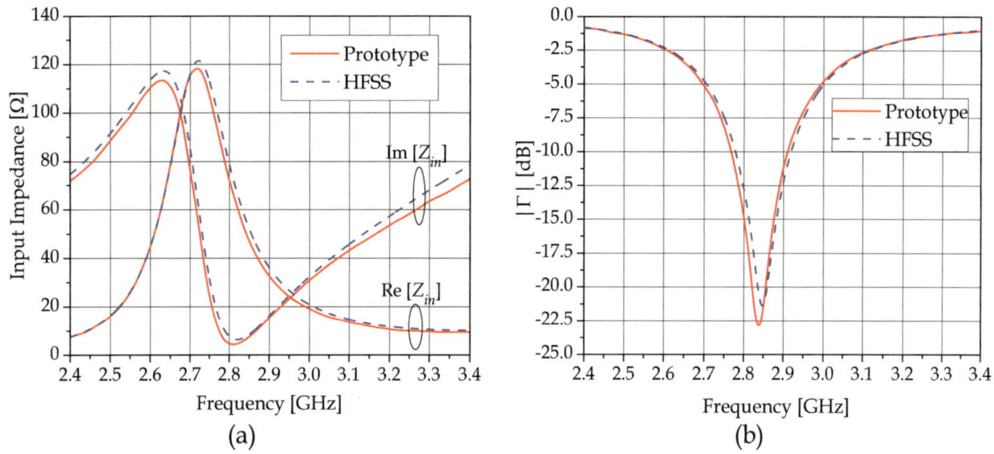

Fig. 25. LP circular patch: (a) input impedance - (b) reflection coefficient magnitude.

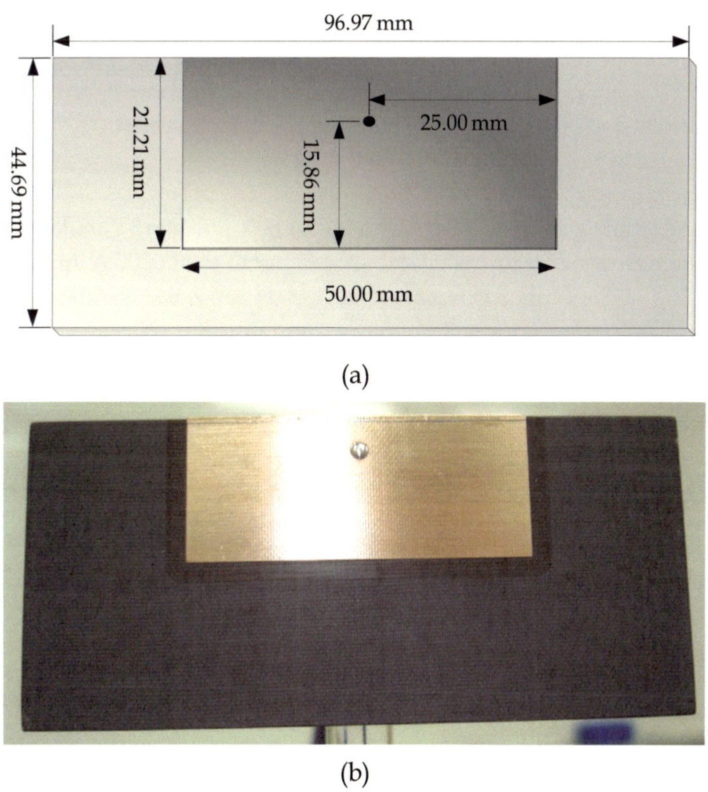

Fig. 26. PIFA antenna: (a) geometry - (b) photo of the prototype.

Results for the input impedance and reflection coefficient magnitude are presented in Figs. 27(a) and (b) respectively. Once again, very good agreement is observed between simulated and measured results.

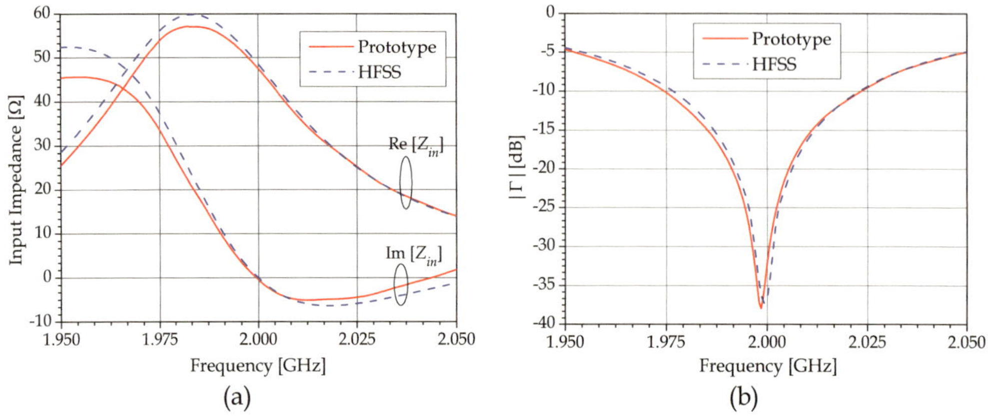

Fig. 27. PIFA antenna: (a) input impedance - (b) reflection coefficient magnitude.

5.2 Circularly-polarized microstrip antennas

Truncated corner probe-fed microstrip antennas can be designed to operate at the zero input reactance $X_{in} = 0$ condition, since their input impedance can be represented by the same equivalent circuit used for nearly-square radiators. To illustrate the applicability of this condition, two CP antennas are analyzed; one designed to meet specific requirements of Glonass applications and other to comply with the Globalstar system.

5.2.1 Glonass antenna

For Glonass applications, receiver antennas have to be right-hand circularly polarized over an operating range from 1.598 to 1.609 GHz (Nascimento et al., 2008). In order to obtain an excellent matching at the same frequency of best axial ratio, the standard truncated-corner square patch will not work, so a rectangular antenna design (Nascimento et al., 2007b) is used.

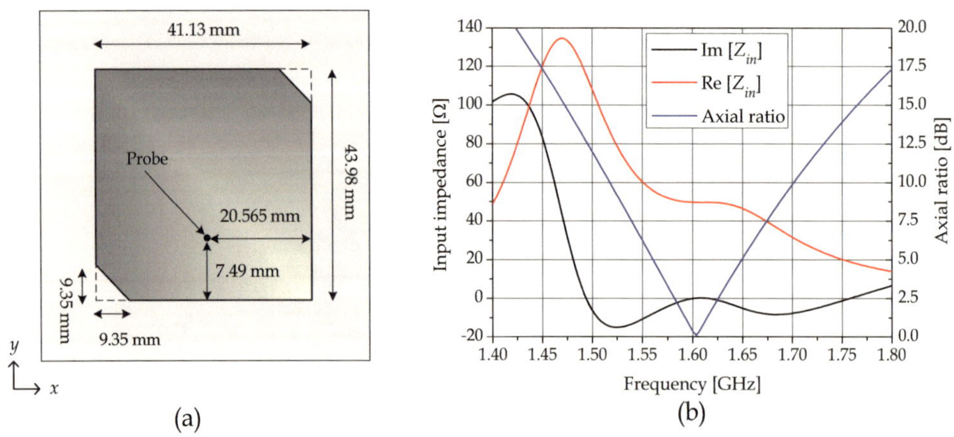

Fig. 28. CP truncated-corner patch: (a) geometry - (b) axial ratio and input impedance.

Following (Nascimento & Lacava, 2009), the truncated-corner rectangular patch (TCRP) illustrated in Fig. 28(a), printed on a finite FR4 substrate ($\varepsilon_r = 4.4$, $\tan\delta = 0.02$ and $h = 6.6$ mm) and a 75-mm square ground plane was designed using the HFSS software for operation at 1.603 GHz. The optimized antenna dimensions are shown in Fig. 28(a), the simulated input impedance and axial ratio results are presented in Fig. 28(b) and the reflection coefficient magnitude in Fig. 29. As expected, the microstrip antenna with the new geometry exhibits very good AR (0.1 dB) and reflection coefficient magnitude (-48 dB) characteristics at 1.603 GHz, without the need for any external matching network.

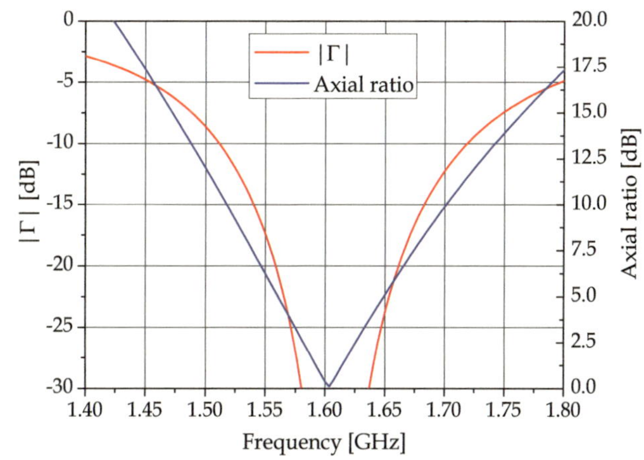

Fig. 29. CP truncated-corner patch: axial ratio and reflection coefficient magnitude.

5.2.2 Globalstar Antenna

Moderately thick microstrip antennas can also be used for bandwidth improvement. To exemplify such application a prototype of a Globalstar antenna was manufactured using a low-loss substrate (ε_r = 2.55, tan δ = 0.0022 and h = 4.572 mm). Left-handed CP Globalstar mobile-terminals require two radiators. The first is designed for uplink frequencies (Tx - 1.61073 to 1.62549 GHz) while the other receives the downlink ones (Rx - 2.48439 to 2.49915 GHz) (Nascimento et al., 2007a). The antenna geometry and a photo of the prototype are shown in Figs. 30(a) and (b), respectively.

The optimized antenna dimensions (using the HFSS software) are presented on Table 2 for the radiators designed on finite ground plane and dielectric (L = 140 mm; W = 85 mm).

Tx		*Rx*	
L_{T1}	54.90 mm	L_{R1}	34.40 mm
L_{T2}	55.90 mm	L_{R2}	35.85 mm
C_T	7.55 mm	C_R	5.75 mm
P_T	15.85 mm	P_R	9.00 mm
D_T	71.00 mm	D_R	15.00 mm

Table 2. Globalstar antenna dimensions.

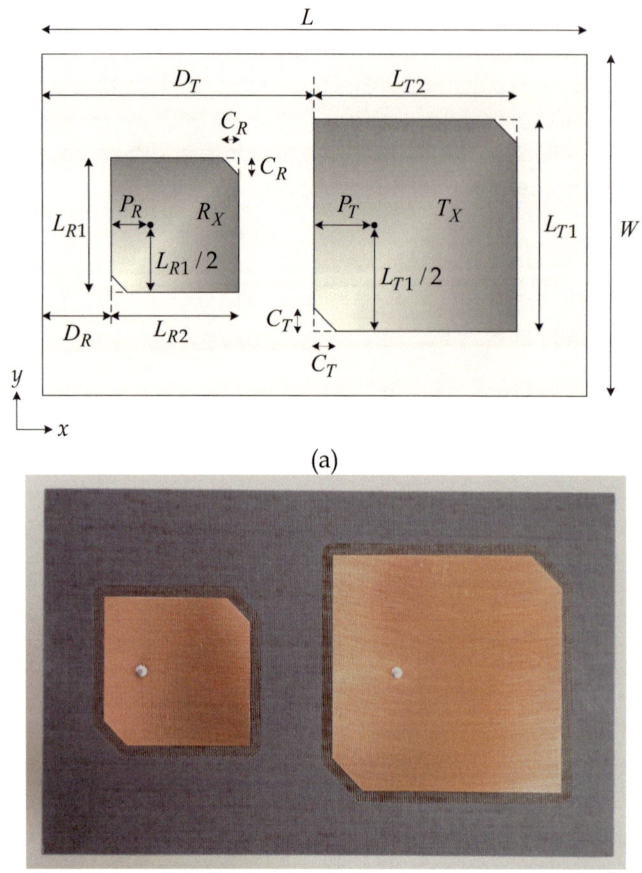

(a)

(b)

Fig. 30. Globalstar antenna: (a) geometry - (b) photo of the prototype.

The axial ratio and reflection coefficient magnitude are presented in Figs. 31 and 32 for the Tx and Rx antennas, respectively.

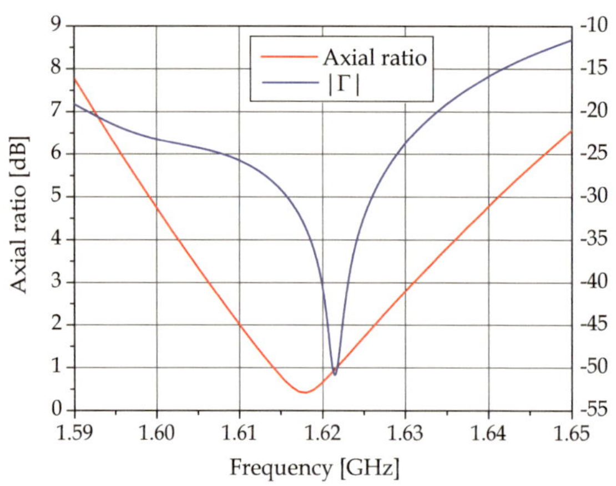

Fig. 31. Globalstar antenna axial ratio and reflection coefficient magnitude: Tx radiator.

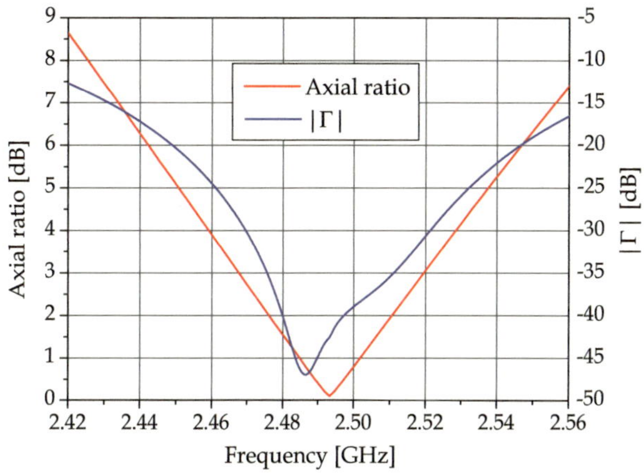

Fig. 32. Globalstar antenna axial ratio and reflection coefficient magnitude: Rx radiator.

Results for the input impedance on the Smith chart are presented in Figs. 33 and 34 for the Tx and Rx antennas, respectively. These results indicate that the antenna meets the Globalstar specifications.

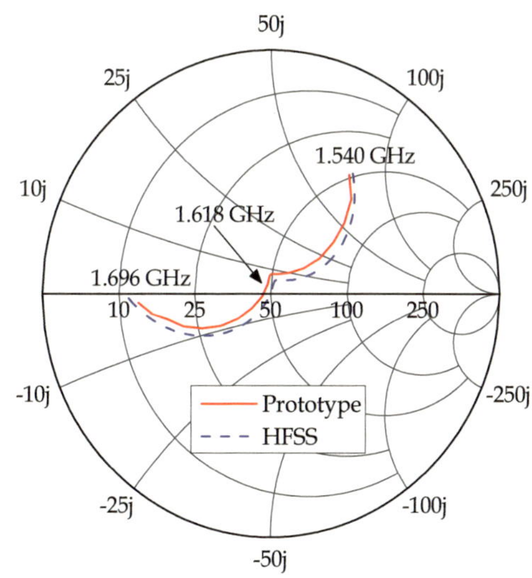

Fig. 33. Globalstar antenna input impedance: Tx radiator.

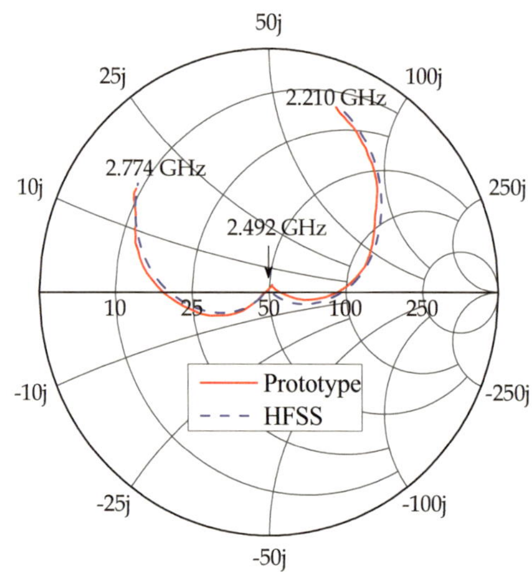

Fig. 34. Globalstar antenna input impedance: Rx radiator.

5.3 CP antenna radiation efficiency measurements

The radiation efficiency of a LP microstrip antenna can be efficiently measured using the Wheeler cap (Choo et al., 2005; Pozar & Kaufman, 1988; Sona & Rahmat-Samii 2006). According to Wheeler, the radiation resistance of an antenna can be separated from its loss resistance by enclosing the antenna with a radiation shield cap placed at a distance greater than $\lambda/(2\pi)$ (Wheeler, 1959). Consequently, since a linearly-polarized microstrip antenna can be modeled as a parallel *RLC* circuit, its efficiency is calculated by

$$\eta = \frac{G_{out} - G_{cap}}{G_{out}}, \tag{11}$$

where G_{cap} is the conductance of the admittance measured with the cap in place and G_{out} is the conductance of the admittance measured with the cap removed.

In the case of a CP microstrip antenna an innovative radiation efficiency analysis using the Wheeler cap method was presented in (Nascimento & Lacava, 2009). This procedure is discussed next, for the case of the Glonass antenna designed in Section 5.2.1.

Differently from the standard design, the two orthogonal resonant modes in the new approach are now asymmetrically positioned in relation to the frequency for optimal axial ratio as presented in Fig. 28 (b). In addition, at the lower resonant frequency (1.468 GHz), its 15.45-dB axial ratio shows the antenna tends to be linearly polarized around this frequency. This result supports the use of the Wheeler cap method for measuring the antenna radiation efficiency at this frequency.

The cap geometry is shown in Fig. 35 where the radiator is positioned inside a cubic cavity of electrically conducting walls of 270-mm internal dimension.

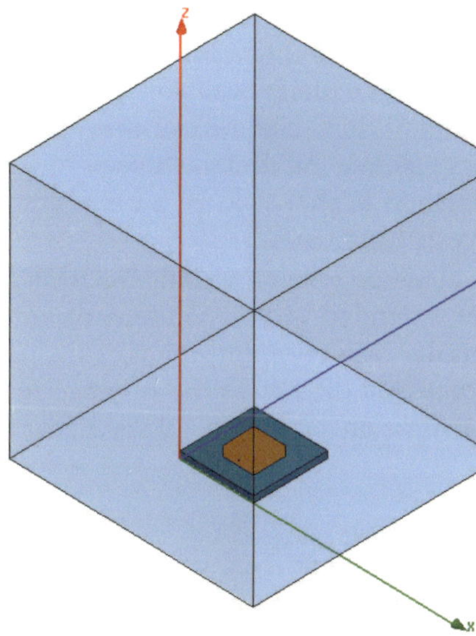

Fig. 35. Geometry of the Wheeler cap simulation through the HFSS package.

HFSS simulation results for the real part of the input impedance are presented in Fig. 36, both with and without the cubic cavity. Making use of equation (11) for the lower resonant mode (G_{cap} = 1.92 mS and G_{out} = 7.43 mS), the radiation efficiency computed from the Wheeler method is 74.16%. The free-space radiation efficiency, computed with the HFSS package is 74.68% at 1.468 GHz, which is only 0.7% off. Consequently, the Wheeler cap method can be used for accurately determining the radiation efficiency of TCRP radiators.

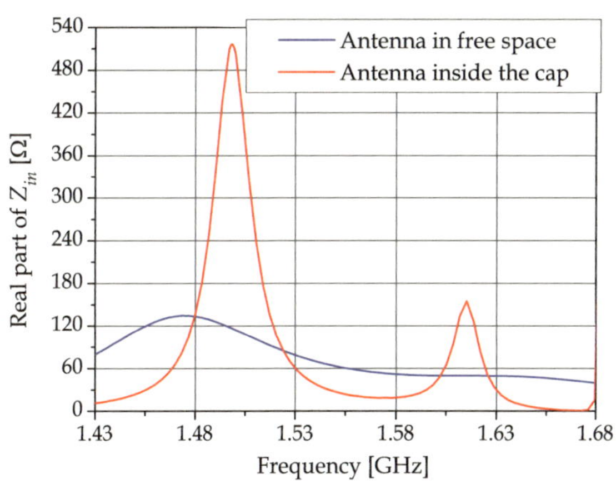

Fig. 36. Low-cost Glonass antenna: real part of its input impedance.

6. Conclusion

In this chapter, new effective strategies for designing probe-fed moderately thick LP and CP microstrip antennas were presented. As the design procedures do not make use of external networks, the antenna construction process is considerably simplified. Besides, as the new methodologies are based on properties of the antenna equivalent circuit, they can be applied to the design of microstrip radiators of arbitrary patch shapes. Moreover, it is not restricted to low-cost substrate thus applying equally well to the design of LP or CP microstrip patches printed on any moderately thick commercial microwave laminates. Experimental results for LP and CP radiators validate the design strategies for both the LP and CP cases. Moreover, the Wheeler cap method is shown to be an effective means for simulating the radiation efficiency of CP microstrip antennas.

The excellent practical results obtained when matching microstrip patch radiators to a 50-Ω SMA connector can be readily extended to the synthesis of inductive or capacitive input impedances, as for example in the case of optimization of the noise figure and stability of low-noise power amplifiers connected directly to the antenna. Another possible application is the design of low-cross-polarization probe-fed microstrip arrays (Marzall, et al., 2009; Marzall et al., 2010).

7. References

Alexander, M. J. (1989). Capacitive matching of microstrip antennas. *IEE Proceedings of Microwaves, Antennas and Propagation,* Vol. 137, No. 2, (Apr. 1989) (172-174), ISSN: 0950-107X.

Chang, F. S. & Wong, K. L. (2001), A broadband probe-fed patch antenna with a thickened probe pin, *Proceedings of Asia-Pacific Microwave Conference,* (1247-1250), ISBN: 0-7803-7138-0, Taipei, China, Dec. 2001

Chen, H. M.; Lin, Y. F.; Cheng, P. S.; Lin, H. H.; Song, C. T. P. & Hall, P. S. (2005), Parametric study on the characteristics of planar inverted-F antenna. *IEE Proceedings of Microwaves, Antennas and Propagation,* (Dec. 2005) (534-538), ISSN: 1350-2417.

Choo, H.; Rogers, R. & Ling, H. (2005), Comparison of three methods for the measurement of printed antennas efficiency, *IEEE Transactions on Antennas and Propagation,* Vol. 53, No. 7, (Jul. 2005) (2328-2332), ISSN: 0018-926X.

Dahele, J. S.; Hall, P. S. & Haskins, P. M. (1989), Microstrip patch antennas on thick substrates, *Proceedings of Antennas and Propagation Society International Symposium,* pp. 458-462, San Jose, CA, USA, Jun. 1989.

Engest, B. & Lo, Y. T. (1985), A study of circularly polarized rectangular microstrip antennas, *Technical Report,* Electromagnetics Laboratory, University of Illinois.

Gardelli, R.; La Cono, G. & Albani, M. (2004), A low-cost suspended patch antenna for WLAN access points and point-to-point links, *IEEE Antennas and Wireless Propagation Letters,* Vol. 3, (2004) (90-93), ISSN: 1536-1225.

Garg, R.; Bhartia, P.; Bahl, I. & Ittipiboon, A. (2001). *Microstrip Antenna Design Handbook,* Artech House, ISBN: 0-89006-513-6, Boston.

Hall, P. S. (1987). Probe compensation in thick microstrip patches. *Electronics Letters,* Vol. 23, No. 11, (May 1987) (606-607), ISSN: 0013-5194.

Haskins, P. M. & Dahele, J. S. (1998), Capacitive coupling to patch antenna by means of modified coaxial connectors, *Electronics Letters,* Vol. 34, No. 23, (Nov. 1998) (2187-2188), ISSN: 0013-5194.

HFSS (2010), Product overview, *Available: http://www.ansoft.com/products/hf/hfss/,* (Sept. 2010).

IEEE Std 145 (1993). *IEEE Standard Definitions of Terms for Antennas*, ISBN: 1-55937-317-2, New York, USA.

James, J. R. & Hall, P. S. (1989). *Handbook of Microstrip Antennas*, Peter Peregrinus, ISBN: 0-86341-150-9, London.

Lee , K. F. & Chen, W. (1997). *Advances in Microstrip and Printed Antennas*, John Wiley, ISBN: 0-471-04421-0, New York.

Lumini, F.; Cividanes, L. & Lacava, J. C. S. (1999), Computer aided design algorithm for singly fed circularly polarized rectangular microstrip patch antennas, *International Journal of RF and Microwave Computer-Aided Engineering*, Vol. 9, No. 1, (Jan. 1999) (32-41), ISBN: 1096-4290.

Marzall, L. F., Schildberg, R. & Lacava, J. C. S. (2009), High-performance, low-cross-polarization suspended patch array for WLAN applications, *Proceedings of Antennas and Propagation Society International Symposium*, pp. 1-4, ISBN: 978-1-4244-3647-7, Charleston, SC, USA, June 2009.

Marzall, L. F., Nascimento D.C., Schildberg, R. & Lacava, J. C. S. (2010), An effective strategy for designing probe-fed linearly-polarized thick microstrip arrays with symmetrical return loss bandwidth, *PIERS Online*, Vol. 6, No. 8, (July 2010) (700-704), ISSN: 1931-7360.

Nascimento, D. C.; Mores Jr., J.A.; Schildberg, R. & Lacava, J. C. S. (2006), Low-cost truncated corner microstrip antenna for GPS application, *Proceedings of Antennas and Propagation Society International Symposium*, pp. 1557-1560, ISBN: 1-4244-0123-2, Albuquerque, NM, USA, July 2006.

Nascimento, D. C.; Bianchi, I.; Schildberg, R. & Lacava, J. C. S. (2007a), Design of probe-fed truncated corner microstrip antennas for Globalstar system, *Proceedings of Antennas and Propagation Society International Symposium*, pp. 3041-3044, ISBN: 978-1-4244-0877-1, Honolulu, HI, USA, June 2007.

Nascimento, D. C.; Schildberg, R. & Lacava, J. C. S. (2007b), New considerations in the design of low-cost probe-fed truncated corner microstrip antennas for GPS applications, *Proceedings of Antennas and Propagation Society International Symposium*, pp. 749-752, ISBN: 978-1-4244-0877-1, Honolulu, HI, USA, June 2007.

Nascimento, D. C.; Schildberg, R. & Lacava, J. C. S. (2008). Design of low-cost microstrip antennas for Glonass applications. *PIERS Online*, Vol. 4, No. 7, (2008) (767-770), ISSN: 1931-7360.

Nascimento, D. C. & Lacava, J. C. S. (2009), Circularly-polarized microstrip antenna radiation efficiency simulation based on the Wheeler cap method, *Proceedings of Antennas and Propagation Society International Symposium*, pp. 1-4, ISBN: 978-1-4244-3647-7, Charleston, SC, USA, June 2009.

Niroojazi, M. & Azarmanesh, M. N. (2004), Practical design of single feed truncated corner microstrip antenna, *Proceedings of Second Annual Conference on Communication Networks and Services Research, 2004*, pp. 25-29, ISBN: 0-7695-2096-0, Fredericton, NB, Canada, May 2004.

Pozar, D. M. & Kaufman, B. (1988), Comparison of three methods for the measurement of printed antennas efficiency, *IEEE Transactions on Antennas and Propagation*, Vol. 36, No. 1, (Jan. 1988) (136-139), ISSN: 0018-926X.

Richards, W. F.; Lo, Y. T. & Harrison, D. D. (1981), An improved theory for microstrip antennas and applications, *IEEE Transactions on Antennas and Propagation*, Vol. 29, No 1, (Jan. 1981) (38-46), ISSN: 0018-926X.

Sona, K. S. & Rahmat-Samii, Y. (2006), On the implementation of Wheeler cap method in FDTD, *Proceedings of Antennas and Propagation Society International Symposium*, pp. 1445-1448, ISBN: 1-4244-0123-2, Albuquerque, NM, USA, July 2006.

Teng, P. L.; Tang, C. L. & Wong, K. L. (2001), A broadband planar patch antenna fed by a short probe feed, *Proceedings of Asia-Pacific Microwave Conference*, pp. 1243-1246, ISBN: 0-7803-7138-0, Taipei, China, Dec. 2001.

Tinoco S., A. F.; Nascimento, D. C. & Lacava, J. C. S. (2008), Rectangular microstrip antenna design suitable for undergraduate courses, *Proceedings of Antennas and Propagation Society International Symposium*, pp. 1-4, ISBN: 978-1-4244-2041-4, San Diego, CA, USA, July 2008.

Tzeng, Y. B.; Su, C. W. & Lee, C. H. (2005), Study of broadband CP patch antenna with its ground plane having an elevated portion, *Proceedings of Asia-Pacific Microwave Conference*, pp. 1-4, ISBN: 0-7803-9433-X, Suzhou, China, Dec. 2005

Vandenbosch, G. A. E. & Van de Capelle, A. R. (1994), Study of the capacitively fed microstrip antenna element, *IEEE Transactions on Antennas and Propagation*, Vol. 42, No. 12, (Dec. 1994) (1648-1652), ISSN: 0018-926X.

Volakis, J. L. (2007). *Antenna Engineering Handbook. 4th ed.*, McGraw-Hill, ISBN: 0-07-147574-5, New York.

Wheeler, H. A. (1959), The radiansphere around a small antenna, *Proceedings of the IRE*, Vol. 47, No. 8, (Aug. 1959) (1325-1331), ISSN: 0096-8390.

2

Analysis of a Rectangular Microstrip Antenna on a Uniaxial Substrate

Amel Boufrioua
Electronics Department University of Constantine,
25000 Constantine,
Algeria

1. Introduction

Over the past years microstrip resonators have been widely used in the range of microwave frequencies. In general these structures are poor radiators, but by proper design the radiation performance can be improved and these structures can be used as antenna elements (Damiano & Papiernik, 1994). In recent years microstrip patch antennas became one of the most popular antenna types for use in aerospace vehicles, telemetry and satellite communication. These antennas consist of a radiating metallic patch on one side of a thin, non conducting, supporting substrate panel with a ground plane on the other side of the panel. For the analysis and the design of microstrip antennas there have been several techniques developed (Damiano & Papiernik, 1994; Mirshekar-Syahkal, 1990). The spectral domain approach is extensively used in microstrip analysis and design (Mirshekar-Syahkal, 1990). In such an approach, the spectral dyadic Green's function relates the tangential electric fields and currents at various conductor planes. It is found that the substrate permittivity is a very important factor to be determined in microstrip antenna designs. Moreover the study of anisotropic substrates is of interest, many practical substrates have a significant amount of anisotropy that can affect the performance of printed circuits and antennas, and thus accurate characterization and design must account for this effect (Bhartia et al. 1991). It is found that the use of such materials may have a beneficial effect on circuit or antenna (Bhartia et al. 1991; Pozar, 1987). For a rigorous solution to the problem of a rectangular microstrip antenna, which is the most widely used configuration because its shape readily allows theoretical analysis, Galerkin's method is employed in the spectral domain with two sets of patch current expansions. One set is based on the complete set of orthogonal modes of the magnetic cavity, and the other employs Chebyshev polynomials with the proper edge condition for the patch currents (Tulintsef et al. 1991).

This chapter describes spectral domain analyses of a rectangular microstrip patch antenna that contains isotropic or anisotropic substrates in which entire domain basis functions are used to model the patch current, we will present the effect of uniaxial anisotropy on the characterization of a rectangular microstrip patch antenna, also because there has been very little work on the scattering radar cross section of printed antennas in the literature, including the effect of a uniaxial anisotropic substrate, a number of results pertaining to this case will be presented in this chapter.

2. Theory

An accurate design of a rectangular patch antenna can be done by using the Galerkin procedure of the moment method (Pozar, 1987; Row & Wong, 1993; Wong et al., 1993). An integral equation can be formulated by using the Green's function on a thick dielectric substrate to determine the electric field at any point.

The patch is assumed to be located on a grounded dielectric slab of infinite extent, and the ground plane is assumed to be perfect electric conductor, the rectangular patch with length *a* and width *b* is embedded in a single substrate, which has a uniform thickness of *h* (see Fig. 1), all the dielectric materials are assumed to be nonmagnetic with permeability μ_0. To simplify the analysis, the antenna feed will not be considered.

The study of anisotropic substrates is of interest, however, the designers should, carefully check for the anisotropic effects in the substrate material with which they will work, and evaluate the effects of anisotropy.

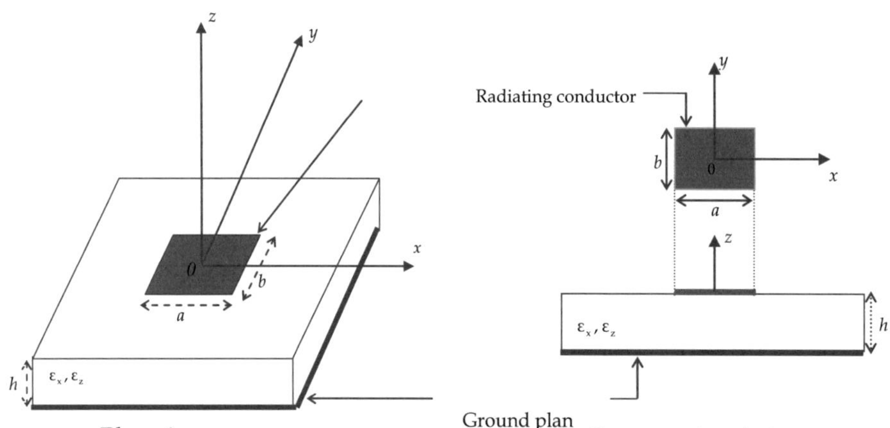

a. Plan view b. Cross sectional view

Fig. 1. Geometry of a rectangular microstrip antenna

Anisotropy is defined as the substrate dielectric constant on the orientation of the applied electric field. Mathematically, the permittivity of an anisotropic substrate can be represented by a tensor or dyadic of this form (Bhartia et al., 1991)

$$\boldsymbol{\varepsilon} = \varepsilon_0 \cdot \begin{bmatrix} \varepsilon_{xx} & \varepsilon_{xy} & \varepsilon_{xz} \\ \varepsilon_{yx} & \varepsilon_{yy} & \varepsilon_{yz} \\ \varepsilon_{zx} & \varepsilon_{zy} & \varepsilon_{zz} \end{bmatrix} \tag{1}$$

For a biaxially anisotropic substrate the permittivity is given by

$$\boldsymbol{\varepsilon} = \varepsilon_0 \cdot \begin{bmatrix} \varepsilon_x & 0 & 0 \\ 0 & \varepsilon_y & 0 \\ 0 & 0 & \varepsilon_z \end{bmatrix} \tag{2}$$

For a uniaxially anisotropic substrate the permittivity is

$$\boldsymbol{\varepsilon} = \varepsilon_0 \cdot \begin{bmatrix} \varepsilon_x & 0 & 0 \\ 0 & \varepsilon_x & 0 \\ 0 & 0 & \varepsilon_z \end{bmatrix} \tag{3}$$

ε_0 is the free-space permittivity.

ε_x is the relative permittivity in the direction perpendicular to the optical axis.

ε_z is the relative permittivity in the direction of the optical axis.

Many substrate materials used for printed circuit antenna exhibit dielectric anisotropy, especially uniaxial anisotropy (Bhartia et al. 1991; Wong et al., 1993).

In the following, the substrate material is taken to be isotropic or uniaxially anisotropic with the optical axis normal to the patch.

The boundary condition on the patch is given by (Pozar, 1987)

$$\mathbf{E}_{scat} + \mathbf{E}_{inc} = 0 \tag{4}$$

$\mathbf{E}_{inc}$ Tangential components of incident electric field.

$\mathbf{E}_{scat}$ Tangential components of scattered electric field.

While it is possible to work with wave equations and the longitudinal components $\tilde{\mathbf{E}}_z$ and $\tilde{\mathbf{H}}_z$, in the Fourier transform domain, it is desired to find the transverse fields in the (TM, TE) representation in terms of the longitudinal components. Assuming an $e^{i\omega t}$ time variation, thus Maxwells equations

$$\nabla \times \mathbf{H} = \varepsilon \frac{\partial \mathbf{E}}{\partial t} = i\,\omega\,\varepsilon\,\mathbf{E} \tag{5}$$

$$\nabla \times \mathbf{E} = \mu_0 \frac{\partial \mathbf{H}}{\partial t} = -i\,\omega\,\mu_0\,\mathbf{H} \tag{6}$$

Applying the divergence condition component

$$\nabla \cdot \mathbf{E} = \frac{\partial \mathbf{E}_x}{\partial x} + \frac{\partial \mathbf{E}_y}{\partial y} + \frac{\partial \mathbf{E}_z}{\partial z} = 0 \tag{7}$$

$$\nabla \cdot \mathbf{H} = \frac{\partial \mathbf{H}_x}{\partial x} + \frac{\partial \mathbf{H}_y}{\partial y} + \frac{\partial \mathbf{H}_z}{\partial z} = 0 \tag{8}$$

$i = \sqrt{-1}$

ω is the angular frequency.

From the above equations and after some algebraic manipulation, the wave equations for $\mathbf{E}_z$ and $\mathbf{H}_z$ are respectively

$$\frac{\partial^2 \mathbf{E}_z}{\partial x^2} + \frac{\partial^2 \mathbf{E}_z}{\partial y^2} + \frac{\varepsilon_x}{\varepsilon_z}\frac{\partial^2 \mathbf{E}_z}{\partial z^2} + \varepsilon_z\, k_0^2\, \mathbf{E}_z = 0 \tag{9}$$

$$\frac{\partial^2 \mathbf{H_z}}{\partial x^2}+\frac{\partial^2 \mathbf{H_z}}{\partial y^2}+\frac{\varepsilon_x}{\varepsilon_z}\frac{\partial^2 \mathbf{H_z}}{\partial z^2}+\varepsilon_z\, k_0^2\, \mathbf{H_z}=0 \tag{10}$$

With

k_0 propagation constant for free space, $k_0 = \omega\sqrt{\varepsilon_0\,\mu_0}$

By assuming plane wave propagation of the form $e^{\pm i k_x x}\, e^{\pm i k_y y}\, e^{\pm i k_z z}$

A Fourier transform pair of the electric field is given by (Pozar, 1987)

$$\widetilde{\mathbf{E}}\left(k_x, k_y, k_z\right)=\iint_{-\infty}\mathbf{E}\left(x, y, z\right)e^{-i k_x x}\, e^{-i k_y y}\, dx\, dy \tag{11}$$

$$\mathbf{E}\left(x, y, z\right)=\frac{1}{4\pi^2}\iint_{-\infty}\widetilde{\mathbf{E}}\left(k_x, k_y, k_z\right)e^{i k_x x}\, e^{i k_y y}\, dk_x\, dk_y \tag{12}$$

A Fourier transform pair of the magnetic field is given by (Pozar, 1987)

$$\widetilde{\mathbf{H}}\left(k_x, k_y, k_z\right)=\iint_{-\infty}\mathbf{H}\left(x, y, z\right)e^{-i k_x x}\, e^{-i k_y y}\, dx\, dy \tag{13}$$

$$\mathbf{H}\left(x, y, z\right)=\frac{1}{4\pi^2}\iint_{-\infty}\widetilde{\mathbf{H}}\left(k_x, k_y, k_z\right)e^{i k_x x}\, e^{i k_y y}\, dk_x\, dk_y \tag{14}$$

It is worth noting that ~ is used to indicate the quantities in spectral domain.

In the spectral domain $\frac{\partial}{\partial x}=i\,k_x$, $\frac{\partial}{\partial y}=i\,k_y$, $\frac{\partial}{\partial z}=i\,k_z$ and $\frac{\partial}{\partial t}=i\,\omega\, t$

After some straightforward algebraic manipulation the transverse field can be written in terms of the longitudinal components $\widetilde{\mathbf{E}}_z$, $\widetilde{\mathbf{H}}_z$

$$\widetilde{\mathbf{E}}_x=\frac{i\,\varepsilon_z\, k_x}{\varepsilon_x\, \mathbf{k}_s^2}\frac{\partial \widetilde{\mathbf{E}}_z}{\partial z}+\frac{\omega\,\mu_0\, k_y}{\mathbf{k}_s^2}\widetilde{\mathbf{H}}_z \tag{15}$$

$$\widetilde{\mathbf{E}}_y=\frac{i\varepsilon_z\, k_y}{\varepsilon_x\, \mathbf{k}_s^2}\frac{\partial \widetilde{\mathbf{E}}_z}{\partial z}-\frac{\omega\,\mu_0\, k_x}{\mathbf{k}_s^2}\widetilde{\mathbf{H}}_z \tag{16}$$

$$\widetilde{\mathbf{H}}_x=-\frac{\omega\,\varepsilon_z\varepsilon_0\, k_y}{\mathbf{k}_s^2}\widetilde{\mathbf{E}}_z+\frac{i\,k_x}{\mathbf{k}_s^2}\frac{\partial \widetilde{\mathbf{H}}_z}{\partial z} \tag{17}$$

$$\widetilde{\mathbf{H}}_y=-\frac{\omega\,\varepsilon_z\varepsilon_0\, k_x}{\mathbf{k}_s^2}\widetilde{\mathbf{E}}_z+\frac{i\,k_y}{\mathbf{k}_s^2}\frac{\partial \widetilde{\mathbf{H}}_z}{\partial z} \tag{18}$$

$\mathbf{k}_s$ is the transverse wave vector, $\mathbf{k}_s=k_x\hat{\mathbf{x}}+k_y\hat{\mathbf{y}}$, $k_s=\left|\mathbf{k}_s\right|$

k_x and k_y are the spectral variables corresponding to x and y respectively.

From the wave equations (9) and (10), the general form of $\widetilde{\mathbf{E}}_z$ and $\widetilde{\mathbf{H}}_z$ is

$$\tilde{E}_z = C_1\, e^{-ik_z z} + D_1\, e^{ik_z z} \tag{19}$$

$$\tilde{H}_z = C_2\, e^{-ik_z z} + D_2\, e^{ik_z z} \tag{20}$$

$\mathbf{C_1}$, $\mathbf{D_1}$, $\mathbf{C_2}$ and $\mathbf{D_2}$ are the unknowns to be determined.
By substitution of (19) and (20) in (15)-(18) and after some algebraic manipulation the transverse field in the (TM, TE) representation can be written by

$$\tilde{\mathbf{E}}_s(\mathbf{k}_s, z) = \begin{bmatrix} \tilde{E}_s^{\,e}(\mathbf{k}_s, z) \\ \tilde{E}_s^{\,h}(\mathbf{k}_s, z) \end{bmatrix} = e^{ik_z z}\, \mathbf{A}(\mathbf{k}_s) + e^{-ik_z z}\, \mathbf{B}(\mathbf{k}_s) \tag{21}$$

$$\tilde{\mathbf{H}}_s(\mathbf{k}_s, z) = \begin{bmatrix} \tilde{H}_s^{\,e}(\mathbf{k}_s, z) \\ \tilde{H}_s^{\,h}(\mathbf{k}_s, z) \end{bmatrix} = \mathbf{g}(\mathbf{k}_s)\left[e^{ik_z z}\, \mathbf{A}(\mathbf{k}_s) - e^{-ik_z z}\, \mathbf{B}(\mathbf{k}_s)\right] \tag{22}$$

The superscripts e and h denote the TM and the TE waves, respectively.
A and **B** are two unknowns vectors to be determined, note that are expressed in terms of $\mathbf{C_1}$, $\mathbf{D_1}$, $\mathbf{C_2}$ and $\mathbf{D_2}$.
Where

$$\mathbf{g}(\mathbf{k}_s) = \begin{bmatrix} \dfrac{\omega \varepsilon_0 \varepsilon_x}{k_z^e} & 0 \\ 0 & \dfrac{k_z^h}{\omega \mu_0} \end{bmatrix} \tag{23}$$

$$\mathbf{k}_z = \begin{bmatrix} k_z^e & 0 \\ 0 & k_z^h \end{bmatrix},\quad k_z^e = \left(\varepsilon_x k_0^2 - \frac{\varepsilon_x}{\varepsilon_z} k_s^2\right)^{\frac{1}{2}} \quad \text{and} \quad k_z^h = \left(\varepsilon_x k_0^2 - k_s^2\right)^{\frac{1}{2}}$$

k_z^e and k_z^h are respectively propagation constants for TM and TE waves in the uniaxial dielectric.
By eliminating the unknowns **A** and **B**, in the equations (21) and (22) we obtain the following matrix which combines the tangential field components on both sides z1 and z2 of the considered layer as input and output quantities

$$\begin{bmatrix} \tilde{\mathbf{E}}^{e(h)}(\mathbf{k}_s, z_2^-) \\ \tilde{\mathbf{H}}^{e(h)}(\mathbf{k}_s, z_2^-) \end{bmatrix} = \begin{bmatrix} \mathbf{I}\cos(k_z^{e(h)} h) & i\mathbf{g}^{-1}\sin(k_z^{e(h)} h) \\ i\mathbf{g}\sin(k_z^{e(h)} h) & \mathbf{I}\cos(k_z^{e(h)} h) \end{bmatrix} \times \begin{bmatrix} \tilde{\mathbf{E}}^{e(h)}(\mathbf{k}_s, z_1^+) \\ \tilde{\mathbf{H}}^{e(h)}(\mathbf{k}_s, z_1^+) \end{bmatrix} - \begin{bmatrix} 0 \\ \tilde{\mathbf{J}}^{e(h)}(\mathbf{k}_s) \end{bmatrix} \tag{24}$$

I is the unit matrix.
$\tilde{\mathbf{J}}^{e(h)}(\mathbf{k}_s)$ is the current on the patch.
In the spectral domain the relationship between the patch current and the electric field on the patch is given by

$$\tilde{\mathbf{E}}_s(\mathbf{k}_s) = \mathbf{G}(\mathbf{k}_s) \cdot \tilde{\mathbf{J}}(\mathbf{k}_s) \tag{25}$$

G is the spectral dyadic Green's function

$$\mathbf{G} = \begin{bmatrix} \mathbf{G}^e & 0 \\ 0 & \mathbf{G}^h \end{bmatrix} \tag{26}$$

$\mathbf{G}^e, \mathbf{G}^h$ are given by

$$\mathbf{G}^e = \frac{1}{i\omega\varepsilon_0} \frac{-k_z^e k_z \sin(k_{z1}h)}{ik_z^e \sin(k_{z1}h) + \varepsilon_x k_z \cos(k_{z1}h)} \tag{26a}$$

$$\mathbf{G}^h = \frac{1}{i\omega\varepsilon_0} \frac{-k_0^2 \sin(k_{z1}h)}{ik_z \sin(k_{z1}h) + k_z^h \cos(k_{z1}h)} \tag{26b}$$

In the case of the isotroipc substrate

$$\mathbf{G}^e = \sqrt{\frac{\mu_0}{\varepsilon_0}} \frac{\cos(k_{z1}\,h)}{\left(1 - i\varepsilon_r\,k_z \cot(k_{z1}\,h)/k_{z1}\right)} \tag{26c}$$

$$\mathbf{G}^h = \sqrt{\frac{\mu_0}{\varepsilon_0}} \frac{1}{\cos(k_{z1}\,h)\left(1 - i\,k_{z1} \cot(k_{z1}\,h)/k_z\right)} \tag{26d}$$

Where

$k_{z1} = k_0 \cos(k_z\,h)$ and $k_z = \left(k_0^2 - k_s^2\right)^{1/2}$

$\tilde{\mathbf{J}}(\mathbf{k}_s)$ is the current on the patch which related to the vector Fourier transform of $J(r_s)$, as (Chew & Liu, 1988)

$$\tilde{\mathbf{J}}(\mathbf{k}_s) = \iint_{-\infty}^{\infty} d\mathbf{k}_s \mathbf{F}(\mathbf{k}_s, -\mathbf{r}_s) \cdot \mathbf{J}(\mathbf{r}_s) \tag{27}$$

Where

$$\mathbf{F}(\mathbf{k}_s, \mathbf{r}_s) = \frac{1}{k_s} \begin{bmatrix} k_x & k_y \\ k_y & -k_x \end{bmatrix} e^{i\mathbf{k}_s \cdot \mathbf{r}_s}, \quad \mathbf{r}_s = x\hat{\mathbf{x}} + y\hat{\mathbf{y}} \tag{28}$$

$\hat{\mathbf{x}}$ unit vector in x direction.

$\hat{\mathbf{y}}$ unit vector in y direction.

The surface current on the patch can be expanded into a series of known basis functions J_{xn} and J_{ym}

$$\mathbf{J}(\mathbf{r}_s) = \sum_{n=1}^{N} a_n \begin{bmatrix} J_{xn}(\mathbf{r}_s) \\ 0 \end{bmatrix} + \sum_{m=1}^{M} b_m \begin{bmatrix} 0 \\ J_{ym}(\mathbf{r}_s) \end{bmatrix} \tag{29}$$

Where a_n and b_m are the unknown coefficients to be determined in the x and y direction respectively.

The latter expression is substituted into equation (27); the results can be given by

$$\tilde{\mathbf{J}}(\mathbf{k}_s) = \frac{1}{\mathbf{k}_s} \begin{bmatrix} k_x \\ k_y \end{bmatrix} \sum_{n=1}^{N} a_n\, \tilde{\mathbf{J}}_{xn}(\mathbf{k}_s) + \frac{1}{\mathbf{k}_s} \begin{bmatrix} k_y \\ -k_x \end{bmatrix} \sum_{m=1}^{M} b_m\, \tilde{\mathbf{J}}_{ym}(\mathbf{k}_s) \tag{30}$$

$\tilde{\mathbf{J}}_{xn}(\mathbf{k}_s)$ and $\tilde{\mathbf{J}}_{ym}(\mathbf{k}_s)$ are the Fourier transforms of $\mathbf{J}_{xn}(\mathbf{r}_s)$ and $\mathbf{J}_{ym}(\mathbf{r}_s)$ respectively.

One of the main problems with the computational procedure is to overcome the complicated time-consuming task of calculating the Green's functions in the procedure of resolution by the moment method. The choice of the basis function is very important for a rapid convergence to the true values (Boufrioua & Benghalia, 2008; Boufrioua, 2009).

Many subsequent analyses involve entire-domain basis functions that are limited to canonical shapes such as rectangles, circles and ellipses. Recently, much work has been published regarding the scattering properties of microstrip antennas on various types of substrate geometries. Virtually all this work has been done with entire domain basis functions for the current on the patch.

For the resonant patch, entire domain expansion currents lead to fast convergence and can be related to a cavity model type of interpretation (Boufrioua, 2009; Pozar & Voda, 1987). The currents can be defined using a sinusoid basis functions defined on the whole domain, without the edge condition (Newman & Forrai, 1987; Row & Wong, 1993), these currents associated with the complete orthogonal modes of the magnetic cavity. Both x and y directed currents were used, with the following forms (Chew & Liu, 1988; Row & Wong, 1993)

$$\mathbf{J}_{xn}(\mathbf{r}_s) = \sin\left[\frac{n_1\,\pi}{a}\left(x+\frac{a}{2}\right)\right]\cos\left[\frac{n_2\,\pi}{b}\left(y+\frac{b}{2}\right)\right] \tag{31a}$$

$$\mathbf{J}_{ym}(\mathbf{r}_s) = \cos\left[\frac{m_1\,\pi}{a}\left(x+\frac{a}{2}\right)\right]\sin\left[\frac{m_2\,\pi}{b}\left(y+\frac{b}{2}\right)\right] \tag{31b}$$

The Fourier transforms of J_{xn} and J_{ym} are obtained from equation (27) and given by

$$\tilde{\mathbf{J}}_{xn}(\mathbf{k}_s) = \int_{-a/2}^{a/2} dx\, e^{-ik_x x} \sin\left(\frac{n_1\pi}{a}\left(x+\frac{a}{2}\right)\right) \times \int_{-b/2}^{b/2} dy\, e^{-ik_y y} \cos\left(\frac{n_2\pi}{b}\left(y+\frac{b}{2}\right)\right) \tag{32a}$$

$$\tilde{\mathbf{J}}_{ym}(\mathbf{k}_s) = \int_{-a/2}^{a/2} dx\, e^{-ik_x x} \cos\left(\frac{m_1\pi}{a}\left(x+\frac{a}{2}\right)\right) \times \int_{-b/2}^{b/2} dy\, e^{-ik_y y} \sin\left(\frac{m_2\pi}{b}\left(y+\frac{b}{2}\right)\right) \tag{32b}$$

Since the chosen basis functions approximate the current on the patch very well for conventional microstrips, only one or two basis functions are used for each current component.

Using the equations (32.a) and (32.b), the integral equation describing the field **E** in the patch can be discretized into the following matrix

$$\begin{bmatrix} (\mathbf{Z}_1)_{N\times N} & (\mathbf{Z}_2)_{N\times M} \\ (\mathbf{Z}_3)_{M\times N} & (\mathbf{Z}_4)_{M\times M} \end{bmatrix} \cdot \begin{bmatrix} (\mathbf{a})_{N\times 1} \\ (\mathbf{b})_{M\times 1} \end{bmatrix} = 0 \tag{33}$$

Where the impedance matrix terms are

$$(\mathbf{Z}_1)_{kn} = \iint_{-\infty}^{\infty} d\mathbf{k}_s \frac{1}{k_s^2}\left[k_x^2 \mathbf{G}^e + k_y^2 \mathbf{G}^h\right] \times \tilde{\mathbf{J}}_{xk}(-\mathbf{k}_s)\,\tilde{\mathbf{J}}_{xn}(\mathbf{k}_s). \tag{34a}$$

$$(\mathbf{Z}_2)_{km} = \iint_{-\infty}^{\infty} d\mathbf{k}_s \frac{k_x k_y}{k_s^2} [\mathbf{G}^e - \mathbf{G}^h] \times \tilde{\mathbf{J}}_{xk}(-\mathbf{k}_s) \tilde{\mathbf{J}}_{ym}(\mathbf{k}_s) \tag{34b}$$

$$(\mathbf{Z}_3)_{ln} = \iint_{-\infty}^{\infty} d\mathbf{k}_s \frac{k_x k_y}{k_s^2} [\mathbf{G}^e - \mathbf{G}^h] \times \tilde{\mathbf{J}}_{yl}(-\mathbf{k}_s) \tilde{\mathbf{J}}_{xn}(\mathbf{k}_s) \tag{34c}$$

$$(\mathbf{Z}_4)_{lm} = \iint_{-\infty}^{\infty} d\mathbf{k}_s \frac{1}{k_s^2} [k_y^2 \mathbf{G}^e + k_x^2 \mathbf{G}^h] \cdot \tilde{\mathbf{J}}_{yl}(-\mathbf{k}_s) \tilde{\mathbf{J}}_{ym}(\mathbf{k}_s) \tag{34d}$$

$\begin{bmatrix} (\mathbf{a})_{N\times 1} \\ (\mathbf{b})_{M\times 1} \end{bmatrix}$ are the unknown current modes on the patch

It should be noted that the roots of the characteristic equation given by (33) are complex, Muller's algorithm has been employed to compute the roots and hence to determine the resonant frequency.

The integration of the matrix elements in equations (34) must be done numerically, but can be simplified by conversion from the (k_x, k_y) coordinates to the polar coordinates (k_ρ, α) with the following change.

$$\iint_{-\infty}^{\infty} dk_s = \int_{-\infty}^{\infty} \int_{-\infty}^{\infty} dk_x \, dk_y = \int_0^{\infty} dk_\rho \, k_\rho \int_0^{2\pi} d\alpha \tag{35}$$

3. Antenna characteristics

Since the resonant frequencies are defined to be the frequencies at which the field and the current can sustain themselves without a driving source. Therefore, for the existence of nontrivial solutions, the determinant of the [z] matrix must be zero, i.e

$$\det(\mathbf{Z}(\omega)) = 0 \tag{36}$$

This condition is satisfied by a complex frequency $f = f_r + i\, f_i$ that gives the resonant frequency f_r, the half power bandwidth $BW = 2f_i / f_r$ and the other antenna characteristics.

Stationary phase evaluation yields convenient and useful results for the calculation of antenna patterns or radar cross section (Pozar, 1987).

The scattered far-zone electric field from the patch can then be found in spherical coordinates with components $\mathbf{E}_\theta$ and $\mathbf{E}_\phi$ and the results are of the form

$$\begin{bmatrix} \mathbf{E}_\theta \\ \mathbf{E}_\varphi \end{bmatrix} = i\, k_0 \frac{\exp(i\, k_0\, r)}{2\pi\, r} \begin{bmatrix} -\mathbf{G}^e & 0 \\ 0 & \mathbf{G}^h \cos\theta \end{bmatrix} \begin{bmatrix} \tilde{\mathbf{J}}^e \\ \tilde{\mathbf{J}}^h \end{bmatrix} \tag{37}$$

In the above equation, k_x and k_y are evaluated at the stationary phase point as

$$k_x = k_0 \sin\theta \cos\varphi \tag{38a}$$

$$k_y = k_0 \sin\theta \sin\varphi \tag{38b}$$

The radar cross section of a microstrip patch has recently been treated (Knott et al., 2004), although, there has been very little work on the radar cross section of patch antennas in the literature. The solution of the electric field integral equation via the method of moments has been a very useful tool for accurately predicting the radar cross section of arbitrarily shaped in the frequency domain (Reddy et al., 1998). In this chapter we will considere only monostatic scattering. The radar cross section computed from (Knott et al., 2004; Reddy et al., 1998), for a unit amplitude incident electric field the typical scattering results are of the form

$$\sigma_{\theta\theta} = 4\,\pi\,r^2 \left|E_\theta^{scat}\right|^2 \tag{39}$$

$\sigma_{\theta\theta}$ is $\hat{\theta}$-polarized backscatter from a unit amplitude $\hat{\theta}$ polarized incident field

$$\mathrm{RCS} = 10\,\log_{10}(\sigma_{\theta\theta}) \tag{40}$$

RCS is the radar cross section.

Computer programs have been written to evaluate the elements of the impedance matrix and then to solve the matrix equation. In Figure 2, comparisons are shown for the calculated and measured data presented by W. C. Chew and Q. Liu, deduced from table. I (Chew & Liu, 1988) and the calculated results from our model, for a perfectly conducting patches of different dimensions a(cm)×b(cm), without dielectric substrates (air) with thickness of 0.317cm. It is important to note that the normalization is with respect to f_0 of the magnetic wall cavity, the mode studied in this work is the dominant mode TM01. Our calculated results agree very well with experimental results, the maximum difference between the experimental and numerical results is less than 7%, this shift may indicate physical tolerances of the patch size or substrate dielectric parameters.

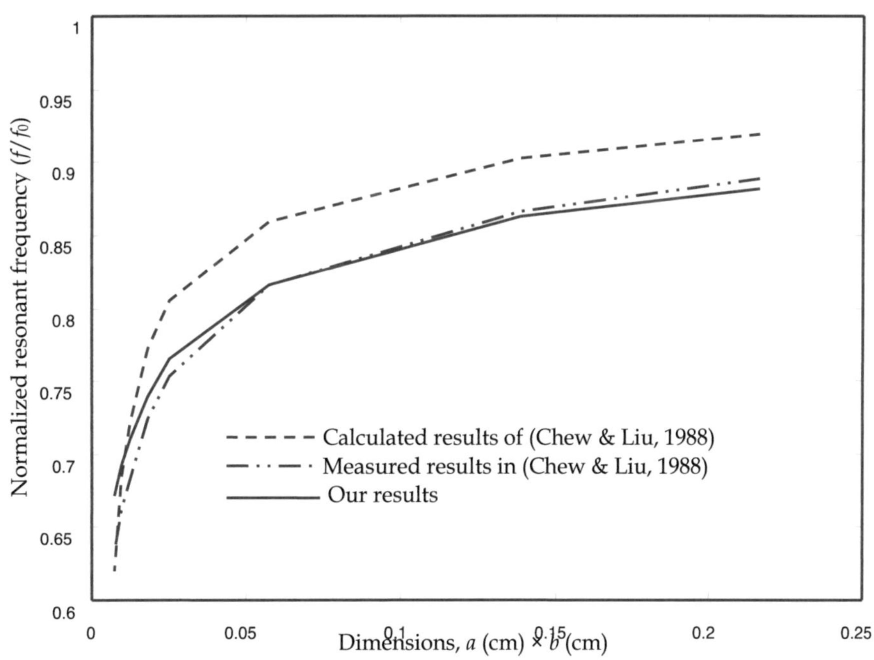

Fig. 2. Comparison between our calculated resonant frequencies and measured results versus the dimensions of the patch.

The influence of uniaxial anisotropy in the substrate on the resonant frequency, the quality factor and the half power band width of a rectangular microstrip patch antenna with dimensions *a*=1.5cm, *b*=1.0cm and the substrate has a thickness *h*=0.159 cm, for different pairs of relative permittivity (ε_x, ε_z) is shown in Table 1. The obtained results show that the positive uniaxial anisotropy slightly increases the resonant frequency and the half power band width, while the negative uniaxial anisotropy slightly decreases both the half power band width and the resonant frequency.

Comparisons are shown in table 2 for the calculated data presented by (Bouttout et al., 1999) and our calculated results for a rectangular patch antenna with dimensions *a*=1.9cm, *b*=2.29 cm and the substrate has a thickness *h*=0.159cm. The obtained results show that when the permittivity along the optical axis ε_z is changed and ε_x remains constant the resonant frequency changes drastically, on the other hand, we found a slight shift in the resonant frequency when the permittivity ε_x is changed and ε_z remains constant, these behaviors agree very well with those obtained by (Bouttout et al., 1999).

Uniaxial anisotropy type	Relative permittivity ε_x	Relative permittivity ε_z	Resonant frequency Ghz	Band width BW %	Quality factor Q
isotropic	2.35	2.35	8.6360194	9.0536891	11.0452213
isotropic	7.0	7.0	5.2253631	3.1806887	31.4397311
positive	1.88	2.35	8.7241626	9.1377564	10.9436053
negative	2.82	2.35	8.5537694	8.9779555	11.1383933
negative	8.4	7.0	5.1688307	3.1550166	31.6955535
positive	5.6	7.0	5.2869433	3.2124019	31.1293545

Table 1. Resonant frequency, band width and quality factor for the isotropic, positive and negative uniaxial anisotropic substrates

ε_x	ε_z	AR	Resonant frequencies (Ghz)	
			(Bouttout et al., 1999)	Our results
2.32	2.32	1	4.123	4.121
4.64	2.32	2	4.042	4.041
2.32	1.16	2	5.476	6.451
1.16	2.32	0.5	4.174	4.171
2.32	4.64	0.5	3.032	3.028

Table 2. Dependence of resonant frequency on relative permittivity (ε_x , ε_z)

The anisortopic ratio $AR = \varepsilon_x / \varepsilon_z$

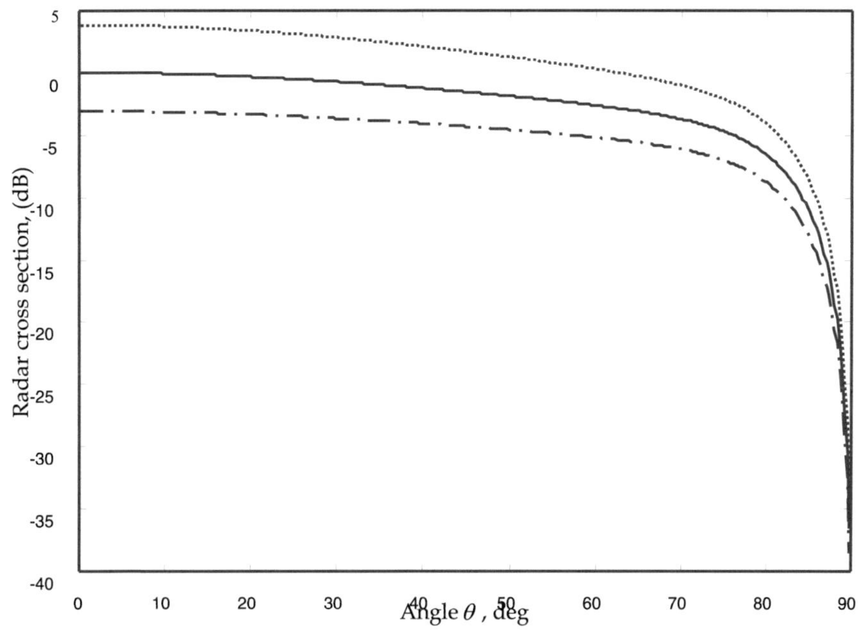

(a) ε_z changed, ——— $\varepsilon_x = \varepsilon_z = 5$, —·—· $\varepsilon_x = 5$, $\varepsilon_z = 6.4$, ············ $\varepsilon_x = 5$, $\varepsilon_z = 3.6$.

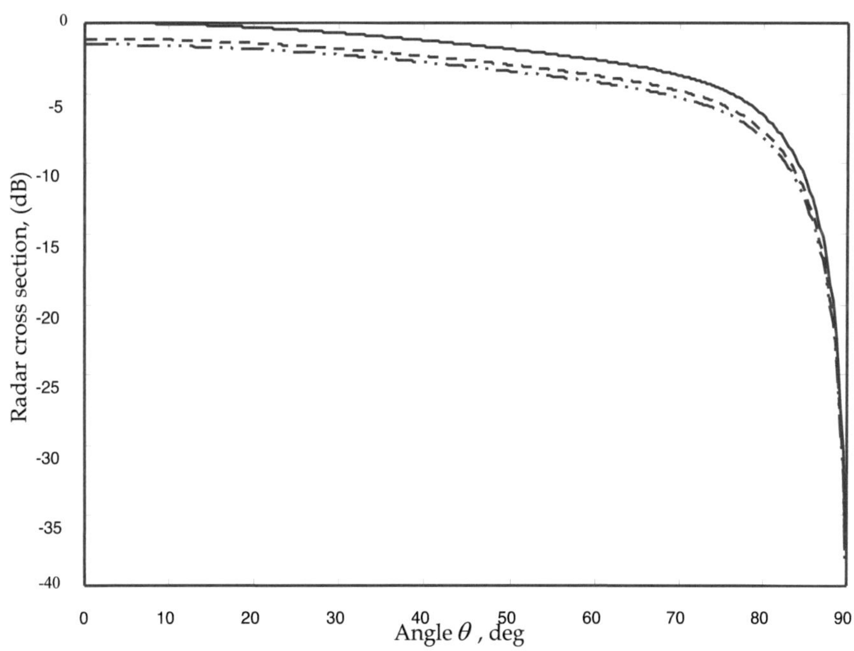

(b) ε_x changed, ——— $\varepsilon_x = \varepsilon_z = 5$, - - - - $\varepsilon_x = 3.6, \varepsilon_z = 5$, —··— $\varepsilon_x = 6.4$, $\varepsilon_z = 5$.

Fig. 3. Normalized radar cross section versus angle θ for the isotropic, positive uniaxial anisotropic and negative uniaxial anisotropic substrates.

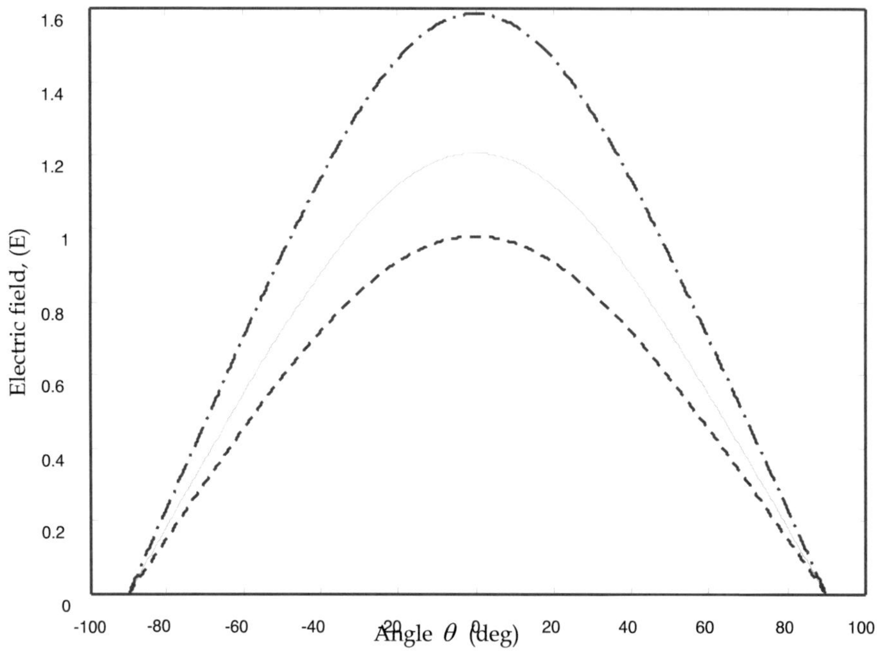

(a) ε_z changed, ——— $\varepsilon_x = \varepsilon_z = 5$, - - - - $\varepsilon_x = 5, \varepsilon_z = 6.4$, —·— $\varepsilon_x = 5, \varepsilon_z = 3.6$

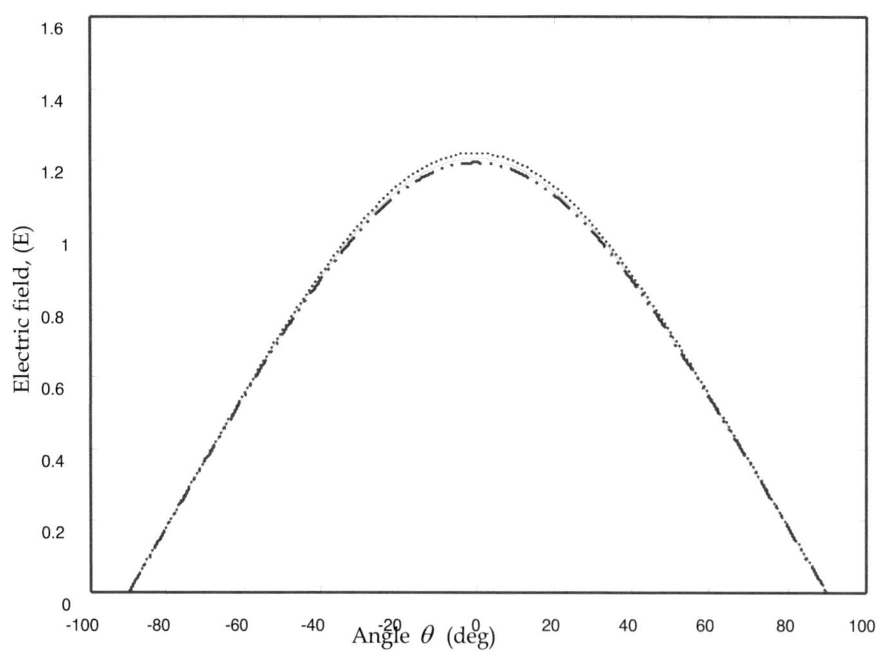

(b) ε_x changed, ——— $\varepsilon_x = \varepsilon_z = 5$, ... $\varepsilon_x = 3.6$, $\varepsilon_z = 5$, —··— $\varepsilon_x = 6.4$, $\varepsilon_z = 5$.

Fig. 4. Radiation pattern versus the angle θ for the isotropic, positive uniaxial anisotropic and negative uniaxial anisotropic substrates.

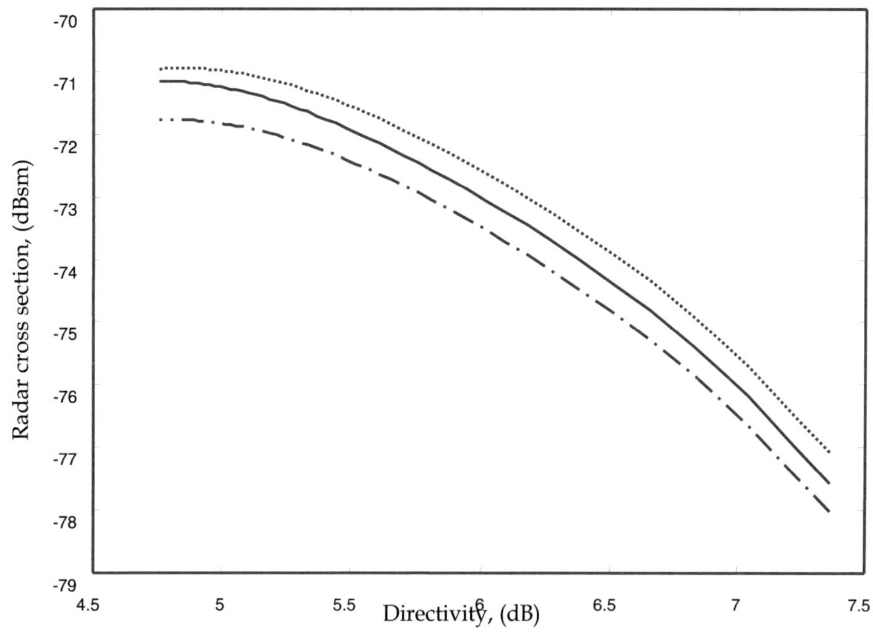

(a) ε_z changed, ——— $\varepsilon_x = \varepsilon_z = 5$, —·— $\varepsilon_x = 5$, $\varepsilon_z = 6.4$, ·············· $\varepsilon_x = 5$, $\varepsilon_z = 3.6$.

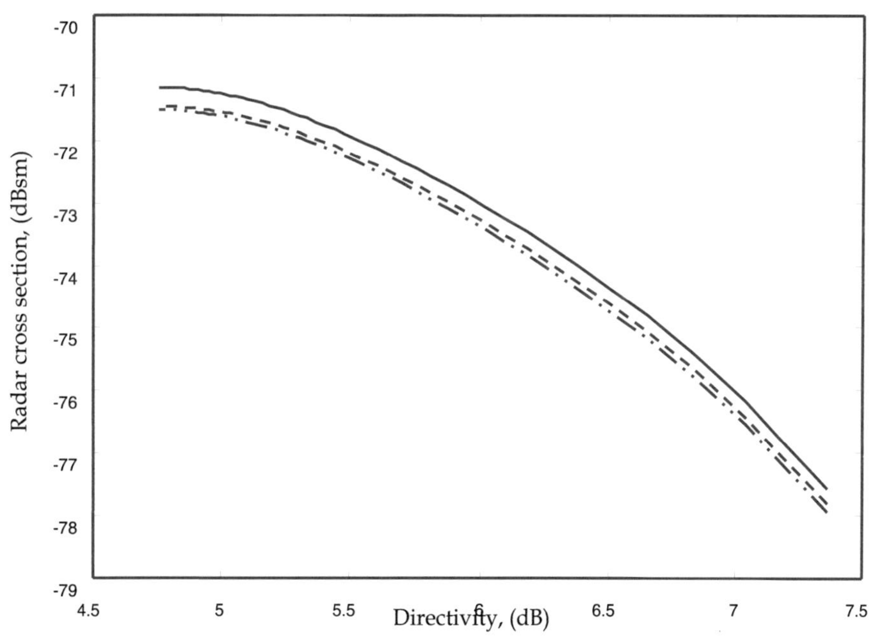

(b) ε_x changed, ——— $\varepsilon_x = \varepsilon_z = 5$, - - - - $\varepsilon_x = 3.6, \varepsilon_z = 5$, —··— $\varepsilon_x = 6.4$, $\varepsilon_z = 5$.

Fig. 5. Radar cross section versus the directivity for the isotropic, positive uniaxial anisotropic and negative uniaxial anisotropic substrates.

Figures 3 and 4 show the influence of uniaxial anisotropy in the substrate on the radiation and the radar cross section displayed as a function of the angle θ at $\varphi = 0°$ plane and at the frequency 5.95 Ghz, where the isotropic($\varepsilon_z = \varepsilon_x$), positive uniaxial anisotropic ($\varepsilon_z > \varepsilon_x$) and negative uniaxial anisotropic substrates ($\varepsilon_z < \varepsilon_x$) are considered, a rectangular patch antenna with dimensions *a*=1.5cm, *b*=1.0 cm is embedded in a single substrate with thickness *h*=0.2 cm. The obtained results can be seen to be the same as discussed previously in the case of the resonant frequency, moreover the permittivity ε_z along the optical axis is the most important factor in determining the resonant frequency, the radiation and the radar cross section when the pair (ε_x, ε_z) changes.
The same remarks hold for the variation of the radar cross section versus the directivity figures (5. a, b).
It is worth noting that the radar cross section in equation (39) is calculated at one frequency. If one needs the radar cross section over a frequency range, this calculation must be repeated for the different frequencies of interest.

4. Conclusion

The moment method technique has been developed to examine the resonant frequency, the radiation, the half power band width, the directivity and the scattering radar cross section of a rectangular microstrip patch antenna. The boundary condition for the electric field was used to derive an integral equation for the electric current, the Galerkin's procedure of the moment method with entire domain sinusoidal basis functions without edge condition was investigated, the resulting system of equations was solved for the unknown current modes on the patch, it is important to note that the dyadic Green's functions of the problem were efficiently determined by the (TM, TE) representation. Since there has been a little work on the scattering radar cross section of patch antennas including the effect of uniaxial anisotropic substrate in the literature, a number of results pertaining to this case were presented in this chapter. The obtained results show that the use of the uniaxial anisotropy substrates significantly affects the characterization of the microstrip patch antennas. The numerical results indicate that the resonant frequency and the half power band width are increased due to the positive uniaxial anisotropy when ε_x change, on the other hand, decreased due to the negative uniaxial anisotropy. Moreover the ε_z permittivity has a stronger effect on the resonant frequency, the radiation and the radar cross section than the permittivity ε_x. Also the effect of the uniaxial substrate on the radar cross section versus the directivity was presented. Accuracy of the computed techniques presented and verified with other calculated results.
A new approach for enhancement circular polarisation output in the rectangular patch antenna based on the formulation presented in this chapter is in progress and will be the subject of a future work, when two chamfer cuts will be used to create the right or the left handed circular polarisation by exciting simultaneously two nearly degenerate patch modes. The analysis presented here can also be extanded to study a biaxially anisotropic substrate and the effect of dielectric cover required for the protection of the antenna from the environment. Also the radar cross section monostatic and bistatic and the other antenna characteristics will be study for this case in our future work.

5. References

Bhartia, P.; Rao, K. V. S.& Tomar, R. S. (1991). *Millimeter Wave Microstrip and Printed Circuit Antennas*, Publisher, Artech House, ISBN 0-89006-333-8, Boston, London

Boufrioua, A. & Benghalia, A. (2008). Radiation and resonant frequency of a resistive patch and uniaxial anisotropic substrate with entire domain and roof top functions, *Elsevier, EABE, Engineering Analysis with Boundary Elements*, Vol., 32, No. 7, (March 2008), (591-596), ISSN 0955-7997

Boufrioua, A. (2009). Resistive Rectangular Patch Antenna with Uniaxial Substrate. In: *Antennas: Parameters, Models and Applications* (Ed. Albert I. Ferrero), (163-190), Publisher, Nova, ISBN 978-1-60692-463-1, New York

Bouttout, F.; Benabdelaziz, F.; Benghalia, A.; Khedrouche, D. & Fortaki, T. (1999), Uniaxially Anisotropic Substrate Effects on Resonance of Rectangular Microstrip Patch Antenna, *Electronics Letters*, Vol. 35, No. 4, (February 1999), (255-256), ISSN 0013-5194

Chew, W. C. & Liu, Q. (1988), Resonance Frequency of a Rectangular Microstrip Patch, *IEEE Transactions on Antennas and Propagation*, Vol. 36, No. 8, (August 1988), (1045-1056), ISSN 0018-926X

Damiano, J. P. & Papiernik, A. (1994), Survey of Analytical and Numerical Models for Probe-Fed Microstrip Antennas, *IEE proceeding. Microwaves, Antennas and Propagation*, Vol. 141, No. 1, (February 1994), (15-22), ISSN 1350-2417

Knott, E. F.; Shaeffer, J. F. & Tuley, M. T. (2004). *Radar Cross Section*, Publisher SciTech, ISBN 1-891121-25-1, Raleigh, NC

Mirshekar-Syahkal, D. (1990). *Spectral Domain Method for Microwave Integrated Circuits*, Publisher, John Wiley & Sons Inc, ISBN 0-86380-099-8, New York

Newman, E. H. & Forrai, D. (1987). Scattering from a Microstrip Patch, *IEEE Transactions on Antennas and Propagation*, Vol. 35, No. 3, (March 1987), (245-251) ISSN 0018-926X

Pozar, D. M. (1987). Radiation and Scattering from a Microstrip Patch on a Uniaxial Substrate, *IEEE Transactions on Antennas and Propagation*, Vol. 35, No. 6, (June 1987), (613-621), ISSN 0018-926X

Pozar, D. M. & Voda, S. M. (1987). A Rigorous Analysis of a Microstripline Fed Patch Antenna, *IEEE Transactions on Antennas and Propagation*, Vol. 35, No. 12, (December 1987), (1343-1350), ISSN 0018-926X

Reddy, V. M.; Deshpand, D.; Cockrell, C. R. & Beck, F. B. (1998). Fast RCS Computation Over a Frequency Band Using Method of Moments in Conjuction with Asymptotic Waveform Evaluation Technique, *IEEE Transactions on Antennas and Propagation*, Vol. 46, No. 8, (August 1998), (1229-1233), ISSN 0018-926X

Row, J. S. & Wong, K. L. (1993). Resonance in a Superstrate-Loaded Rectangular Microstrip Structure, *IEEE Transactions on Antennas and Propagation*, Vol. 41, No. 8, (August 1993), (1349-1355), ISSN 0018-9480

Tulintsef, A. N.; Ali, S. M. & Kong, J. A. (1991). Input Impedance of a Probe-Fed Stacked Circular Microstrip Antenna, *IEEE Transactions on Antennas and Propagation*, Vol. 39, No. 3, (March 1991), (381-390), ISSN 0018-926X

Wong, K. L.; Row, J. S.; Kuo, C. W. & Huang, K. C. (1993). Resonance of a Rectangular Microstrip Patch on a Uniaxial Substrate, *IEEE Transactions on Microwave Theory and Techniques*, Vol., 41 No. 4, (April 1993), (698-701), ISSN 0018-9480

3

Artificial Materials based Microstrip Antenna Design

Merih Palandöken
Berlin Institute of Technology
Germany

1. Introduction

The demand on the portable mobile devices is increasing progressively with the development of novel wireless communication techniques. In that respect, novel design methods of wireless components have to be introduced to fulfill many performance criteria simultaneously. Compact size, light weight, low profile and low cost are now quite important challenges for a system designer to accomplish in the design and performance enhancement of every wireless mobile component. One of these wireless components to be enhanced inevitably and matched with these new challenges in any communication system is the antenna. Therefore, this chapter is mainly dealing with the novel antenna design method, which is based on artificial materials. How to engineer artificially the ground plane, substrate or radiating part of the antenna as a solution for better antenna designs are explained from basic electrical limitations upto proposed design solutions.
In this chapter, basic concepts in electrically small antennas are introduced in Section 2 to determine the fundamental performance limitations of small antennas. The minimum Q and maximum gain of electrically small antennas, which are the main targets in the antenna design, are addressed to point out the effect of electromagnetic material parameters and physical dimensions on the antenna radiation performance. In Section 3, the concept of electromagnetic parameter engineering and how to engineer substrate effective parameters are introduced along with the design of broadband artificial material loaded dipole antenna in detail as a design example to understand how to combine these artificial materials with the radiating sections for better antenna performance than conventional alternative designs. In Section 4, rather than manipulation of effective parameters of substrate, the concept of artificial ground plane, like high impedance surface (HIS), electromagnetic bandgap structures (EBG) for more directive and high gain antennas is explained with a numerical example of the design of electrically small artificial magnetic conductor (AMC) ground plane. In the last section, in the microstrip antenna design with artificial materials, the radiating part of the antenna is structured with electrically small self-resonant cells to design electrically smaller, higher gain antennas in a certain physical dimension than the conventional antennas. Each above mentioned concepts are detailed with pioneering references for more interested readers.

2. Fundamental limits of small antennas

Electrically small antenna design has long been the current trend as one of the most significant and interested research topics in microwave community. The large physical dimensions of conventional $\lambda/2$ dipole or $\lambda/4$ monopole antennas in the short-, or long-wave radio communication has further triggered to miniaturize the antenna dimensions in new generation applications. Especially, nowadays the demand on multifunctional complex systems imposes to design miniaturized mobile terminals with a predetermined bandwidth. One of the challenging issue in the design of these mobile systems is to miniaturize their radiating parts with a prerequired gain for data transmission and navigation. The small antennas have an additional advantage, especially in wireless/contactless measurement systems,of not to influence the measured field for more reliable measurements. Hence, to design compact, high efficient/gain, broadband antennas is the current challenge to be overcome for any high performance wireless systems.

An electrically small antenna is commonly defined as an antenna occupying a small fraction of one radiansphere, which circumscribes the maximum dimension of the antenna (Wheeler, 1975;Wheeler, 1959;Wheeler, 1947;Johnson,1993). The radianphere is by definition the spherical volume having the radius of $\lambda/2\pi$. This defines logically the maximum dimension to be smaller than λ/π, 0.314λ where λ is the free space wavelength. A conventional upperlimit for the greatest dimension is less than one-quarter wavelength (including any image resulting from the ground plane), which is approximately 0.785 times radius of the radiansphere. This results the electrically small antenna to have maximum ka value of 0.78, where k is the free space wave number and a is the radius of the imaginary enclosing sphere. The main reason to have the radiansphere as a logical reference in the above definition is that because around a small antenna, this volume is the space occupied mainly by the stored energy of its electrical or magnetic field for the electrical dipole and magnetic dipole respectively. This region is also termed as reactive near field region. The radius of the enclosing radiansphere is also the distance at which the reactive part of the input impedance is equal to the real part for infinitesimally small electric and magnetic dipole. Far distances away from this radius, the real part is more dominating than the imaginary part, leading to have more plane wave formed wave propagation like in the transition from Fresnel to Fraunhofer regime. Thus, an electrically small antenna is essentially a capacitive electric or inductive magnetic dipole or possibly a self resonant antenna, which is the combination of both types.

The operation bandwidth of any antenna type can be determined by the frequency dependence of feed point impedance, which is calculated by the surface integration of transversal field components over the spherical surface of radius a. The feed point impedance is given by $Z(\omega) = R(\omega) + j X(\omega)$, where ω is the radian frequency. The resistive part of input impedance includes both the radiation and conduction loss. The imaginary part is proportional to the difference of time averaged stored magnetic and electric energy outside the spherical volume of radius a. This could be easily deduced from the following Poynting formulation,

$$-\frac{1}{2}\oint_S \left(\vec{E}x\overline{H^*}\right)\bullet dA = j2\omega\int_V\left(\frac{1}{4}\mu|H|^2-\frac{1}{4}\varepsilon|E|^2\right)dV+\frac{1}{2}\int_V\left(\vec{E}\bullet\overline{J^*}\right)dV=\frac{1}{2}I^2Z(\omega) \qquad (1)$$

The impedance bandwidth of matched small antennas is often characterized by its Q at the resonant frequency, which has an achievable theoretical lower bound and inversely proportional to the respective fractional bandwidth. Therefore, no electrically small antenna can exhibit a Q less than its theoretical lower bound or bandwidth larger than a specific upper bound. The antenna Q is generally defined as, in (2), where We and Wm are time averaged stored electric and magnetic energy outside the enclosing sphere of radius a, respectively and P_{loss} is the sum of total radiated and metallic loss power at the resonance frequency, ω_0 .

$$Q = \begin{cases} 2\omega_0 \dfrac{W_e}{P_{loss}}, & W_e > W_m \\ 2\omega_0 \dfrac{W_m}{P_{loss}}, & W_m > W_e \end{cases} \tag{2}$$

This formulation is actually the same formulation, which could be derived from Poynting formulation in (1) by taking the ratio of imaginary and real part of complex power for high Q antennas. In the theoretical lower bound calculation of Q, an arbitrary current distribution is optimized to excite all possible higher order orthogonal spherical TM modes with respective rotation symmetrical modal near fields for a vertical wire antenna. The modal Q_n of vertically oriented omnidirectional antenna can be calculated in the terms of modal tangential field quantities on the spherical surface of the reactive near field region of radius a as (Chu,1948),

$$\begin{aligned} Q_n &= \frac{2\omega W_n}{P_n} = \frac{1}{2} \mid \rho h_n \mid^2 \left[\rho \frac{dX_n}{d\rho} - X_n \right] \\ X_n &= \left[\rho j_n \left(\rho j_n\right)' + \rho n_n \left(\rho n_n\right)' \right] \left|\rho h_n\right|^{-2} \end{aligned} \tag{3}$$

where h_n is the nth order spherical Hankel function of second kind, which is defined as $h_n = j_n - jn_n$, ρ is ka, which is calculated on a spherical closed surface. The antenna Q resulting from the spherical wave function expansion method is lower than the real value since the stored energy inside the sphere has been totally ignored in the calculation. The above relation is plotted in Figure 1. As it could be understood from Figure 1, for a specific operation frequency and excited mode of electrically small antenna, to increase the antenna's maximum dimension 2a or its physical volume, reduces the total stored energy outside the antenna volume along with the increase in the radiation resistance, hence Q_n. .The reason of reduction in the amount of stored energy or increase in the radiation resistance could be deduced by the introduction of additional radiating sections which enhances the specific mode to propagate with real impedance and less reactive part. Especially, for electrically small antennas, to increase the antenna length leads the excited near field to be more bounded in the antenna volume due to the small ratio of reactive near field region radius, $0.62\sqrt{\frac{D^3}{\lambda}}$ to antenna size. This results modal stored energy, which is calculated in the spherical volume outside the radius a, to be reduced.

On the other hand, higher order modes result the antenna to have higher Q for the specific size. Thus, to excite higher order modes is in practice quite challenging due to the difficulty in matching high reactive input impedance. The main conclusion to be conducted in minimum Q calculation is that Q is only to be minimized under one mode excitation of the

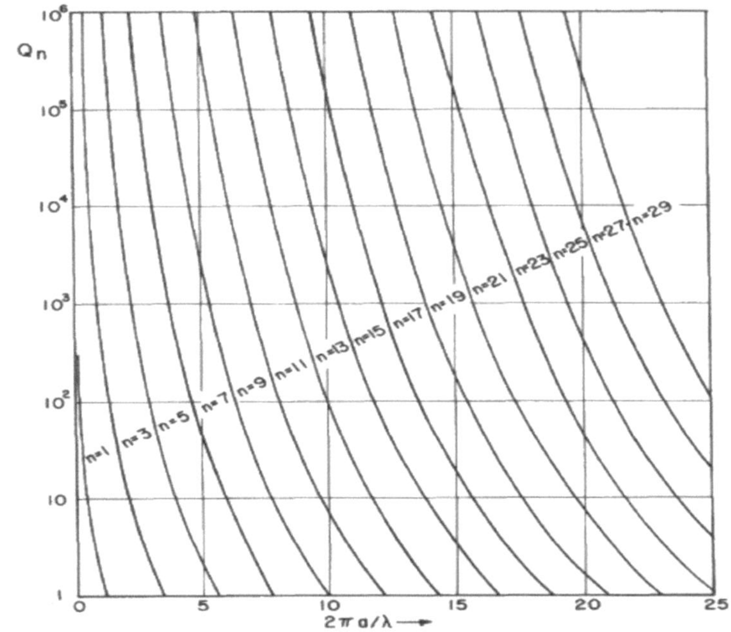

Fig. 2.1. Q_n of the equivalent circuit of TM_n or TE_n wave (Chu,1948)

antenna without multimodal operation. Since Q1 has the lowest amplitude, the antenna which generates a field outside the sphere corresponding to that of an infinitesimally small dipole has potentially the broadest bandwidth of all antennas (Chu, 1948). Therefore, the optimum method to obtain broadbandwidth operation for a specific antenna size is to operate the antenna with lowest of all possible modes. The theoretical lower bound of Q for small directional and omnidirectional antennas is expressed as (Geyi, 2003)

$$Q_{small}^{min} = \left(\frac{1}{2(ka)^3} + \frac{1}{ka} \right) \tag{4}$$

In general, for a tuned antenna of any size exhibiting a single resonance frequency, Q can be calculated from the input feeding impedance as (Yaghjian & Best,2005)

$$Q(\omega_0) = \frac{\omega_0}{2R(\omega_0)} \sqrt{R'(\omega_0)^2 + \left(X'(\omega_0) + \frac{|X(\omega_0)|}{\omega_0} \right)^2} \tag{5}$$

where $R'(\omega_0)$ and $X'(\omega_0)$ are the frequency derivatives of real and imaginary parts of feed point impedance of unmatched antenna at the resonance frequency. As it can be deduced from the formulation, Q can be minimized with higher radiation resistance, for instance, by the design of self-resonating elements, which has an additional advantage of compensated reactive part, $X(\omega_0) = 0$ to minimize the numerator term and increase the antenna overall gain and efficiency without introducing any conduction loss due to the matching network. It is an important issue because it is generally known that in electrically small antennas the loss resistance within the matching network often exceeds the radiation resistance, resulting in low overall efficiency. The efficiency reduction in mismatch loss achieved with the impedance match often exceeds the increase in loss due to the matching components and better overall performance is achieved relative to the isolated mismatched antenna. Thus, one strategy in the antenna design could be to design self resonating structures to increase the radiation resistance and enhance the operation bandwidth in comparison to the non resonating capacitive or inductive antennas. $R'(\omega_0)$ in (5) is mostly quite small in comparison to the other terms.

However for the case of gain optimization, it is well known that there is no mathematical limit to the gain to be obtained from currents confined to an arbitrary small volume (Chu,1948). But a small sized antenna with extremely high gain produces high field intensity in the vicinity of the antenna, which results in high heat loss or high stored energy. This leads the antenna gain to be limited with an upper bound. The maximum gain obtainable is calculated as Gmax= N(N+2) by artificially truncating the spherical wave function expansions of the fields to the order N (Harrington, 1960). Hence as N increases (equivalently the antenna complexity increases) the maximum gain and practical difficulty in the matching network of high reactive antenna increases. Therefore, in principle, the main target is to excite higher number of modes in the antenna by imposing the specified optimum current distribution, which increases the radiation resistance and hence gain. However, as it is stated, this leads in turn these higher order modes to attenuate more in comparison to the lower order modes as the evanescent fields, leading large stored energy in comparison to the radiation loss. This could be quite easily deduced in analogy to the reduction of propagation constant with the higher order modes in dielectric waveguides, which increases the leaking power of the guided field from the substrate to the free space, hence radiated power. In other words to express, the gain enhancement with the account of high number of in-phase excited modes and hence the superposition of radiated field of each mode could counteract to the amount of stored energy in the near field, because they could not propagate effectively into the free space. It is because that these evanescent waves have pure capacitive or inductive impedance depending on the excitation of TM or TE modes, respectively.

Therefore, due to the trade-off between gain and Q of an antenna, there is one additional parameter to be maximized for optimum antenna design. This parameter is the ratio of gain (G) to Q. The maximum G/Q ratio for the small antennas is expressed as,

$$\max \frac{G}{Q}\Big|_{dir}^{small} \approx \frac{6(ka)^3}{2(ka)^2+1}$$
$$\max \frac{G}{Q}\Big|_{omni}^{small} \approx \frac{3(ka)^3}{2(ka)^2+1} \tag{6}$$

Antenna gain can be alternatively expressed with an additional parameter called radiation power factor (PF) in terms of antenna input impedance (Wheeler, 1947). It is descriptive of the radiated power from an electrically small antenna taking a much larger value of reactive power. This term is applicable to either kind of electrically small electric or magnetic reactor, and its small value is limited to some measure of the physical size in either kind. It is equal to 1/Q and defined as the ratio of input resistance to reactance at the resonant frequency. For any shape of electrically small antenna, the radiation PF at one frequency is proportional to its volume. Moreover, it is nearly equal for electric and magnetic dipole type radiators of nearly equal volume. However, radiation PF is somewhat different from the radiation parameters, which have been calculated before. The main difference is that in principle the effective volume of any small electric or magnetic antenna is determined by mapping the near field distribution of the antenna around the radiansphere to that of the equivalent circular capacitor shaped electric and spiral inductor formed magnetic antenna, which has not been done in previous calculations. The integration volume in the calculation of previously described parameters is the volume outside the smallest sphere enclosing the antenna, which is unbounded, in contrast to that, for radiation PF calculation, the reference volume is the volume inside the radiansphere, which is clearly bounded. The radiation PF can be alternatively defined in terms of effective volume V' and volume of radiansphere V as (Wheeler, 1947; Johnson,1993)

$$p=\frac{2}{9}\frac{V'}{Vs}=\frac{2}{9}\left(\frac{2\pi a'}{\lambda}\right)^3 \tag{7}$$

Thus, one method to enhance the radiation PF of small antenna is to increase the effective volume of electric or magnetic dipoles by decreasing electric permittivity of dielectric or increasing magnetic permeability of inductive core, respectively. The logical approach of this method is to eliminate or minimize the avoidable stored energy inside the magnetic or electric core by leaving unavoidable amount of stored energy outside the capacitor or inductor core but mostly inside the radiansphere. The accompanying enhancement of radiation resistance with decreasing permittivity and increasing permeability is also consistent with the medium parameter dependence of radiation resistance of Hertzian electric and magnetic dipoles.

After having pointed out the influence of material parameters on the antenna performance, how to engineer the substrate material to design broadband microstrip dipole antenna is explained with numerical simulations in the next section.

3. Artificial substrate design

Metamaterials are artificially structured materials providing electromagnetic properties not encountered in nature (Veselago, 1968; Engheta & Ziolkowski, 2006, Caloz & Itoh, 2005, Eleftheriades & Balmain, 2005). The electrodynamics of hypothetical materials having simultaneously negative permittivity and permeability was first theoretically predicted by Veselago (Veselago, 1968). These materials are termed as "left handed materials (LHM) " due to the left-handedness of electric, magnetic field and wave vector. A left handed material was first implemented in a two dimensional periodic array of split ring resonators and long wire strips by Smith (Smith et al., 2000). The logical approach was to excite the split ring resonators and wire strips in order to force the structure to behave like magnetic and electric dipoles, respectively. Since then, there have been large numbers of experimental investigations on the observation of this phenomenon. The effective electromagnetic parameters were also retrieved experimentally and numerically from the transmission and reflection data (Chen et al., 2006; Smith et al., 2005; Alexopoulos et al., 2007; Smith et al., 2000). Rather than split ring resonators and wire strips, the left handed feature can also be

realized with periodic loading of conventional microstrip transmission lines with series capacitors and shunt inductors [12],[20]. Many microwave circuits have been implemented by using this strategy such as compact broadband couplers, broadband phase shifters, compact wideband filters, compact resonator antennas, LH leaky wave antennas, which have a very unique property of backfire-to-endfire frequency scanning capability with broadside radiation, which is not possible for RH leaky wave antennas (Caloz & Itoh, 2005, Eleftheriades & Balmain, 2005).
In this section, the design of a novel microstrip dipole antenna by artificially engineering the substrate material with left-handed metamaterials (LHM) is explained for compact wideband wireless applications. The broadband microstrip antenna is composed of a dipole and six LHM unit cells. The antenna is matched to 50Ω with the stepped impedance transformer and rectangular slot in the truncated ground plane. By the utilization of phase compensation and coupled resonance feature of LHMs, the narrowband dipole antenna is operated at broader bandwidth. First in Section 3.1, the structure of the electrically small LHM unit cell is described. A one dimensional dispersion diagram is numerically calculated by Finite Element Method (FEM) to prove the lefthandedness and respective negative refractive index of the proposed unit cell. The effective permittivity and permeability are also retrieved from the reflection and transmission data of one unit cell. In Section 3.2, the configuration and operation principle of the proposed antenna are explained. The simulated and measured return loss, radiation pattern and numerically computed radiation parameters are presented.

3.1 LHM unit cell design

The negative material parameters are synthesized by the simultaneous excitation of electric and magnetic dipoles in the LHM unit cell. The original structure proposed in (Smith et al., 2000) consists of a bulky combination of metal wires and split ring resonators (SRR) disposed in alternating rows. The excited wires and SRRs are electric and magnetic dipoles, thus creating the left-handed behavior. Because the typical LHM designs are inherently inhomogeneous, novel strategies to miniaturize the unit cell with different topological and geometrical methods are important.

3.1.1 Description of the structure

LHM behavior implies small unit cells as compared to the free space wavelength λo. The upper limit of the unit cell size is one fourth of the guided wavelength (Caloz & Itoh, 2005). One well-known method of miniaturization is to increase the coupling between the resonators. This strategy was chosen for the proposed LHM unit cell, Figure 3.1 with geometrical parameters in (Palandöken et al. 2009), in which wire strips and spiral resonators (SR) are directly connected with each other, on both sides of the substrate. Further, instead of SRRs as in the original proposals, SRs are used, which have half the resonance frequency of SRRs (Baena et al., 2004). In the design, the geometrical parameters of the front and back side unit cells are the same, except shorter wire strip length on the front side. Different strip wire lengths lead to a smaller resonance frequency and larger bandwidth. The substrate material is nonmagnetic FR4-Epoxy with a relative permittivity of 4.4 and loss tangent of 0.02.
The unit cell size is 3x3.5 mm. The validity of the model is shown by retrieving the effective constitutive parameters from S parameters and by the opposite direction of group and phase velocity.

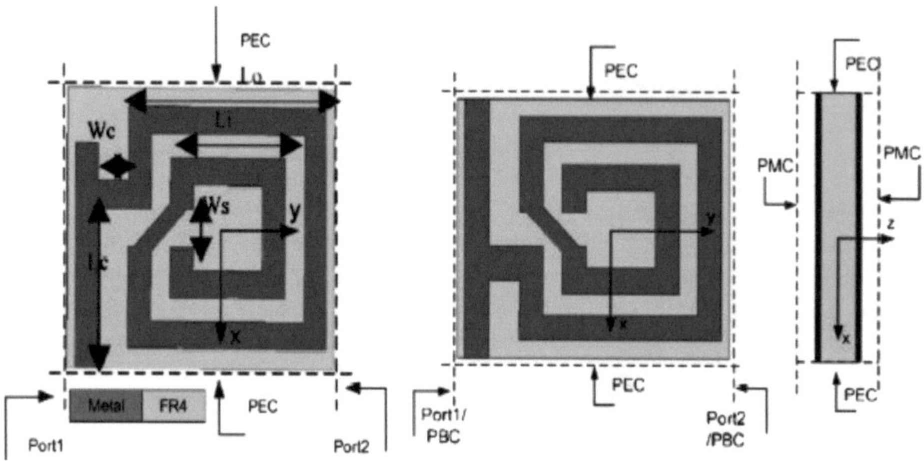

Fig. 3.1. LHM unit cell geometry. (a) Front and (b) back side of one LHM unit cell

3.1.2 Simulation results

To determine the frequency interval of left-handedness, a one dimensional Brillouin diagram is studied at first. In order to obtain the dispersion relation of the infinite periodic structure, the cells must be excited with the magnetic field perpendicular to the SR plane (z-direction), and the electric field in the direction of strip wires (x-direction), Figure 3.1. Therefore, the eigenfrequencies of a unit cell are calculated with perfect magnetic boundaries (PMC) in z-direction and perfect electric boundaries (PEC) in x-direction. Periodic boundary conditions (PBC) are imposed in y-direction. The simulation was done with the FEM based commercial software HFSS and is shown in Figure 3.2. Oppositely directed phase and group velocities are observed in the LH band between 2.15-2.56 GHz with 410MHz bandwidth, which proves additionally the negative refractive index of the proposed unit cell. Alternatively, the same unit cell structure but with longer strip wires on the front side leads to higher cutoff frequencies (2.58-2.65 GHz) and a narrow LH passband

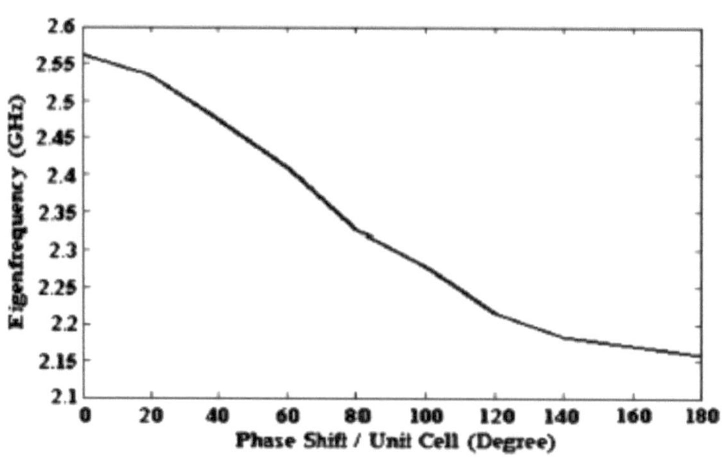

Fig. 3.2. Dispersion diagram of the proposed LHM structure

(69.7MHz). Also, if the front and back side are chosen identically, the LH passband is between 3.45-3.51 GHz, which is relatively narrow and which is at higher frequencies than for the proposed design. This explains the use of a shorter wire strip on the front side of the substrate, which reduces the resonance frequency and increases the bandwidth.

In addition to the dispersion diagram, the effective constitutive parameters are retrieved from the scattering parameters of a one cell thick LHM sample. Therefore, the lefthandedness of the unit cell is not only proved with the opposite phase and group velocities as in Figure 3.2, but also with the values and the sign of the retrieved parameters.

The reflection and transmission parameters are numerically calculated for x polarized and in y-direction propagating plane waves. PEC and PMC boundary conditions are imposed in x- and z- direction. The effective permittivity and permeability are retrieved from the simulated S parameters and shown in Figure 3.3.

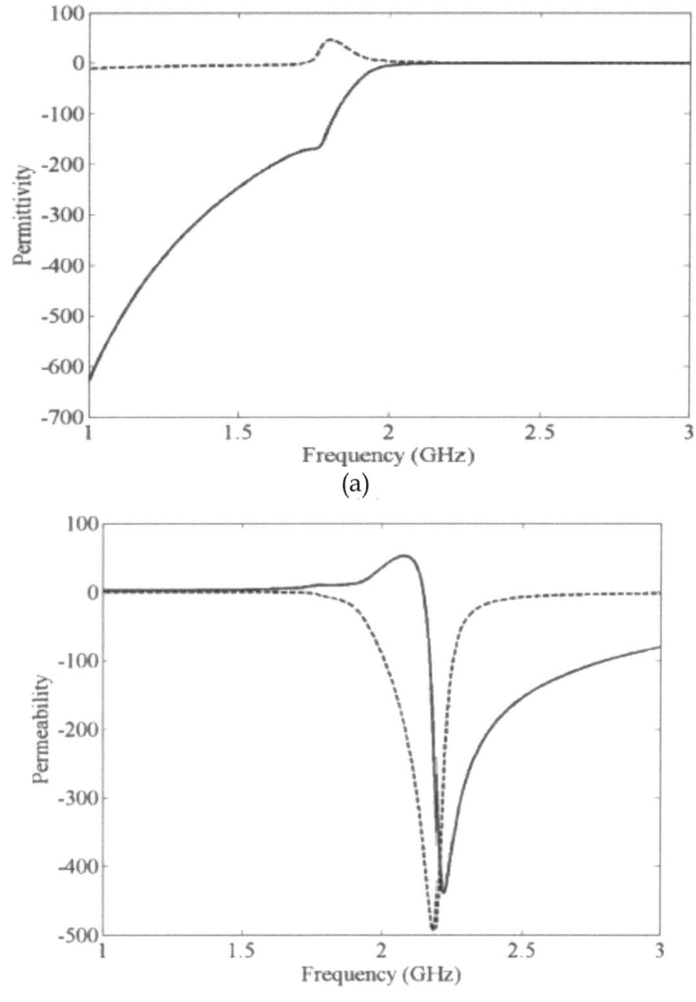

Fig. 3.3. Real (solid) and imaginary (dashed) part of retrieved effective parameters of LHM: (a) complex permittivity, (b) complex permeability

Therefore, by the introduction of metallic inclusions as wire strips and SRs on the substrate, permittivity and permeability of the material, composed of periodical arrangement of these cells can be engineered. This is the main motivation in the performance enhancement of antennas due to the controllable manipulation of substrate parameters. There are important issues to be discussed about the frequency dispersion of retrieved parameters. First of all, the retrieval procedure leads in general to satisfying results - an expected Lorentzian type magnetic resonance for μ - but unphysical artifacts occur such as a positive imaginary part of ε. The reason is that the homogenization limit has not been reached (Smith et al., 2005), although the unit cell is approximately 1/23 of the guided wavelength in the substrate. The anti-resonance of the real part of ε near 1.8 GHz leads consistently to a positive imaginary part. This is an inherent artifact for inhomogeneous, periodic structures because of the finite unit cell size. Secondly, there is a LH resonance near 1.94 GHz, which is smaller than the lower cutoff frequency in the Brillouin diagram and is attributed to the single cell simulation. The Bloch impedance of the infinitely periodic LHM is no longer valid for an isolated single cell. Recently, a new parameter retrieval procedure, which is based on two-port network formulation of one unit cell thick sample and virtual continuation of one cell periodically into infinite number of unit cells in the propagation direction by Bloch Theorem is introduced (Palandöken & Henke, 2009). This method will be detailed in Section 4. The LH band for retrieved parameters extends from 1.75 up to 2.55 GHz. It is in good correspondence with the simulated band in the range from 2.17 to 2.53 GHz in terms of the refractive indices calculated directly from the dispersion diagram in Figure 3.2.
The size of a unit cell is approximately 1/43 of λ_o at 2 GHz, which is directly connected, in first approximation and neglecting all coupling, to the total metallic length from the open circuited SR to the short circuited wire strip. The varying degree of coupling between the resonators shifts and broadens the transmission band. If the electrically small unit cells are excited by their eigencurrents, they represent effective radiating elements and are key elements for the future aspects in the antenna miniaturization.

3.2 Antenna design

3.2.1 Operation principle

The operation principle of the antenna depends on the radiation of the dipole antenna and the excitation of LHM unit cells with the dipole field. The excitation of LH cells in their eigenmodes causes the individual electric and magnetic dipoles to be coupled in the same way as in the eigenmode simulation. These unit cell dipoles are also radiation sources in addition to the exciting dipole antenna even though they are designed as loads for the dipole. The magnetic and electric dipole moments are expressed by the surface current density as in (Li et al., 2006). For each unit cell, the electric and magnetic dipoles are simultaneously excited in principle. However, the magnetic dipoles are more effective than the electric ones. At first, magnetic dipole fields do not cancel in the far field because of inplane electrical coupling among the cells on the front and back side. The second reason is that the current on the back side strip wire has partially opposite directions and do not excite the electric dipole as effectively as the magnetic dipole. As a last reason, the surface current on the back side unit cell spirals in the same direction as the surface current on the front side unit cell, thus doubling the magnetic dipole moment. In that respect, front and back side cells are mainly magnetically coupled and the back side cells can be considered as the artificial magnetic ground plane for the front side cells, which will be discussed in

Section 4. It also follows from the Lorentzian type magnetic resonance in Figure 3.3.b, which is the dominating resonance in the retrieved effective parameters. However, the antenna radiates mainly in the dipole mode, which is the reason why we call it as an LHM loaded dipole antenna.

3.2.2 Antenna design

As a first step in the antenna design, the front and back side unit cells were connected symmetrically with adjacent cells in x-direction and periodically in y-direction, see Figure 3.1. These requirements follow from the boundary condition in the eigenmode simulation. Six unit cells were used without vertical stacking and arranged in a 2x3 array, Figure 3.4. The front sides of unit cells are directly connected to the dipole in order to increase the coupling from the dipole to the LH load. In that way, the impedance of the LH load is transformed by the dipole. The truncated ground plane leads to a decreased stored energy because of lower field components near the metallic interfaces (decreased effective permittivity).The effect of the slot can be modeled by a shunt element consisting of a parallel LC resonator in series with the capacitance. The width of the slot is appreciably smaller than half a wavelength in the substrate and is optimized together with the length. Geometrical parameters are given in (Palandöken et al., 2009). The overall size of the antenna is 55x14 mm, while the size of main radiating section of the loaded dipole is 30x14 mm.

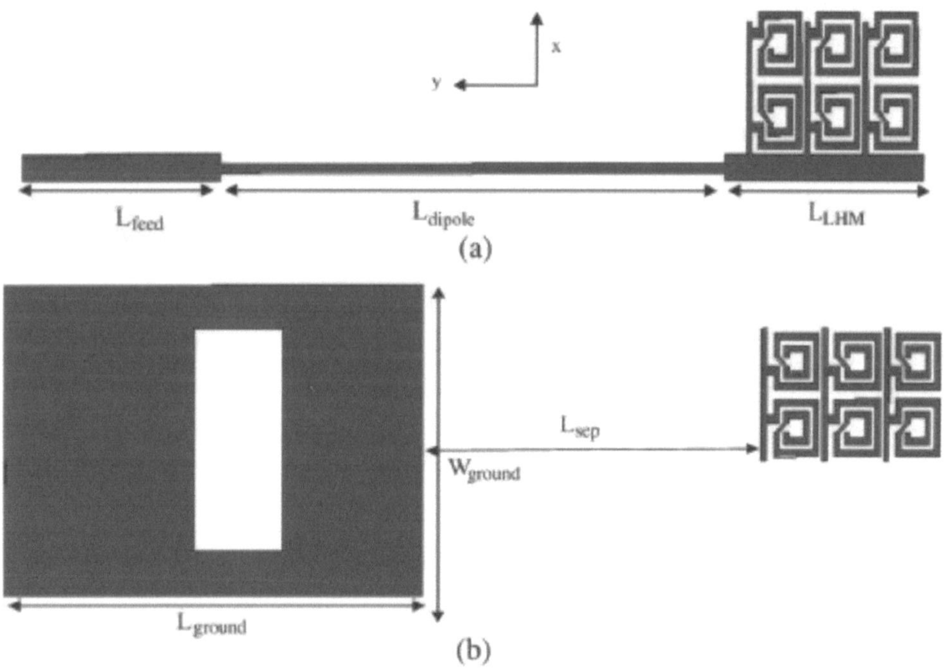

Fig. 3.4. (a) Top, (b) bottom geometry of the proposed antenna.

3.2.3 Experimental and simulation results

The return loss of the antenna was measured with the vector network analyzer HP 8722C and is shown in Figure 3.5 together with the simulation result.

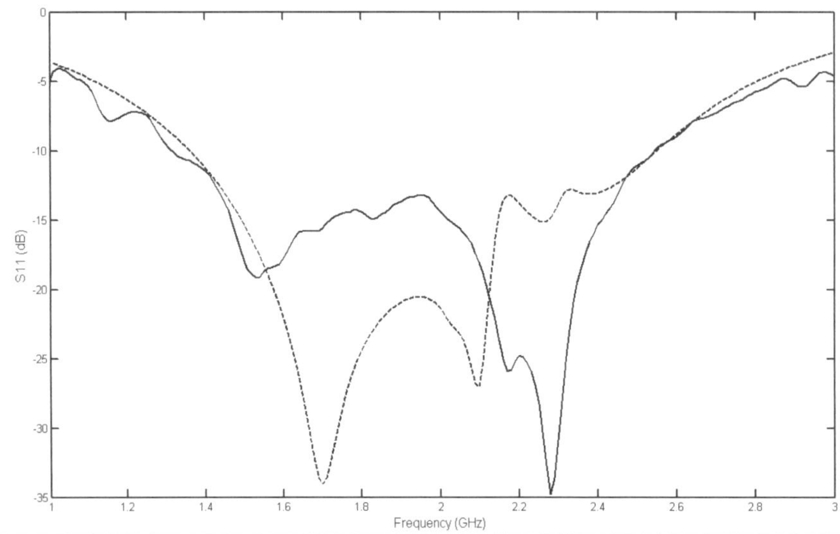

Fig. 3.5. Measured (solid line) and simulated (dashed line) reflection coefficient of the proposed antenna

The bandwidth of 63.16 % extends from approximately 1.3 GHz to 2.5 GHz with the center frequency of 1.9 GHz. Two unit cell resonances can be clearly observed in the passband. The low frequency ripples are attributed to the inaccurate modeling of the coax-microstrip line transition due to the inherent uncertainty of substrate epsilon. In summary, the measured and simulated return losses are in good agreement.

There are nevertheless some issues to be discussed from the measured and simulated results. First of all, in the experimental result, there are lower resonance frequencies than those of the LH passband in Figure 3.2, which is also the case in the simulated return loss. These lower resonance frequencies are due to the direct coupling between the dipole antenna and LHM unit cells and are not emerging from the LHM resonances. In order to prove this reasoning, the current distribution in LHM cells and the dipole is examined. At 1.7GHz, the dipole is stronger excited than the LHM cells, which is obvious because the resonance of the LH load is out-of-band. In other words, the LH load impedance is transformed by the dipole to match at this lower frequency. Secondly, the bandwidth is enhanced by the fact that different sections of LHM cells and dipole are excited at different frequencies. Still, the effect of the LH load is quite important for broadband operation. It is because the unit cell resonances are closer to each other at the lower frequencies than at higher frequencies. This unique property results in a broadband behavior at low frequencies, which is not the case for RH operation. The same reasoning can also be deduced from the dispersion diagram in Figure 3.2. Therefore, the coupled resonance feature of LHM cells results in an antenna input impedance as smooth as in the case of tapering. It is the main reason why the antenna is broadband (Geyi et al., 2000). The topology of the matching network is as important as the broadband load for the wideband operation. The third important issue is the radiation of electrically small LHM cells. It could be verified not only from the current distribution and the return loss but also from the radiation pattern, which is explained next. The antenna matching can be explained by the phase compensation feature of LHM as for instance in the case for the length independent subwavelength resonators (Engheta & Ziolkowski, 2006) and antennas (Jiang et al., 2007).

The normalized radiation patterns of the antenna in y-z and x-z planes at 1.7 GHz and 2.3 GHz are shown in Figure 3.6. They are mainly dipole-like radiation patterns in E and H planes, which is the reason to call the antenna an LHM loaded dipole antenna. The radiation of the electrically small LHM cells is also observed from the radiation pattern at 2.3 GHz. As it is shown in Figure 3.6.b, the more effective excitation of the LHM cells at 2.3 GHz than at 1.7 GHz results in an asymmetric radiation pattern because of the structure asymmetry along the y axis. The cross polarization in the y-z plane is 8 dB higher at 2.3 GHz than that at 1.7 GHz, see Figure 3.2, because of LH passband resonance.

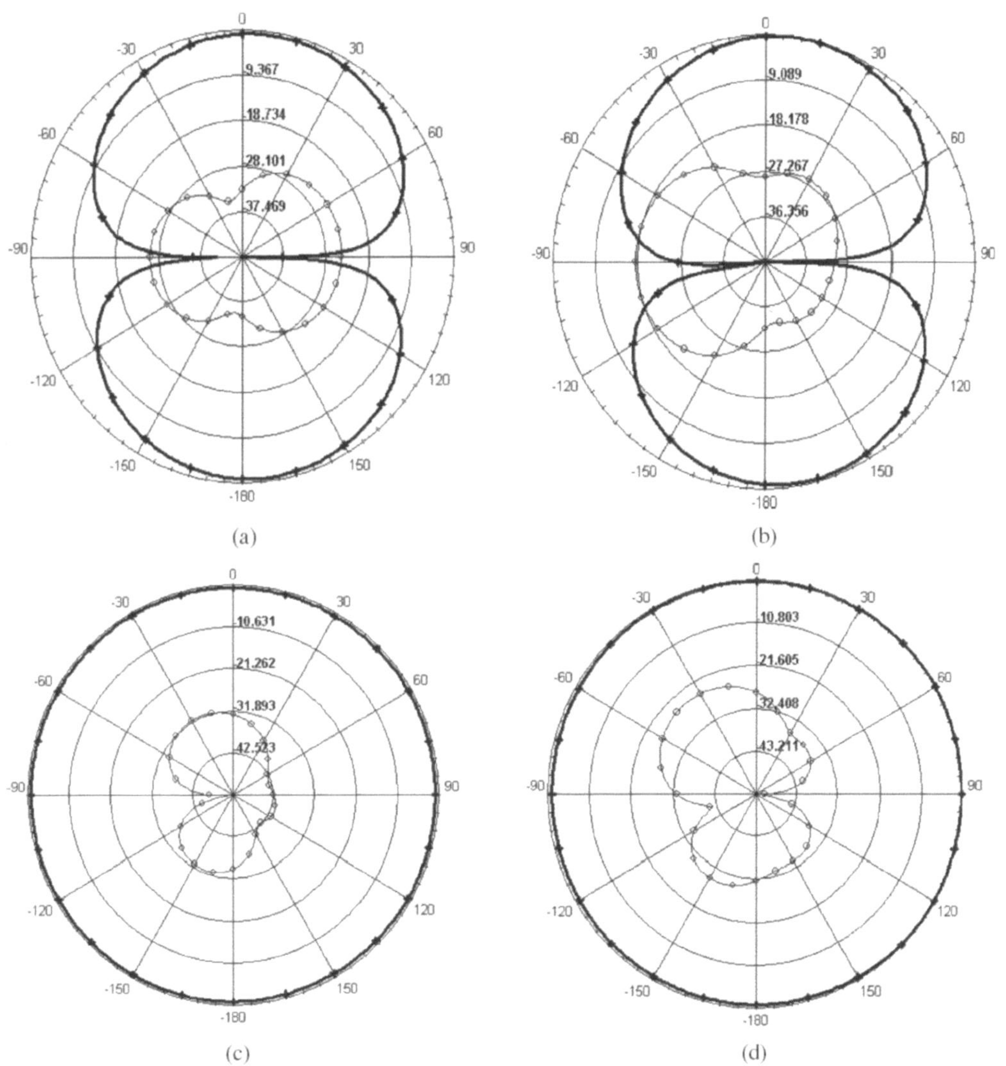

Fig. 3.6. Normalized radiation patterns cross-polarization (o-light line) and co-polarization (+ - dark line) at 1.7 GHz in (a) y-z and (c) x-z plane, and at 2.3 GHz in (b) y-z and (d) x-z plane

The gain of the broadband antenna is unfortunately small. The maximum gain and directivity are -1 dBi and 3 dB with 40% efficiency at 2.5 GHz, respectively. For the comparison of the overall size and radiation parameters of the proposed design with conventional microstrip dipole antennas, two edge excited $\lambda/4$ and $\lambda/2$ dipole antennas are designed and radiation parameters are tabulated along with the frequency dependent efficiency and gain of the proposed LHM loaded dipole in (Palandöken et al., 2009). The proposed antenna has relatively better radiation performance than these conventional dipole antennas. In addition, the gain of the proposed antenna is higher than different kinds of miniaturized and narrow band antennas in literature (Skrivervik et. al, 2001; Iizuka & Hall, 2007; Lee et al., 2006; Lee et al., 2005).

On the other hand, instead of loading a narrow-band dipole with a number of LHM unit cells to broaden the bandwidth, there are well-known alternative design techniques, some of which are increasing the thickness of the substrate, using different shaped slots or radiating patches (Lau et al., 2007), stacking different radiating elements or loading of the antenna laterally or vertically (Matin et al., 2007; Ooi et al., 2002), utilizing magnetodielectric substrates (Sarabandi et al., 2002) and engineering the ground plane as in the case of EBG metamaterials (Engheta & Ziolkowski, 2006).

The main reasons of low antenna gain are substrate/copper loss and horizontal orientation of the radiating section over the ground plane. It is like in the case of gain reduction of the dipole antenna with the smaller aperture (angle) between two excited lines. However, the gain can be increased by orienting the radiating element vertically to the ground plane to have same direction directed electric dipoles, unfortunately with the cost of high profile. Hence, a frequently addressed solution to decrease the antenna profile with the advantage of higher gain is to design artificial magnetic ground plane, on which the electric dipole can be oriented horizontally with the simultaneous gain enhancement, whose design is the main task of the next section.

4. Artificial ground plane design

In general, the performance of low profile wire antennas is degraded by their ground plane backings due to out-off phase image current distribution especially when the antenna is in close proximity to the ground plane. If the separation distance between the radiating section of the antenna and ground plane is $\lambda/4$, the ground plane reflects the exciting antenna radiation in phase with approximately 3 dB increase in gain perpendicular to ground. The problem, however, is that if the ground plane-antenna separation distance is smaller than $\lambda/4$, it cannot provide 3 dB increase, because the reflected antenna back-radiation interferes destructively with the antenna forward-radiation. Therefore, the antenna can be attributed in this case to be partially "short circuited". A second problem in microstrip antenna design is the generation of surface waves due to the dielectric layer. In surface wave excitation, the field distribution on the feeding line and the near field distribution of the antenna excite the propagating surface wave modes of ground-substrate-air system. This results the radiation efficiency degradation due to the near field coupling of antenna to the guided wave along the substrate, which does not actually contribute to the antenna radiation in the desired manner. Additionally, the guided waves can deteriorate the antenna radiation pattern by reflecting from and diffracting at the substrate edges and other metallic parts on the substrate. To solve these problems a Perfect Magnetic Conductor (PMC) would be an ideal solution for low profile antennas on which the input radiation reflects without a phase-shift due to high surface impedance. A PMC can be designed by introducing certain shaped metallic inclusions on the substrate surface to have resonances at the operation frequency. These surfaces are called EBG surfaces or Artificial Magnetic Conductors (AMC) (Goussetis et al., 2006; Engheta & Ziolkowski, 2006).

There are two bandgap regions in EBG structures. The first one is caused as a result of EBGs array resonance and array periodicity. This is the region where surface waves are suppressed and reflected due reactive Bloch impedance and complex propagation constant of the periodic array. The second region is caused by the cavity resonance between the ground plane and high impedance surface (HIS) on which radiating waves are reflected with no phase shift as in the case of PMC. The most commonly known EBG surface is the mushroom EBG (Sievenpiper et al., 1999). It consists of an array of metal patches, each patch connected with a via to ground through a substrate. The capacitively-coupled metal patches and inductive vias create a grid of LC resonators. A planar EBG can also be designed, which does not have vias and acts as a periodic frequency selective surface (FSS). A widely used EBG surface of this kind is the Jerusalem-cross (Yang et al, 1999), which consists of metal pads connected with narrow lines to create a LC network. Advanced structures without vias, consisting of square pads and narrow lines with insets, have also been proposed which are simpler to fabricate (Yang et al, 1999). On the other hand, split-ring resonators have also been frequently used in AMC design (Oh & Shafai, 2006). When the exciting magnetic field (H) is directed perpendicular to the SRR surface, strong magnetic material-like responses are produced around its resonant frequencies, thus resulting its effective permeability to be negative. However, another possibility is to excite SRRs with the magnetic field parallel to the SRRs, which results the effective permittivity to be negative rather than effective permeability. The possibility of using SRRs for the PMC surface where the magnetic vector *H* was normal to the rings surface or the propagation vector *k* perpendicular to the rings surface with the magnetic field vector *H* parallel to the surface was investigated (Oh & Shafai, 2006).

In this section, the design of an electrically small fractal spiral resonator is explained as a basic unit cell of an AMC. In Section 4.1, the geometry of one unit cell of periodic artificial magnetic material is introduced. In Section 4.2, the magnetic resonance from the numerically calculated field pattern is illustrated along with the effective permeability, which is analytically calculated from the numerical data in addition to the dispersion diagram. The negative permeability in the vicinity of resonance frequency validates the proposed design to be an artificial magnetic material.

4.1 Structural description

The topology of the artificial magnetic material is shown in Figure 4.1. Each of the outer and inner rings are the mirrored image of first order Hilbert fractal to form the ring shape. They are then connected at one end to obtain the spiral form from these two concentric Hilbert fractal curves. The marked inner section is the extension of the inner Hilbert curve so as to increase the resonant length due to the increased inductive and capacitive coupling between the different sections. The substrate is 0.5 mm thick FR4 with dielectric constant 4.4 and tan(δ) 0.02. The metallization is copper. The geometrical parameters are L_1= 2.2mm, L_2= 0.8mm and L_3 = 1mm. The unit cell size is a_x = 5mm, a_y = 2mm, a_z = 5mm. Only one side of the substrate is structured with the prescribed fractal geometry while leaving the other side without any metal layer.

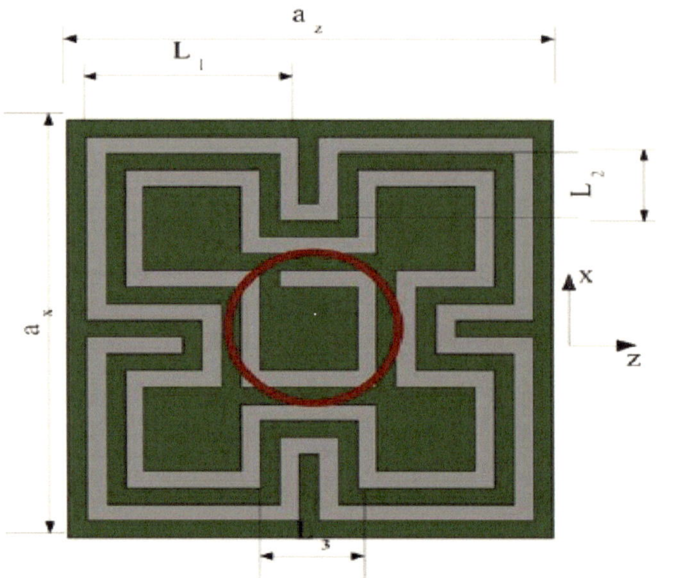

Fig. 4.1. Magnetic metamaterial geometry

4.2 Numerical simulations

In order to induce the magnetic resonance for the negative permeability, the structure has to be excited with out-of-plane directed magnetic field. Thus, in the numerical model, the structure is excited by z-direction propagating, x-direction polarized plane wave. Perfect Electric Conductor (PEC) at two x planes and Perfect Magnetic Conductor (PMC) at two y planes are assigned as boundary conditions. The numerical model was simulated with HFSS. The simulated S-parameters are shown in Figure 4.2. The resonance frequency is

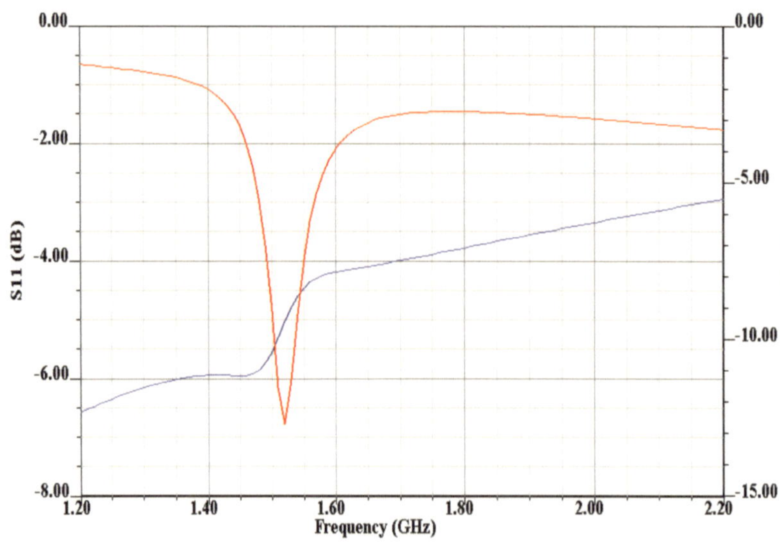

Fig. 4.2. Transmission (red) and reflection (blue) parameters

Fig. 4.3. Surface current distribution at the resonant frequency

1.52 GHz. The surface current distribution at the resonance frequency is shown in Figure 4.3. Because the surface current spirals with the result of out-of-plane directed magnetic field, this electrically small structure can be considered as a resonant magnetic dipole. The transmission deep in the S-parameter is effectively due to the depolarization effect of this magnetic dipole for the incoming field.

This is the reason why this artificial magnetic material is regarded as a negative permeability material in a certain frequency band. As a next step, the effective permeability of the structure is retrieved to confirm the negative permeability and justify the above-mentioned remarks. In principle, the effective parameters of such materials have to be calculated to characterise them as artificial magnetic or dielectric materials.

They are conventionally retrieved from S parameters of one unit cell thick sample under the plane wave excitation. However, the resulted effective parameters are only assigned to this one sample. An alternative procedure which has been recently introduced is to calculate the dispersion diagram of infinite number of proposed unit cell in the propagation direction with certain phase shifts (Palandöken&Henke, 2009). As a first step in this method, the numerically calculated Z parameters, which are deembedded upto the cell interfaces, are transformed to ABCD parameters,

$$A=\frac{Z_{11}}{Z_{21}},\quad B=\frac{Z_{11}Z_{22}-Z_{21}^2}{Z_{21}},\quad C=\frac{1}{Z_{21}},\quad D=\frac{Z_{22}}{Z_{21}}, \tag{8}$$

Then, the Bloch-Floquet theorem is utilized to calculate Bloch impedance and 1D Brillioun diagram from ABCD parameters with complex propagation constant γ and period d.

$$\gamma=\frac{\operatorname{arccosh}\left(\frac{Z_{11}+Z_{22}}{2Z_{21}}\right)}{d}, \tag{9}$$

$$Z_{Bloch}=\frac{B}{\exp(\gamma d)-A}. \tag{10}$$

The effective permeability can then be easily calculated from Bloch impedance and propagation constant with free space wave number k_o, and line impedance Z_o

$$\mu_{eff.} = \frac{-j\gamma Z_{Bloch}}{k_0 Z_0}. \quad (11)$$

The propagation constant and effective permeability are shown in Figure 4.4.

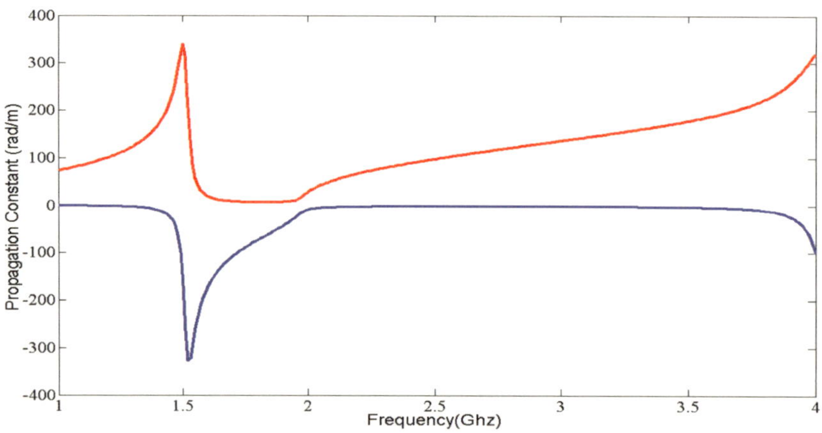

Fig. 4.4.a. Real (blue) and imaginary (red) part of complex propagation constant

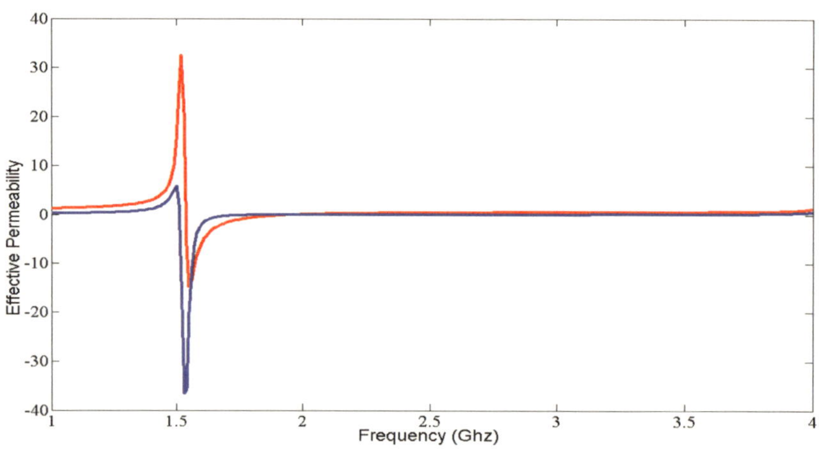

Fig. 4.4.b. Real (red) and imaginary (blue) part of effective relative permeability

As it can be deduced from Figure 4.4.a, the exciting plane wave is exponentially attenuated in the frequency band of 1.45-1.96 GHz due to nonzero attenuation parameter. This results to have pure capacitive Bloch line impedance through the periodic unit cells. This reactive Bloch impedance could be modeled as series and shunt capacitances, which is the transmission line model of negative permeability metamaterials (Caloz & Itoh, 2005, Eleftheriades & Balmain, 2005). From Figure 4.4.b, the magnetic resonance frequency, 1.52 GHz, can be identified at which EBG, which is composed of periodically oriented fractal spiral resonators, can be operated as AMC due to high Bloch impedance and relative permeability. In addition, there is also a broader bandwidth of negative permeability, which is obtained between the magnetic resonance and plasma frequency at which surface wave propagation through the substrate could be suppressed. As a result, the proposed fractal spiral resonator, which is electrically small with the unit cell size of approximately 1/40 of resonant wavelength can be effectively utilized in artificial ground plane design. In the antenna design, the radiating element of either magnetic or electric dipole antenna is mostly located on the EBG surface with a substrate of certain distance inbetween. In addition, rather than artificially structuring the substrate and ground plane of microstrip antennas,which was explained in the previous sections, the radiating element itself can also be designed with electrically small self-resonating structures, which is explained in the next section.

5. Metamaterials based antenna design

As it was pointed out in the previous sections, metamaterial structures are able to sustain strong subwavelength resonances in the form of magnetic or electric dipole. These two types of dipoles can also be coupled in the same unit cell to excite both radiators in smaller size. In other words, the inductive impedance of negative permittivity material (ENG) can be compensated with the capacitive impedance of negative permeability material (MNG) as in double negative material (DNG). This leads these electrically small structures to be utilized as the resonators in miniturized antennas. Rather than designing self-resonating structures for new antenna topologies, there is a great deal of interest in enhancing the performance of conventional electrically small non-resonant antennas (ESA). As an attempt, the performance of ESAs surrounded by metamaterial shells was originally shown in papers (Ziolkowski et al.,2006; Ziolkowski et al.,2005; Ziolkowski et al.,2003; Li et al.,2001), in which significant gain enhancement of an electrically small dipole can be accomplished by surrounding it with a (DNG) shell. It is because high capacitive impedance of dipole is compensated with inherent inductance and capacitance of DNG, which results the resonance frequency to shift from the eigenfrequencies in the passband of DNG due to capacitive loading. In other words, the whole system comprised of electric dipole and DNG shell resonates rather than dipole itself. The gain is therefore higher due to not only matching of non-resonant dipole but also the contribution of metamaterial shell into the radiation as in the case of phased antenna arrays. A similar configuration for an infinitesimal dipole surrounded by an ENG shell in (Ziolkowski et al.,2007) was shown to demonstrate a very large power gain, due to the resonance between the inductive load offered by the ENG shell and the capacitive impedance of the dipole in the inner medium. A multilayer spherical configuration was presented in (Kim et al, 2007) to achieve gain enhancement for an electrically small antenna. And the radiated power gain of the DNG/MNG shell was also compared with respect to a loop antenna of the same radius as the outer radius of the shell and reasonably good power gains were obtained (Ghosh et al.,2008). In analogy with the electrically small dipole, the inductive impedance of electrically small loop antenna can be matched with the capacitive impedance of MNG shell. The resulting shell/magnetic loop system couples to DNG material effectively due to enhanced near field, resulting the whole system to resonate and cumulatively radiate in large volume.

In this section, a metamaterial based antenna, which is composed of self-resonating slots and an exciting small microstrip monopole, is explained. The electrically small monopole is coupled to the slot radiators capacitively to excite the compact resonators for subwavelength operation. The radiating slots are located horizontally over the large ground plane while the exciting microstrip monopole is located vertically on the ground plane and connected to the inner conductor of SMA connector. The main radiating section of the antenna is composed of four slotted array of the unit cell shown in Figure 3.1 in Section 3. In Section 5.1, the geometrical model of the metamaterial inspired radiator is explained. In Section 5.2, the antenna geometry is explained along with the design and operation principle. In Section 5.3, the numerically calculated return loss and radiation patterns are presented.

5.1 Metamaterial inspired radiating structure

The main radiating section of the antenna is composed of four slots of the unit cell proposed in (Palandöken et al. 2009). Because the electrical length of each resonator can be increased by the direct connection with the neighboring resonator antisymmetrically, the radiator is structured as shown in Figure 5.1. This perforated structure is located perpendicular to the substrate of exciting monopole. The overall size is 14 mm x 6 mm. The separation distance between each pair of resonators is 0.4 mm.

Fig. 5.1. Metamaterial slot radiator geometry

5.2 Antenna design

The metamaterial resonator based slot antenna is shown in Figure 5.2. In the antenna model, the inner conductor of SMA is extended to be connected with the microstrip line of monopole. The length of extended inner conductor from the ground plane surface is 2 mm. The substrate material is 0.5 mm thick FR4 with the ground plane length L_{grn} = 6mm. There is a small gap between the slot resonators and microstrip line with the arm width, W_{mat} =6.5mm and length L_{mat}=6mm, which enhances the impedance matching and capacitive field coupling from the feeding line to the slots. The monopole feeding line is situated exactly in the middle of the resonator in case two slot resonators are excited to be coupled magnetically. The microstrip monopole has two main design advantages. Firstly, it is a supporting material for the radiating slot resonator to be located on due to no substrate under the radiator. Secondly, it results matching network to be designed on the feeding monopole without increasing antenna size. The microstrip monopole with T-formed matching section is shown in Figure 5.3.

(a) (b)

Fig. 5.2. (a) Top and (b) side view of metamaterial based microstrip antenna

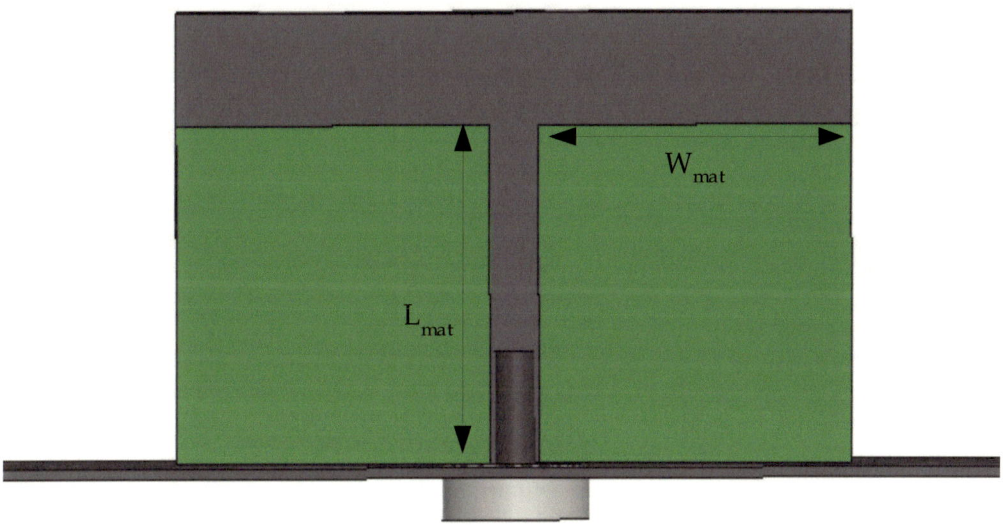

Fig. 5.3. Microstrip feeding monopole antenna

On the other hand, the spiral nature of SRs and linear form of the slotted thin wires in the radiating section lead the excited virtual magnetic currents to react the structure as a combination of electric and magnetic dipole, respectively. Therefore , the operation principle of the antenna is based on the excitation of the horizontally oriented magnetic and vertically oriented electric dipoles. Because these dipoles have the same direction directed image currents due to perfect electric ground plane, this radiator topology results both dipole types to radiate effectively.

5.3 Simulation results

The return loss of metamaterial inspired antenna is numerically calculated with HFSS and shown in Figure 5.4. The resonance frequency is 5.25 GHz. The resonance frequency is higher than the eigenfrequencies in the passband of unit cell due to the slotted form and no existence of substrate. This proposed topology increases the radiation efficiency and gain of the antenna in comparison to the alternative design with substrate. The truncated ground of microstrip monopole (Figure 5.2.b) increases the field coupling from the guided line to the slot resonators, which results the antenna to be better impedance matched. On the other hand, another advantage of the feeding line ground plane is to couple the incoming field from the input port more effectively to the antenna without any leaking fields to the surrounding large ground plane.

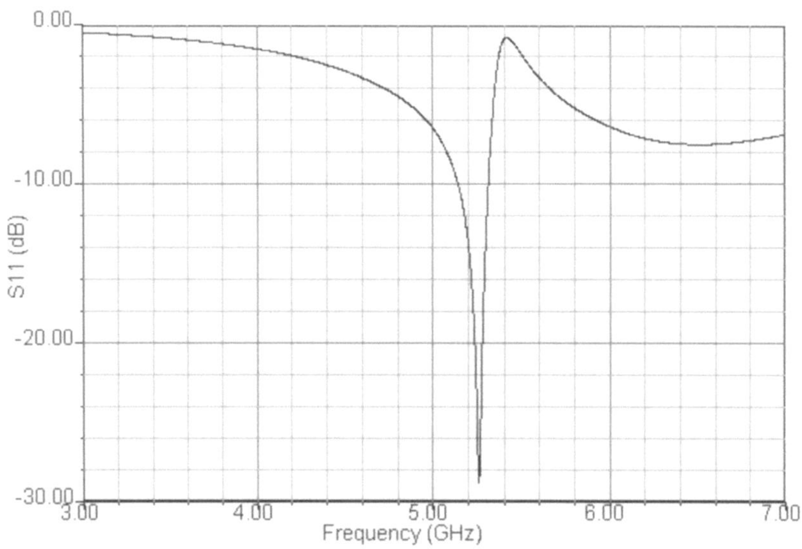

Fig. 5.4. Return loss of the metamaterial based slot antenna

The normalized radiation patterns on H and E planes at 5.25 GHz are shown in Figure 5.5. As deduced from Figure 5.5.b, the co-polarization radiation pattern on H plane has quite similar form of H-plane radiation pattern of electric dipole. This is because of the excitation of the spiral resonator with the eigencurrents in the form of virtual magnetic current. Spiraling magnetic current generates the near field distribution as in the form of electric dipole excitation due to Babinet's principle and duality. However, the radiation pattern on E-plane is not similar to the dipole radiation because of the lack of radiation null on the dipole axis. The reason of the enhanced radiation on the dipole axis is due to the superposition of the dipole fields in four element dipole array on H plane and shift of the antenna phase center. However, the radiation intensity along the dipole axis is minimum. On the other hand, the cross-polarization radiation pattern on H plane is quite similar to the radiation pattern of horizontally directed magnetic dipole. It is due to the excitation of slotted connection lines of the spiral resonators in the slot radiator. There is another possible magnetic radiation source, which originates from the electric coupling in the gap separation between the feeding line and slot radiators. However, it has quite small effect on the radiation, which could be proved due to small cross-polarization on H-plane. The cross-polarization level is better than -90dB on E and -80dB on H plane.

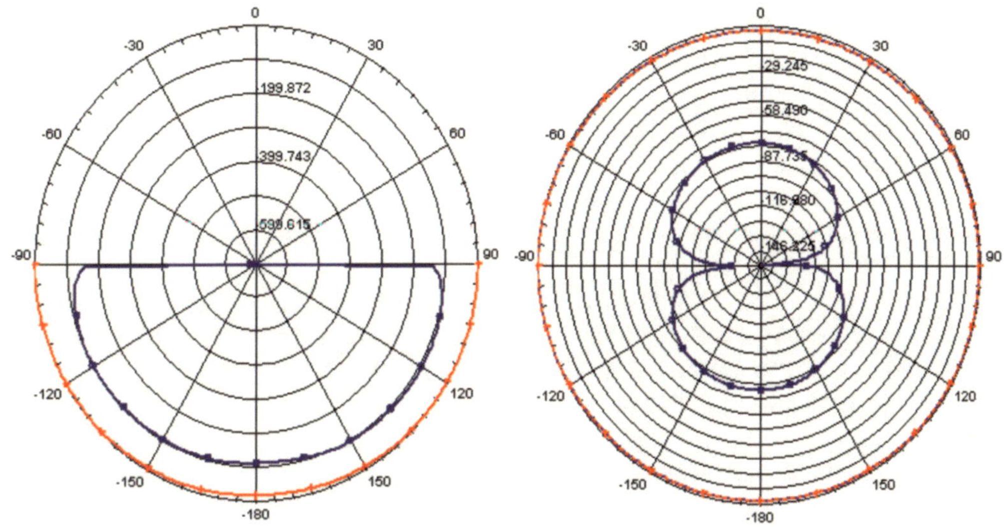

(a) (b)

Fig. 5.5. Co-polarization (red) / cross-polarization (blue) radiation patterns on (a) E- and (b) H-plane at 5.25GHz

The antenna gain is 5.5dBi with the overall efficiency of more than 90%. It is quite efficient radiator due to the cumulative excitation of the vertical electric and horizontal magnetic dipole over the metallic ground plane and reduced level of field leaking from the input port. Vertical orientation of the slot resonators or horizontal orientation of the proposed unit cells without slotted form would result the field cancellation of the radiation in the far field due the ground plane. It is an important issue in the high performance antenna design. The overall antenna size is 0.24λ x 0.1λ, which is electrically small and therefore, well suited for the modern wireless communication systems.

6. References

N. G. Alexopoulos, C. A. Kyriazidou, and H. F. Contopanagos, "Effective parameters for metamorphic materials and metamaterials through a resonant inverse scattering approach," *IEEE Trans. Microw. Theory Tech.*, vol. 55, no. 2, pp. 254-267, Feb. 2007.

J. Baena, R. Marqués, and F. Medina, "Artificial magnetic metamaterial design by using spiral resonators," *Phy. Rev. B*, vol. 69, pp. 0144021-0144025, 2004

Caloz and T. Itoh, *Electromagnetic Metamaterials: Transmission Line Theory and Microwave Applications*. Piscataway, NJ: Wiley- IEEE, 2005.

H. Chen, J. Zhang, Y. Bai, Y. Luo, L. Ran, Q. Jiang, and J. A. Kong, "Experimental retrieval of the effective parameters of metamaterials based on a waveguide method," *Opt. Express*, vol. 14, no. 26, pp. 12944-12949, Dec. 2006.

Chen, T. M. Grzegorczyk, B.-I.Wu, J. Pacheco, Jr., and J. A. Kong, "Robust method to retrieve the constitutive effective parameters of metamaterials," *Phys. Rev. Lett. E*, vol. 70, pp. 0166081-0166087, 2004.

J. Chu, "Physical limitations on omnidirectional antennas," *J. Appl. Phys.*, vol. 19, pp. 1163–1175, Dec. 1948

G. V. Eleftheriades and K. G. Balmain, *Negative Refraction Metamaterials: Fundamental Principles and Applications*. New York:Wiley Interscience, 2005

N. Engheta and R. W. Ziolkowski, *Metamaterials Physics and Engineering Explorations*, , Eds. New York: Wiley/IEEE, 2006.

W. Geyi, P. Jarmuszewski, and Y. Qi, "The Foster reactance theorem for antenna radiation Q," *IEEE Trans. Antennas Propag.*, vol. 48, no. 3, pp. 401–408, Mar. 2000.

Geyi, W. , "Physical limitations of antenna,", *Antennas and Propagation, IEEE Transactions on* pp. 2116 - 2123 , Volume: 51 Issue: 8, Aug. 2003

Ghosh, B., S. Ghosh, and A. B. Kakade, "Investigation of gain enhancement of electrically small antennas using double-negative, single-negative, and double-positive materials," *Physical Review E*, Vol. 78, 026611, 2008.

G. Goussetis, A.P. Feresidis, J.C. Vardaxoglou, Tailoring the AMC and EBG Characteristics of Periodic Metallic Arrays Printed on Grounded Dielectric Substrate. IEEE Transactions on Antennas and Propagation, vol. 54, no. 1, January 2006.

R. F. Harrington, "Effect of antenna size on gain, bandwidth, and efficiency", *J. Res. Nat. Bureau Stand.*, vol. 64D, pp. 1 - 12, 1960.

H. Iizuka and P. S. Hall, "Left-handed dipole antennas and their implementations," *IEEE Trans. Antennas Propag.*, vol. 55, no. 5, pp. 1246–1253, May 2007

T. Jiang, Y. Yuan, D. Wang, L. Ran, and J. A. Kong, "High directive cavity antenna based on 1D LHM-RHM resonator," in *PIERS*, 2007, vol. 3, no. 3, pp. 2054–2056.

R. C. Johnson, *Antenna Engineering Handbook*, 3rd ed. New York: McGraw-Hill.

Kim, H. Y., J. K. Kim, J. H. Kim, Y. J. Kim, and H. M. Lee, "Design of metamaterial structure based electrically small monopole antenna," *2007 Autumn Microwave & Radio Wave Conference*, Vol. 30, 577–580, September 2007

K. L. Lau, K. C. Kong, and K. M. Luk, "A miniature folded shorted patch antenna for dual-band operation," *IEEE Trans. Antennas. Prop.*,vol. 55, no.8, pp. 2391–2398, Aug. 2007

C.-J. Lee, K. M. K. H. Leong, and T. Itoh, "Composite right/left-handed transmission line based compact resonant antennas for RF module integration," *IEEE Trans. Antennas Propag.*, vol. 54, no. 8, pp. 2283–2291, Aug. 2006.

C.-J. Lee, K. M. K. H. Leong, and T. Itoh, "Design of resonant small antenna using composite right/left handed transmission line," in *Antennas Prop. Soc. Int. Symp.*, 2005, vol. 2B.

Li, H. Yao, Q. Wu, and Z. Chen, "Broad-bandwidth and low-loss metamaterials: Theory, design and realization," *J. Zhejiang Univ. Science A*, vol. 7, no. 1, pp. 5–23, Jan. 2006.

A. Matin, B. S. Sharif, and C. C. Tsimenidis, "Probe fed stacked patch antenna for wideband applications," *IEEE Antennas Prop.*, vol. 55, no. 8, pp. 2385–2388, Aug. 2007.

Soon-Soo Oh, Lotfollah Shafai, "Artificial magnetic conductor using split ring resonators and its applications to antennas," *Microwave and Techn. Letters*, vol. 48, no.2, pp.329–334, Feb. 2006.

B.-L. Ooi, S. Qin, and M.-S. Leong, "Novel design of broad-band stacked patch antenna," *IEEE Trans. Antennas Propag.*, vol. 50, no. 10, pp. 1391–1395, Oct. 2002

K. Sarabandi, R. Azadegan, H. Mosallaei, and J. Harvey, "Antenna miniaturization techniques for applications in compact wireless transceivers," in *Proc. URSI*, 2002, pp. 2037–2040.

M. Palandöken, A. Grede, H. Henke, "Broadband microstrip antenna with Left-handed Metamaterials, "*IEEE Trans. Antennas Propag.*, vol. 57, no.2, pp. 331–338, Feb. 2009

M. Palandoken, H. Henke, "Fractal Spiral Resonator as Magnetic Metamaterial", Applied Electromagnetics Conference (AEMC),pages 1-4, 2009

Sievenpiper, L. Zhang, R. Broas, N.G. Alexopolous, and E. Yablonovitch, High-impedanceelectromagnetic surfaces with a forbidden frequency band, IEEE Trans Microwave Theory Tech MTT-47 (1999), 2059–2074.

A. K. Skrivervik, J. -F. Zürcher, O. Staub, and J. R. Mosig, "PCS antenna design: The challenge of miniaturization," *IEEE Antennas Propag. Mag.*, vol. 43, no. 4, pp. 12–27, Aug. 2001.

R. Smith, D. C. Vier, T. Koschny, and C. M. Soukoulis, "Electromagnetic parameter retrieval from inhomogeneous metamaterials," *Phys. Rev. E*, vol. 71, pp. 0366171–03661711, 2005.

R. Smith, W. J. Padilla, D. C. Vier, S. C. Nemat-Nasser, and S. Schultz, "Composite medium with simultaneously negative permeability and permittivity," *Phys. Rev. Lett.*, vol. 84, no. 18, pp. 4184–4187, 2000.

Veselago G., "The electrodynamics of substance with simultaneously negative values of ε and μ" *Sov. Phys.-Usp.*, vol. 10, no. 4, pp.509–514, Jan.–Feb. 1968.

Wheeler, H.A, "Small Antennas," IEEE Trans. Antennas Propagat. , vol . AP-23, July 1975, pp.462-469

Wheeler, H.A, "The radiansphere around a small antenna," Proc. Of the IRE, vol.47, pp.1325-1331, Aug. 1959.

Wheller , H.A.: 'Fundamental limitations of small antennas', Proc IRE, December 1947, pp. 1479–1488

A.D. Yaghjian and S.R. Best, Impedance, bandwidth, and *Q* of antennas, Trans IEEE, AP-53 (2005), 1298–1324.

F-R. Yang, K-P. Ma, Y. Qian, and T. Itoh, A novel TEM waveguide using uniplanar compact photonic-bandgap (UC-PBG) structure, IEEE Trans Microwave Theory Tech MTT-47 (1999), 2092–2098

F. Yang, K. Ma, Y. Qian, T. Itoh, A Uni-planar Compact Photonic-Bandgap (UCPBG) Structure and Its Applications for Microwave Circuits. IEEE Transactions on Microwave Theory and Techniques, vol. 47, no. 8, August 1999.

Ziolkowski, R. W. and A. Erentok, "At and beyond the chu limit: Passive and active broad bandwidth metamaterial-based efficient electrically small antennas," *IET Microw., Antennas Propag.*, Vol. 1, 116–128, February 2007.

Ziolkowski, R. W. and A. Erentok, "Metamaterial-based efficient electrically small antennas, " *IEEE Trans. Antennas Propag.*, Vol. 54, 2113–2130, July 2006.

Ziolkowski, R. W. and A. D. Kipple, "Reciprocity between the effects of resonant scattering and enhanced radiated power by electrically emall antennas in the presence of nested metamaterial shells," *Physical Review E*, Vol. 72, 036602, September 2005.

Ziolkowski, R. W. and A. Kipple, "Application of double negative metamaterials to increase the power radiated by electrically small antennas," *IEEE Trans. Antennas Propag.*, Vol. 51, 2626–2640, October 2003.

4

Particle-Swarm-Optimization-Based Selective Neural Network Ensemble and Its Application to Modeling Resonant Frequency of Microstrip Antenna

Tian Yu-Bo and Xie Zhi-Bin
School of Electronics and Information, Jiangsu University of Science and Technology, Zhenjiang, Jiangsu province 212003, P. R. China

1. Introduction

From communication systems to biomedical systems, microstrip antennas (MSAs) are used in a broad range of applications, and this primarily due to their simplicity, conformability, low manufacturing cost, light weight, low profile, reproducibility, reliability, and ease in fabrication and integration with solid-state devices [1][2]. Recently, these attractive features have increased the applications of MSAs and stimulated greater effort to investigate their performance. In designing MSA, it is very important to determine its resonant frequencies accurately, because MSA has narrow bandwidths and can only operate effectively in the vicinity of the resonant frequency. So, a model to determine the resonant frequency is helpful in antenna designs. Several methods, varying in accuracy and computational effort, have been proposed and used to calculate the resonant frequency of rectangular MSA [3]-[13]. These methods can be broadly classified into two categories: analytical and numerical methods. Based on some fundamental simplifying physical assumptions regarding the radiation mechanism of antennas, the analytical methods are the most useful for practical design as well as providing a good intuitive explanation of the operation of MSAs. However, these methods are not suitable for many structures, in particular, if the thickness of the substrate is not very thin. The numerical methods provide accurate results but usually require tremendous computational effort and numerical procedures, resulting in roundoff errors, and may also need final experimental adjustment to the theoretical results. They suffer from a lack of computational efficiency, which in practice can restrict their usefulness due to high computational time and costs. The numerical methods also suffer from the fact that any change in the geometry, including patch shape, feeding method, addition of a cover layer, etc., requires the development of a new solution.
During the last decade, artificial neural network (NN) models have been increasingly used in the design of antennas, microwave devices, and circuits due to their ability and adaptability to learn, generalization, smaller information requirement, fast real-time operation, and ease of implementation features [14][15]. Through training process, a NN model can be developed by learning from measured/simulated data. The aim of the training

process is to minimize error between target output and actual output of the NN. The trained NN model can be used during electromagnetic design to provide instant answers to the task it learned. Due to their attractive features, NN was used in computing the resonant frequencies of rectangular MSAs [16][17]. Reference [16] presented a NN model trained with the backpropagation (BP), delta-bar-delta (DBD), and extended delta-bar-delta (EDBD) algorithms for calculating the resonant frequencies of MSAs. The performance of this NN model was improved in reference [17] by using a parallel tabu search (PTS) algorithm for the training process. The results in [16] and [17] are not in very good agreement with the experimental results in literature [12] and [18] for the rectangular MSAs.

At present, neural network ensemble (NNE) has gradually become the hotspot in the field of machine learning and neural computation [19]. Through independently training several NNs and ensembling their computing results together, generalization ability of NNE modeling complex problems can be improved remarkably. In this chapter, selective NNE methods based on decimal particle swarm optimization (DePSO) algorithm and binary particle swarm optimization (BiPSO) algorithm are proposed. The NNs for constructing NNE are optimally selected in terms of particle swarm optimization (PSO) algorithm. This can maintain the diversity of NNs and decrease the effects of collinearity and noise of samples. At the same time, chaos mutation is adopted to increase the particles diversity of PSO algorithm. The next section briefly describes the computing method of resonant frequency of rectangular MSAs. The basic principles of PSO-based selective NNE are presented in the following section. Subsequently, the selective NNE is applied to the calculation of resonant frequency, and the computing results are better than available ones. The conclusions are then made.

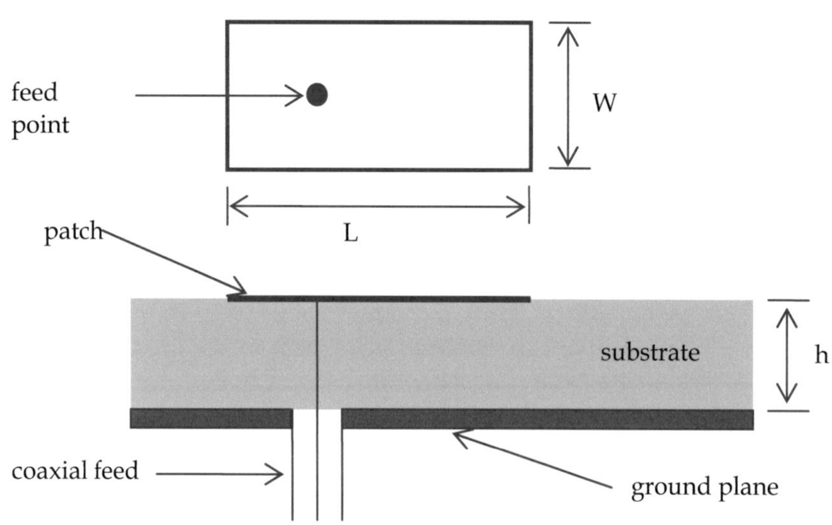

Fig. 1. Geometry of rectangular MSA

2. Resonant frequency of rectangular MSA

Fig. 1 illustrates a rectangular patch of width W and length L over a ground plane with a substrate of thickness h and a relative dielectric constant ε_r. The resonant frequency f_{mn} of the antenna can be calculated from (1) - (3) [1] [2].

$$f_{mn} = \frac{c}{2\sqrt{\varepsilon_e}}\left[\left(\frac{m}{L_e}\right)^2 + \left(\frac{n}{W_e}\right)^2\right]^{1/2} \tag{1}$$

Where ε_e is effective dielectric constant for the patch, c is velocity of electromagnetic waves in free space, m and n take integer values, and L_e and W_e are effective dimensions. To compute the resonant frequency of a rectangular patch antenna driven at its fundamental TM_{10} mode, expression (1) is written as

$$f_{10} = \frac{c}{2L_e\sqrt{\varepsilon_e}} \tag{2}$$

The effective length L_e can be defined as follows:

$$L_e = L + 2\Delta L \tag{3}$$

Where ΔL is edge extension, and its value is relative to the thickness h of the dielectric substrate.
Obviously, resonant frequency of the rectangular MSA is decided by h, ε_r, m, n and the size of patch W and L.

3. Particle swarm optimization algorithm

PSO, which was first developed by Kennedy and Eberhart[20], is a kind of evolutionary computational technology based on intelligent behavior of organisms, and its basic idea is originally from artificial life and evolutionary computation[21][22]. The main feature of PSO is to solve problems in real number field, the adjusted parameters are few, and it is a kind of general global research algorithm. Therefore, the method has been widely used in many fields, such as NN training, function optimization, fuzzy control system, etc.[23]. The advantages of PSO are characterized as simple, easy to implement, and efficient to compute. Unlike the other heuristic techniques, PSO has a flexible and well-balanced mechanism to enhance the global and local exploration abilities. At present, as a robust global optimal method, PSO is also utilized in electromagnetic field [24][25], such as the design of absorbing material, antenna design and so forth.
PSO simulates the behaviors of bird flock [21][22]. Suppose the following scenario: a group of birds are randomly searching food in an area. There is only one piece of food being searched in the area. All the birds do not know where the food is. But they know how far the food is during each search iteration. So what's the best strategy to find the food? The effective one is to follow the bird that is nearest to the food. PSO learns from the scenario and uses it to solve the optimization problems. In PSO, each single solution is a "bird" in the search space. We call it "particle". All of particles have fitness values that are evaluated by the fitness function to be optimized, and have velocities that direct the flying of the particles. The

particles are "flown" through the problem space by following the current optimum particles. PSO is initialized with a group of random particles (solutions) and then searches for optima by updating generations. In every iteration, each particle is updated by following two "best" values. The first one is the best solution (fitness) it has achieved so far. This value is called *pbest*. The other that is tracked by the particle swarm optimizer is the best value, which is obtained so far by any particle in the swarm. This best value is a global best and called *gbest*. After finding the two best values, the particle updates its velocity and position with following formulas:

$$v_{i,d}^{k+1} = \omega \cdot v_{i,d}^{k} + c_1 \cdot rand(\ ) \cdot \left(pbest_{i,d}^{k} - x_{i,d}^{k}\right) + c_2 \cdot rand(\ ) \cdot \left(gbest_{d}^{k} - x_{i,d}^{k}\right) \quad (4)$$

$$x_{i,d}^{k+1} = x_{i,d}^{k} + v_{i,d}^{k+1} \quad (5)$$

Where ω is inertia weight and control the PSO's exploitation ability and exploration ability. c_1 and c_2 are learning factors, and usually $c_1 = c_2 = 2$. $rand(\ )$ is a random number between (0,1). $v_{i,d}^{k}$ and $x_{i,d}^{k}$ are, respectively, velocity and position of particle i in dth dimension and kth iteration, and they are limited to a scope.

4. Neural Network and Neural Network ensemble

4.1 Neural Network

Neural Network (NN), which is also called Artificial Neural Network (ANN) in this study, is a simplified model to descript and abstract biological neural system of human brain according to mathematical and physical method and from the angle of information processing based on the understanding of human brain [26][27]. NN is a kind of description of characteristics of human brain system. Also, it is considered a computer system including many very simple processing units that connect each other in accordance with some manners. The system processes information on the basis of dynamic response of its state to external inputs. Simply speaking, NN is a mathematical model, and it can be implemented by electronic circuits or simulated by computer program. It is a kind of method in artificial intelligent (AI). Because NN can solve practical problems, its applying fields are expanded continuously, including not only engineering, science and mathematics but also iatrology, business, finance, and literature. To complex microwave engineering, it is very difficult to design by traditional manual methods, even not satisfy the demand. Therefore, rapid and efficient CAD (Computer aided design) method is urgently needed. From mathematical point of view, CAD model of electromagnetic problem is a kind of relationship of mapping, and NN can descript the relation effectively and accurately. Moreover, the computation of the mapping is convenient and fast. Because NN is very suitable for modeling and optimization of complex electromagnetic systems that face CAD optimized process, it is widely used in electromagnetic field [14][15].

4.2 Neural Network ensemble

NN has been successfully used in many fields. However, because of lack of instruction of rigorous theory, its applying effectiveness mainly depends on users' experience. Although feed fordward network with only one nonlinear hidden layer can approach functions with arbitrary complexity by arbitrary precision, the configuration and training of NN are NP

(Non-deterministic Polynomial) problem. In practical applications, model, algorithm and parameters of NN are determined only by many time-consuming experiments. Furthermore, the users of NN are ordinary engineers, and they are usually lack in applying experiences of NN. If there isn't easily used NN in engineering, the applying effectiveness of NN will be very difficult to be ensured. In 1990, Hansen and Salamon creationarily put forward the NNE (Neural network ensemble) method [19]. They proved that generalization ability of the neural computing system can be improved obviously by the way of training several NNs and then ensembling the results according to the rule of relative plurality voting or absolute plurality voting. Because the method is easy to use and its effectiveness is obvious, users without applying experiences of NN can get some benefits from it. Therefore, the NNE seems to be an effective neural computing method in engineering [28].

5. PSO-based selective NNE

5.1 Ensemble methods of NNs

An important research topic about NNE is its ensemble methods. This directly concerns its generalization ability. At present, the studies of ensemble methods mainly focus on two aspects: one is how to build/select every individual network; and the other is how to ensemble the outputs of every built/selected individual network. Taking the regression problem as an example, matrix inversion has to be carried out for getting combination weights of some conventional methods, and it is affected easily by multi-dimensional collinearity and noise in the data, which may decrease the generalization ability of the NNE [29][30]. In order to solve the problem of multi-dimensional collinearity, we can adopt some methods, such as, avoiding matrix inversion and restricting combination weights [29], selecting ensemble method [31], extracting principal components [29], and so on. In order to decrease the influence of noise, we can also adopt some methods, such as, restricting combination weights [32], adjusting objective function [32], and so forth. From the angles of avoiding matrix inversion and restricting combination weights, selective NNE methods based on DePSO and BiPSO are, respectively, adopted in this chapter. The basic idea of the methods is to optimally select NNs to construct NNE with the aid of PSO algorithm. This can maintain the diversity of individuals and decrease the effects of collinearity and noise of samples. Simultaneously, chaos mutation is injected into the iterative process to increase the particles diversity of PSO algorithm.

5.2 Implementation of PSO-based selective NNE

Assume that n NNs $f_1, f_2, \cdots, f_n$ have been trained separately, and apply them to construct NNE that approximate the mapping $f : \boldsymbol{R}^m \to \boldsymbol{R}^n$. In order to discuss the problem simply, assume every NN has only one output variable. Namely, the approximated mapping is $f : \boldsymbol{R}^m \to \boldsymbol{R}$. Obviously, the conclusion in this chapter may be extended easily to the situation with many output variables. The ensemble process can be achieved by using above-mentioned PSO algorithm. Every swarm of particles represent a kind of ensemble of $\{f_1, f_2, \cdots, f_n\}$, and particles length (dimension of particles space) equals to n that is the quantity of the NNs. We can adopt DePSO algorithm and BiPSO algorithm to achieve the network's selection.

To selective NNE based on DePSO algorithm, its actual output corresponding to the input $\boldsymbol{x}$ is given by

$$\bar{f}(\boldsymbol{x})=\sum_{i=1}^{n}\bar{\beta}_i f_i(\boldsymbol{x}),\ \bar{\beta}_i=\frac{\beta_i}{\sum_{i=1}^{n}\beta_i} \tag{6}$$

where $\bar{\boldsymbol{\beta}}$ represents the importance (weight) of every NN in the ensemble, and it corresponds to the position vector in formula (5). Furthermore, it is optimized by above-mentioned DePSO algorithm.

To selective NNE based on BiPSO algorithm, the value of every particle in every dimension is discrete 0 or 1. If the value is 1, it indicates that the corresponding individual NN does participate the ensemble; if the value is 0, it indicates that the corresponding individual NN doesn't participate the ensemble. Therefore, the problem of selecting the individual NN to construct NNE can be transformed to the PSO problem that selects optimal particles in n-dimensional 0 and -1 space. The above-mentioned PSO algorithm can only be used in continuous space. In order to solve the optimal problems of discrete space, Kennedy proposed separate binary PSO algorithm [33]. To binary PSO algorithm, position of particle in every dimension has only two situations 0 or 1. The updating method of velocity is same as continuous PSO, but the updating method of position depends on the situation transition probability determined by the velocity of particle. The bigger the velocity is, the bigger the possibility of particle value being 1 becomes, and vise versa. The iterative formula is given by

$$x_{i,d}^{k+1}=\begin{cases}1 & \rho_{i,d}^{k+1}\le sig\left(v_{i,d}^{k+1}\right)\\ 0 & \rho_{i,d}^{k+1}> sig\left(v_{i,d}^{k+1}\right)\end{cases} \tag{7}$$

where $\rho_{i,d}^{k+1}\in[0,1]$ is a random number. $sig(\cdot)$ is Sigmoid function, which is the velocity transform function. When adopting BiPSO algorithm constructs selective NNE, its actual output corresponding to the input $\boldsymbol{x}$ is given by

$$\bar{f}(\boldsymbol{x})=\frac{\sum_{i=1}^{n}\beta_i f_i(\boldsymbol{x})}{Number} \tag{8}$$

Similarly, $\boldsymbol{\beta}$ represents to the position vector in formula (7), and *Number* represents the count of value 1 in $\boldsymbol{\beta}$.

In order to ensure the diversity of particles in the evolution process, velocity vector is mutated by chaos theory. Chaos is a significant concept in modern science, and it is also a very important content in nonlinear science [34]. Although it seems chaotic, its internal structure is exquisite. It is sensitive to initial conditions and has several advantages, such as randomicity, ergodicity, and regularity. Compared with general random search methods, its local search ability in a small area is excellent, and the effectiveness of detailed search is good. In this chapter, Logistic chaotic mapping is used to generate chaotic signals, and the iterative formula is as follows:

$$c^{k+1}=\mu c^k\left(1-c^k\right) \tag{9}$$

where μ is control parameter. The Logistic chaos is complete when $\mu=4$ and $0<c^0<1$, $c^0\ne 0.25$, $c^0\ne 0.75$. After chaotic signal is generated, it is transformed to the optimized variables by the method of signal carrier, which makes them present chaotic state. Simultaneously, enlarge the ergodic scope of chaotic movement to the scope of optimized variables, and then directly search by using chaotic variables. After Chaotic signal is generated, $v_{i,d}^{k+1}$ in formula (7) is mutated by the following formula:

$$v_{i,d}^{k+1} = v_{i,d}^{k+1} + 2\alpha c^{k+1} - \alpha \tag{10}$$

where parameter α controls the extent of chaos mutation.

5.3 Numerical experiments

Experimental data is generated by Friedman # 1 function and Friedman # 3 function[29]-[31]. These two functions are representative and typical, and they are widely researched in regression analysis problem. The expressions of two functions are shown in table 1. Seven ensemble methods, which are simple evenness method (BEM) [29], generalized ensemble method (GEM) [20][21], linear regression (LR) method [29][30], principal components regression (PCR) method [29], genetic algorithm selective ensemble (GASEN) method [31], chaos DePSO method and chaos BiPSO method, are compared in this chapter.

According to every function's expression, 2200 data are generated, including 200 for training and the other 2000 for testing. By using NNs toolbox in Matlab, the variable gradient BP algorithm is used to train 20 networks. In order to ensure the differences of individual networks, the initial weights are generated randomly according to chaos theory. MSE_{train} and MSE_{test}, which respectively represent the mean square error of training set and testing set, show the generalization ability of NNE.

The computing results of Friedman # 1 function and Friedman # 3 function based on previously known methods are given in table 2, and computing results based on chaos DePSO method and chaos BiPSO method are given in table 3. In the computing process, the particles number is 30, the learning factors are selected according to the literature [35], namely $c_1 = 2.8$, $c_2 = 1.3$, the inertia weight ω in expression (4) changes linearly from 1 to 0.4 [36], and the evolutionary generation is 1000. In order to decrease the randomness, all of the ensemble methods are run 20 times repeatedly. The data in brackets of table 3 represent the count of NNs selected by the BiPSO algorithm when the corresponding computing result is gotten. Table 4 gives the statistic of NNs that participate the ensemble in all 20 times repeat computation in the BiPSO algorithm.

	Function expression	Variables range
Friedman # 1	$y = 10\sin(\pi x_1 x_2) + 20(x_3 - 0.5)^2 + 10x_4 + 5x_5$	$x_i \in U[0,1]$, $\varepsilon \sim N(0,1)$
Friedman # 3	$y = \tan^{-1}\left(\dfrac{x_2 x_3 - \dfrac{1}{x_2 x_4}}{x_1}\right) + \varepsilon$	$x_1 \in U[0,100]$, $x_2 \in U[40\pi, 560\pi$ $x_3 \in U[0,1]$, $x_4 \in U[0,11]$ $\varepsilon \sim N(0,0.01)$

Table 1. Experimental functions

Ensemble methods	Friedman # 1		Friedman # 3 ($\times 10^{-2}$)	
	MSE_{train}	MSE_{test}	MSE_{train}	MSE_{test}
BEM	0.207	0.341	0.258	0.820
GEM	0.270	0.467	0.368	1.624
LR	0.270	0.469	0.370	1.638
PCR	0.184	0.280	0.288	0.960
GASEN	0.183	0.281	0.252	0.806

Table 2. Experimental results of known ensemble methods

Friedman # 1 function

Ensemble methods	MSE_{train} ($\times 10^{-2}$)			MSE_{test} ($\times 10^{-2}$)		
	Best	Worst	Average	Best	Worst	Average
BiPSO	0.5195(9)	1.1867(7)	0.7699	1.1189(8)	4.6976(13)	2.1402
DePSO	0.5516	1.0008	0.7334	1.3102	4.1716	2.3651

Friedman # 3 function

Ensemble methods	MSE_{train} ($\times 10^{-2}$)			MSE_{test} ($\times 10^{-2}$)		
	Best	Worst	Average	Best	Worst	Average
BiPSO	0.0815(11)	0.9503(7)	0.2584	0.1889(11)	1.3688(10)	0.8287
DePSO	0.0884	0.9799	0.2617	0.1895	1.5768	0.8346

Table 3. Experimental results based on ensemble methods in this study

	Friedman # 1							Friedman # 3							
Number of NNs	5	7	8	9	10	11	13	5	6	7	8	9	10	11	12
Statistic of appearance number	1	1	6	5	4	2	1	1	2	2	3	4	4	3	1

Table 4. Statistic of NNs that participate ensemble based on the BiPSO algorithm (The total repeat times is 20)

From the experimental results, we can safely draw the conclusions as follows.① Friedman # 1 function and Friedman # 3 function exist multi-dimension collinearity and noise problems. Due to low restraint to weights, existing matrix inverse and sensitiveness to noise, the ensemble performances by GEM and LR methods obviously are inferior to other methods. BEM highly restricts the weights, and it is insensitive to noise. PCR can eliminate some noise by extracting the principal components. GASEN adopts simple average method, and it can also suppress the noise. The BiPSO algorithm and the DePSO algorithm have high restraint to weights, and they can suppress the noise, too. Because all the last 5 methods can restrict weights, the performances of NNE decided by them are good. ② The influences of multi-dimension collinearity and noise to the BiPSO algorithm and DePSO algorithm are least, so the performances of NNE decided by these two algorithms are obviously better than these of GEM and LR methods as well as PCR. Compared with BEM and GASEN, to Friedman # 1 function, the BiPSO algorithm and the DePSO algorithm are obvious better than BEM and GASEN; to Friedman # 3 function, the BiPSO algorithm and the DePSO algorithm are almost same with BEM and GASEN. If we increase the iteration number of PSO, the computing results would get better. In addition, with the same evolutionary generation, the computing time of PSO algorithm is shorter than that of GASEN because of its simplicity.③ Table 3 shows that the computing results based on the BiPSO algorithm are a little better than that based on the DePSO algorithm. ④ Table 4 shows that, to the selective NNE based on the BiPSO algorithm, the statistic of NNs selected is around 8 to 10 generally.

6. Modeling resonant frequency of rectangular MSA by using BiPSO-Based selective NNE

Applying the above-discussed selective NNE based on the BiPSO algorithm models the resonant frequency of rectangular MSA at its fundamental TM_{10} mode. The input sample sets is (W,L,h, ε_r), and the output is corresponding measured resonant frequency f_{ME}. The trained NNE can give the mapping between the relative parameters of the rectangular MSA and its measured resonant frequency. The training and testing data sets used in this study have been obtained from previous experimental works [12] [18]. The relative data of the MSA are given from column 2 to column 5 of table 5, and the measured resonant frequency for TM_{10} mode are given in column 6 of table 5. Total 33 data sets are listed in table 5. Twenty-six data sets are used to train the NNE, and the remaining seven data sets, marked with an asterisk in table 5, are used for testing the NNE. Considering the influence caused by randomness, the program is run 20 times repeatedly, and the result is their average value. The computing results by using the method proposed in this chapter are given in column 7 of table 5, and the results obtained by using NN proposed by Guney [16] and by Sagiroglu and Kalinli [17] are given from column 8 to column 10 in table 5. The f_{EDBD}, f_{DBD}, f_{BP} and f_{PTS} in table 5 represent, respectively, the values calculated by using the NNs trained with EDBD(Extended delta-bar-delta), DBD(Delta-bar-delta), BP(Back propagation), and PTS(Parallel tabu search) algorithms. Simultaneously, the sum of absolute errors between theoretical and experimental results for every method is listed in table 5, too. Table 5 shows that the results computed by using the chaos BiPSO-based selective NNE are better agreement with the experimental results as compared with these of the previous NNs in [16] and [17], which means that the model of this chapter is better than previous ones. In order to make a further comparison, the resonant frequency of rectangular MSA obtained by

No	W (cm)	L (cm)	h (cm)	ε_r	Measured f_{ME}	$f_{BiPSO\text{-}NNE}$	f_{EDBD}[16]	f_{DBD}[16]	f_{BP}[16]	f_{PTS}[17]
1	0.850	1.290	0.017	2.22	7740	7764	7935.5	7890.1	7858.6	7847.4
2*	0.790	1.185	0.017	2.22	8450	8169	8328.2	8226.0	8233.1	8148.6
3	2.000	2.500	0.079	2.22	3970	3980	4046.4	4023.0	4075.4	3971.5
4	1.063	1.183	0.079	2.25	7730	7698	7590.1	7567.3	7616.8	7881.6
5	0.910	1.000	0.127	10.20	4600	4601	4604.8	4573.9	4592.4	4603.4
6	1.720	1.860	0.157	2.33	5060	5036	4934.2	4914.0	4930.3	4969.4
7*	1.810	1.960	0.157	2.33	4805	4796	4699.2	4684.8	4703.3	4879.0
8	1.270	1.350	0.163	2.55	6560	6559	6528.6	6502.8	6516.5	6635.8
9	1.500	1.621	0.163	2.55	5600	5605	5503.2	5473.3	5449.0	5516.3
10*	1.337	1.412	0.200	2.55	6200	6196	6176.6	6142.6	6147.2	6205.7
11	1.120	1.200	0.242	2.55	7050	7064	7099.6	7064.3	7132.9	7113.8
12	1.403	1.485	0.252	2.55	5800	5803	5805.6	5768.8	5765.7	5794.3
13	1.530	1.630	0.300	2.50	5270	5279	5287.7	5260.3	5254.0	5313.0
14	0.905	1.018	0.300	2.50	7990	7983	7975.5	7881.8	8002.2	7776.6
15	1.170	1.280	0.300	2.50	6570	6577	6674.8	6632.8	6682.7	6481.9
16*	1.375	1.580	0.476	2.55	5100	5182	5311.8	5293.2	5291.4	5191.4
17	0.776	1.080	0.330	2.55	8000	7948	7911.1	7841.6	7942.5	7893.0
18	0.790	1.255	0.400	2.55	7134	7176	7183.2	7162.1	7215.9	7267.0
19	0.987	1.450	0.450	2.55	6070	6092	6173.0	6155.1	6170.2	6030.4
20*	1.000	1.520	0.476	2.55	5820	5853	5931.0	5918.0	5924.5	5780.3
21	0.814	1.440	0.476	2.55	6380	6425	6424.0	6417.5	6430.7	6500.0
22	0.790	1.620	0.550	2.55	5990	5925	5866.1	5873.9	5870.5	6004.0
23	1.200	1.970	0.626	2.55	4660	4641	4699.0	4728.0	4718.9	4562.8
24	0.783	2.300	0.854	2.55	4600	4603	4459.1	4517.1	4519.2	4591.2
25*	1.256	2.756	0.952	2.55	3580	3614	3659.8	3655.7	3644.6	3685.2
26	0.974	2.620	0.952	2.55	3980	3977	3952.9	3982.6	3975.9	3948.5
27	1.020	2.640	0.952	2.55	3900	3912	3905.4	3930.0	3922.2	3891.4
28	0.883	2.676	1.000	2.55	3980	3986	3938.8	3970.7	3965.3	3969.4
29	0.777	2.835	1.100	2.55	3900	3895	3825.5	3851.1	3845.9	3893.0
30	0.920	3.130	1.200	2.55	3470	3472	3481.4	3466.2	3458.4	3456.9
31*	1.030	3.380	1.281	2.55	3200	3196	3230.3	3184.7	3178.0	3167.0
32	1.265	3.500	1.281	2.55	2980	2982	3036.1	2965.6	2961.2	3035.5
33	1.080	3.400	1.281	2.55	3150	3149	3191.2	3140.4	3134.0	3135.3
		Errors				863	2392	2427	2372	2239

*Test data sets. Resonant frequencies and errors are in MHz.

Table 5. Resonant frequency of rectangular MSAs for TM_{10} mode

No	f_{ME}	[3]	[4]	[5]	[6]	[7]	[8]	[9]	[10]	[11]	[12]	[13]
1	7740	7804	7697	7750	7791	7635	7737	7763	7720	7717	412	7765
2	8450	8496	8369	8431	8478	8298	8417	8446	8396	8389	488	8451
3	3970	4027	3898	3949	3983	3838	3951	3950	3917	3887	510	3977
4	7730	7940	7442	7605	7733	7322	7763	7639	7551	7376	1610	7730
5	4600	4697	4254	4407	4641	4455	4979	4759	4614	4430	113	4618
6	5060	5283	4865	4989	5070	4741	5101	4958	4924	4797	1621	5077
7	4805	5014	4635	4749	4824	4520	4846	4724	4688	4573	1460	4830
8	6560	6958	6220	6421	6566	6067	6729	6382	6357	6114	2550	6563
9	5600	5795	5270	5424	5535	5158	5625	5414	5374	5194	1769	5535
10	6200	6653	5845	6053	6201	5682	6413	5987	5988	5735	2860	6193
11	7050	7828	6566	6867	7052	6320	7504	6682	6769	6433	4792	7030
12	5800	6325	5435	5653	5801	5259	6078	5552	5586	5326	3259	5787
13	5270	5820	4943	5155	5287	4762	5572	5030	5081	4842	3383	5273
14	7990	9319	7334	7813	7981	6917	8885	7339	7570	6822	8674	8101
15	6570	7412	6070	6390	6550	5794	7076	6135	6264	5951	5486	6543
16	5100	5945	4667	4993	5092	4407	5693	4678	4830	4338	5437	5193
17	8000	8698	6845	7546	7519	6464	8447	6889	7160	6367	8067	7948
18	7134	7485	5870	6601	6484	5525	7342	5904	6179	5452	7242	7169
19	6070	6478	5092	5660	5606	4803	6317	5125	5341	4735	6103	6026
20	5820	6180	4855	5423	5352	4576	6042	4886	5100	4513	5875	5817
21	6380	6523	5101	5823	5660	4784	6453	5122	5396	4729	6546	6515
22	5990	5798	4539	5264	5063	4239	5804	4550	4830	4196	5976	6064
23	4660	4768	3746	4227	4141	3526	4689	3770	3949	3479	4600	4613
24	4600	4084	3201	3824	3615	2938	4209	3168	3446	2921	4603	4550
25	3580	3408	2668	3115	2983	2485	3430	2670	2845	2461	3574	3628
26	3980	3585	2808	3335	3162	2590	3668	2790	3015	2572	3955	3956
27	3900	3558	2785	3299	3133	2573	3629	2771	2987	2555	3895	3907
28	3980	3510	2753	3294	3112	2522	3626	2721	2966	2509	3982	3922
29	3900	3313	2608	3147	2964	2364	3473	2554	2823	2356	3903	3747
30	3470	3001	2358	2838	2675	2146	3129	2317	2549	2137	3493	3381
31	3200	2779	2183	2623	2474	1992	2889	2151	2357	1983	3197	3123
32	2980	2684	2102	2502	2370	1936	2752	2086	2259	1924	2982	2972
33	3150	2763	2168	2600	2453	1982	2863	2139	2338	1972	3160	3096
Errors		13136	24097	11539	12322	30669	8468	22572	18148	30504	56698	1393

Resonant frequencies and errors are in MHz.

Table 6. Resonant frequency obtained from traditional methods for rectangular MSAs and sum of absolute errors between experimental results and theoretical results conventional methods [3]-[13] are listed in table 6. The sum of absolute errors between experimental and theoretical results for every method is also listed in the last row of table 6. It is clear from table 5 and table 6 that the computing results of the chaos BiPSO-based selective NNE are better than these of previously proposed methods, which proves the validity of the algorithm further.

7. Conclusion

Selective neural network ensemble (NNE) methods based on decimal particle swarm optimization (DePSO) algorithm and binary particle swarm optimization (BiPSO) algorithm are proposed in this study. In these algorithms, optimally select neural networks (NNs) to construct NNE with the aid of particle swarm optimization (PSO) algorithm, which can maintain the diversity of NNs. In the process of ensemble, the performance of NNE may be improved by appropriate restriction on combination weights based on BiPSO algorithm. And this may avoid calculating the matrix inversion and decrease the multi-dimensional collinearity and the over-fitting problem of noise. In order to effectively ensure the particles diversity of PSO algorithm, chaos mutation is adopted during the iteration process according to randomicity, ergodicity and regularity in chaos theory. Experimental results show that the chaos BiPSO algorithm can improve the generalization ability of NNE. By using the chaos BiPSO-based selective NNE, resonant frequency of rectangular microstrip antenna (MSA) is modeled, and the computing results are better than available ones, which mean that the proposed NNE in this study is effective. The method of NNE proposed in this study may be conveniently extended to other microwave engineering and designs.

8. Acknowledge

This work is supported by Pre-research foundation of shipping industry of China under grant No. 10J3.5.2, and Natural Science Foundation of Higher Education of Jiangsu Province of China under grant No. 07KJB510032.

9. Reference

[1] Wong K L, "*Compact and broadband microstrip antennas*", New York: John Wiley & Sons Inc., 2002.
[2] Kumar G, and Ray K P, "*Broadband microstrip antennas*", MA: Artech House, 2003.
[3] Howell J Q, "Microstrip antennas", *IEEE Transactions on Antennas and Propagation*, 1975, 23(1): 90-93.
[4] Hammerstad E O, "Equations for microstrip circuits design", Proc. 5th Eur. Microw. Conf., Hamburg, Germany, Sep. 1975, pp. 268–272.
[5] Carver K R, "Practical analytical techniques for the microstrip antenna", Proc. Workshop Printed Circuit Antenna Tech., New Mexico State Univ., Las Cruces, NM, Oct. 1979, pp. 7.1-7.20.
[6] Bahl I J, and Bhartia P, "*Microstrip Antennas*", MA: Artech House, 1980.
[7] James J R, Hall P S, and Wood C, "*Microstrip antennas-theory and design*", London: Peregrinus, 1981.
[8] Sengupta D L, "Approximate expression for the resonant frequency of a rectangular patch antenna", *Electronics Letters*, 1983, 19: 834-835.

[9] Garg R, and Long S A, "Resonant frequency of electrically thick rectangular microstrip antennas", *Electronics Letters*, 1987, 23: 1149-1151.
[10] Chew W C, and Liu Q, "Resonance frequency of a rectangular microstrip patch", *IEEE Transactions on Antennas Propagation*, 1988, 36: 1045-1056.
[11] Guney K, "A new edge extension expression for the resonant frequency of electrically thick rectangular microstrip antennas", *Int. J. Electron.*, 1993, 75: 767-770.
[12] Kara M, "Closed-form expressions for the resonant frequency of rectangular microstrip antenna elements with thick substrates", *Microwave and Optical Technology Letters*,1996, 12: 131-136.
[13] Guney K, "A new edge extension expression for the resonant frequency of rectangular microstrip antennas with thin and thick substrates", *J. Commun. Tech. Electron.*, 2004, 49: 49 53.
[14] Zhang Q J, and Gupta K C, "*Neural networks for RF and microwave design*", MA: Artech House, 2000.
[15] Christodoulou C, and Georgiopoulos M, "*Applications of Neural Networks in Electromagnetic*", MA: Artech House, 2001.
[16] Guney K, Sagiroglu S, and Erler M, "Generalized neural method to determine resonant frequencies of various microstrip antennas", *International Journal of RF and Microwave Computer-Aided Engineering*, 2002, 12(1): 131-139.
[17] Sagiroglu S, and Kalinli A, "Determining resonant frequencies of various microstrip antennas within a single neural model trained using parallel tabu search algorithm", *Electromagnetics*, 2005, 25(6): 551-565.
[18] Kara M, "The resonant frequency of rectangular microstrip antenna elements with various substrate thicknesses", *Microwave and Optical Technology Letters*, 1996, 11: 55-59.
[19] Hansen L K, and Salamon P, "Neural network ensembles", *IEEE Transactions on Pattern Analysis and Machine Intelligence*, 1990, 12(10): 993-1001.
[20] Kennedy J, and Eberhart R C, "Particle Swarm Optimization", IEEE International Conference on Neural Networks, Piscataway, NJ: IEEE Press, 1995, 1942-1948.
[21] Zeng J C, Jie J, and Cui Z H, "*Particle swarm optimization*", Beijing: Science Press, 2004.
[22] Clerc M, "*Particle Swarm Optimization*", ISTE Publishing Company, 2006.
[23] R. Poli. Analysis of the publications on the applications of particle swarm optimization. Journal of Artificial Evolution and Applications, 2008, Article No. 4.
[24] Robinson J, and Rahmat-Samii Y, "Particle swarm optimization in electromagnetics", *IEEE Transactions on Antennas and Propagation*, 2004, 52(2): 397-407.
[25] Mussetta M, Selleri S, Pirinoli P, et al., "Improved Particle Swarm Optimization algorithms for electromagnetic optimization", *Journal of Intelligent and Fuzzy Systems*, 2008, 19(1): 75-84.
[26] M. T. Hagan, H. B. Demuth, and M. H. Beale, Neural Network Design, Boston: PWS Pub. Co., 1995.
[27] S. Haykin. Neural Networks: A Comprehensive Foundation (2nd Edition), Prentice Hall, 1999.
[28] Y. B. Tian, Hybrid neural network techniques, Beijing: Science Press, 2010.
[29] Merz C J, and Pazzani M J, "A principal components approach to combining regression estimates", *Machine Learning*, 1999, 36 (1-2): 9-32.

[30] Hashem S, "Treating harmful collinearity in neural network ensembles", In: Sharkey A J C, ed. Combining artificial neural nets: Ensemble and modular multi2net systems, Great Britain: Springer-Verlag London Limited, 1999. 101-123.

[31] Zhou Zhihua, Wu Jianxin, and Tang Wei, "Ensembling neural networks: Many could be better than all", *Artificial Intelligence*, 2002, 137 (1-2): 239-263.

[32] Dietterich T G, "An experimental comparison of three methods for constructing ensembles of decision trees: Bagging, boosting, and randomization", *Machine Learning*, 2000, 40: 139-157.

[33] Kennedy J, and Eberhart R, "A discrete binary version of the particle swarm optimization", Proceedings IEEE International Conference on Computational Cybernetics and Simulation. Piscataway, NJ: IEEE, 1997: 4104-4108.

[34] Huang R S, and Huang H, "*Chaos and its applications*", China Wuhan: Wuhan University press, 2005.

[35] Zhang L P, "*Theory and applications of particle swarm optimization*", Ph. D. dissertation, Zhejiang University, China, 2005.

[36] Shi Y, and Eberhart R, "Empirical study of particle swarm optimization", Proceedings of the 1999 Congress on Evolutionary Computation, 1999: 1945-1950.

5

Microstrip Antennas Conformed onto Spherical Surfaces

Daniel B. Ferreira and J. C. da S. Lacava
Technological Institute of Aeronautics
Brazil

1. Introduction

Microstrip antennas are customary components in modern communications systems, since they are low-profile, low-weight, low-cost, and well suited for integration with microwave circuits. Antennas printed on planar surfaces or conformed onto cylindrical bodies have been discussed in many publications, being the subject of a variety of analytical and numerical methods developed for their investigation (Josefsson & Persson, 2006; Garg et al., 2001; Wong, 1999). However, such is not the case of spherical microstrip antennas and arrays composed of these radiators. Even commercial electromagnetic software, like *HFSS*® and *CST*®, do not provide a tool to assist the design of spherical antennas and arrays, i.e., electromagnetic simulators do not have an estimator tool for establishing the initial dimensions of a spherical microstrip antenna for further numerical analysis, as available for planar geometries. Moreover, this software is time-consuming when utilized to simulate spherical radiators, hence it is desirable that the antenna geometry to be analyzed is not too far off from the final optimized one, otherwise the project cost will likely be affected. Nonetheless, spherical microstrip antenna arrays have a great practical interest because they can direct a beam in an arbitrary direction throughout the space, i.e., without limiting the scan angles, differently from the planar antenna behaviour. This characteristic makes them feasible for use in communication satellites and telemetry (Sipus et al., 2006), for example.

Rigorous analysis of spherical microstrip antennas and their respective arrays has been conducted through the Method of Moments (MoM) (Tam et al., 1995; Wong, 1999; Sipus et al., 2006). But the MoM involves highly complex and time-consuming calculations. On the other hand, whenever the objective is the analysis of spherical thin radiators, the cavity model (Lima et al., 1991) can be applied, instead of the MoM. However, for both MoM and cavity model, the behaviour of the antenna input impedance and radiated electric field is described by the associated Legendre functions, hence efficient numerical routines for their evaluation are required, otherwise the scope of the antennas analyzed is restricted.

In order to overcome the drawbacks described above, a *Mathematica*®-based CAD software capable of performing the analysis and synthesis of spherical-annular and -circular thin microstrip antennas and their respective arrays with high computational efficiency is presented in this chapter. It is worth mentioning that the theoretical model utilized in the CAD can be extended to other canonical spherical patch geometries such as rectangular or triangular ones. The *Mathematica*® package, an integrated scientifical computing software with a vast collection of built-in functions, was chosen mainly due to its powerful algorithms for calculating cylindrical and spherical harmonics functions what makes it suitable for the analysis of conformed antennas. *Mathematica®* permits the analysis and synthesis of various spherical microstrip radiators, thus avoiding the use of the normalized Legendre functions that are sometimes employed to overcome numerical difficulties (Sipus et al., 2006). Furthermore, it is important to point out that the developed CAD does not require a powerful computer to run on, working well and quickly in a regular classroom PC, since its code does not utilize complex numerical techniques, like MoM or finite element method (FEM). In Section 2, the theoretical model implemented in the developed CAD to evaluate the antenna input impedance, quality factor, radiation pattern and directivity is discussed. Furthermore, comparisons between the CAD results and the *HFSS®* full wave solver data are presented in order to validate the accuracy of the utilized technique.

An effective procedure, based on the global coordinate system technique (Sengupta, 1968), to determine the radiation patterns of thin spherical meridian and circumferential arrays is utilized in the special-purpose CAD, as addressed in Section 3. The array radiation patterns so obtained with the CAD are also compared to those from the *HFSS®* software. Section 4 is devoted to present an alternative strategy for fabricating a low-cost spherical-circular microstrip antenna along with the respective experimental results supporting the proposed antenna fabrication approach.

2. Analysis and synthesis of spherical thin microstrip antennas

The geometry of a probe-fed spherical-annular microstrip antenna embedded in free space (electric permittivity ε_0 and magnetic permeability μ_0) is shown in Fig. 1. It is composed of a metallic sphere of radius a, called ground sphere, covered by a dielectric layer (ε and μ_0) of thickness $h = b - a$.

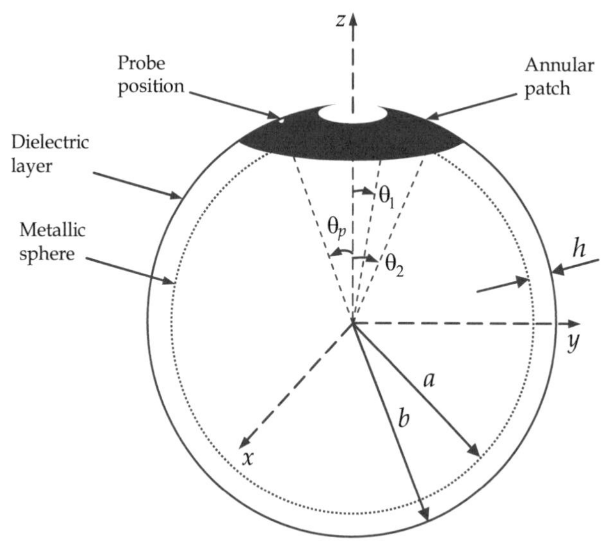

Fig. 1. Geometry of a probe-fed spherical-annular microstrip antenna.

A symmetrical annular metallic patch, defined by the angles θ_1 and θ_2 ($\theta_2 > \theta_1 > 0$), is fed by a coaxial probe positioned at (θ_p, ϕ_p). The radiators treated in this chapter are electrically thin, i.e., $h << \lambda$ (λ is the wavelength in the dielectric layer), so the cavity model (Lo et al.,

1979) is well suited for the analysis of such antennas. Based on this model it is possible to develop expressions for computing the antenna input impedance and for estimating the electric surface current density on the patch without employing any complex numerical method such as the MoM.

Before starting the input impedance calculation, the expression for computing the resonant frequencies of the modes established in a lossless equivalent cavity is determined. In the case of electrically thin radiators, the electric field within the cavity can be considered to have a radial component only, which is r-independent. Therefore, applying Maxwell's equations to the dielectric layer region, and disregarding the feeder presence, the following equation for the r-component of the electric field is obtained

$$\frac{1}{a^2 \sin\theta}\frac{\partial}{\partial\theta}\left(\sin\theta\frac{\partial E_r}{\partial\theta}\right)+\frac{1}{a^2\sin^2\theta}\frac{\partial^2 E_r}{\partial\phi^2}+k^2E_r=0\,, \tag{1}$$

where $k^2 = \omega^2\mu_0\varepsilon$ and ω denotes the angular frequency. Consequently, only TM^r modes can be established in the equivalent cavity.

Solving the wave equation (1) via the method of separation of variables (Balanis, 1989), results in the electric field

$$E_r(\theta,\phi)=[A\mathrm{P}_\ell^m(\cos\theta)+B\mathrm{Q}_\ell^m(\cos\theta)][C\cos(m\phi)+D\sin(m\phi)]\,, \tag{2}$$

where $\mathrm{P}_\ell^m(.)$ and $\mathrm{Q}_\ell^m(.)$ are the associated Legendre functions of ℓ-th degree and m-th order of the first and the second kinds, respectively, $\ell(\ell+1) = k^2a^2$ and A, B, C and D are constants dependent on the boundary conditions.

Enforcing the boundary conditions related to the equivalent cavity of annular geometry and taking into account that it is symmetrical in relation to the z-axis, the electric field (2) reduces to

$$E_r(\theta,\phi)=E_{\ell m}\,\mathrm{R}_\ell^m(\cos\theta)\cos(m\phi)\,, \tag{3}$$

where

$$\mathrm{R}_\ell^m(\cos\theta)=\sin\theta_{1c}[\mathrm{P}_\ell^{m\prime}(\cos\theta_{1c})\mathrm{Q}_\ell^m(\cos\theta)-\mathrm{Q}_\ell^{m\prime}(\cos\theta_{1c})\mathrm{P}_\ell^m(\cos\theta)]\,, \tag{4}$$

$m = 0, 1, 2,\ldots$ and the index ℓ must satisfy the transcendental equation

$$\mathrm{P}_\ell^{m\prime}(\cos\theta_{1c})\mathrm{Q}_\ell^{m\prime}(\cos\theta_{2c})-\mathrm{Q}_\ell^{m\prime}(\cos\theta_{1c})\mathrm{P}_\ell^{m\prime}(\cos\theta_{2c})=0\,, \tag{5}$$

with the angles θ_{1c} and θ_{2c} ($\theta_{2c} > \theta_{1c}$) indicating the equivalent cavity borders in the θ direction, i.e., $\theta_{1c} \le \theta \le \theta_{2c}$, $0 \le \phi < 2\pi$, $E_{\ell m}$ are the coefficients of the natural modes and the prime denotes a derivative. Hence, once the indexes ℓ and m are determined it is possible to evaluate the $TM_{\ell m}$ mode resonant frequency from the following expression

$$f_{\ell m}=\frac{\sqrt{\ell(\ell+1)}}{2\pi a\sqrt{\mu_0\varepsilon}}\,. \tag{6}$$

Before solving the transcendental equation (5) it is necessary to determine the equivalent cavity dimensions θ_{1c} and θ_{2c}, which correspond to the actual patch dimensions with the addition of the fringe field extension. However, differently from planar microstrip antennas, the literature does not currently present expressions for estimating the dimensions of spherical equivalent cavities based on the physical antenna dimensions and the dielectric substrate characteristics. Therefore, in this chapter, the expressions used for estimating the equivalent cavity dimensions of a planar-annular microstrip antenna are extended to the spherical-annular case (Kishk, 1993), i.e., the spherical-annular equivalent cavity arc lengths were considered equal to the respective linear dimensions of the planar-annular equivalent cavity. The proposed expressions are given below; equations (7.a) and (7.b),

$$\theta_{1c} = \theta_1 \sqrt{1 - \frac{2h\,\mathrm{F}(\theta_1)}{\pi b \theta_1 \varepsilon_r}}\,, \tag{7.a}$$

$$\theta_{2c} = \theta_2 \sqrt{1 + \frac{2h\,\mathrm{F}(\theta_2)}{\pi b \theta_2 \varepsilon_r}}\,, \tag{7.b}$$

where $\mathrm{F}(\theta) = \ell\mathrm{n}(b\theta / 2h) + 1.41\varepsilon_r + 1.77 + h(0.268\varepsilon_r + 1.65) / b\theta$ and ε_r is the relative electric permittivity of the dielectric substrate.

2.1 Input impedance

The input impedance of the spherical-annular microstrip antenna illustrated in Fig. 1 fed by a coaxial probe can be evaluated following the procedure proposed in (Richards et al., 1981), i.e., the coaxial probe is modelled by a strip of current whose electric current density is given by,

$$\vec{J}_f(\theta,\phi) = \frac{1}{a^2 \sin\theta_p} J(\phi)\delta(\theta - \theta_p)\hat{r}\,, \tag{8}$$

where $\delta(.)$ indicates the Dirac's delta function and

$$J(\phi) = \begin{cases} J_0, \text{ if } \phi_p - \Delta\phi / 2 \le \phi \le \phi_p + \Delta\phi / 2 \\ 0, \text{ otherwise} \end{cases} \tag{9}$$

with $\Delta\phi$ denoting the strip angular length relative to the ϕ–direction. In our analysis, also following the procedure established in (Richards et al., 1981) for planar microstrip antennas, the strip arc length has been assumed to be five times the coaxial probe diameter d, expressed as

$$\Delta\phi = 5d / a\sin\theta_p\,. \tag{10}$$

It is important to point out that the electric current density (8) is an r-independent function since the antenna under analysis is electrically thin. Thus, to take into account the current strip, the wave equation (1) for the electric field is modified to

$$\frac{1}{a^2 \sin\theta}\frac{\partial}{\partial\theta}\left(\sin\theta \frac{\partial E_r}{\partial\theta}\right) + \frac{1}{a^2 \sin^2\theta}\frac{\partial^2 E_r}{\partial\phi^2} + k^2 E_r = j\omega\mu_0 \vec{J}_f(\theta,\phi)\cdot\hat{r}\,. \tag{11}$$

Expanding the r-component of the electric field into its eigenmodes (3), the solution for equation (11) is given by

$$E_r(\theta,\phi) = j\frac{\omega\mu_0 J_0 \Delta\phi}{\pi a^2}\sum_m\sum_\ell \frac{R_\ell^m(\cos\theta_p)\,\mathrm{sinc}(m\Delta\phi/2)\cos(m\phi_p)}{\xi_m[k^2-k_{\ell m}^2]\displaystyle\int_{\nu=\cos\theta_{2c}}^{\cos\theta_{1c}}[R_\ell^m(\nu)]^2 d\nu} R_\ell^m(\cos\theta)\cos(m\phi), \quad (12)$$

where

$$\xi_m = \begin{cases} 2, \text{ if } m=0 \\ 1, \text{ otherwise} \end{cases},$$

$k_{\ell m} = \sqrt{\ell(\ell+1)}/a$ and $\mathrm{sinc}(x)=\sin(x)/x$.

Since the procedure just described has been developed for ideal cavities, equation (12) is purely imaginary. So, for incorporating the radiated power and the dielectric and metallic losses into the cavity model, the concept of effective loss tangent, $\tan\delta_{eff}$, (Richards et al., 1979) is employed. Based on this approach, the wave number k is replaced by an effective wave number

$$k_{eff} = k\sqrt{1 - j\tan\delta_{eff}}. \quad (13)$$

Once the electric field inside the equivalent cavity has been determined, the antenna input voltage V_{in} can be computed from the expression,

$$V_{in} = -\bar{E}_r h, \quad (14)$$

where $\bar{E}_r$ denotes the average value of $E_r(\theta_p,\phi)$ over the strip of current. Consequently, the input impedance Z_{in} of the spherical-annular microstrip antenna is given by

$$Z_{in} = \frac{V_{in}}{J_0\Delta\phi} = j\frac{\omega\mu_0 h}{\pi a^2}\sum_m\sum_\ell \frac{[R_\ell^m(\cos\theta_p)]^2\,\mathrm{sinc}^2(m\Delta\phi/2)\cos^2(m\phi_p)}{\xi_m[k_{\ell m}^2 - k^2(1-j\tan\delta_{eff})]\displaystyle\int_{\nu=\cos\theta_{2c}}^{\cos\theta_{1c}}[R_\ell^m(\nu)]^2 d\nu}. \quad (15)$$

An alternative representation for frequencies close to the TM_{LM} resonant mode but sufficiently apart from the other modes can be obtained by rewriting the antenna input impedance as

$$Z_{in} \cong \frac{j\omega\alpha_{LM}}{\omega_{LM}^2 - (1-j\tan\delta_{eff})\omega^2} + j\omega\sum_m\sum_{\substack{\ell \\ (\ell,m)\neq(L,M)}} \frac{\alpha_{\ell m}}{\omega_{\ell m}^2 - \omega^2}, \quad (16)$$

where $\alpha_{\ell m} = h[R_\ell^m(\cos\theta_p)]^2\,\mathrm{sinc}^2(m\Delta\phi/2)\cos^2(m\phi_p)/\pi a^2\varepsilon\xi_m \displaystyle\int_{\nu=\cos\theta_{2c}}^{\cos\theta_{1c}}[R_\ell^m(\nu)]^2 d\nu$.

The expression (16) corresponds to the equivalent circuit shown in Fig. 2, i.e., a parallel RLC circuit with a series inductance L_p. In this case, the series inductance represents the probe effect and its value is that of the double sum in (16). However, as this is a slowly convergent series, the developed CAD utilizes, alternatively, the equation due to (Damiano & Papiernik, 1994) for calculating the probe reactance X_p, given by

$$X_p = -j60k_0 h\ell n(kd/2), \quad (17)$$

where $k_0 = \omega\sqrt{\mu_0\varepsilon_0}$ and provided $k_0 d << 0.2$.

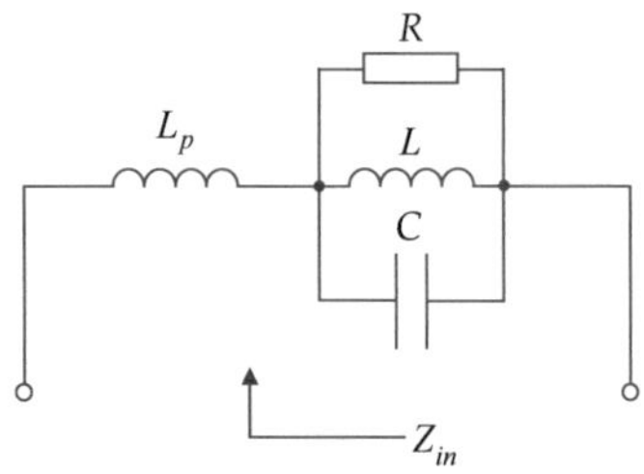

Fig. 2. Simplified equivalent circuit for thin microstrip antennas.

The previous expressions developed for computing the resonant frequencies and the input impedance of spherical-annular microstrip antennas can also be used for analysing wraparound radiators. However, in the limit case when $\theta_1 \to 0$, i.e., the antenna patch corresponding to a circular one (Fig. 3), the associated Legendre function of the second kind becomes unbounded for $\theta \to 0$, so it is no longer part of the function that describes the electromagnetic field within the equivalent cavity. So, to obtain the expressions for spherical-circular microstrip antennas it is enough to eliminate the Legendre function of the second kind from the previously developed solution for spherical-annular antennas. These expressions are presented in Table 1. In Section 2.3 examples are given for spherical-annular and -circular microstrip antennas.

Resonant frequency	$f_{\ell m} = \dfrac{\sqrt{\ell(\ell+1)}}{2\pi a\sqrt{\mu_0\varepsilon}}$, the index ℓ is obtained from $\mathrm{P}_\ell^{m\prime}(\cos\theta_{2c}) = 0$
Input impedance	$Z_{in} = j\dfrac{\omega\mu_0 h}{\pi a^2}\sum_m\sum_\ell \dfrac{[\mathrm{P}_\ell^m(\cos\theta_p)]^2\,\mathrm{sinc}^2(m\Delta\phi/2)\cos^2(m\phi_p)}{\xi_m[k_{\ell m}^2 - k^2(1-j\tan\delta_{eff})]\int_{\nu=\cos\theta_{2c}}^{\cos\theta_{1c}}[\mathrm{P}_\ell^m(\nu)]^2 d\nu}$

Table 1. Spherical-circular microstrip antenna expressions.

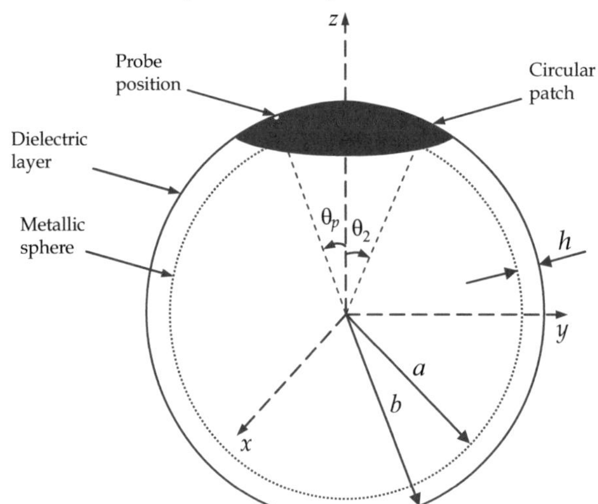

Fig. 3. Geometry of a probe-fed spherical-circular microstrip antenna.

2.2 Radiated far electric field

In the developed CAD, the electric surface current method (Tam & Luk, 1991) is used to determine the far electric field radiated by the thin spherical-annular and -circular microstrip antennas. This method is very convenient in the case of spherical-annular and -circular patches since both are electrically symmetrical. As a result, no numerical integration is required for the calculation of the spectral current density and the radiated power. Moreover, differently from planar and cylindrical geometries, where truncation of the ground layer and diffraction at the edges of the conducting surfaces affect the radiation patterns, thin spherical microstrip patches of canonical geometries can be efficiently analyzed by combining the cavity model with the electric surface current method.

The procedure proposed here starts from observing that the geometry shown in Fig 1 (or in Fig. 3) can be treated as a three-layer structure, made out of ground, dielectric substrate and free space. Consequently, its spectral dyadic Green's function, necessary for calculating the far electric field via the electric surface current method, can be easily evaluated using an equivalent circuital model (Giang et al., 2005). As it is based on the (*ABCD*) matrix (transmission matrix) concept, *Mathematica®*'s symbolic capability can be used for calculating the matrices involved. The technique for establishing the structure's equivalent circuital model and, consequently, its spectral dyadic Green's function is presented next.

The fields within the dielectric layer can be written as the sum of TE^r and TM^r modes with the aid of the vector auxiliary potential approach (Balanis, 1989). In this case, the expressions for the transversal components of the electromagnetic field are given by

$$E_\theta(r,\theta,\phi)=\sum_{m=-\infty}^{+\infty}\sum_{n=|m|}^{+\infty}\frac{1}{r}\left\{\frac{1}{j\sqrt{\mu_0\varepsilon}}\left[A_n^m\hat{H}_n^{(1)\prime}(kr)+B_n^m\hat{H}_n^{(2)\prime}(kr)\right]\frac{d}{d\theta}P_n^{|m|}(\cos\theta)\right.$$

$$\left.+\frac{m}{j\varepsilon\sin\theta}\left[C_n^m\hat{H}_n^{(1)}(kr)+D_n^m\hat{H}_n^{(2)}(kr)\right]P_n^{|m|}(\cos\theta)\right\}e^{jm\phi},\tag{18}$$

$$E_\phi(r,\theta,\phi)=\sum_{m=-\infty}^{+\infty}\sum_{n=|m|}^{+\infty}\frac{1}{r}\left\{\frac{m}{\sqrt{\mu_0\varepsilon}\sin\theta}\left[A_n^m\hat{H}_n^{(1)\prime}(kr)+B_n^m\hat{H}_n^{(2)\prime}(kr)\right]P_n^{|m|}(\cos\theta)\right.$$

$$\left.+\frac{1}{\varepsilon}\left[C_n^m\hat{H}_n^{(1)}(kr)+D_n^m\hat{H}_n^{(2)}(kr)\right]\frac{d}{d\theta}P_n^{|m|}(\cos\theta)\right\}e^{jm\phi},\tag{19}$$

$$H_\theta(r,\theta,\phi)=\sum_{m=-\infty}^{+\infty}\sum_{n=|m|}^{+\infty}\frac{1}{r}\left\{\frac{jm}{\mu_0\sin\theta}\left[A_n^m\hat{H}_n^{(1)}(kr)+B_n^m\hat{H}_n^{(2)}(kr)\right]P_n^{|m|}(\cos\theta)\right.$$

$$\left.+\frac{1}{j\sqrt{\mu_0\varepsilon}}\left[C_n^m\hat{H}_n^{(1)\prime}(kr)+D_n^m\hat{H}_n^{(2)\prime}(kr)\right]\frac{d}{d\theta}P_n^{|m|}(\cos\theta)\right\}e^{jm\phi},\tag{20}$$

$$H_\phi(r,\theta,\phi)=\sum_{m=-\infty}^{+\infty}\sum_{n=|m|}^{+\infty}\frac{1}{r}\left\{-\frac{1}{\mu_0}\left[A_n^m\hat{H}_n^{(1)}(kr)+B_n^m\hat{H}_n^{(2)}(kr)\right]\frac{d}{d\theta}P_n^{|m|}(\cos\theta)\right.$$

$$+\frac{m}{\sqrt{\mu_0\varepsilon}\sin\theta}\left[C_n^m\hat{H}_n^{(1)\prime}(kr)+D_n^m\hat{H}_n^{(2)\prime}(kr)\right]P_n^{|m|}(\cos\theta)\Bigg\}e^{jm\phi}, \tag{21}$$

where the coefficients A_n^m, B_n^m, C_n^m and D_n^m are dependent on the boundary conditions at the interfaces $r=a$ and $r=b$, and $\hat{H}_n^{(\tau)}(.)$ is the Schelkunoff spherical Hankel function of n-th order and τ-th kind (τ = 1 or 2). The fields (18) to (21) can be rewritten in a more adequate form as

$$\begin{bmatrix}E_\theta\\E_\phi\end{bmatrix}=\sum_{m=-\infty}^{+\infty}\sum_{n=|m|}^{+\infty}\tilde{L}(n,m,\theta)\cdot\vec{\mathsf{E}}(r,n)e^{jm\phi}, \tag{22}$$

$$\begin{bmatrix}H_\theta\\H_\phi\end{bmatrix}=\sum_{m=-\infty}^{+\infty}\sum_{n=|m|}^{+\infty}\tilde{L}(n,m,\theta)\cdot\vec{\mathsf{H}}(r,n)e^{jm\phi}, \tag{23}$$

where

$$\tilde{L}(n,m,\theta)=\begin{bmatrix}\frac{d}{d\theta}P_n^{|m|}(\cos\theta) & -jm\frac{P_n^{|m|}(\cos\theta)}{\sin\theta}\\ jm\frac{P_n^{|m|}(\cos\theta)}{\sin\theta} & \frac{d}{d\theta}P_n^{|m|}(\cos\theta)\end{bmatrix}, \tag{24}$$

$$\vec{\mathsf{E}}(r,n)=\begin{bmatrix}\mathsf{E}_\theta\\\mathsf{E}_\phi\end{bmatrix}=\begin{bmatrix}\frac{\omega}{jkr}\left[A_n^m\hat{H}_n^{(1)\prime}(kr)+B_n^m\hat{H}_n^{(2)\prime}(kr)\right]\\ \frac{1}{r\varepsilon}\left[C_n^m\hat{H}_n^{(1)}(kr)+D_n^m\hat{H}_n^{(2)}(kr)\right]\end{bmatrix}, \tag{25}$$

$$\vec{\mathsf{H}}(r,n)=\begin{bmatrix}\mathsf{H}_\theta\\\mathsf{H}_\phi\end{bmatrix}=\begin{bmatrix}\frac{\omega}{jkr}\left[C_n^m\hat{H}_n^{(1)\prime}(kr)+D_n^m\hat{H}_n^{(2)\prime}(kr)\right]\\ -\frac{1}{r\mu_0}\left[A_n^m\hat{H}_n^{(1)}(kr)+B_n^m\hat{H}_n^{(2)}(kr)\right]\end{bmatrix}, \tag{26}$$

and the argument (r,θ,ϕ) was omitted in (22) and (23) only for simplifying the notation. The vectors $\vec{\mathsf{E}}(r,n)$ and $\vec{\mathsf{H}}(r,n)$ are the transversal electric and magnetic fields in the spectral domain, respectively. In this chapter, the pair of vector-Legendre transforms (Sipus et al., 2006) is defined as follows,

$$\vec{\mathsf{X}}(r,n)=\frac{1}{2\pi S(n,m)}\int_{\phi=0}^{2\pi}\int_{\theta=0}^{\pi}\tilde{L}(n,m,\theta)\cdot\vec{X}(r,\theta,\phi)\sin\theta e^{-jm\phi}d\theta d\phi \tag{27}$$

and

$$\vec{X}(r,\theta,\phi)=\sum_{m=-\infty}^{+\infty}\sum_{n=|m|}^{+\infty}\tilde{L}(n,m,\theta)\cdot\vec{\mathsf{X}}(r,n)e^{jm\phi}, \tag{28}$$

where $S(n,m) = 2n(n+1)(n+|m|)!/(2n+1)(n-|m|)!$ and $\vec{\mathsf{X}}(r,n)$, the vector-Legendre transform of $\vec{X}(r,\theta,\phi)$, has only the θ and/or ϕ components.

From evaluating the expressions (25) and (26) at the lower ($r=a$) and upper ($r=b$) interfaces it is possible to establish the following relation

$$\begin{bmatrix} \vec{\mathsf{E}}(a,n) \\ \vec{\mathsf{H}}(a,n) \end{bmatrix} = \begin{bmatrix} \tilde{V} & \tilde{Z} \\ \tilde{Y} & \tilde{B} \end{bmatrix} \cdot \begin{bmatrix} \vec{\mathsf{E}}(b,n) \\ \vec{\mathsf{H}}(b,n) \end{bmatrix}, \tag{29}$$

and the matrices $\tilde{V}$, $\tilde{Z}$, $\tilde{Y}$ and $\tilde{B}$ can be found in (Ferreira, 2009).

Based on equation (29), the two-port network illustrated in Fig. 4, representing the dielectric layer, can be defined. The related transmission (*ABCD*) matrix is given in (30).

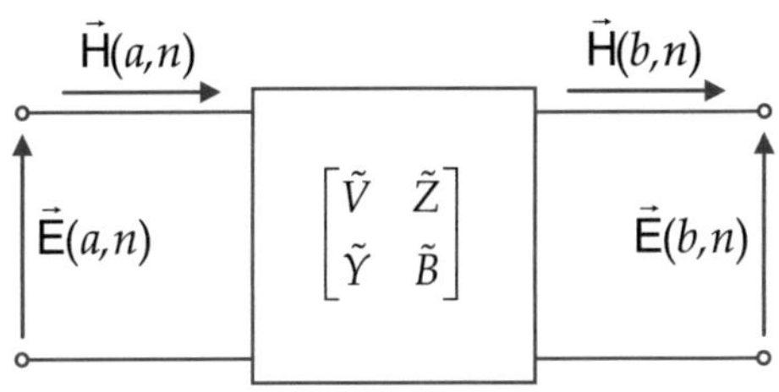

$$(ABCD) = \begin{bmatrix} \tilde{V} & \tilde{Z} \\ \tilde{Y} & \tilde{B} \end{bmatrix}. \tag{30}$$

Fig. 4. Transmission (*ABCD*) network.

In a similar way, the following relation between the free-space spectral electric $\vec{\mathsf{E}}_0$ and magnetic $\vec{\mathsf{H}}_0$ transversal field components can be determined,

$$\vec{\mathsf{H}}_0(b,n) = \tilde{Y}_0 \cdot \vec{\mathsf{E}}_0(b,n), \tag{31}$$

where

$$\tilde{Y}_0 = \begin{bmatrix} 0 & \hat{\mathrm{H}}_n^{(2)\prime}(k_0 b) / j\eta_0 \hat{\mathrm{H}}_n^{(2)}(k_0 b) \\ \hat{\mathrm{H}}_n^{(2)}(k_0 b) / j\eta_0 \hat{\mathrm{H}}_n^{(2)\prime}(k_0 b) & 0 \end{bmatrix}, \tag{32}$$

and η_0 denotes the free space intrinsic impedance. Consequently, free space can be represented by the admittance load $\tilde{Y}_0$ in the circuital model. It is worth mentioning that the matrices $\tilde{V}$, $\tilde{Z}$, $\tilde{Y}$, $\tilde{B}$ and $\tilde{Y}_0$ can be evaluated in a straightforward manner utilizing the *Mathematica®*'s symbolic capability.

As the ground layer is considered a perfect electric conductor, it is well represented by a short circuit that corresponds to null electric field ($\vec{\mathsf{E}}_g = 0$). On the other hand, the spectral electric surface current density $\vec{\mathsf{J}}_s$ located on the metallic patch is modelled by an ideal current source. The circuital representations for both short circuit and ideal current source are given in Fig. 5.

Finally, by properly combining the circuit elements, the three-layer structure model is the equivalent circuit illustrated in Fig. 6, whose resolution produces the transversal dyadic Green's function $\tilde{G}$ in the spectral domain. Notice that the Green's function, calculated according to this approach, is evaluated at the dielectric substrate – free space interface ($r=b$).

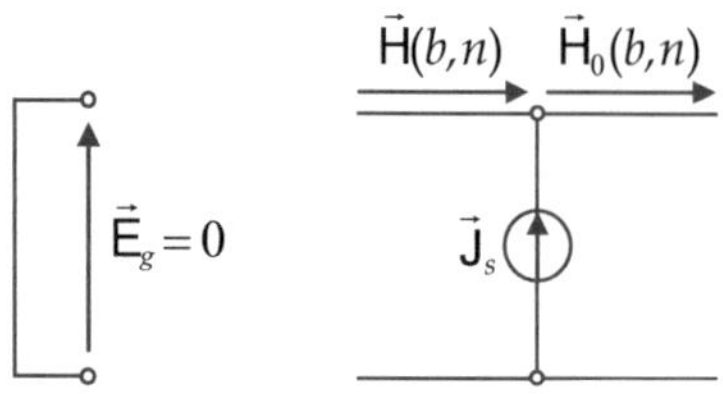

(a) (b)

Fig. 5. Short circuit (a) and ideal current source (b) circuital representations.

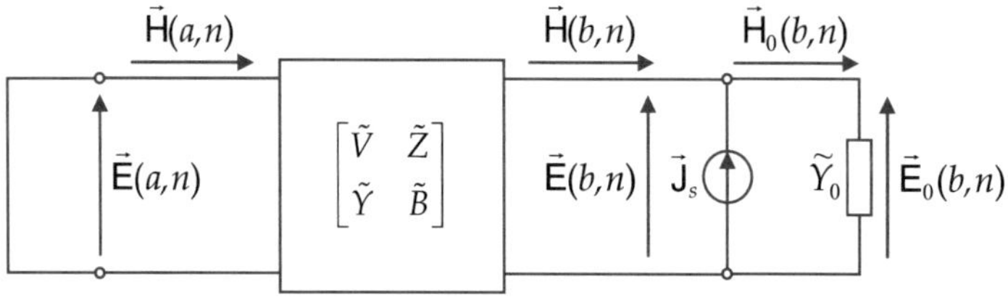

Fig. 6. Circuital representation for the spherical three-layer structure.

It is important to point out that the *Mathematica®*'s symbolic capability is also helpful for the circuit resolution and allows writing the related functions in a compact form, as shown:

$$\vec{\mathsf{E}}_0(b,n) = \tilde{G} \cdot \vec{\mathsf{J}}_s, \tag{33}$$

where

$$\tilde{G} = \tilde{G}(b,n) = \begin{bmatrix} 0 & -G_{\theta\theta} \\ G_{\phi\phi} & 0 \end{bmatrix}, \tag{34}$$

$$G_{\theta\theta} = \frac{\eta_0\, q_n \hat{\mathrm{H}}_n^{(2)\prime}(k_0 b)}{j(\sqrt{\varepsilon_r}\, p_n \hat{\mathrm{H}}_n^{(2)\prime}(k_0 b) + q_n \hat{\mathrm{H}}_n^{(2)}(k_0 b))}, \tag{35}$$

$$G_{\phi\phi} = \frac{j\eta_0 s_n \hat{\mathrm{H}}_n^{(2)}(k_0 b)}{\sqrt{\varepsilon_r}\, r_n \hat{\mathrm{H}}_n^{(2)}(k_0 b) + s_n \hat{\mathrm{H}}_n^{(2)\prime}(k_0 b)}, \tag{36}$$

with

$$p_n = \hat{\mathrm{H}}_n^{(2)}(kb)\hat{\mathrm{H}}_n^{(1)\prime}(ka) - \hat{\mathrm{H}}_n^{(1)}(kb)\hat{\mathrm{H}}_n^{(2)\prime}(ka), \tag{37.a}$$

$$q_n = \hat{\mathrm{H}}_n^{(2)\prime}(ka)\hat{\mathrm{H}}_n^{(1)\prime}(kb) - \hat{\mathrm{H}}_n^{(1)\prime}(ka)\hat{\mathrm{H}}_n^{(2)\prime}(kb), \tag{37.b}$$

$$r_n = \hat{\mathrm{H}}_n^{(2)}(ka)\hat{\mathrm{H}}_n^{(1)\prime}(kb) - \hat{\mathrm{H}}_n^{(1)}(ka)\hat{\mathrm{H}}_n^{(2)\prime}(kb), \tag{37.c}$$

$$s_n = \hat{\mathrm{H}}_n^{(2)}(kb)\hat{\mathrm{H}}_n^{(1)}(ka) - \hat{\mathrm{H}}_n^{(1)}(kb)\hat{\mathrm{H}}_n^{(2)}(ka) \tag{37.d}$$

and $\vec{\mathsf{J}}_s = \begin{bmatrix} \mathsf{J}_{s\phi} & -\mathsf{J}_{s\theta} \end{bmatrix}^T$ whose superscript T indicates the transpose operator.

Writing the free-space spectral electric field ($r > b$) as a function of the field evaluated at the dielectric substrate - free space interface ($r = b$) and taking the asymptotic expression ($r \to \infty$) for the Schelkunoff spherical Hankel function of second kind and n-th order (Olver, 1972), the spectral far electric field is derived as

$$\tilde{\mathbf{E}}_0(r,n) \cong j^n b\tilde{A}\cdot\tilde{G}\cdot\tilde{\mathbf{J}}_s \frac{e^{-jk_0 r}}{r}, \tag{38}$$

where

$$\tilde{A} = \begin{bmatrix} [\hat{\mathrm{H}}_n^{(2)\prime}(k_0 b)]^{-1} & 0 \\ 0 & j[\hat{\mathrm{H}}_n^{(2)}(k_0 b)]^{-1} \end{bmatrix}.$$

So, applying (28) to the spectral field (38), the spatial far electric field radiated from the spherical microstrip antenna is determined,

$$\begin{bmatrix} E_\theta \\ E_\phi \end{bmatrix} = \sum_{m=-\infty}^{+\infty}\sum_{n=|m|}^{+\infty} j^n b\tilde{L}(n,m,\theta)\cdot\tilde{A}\cdot\tilde{G}\cdot\tilde{\mathbf{J}}_s e^{jm\phi}\frac{e^{-jk_0 r}}{r}. \tag{39}$$

Notice that the present development did not take into account the patch geometry, since the electric surface current density has not been specified yet. Hence, expression (39) can be applied to any arbitrary patch geometry conformed onto the structure of Fig. 1, and not only to the annular or circular ones. However, as this chapter's purpose is to develop a computationally efficient CAD for the analysis of thin spherical-annular and -circular microstrip antennas, instead of employing a complex numerical method, such as the MoM, for determining the electric surface current densities on the patches, the cavity model was enough for their accurate estimation. Following this approach for the case of the spherical-annular patch operating in the TM_{LM} mode, the expressions below are obtained

$$J_{s\theta} = -j\frac{E_{LM}}{\omega\mu_0 a}\frac{d}{d\theta}\mathrm{R}_L^M(\cos\theta)\cos(M\phi), \tag{40.a}$$

$$J_{s\phi} = j\frac{ME_{LM}}{\omega\mu_0 a\sin\theta}\mathrm{R}_L^M(\cos\theta)\sin(M\phi). \tag{40.b}$$

So, after the vector-Legendre transform, the spectral components of the surface current density can be evaluated in closed form as,

$$\begin{aligned}\mathbf{J}_{s\theta} = j\frac{E_{LM}}{2\omega\mu_0 aS(n,M)}\frac{L(L+1)}{n(n+1)-L(L+1)}[&\sin^2\theta_{1c}\,\mathrm{P}_n^{M\prime}(\cos\theta_{1c})\mathrm{R}_L^M(\cos\theta_{1c}) \\ &-\sin^2\theta_{2c}\,\mathrm{P}_n^{M\prime}(\cos\theta_{2c})\mathrm{R}_L^M(\cos\theta_{2c})],\end{aligned} \tag{41.a}$$

$$\mathbf{J}_{s\phi} = \frac{mE_{LM}}{2\omega\mu_0 aS(n,M)}[\mathrm{P}_n^M(\cos\theta_{2c})\mathrm{R}_L^M(\cos\theta_{2c}) - \mathrm{P}_n^M(\cos\theta_{1c})\mathrm{R}_L^M(\cos\theta_{1c})], \tag{41.b}$$

if $m = M$ or $m = -M$. Otherwise, $\tilde{\mathbf{J}}_s = 0$. Consequently the expression for the far electric field radiated by this radiator is also determined in closed form.

In a similar way, expressions for the spatial and spectral electric surface current densities are derived for the spherical-circular microstrip antenna (Fig. 3) operating in the TM_{LM} mode

$$J_{s\theta} = -j\frac{E_{LM}}{\omega\mu_0 a}\frac{d}{d\theta}P_L^M(\cos\theta)\cos(M\phi), \tag{42.a}$$

$$J_{s\phi} = j\frac{ME_{LM}}{\omega\mu_0 a\sin\theta}P_L^M(\cos\theta)\sin(M\phi), \tag{42.b}$$

$$\mathrm{J}_{s\theta} = \frac{-jE_{LM}}{2\omega\mu_0 aS(n,M)}\frac{L(L+1)}{n(n+1)-L(L+1)}\sin^2\theta_{2c}\,P_n^{M\prime}(\cos\theta_{2c})P_L^M(\cos\theta_{2c}), \tag{43.a}$$

$$\mathrm{J}_{s\phi} = \frac{mE_{LM}}{2\omega\mu_0 aS(n,M)}P_n^M(\cos\theta_{2c})P_L^M(\cos\theta_{2c}), \tag{43.b}$$

if $m = M$ or $m = -M$. Otherwise, $\vec{\mathrm{J}}_s = 0$.
Once the far electric field radiated by spherical antennas is known, an expression for its average radiated power can be established. Starting from equation (44) (Balanis, 2005),

$$P_0 = \frac{1}{2\eta_0}\int_{\phi=0}^{2\pi}\int_{\theta=0}^{\pi}\vec{E}\cdot\vec{E}^* r^2\sin\theta d\theta d\phi, \tag{44}$$

where $\vec{E}$ denotes the far electric field determined by (39) and the superscript * indicates the complex conjugate operator. After the double integration, the following expression can be obtained

$$P_0 = \frac{\pi b^2}{\eta_0}\sum_{m=-\infty}^{+\infty}\sum_{n=|m|}^{+\infty}S(n,m)\left[\left|\frac{\mathrm{J}_{s\theta}G_{\theta\theta}}{\hat{\mathrm{H}}_n^{(2)\prime}(k_0 b)}\right|^2 + \left|\frac{\mathrm{J}_{s\phi}G_{\phi\phi}}{\hat{\mathrm{H}}_n^{(2)}(k_0 b)}\right|^2\right]. \tag{45}$$

In order to calculate the directivity of thin spherical-annular and -circular microstrip antennas, the developed CAD employs (45), since its evaluation requires no numerical integrations, which, as previously mentioned, is an advantage. In addition, equation (45) is employed in the CAD for computing the quality factor associated to the radiated power, from which the effective loss tangent $\tan\delta_{eff}$ and, consequently, the antenna input impedance (15) are estimated.

2.3 CAD results

Before presenting some CAD results and comparing them with *HFSS®* output data, a brief overview of the CAD structure will be given. The current version of the CAD contains two independent sections: the synthesis and the analysis modules that can be accessed from their respective tabs. By selecting the *Synthesis* option, the design interface (Fig. 7) is presented. The inputs required for the synthesis procedure to start are the desired frequency, the ground sphere radius a and the substrate parameters, such as relative permittivity ε_r, loss tangent $\tan\delta$ and thickness h. As results, the CAD returns the patch physical dimensions (θ_1 and θ_2 for the annular patch and only θ_2 for the circular one) and the probe position θ_p

considering the antenna fed by a 50-ohm SMA connector at $\phi_p = 0°$. The *Analysis* module evaluates the electromagnetic characteristics of a synthesized antenna. The inputs required for the analysis procedure to start are the metallic sphere radius *a*, the substrate parameters, the patch angular dimensions and the probe position. As outputs, the antenna input impedance (rectangular plot), quality factor, radiation patterns (polar plots) and directivity are calculated. Notice that the window illustrated in Fig. 7 is relative to the spherical-circular microstrip antenna; another similar one was developed for the spherical-annular radiator.

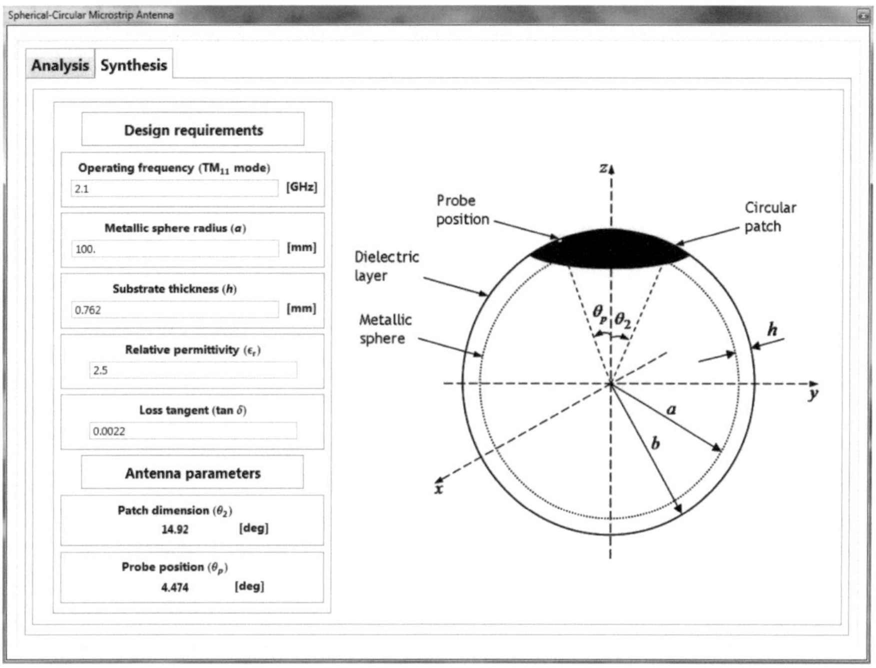

Fig. 7. The Synthesis module interface.

The CAD algorithm was implemented in *Mathematica®* mainly due to the efficient numerical routines for the computation of the associated Legendre functions. Besides, *Mathematica®* has a vast collection of built-in functions that permit implementing the respective code in a short number of lines, plus its many graphical resources.

In order to solve the transcendental equation and to calculate the equivalent cavity resonant frequencies in a fast and accurate way, the CAD utilizes a set of interpolation polynomials specially developed to provide seed values for the *Mathematica®*'s numerical routine that searches the transcendental equation root in a given operation mode. The interpolation polynomials were calculated based on graphical analysis, so the CAD can determine the resonant frequency of a specific mode without any further graphical inspection. For example, the interpolation polynomial associated to the TM_{11} mode of a spherical-circular cavity which is employed by the CAD is:

$$\begin{aligned}\ell_{11}(\theta_{2c}) = 54.46 - 11.06\,\theta_{2c} + 1.13\theta_{2c}^2 - 6.21\times10^{-2}\theta_{2c}^3 + 1.67\times10^{-3}\theta_{2c}^4 \\ - 5.36\times10^{-6}\theta_{2c}^5 - 8.71\times10^{-7}\theta_{2c}^6 + 2.07\times10^{-8}\theta_{2c}^7 - 1.53\times10^{-10}\theta_{2c}^8,\end{aligned} \quad (46)$$

where θ_{2c} is given in degrees.

To illustrate the CAD synthesis procedure, a spherical-circular antenna conformed onto a typical microwave laminate ($\varepsilon_r = 2.5$, loss tangent = 0.0022 and $h = 0.762$ mm) fed by a 50-ohm SMA coaxial connector of 0.65-mm radius, was designed to operate at 2.1 GHz. The radius of the metallic sphere is a =100 mm. After entering these parameters, the CAD outputs $\theta_2 = 14.92$ degrees and $\theta_p = 4.47$ degrees, in a few minutes of computational time, even running on a regular classroom desktop computer. The input data and the results are automatically saved for use in the analysis module. In Fig. 8, the comparison is shown between the radiation patterns, at the operating frequency, obtained from the developed CAD and from the *HFSS*® package for the spherical-circular microstrip antenna so designed. It is seen that the radiation patterns exhibit excellent agreement, thus validating our procedure to calculate the radiated far electric field based on the combination of the cavity model with the electric surface current method. It is important to point out that *HFSS*® employs the FEM (finite element method) for solving high frequency structures, so it takes considerable time to determine the structure solution. In addition, it does not provide an estimator tool to establish the initial geometry of the spherical radiator as the developed CAD does.
Results for the real and imaginary parts of the antenna input impedance are presented in Fig. 9; once again, the curves are very similar.

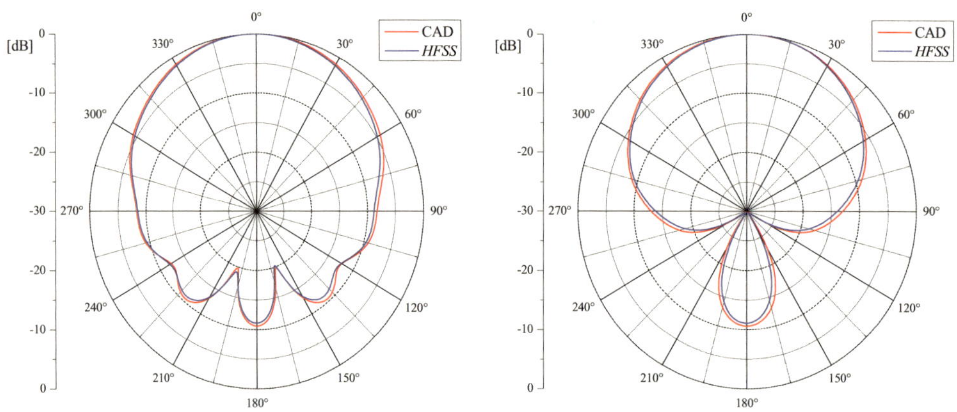

E_θ radiation pattern: *xz* plane. E_ϕ radiation pattern: *yz* plane.

Fig. 8. Comparison between the radiation patterns.

The half power quality factor Q and the antenna directivity D shown in Table 2 are also in very good agreement.

	CAD	*HFSS*®	Deviation
Q	78.8	80.8	2.5%
D (dB)	6.6	6.9	0.3 dB

Table 2. Quality factor and directivity.

As another illustrative example, a spherical-annular antenna fed by a 50-ohm SMA coaxial connector of 0.65-mm radius and conformed onto the same typical microwave laminate used before was designed to operate at 1.364 GHz in the TM_{10} mode. In this case, $\theta_1 = 10.0$ degrees, $\theta_2 = 30.0$ degrees, $\theta_p = 13.21$ degrees and the ground sphere has a radius $a = 200$ mm. The radiation pattern in the E-plane, at 1.364 GHz, and the input impedance curve evaluated in the CAD and *HFSS*® are presented in Figs. 10 and 11, respectively. It is clear from these figures that, once again, the results are very similar.

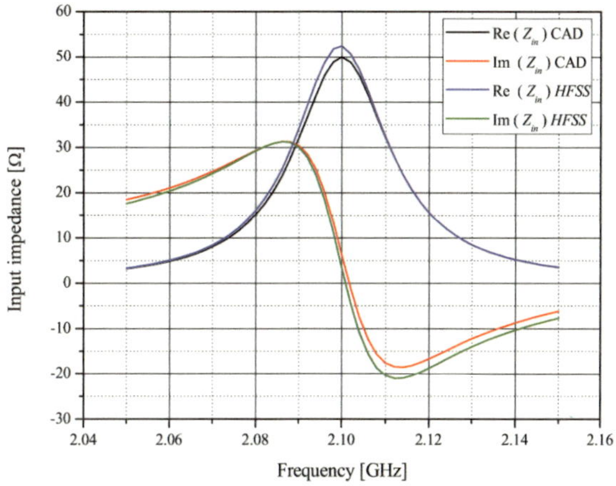

Fig. 9. Spherical-circular microstrip antenna input impedance.

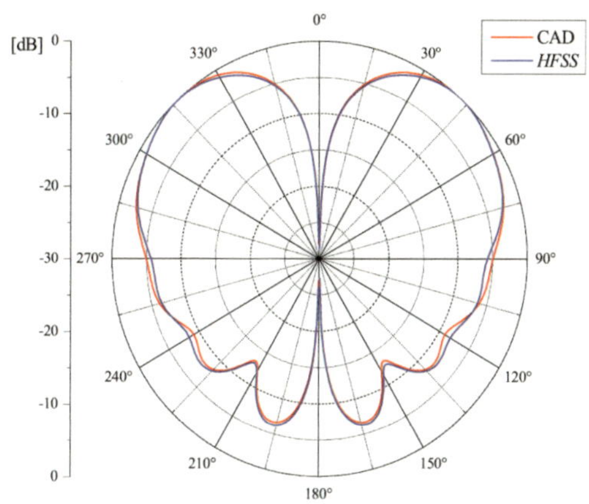

Fig. 10. E_θ radiation pattern: E-plane.

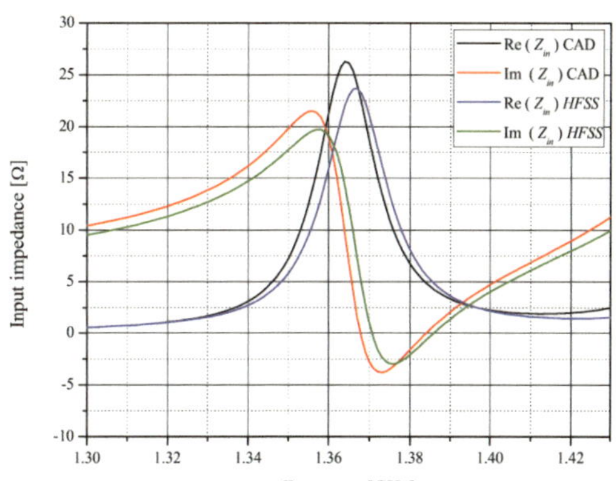

Fig. 11. Spherical-annular antenna input impedance (TM_{10} mode).

3. Radiation patterns of spherical arrays

As aforementioned, a great advantage of using spherical arrays is the possibility of 360° coverage in any radial direction. So, they have potential application in tracking, telemetry and command services for low-earth and medium-earth orbit satellites (Sipus et al., 2008). Rigorous analysis of spherical microstrip antenna arrays has been carried out using the MoM (Sipus et al., 2006). However, the MoM involves highly-complex and time-consuming calculations even considering the far-field evaluation alone. On the other hand, when spherical-annular or –circular patches of thin radiators are positioned symmetrically in relation to the z-axis, they can be effectively analyzed through the electric surface current method in association with the cavity model, as shown in Section 2. In case of spherical arrays, not all array elements can be positioned symmetrically with respect to the z-axis. Hence, in this chapter, the global coordinate system technique (Sengupta et al., 1968) is employed to evaluate the far electric field radiated by each one of the array elements.
To illustrate the proposed technique, let's analyze the spherical-circular microstrip antenna shown in Fig. 12, which represents a generic spherical array element whose centre is located at (α_n, β_n).

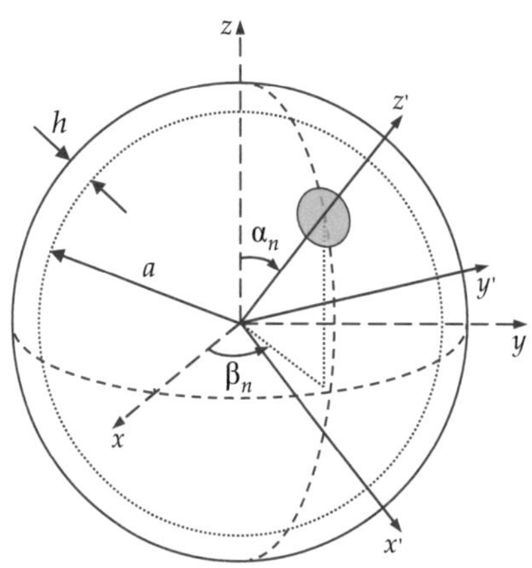

Fig. 12. Geometry of a spherical-circular array element.

Starting from the expressions for the far electric field components $E_{\theta'}(.)$ and $E_{\phi'}(.)$ of a patch that is symmetrically positioned around the z'-axis, as calculated in Section 2.2, and using the global coordinate system, the following expressions for the radiated far electric field components in the reference (r,θ,ϕ) coordinate system are obtained

$$E_{rot\theta}(\theta,\phi) = A_n E_{\theta'}(\theta',\phi') - B_n E_{\phi'}(\theta',\phi'), \tag{47}$$

$$E_{rot\phi}(\theta,\phi) = B_n E_{\theta'}(\theta',\phi') + A_n E_{\phi'}(\theta',\phi'), \tag{48}$$

where

$$A_n = [-\cos\theta \sin\alpha_n \cos(\phi - \beta_n) + \sin\theta \cos\alpha_n] / \sin\theta', \tag{49}$$

$$B_n = [\sin\alpha_n \sin(\phi - \beta_n)] / \sin\theta', \tag{50}$$

with

$$\cos\theta' = \sin\alpha_n \sin\theta \cos(\phi - \beta_n) + \cos\alpha_n \cos\theta \,, \tag{51}$$

and

$$\cot\phi' = \frac{\cos\alpha_n \sin\theta \cos(\phi - \beta_n) - \sin\alpha_n \cos\theta}{\sin\theta \sin(\phi - \beta_n)}. \tag{52}$$

To verify this approach, a spherical-circular single-element antenna, such as the one illustrated in Fig. 12, whose centre is positioned at ($\alpha_n = 30°$, $\beta_n = 0°$), was designed in our CAD to operate at 3.1 GHz (TM_{11} mode). The spherical-circular patch, fed at ($\theta_{pn} = 27.1°$, $\phi_{pn} = 0°$) by a 50-ohm SMA coaxial connector of 0.65-mm radius, is conformed onto a microwave laminate with $\varepsilon_r = 2.5$, loss tangent = 0.0022 and $h = 0.762$ mm. The radius of the metallic sphere is a =100 mm. The designed antenna was also simulated in *HFSS®* package for comparison purposes. Fig. 13 shows the results obtained from the CAD for the radiation patterns in *xz*- and *yz*-planes compared to those simulated in *HFSS®*. As observed, they exhibit excellent agreement.

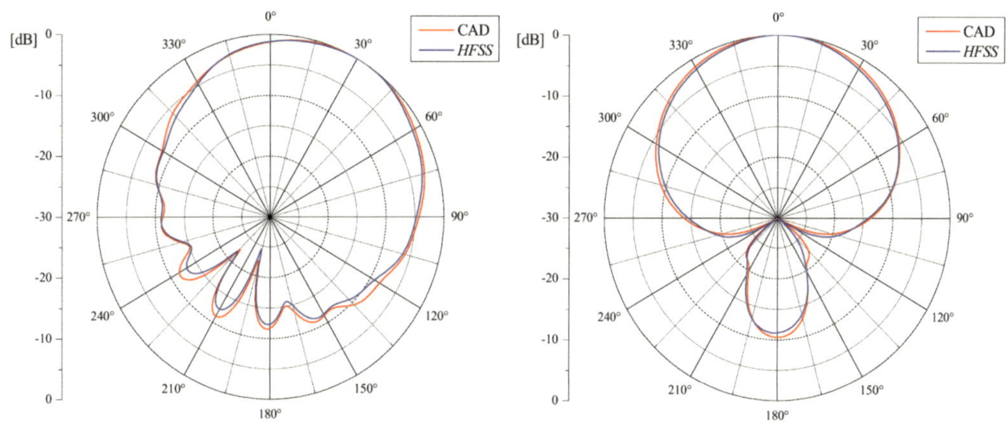

$E_{rot\theta}$ radiation pattern: *xz* plane. $E_{rot\phi}$ radiation pattern: *yz* plane.

Fig. 13. Radiation patterns of the designed rotated element.

After validating the adopted procedure for calculating the radiation pattern of a generic spherical array element, the array analysis can be carried out. Since for spherical arrays there is no diffraction at the edges of the conducting surfaces and considering that coupling among the array elements can be neglected for radiation pattern purposes, the components of the far electric field radiated by an spherical array can be calculated by superposing the fields radiated by each element individually. Following this approach, the components of the far electric field radiated by a spherical array of N elements can be evaluated from

$$E_{R\theta}(\theta,\phi) = \sum_{n=1}^{N} A_n E_{\theta'}(\theta',\phi') - B_n E_{\phi'}(\theta',\phi'), \tag{53}$$

$$E_{R\phi}(\theta,\phi) = \sum_{n=1}^{N} B_n E_{\theta'}(\theta',\phi') + A_n E_{\phi'}(\theta',\phi'). \tag{54}$$

To illustrate the proposed procedure, an array consisting of two spherical-circular elements, as shown in Fig. 14, was designed to operate at 3.1 GHz (TM_{11} mode, $\varepsilon_r = 2.5$, $\tan\delta = 0.0022$, $h = 0.762$ mm and $a = 100$ mm). The antennas are fed by identical currents, $\beta_1 = 0°$ and $\beta_2 = 180°$. The patch spacing α was chosen to be 15° and 90°, one at time, in order to analyze the developed approach for a wide range of α. Figs. 15 and 16 show the radiation patterns in the *xz*- and *yz*-planes evaluated both with the CAD and *HFSS*®. As seen, they are in excellent agreement, even in the case when the patches are closer together ($\alpha = 15°$), thus validating the adopted technique. In the next sections, two spherical arrays configurations are discussed: the meridian-spherical and circumferential-spherical arrays whose radiation patterns will be evaluated following this approach.

3.1 Meridian-spherical arrays

The geometry of the spherical-circular meridian array, i.e. one whose patches are all centred along a constant-ϕ plane, is shown in Fig. 17. In this particular configuration, the array is positioned along the $\phi = \beta$ plane and the patch centres are located at α_i, where $i = 1, 2, \ldots, N$. Note the maximum number of elements N is a function of the sphere radius, the dielectric permittivity and the operating frequency, in a way to avoid the superposition of patches.

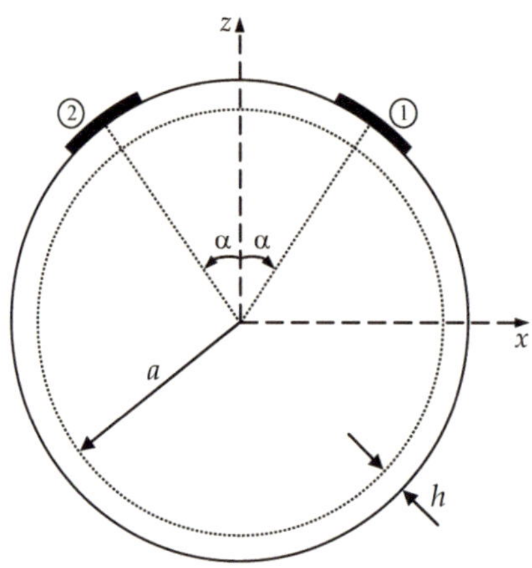

Fig. 14. Two-element array: cut in *xz*-plane.

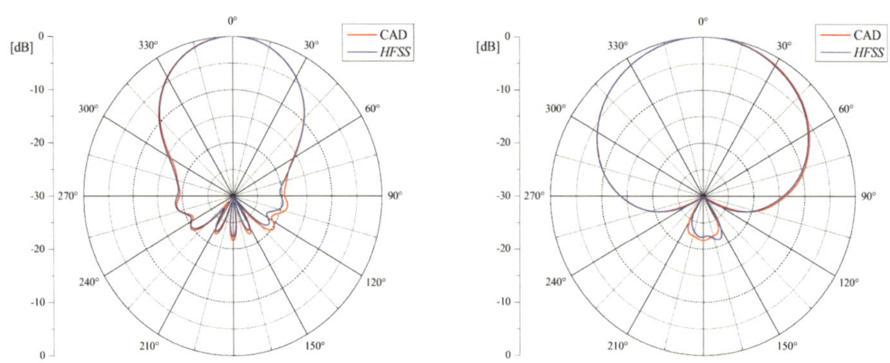

$E_{R\theta}$ radiation pattern: *xz* plane. $E_{R\phi}$ radiation pattern: *yz* plane.

Fig. 15. Two-element array radiation patterns: $\alpha = 15°$.

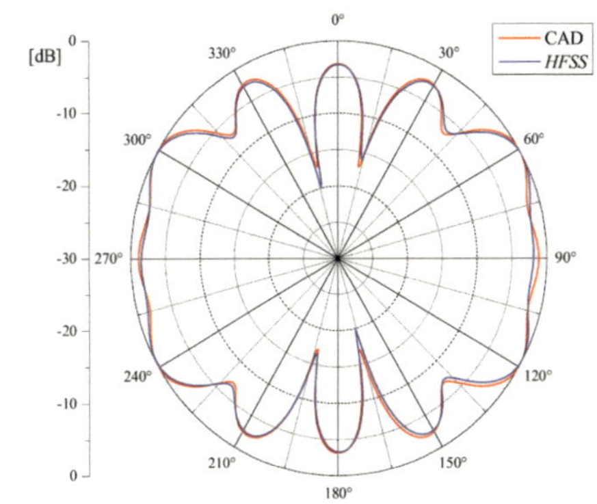

$E_{R\theta}$ radiation pattern: xz plane.

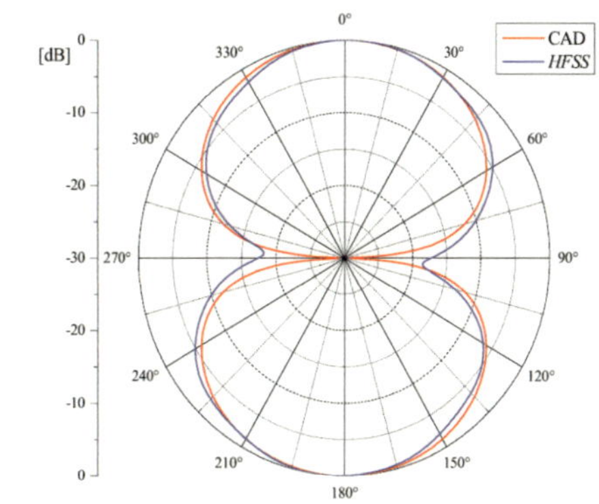

$E_{R\phi}$ radiation pattern: yz plane.

Fig. 16. Two-element array radiation patterns: $\alpha = 90^o$.

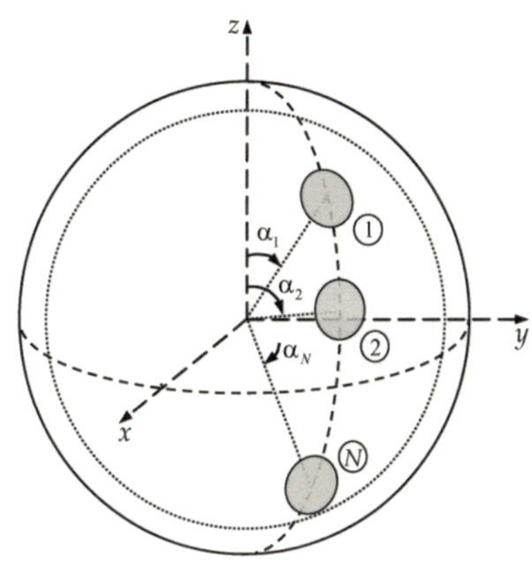

Fig. 17. Meridian-spherical array.

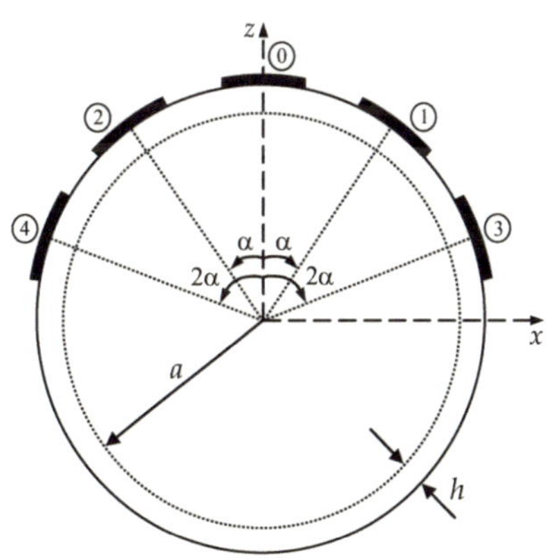

Fig. 18. Five-element array: cut in xz plane.

As an example of a spherical-circular meridian array, consider the five-element array shown in Fig. 18. This array was also designed to operate at 3.1 GHz (TM_{11} mode, $\varepsilon_r = 2.5$, $\tan\delta = 0.0022$, $h = 0.762$ mm and $a = 100$ mm) and its elements are fed by identical currents. The uniform patch spacing α was chosen to be 27.5°. Results for the corresponding radiation patterns, evaluated with both our CAD and *HFSS®* are illustrated in Fig. 19. Once again, the radiation patterns are in very good agreement, thus demonstrating that the coupling between the array elements can be neglected in the calculation of the far electric field the array radiates.

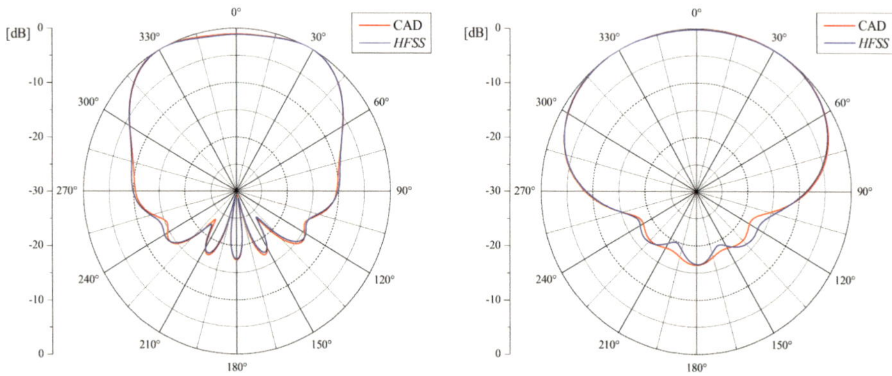

$E_{R\theta}$ radiation pattern: *xz* plane. $E_{R\phi}$ radiation pattern: *yz* plane.

Fig. 19. Five-element array radiation patterns.

3.2 Circumferential-spherical arrays

A circumferential-spherical array of *N*-element is shown in Fig. 20. In this case, the patches are centred along a θ-constant cone and the maximum number of elements *N* is a function of θ, the sphere radius *a*, the dielectric permittivity and the operating frequency, in a way to avoid the superposition of patches.

To illustrate the analysis technique, let's consider the four-element array presented in Fig. 21. This array was also designed to operate at 3.1 GHz (TM_{11} mode, $\varepsilon_r = 2.5$, $\tan\delta = 0.0022$, $h = 0.762$ mm and $a = 100$ mm) and its elements are fed by identical currents, but $\alpha = 35°$. Results for the radiation patterns in the *xz*- and *yz*-planes evaluated with both the CAD and *HFSS®* are shown in Fig. 22. As seen from these results, the radiation patterns are in excellent agreement, thus supporting the validation of the superposition procedure presented in this chapter for the calculation of the far electric field radiated by spherical microstrip antenna arrays.

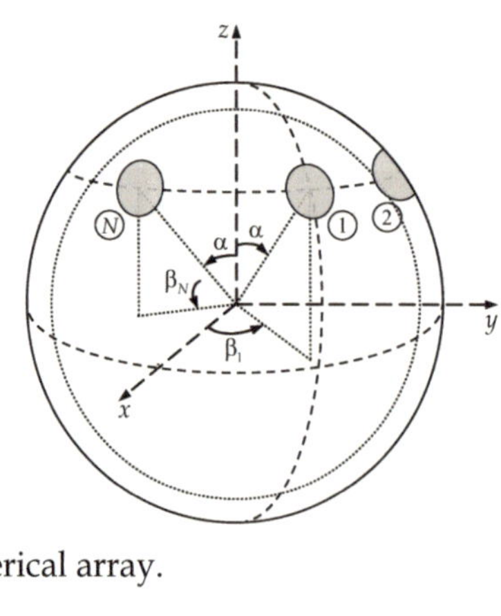

Fig. 20. Circumferential-spherical array.

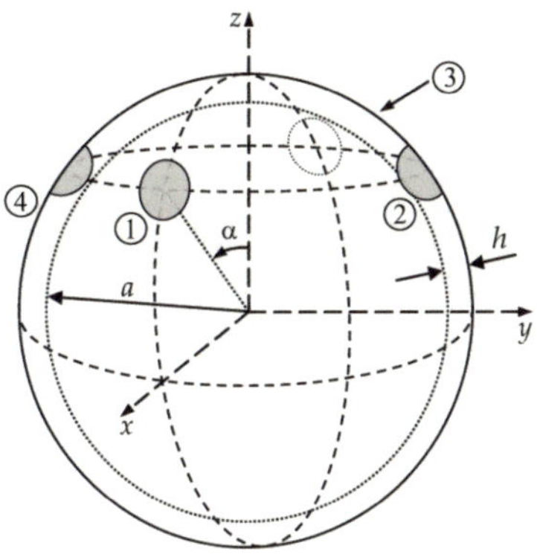

Fig. 21. Four-element circumferential array.

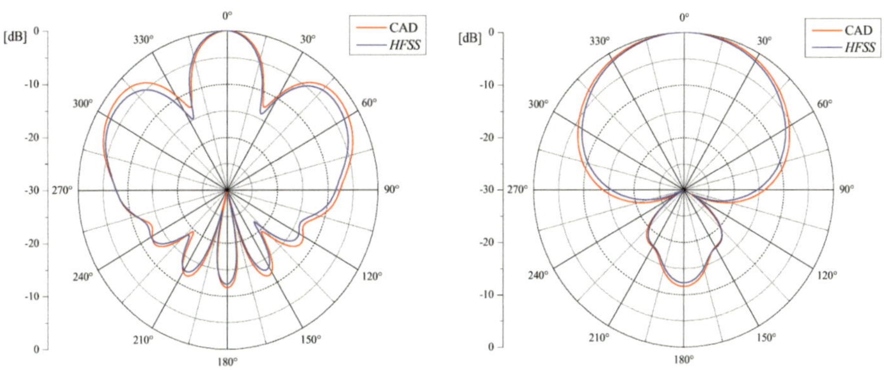

$E_{R\theta}$ radiation pattern: yz plane. $E_{R\phi}$ radiation pattern: xz plane.

Fig. 22. Four-element array radiation patterns.

Although the examples given in this section involve spherical arrays whose patches are circular, the proposed technique can be applied in the same manner to spherical arrays whose patches are annular.

4. Prototype design and experimental results

The theoretical model developed in the previous sections considers the dielectric substrate and the patch are both conformed onto the metallic ground sphere. Although the fabrication of spherical-microstrip antennas starting from planar radiators is a very challenging task (Piper & Bialkowski, 2004), the procedure can be eased if the geometry is slightly modified, i.e., if a facet is cut on the metallic spherical layer for mounting a planar antenna. An example of such modified geometry is illustrated in Fig. 23 where a planar circular patch is mounted onto the facet. The same adaptation could be made for other patch geometries, as the annular or rectangular, for instance. But, for this modified geometry, an essential question is posed: how well can its electromagnetic behavior be predicted from the theoretical model previously developed?

When the dimensions of the planar patch are much smaller than the metallic sphere radius, the electrical characteristics of the hybrid geometry tend to those of an equivalent antenna whose patch and dielectric substrate are conformed onto the ground sphere. So, the

electromagnetic behavior of the modified geometry under this condition can in a first step be predicted using the special-purpose CAD. To validate this statement, a spherical-circular patch conformed onto a typical microwave laminate ($\varepsilon_r = 2.55$, $\tan\delta = 0.0022$, $h = 0.762$ mm) fed by a 0.65-mm radius, 50-ohm SMA coaxial connector, was designed to operate at 3.1 GHz (TM_{11} mode). The available ground sphere has a radius of 98 mm. Using the Synthesis module of the developed CAD the following dimensions result: $\theta_2 = 10.11$ degrees and $\theta_p = 2.91$ degrees.
To generate the initial geometry of the faceted antenna, the arc lengths of the spherical-circular antenna were used to establish the dimensions of the planar circular patch. The dimensions so obtained are depicted in Table 3, where r_F denotes the radius of the metallic patch, r_p the probe position (relative to the patch centre) and r_d the planar substrate layer radius. Notice that r_d was chosen to be equal to the patch radius extended by four times the dielectric substrate thickness, in order not to affect the fringe fields.

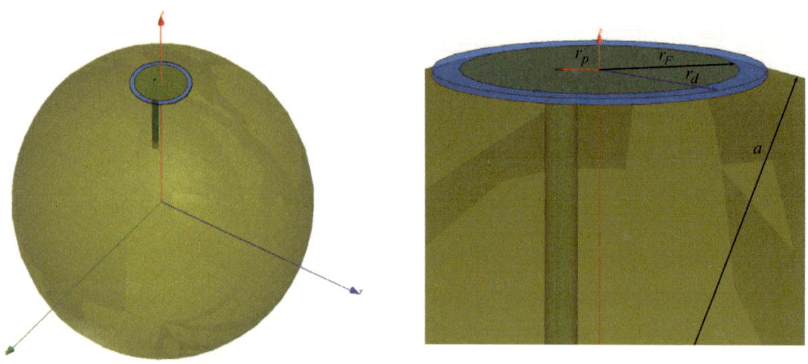

Fig. 23. Faceted antenna geometry.

r_F	17.43
r_p	5.02
r_d	20.47

Table 3. Initial dimensions in mm.

With the dimensions presented in Table 3, the faceted antenna was analyzed in the *HFSS®* software and the results for input impedance and radiation patterns compared to those evaluated with our CAD are shown in Figs. 24 and 25, respectively. It is worth mentioning that the radiation patterns were calculated in the frequencies where the antenna input resistance assumes its maximum value, i.e., 3.002 GHz for faceted antenna and 3.1 GHz for the corresponding conformed antenna. As seen in Figs. 24 and 25, the shapes of the curves so obtained are in good agreement despite a small frequency shift. Therefore, in the optimization of the modified geometry dimensions, the operating frequency will be shifted from 3.002 to 3.1 GHz, hence justifying it as an alternative topology for spherical-circular antennas.
After the *HFSS®* optimization, the resulting faceted antenna dimensions are given in Table 4. This same table presents the deviation between the initial dimensions and the final ones. It is important to point out that the optimized antenna dimensions are not far off the initial ones, thus validating the developed CAD ability to generate the initial modified antenna geometry.

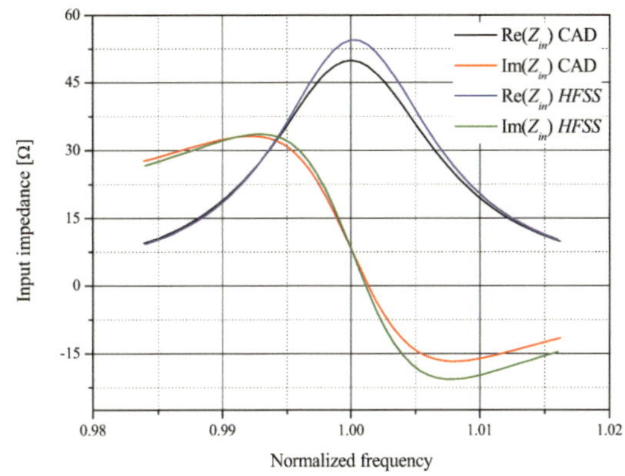

Fig. 24. Input impedance.

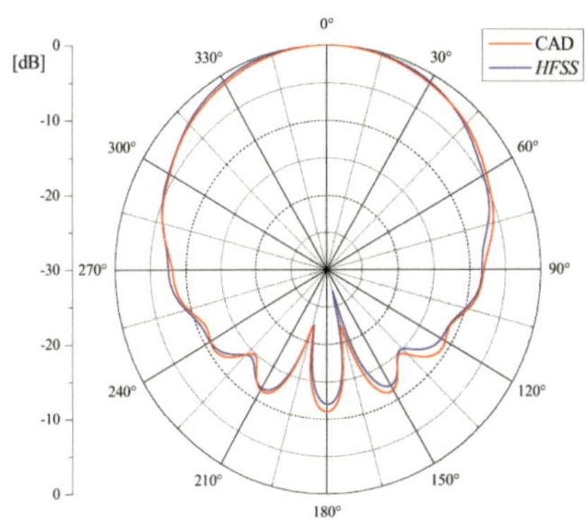

E_θ radiation pattern: *xz* plane.

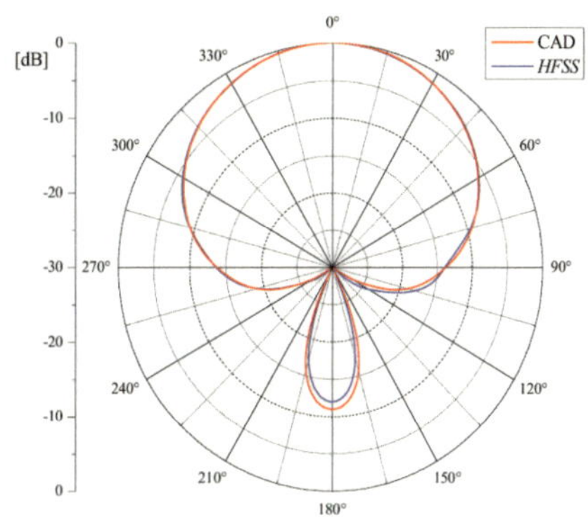

E_ϕ radiation pattern: *yz* plane.

Fig. 25. Radiation patterns.

r_F	17.20	1.3%
r_p	4.68	6.8%
r_d	20.47	0%

Table 4. Final dimensions in mm and percent deviation.

For the modified antenna geometry under consideration, although the dielectric layer and the metallic patch are not conformed onto the ground sphere, it is still necessary to produce the metallic sphere, which is in general an expensive and complex structure to fabricate. However, if the mechanical considerations can be relaxed, as, for instance, for educational purposes, a styrofoam sphere covered with aluminium foil, easily found in any hardware store, can serve well the fabrication of the ground sphere, as illustrated in Fig 26. Measurements were conducted for validating the fabrication of the antenna prototype following this simplified approach. Fig. 27 presents the comparison between the measured input impedance and the one calculated in *HFSS*® software. As seen, the results agree very well, the 3.1-MHz shift between the curves representing just 0.1% of the antenna operating frequency. This deviation is less than the one allowed due to the substrate relative permittivity tolerance. So, the good agreement between the curves validates the procedure used for manufacturing the proposed antenna prototype.

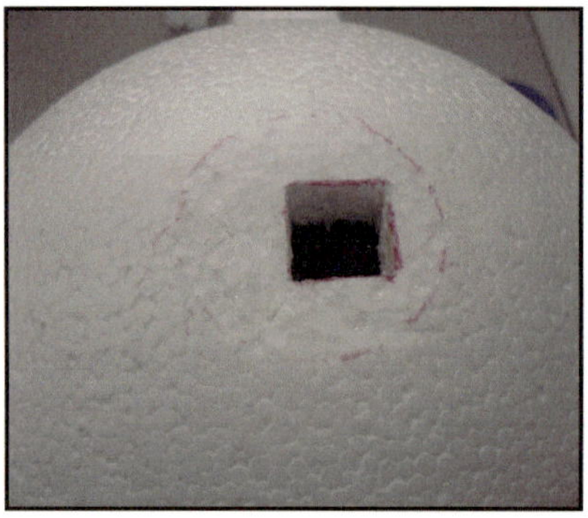

(a) Styrofoam sphere.

(b) Planar circular patch.

(c) Antenna prototype.

Fig. 26. Photos of the faceted antenna.

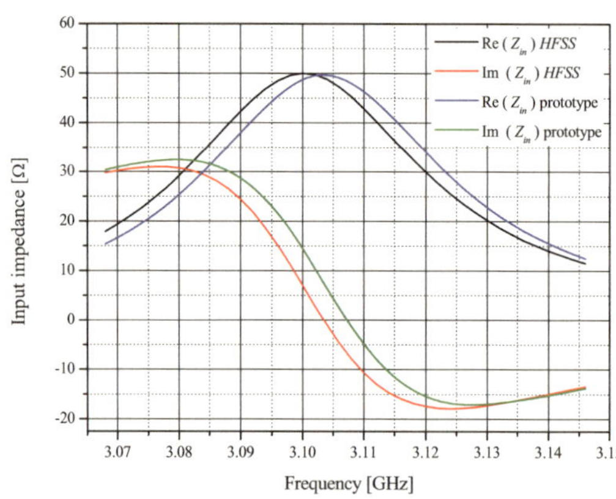

Fig. 27. Input impedance comparison.

5. Conclusions

In summary, an effective CAD algorithm capable of synthesizing and analyzing probe-fed spherical thin radiators was discussed. Some of its intrinsic advantages in terms of fine accuracy, short computational time and low cost were described; besides, quite conveniently, the CAD does not require a powerful computer to run on. Such characteristics make this program adequate not only for educational purposes but also for practical antenna design. The CAD's ability to generate a precise initial geometry before further optimization with commercial software such as *HFSS®*, can effect significant reductions in design time and cost. Furthermore, through the use of the global coordinate system technique, radiation patterns of meridian- and circumferential-spherical arrays can also be generated. Finally, a prototype was designed and built using a simplified fabrication technique, resulting in an operating frequency shift of only 0.1%, which is less than the deviation due to the substrate relative permittivity tolerance itself, thus validating the fabrication technique proposed.

6. References

Balanis, C. A. (1989). *Advanced engineering electromagnetics*, John Wiley & Sons, ISBN: 0-471-62194-3, New York.

Balanis, C. A. (2005). *Antenna theory: analysis and design*, Wiley-Interscience, ISBN: 0-471-66782-X, New York.

Damiano, J. P. & Papiernik, A. (1994). Survey of analytical and numerical models for probe-fed microstrip antennas. *IEE Proceedings Microwaves, Antennas and Propagation*, Vol. 141, No. 1, pp. 15-22, ISSN: 1350-2417.

Ferreira, D. B. (2009). *Microstrip antennas conformed onto spherical surfaces*, Graduation thesis, Technological Institute of Aeronautics, São José dos Campos, Brazil (in Portuguese).

Garg, R.; Bhartia, P.; Bahl, I. & Ittipiboon, A. (2001). *Microstrip Antenna Design Handbook*, Artech House, ISBN: 0-89006-513-6, Massachusetts.

Giang, T. V. B.; Thiel, M. & Dreher, A. (2005). A unified approach to the analysis of radial waveguides, dielectric resonators, and microstrip antennas on spherical multilayer structures. *IEEE Transactions on Microwave Theory and Techniques*, Vol. 53, No. 1, pp. 404-409, ISSN: 0018-9480.

Josefsson, L. & Persson, P. (2006). *Conformal Array Antenna Theory and Design*, Wiley-Interscience, ISBN: 0-471-46584-4, New Jersey.

Kishk, A. A. (1993). Analysis of spherical annular microstrip antennas. *IEEE Transactions on Antennas and Propagation*, Vol. 41, No. 3, pp. 338-343, ISSN: 0018-926X.

Lima, A. C. C.; Descardeci, J. R. & Giarola, A. J. (1991). Microstrip antenna on a spherical surface, *Proceedings of Antennas and Propagation Society International Symposium*, pp. 820-823, ISBN: 0-7803-0144-7, London, June 1991, New York.

Lo, Y. T.; Solomon, D. & Richards, W. F. (1979). Theory and experiment on microstrip antennas. *IEEE Transactions on Antennas and Propagation*, Vol. 27, No. 2, pp. 137-145, ISSN: 0018-926X.

Olver, F. W. J. (1972). Bessel functions of integer order, In: *Handbook of Mathematical Functions with Formulas, Graphs, and Mathematical Tables*, Abramowitz, M. & Stegun, I. A., pp. 355-389, Dover Publications, ISBN: 978-0-486-61272-0, New York.

Piper, B. R. & Bialkowski, M. E. (2004). Modelling the distortions to manufacture spherical conformal microstrip antennas, *Proceedings of Antennas and Propagation Society International Symposium*, pp. 3525-3528, ISBN: 0-7803-8302-8, Monterey, June 2004, IEEE, New York.

Richards, W. F.; Lo, Y. T. & Harrison, D. D. (1979). Improved theory for microstrip antennas. *Electronics Letters*, Vol. 15, No. 2, pp. 42-44, ISSN: 0013-5194.

Richards, W. F.; Lo, Y. T. & Harrison, D. D. (1981). An improved theory for microstrip antennas and applications. *IEEE Transactions on Antennas and Propagation*, Vol. 29, No. 1, pp. 38-46, ISSN: 0018-926X.

Sengupta, D. L.; Smith, T. M. & Larson, R. W. (1968). Radiation characteristics of a spherical array of circularly polarized elements. *IEEE Transactions on Antennas and Propagation*, Vol. 16, No. 1, pp. 2-7, ISSN: 0018-926X.

Sipus, Z.; Burum, N.; Skokic, S. & Kildal, P. –S. (2006). Analysis of spherical arrays of microstrip antennas using moment method in spectral domain. *IEE Proceedings Microwaves, Antennas and Propagation*, Vol. 153, No. 6, pp. 533-543, ISSN: 1350-2417.

Sipus, Z.; Skokic, S.; Bosiljevac, M. & Burum, N. (2008). Study of mutual coupling between circular stacked-patch antennas on a sphere. *IEEE Transactions on Antennas and Propagation*, Vol. 56, No. 7, pp. 1834-1844, ISSN: 0018-926X.

Tam, W. -Y. & Luk, K. -M. (1991). Far field analysis of spherical-circular microstrip antennas by electric surface current models. *IEE Proceedings H Microwaves, Antennas and Propagation*, Vol. 138, No. 1, pp. 98-102, ISSN: 0950-107X.

Tam, W. Y.; Lai, A. K. Y. & Luk, K. M. (1995). Input impedance of spherical microstrip antenna. *IEE Proceedings Microwaves, Antennas and Propagation*, Vol. 142, No. 3, pp. 285-288, ISSN: 1350-2417.

Wong, K. –L. (1999). *Design of Nonplanar Microstrip Antennas and Transmission Lines*, John Wiley & Sons, ISBN: 0-471-18244-3, New York.

6

Mathematical Modeling of Spherical Microstrip Antennas and Applications

Nikolaos L. Tsitsas and Constantinos A. Valagiannopoulos
National Technical University of Athens
Greece

1. Introduction

Microstrip antennas offer many attractive features and hence it is very challenging to model efficiently their resonance and radiation characteristics (Pozar & Schaubert, 1995). In particular, microstrip antennas composed of perfectly electric conducting (PEC) patches mounted on spherical surfaces possess two basic advantages against the planar ones. First, their curved shape makes them more applicable to real-world moving configurations such as missiles, satellites and aircrafts, where large plane boundaries are usually absent. Second, the frequency scanning problem encountered in case of planar microstrips at low elevations may be overcome due to the conformability of spherical antennas. Apart from these two advantages, spherical microstrip antennas offer (due to their relatively simple geometry) additional attractive features like low cost, light weight, easy fabrication as well as integrability with microwave and millimeter-wave circuits. As far as the metallic patch is concerned, circular or annular-ring ones are preferable for mounting on a spherical body. Interesting design characteristics of microstrip antennas are mainly the complex resonant frequencies, the radiation pattern and the input impedance. An extensive literature survey concerning the analysis of non-planar radiating microstrip structures, as well as the investigations of the aforementioned design characteristics is included in the classic book (Wong, 1999).

The mathematical methods, developed for analyzing the behavior of a spherical microstrip, may be categorized as follows: (i) the *full-wave approach* implements the method of moments (MoM) to approximate the surface current on the PEC patch by means of suitable basis functions (Uwaro & Itoh, 1989; Tam & Luk, 1991; Wong, 1999, Sipus et al., 2003; Giang et al. 2005), (ii) the *cavity model analysis* assumes that the distance between the substrate and the radiating metallic patch is electrically small, forming a cavity into which the vector of the electric field is independent of the longitudinal coordinate (Lo et al., 1979; Luk & Tam 1991; Chen & Wong, 1994), (iii) the *generalized transmission-line model* (GTLM) extracts an equivalent circuit for the spherical microstrip, where the line parameters are the electromagnetic fields under the patch (Ke & Kishk, 1991; Kishk, 1993).

Regarding the analysis of spherical microstrips, whose circular metallic patch is located inside a coating, composed of an arbitrary number of concentric spherical substrate and superstrate layers, an efficient approach for the analysis of the related resonance and radiation characteristics has been proposed in (Valagiannopoulos & Tsitsas, 2008a; Valagiannopoulos & Tsitsas, 2008b). The generic structure of that microstrip configuration offers in particular additional degrees of freedom to the designer in order to achieve the required operational characteristics. More precisely the method, presented in (Valagiannopoulos & Tsitsas, 2008a), utilizes appropriately the *Legendre transform*, since the microstrip structure is non-entire with respect to the spherical coordinates variable θ, while the effect of the layers above and below the patch is handled efficiently by using a *T-matrix method* (Chew, 1995; Tsitsas & Athanasiadis, 2006). Suitable basis functions for describing the surface current on the patch are chosen by applying the cavity model analysis (Lo et al., 1979). Subsequently, Galerkin's method is employed to formulate a homogeneous linear system, the singular points of which are the complex resonant frequencies. Finally, the inverse Legendre transform and far-field asymptotics lead to the computation of the radiation patterns. Importantly, structures, involving microstrips with a fixed small number of substrate and superstrate layers (Tam & Luk, 1991; Wong et al., 1993a; Wong & Chen, 1993), are incorporated as special cases of this method.

Several spherical microstrip configurations have been reported in the literature over the past two decades, exhibiting various interesting radiation features and demonstrating useful design characteristics. For example, the resonance properties of a simple spherical microstrip, possessing a single dielectric substrate between the metallic sphere and the circular patch are investigated in (Tam & Luk 1991). Moreover, in (Wong et al., 1993a) the authors studied the influence of a dielectric superstrate on the resonance and radiation properties of the spherical circular patch antenna proposed in (Tam & Luk 1991). A spherical microstrip with a substrate of two layers, one of which is an airgap, was treated in (Wong & Chen, 1993). In a similar context, a spherical microstrip with an annular-ring patch and an air gap between the metallic sphere and the dielectric was treated in (Ribero et al., 1999). On the other hand, a 4-layered microstrip with two airgaps surrounding an amplifying layer (constituting actually a φ- and θ-entire excitation) was proposed in (Valagiannopoulos & Tsitsas, 2008a), where the control mechanism of the amplifying capability via the layer's thickness and dielectric permittivity was also reported. Moreover, a spherical microstrip with two airgaps, surrounded by a zero-index material, expected to find applications in receiving antennas, was proposed in (Valagiannopoulos & Tsitsas, 2008a). On the other hand, a coating's continuous distribution, following a "shifted" Luneburg law may be treated by a step approximation of the radial function of its dielectric permittivity (Nikolic et al., 2007; Sakurai et al. 1998; Liang et al. 2005; Valagiannopoulos & Tsitsas, 2008a). Such a configuration exhibits larger quality factor values than respective single-layered spherical microstrips, which especially become extremely large as the patch approaches the metallic sphere (Valagiannopoulos & Tsitsas, 2008a). Besides, a PEC patch (implant) radiating inside a 6-layered spherical human head model was investigated in (Kim & Rahmat-Samii, 2004).

The present chapter is organized as follows. Section 2 contains an overview of existing mathematical methods, mainly treating single-layered spherical microstrip antennas. Section 3 discusses the motivation to model multi-layered spherical microstrips, where the patch is arbitrarily located inside the multi-layered sphere, and summarizes the basic features of the combined Legendre transform and T-matrix methodology for the analysis of such microstrip configurations. Furthermore, in Section 4 potential applications of multi-layered spherical microstrips are pointed out, dealing with coating's materials characterized by (i) zero refractive index, (ii) amplifying properties, and (iii) continuous dielectric permittivity distribution.

2. Overview of mathematical methods treating single-layered microstrips

In this Section we give a brief report of the existing mathematical methodologies, which provide the essential tools for the analysis of single-layered spherical microstrips. Hereinafter, an $\exp(+j\omega t)$ time dependence of the field quantities is assumed and suppressed.

2.1 Full-wave approach

In full wave analysis, one seeks the unique field solution satisfying the Helmholtz differential equation as well as the appropriate boundary conditions (see Wong, 1999 for an excellent survey on full wave methods). Into this context, we consider the microstrip configuration of Fig. 1, where a PEC sphere of radius a_1 is coated within a dielectric substrate of thickness d. A circular PEC radiating patch, located at $r=a_2$ and $\theta<\theta_p$ and assuming to possess negligible thickness, is mounted on the dielectric substrate. In order to fulfill the boundary conditions, the tangential electric field is required to be equal to zero on the PEC patch.

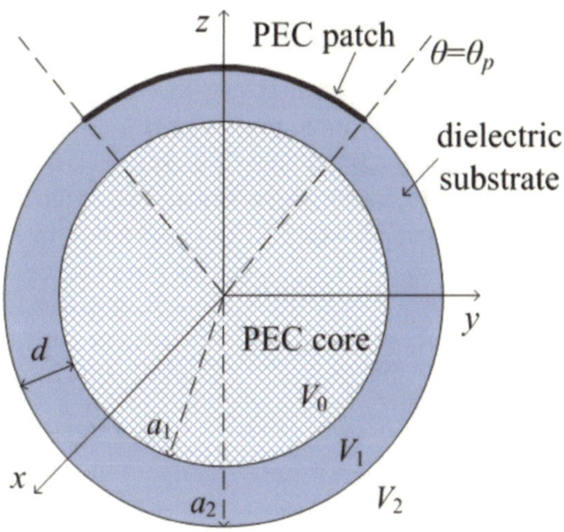

Fig. 1. Geometry of the spherical microstrip composed of a PEC sphere of radius a_1 coated with a dielectric substrate of thickness d; the circular PEC patch is located at $r=a_2$ and $\theta<\theta_p$.

The solution of the radiation problem is actually reduced to determining the surface current distribution on the patch. Since the patch is not entire with respect to the inclination angle θ (although it is entire with respect to the azimuthal angle φ), it is convenient to utilize the *vector Legendre Transform pair*, defined in (Tam & Luk, 1991; Wong, 1999; Giang et al. 2005) by

$$\mathbf{F}(r,\theta) = \sum_{n=|m|}^{+\infty} \overline{\overline{\mathbf{L}}}(n,m,\theta) \cdot \tilde{\mathbf{F}}(r,n)\,, \tag{1a}$$

$$\tilde{\mathbf{F}}(r,n) = \frac{1}{S(n,m)} \int_0^{\pi} \overline{\overline{\mathbf{L}}}(n,m,\theta) \cdot \mathbf{F}(r,\theta) \sin\theta d\theta\,, \tag{1b}$$

where $\tilde{\mathbf{F}}(r,n)$ constitutes the Legendre Transform of $\mathbf{F}(r,\theta)$. Both functions are 2×1 vectors, containing the transverse (with respect to $\hat{\mathbf{r}}$) components of either the electric or the magnetic field or the surface current, namely $\mathbf{F}(r,\theta)=[\mathbf{F}_\theta(r,\theta),\mathbf{F}_\varphi(r,\theta)]^T$ and $\tilde{\mathbf{F}}(r,n)=[\tilde{\mathbf{F}}_\theta(r,n),\tilde{\mathbf{F}}_\varphi(r,n)]^T$. The remaining participating quantities in (1a) and (1b) are given by

$$\overline{\overline{\mathbf{L}}}(n,m,\theta)=\begin{bmatrix} \dfrac{\partial P_n^m(\cos\theta)}{\partial\theta} & -jm\dfrac{P_n^m(\cos\theta)}{\sin\theta} \\ jm\dfrac{P_n^m(\cos\theta)}{\sin\theta} & \dfrac{\partial P_n^m(\cos\theta)}{\partial\theta} \end{bmatrix}, \tag{2a}$$

$$S(n,m)=\frac{2n(n+1)(n+m)!}{(2n+1)(n-m)!}. \tag{2b}$$

Note that the use of the vector Legendre Transform in the spherical microstrip is the analog situation to the utilization of the Fourier (Tsalamengas et al., 1985; Wong et al., 1993b) and the Hankel (Ali et al., 1982; Nie et al., 1990) transform in the planar one.
The electric and magnetic fields $\mathbf{E}^1$ and $\mathbf{H}^1$ in the substrate region V_1 as well as the respective ones $\mathbf{E}^2$ and $\mathbf{H}^2$ in the outer (air) region V_2 are expressed by means of the corresponding electric and magnetic potentials. Subsequently, these potentials are expanded with respect to an appropriate basis of spherical harmonic functions. The unknown weighting coefficients in these expansions will be determined by imposing the appropriate boundary conditions.
To this end, the vector Legendre Transform is utilized in order to overcome the difficulty concerning the limited θ extent of the patch. Specifically, the boundary conditions are transformed from the spatial into the spectral (Legendre) domain as

$$\tilde{\mathbf{E}}^1(r,n)=\begin{cases} \mathbf{0}, & r=a_1 \\ \tilde{\mathbf{E}}^2(r,n), & r=a_2 \end{cases}, \tag{3a}$$

$$\tilde{\mathbf{J}}_s(a_2,n)=\begin{bmatrix} 0 & -1 \\ 1 & 0 \end{bmatrix}\left[\tilde{\mathbf{H}}^2(a_2,n)-\tilde{\mathbf{H}}^1(a_2,n)\right], \tag{3b}$$

where $\tilde{\mathbf{J}}_s(a_2,n)$ is the spectral amplitude of the surface current density on the circular patch. By transforming the electric and magnetic field expressions in the spectral domain (resulting by considering the corresponding expansions of the potentials, as described above) and utilizing the transformed boundary conditions (3a) and (3b), the following spectral equation, connecting the surface current density with the tangential electric field on the patch, is derived

$$\tilde{\mathbf{E}}(a_2,n)=\overline{\overline{\mathbf{Z}}}(n)\cdot\tilde{\mathbf{J}}_s(a_2,n), \tag{4}$$

where $\overline{\overline{\mathbf{Z}}}(n)$ is a determined "impedance" matrix, while $\tilde{\mathbf{E}}(a_2,n)$ denotes the common value of the tangential electric fields $\tilde{\mathbf{E}}^1(r,n)$ and $\tilde{\mathbf{E}}^2(r,n)$ on $r=a_2$ (see the boundary condition (3a)).

Next, the current distribution on the PEC patch is expanded into a suitable set of orthogonal functions and the unknown weighting coefficients of these expansions are finally determined by using a Galerkin's moment method. More precisely, the surface current distribution is expressed by the following weighted finite sum of suitably chosen basis functions, as

$$\tilde{\mathbf{J}}_s(a_2,n)=\sum_{\nu=1}^{N} c_\nu \tilde{\mathbf{J}}_{s,\nu}(a_2,n), \tag{5}$$

where c_ν the unknown coefficients and $\tilde{\mathbf{J}}_{s,\nu}(a_2,n)$ the spectral amplitude of the ν-th expansion function. Then, by applying a Galerkin scheme, namely utilizing as testing functions the same set of basis (expansion) functions, the following homogeneous linear system is formulated

$$\overline{\overline{\mathbf{A}}}(\omega)\cdot\mathbf{c}=\mathbf{0}, \tag{6}$$

where the vector **c** incorporates the coefficients c_ν, while the elements of matrix $\overline{\overline{\mathbf{A}}}$ are given by (the * denotes the complex conjugate transpose)

$$A_{\mu\nu}(\omega)=\sum_{n=|m|}^{+\infty} S(n,m)\tilde{\mathbf{J}}^{*}_{s,\mu}(a_2,n)\cdot\overline{\overline{\mathbf{Z}}}(n)\cdot\tilde{\mathbf{J}}_{s,\nu}(a_2,n) \quad (\mu,\nu=1,\ldots,\mathrm{N}). \tag{7}$$

In view of (6), the complex resonant frequencies $\omega_0=\mathrm{Re}(\omega_0)+j\mathrm{Im}(\omega_0)$ of the spherical microstrip structure are calculated by solving numerically the equation

$$\det\left[\overline{\overline{\mathbf{A}}}(\omega_0)\right]=0. \tag{8}$$

Note that one should retain only the roots of (8) with $\mathrm{Re}(\omega_0)>0$ and $\mathrm{Im}(\omega_0)>0$ in order to avoid unbounded oscillations as time advances. In particular, $\mathrm{Re}(\omega_0)$ is the resonant frequency of the spherical microstrip and $\mathrm{Im}(\omega_0)$ expresses the radiation loss, defining the quality factor Q of the structure by

$$Q=\frac{\mathrm{Re}(\omega_0)}{2\mathrm{Im}(\omega_0)}. \tag{9}$$

2.2 Cavity model analysis

The appropriate basis functions for the expansion of the surface current on the patch are selected according to the cavity model theory (Lo et al., 1979; Luk & Tam 1991; Chen & Wong, 1994). The basic requirements for these basis functions concern the facts that they should be orthogonal and also be transformed into the spectral domain into closed form (Wong, 1999). In particular, the cavity model theory dictates that the region between the patch and the PEC core forms a cavity which is bounded by a magnetic wall along the edge of the patch and by electric walls from above and below, and that the field in the cavity is independent of r (Wong, 1999).

Subsequently, the Helmholtz equations for the electric and magnetic potentials are solved locally in an area just enclosing the patch, on which flows the unknown current. Because the patch is non-entire with respect to θ and due to the principle of charge preservation (vanishing current component normal to the edge of the metallic patch), only specific degrees of the Legendre functions provide solutions with physical meaning. More specifically, these suitable degrees for the TM$^{m\ell}$ and TE$^{m\ell}$ cases are the roots $\ell^{TM}(i)$ and $\ell^{TE}(i)$ respectively of the equations

$$P'^{m}_{\ell^{TM}(i)}(\cos\theta_0) = 0 \quad \text{and} \quad P^{m}_{\ell^{TE}(i)}(\cos\theta_0) = 0\,, \tag{10}$$

where the prime indicates differentiation with respect to θ and i is an integer denoting the ith root of the corresponding equation. The roots of (10) lie on the real axis (Abramowitz & Stegun, 1972). Each value of $\ell^{TM}(i)$ and $\ell^{TE}(i)$ corresponds to a specific operation TM$^{m\ell}$ and TE$^{m\ell}$ mode. Furthermore, by imposing the discontinuity boundary condition for the magnetic field, one obtains the explicit forms of the current basis functions for the TM$^{m\ell}$ and TE$^{m\ell}$ modes

$$\mathbf{J}_s^{TM_{m\ell(i)}}(\theta) = \begin{cases} \begin{bmatrix} P'^{m}_{\ell^{TM}(i)}(\cos\theta) \\ \frac{jm}{\sin\theta} P^{m}_{\ell^{TM}(i)}(\cos\theta) \end{bmatrix}, & \theta < \theta_p \\ \mathbf{0} & \theta > \theta_p \end{cases}, \tag{11a}$$

$$\mathbf{J}_s^{TE_{m\ell(i)}}(\theta) = \begin{cases} \begin{bmatrix} \frac{jm}{\sin\theta} P^{m}_{\ell^{TE}(i)}(\cos\theta) \\ -P'^{m}_{\ell^{TE}(i)}(\cos\theta) \end{bmatrix}, & \theta < \theta_p \\ \mathbf{0}, & \theta > \theta_p \end{cases}. \tag{11b}$$

Both current basis functions are zero for $\theta_p<\theta<\pi$, since the patch there is absent.
Importantly, the elements $A_{\mu\nu}(\omega)$ of (7) involve the current basis functions in the spectral (Legendre) domain, which are derived by imposing the Legendre transformations on the currents (11), yielding

$$\tilde{\mathbf{J}}_s^{TM_{m\ell(i)}}(n) = \frac{1}{S(n,m)} \begin{bmatrix} \dfrac{\ell^{TM}(i)\left(\ell^{TM}(i)+1\right)}{\ell^{TM}(i)\left(\ell^{TM}(i)+1\right)-n(n+1)} P^{m}_{\ell^{TM}(i)}(\cos\theta_p) P'^{m}_{n}(\cos\theta_p)\sin\theta_p \\ jm P^{m}_{\ell^{TM}(i)}(\cos\theta_p) P^{m}_{n}(\cos\theta_p) \end{bmatrix}, \tag{12a}$$

$$\tilde{\mathbf{J}}_s^{TE_{m\ell(i)}}(n) = \frac{1}{S(n,m)} \begin{bmatrix} 0 \\ \dfrac{n(n+1)}{\ell^{TE}(i)\left(\ell^{TE}(i)+1\right)-n(n+1)} P'^{m}_{\ell^{TE}(i)}(\cos\theta_p) P^{m}_{n}(\cos\theta_p)\sin\theta_p \end{bmatrix}. \tag{12b}$$

2.3 Generalized transmission line model

The transmission-line model (TLM) was one of the first and simplest methods for the analysis of microstrip antennas with relatively fair accuracy compared to other more complicated methods (Pues & Van de Capelle, 1984). The TLM method in its original form incorporates as line parameters the characteristic impedance and the effective propagation constant and is applicable only for planar rectangular or square microstrip antennas (Wong, 1999). To this end, the generalized transmission line model (GTLM) theory was proposed (Bhattacharyya & Garg, 1985), where the line parameters are the electromagnetic fields under the patch. Under this consideration, as long as the separation of variables is feasible for the wave equation expressed in a particular coordinate system, GTLM theory is applicable to microstrip antennas of any patch shape.

Focusing on the modeling of spherical microstrips by the GTLM theory, the patch is considered as a transmission line in the θ-direction, which is loaded with a wall admittance evaluated at the radiation apertures (Ke & Kishk, 1991; Kishk, 1993). Besides, the effect of other apertures is considered as leakage of the transmission line. The equivalent transmission line of the circular patch can be replaced by a Π-network, while the wall admittance at the patch edge is considered constant. Fig. 2 depicts the equivalent circuit of the spherical microstrip with a circular patch of Fig. 1 and a probe feeding at θ_f and φ_f (Wong, 1999). The network of the circuit elements, g_1, g_2, and g_3, represents the transmission-line section between the feeding and the radiation aperture at the patch edge. The shorted transmission-line section between the feeding and the patch center is replaced by an equivalent admittance y_f.

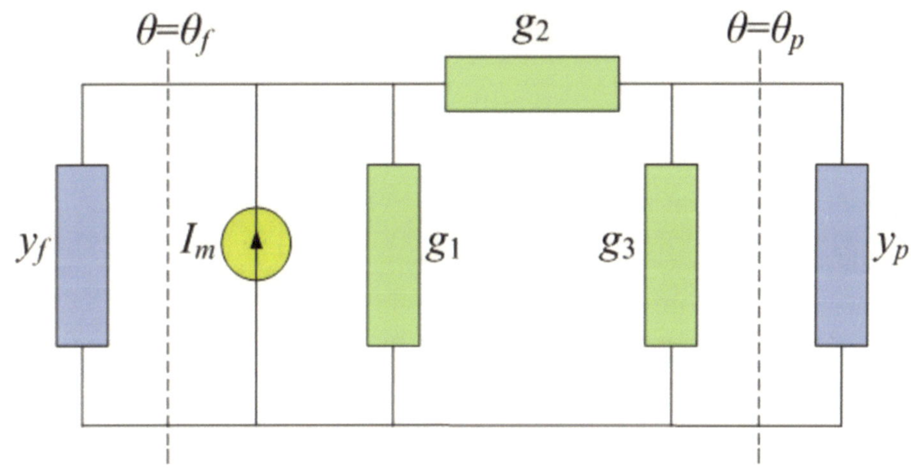

Fig. 2. Equivalent GTLM circuit for the spherical microstrip with circular patch of angular extent θ_p (depicted in Fig. 1), which is subject to probe feeding at θ_f and φ_f.

By deriving expressions for the corresponding circuit elements, the input impedance of the patch antenna at the feed position is readily obtained. More precisely, by following the GTLM formulation (Ke & Kishk, 1991; Kishk, 1993), the microstrip antenna is modeled as a transmission line in the θ-direction, while the modal voltage and the modal current are, respectively, defined by E_{rm}, and $\pm a_1 \sin\theta\, H_{\varphi m}$ for wave propagation in the $\mp\hat{\boldsymbol{\theta}}$ direction, where E_{rm}, and $H_{\varphi m}$ are the electric and magnetic fields inside the substrate layer under the patch (Wong, 1999), which for the TM_{mn} mode are given by

$$E_{rm}(\theta,\varphi) = E_0 \cos(m\varphi) \begin{cases} P_v^m(\cos\theta), & 0 \le \theta \le \theta_f \\ \left[P_v^m(\cos\theta) + C_m Q_v^m(\cos\theta) \right], & \theta_f \le \theta \le \theta_p \end{cases}, \tag{13a}$$

$$H_{\varphi m}(\theta,\varphi) = -E_0 \tfrac{\sin\theta}{j\omega\mu_0 a_1} \cos(m\varphi) \begin{cases} P_v'^m(\cos\theta), & 0 \le \theta \le \theta_f \\ \left[P_v'^m(\cos\theta) + C_m Q_v'^m(\cos\theta) \right], & \theta_f \le \theta \le \theta_p \end{cases}, \tag{13b}$$

where P_v^m and Q_v^m are the associated Legendre functions of the first and second kind, respectively, while C_m are unknown coefficients. Now, referring to Fig. 2, the circular patch is modeled as a Π-network (g_1, g_2, and g_3); y_f is the wall admittance at $\theta = \theta_f^-$, y_p is the wall admittance at the patch edge, and I_m is the feed current corresponding to the TM$_m$ mode excitation. The explicit expressions of the equivalent-circuit elements for the Π-network and the wall admittances at θ_f and θ_p are given in (Wong, 1999, p. 231-232).
Finally, having determined all the elements in the equivalent circuit, the input impedance of the antenna, seen by the probe feed at the TM$_{11}$ mode, is given by

$$Z_{in} = \frac{d}{\pi} \left[y_f + g_1 + \frac{g_2(g_3 + y_p)}{g_2 + g_3 + y_p} \right]^{-1}. \tag{14}$$

3. A combined Legendre transform and T-matrix method for the analysis of multi-layered microstrips

The techniques of Section 2 refer to the mathematical analysis of a single-layered spherical microstrip. In the present Section we summarize certain appropriate extensions of the aforementioned mathematical techniques in order to model the resonance and radiation characteristics of spherical microstrips composed of an arbitrary number of concentric spherical layers. For further details the reader is referred to (Valagiannopoulos & Tsitsas, 2008a; Valagiannopoulos & Tsitsas, 2008b). These extensions mainly refer to the suitable implementation of a T-matrix method, which handles effectively the effect of the dielectric layers above and below the PEC patch.

3.1 Geometrical configuration of the multi-layered spherical microstrip

The geometry of the under consideration multi-layered spherical microstrip is depicted in Fig. 3. A layered sphere of radius a_{U+1} is divided by U concentric spherical surfaces, defined by $r=a_u$ ($u=1,\ldots,U$), into $U+1$ layers V_u ($u=0,\ldots,U$). The sphere's core V_0 ($0\le r\le a_1$) is a perfect electric conductor. The layers V_u, defined by $a_u\le r\le a_{u+1}$ ($u=1,\ldots,U$), are filled with materials of complex dielectric permittivity ε_u and wavenumber k_u. A PEC circular patch is printed on $r=a_p$ for $0\le\theta\le\theta_p$ (between layers V_{p-1} and V_p). The exterior V_{U+1} ($r>a_{U+1}$) of the sphere is homogeneous with free-space dielectric permittivity ε_0, wavenumber k_0 and intrinsic impedance ζ_0. The entire space is magnetically inert with permeability μ_0. The feeding of the microstrip is not taken into account, since we focus on the inherent resonance and radiation properties of the microstrip structure and not the influence on them by an external excitation. However, we note that with the incorporation of certain appropriate modifications, a probe feeding of the microstrip may be also taken into account in the analysis (see for example the discussions in Chen et al., 1997).

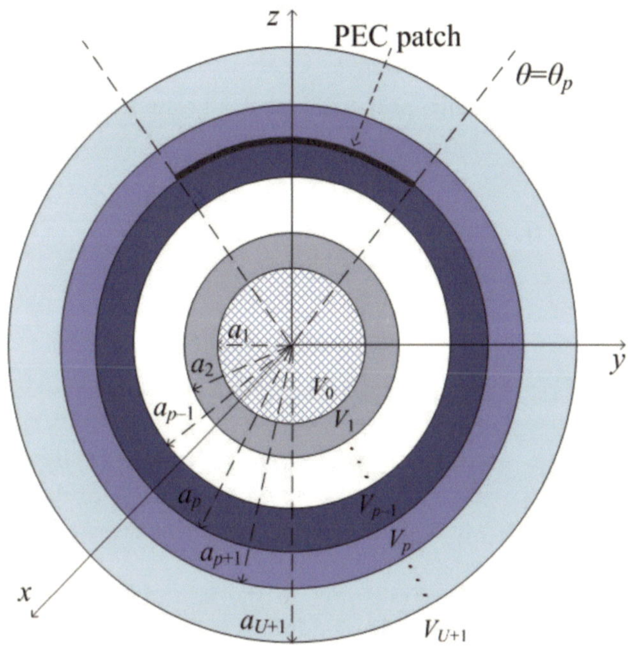

Fig. 3. The under consideration multi-layered spherical microstrip composed of a PEC spherical core of radius a_1 covered by an arbitrary number U of dielectric layers; the circular PEC patch is located at $r=a_p$ and $\theta<\theta_p$ (between layers V_{p-1} and V_p).

3.2 T-matrix method

The electric potential $\mathbf{A}_u$ and magnetic potential $\mathbf{F}_u$ in layer V_u (u=1,...,U+1), expressing the transverse magnetic TM^r and transverse electric TE^r modes respectively, admit the following expansions

$$\mathbf{A}_u(\mathbf{r}) = \hat{\mathbf{r}}A_u(\mathbf{r}) = \hat{\mathbf{r}}\, e^{jm\varphi} \sum_{n=|m|}^{+\infty} \left[\alpha_u(n)j_n(k_u r) + \beta_u(n)h_n(k_u r)\right] P_n^m(\cos\theta) \quad (a_u \le r \le a_{u+1}), \quad (15a)$$

$$\mathbf{F}_u(\mathbf{r}) = \hat{\mathbf{r}}F_u(\mathbf{r}) = \hat{\mathbf{r}}\, e^{jm\varphi} \sum_{n=|m|}^{+\infty} \left[\gamma_u(n)j_n(k_u r) + \delta_u(n)h_n(k_u r)\right] P_n^m(\cos\theta) \quad (a_u \le r \le a_{u+1}), \quad (15b)$$

where j_n and h_n are the n-th order spherical Bessel and Hankel functions of the second kind, P_n^m is the associated Legendre function of the first kind of order m and degree n and a_u, β_u, γ_u, and δ_u are under determination coefficients. The electric and magnetic potentials in the exterior V_{U+1} of the microstrip have the expansions (15) with $u=U+1$ and $a_{U+1}(n)=\gamma_{U+1}(n)=0$, valid for $r>a_{U+1}$, in order that the radiation condition is satisfied. The transverse electric and magnetic field components in layer V_u are then readily expressed in terms of the corresponding potentials (see Eq. (2) of Valagiannopoulos & Tsitsas, 2008a).

Next, since the spherical microstrip is non-entire with respect to θ, we utilize the vector Legendre Transform, given by (1) and (2) (as analyzed in details in Section 2). The Legendre Transforms $\tilde{\mathbf{E}}_u(r,n)=[\tilde{\mathbf{E}}_{u,\theta}(r,n),\tilde{\mathbf{E}}_{u,\varphi}(r,n)]^T$ and $\tilde{\mathbf{H}}_u(r,n)=[\tilde{\mathbf{H}}_{u,\theta}(r,n),\tilde{\mathbf{H}}_{u,\varphi}(r,n)]^T$ of the transverse electric $\mathbf{E}_u(r,n)=[\mathbf{E}_{u,\theta}(r,n),\mathbf{E}_{u,\varphi}(r,n)]^T$ and magnetic $\mathbf{H}_u(r,n)=[\mathbf{H}_{u,\theta}(r,n),\mathbf{H}_{u,\varphi}(r,n)]^T$ fields in layer V_u are determined by means of (15) and (1b) and by imposing the orthogonality properties of the Legendre functions (Gradshteyn & Ryzhik, 2000)

$$\tilde{\mathbf{E}}_u(r,n)=k_0\begin{bmatrix}\sqrt{\varepsilon_u}\left(\gamma_u(n)j_n^d(k_u r)+\delta_u(n)h_n^d(k_u r)\right)\\ j\zeta_0\left(\alpha_u(n)j_n(k_u r)+\beta_u(n)h_n(k_u r)\right)\end{bmatrix}, \tag{16a}$$

$$\tilde{\mathbf{H}}_u(r,n)=k_0\begin{bmatrix}\sqrt{\varepsilon_u}\left(\alpha_u(n)j_n^d(k_u r)+\beta_u(n)h_n^d(k_u r)\right)\\ (j\zeta_0)^{-1}\varepsilon_u\left(\gamma_u(n)j_n(k_u r)+\delta_u(n)h_n(k_u r)\right)\end{bmatrix}, \tag{16b}$$

where $zj_n^d(z)=d(zj_n(z))/dz$ and $zh_n^d(z)=d(zh_n(z))/dz$. Notice that by using the Legendre Transform, the coefficient pairs a_u, β_u and γ_u, δ_u, corresponding to the TM^r and TE^r fields respectively, appear exclusively in only one component of the transformed vectors $\tilde{\mathbf{E}}_u$, $\tilde{\mathbf{H}}_u$, and thus a decoupling is achieved.

By imposing the vector Legendre Transform to the boundary conditions in the spatial domain, we obtain the following boundary conditions in the spectral domain

$$\tilde{\mathbf{E}}_1(a_1,n)=\mathbf{0}, \tag{17a}$$

$$\tilde{\mathbf{E}}_u(a_u,n)-\tilde{\mathbf{E}}_{u-1}(a_u,n)=\mathbf{0}\quad(u=2,\dots,U+1), \tag{17b}$$

$$\tilde{\mathbf{H}}_u(a_u,n)-\tilde{\mathbf{H}}_{u-1}(a_u,n)=\mathbf{0}\quad(u=2,\dots,U+1\ ,\ u\neq p), \tag{17c}$$

$$\tilde{\mathbf{J}}_s(n)=\begin{bmatrix}0 & -1\\ 1 & 0\end{bmatrix}\left(\tilde{\mathbf{H}}_p(a_p,n)-\tilde{\mathbf{H}}_{p-1}(a_p,n)\right), \tag{17d}$$

where $\mathbf{J}_s$ is the surface current distribution at $r=a_p$, due to the presence of the patch.

Now, we describe a T-matrix scheme, leading to the successive connection of the fields coefficients (connections between the fields coefficients of spherically layered media are analyzed in details in Tsitsas & Athanasiadis, 2006). Specifically, successive applications of (17b) and (17c) imply the following transformations of the electric field coefficients in layer V_{u-1} to those in V_u

$$\begin{bmatrix}\alpha_u(n)\\ \beta_u(n)\end{bmatrix}=\bar{\bar{\mathbf{T}}}_u^{TM}(n)\begin{bmatrix}\alpha_{u-1}(n)\\ \beta_{u-1}(n)\end{bmatrix}\quad(u=2,\dots,U\ ,\ u\neq p), \tag{18a}$$

$$\begin{bmatrix}\gamma_u(n)\\ \delta_u(n)\end{bmatrix}=\bar{\bar{\mathbf{T}}}_u^{TE}(n)\begin{bmatrix}\gamma_{u-1}(n)\\ \delta_{u-1}(n)\end{bmatrix}\quad(u=2,\dots,U\ ,\ u\neq p), \tag{18b}$$

where the explicit expressions of the elements of the 2×2 transition matrices from V_{u-1} to V_u, appearing above, are given in (Valagiannopoulos & Tsitsas, 2008a).

Furthermore, application of the boundary condition (17a) on the PEC core implies

$$\begin{bmatrix}\alpha_1(n)\\ \beta_1(n)\end{bmatrix}=\mathbf{t}_1^{TM}(n)\alpha_1(n)\,, \tag{19a}$$

$$\begin{bmatrix}\gamma_1(n)\\ \delta_1(n)\end{bmatrix}=\mathbf{t}_1^{TE}(n)\gamma_1(n)\,, \tag{19b}$$

while application of (17b) and (17c) for $u=U$ yields

$$\begin{bmatrix}\alpha_U(n)\\ \beta_U(n)\end{bmatrix}=\mathbf{t}_{U+1}^{TM}(n)\beta_{U+1}(n)\,, \tag{20a}$$

$$\begin{bmatrix}\gamma_U(n)\\ \delta_U(n)\end{bmatrix}=\mathbf{t}_{U+1}^{TE}(n)\delta_{U+1}(n)\,, \tag{20b}$$

where the 2×1 vectors $\mathbf{t}_1^{TM}$, $\mathbf{t}_1^{TE}(n)$, $\mathbf{t}_{U+1}^{TM}(n)$, and $\mathbf{t}_{U+1}^{TE}(n)$ are given in (Valagiannopoulos & Tsitsas, 2008a).

Now, by applying successively the transformations (18) for u=2,…,p–1 and using (19), we see that the field coefficients in layer V_{p-1} are related to a_1 and γ_1 of V_1 by

$$\begin{bmatrix}\alpha_{p-1}(n)\\ \beta_{p-1}(n)\end{bmatrix}=\mathbf{t}_-^{TM}(n)\alpha_1(n)\,, \tag{21a}$$

$$\begin{bmatrix}\gamma_{p-1}(n)\\ \delta_{p-1}(n)\end{bmatrix}=\mathbf{t}_-^{TE}(n)\gamma_1(n)\,, \tag{21b}$$

where

$$\mathbf{t}_-^{TM}(n)=\begin{bmatrix}f_-^{TM}(n)\\ g_-^{TM}(n)\end{bmatrix}=\overline{\overline{\mathbf{T}}}_{w-1}^{TM}(n)\overline{\overline{\mathbf{T}}}_{w-2}^{TM}(n)\cdots\overline{\overline{\mathbf{T}}}_2^{TM}(n)\mathbf{t}_1^{TM}(n)\,, \tag{22a}$$

$$\mathbf{t}_-^{TE}(n)=\begin{bmatrix}f_-^{TE}(n)\\ g_-^{TE}(n)\end{bmatrix}=\overline{\overline{\mathbf{T}}}_{w-1}^{TE}(n)\overline{\overline{\mathbf{T}}}_{w-2}^{TE}(n)\cdots\overline{\overline{\mathbf{T}}}_2^{TE}(n)\mathbf{t}_1^{TE}(n)\,. \tag{22b}$$

In a similar way by using (18) for $u=p$+1,…,U and (20), we obtain

$$\begin{bmatrix}\alpha_p(n)\\ \beta_p(n)\end{bmatrix}=\mathbf{t}_+^{TM}(n)\beta_{U+1}(n)\,, \tag{23a}$$

$$\begin{bmatrix}\gamma_p(n)\\ \delta_p(n)\end{bmatrix}=\mathbf{t}_+^{TE}(n)\delta_{U+1}(n)\,, \tag{23b}$$

where

$$\mathbf{t}_{+}^{TM}(n)=\begin{bmatrix} f_{+}^{TM}(n) \\ g_{+}^{TM}(n) \end{bmatrix}=\left[\overline{\overline{\mathbf{T}}}_{w+1}^{TM}(n)\right]^{-1}\left[\overline{\overline{\mathbf{T}}}_{w+2}^{TM}(n)\right]^{-1}\cdots\left[\overline{\overline{\mathbf{T}}}_{U}^{TM}(n)\right]^{-1}\mathbf{t}_{U+1}^{TM}(n), \tag{24a}$$

$$\mathbf{t}_{+}^{TE}(n)=\begin{bmatrix} f_{+}^{TE}(n) \\ g_{+}^{TE}(n) \end{bmatrix}=\left[\overline{\overline{\mathbf{T}}}_{w+1}^{TE}(n)\right]^{-1}\left[\overline{\overline{\mathbf{T}}}_{w+2}^{TE}(n)\right]^{-1}\cdots\left[\overline{\overline{\mathbf{T}}}_{U}^{TE}(n)\right]^{-1}\mathbf{t}_{U+1}^{TE}(n). \tag{24b}$$

The subscripts - in (21) and + in (23) indicate approach of the patch surface $r=a_p$ from the layers below and above respectively.

3.3 Determination of the complex resonant frequencies

By the continuity boundary condition (17b) for $u=p$, the tangential electric field of layer V_{p-1} coincides with that of V_p; their common value will be denoted hereafter by $\tilde{\mathbf{E}}(a_p,n)$. Thus, by combining (17d), (21) and (23), we result to the following spectral equation for the surface current on the patch

$$\tilde{\mathbf{E}}(\alpha_p,n)=\left[\overline{\overline{\mathbf{Y}}}_{+}(n)-\overline{\overline{\mathbf{Y}}}_{-}(n)\right]^{-1}\tilde{\mathbf{J}}_{s}(n), \tag{25}$$

where

$$\overline{\overline{\mathbf{Y}}}_{+}(n)=j\frac{\sqrt{\varepsilon_p}}{\zeta_0}\begin{bmatrix} \dfrac{f_{+}^{TE}(n)j_n(x_p)+g_{+}^{TE}(n)h_n(x_p)}{f_{+}^{TE}(n)j_n^d(x_p)+g_{+}^{TE}(n)h_n^d(x_p)} & 0 \\ 0 & -\dfrac{f_{+}^{TM}(n)j_n^d(x_p)+g_{+}^{TM}(n)h_n^d(x_p)}{f_{+}^{TM}(n)j_n(x_p)+g_{+}^{TM}(n)h_n(x_p)} \end{bmatrix}, \tag{26a}$$

$$\overline{\overline{\mathbf{Y}}}_{-}(n)=j\frac{\sqrt{\varepsilon_{p-1}}}{\zeta_0}\begin{bmatrix} \dfrac{f_{-}^{TE}(n)j_n(y_p)+g_{-}^{TE}(n)h_n(y_p)}{f_{-}^{TE}(n)j_n^d(y_p)+g_{-}^{TE}(n)h_n^d(y_p)} & 0 \\ 0 & -\dfrac{f_{-}^{TM}(n)j_n^d(y_p)+g_{-}^{TM}(n)h_n^d(y_p)}{f_{-}^{TM}(n)j_n(y_p)+g_{-}^{TM}(n)h_n(y_p)} \end{bmatrix}, \tag{26b}$$

while $x_p=k_pa_p$, $y_p=k_{p-1}a_p$.

Subsequently, we follow the general procedure, analyzed for the single-layered microstrip case in Sections 2.1 and 2.2. More precisely, the surface current density on the patch is expanded into a linear combination of the form (5) with respect to appropriate basis functions, which are specifically chosen by using the cavity model theory (Lo et al., 1979). Then, we apply a Galerkin's procedure for the computation of the linear combinations unknown weighting coefficients, finally resulting to the homogeneous linear system

$$\overline{\overline{\mathbf{B}}}(\omega)\mathbf{c}=\mathbf{0}, \tag{27}$$

where

$$B_{\mu\nu}(\omega)=\sum_{n=|m|}^{+\infty}S(n,m)\tilde{\mathbf{J}}_{s,\mu}^{*}(n)\left[\overline{\overline{\mathbf{Y}}}_{+}(n)-\overline{\overline{\mathbf{Y}}}_{-}(n)\right]^{-1}\tilde{\mathbf{J}}_{s,\nu}(n) \quad (\mu,\nu=1,\ldots,\mathrm{N}). \tag{28}$$

The possible complex resonant frequencies correspond to the non-trivial solutions of (27), namely to the values of the parameter ω for which the corresponding system matrix is singular. Hence, the complex resonant frequency $\omega_0=\mathrm{Re}(\omega_0)+j\mathrm{Im}(\omega_0)$ is the root of the determinant of (27); for further details see also the related discussion in Section 2.1.
It is worth to point-out that the elements of the matrix **B** in (27) are computed numerically by truncating the infinite sum. Thus, it is required to investigate the variation of the sufficient truncation order with respect to the microstrip characteristics, and in particular with respect to the angular extent θ_p of the patch. A typical convergence pattern of the resonant frequency of a three-layered spherical microstrip is depicted in Fig. 4. It is evident that the required truncation order is large for $\theta_p \rightarrow 0$ and decreases rapidly as θ_p increases. The truncation order has also been tested against the variation of the other geometrical and physical parameters of the microstrip and it has been concluded that θ_p is the most significant parameter. For further details on the influence of the microstrip's parameters on the convergence pattern see (Valagiannopoulos & Tsitsas, 2008a).

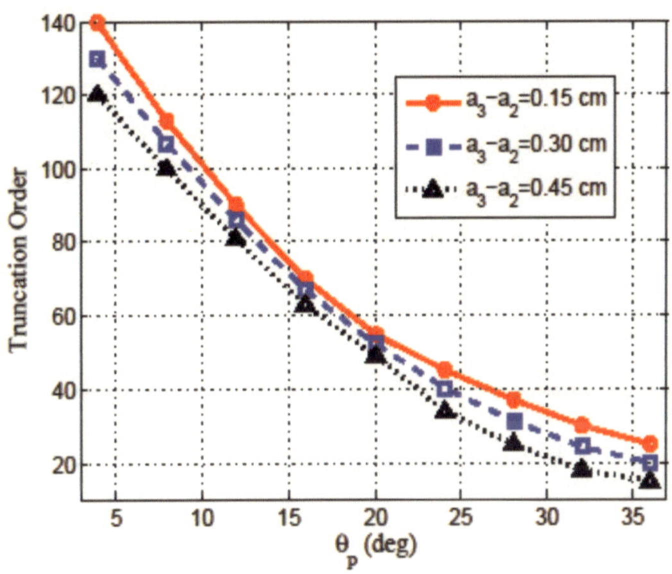

Fig. 4. Sufficient truncation order for convergence of the resonant frequency (as computed by solving the homogeneous linear system (27)) as a function of the angular extent θ_p of the patch for different superstrate thicknesses a_3-a_2 with a_1=5 cm, a_2=5.15 cm, $\varepsilon_1=\varepsilon_2=2.5\varepsilon_0$ and p=2.

3.4 Far-field radiation pattern

The determined resonant frequency of (27) may be utilized for the computation of the far-field radiation pattern by means of the inverse Legendre Transform of (16). The patch current near the resonant frequency is well described by the cavity-model modes of (11). We consider only the fundamental TM_{11} mode and hence $\tilde{\mathbf{J}}_s(n) = \tilde{\mathbf{J}}_s^{TM_{11}}(n)$.
By following the procedure analyzed in (Valagiannopoulos & Tsitsas, 2008a), we obtain the tangential electric field vector in the infinite exterior region V_{U+1}

$$\mathbf{E}_{U+1}(r,\theta)=\begin{bmatrix}\mathrm{E}_{\theta,U+1}(r,\theta)\\ \mathrm{E}_{\varphi,U+1}(r,\theta)\end{bmatrix}$$
$$=k_0\begin{bmatrix}\sum_{n=1}^{+\infty}\left(\delta_{U+1}(n)h_n^d(k_0r)P_n'^1(\cos\theta)+\zeta_0\beta_{U+1}(n)h_n(k_0r)\dfrac{P_n^1(\cos\theta)}{\sin\theta}\right)\\ \sum_{n=1}^{+\infty}\left(j\delta_{U+1}(n)h_n^d(k_0r)\dfrac{P_n^1(\cos\theta)}{\sin\theta}+j\zeta_0\beta_{U+1}(n)h_n(k_0r)P_n'^1(\cos\theta)\right)\end{bmatrix}, \tag{29}$$

where the coefficients δ_{U+1}, β_{U+1} are determined by means of the computed tangential electric field on the patch after combining (23) with (25).
Next, in order to obtain the far-field pattern expressions, we impose in (29) suitable asymptotic expansions of the spherical Hankel functions and their derivatives (Abramowitz & Stegun, 1972) and finally result to

$$\mathbf{E}_{U+1}(r,\theta)\sim h_0(k_0r)\begin{bmatrix}P_\theta(\theta)\\ P_\varphi(\theta)\end{bmatrix},\ k_0r\to+\infty, \tag{30}$$

where the far-field pattern components are given by

$$\begin{bmatrix}P_\theta(\theta)\\ P_\varphi(\theta)\end{bmatrix}=e^{-j\frac{\pi}{4}}k_0\begin{bmatrix}\sum_{n=1}^{+\infty}j^n\left(-j\delta_{U+1}(n)P_n'^1(\cos\theta)+\zeta_0\beta_{U+1}(n)\dfrac{P_n^1(\cos\theta)}{\sin\theta}\right)\\ \sum_{n=1}^{+\infty}j^n\left(\delta_{U+1}(n)\dfrac{P_n^1(\cos\theta)}{\sin\theta}+j\zeta_0\beta_{U+1}(n)P_n'^1(\cos\theta)\right)\end{bmatrix}. \tag{31}$$

4. Applications

In this Section various interesting multi-layered spherical microstrip configurations will be presented and their operational characteristics will be briefly discussed. The complex resonant frequencies of these configurations as well as their far-field radiation patterns will be computed by the combined vector Legendre Transform and T-matrix technique, outlined in Section 3. For simplicity, we take into account hereinafter only the fundamental TM_{11} mode, in order to adopt a qualitative approach for the investigation of certain novel spherical microstrip configurations. However, by following the general analysis of Section 3, the respective numerical results corresponding to higher order modes can be also readily computed.

4.1 Zero refractive index metamaterial coating

The merits of using zero refractive index materials as substrates in planar antennas have been highlighted in (Wu et al. 2005; Caloz & Itoh, 2006). Such materials are usually realized by periodic collections of rods and rings (Grzegorczyk et al. 2005). Particularly, in (Pendry et al. 1998) it was shown that by using a metallic mesh of thin wires, a plasma-like metamaterial medium is obtained, which is characterized by zero effective permittivity (and hence zero refractive index) at the plasma frequency. Zero index materials possess (according to Snell's law) the property of aligning the traveling rays from inside such a medium to free space, regardless of the angle of incidence, hence achieving high directivity by controlling the direction of emission (see Fig. 1 of Wu et al. 2005).

The spherical analog of such an emitting antenna does not lead to equally interesting results, since its operation might be expected to be nearly omnidirectional. However, the same configuration, utilized as a receiving antenna, could be potentially more promising. More precisely, suppose a spherical shell of zero refractive index enclosing the whole spherical microstrip. Then, by local application of Snell's law (similarly to the planar case), one would expect that all the ingoing rays traveling inside the zero index medium would enter the

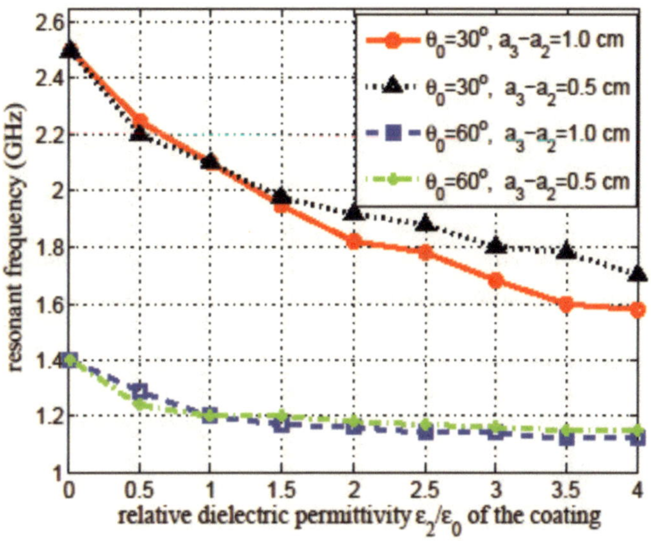

(a)

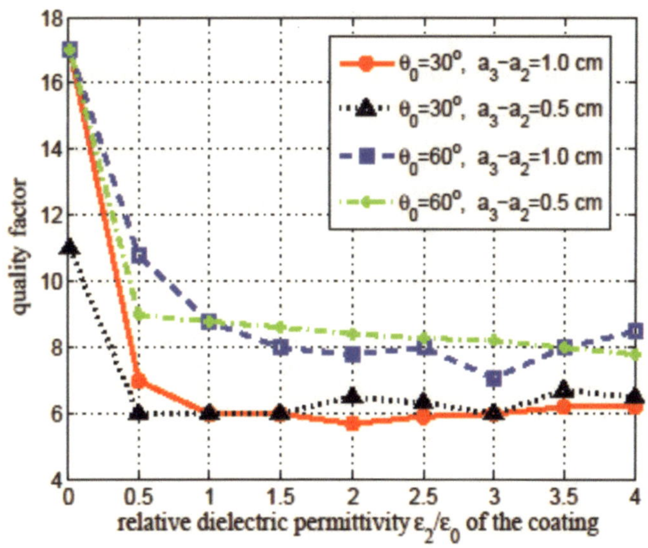

(b)

Fig. 5. Variations of (a) resonant frequency and (b) quality factor with respect to the coating's relative permittivity $\varepsilon_2/\varepsilon_0$ for different coating's thicknesses a_3-a_2 and angular patch extents θ_p in the case of a 3-layered microstrip with a_1=6 cm, a_2=7 cm, p=2 and $\varepsilon_1=\varepsilon_0$.

microstrip aligned perpendicular to the local tangential planes. Thus, the rays arrive with minimum path losses either on the patch or on the surface of the PEC sphere, therefore improving the receiving procedure. A representative sketch for the clarification of this physical situation is depicted in (Fig. 5 of Valagiannopoulos & Tsitsas, 2008a).

The preceding considerations motivate the investigation of the resonant properties of a spherical microstrip surrounded by a zero refractive index material. To this end, we consider a 3-layered microstrip with an airgap between the PEC sphere and the patch and a material with varying dielectric permittivity and thickness between the patch and the free space. Figs. 5a and 5b show the variation of the resonant frequency and quality factor of the aforementioned microstrip with respect to the coating's dielectric permittivity for two different coating's thicknesses a_3-a_2 and two different angular extents θ_p of the patch. The resonant frequency decreases and the quality factor increases with increasing angular extent θ_p for fixed coating's thickness and dielectric permittivity. Especially, Q attains a global maximum value for $\varepsilon_2 \to 0$, demonstrating that such a microstrip also possesses a high quality factor, when a material with nearly zero refractive index is utilized as coating. Furthermore, by Fig. 5a it is evident that for fixed θ_p the influence of the coating's thickness on the resonant frequency becomes weaker as its dielectric permittivity ε_2 decreases.

4.2 Amplifying excitation layer

The feeding of the spherical microstrip is usually modeled as a source located between the PEC core and the patch (Chen et al., 1997; Sipus et al., 2008; Khamas, 2009). A common type of source utilized in such cases is an axial probe connecting the two PEC surfaces. However, the antenna's operation is strongly dependent on the position of the feeding.

To this direction, a new type of φ- and θ- entire excitation, not affecting considerably the inherent resonance and radiation properties of the antenna, was proposed in (Valagiannopoulos & Tsitsas, 2008a). This excitation is an active (plasma) layer, located between core and patch, and filled with material possessing dielectric permittivity of positive imaginary part. Note that the plasma behavior of the material should be achieved close to the microstrip's resonant frequency.

In particular, a 4-layered microstrip with two airgaps surrounding the active layer was considered with the amplifying capability being controllable via both the layer's thickness a_3–a_2 as well as its dielectric permittivity $\varepsilon_2=\mathrm{Re}[\varepsilon_2]+j\mathrm{Im}[\varepsilon_2]$. Figs 6a and 6b depict the variation of the resonant frequency and the quality factor with respect to the active layer's thickness for different imaginary parts of its permittivity. The resonant frequency seems to be rather insensitive to the increase of $\mathrm{Im}[\varepsilon_2]$, indicating a rather weak dependence of the resonant frequency on the amplifying capability of the active layer. As far as the quality factor is concerned, it is indeed an increasing function of both a_3–a_2 and $\mathrm{Im}[\varepsilon_2]$. Specifically, in applications demanding large values of Q, this configuration may be used with $a_3 \to a_4$ and $\mathrm{Im}[\varepsilon_2]$ suitably large.

4.3 Shifted Luneburg lens material

In the literature are usually examined microstrips with a fixed small number of concentric layers (Tam & Luk, 1991; Wong et al., 1993a; Wong & Chen, 1993). However, it is particularly interesting to investigate also the case of the coating's region $a_1<r<a_{U+1}$ having continuously varying dielectric permittivity with radial dependence because such a configuration would offer more degrees of freedom to the designer in order to achieve the

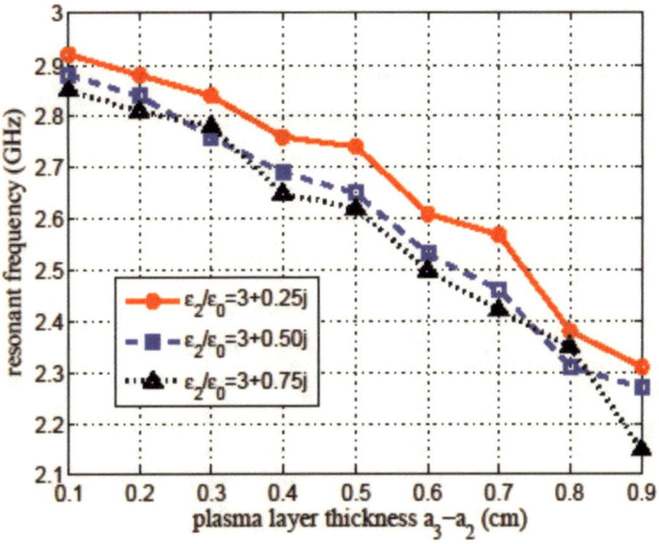

(a)

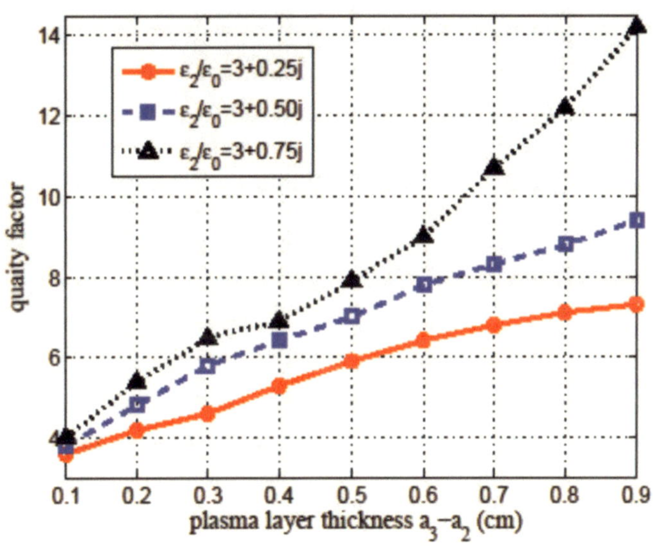

(b)

Fig. 6. (a) Resonant frequency and (b) quality factor as functions of the active (plasma) layer's thickness for different imaginary parts of its dielectric permittivity for a 4-layered microstrip with a_1=6 cm, a_2=6.5 cm, a_4=7.5 cm, p=4, θ_p=15°, and $\varepsilon_1=\varepsilon_3=\varepsilon_0$.

required operational characteristics. Since the method, summarized in Section 3, is suitable for handling efficiently microstrips with arbitrary number U of spherical layers, a continuous distribution of the radial function of the dielectric permittivity may be effectively treated by the corresponding step approximation.

In (Valagiannopoulos & Tsitsas, 2008a) a spherical microstrip with discrete shifted-Luneburg coating was investigated. More precisely, since the microstrip is composed of a PEC core, its coating is assumed to obey a continuous distribution with radial dependence following a shifted Luneburg lens law (Luneburg, 1941)

$$\varepsilon(r) = 2 - \left(\frac{r - a_1}{a_{U+1} - a_1} \right)^2 \quad (a_1 < r < a_{U+1}) . \tag{32}$$

This continuous distribution is approximated by the analyzed model of U spherical layers. The patch is still located inside the coating on $r=a_p$.

Typical variations of the resonant frequency and the quality factor as functions of the patch location a_p inside the discrete Luneburg coating for different angular extents θ_p of the patch are depicted in Fig. 7 of (Valagiannopoulos & Tsitsas, 2008a). The resonant frequency decreases with increasing θ_p for fixed a_p and with increasing a_p for fixed θ_p. On the other hand, the quality factor decreases rapidly as the patch gets distant from the PEC sphere for fixed θ_p and increases with θ_p for fixed a_p (as opposed to the respective behavior of the resonant frequency for increasing θ_p). The values of the quality factor become extremely large as the patch approaches the PEC sphere, a fact which may be exploited appropriately in various applications.

Finally, regarding the electric far-field pattern components $P_\theta(\theta)$ and $P_\varphi(\theta)$, which are computed by means of (31), Figs 7a and 7b depict the variation of the normalized components $|P_\theta(\theta)|$ and $|P_\varphi(\theta)|$ for three different patch locations a_p inside the Luneburg coating with fixed patch angular extent θ_p=30°. For all three patch locations the curves of $|P_\theta(\theta)|$ attain three local maxima at 0°, 95°, and 180°, and two local minima at 50°, and 135°. On the other hand, the curves of $|P_\varphi(\theta)|$ attain two local maxima at 0°, and 180°, and one minimum at 110°. The locations of these maxima and minima do not seem to be strongly dependent on the patch location. However, their values increase as a_p increases, namely as patch gets distant from the PEC sphere V_0 and gets closer to the free-space region V_{U+1}.

5. Conclusions

Spherical microstrip antennas have certain important advantages, which have been meticulously highlighted in the literature. Among the mathematical methods, which have been developed for analyzing the behavior of a spherical microstrip, the most widely used and effective ones are the full-wave approach, the cavity model analysis, and the generalized transmission-line model. A brief overview of each of these methods has been given. On the other hand, we summarized the basic features of an efficient method which combines the vector Legendre Transform with the T-matrix method and is utilized to analyze and model the resonance and radiation phenomena by a multi-layered spherical microstrip, whose circular metallic patch is located inside a coating, composed of an arbitrary number of concentric spherical substrate and superstrate layers. Finally, certain potential applications of multi-layered spherical microstrips were pointed out, where emphasis was given on the variation of the resonant frequency with respect to the microstrip's physical and geometrical characteristics as well as on the conditions under which high values of the quality factor may be achieved.

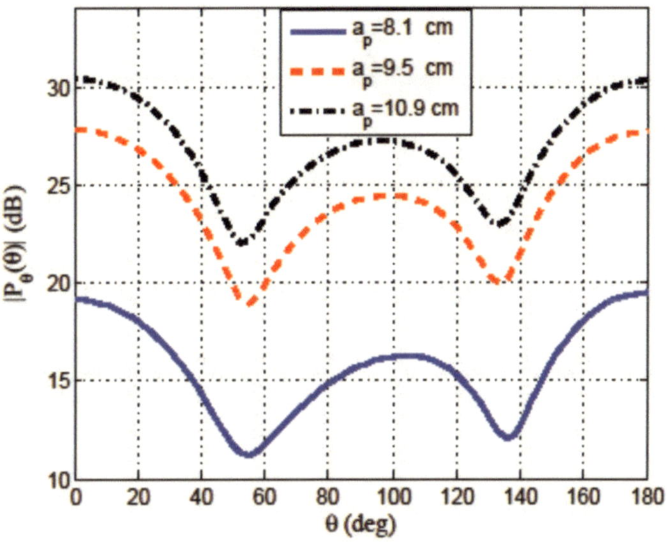

(a)

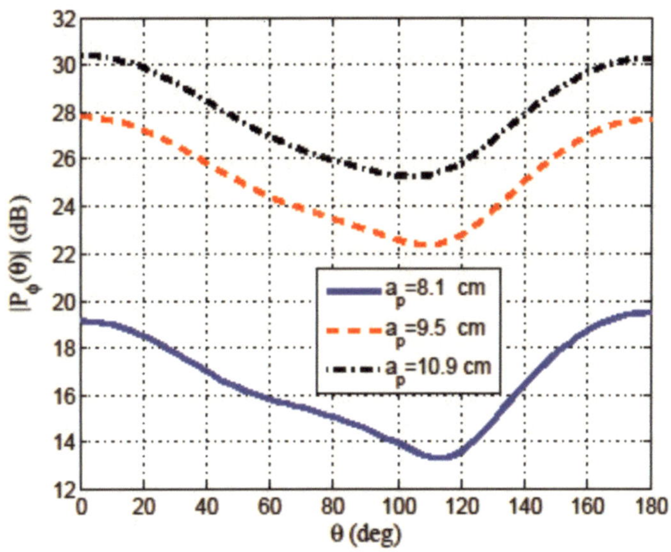

(b)

Fig. 7. Variation of the normalized electric far-field components $|P_\theta(\theta)|$ and $|P_\varphi(\theta)|$ for three different patch locations a_p with fixed patch angular extent θ_p=30°, inside a discrete Luneburg coating with constant parameters a_1=8 cm, a_{21}=11 cm, U=20.

6. References

Abramowitz M. & Stegun I. (1972). *Handbook of Mathematical Functions*, Dover, New York

Ali S. M., Chew W. C. & Kong J. A. (1982). Vector Hankel Transform Analysis of Annular-Ring Microstrip Antenna. *IEEE Trans. Antennas Propag.*, Vol. AP-30, No. 4, pp. 637-644

Bhattacharyya A. K. & Garg R. (1985). Generalised transmission line model for microstrip patches. *IEE Proc. pt. H.*, Vol. 132, pp. 93-98.

Caloz C. & Itoh T. (2006). *Electromagnetic Metamaterials: Transmission Line Theory and Microwave*, John Wiley & Sons, New Jersey

Chen H. T. & Wong K. L. (1994). Analysis of probe-fed spherical circular microstrip antennas using cavity-model theory. *Microwave Opt. Technol. Lett.*, Vol. 25, pp. 309-312

Chen H.-T., Chen H.-D & Cheng Y.-T. (1997). Full-Wave Analysis of the Annular-Ring Loaded Spherical-Circular Microstrip Antenna. *IEEE Trans. Antennas Propag.*, Vol. 45, No. 11, pp. 1581-1583

Chew W. C. (1995). *Waves and Fields in Inhomogeneous Media*, IEEE Press, New York

Giang T. V. B., Thiel M. & Dreher A. (2005). A Unified Approach to the Analysis of Radial Waveguides, Dielectric Resonators, and Microstrip Antennas on Spherical Multilayer Structures. *IEEE Trans. Microw. Theory Tech.*, Vol. 53, No. 1, pp. 404-409

Gradshteyn I. S. & Ryzhik I. M. (2000). *Table of Integrals, Series, and Products*, Academic Press, San Diego

Grzegorczyk T. M., Moss C. D., Lu J., Chen X., Pacheco J. Jr. & Kong J. A. (2005). Properties of Left-Handed Metamaterials: Transmission, Backward Phase, Negative Refraction, and Focusing. *IEEE Trans. Microw. Theory Tech.*, Vol. 53, No. 9, pp. 2956-2967

Ke B., & Kishk A. A. (1991). Analysis of spherical circular microstrip antennas. *IEE Proc. pt. H*, Vol. 138, pp. 542-548

Khamas S. K. (2009). Electromagnetic Radiation by Antennas of Arbitrary Shape in a Layered Spherical Media. *IEEE Trans. Antennas Propag.*, Vol. 57, No. 12, pp. 3827-3834

Kim J. & Rahmat-Samii Y. (2004). Implanted Antennas Inside a Human Body: Simulations, Designs, and Characterizations. *IEEE Trans. Microw. Theory Tech.*, Vol. 52, No. 8, pp. 1934-1943

Kishk A. A. (1993). Analysis of spherical annular-ring microstrip antennas. *IEEE Trans. Antennas Propagat.*, Vol. 41, pp. 338-343

Liang C. S., Streater D. A., Jin J.-M., Dunn E., & Rozendal T. (2005). A quantitative study of Luneberg-lens reflectors. *IEEE Antenna Propag. Mag.*, Vol. 47, No. 2, pp. 30-41

Lo Y. T., Solomon D., & Richards W. F. (1979). Theory and Experiment on Microstrip Antennas. *IEEE Trans. Antennas Propag.*, Vol. AP-27, No. 2, pp. 137-145

Luk K. M. & Tam W. Y. (1991). Patch antennas on spherical body. *Proc. Inst. Elect. Eng., pt. H*, Vol. 138, pp. 103-108

Luneburg R. K. (1941). *The Mathematical Theory of Optics*, Brown Univ. Press: Providence, RI.

Nie Z., Chew W. C. & Lo Y. T. (1990). Analysis of the Annular-Ring-Loaded Circular-Disk Microstrip Antenna. *IEEE Trans. Antennas Propag.*, Vol. 38, No. 6, pp. 806-813

Nikolic N., Kot J. S., & Vinogradov S. (2007). Scattering by a Luneberg lens partially covered by a metallic cap. *J. of Electromagn. Waves and Appl.*, Vol. 21, No. 4, pp. 549-563

Pendry J. B., Holden A. J., Robbins D. J. & Stewart W. J. (1998). Low frequency plasmons in thin-wire structures. *J. Physics-Condensed Matter*, Vol. 10, pp. 4785-4809

Pozar D. & Schaubert D. (1995). *Microstrip Antennas: The Analysis and Design of Microstrip Antennas and Arrays*, Wiley-IEEE Press: New York

Pues H. & Van de Capelle A. (1984). Accurate transmission-line model for the rectangular microstrip antenna. *IEE Proc. pt. H*, Vol. 131, pp. 334-340

Ribero J.-M., Damiano J.-P. & Pirinoli P. (1999). Analysis and synthesis of spherical circular and annular-ring microstrip structures using a new algebraic tool. *Microw. and Opt. Tech. Lett.*, Vol. 20, No. 4, pp. 274-280

Sakurai H., Hashidate T., Ohki M., Motojima K., & Kozaki S. (1998). Electromagnetic scattering by the Luneberg lens with reflecting cap. *IEEE Trans. Electromagn. Compat.*, Vol. 40, No. 2, pp. 94-96

Sipus Z., Burum N. & Bartolic J. (2003). Analysis of rectangular microstrip patch antennas on spherical structures. *Microw. and Opt. Tech. Lett.*, Vol. 36, No. 4, pp. 276-280

Sipus Z., Skokic S., Bosiljevac M. & Burum N. (2008). Study of Mutual Coupling Between Circular Stacked-Patch Antennas on a Sphere. *IEEE Trans. Antennas Propag.*, Vol. 56, No. 7, pp. 1834-1844

Tam W. Y. & Luk K. M. (1991). Resonance in spherical-circular microstrip structure. *IEEE Trans. Microwave Theory Tech.*, Vol. 39, pp. 700-704

Tsalamengas J. L., Uzunoglu N. K. & Alexopoulos N. G. (1985). Propagation Characteristics of a Microstrip Line Printed on a General Anisotropic Substrate. *IEEE Trans. Microw. Theory Tech.*, Vol. MTT-33, No. 10, pp. 941-945

Tsitsas N. L. & Athanasiadis C. (2006). On the scattering of spherical electromagnetic waves by a layered sphere. *Quart. J. Mech. Appl. Math.*, Vol. 59, No. 1, pp. 55-74

Uwaro, T. & Itoh, T. (1989). Chapter 5, In: *Numerical Techniques for Microwave and Millimeter-Wave Passive Structures*, Itoh T., (Ed.), John Wiley & Sons, New York

Valagiannopoulos C. A. & Tsitsas N. L. (2008a). On the Resonance and Radiation Characteristics of Multi-Layered Spherical Microstrip Antennas. *Electromagnetics*, Vol. 28, pp. 243-264

Valagiannopoulos C. A. & Tsitsas N. L. (2008b). Complex resonant frequencies of multi-layered spherical microstrip antennas, *Proceedings of 12th International Conference on Mathematical Methods in Electromagnetic Theory, 2008 (MMET 2008)*, pp. 213-215, Odessa, Ukraine, Jun. 2008

Wong, K.-L. (1999). *Design of Nonplanar Microstrip Antennas and Transmission Lines*, Wiley, New York

Wong K.-L. & Chen H.-T. (1993). Resonance in a Spherical-Circular Microstrip Structure with an Airgap. *IEEE Trans. Microw. Theory Tech.*, Vol. 41, No. 8, pp. 1466-1468

Wong K.-L., Hsiao S.-F. & Chen H.-T. (1993a). Resonance and Radiation of a Superstrate-Loaded Spherical-Circular Microstrip Patch Antenna. *IEEE Trans. Antennas Propag.*, Vol. 41, No. 5, pp. 686-690

Wong K.-L., Row J.-S., Kuo C.-W. & Huang K.-C. (1993b). Resonance of a Rectangular Microstrip Patch on a Uniaxial Substrate. *IEEE Trans. Microw. Theory Tech.*, Vol. 41, No. 4, pp. 698-701

Wu B.-I., Wang W., Pacheco J., Chen X., Grzegorczyk T. & Kong J. A. (2005). A study of using metamaterials as antenna substrate to enhance gain. *Progress in Electromagnetic Research*, Vol. PIER 51, pp. 295-328

7

Cavity-Backed Cylindrical Wraparound Antennas

O. M. C. Pereira-Filho[1], T. B. Ventura[2], C. G. Rego[3],
A. F. Tinoco-S.[4], and J. C. da S. Lacava[4]
[1]*Federal University of Pernambuco,*
[2]*UNA College,*
[3]*Federal University of Minas Gerais,*
[4]*Technological Institute of Aeronautics*
Brazil

1. Introduction

One of the most important advantages of microstrip antennas is their ability to conform to various surfaces, accordingly to mechanical or aerodynamical needs. Cavity-backed cylindrical wraparound antennas are microstrip-ring radiators that are embedded in a circular cylinder, like a missile for example. The purpose of this chapter is to present a detailed and accurate analysis of such antennas that fully takes into account its geometry, as well as overcomes some difficulties encountered for cylindrical microstrip antennas.

Most usually cylindrical microstrip antennas refer to metallic patches printed onto dielectric-coated infinite cylinders. Various methods have been applied to analyze such antennas (Wong,1999), as cavity method, electric surface current method, method of moments, and hybrid methods. The cavity method (Krowne,1983; Luk et al.,1989; Lumini,1991; Heckler et al.,2003; Yang & Ruan,1993; Wong & Ke,1993) is based on the fact that the electromagnetic fields are concentrated between the patch and the cylinder, and can be approximated by a modal expansion within this cavity. The dielectric layer external to the cavity is usually disregarded and equivalent magnetic surface currents are used as radiation sources. Although limited to thin dielectrics the cavity method provides valuable insight information about the field distribution. The dielectric effect may be included by the Green's functions as in (Fonseca & Giarola,1983). The electric surface current method (Ashkenazy et al.,1985) uses an approximation of the electric surface current on the antenna and the Green's functions to calculate the electrical characteristics of the antenna. It presents more accurate results than cavity method as the dielectric layer is taken into account through the Green's functions. Full-wave analysis have also been extensively employed to analyze a variety of microstrip antennas through the method of moments (Harrington,1968). This procedure expands the surface current distribution on the antenna in a set of functions known as basis-functions. The fields are obtained using Green's functions, and the expansion coefficients are determined by imposing the boundary conditions. The complexity of such method depends on the choice of these functions. It was used in (Ali et al.,1989) and (Habashy et al.,1990) to determine the resonance frequencies, radiation patterns and input impedances of rectangular and wraparound antennas printed on a single dielectric layer over a cylindrical conductor, when fed by a coaxial cable. The effect of parasitic rectangular patches are analyzed in (Tam et al.,1995) and cylindrical arrays of rectangular patches were analyzed in (Silva et al.,1991). Superstrate-loaded cylindrical microstrip antennas were also analyzed. In (Silva et al.,1991b) the rectangular patch was placed between two dielectric layers over a cylindrical conductor, and fed by a transmission line. The superstrate effect on the resonance frequency of rectangular patches was described in (Wong et al.,1993b), and on the input impedance of a probe-fed rectangular patch was shown in (Ke & Wong,1994). Different configurations of arrays of cylindrical-rectangular patches in a multilayer circular cylindrical structure were shown in (Raffaelli et al.,2005). Mutual impedance of superstrate-loaded wraparound antennas was presented in (Silva & Lacava,1995). A hybrid solution combining finite-difference time-domain and Green's functions was used in (Mang & Xiaowen,2004) to obtain the input impedance and radiation pattern of probe-fed multilayered cylindrical-rectangular antennas.

Many practical situations require the use of embedded antennas for which the dielectric-coated infinite cylinder is no longer an accurate model. These include cavity-backed rectangular antennas (Kempel & Volakis,1995) analyzed by hybrid finite element-boundary integral method, and flush-mounted cylindrical-rectangular microstrip antennas (Pereira Filho,2009) by method of moments. These geometries avoid the surface waves at the infinite dielectric layers, diminishing the coupling between array elements, and increasing the antenna efficiency.

The purpose of this chapter is to present a detailed analysis of a cavity-backed wraparound cylindrical antennas, using equivalence principle and method of moments. A cavity is built within a cylindrical perfect electric conductor, and filled with a dielectric. A wraparound patch is printed onto the dielectric interface. The antenna is fed by a parallel network of equally spaced coaxial cables along a circumference, allowing the antenna to radiate an almost omni-directional pattern. The procedure is based on equivalence principle to split the region in two, a cylindrical cavity and an external region, with equivalent magnetic surface currents in the interface. Method of moments is applied to determine the equivalent currents, using full-domain basis-functions in ϕ-direction and sub-domain basis-functions in z-direction.

Method of moments analyses of cylindrical microstrip antennas are known to present limitations when dealing with large cylinders. This is due to numerical problems in evaluating Bessel functions of large complex arguments, that arise from the wave equation solution in cylindrical coordinates. An analytical procedure is presented that mitigate this problem and allows the analysis to be performed for larger cylinders. The procedure proposed here starts from writing the Green's function in an appropriate form, where the Bessel functions appear in specific combinations which are evaluated differently according to the argument range. For very small arguments analytical expressions of such combinations are obtained from analytical limits of the Bessel functions. These analytical limits of the combinations can be evaluated for arguments much smaller than the Bessel functions individually, due to opposite behavior of the Bessel functions in these combinations. Similarly, for very large arguments asymptotic expressions of Bessel functions are used to obtain analytical expressions of the combinations that can be calculated much further than single Bessel functions. Numerical routines of Bessel functions are only used for an argument range without numerical issues.

Results for radiation pattern and input impedance are compared to those from a commercial software or from other methods. The input impedance is obtained from a variational expression, based on reciprocity theorem. It includes the reaction of the fields due to the basis-functions on the feeding network, as well as the reaction of the fields due to the feeding network on itself, which leads to the self-impedance of the feed. Unlike for rectangular or spherical geometries, coaxial cables in cylindrical layers generate both transverse electric

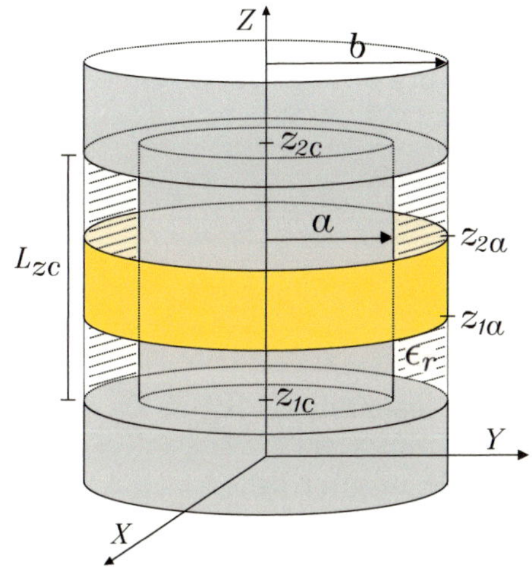

Fig. 1. Geometry of cavity-backed cylindrical wraparound antenna.

and transverse magnetic fields, and the self-impedance of the feed is not straightforward. This chapter introduces an approximation for the self-impedance of probes in cylindrically layered medium, by disregarding its radiation contribution. Comparison of the results with those obtained using commercial software validates the analysis. The effects of the cavity on radiation fields and input impedance are also presented.

2. Application of equivalence principle

The geometry of the cavity-backed cylindrical wraparound antenna is shown in Fig. 1. A cylindrical perfect electric conductor has radius b. A cavity is built within the conductor, from $z = z_{1c}$ to $z = z_{2c}$, with inner radius a, and filled with a dielectric of permeability $\mu_d = \mu_o$ and permittivity $\varepsilon_d = \varepsilon_r \varepsilon_o(1 - j\tan\delta)$, where ε_r and $\tan\delta$ are the relative permittivity and loss tangent of the dielectric, and μ_o and ε_o are free-space permeability and permittivity, respectively. A wraparound antenna is printed onto the dielectric surface ($\rho = b$), from $z = z_{1a}$ to $z = z_{2a}$. The antenna is fed by a parallel network of equally spaced N_p coaxial cables along a circumference ($z = z_f$, $\rho = b$) of the antenna. All conductors are assumed to be perfect, and a time-variation of e^{jwt} is assumed and suppressed.

The antenna is analyzed with the aid of the equivalence principle. The original geometry shown in Fig. 1 is split into 2 regions, by adding perfect electric conductors at $\rho = b$ from z_{1c} to z_{1a} and from z_{2a} to z_{2c}, and equivalent magnetic surface currents. The internal region forms a cavity from $\rho = a$ to $\rho = b$, and from $z = z_{1c}$ to $z = z_{2c}$. The fields within the cavity are due to the excitation probes and to the equivalent magnetic surface current ($\bar{M}_s$) at $\rho = b$. The fields in the external region ($\rho > b$) are due to the equivalent magnetic surface current (-$\bar{M}_s$) at $\rho = b$ in the presence of an infinite perfect electric cylindrical conductor. The problem is solved using method of moments in spectral domain, as described below.

2.1 Internal region

The fields within the cavity are written as the sum of TE_z and TM_z components, with the aid of vector potential components F_{zd} and A_{zd}, respectively. The boundary conditions in

z-direction require that $\partial A_{zd}/\partial z = 0$ and $F_{zd} = 0$ at $z = z_{1c}$ and $z = z_{2c}$. For both cases the boundary conditions require the continuity of the potentials in ϕ-direction. As a result the vector potentials can be expressed in a double-Fourier series already satisfying the boundary conditions in z and ϕ-directions as:

$$A_{zd}(\rho,\phi,z) = \sum_{n=-\infty}^{\infty}\sum_{q=0}^{\infty} \widetilde{A}_{zd}^{ec}(\rho,n,q)e^{-jn\phi}\cos\left[\frac{q\pi}{L_{zc}}(z-z_{1c})\right] \tag{1}$$

$$F_{zd}(\rho,\phi,z) = \sum_{n=-\infty}^{\infty}\sum_{q=1}^{\infty} \widetilde{F}_{zd}^{es}(\rho,n,q)e^{-jn\phi}\sin\left[\frac{q\pi}{L_{zc}}(z-z_{1c})\right] \tag{2}$$

where $L_{zc} = (z_{2c} - z_{1c})$, the first superscript in $\widetilde{A}_{zd}^{ec}$ or $\widetilde{F}_{zd}^{es}$ refers to an exponential Fourier series in ϕ, and the second superscript refers to a sine or cosine Fourier series in z. And the vector potentials in spectral domain are given by:

$$\widetilde{A}_{zd}^{ec}(\rho,n,q) = \frac{1}{2\pi}\frac{\epsilon_q}{L_{zc}}\int_0^{2\pi}\int_{z_{1c}}^{z_{2c}} A_{zd}(\rho,\phi,z)e^{jn\phi}\cos\left[\frac{q\pi}{L_{zc}}(z-z_{1c})\right]dz\,d\phi \tag{3}$$

$$\widetilde{F}_{zd}^{es}(\rho,n,q) = \frac{1}{2\pi}\frac{2}{L_{zc}}\int_0^{2\pi}\int_{z_{1c}}^{z_{2c}} F_{zd}(\rho,\phi,z)e^{jn\phi}\sin\left[\frac{q\pi}{L_{zc}}(z-z_{1c})\right]dz\,d\phi \tag{4}$$

where $\epsilon_q = 1$ if $q = 0$, and $\epsilon_q = 2$ if $q > 0$.
In space-domain the electromagnetic fields are obtained from the vector potentials by (Harrington,1961):

$$\begin{aligned}
E_\rho &= \frac{1}{j\omega\varepsilon_d}\frac{\partial^2 A_{zd}}{\partial\rho\partial z} - \frac{1}{\rho}\frac{\partial F_{zd}}{\partial\phi} & H_\rho &= \frac{1}{j\omega\mu_d}\frac{\partial^2 F_{zd}}{\partial\rho\partial z} + \frac{1}{\rho}\frac{\partial A_{zd}}{\partial\phi} \\
E_\phi &= \frac{1}{j\omega\varepsilon_d\rho}\frac{\partial^2 A_{zd}}{\partial\phi\partial z} + \frac{\partial F_{zd}}{\partial\rho} & H_\phi &= \frac{1}{j\omega\mu_d\rho}\frac{\partial^2 F_{zd}}{\partial\phi\partial z} - \frac{\partial A_{zd}}{\partial\rho} \\
E_z &= \frac{1}{j\omega\varepsilon_d}\left(\frac{\partial^2}{\partial z^2} + k_d^2\right)A_{zd} & H_z &= \frac{1}{j\omega\mu_d}\left(\frac{\partial^2}{\partial z^2} + k_d^2\right)F_{zd}
\end{aligned} \tag{5}$$

where $k_d = \omega\sqrt{\varepsilon_d\mu_d}$ is the wavenumber in the dielectric, including losses. Using the double-Fourier series (1) and (2) the components of the electromagnetic fields in spectral domain are given by:

$$\widetilde{E}_\rho^{es}(\rho,n,q) = -\frac{1}{j\omega\varepsilon_d}\frac{q\pi}{L_{zc}}\frac{\partial}{\partial\rho}\widetilde{A}_{zd}^{ec}(\rho,n,q) + \frac{jn}{\rho}\widetilde{F}_{zd}^{es}(\rho,n,q) \tag{6}$$

$$\widetilde{E}_\phi^{es}(\rho,n,q) = \frac{n}{\omega\varepsilon_d\rho}\frac{q\pi}{L_{zc}}\widetilde{A}_{zd}^{ec}(\rho,n,q) + \frac{\partial}{\partial\rho}\widetilde{F}_{zd}^{es}(\rho,n,q) \tag{7}$$

$$\widetilde{E}_z^{ec}(\rho,n,q) = \frac{1}{j\omega\varepsilon_d}k_{\rho_d}^2\widetilde{A}_{zd}^{ec}(\rho,n,q) \tag{8}$$

$$\widetilde{H}_\rho^{ec}(\rho,n,q) = -\frac{jn}{\rho}\widetilde{A}_{zd}^{ec}(\rho,n,q) + \frac{1}{j\omega\mu_d}\frac{q\pi}{L_{zc}}\frac{\partial}{\partial\rho}\widetilde{F}_{zd}^{es}(\rho,n,q) \tag{9}$$

$$\widetilde{H}_\phi^{ec}(\rho,n,q) = -\frac{\partial}{\partial\rho}\widetilde{A}_{zd}^{ec}(\rho,n,q) - \frac{n}{\omega\mu_d\rho}\frac{q\pi}{L_{zc}}\widetilde{F}_{zd}^{es}(\rho,n,q) \tag{10}$$

$$\widetilde{H}_z^{es}(\rho,n,q) = \frac{1}{j\omega\mu_d}k_{\rho_d}^2\widetilde{F}_{zd}^{es}(\rho,n,q) \tag{11}$$

where $k_{\rho_d}^2 = k_d^2 - \left(\frac{q\pi}{L_{zc}}\right)^2$.

The vector potentials in spectral domain ($\tilde{A}_z^{ec}(\rho,n,q)$ and $\tilde{F}_z^{es}(\rho,n,q)$) are combinations of Bessel functions of order n satisfying the boundary conditions at $\rho = a$ and $\rho = b$. At the inner cylindrical conductor ($\rho = a$) the zero tangential electric field requires that $A_z^d = 0$ and $\partial F_z^d/\partial\rho = 0$, and the potentials can be written as:

$$\widetilde{A}_{zd}^{ec}(\rho,n,q) = c_1\left[J_n(k_{\rho_d}\rho)H_n^{(2)}(k_{\rho_d}a) - J_n(k_{\rho_d}a)H_n^{(2)}(k_{\rho_d}\rho)\right] \tag{12}$$

$$\widetilde{F}_{zd}^{es}(\rho,n,q) = c_2\left[J_n(k_{\rho_d}\rho)H_n^{(2)'}(k_{\rho_d}a) - J_n'(k_{\rho_d}a)H_n^{(2)}(k_{\rho_d}\rho)\right] \tag{13}$$

where $B_n'(.)$ is the derivative of the Bessel function regarding to the argument. The boundary conditions at $\rho = b$ require that the equivalent surface magnetic current $\bar{M}_s(\phi,z) = -\hat{n}\times\bar{E}(b,\phi,z) = \hat{a}_\rho\times\bar{E}(b,\phi,z)$, where $\hat{n}$ is the unit vector normal to the surface $\rho = b$ pointing inwards. Consequently $M_z(\phi,z) = E_\phi(b,\phi,z)$ and $M_\phi(\phi,z) = -E_z(b,\phi,z)$. In spectral domain it reads:

$$\widetilde{M}_z^{es}(n,q) = \widetilde{E}_\phi^{es}(b,n,q) = \frac{n}{\omega\varepsilon_d b}\frac{q\pi}{L_{zc}}\widetilde{A}_{zd}^{ec}(b,n,q) + \frac{\partial}{\partial\rho}\widetilde{F}_{zd}^{es}(b,n,q) \tag{14}$$

$$\widetilde{M}_\phi^{ec}(n,q) = -\widetilde{E}_z^{ec}(b,n,q) = -\frac{1}{j\omega\varepsilon_d}k_{\rho_d}^2\widetilde{A}_{zd}^{ec}(b,n,q) \tag{15}$$

For a given equivalent surface magnetic current the coefficients c_1 and c_2 are determined by substituting (12) and (13) into (14) and (15). And the vector potentials can be written as:

$$\widetilde{A}_{zd}^{ec}(\rho,n,q) = \widetilde{G}_{M\phi}^{Ad}(\rho,n,q)\widetilde{M}_\phi^{ec}(n,q), \tag{16}$$

$$\widetilde{F}_{zd}^{es}(\rho,n,q) = \widetilde{G}_{M\phi}^{Fd}(\rho,n,q)\widetilde{M}_\phi^{ec}(n,q) + \widetilde{G}_{Mz}^{Fd}(\rho,n,q)\widetilde{M}_z^{es}(n,q) \tag{17}$$

where $\tilde{G}_{M\phi}^{Ad}(\rho,n,q)$ corresponds to the transform of the Green's function for the potential $A_z^d(\rho,\phi,z)$ due to a ϕ-directed magnetic surface current at $\rho = b$. And similarly for the other terms, which are given by:

$$\widetilde{G}_{M\phi}^{Ad}(\rho,n,q) = -\frac{j\omega\varepsilon_d}{(k_{\rho_d})^2\vartheta_5(n,q)}\left[J_n(k_{\rho_d}\rho)H_n^{(2)}(k_{\rho_d}a) - J_n(k_{\rho_d}a)H_n^{(2)}(k_{\rho_d}\rho)\right] \tag{18}$$

$$\widetilde{G}_{M\phi}^{Fd}(\rho,n,q) = \frac{jn}{b\left(k_{\rho_d}\right)^3\vartheta_1(n,q)}\frac{q\pi}{L_{zc}}\left[J_n(k_{\rho_d}\rho)H_n^{(2)'}(k_{\rho_d}a) - J_n'(k_{\rho_d}a)H_n^{(2)}(k_{\rho_d}\rho)\right] \tag{19}$$

$$\widetilde{G}_{Mz}^{Fd}(\rho,n,q) = \frac{1}{\left(k_{\rho_d}\right)\vartheta_1(n,q)}\left[J_n(k_{\rho_d}\rho)H_n^{(2)'}(k_{\rho_d}a) - J_n'(k_{\rho_d}a)H_n^{(2)}(k_{\rho_d}\rho)\right] \tag{20}$$

where:

$$\vartheta_1(n,q) = \left[J_n'(k_{\rho_d}b)H_n^{(2)'}(k_{\rho_d}a) - J_n'(k_{\rho_d}a)H_n^{(2)'}(k_{\rho_d}b)\right] \tag{21}$$

$$\vartheta_5(n,q) = \left[J_n(k_{\rho_d}b)H_n^{(2)}(k_{\rho_d}a) - J_n(k_{\rho_d}a)H_n^{(2)}(k_{\rho_d}b)\right] \tag{22}$$

The electromagnetic fields within the cavity are obtained substituting (16), (17) into (6)-(11), and taking the appropriate inverse transforms.

2.2 External region

The external region consists of an equivalent surface magnetic current ($-\bar{M}_s$) over an infinite and perfect electric conductor cylinder of radius b. The external electromagnetic fields can also be written as a sum of TE_z and TM_z components (Harrington,1961), with vector potentials expanded in cylindrical waves, satisfying the radiation condition:

$$A_{zo}(\rho,\phi,z) = \sum_{n=-\infty}^{\infty} \int_{-\infty}^{\infty} \widetilde{A}_{zo}^{ef}(\rho,n,k_z)e^{-jk_z z}e^{-jn\phi}dk_z \tag{23}$$

$$F_{zo}(\rho,\phi,z) = \sum_{n=-\infty}^{\infty} \int_{-\infty}^{\infty} \widetilde{F}_{zo}^{ef}(\rho,n,k_z)e^{-jk_z z}e^{-jn\phi}dk_z \tag{24}$$

where the superscripts ef refer to an exponential series in ϕ and a Fourier transform in z. And the vector potentials in spectral domain are given by:

$$\widetilde{A}_{zo}^{ef}(\rho,n,k_z) = \frac{1}{(2\pi)^2}\int_{-\infty}^{\infty}\int_{-\pi}^{\pi} A_{zo}(\rho,\phi,z)e^{jk_z z}e^{jn\phi}d\phi\, dk_z \tag{25}$$

$$\widetilde{F}_{zo}^{ef}(\rho,n,k_z) = \frac{1}{(2\pi)^2}\int_{-\infty}^{\infty}\int_{-\pi}^{\pi} F_{zo}(\rho,\phi,z)e^{jk_z z}e^{jn\phi}d\phi\, dk_z \tag{26}$$

In space domain the fields are given by (5) with ε_d, μ_d and k_d substituted by ε_o, μ_o and $k_o = \omega\sqrt{\varepsilon_o\mu_o}$, respectively. In the transform domain the components of the electromagnetic fields are given by:

$$\widetilde{E}_{\rho}^{ef}(\rho,n,k_z) = -\frac{k_z}{\omega\varepsilon_o}\frac{\partial\widetilde{A}_{zo}^{ef}}{\partial\rho} + \frac{jn}{\rho}\widetilde{F}_{zo}^{ef} \tag{27}$$

$$\widetilde{E}_{\phi}^{ef}(\rho,n,k_z) = -\frac{nk_z}{j\omega\varepsilon_o\rho}\widetilde{A}_{zo}^{ef} + \frac{\partial\widetilde{F}_{zo}^{ef}}{\partial\rho} \tag{28}$$

$$\widetilde{E}_{z}^{ef}(\rho,n,k_z) = \frac{k_{\rho_o}^2}{j\omega\varepsilon_o}\widetilde{A}_{zo}^{ef} \tag{29}$$

$$\widetilde{H}_{\rho}^{ef}(\rho,n,k_z) = -\frac{jn}{\rho}\widetilde{A}_{zo}^{ef} - \frac{k_z}{\omega\mu_o}\frac{\partial\widetilde{F}_{zo}^{ef}}{\partial\rho} \tag{30}$$

$$\widetilde{H}_{\phi}^{ef}(\rho,n,k_z) = -\frac{\partial\widetilde{A}_{zo}^{ef}}{\partial\rho} - \frac{nk_z}{j\omega\mu_o\rho}\widetilde{F}_{zo}^{ef} \tag{31}$$

$$\widetilde{H}_{z}^{ef}(\rho,n,k_z) = \frac{k_{\rho_o}^2}{j\omega\mu_o}\widetilde{F}_{zo}^{ef} \tag{32}$$

where $k_{\rho_o}^2 = (k_o^2 - k_z^2)$. In spectral domain:

$$\widetilde{A}_{zo}^{ef} = c_3 H_n^{(2)}(k_{\rho_o}\rho) \tag{33}$$

$$\widetilde{F}_{zo}^{ef} = c_4 H_n^{(2)}(k_{\rho_o}\rho) \tag{34}$$

The boundary conditions at $\rho = b$ require that the equivalent surface magnetic current $-\bar{M}_s(\phi,z) = -\hat{n}\times\bar{E}(b,\phi,z) = -\hat{a}_\rho\times\bar{E}(b,\phi,z)$, where $\hat{n}$ is the unit vector normal to the surface $\rho = b$ pointing outwards. Consequently $M_z(\phi, z) = E_\phi(b, \phi, z)$ and $M_\phi(\phi, z) = -E_z(b, \phi, z)$. In spectral domain it reads:

$$\widetilde{M}_z^{ef}(n,q) = \widetilde{E}_\phi^{ef}(\rho, n, k_z) = -\frac{nk_z}{j\omega\varepsilon_o b}\widetilde{A}_{zo}^{ef} + \frac{\partial \widetilde{F}_{zo}^{ef}}{\partial \rho} \tag{35}$$

$$\widetilde{M}_\phi^{ef}(n,q) = -\widetilde{E}_z^{ef}(\rho, n, k_z) = -\frac{k_{\rho o}^2}{j\omega\varepsilon_o}\widetilde{A}_{zo}^{ef} \tag{36}$$

For a given magnetic surface current the coefficients c_3 and c_4 are determined substituting (33) and (34) into (35) and (36), and the potentials are given by:

$$\widetilde{A}_{zo}^{ef}(\rho, n, k_z) = \widetilde{G}_{M\phi}^{Ao}(\rho, n, k_z)\widetilde{M}_\phi^{ef}(n, k_z) \tag{37}$$

$$\widetilde{F}_{zo}^{ef}(\rho, n, k_z) = \widetilde{G}_{M\phi}^{Fo}(\rho, n, k_z)\widetilde{M}_\phi^{ef}(n, k_z) + \widetilde{G}_{Mz}^{Fo}(\rho, n, k_z)\widetilde{M}_z^{ef}(n, k_z) \tag{38}$$

where:

$$\widetilde{G}_{M\phi}^{Ao}(\rho, n, k_z) = -\frac{j\omega\varepsilon_o}{k_{\rho_o}^2}\frac{H_n^{(2)}(k_{\rho_o}\rho)}{H_n^{(2)}(k_{\rho_o}b)} \tag{39}$$

$$\widetilde{G}_{M\phi}^{Fo}(\rho, n, k_z) = -\frac{nk_z}{k_{\rho_o}^3 b}\frac{H_n^{(2)}(k_{\rho_o}\rho)}{H_n^{(2)\prime}(k_{\rho_o}b)} \tag{40}$$

$$\widetilde{G}_{Mz}^{Fo}(\rho, n, k_z) = \frac{1}{k_{\rho_o}}\frac{H_n^{(2)}(k_{\rho_o}\rho)}{H_n^{(2)\prime}(k_{\rho_o}b)} \tag{41}$$

The electromagnetic fields in the external region are obtained substituting (37)-(38) into (27)-(32), and taking the inverse transform.

3. Feeding network

The wraparound antenna is fed by a parallel network of equally spaced N_p coaxial cables along a circumference ($z = z_f$, $\rho = b$) of the antenna. An example of the feeding network, with 6 coaxial cables is shown in Fig. 2. Each probe is modeled as a strip at $z = z_f$, from $\rho = a$ to $\rho = b$, centered at $\phi_f(i) = [\phi_{fo} + (i-1)2\pi/N_p]$, $i = 1, ..., N_p$, and with angular width $\Delta\phi_f$. The strip i is limited from $\phi_{1f}(i)$ to $\phi_{2f}(i)$:

$$\phi_{1f}(i) = \phi_f(i) - \frac{\Delta\phi_f}{2} \tag{42}$$

$$\phi_{2f}(i) = \phi_f(i) + \frac{\Delta\phi_f}{2} \tag{43}$$

It should be observed that the widths of the strips ($= \rho\Delta\phi_f$) vary with the radius ρ, but for practical cases of thin dielectrics they become almost constant. A mean width of the strip values $W_f = (a+b)\Delta\phi_f/2$. The total feeding volume current distribution, in ρ-direction, is given by:

$$J_f(\rho, \phi, z) = \sum_{i=1}^{N_p} J_{fi}(\rho, \phi, z) \tag{44}$$

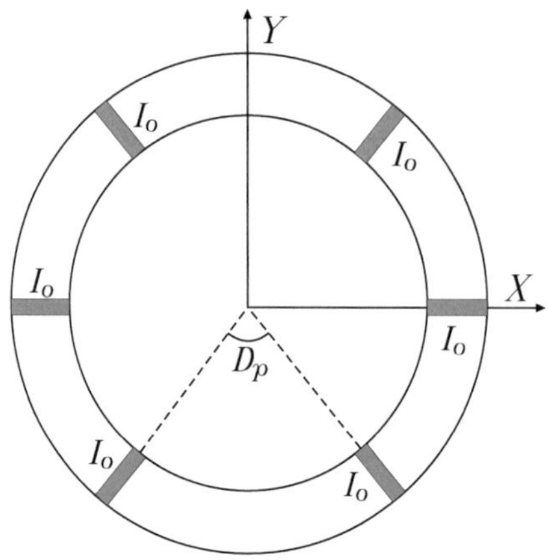

Fig. 2. Example of feeding network with 6 probes.

where

$$J_{fi}(\rho,\phi,z) = \frac{I_o}{\rho\Delta\phi_f}\delta(z - z_f) \qquad \phi_{1f}(i) < \phi < \phi_{2f}(i) \quad a < \rho < b \tag{45}$$

and I_o is the amplitude of the current on each probe. For each probe the coefficients of the exponential series in ϕ and sine series in z are equal to:

$$\widetilde{J}^{es}_{fi}(\rho,n,q) = \frac{I_o}{\pi L_{zc}\rho}\sin\left[\frac{q\pi}{L_{zc}}\left(z_f - z_{1c}\right)\right]\operatorname{sinc}\left(\frac{n\Delta\phi_f}{2\pi}\right)e^{jn\phi_f(i)} \tag{46}$$

where $\operatorname{sinc}(x) = \sin(\pi x)/(\pi x)$.
As each feed point carries a current I_o, the total current in the network is $N_p I_o$.

4. Basis-functions

Following the method of moments procedure the equivalent surface magnetic currents are expanded in a set of basis-functions. These currents extend from z_{1c} to z_{1a}, and from z_{2a} to z_{2c}, over the electric conductor introduced by the equivalence principle. Each component of the magnetic current is expanded in exponential Fourier series in ϕ-direction, and in sub-domain basis-functions in z-direction, forming a piecewise-linear approximation for M_z and a piecewise-constant one for M_ϕ.
The boundary conditions require that ϕ-component of the electric field be zero at the cavity limits (z_{1c} and z_{2c}), which implies that $M_z = 0$ at these points. The piecewise-linear basis-functions used for M_z are shown in Fig. 3, and are composed by basis-functions at the edges of the antenna (z_{1a} and z_{2a}) M_{zu1} and M_{zu2}, and *rooftop* basis-functions M_{zut} ($3 \le t \le T_z + 2$):

$$M_z(\phi,z) = \sum_{u=-U}^{U}\sum_{t=1}^{T_z+2} c_{ut}M_{zut}(\phi,z) \tag{47}$$

where U is the maximum number of harmonics in ϕ, T_z is the number of *rooftop* basis-functions, and $c's$ are the coefficients to be determined. These basis-functions are shown in Fig. 4. The basis-functions at the edges of the antenna are given by:

$$M_{zu1}(\phi,z) = f_1(z)\,e^{-ju\phi} \tag{48}$$

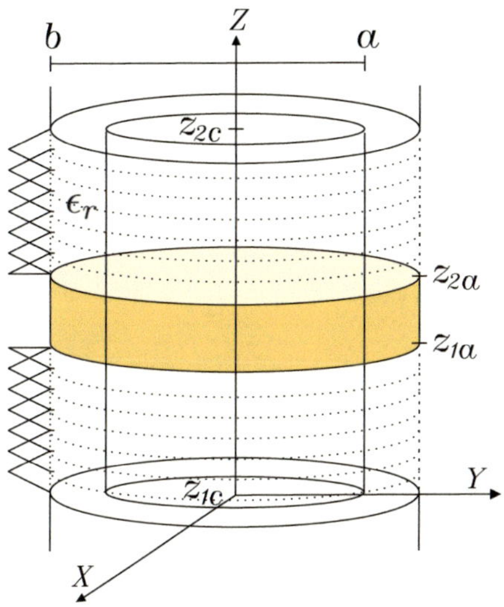

Fig. 3. Set of basis-functions for M_z.

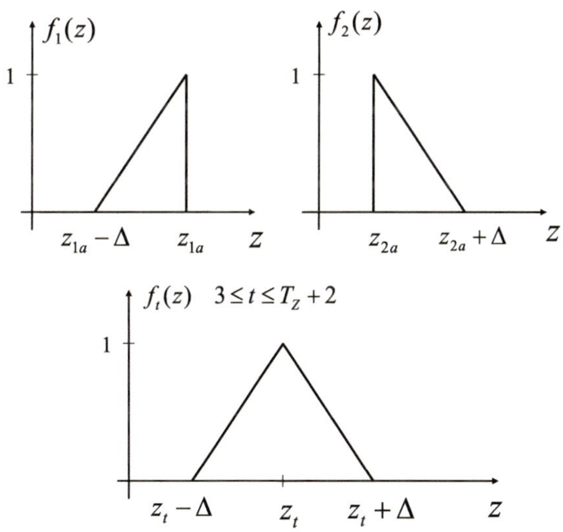

Fig. 4. Basis-functions for M_z.

$$M_{zu2}(\phi, z) = f_2(z)\, e^{-ju\phi} \tag{49}$$

where:

$$f_1(z) = \begin{cases} \frac{z-z_{1a}}{\Delta} + 1 & (z_{1a} - \Delta) < z < z_{1a} \\ 0 & \text{otherwise} \end{cases} \tag{50}$$

$$f_2(z) = \begin{cases} \frac{z_{2a}-z}{\Delta} + 1 & z_{2a} < z < (z_{2a} + \Delta) \\ 0 & \text{otherwise} \end{cases} \tag{51}$$

and Δ is the segment length in z-direction. The *rooftop* functions ($3 \leq t \leq T_z + 2$) are given by:

$$M_{zut}(\phi, z) = f_t(z)\, e^{-ju\phi} \tag{52}$$

where

$$f_t(z) = \begin{cases} 1 + \frac{z-z_t}{\Delta} & z_t - \Delta < z < z_t \\ 1 + \frac{z_t-z}{\Delta} & z_t < z < z_t + \Delta \\ 0 & \text{otherwise} \end{cases} \tag{53}$$

and z_t is the center coordinate of the *rooftop* basis-function.
Some transforms of these z-directed basis-functions will be needed throughout the chapter, and are given below. The exponential series in ϕ and Fourier transform in z are given by:

$$\widetilde{M}^{ef}_{zu1}(n, k_z) = \delta_{n,u} \widetilde{f}^{f}_1(k_z) \tag{54}$$

$$\widetilde{M}^{ef}_{zu2}(n, k_z) = \delta_{n,u} \widetilde{f}^{f}_2(k_z) \tag{55}$$

$$\widetilde{M}^{ef}_{zut}(n, k_z) = \delta_{n,u} \widetilde{f}^{f}_t(k_z) \tag{56}$$

where $\delta_{n,u}$ is the Kronecker's delta function given by:

$$\delta_{n,u} = \begin{cases} 1 & n = u \\ 0 & n \neq u \end{cases} \tag{57}$$

and

$$\widetilde{f}^{f}_1(k_z) = \frac{e^{jk_z z_{1a}}}{2\pi} \left(\frac{1}{jk_z} + \frac{1}{\Delta\, k_z^2} \right) - \frac{e^{jk_z (z_{1a} - \Delta)}}{2\pi\Delta\, k_z^2} \tag{58}$$

$$\widetilde{f}^{f}_2(k_z) = \frac{e^{jk_z z_{2a}}}{2\pi} \left(-\frac{1}{jk_z} + \frac{1}{\Delta\, k_z^2} \right) - \frac{e^{jk_z (z_{2a} + \Delta)}}{2\pi\Delta\, k_z^2} \tag{59}$$

$$\widetilde{f}^{f}_t(k_z) = \frac{e^{jk_z z_t}}{\pi\Delta\, k_z^2} \left[1 - \cos(k_z \Delta)\right] \tag{60}$$

Similarly the exponential series in ϕ and sine series in z are given by:

$$\widetilde{M}^{es}_{zu1}(n, k_z) = \delta_{n,u} \widetilde{f}^{s}_1(k_z) \tag{61}$$

$$\widetilde{M}^{es}_{zu2}(n, k_z) = \delta_{n,u} \widetilde{f}^{s}_2(k_z) \tag{62}$$

$$\widetilde{M}^{es}_{zut}(n, k_z) = \delta_{n,u} \widetilde{f}^{s}_t(k_z) \tag{63}$$

where

$$\widetilde{f}_1^s(k_z) = \frac{2}{q\pi}\left\{-\cos\left(\frac{q\pi}{L_{zc}}(z_{1a}-z_{1c})\right)+\right.$$
$$\left.\frac{L_{zc}}{\Delta q\pi}\left[\sin\left(\frac{q\pi}{L_{zc}}(z_{1a}-z_{1c})\right)-\sin\left(\frac{q\pi}{L_{zc}}(z_{1a}-\Delta-z_{1c})\right)\right]\right\} \tag{64}$$

$$\widetilde{f}_2^s(k_z) = \frac{2}{q\pi}\left\{\cos\left(\frac{q\pi}{L_{zc}}(z_{2a}-z_{1c})\right)-\right.$$
$$\left.\frac{L_{zc}}{\Delta q\pi}\left[\sin\left(\frac{q\pi}{L_{zc}}(z_{2a}+\Delta-z_{1c})\right)-\sin\left(\frac{q\pi}{L_{zc}}(z_{2a}-z_{1c})\right)\right]\right\} \tag{65}$$

$$\widetilde{f}_t^s(k_z) = \frac{4}{L_{zc}\Delta\,(q\pi/L_{zc})^2}\sin\left[\frac{q\pi}{L_{zc}}(z_t-z_{1c})\right]\left[1-\cos\left(\frac{q\pi\Delta}{L_{zc}}\right)\right] \tag{66}$$

On the other hand the ϕ-directed equivalent magnetic surface currents are expanded in piecewise-constant basis functions as:

$$M_\phi(\phi,z) = \sum_{u=-U}^{U}\sum_{t=1}^{T_\phi} d_{ut}M_{\phi ut}(\phi,z) \tag{67}$$

where U is the maximum number of harmonics in ϕ, $d's$ are the coefficients to be determined, and T_ϕ is the number of piecewise-constant basis-functions $M_{\phi ut}$ given by:

$$M_{\phi ut}(\phi,z) = g_t(z)\,e^{-ju\phi} \tag{68}$$

where:

$$g_t(z) = \begin{cases} 1 & z_{t-1} < z < z_t \\ 0 & \text{otherwise} \end{cases} \tag{69}$$

The piecewise-constant basis-functions used for M_ϕ are shown in Fig. 5, and a single basis is shown in Fig. 6. Again we will introduce some transforms of $M_{\phi ut}$ that will be required later. The exponential series in ϕ and Fourier transform in z is given by:

$$\widetilde{M}_{\phi ut}^{ef}(n,k_z) = \delta_{n,u}\widetilde{g}_t^f(k_z) \tag{70}$$

where

$$\widetilde{g}_t^f(k_z) = \frac{e^{jk_z z_t}}{j2\pi k_z}(1-e^{jk_z\Delta}) \tag{71}$$

And the exponential series in ϕ and cosine series in z is given by:

$$\widetilde{M}_{\phi ut}^{ec}(n,k_z) = \delta_{n,u}\widetilde{g}_t^c(k_z) \tag{72}$$

where

$$\widetilde{g}_t^c(k_z) = \frac{\epsilon_q}{q\pi}\left\{\sin\left[\frac{q\pi}{L_{zc}}(z_t-z_{1c})\right]-\sin\left[\frac{q\pi}{L_{zc}}(z_t-\Delta-z_{1c})\right]\right\} \tag{73}$$

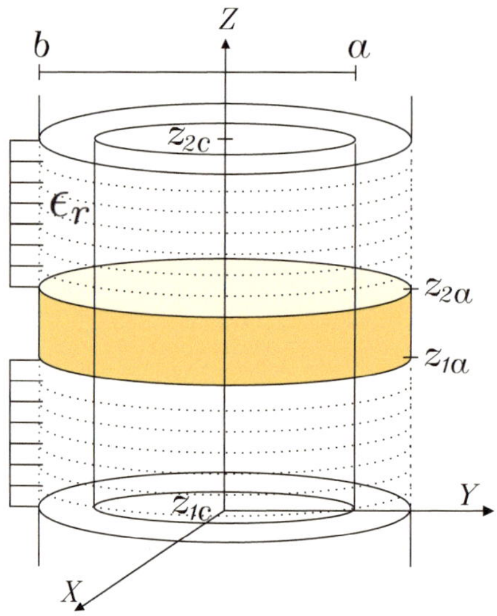

Fig. 5. Set of basis-functions for M_{ϕ}.

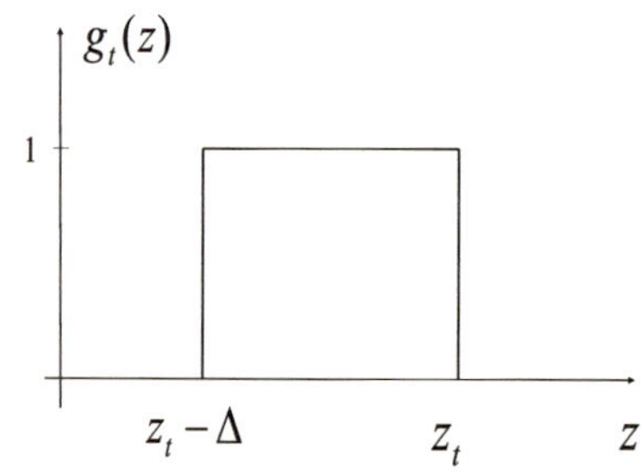

Fig. 6. Basis-function for M_{ϕ}.

5. Method of moments linear system

The coefficients c_{ut} and d_{ut} of the expansions of magnetic surface currents (47) and (67) are determined in order to satisfy the boundary conditions. The continuity of the tangential electric field is already guaranteed as the equivalent magnetic surface currents in the inner and outer regions have opposite signs. And the continuity of the tangential components of the magnetic field over the interface between the inner and outer regions imply:

$$\underbrace{H^{<}(M_z) + H^{<}(M_{\phi}) + H^{<}(J_f)}_{H^{<}} = \underbrace{H^{>}(M_z) + H^{>}(M_{\phi})}_{H^{>}} \tag{74}$$

for the z and ϕ-components, over the surface $\rho = b$, $z_{1c} < z < z_{1a}$ and $z_{2a} < z < z_{2c}$. The superscript $<$ refers to the cavity region, and $>$ refers to the outer region. For H_z, using the current expansions (47) and (67):

$$\sum_{u=-U}^{U}\sum_{t=1}^{T_z+2} c_{ut}\left[H_z^{<}(M_{zut}) - H_z^{>}(M_{zut})\right] + \sum_{u=-U}^{U}\sum_{t=1}^{T_\phi} d_{ut}\left[H_z^{<}(M_{\phi ut}) - H_z^{>}(M_{\phi ut})\right] = -H_z^{<}(J_f) \tag{75}$$

Defining the symmetric product:

$$< f, g >= \int_0^{2\pi}\int_{-\infty}^{+\infty} f(\phi, z)\, g(\phi, z)\, b\, d\phi\, dz \tag{76}$$

and taking the symmetric product of (75) by testing functions which are chosen as the same set of basis-functions $M_{zms}(\phi, z),\ -U \leq m \leq +U,\ s = 1, ..., T_z + 2$ results:

$$\begin{aligned}&\sum_{u=-U}^{U}\sum_{t=1}^{T_z+2} c_{ut} < H_z^{<}(M_{zut}) - H_z^{>}(M_{zut}), M_{zms} > +\\ &\sum_{u=-U}^{U}\sum_{t=1}^{T_\phi} d_{ut} < H_z^{<}(M_{\phi ut}) - H_z^{>}(M_{\phi ut}), M_{zms} > = - < H_z^{<}(J_f), M_{zms} >\\ &\qquad -U \leq m \leq +U,\ s = 1, ..., T_z + 2\end{aligned} \tag{77}$$

Similarly for H_ϕ, using the current expansions (47) and (67) and taking the symmetric product of (75) by testing functions which are chosen as the same set of basis-functions $M_{\phi ms}(\phi, z),\ -U \leq m \leq +U,\ s = 1, ..., T_\phi$ results:

$$\begin{aligned}&\sum_{u=-U}^{U}\sum_{t=1}^{T_z+2} c_{ut} < H_\phi^{<}(M_{zut}) - H_\phi^{>}(M_{zut}), M_{\phi ms} > +\\ &\sum_{u=-U}^{U}\sum_{t=1}^{T_\phi} d_{ut} < H_\phi^{<}(M_{\phi ut}) - H_\phi^{>}(M_{\phi ut}), M_{\phi ms} > = - < H_\phi^{<}(J_f), M_{\phi ms} >\\ &\qquad -U \leq m \leq +U,\ s = 1, ..., T_\phi\end{aligned} \tag{78}$$

Re-arranging (77) and (78) in a matricial form:

$$\begin{bmatrix} [Z^{zz}] & \vdots & [Z^{z\phi}] \\ \cdots & \cdots & \cdots \\ [Z^{\phi z}] & \vdots & [Z^{\phi\phi}] \end{bmatrix} \begin{bmatrix} [c] \\ \cdots \\ [d] \end{bmatrix} = \begin{bmatrix} [V^z] \\ \cdots \\ [V^\phi] \end{bmatrix} \tag{79}$$

where $[c]$ and $[d]$ are arrays which elements are the coefficients c_{ut} and d_{ut}, respectively. And the elements of the submatrices are given by:

$$Z_{ms,ut}^{zz} = < H_z^{<}(M_{zut}), M_{zms} > - < H_z^{>}(M_{zut}), M_{zms} > \tag{80}$$

$$Z_{ms,ut}^{z\phi} = < H_z^{<}(M_{\phi ut}), M_{zms} > - < H_z^{>}(M_{\phi ut}), M_{zms} > \tag{81}$$

$$Z_{ms,ut}^{\phi z} = < H_\phi^{<}(M_{zut}), M_{\phi ms} > - < H_\phi^{>}(M_{zut}), M_{\phi ms} > \tag{82}$$

$$Z_{ms,ut}^{\phi\phi} = < H_\phi^{<}(M_{\phi ut}), M_{\phi ms} > - < H_\phi^{>}(M_{\phi ut}), M_{\phi ms} > \tag{83}$$

$$V_{ms}^{z} = - < H_z^{<}(J_f), M_{zms} > \tag{84}$$

$$V_{ms}^{\phi} = - < H_\phi^{<}(J_f), M_{\phi ms} > \tag{85}$$

Each symmetric product above represents the reaction of the field due to a basis-function, or due to the feed, on a testing function. The procedure will be illustrated for the elements of the submatrix $[Z^{zz}]$. The field H_z due to M_{zut} in the cavity is given by the inverse transform of (11), with the vector potential given by (17):

$$H_z^{<}(M_{zut})(\rho,\phi,z) = \sum_{n=-\infty}^{\infty}\sum_{q=1}^{\infty}\left[\frac{(k_{\rho_d})^2}{j\omega\mu_d}\widetilde{G}_{Mz}^{Fd}(\rho,n,q)\,\widetilde{M}_{zut}^{es}(n,q)\right]e^{-jn\phi}\sin\left[\frac{q\pi}{L_{zc}}(z-z_{1c})\right] \quad (86)$$

And the field H_z due to M_{zut} outside the cylinder is given by the inverse transform of (32) with the vector potential given by (38):

$$H_z^{>}(M_{zut})(\rho,\phi,z) = \sum_{n=-\infty}^{\infty}\int_{-\infty}^{\infty}\left[\frac{(k_{\rho_o})^2}{j\omega\mu_o}\widetilde{G}_{Mz}^{Fo}(\rho,n,k_z)\,\widetilde{M}_{zut}^{ef}(n,k_z)\right]e^{-jk_z z}\,e^{-jn\phi}dk_z \quad (87)$$

Substituting (86) and (87) into (80) after some manipulations we obtain:

$$\begin{aligned} Z_{ms,ut}^{zz} &= \delta_{-u,m}\frac{b\pi L_{zc}}{j\omega\mu_d}\sum_{q=1}^{\infty}(k_{\rho_d})^2\widetilde{G}_{Mz}^{Fd}(b,u,q)\,\widetilde{f}_t^{s}(q)\widetilde{f}_s^{s}(q) - \\ &\quad \delta_{-u,m}\frac{(2\pi)^2 b}{j\omega\mu_o}\int_{-\infty}^{\infty}(k_{\rho_o})^2\widetilde{G}_{Mz}^{Fo}(b,u,k_z)\,\widetilde{f}_t^{f}(k_z)\widetilde{f}_s^{f}(-k_z)dk_z \end{aligned} \quad (88)$$

Similarly for the elements of submatrices $Z^{z\phi}$, $Z^{\phi z}$ and $Z^{\phi\phi}$:

$$\begin{aligned} Z_{ms,ut}^{z\phi} &= \delta_{-u,m}\frac{b\pi L_{zc}}{j\omega\mu_d}\sum_{q=1}^{\infty}(k_{\rho_d})^2\widetilde{G}_{M\phi}^{Fd}(b,u,q)\,\widetilde{g}_t^{c}(q)\widetilde{f}_s^{s}(q) - \\ &\quad \delta_{-u,m}\frac{(2\pi)^2 b}{j\omega\mu_o}\int_{-\infty}^{\infty}(k_{\rho_o})^2\widetilde{G}_{M\phi}^{Fo}(b,u,k_z)\,\widetilde{g}_t^{f}(k_z)\widetilde{f}_s^{f}(-k_z)dk_z \end{aligned} \quad (89)$$

$$\begin{aligned} Z_{ms,ut}^{\phi z} &= -\delta_{-u,m}\frac{u\pi^2}{\omega\mu_d}\sum_{q=1}^{\infty}q\widetilde{G}_{Mz}^{Fd}(b,u,q)\,\widetilde{f}_t^{s}(q)\,\widetilde{g}_s^{c}(q) + \\ &\quad \delta_{-u,m}\frac{(2\pi)^2 u}{j\omega\mu_o}\int_{-\infty}^{\infty}k_z\widetilde{G}_{Mz}^{Fo}(b,u,k_z)\,\widetilde{f}_t^{f}(k_z)\,\widetilde{g}_s^{f}(-k_z)\,dk_z \end{aligned} \quad (90)$$

$$\begin{aligned} Z_{ms,ut}^{\phi\phi} &= \delta_{-u,m}2\pi bL_{zc}\sum_{q=0}^{\infty}\frac{1}{\epsilon_q}\left[-\frac{\partial}{\partial\rho}\widetilde{G}_{M\phi}^{Ad}(b,u,q) - \frac{u}{\omega\mu_d b}\frac{q\pi}{L_{zc}}\widetilde{G}_{M\phi}^{Fd}(b,u,q)\right]\widetilde{g}_t^{c}(q)\,\widetilde{g}_s^{c}(q) - \\ &\quad \delta_{-u,m}(2\pi)^2 b\int_{-\infty}^{\infty}\left[-\frac{\partial}{\partial\rho}\widetilde{G}_{M\phi}^{Ao}(b,u,k_z) - \frac{uk_z}{j\omega\mu_o b}\widetilde{G}_{M\phi}^{Fo}(b,u,k_z)\right]\widetilde{g}_t^{f}(k_z)\,\widetilde{g}_s^{f}(-k_z)\,dk_z \end{aligned} \quad (91)$$

In fact the elements of submatrix $Z^{z\phi}$ (81) are related to those of $Z^{\phi z}$ (82) by the reciprocity theorem (Harrington,1961):

$$\begin{aligned} Z_{ms,ut}^{z\phi} &= < H_z^{<}(M_{\phi ut}), M_{zms} > - < H_z^{>}(M_{\phi ut}), M_{zms} > \\ &= < M_{\phi ut}, H_\phi^{<}(M_{zms}) > - < M_{\phi ut}, H_\phi^{>}(M_{zms}) > \\ &= < H_\phi^{<}(M_{zms}), M_{\phi ut} > - < H_\phi^{>}(M_{zms}), M_{\phi ut} > \\ &= Z_{ut,ms}^{\phi z} \end{aligned} \quad (92)$$

Reciprocity theorem can also be used to calculate the elements of the voltage vector $[V]$. For example the elements of subvector $[V^z]$ given by (84) can be written as:

$$\begin{aligned} V^z_{ms} &= - < H^<_z(J_f), M_{zms} > \\ &= - \int_0^{2\pi} \int_{-\infty}^{\infty} H^<_z(J_f)\,(b,\phi,z)\, M_{zms}(\phi,z)\; b\; d\phi\; dz \\ &= \iiint_V E^<_\rho(M_{zms})(\rho,\phi,z)\; J_f(b,\phi,z)\; dv \end{aligned} \tag{93}$$

where the volume V refers to the cavity. $E^<_\rho(M_{zms})(\rho,\phi,z)$ is given by the inverse transform of (6), using (16) and (17):

$$E^<_\rho(M_{zms})\,(\rho,\phi,z) = \sum_{n=-\infty}^{\infty} \sum_{q=1}^{\infty} \frac{jn}{\rho} \widetilde{G}^{Fd}_{Mz}(\rho,n,q)\, \widetilde{M}^{es}_{zms}(n,q)\, e^{-jn\phi} \sin\left[\frac{q\pi}{L_{zc}}(z - z_{1c})\right] \tag{94}$$

Substituting (94) and (44) into (93) after some manipulations results:

$$V^z_{ms} = jm\pi L_{zc} \int_a^b \sum_{q=1}^{\infty} \widetilde{G}^{Fd}_{Mz}(\rho,m,q)\, \widetilde{f}^s_s(q) \widetilde{J}^{es}_f(\rho,-m,q)\, d\rho \tag{95}$$

Proceeding similarly for V^ϕ_{ms} (85):

$$V^\phi_{ms} = \iiint_V E^<_\rho(M_{\phi ms})(\rho,\phi,z)\; J_f(b,\phi,z)\; dv \tag{96}$$

$$= \pi L_{zc} \int_a^b \sum_{q=1}^{\infty} \left[\frac{-1}{j\omega\varepsilon_d} \frac{q\pi}{L_{zc}} \frac{\partial}{\partial\rho} \widetilde{G}^{Ad}_{M\phi}(\rho,m,q) + \frac{jm}{b} \widetilde{G}^{Fd}_{M\phi}(\rho,m,q) \right] \widetilde{g}^c_s(q) \widetilde{J}^{es}_f(\rho,-m,q)\, \rho\, d\rho \tag{97}$$

6. Numerical considerations

The method of moments linear system (79) requires the evaluation of Bessel functions of a wide range of arguments and orders, which is limited by the accuracy of the numerical routines, or by the intrinsic behavior of these functions. This limitation is enhanced for larger radius (Erturk & Rojas,2002; Erturk & Rojas,2003), as the arguments of the Bessel functions are proportional to either a or b. In order to mitigate this problem the Green's functions have been written in such a way that the Bessel functions appear in specific combinations as (21), (22) and:

$$\vartheta_3(n,q) = \left[J_n(k_{\rho_d} b)\, H^{(2)'}_n(k_{\rho_d} a) - J'_n(k_{\rho_d} a)\, H^{(2)}_n(k_{\rho_d} b) \right] \tag{98}$$

$$\vartheta_7(n,q) = \left[J'_n(k_{\rho_d} b)\, H^{(2)}_n(k_{\rho_d} a) - J_n(k_{\rho_d} a)\, H^{(2)'}_n(k_{\rho_d} b) \right] \tag{99}$$

$$\vartheta_9(n,q) = \frac{H^{(2)}_n(k_{\rho_d} b)}{H^{(2)'}_n(k_{\rho_d} b)} \tag{100}$$

These combinations are evaluated differently according to the argument range:
(i) For very small arguments analytical expressions of such combinations are obtained from

the limits of the Bessel functions (Harrington,1961):

$$\lim_{|z|\to 0} J_n(z) = \begin{cases} 1-\frac{z^2}{4} & n=0 \\ \frac{1}{n!}\left(\frac{z}{2}\right)^n & n>0 \end{cases} \tag{101}$$

$$\lim_{|z|\to 0} Y_n(z) = \begin{cases} \frac{2}{\pi}\ln\left(\frac{\gamma z}{2}\right) & n=0 \\ -\frac{(n-1)!}{\pi}\left(\frac{2}{z}\right)^n & n>0 \end{cases} \tag{102}$$

where $\gamma = 1.781072418....$ Substituting into (21), (22), (98)-(100):

$$\lim_{|k_{\rho_d}a|\to 0} \vartheta_1(n,q) = \begin{cases} \frac{j}{\pi}\left(\frac{b}{a}-\frac{a}{b}\right) & n=0 \\ \frac{jn}{\pi k_{\rho_d}^2}\left[-\frac{1}{a^2}\left(\frac{b}{a}\right)^{n-1}+\frac{1}{b^2}\left(\frac{a}{b}\right)^{n-1}\right] & n>0 \end{cases} \tag{103}$$

$$\lim_{|k_{\rho_d}a|\to 0} \vartheta_3(n,q) = \begin{cases} -\frac{2j}{\pi}\left[\frac{1}{k_{\rho_d}a}+\frac{k_{\rho_d}a}{2}\ln\left(\frac{\gamma}{2}k_{\rho_d}b\right)\right] & n=0 \\ -\frac{j}{\pi k_{\rho_d}}\left[\frac{1}{a}\left(\frac{b}{a}\right)^n+\frac{1}{b}\left(\frac{a}{b}\right)^{n-1}\right] & n>0 \end{cases} \tag{104}$$

$$\lim_{|k_{\rho_d}a|\to 0} \vartheta_5(n,q) = \begin{cases} \frac{2j}{\pi}\ln\left(\frac{b}{a}\right) & n=0 \\ \frac{j}{n\pi}\left[\left(\frac{b}{a}\right)^n-\left(\frac{a}{b}\right)^n\right] & n>0 \end{cases} \tag{105}$$

$$\lim_{|k_{\rho_d}a|\to 0} \vartheta_7(n,q) = \begin{cases} \frac{2j}{\pi}\left[\frac{1}{k_{\rho_d}b}+\frac{k_{\rho_d}b}{2}\ln\left(\frac{\gamma}{2}k_{\rho_d}a\right)\right] & n=0 \\ \frac{j}{\pi k_{\rho_d}}\left[\frac{1}{a}\left(\frac{b}{a}\right)^{n-1}+\frac{1}{b}\left(\frac{a}{b}\right)^n\right] & n>0 \end{cases} \tag{106}$$

$$\lim_{|k_{\rho_d}b|\to 0} \vartheta_9(n,q) = \begin{cases} \frac{j\pi}{2}k_{\rho_d}b+k_{\rho_d}b\ln\left(\frac{\gamma}{2}k_{\rho_d}b\right) & n=0 \\ -\frac{k_{\rho_d}b}{n} & n>0 \end{cases} \tag{107}$$

These analytical limits can be evaluated for arguments much smaller than the Bessel functions individually, due to opposite behavior of the Bessel functions in these combinations. The maximum magnitude of the argument $|z_1|$ for which the limits above should be used depends on n. A simple expression for this threshold can be found to ensure that the magnitude of the Bessel functions are larger than 10^{-Exp1} and smaller than 10^{+Exp1}, where $Exp1$ depends on the precision used:

$$|z_1| = \begin{cases} 2\times 10^{-Exp1} & n=0 \\ 2\times 10^{\frac{1}{n+1}\left[\log_{10}\left(\frac{n!}{2\pi}\right)-Exp1\right]} & n>0 \end{cases} \tag{108}$$

(ii) For very large arguments asymptotic expressions of Bessel functions (Harrington,1961) are used to obtain analytical expressions of the combinations.

$$\lim_{|z|\to\infty} J_n(z) = \sqrt{\frac{2}{\pi z}}\cos\left(z-n\pi/2-\pi/4\right) \tag{109}$$

$$\lim_{|z|\to\infty} Y_n(z) = \sqrt{\frac{2}{\pi z}}\sin\left(z-n\pi/2-\pi/4\right) \tag{110}$$

Substituting into (21), (22), (98)-(100):

$$\lim_{|k_{\rho_d}b|\to\infty} \vartheta_1(n,q) = \frac{2j}{\pi k_{\rho_d}\sqrt{ab}} \left\{ \left(1+\frac{1}{4k_{\rho_d}^2 ab}\right) \sin\left[k_{\rho_d}(b-a)\right] + \frac{1}{2k_{\rho_d}}\left(\frac{1}{b}-\frac{1}{a}\right) \cos\left[k_{\rho_d}(b-a)\right] \right\} \quad (111)$$

$$\lim_{|k_{\rho_d}b|\to\infty} \vartheta_3(n,q) = -\frac{2j}{\pi k_{\rho_d}\sqrt{ab}} \left\{ \frac{1}{2k_{\rho_d}a} \sin\left[k_{\rho_d}(b-a)\right] + \cos\left[k_{\rho_d}(b-a)\right] \right\} \quad (112)$$

$$\lim_{|k_{\rho_d}b|\to\infty} \vartheta_5(n,q) = \frac{2j}{\pi k_{\rho_d}\sqrt{ab}} \sin\left[k_{\rho_d}(b-a)\right] \quad (113)$$

$$\lim_{|k_{\rho_d}b|\to\infty} \vartheta_7(n,q) = \frac{2j}{\pi k_{\rho_d}\sqrt{ab}} \left\{ \cos\left[k_{\rho_d}(b-a)\right] - \frac{1}{2k_{\rho_d}b} \sin\left[k_{\rho_d}(b-a)\right] \right\} \quad (114)$$

$$\lim_{|k_{\rho_d}b|\to\infty} \vartheta_9(n,q) = j \quad (115)$$

The use of these asymptotic limits are specially important for the case when the imaginary part of the arguments are large and the Bessel functions present an exponential behavior. In this case asymptotic limits of the combinations also present an exponentially behavior, but the exponent is no longer proportional to the cylinder radius, but to the dielectric thickness, allowing the combinations to be calculated for much larger arguments. The minimum value of the magnitude of the argument ($|z_2|$) from which the asymptotic limits are used is specified depending on the precision used.
(iii) Numerical routines of Bessel functions are only used if the magnitude of the arguments lies between $|z_1|$ and $|z_2|$, where they present good convergence.
This procedure avoids that the numerical routines result in overflow or underflow.

7. Input impedance

The input impedance observed in a single coaxial cable, for example cable 1, is given by the variational expression (Harrington,1961):

$$Z_{in} = -\frac{1}{I_o^2} \iiint_V \left[\vec{E}(\vec{M}) + \vec{E}(\vec{J}_f)\right] \cdot \vec{J}_{f1} dv \quad (116)$$

where $\bar{M}$ represents the equivalent magnetic surface current distribution, including z and ϕ components, and $\bar{J}_f$ is the total network feed current distribution. Isolating initially the component due to $\bar{M}$, and using the basis expansions (47) and (67):

$$Z_{in}^M = -\frac{1}{I_o^2} \sum_{u=-U}^{U} \sum_{t=1}^{T_z+2} c_{ut} \iiint_V E_\rho(M_{zut})(\rho,\phi,z) J_{f1}(\rho,\phi,z) dv - \frac{1}{I_o^2} \sum_{u=-U}^{U} \sum_{t=1}^{T_\phi} d_{ut} \iiint_V E_\rho(M_{\phi ut})(\rho,\phi,z) J_{f1}(\rho,\phi,z) dv \quad (117)$$

As the coaxial cables are equally spaced and fed with the same current I_o, the same input impedance is observed at any of the coaxial cables, and we can write:

$$N_p Z_{in}^M = -\frac{1}{I_o^2}\sum_{u=-U}^{U}\sum_{t=1}^{T_z+2} c_{ut}\iiint_V E_\rho(M_{zut})(\rho,\phi,z)J_f(\rho,\phi,z)dv - \frac{1}{I_o^2}\sum_{u=-U}^{U}\sum_{t=1}^{T_\phi} d_{ut}\iiint_V E_\rho(M_{\phi ut})(\rho,\phi,z)J_f(\rho,\phi,z)dv \quad (118)$$

where J_f (44) includes all N_p coaxial cables. From (93) and (96) the integrals above can be identified as elements of the voltage vector of the method of moments linear system:

$$Z_{in}^M = \frac{-1}{N_p I_o^2}\sum_{u=-U}^{U}\left[\sum_{t=1}^{T_z+2} c_{ut}V_{ut}^z + \sum_{t=1}^{T_\phi} d_{ut}V_{ut}^\phi\right] \quad (119)$$

The second portion of Z_{in} (116) includes the reaction of the fields due to the feeding network on a single coaxial cable and will be called Z_{in}^f. An approximation of this impedance will be obtained by a similar procedure to those presented at (Zheng & Chang,1991) and (Tulintseff et. al.,1991) for self-impedance of the probe in planar dielectric layers. This approximation disregards the radiation contribution of the feed network, as if the feed is enclosed in the cavity limited by $\rho = b$, and assumes that the dielectric is thin and the fields have no variation with ρ. Within this cavity the ρ-component of the electric field is given by:

$$\frac{1}{d^2}\frac{\partial^2}{\partial\phi^2}E_\rho + \frac{\partial^2}{\partial z^2}E_\rho + k_d^2 E_\rho = j\omega\mu_d J_f \quad (120)$$

where $d = (a+b)/2$. The eigenmodes of the cavity are given by:

$$E_{\rho mn}(\phi,z) = E_{mn}e^{jm\phi}\sin\left[\frac{n\pi}{L_{zc}}(z - z_{1c})\right] \quad (121)$$

which is the solution of:

$$\frac{1}{d^2}\frac{\partial^2}{\partial\phi^2}E_{\rho mn} + \frac{\partial^2}{\partial z^2}E_{\rho mn} + k_{mn}^2 E_{\rho mn} = 0 \quad (122)$$

where $k_{mn}^2 = (m/d)^2 + (n\pi/L_{zc})^2$. And the electric field E_ρ (120) can be expanded into eigenmodes of the cavity given by:

$$E_\rho(\phi,z) = \sum_{m=-\infty}^{\infty}\sum_{n=1}^{\infty} E_{mn}\, e^{jm\phi}\sin\left[\frac{n\pi}{L_{zc}}(z - z_{1c})\right] \quad (123)$$

The coefficients of the modal expansion E_{mn} are determined substituting (123) into (120) and using the network feeding current (44):

$$E_{mn} = \frac{j\omega\mu_d I_o}{\pi d L_{zc}\left[k_d^2 - k_{mn}^2\right]}\sin\left[\frac{n\pi}{L_{zc}}(z_f - z_{1c})\right]\text{sinc}\left[\frac{m\Delta_{\phi_f}}{2\pi}\right]e^{-jm\phi_{fo}}\left(\sum_{i=1}^{N_p} e^{-jm(i-1)2\pi/N_p}\right) \quad (124)$$

The approximation for Z_{in}^f is obtained substituting the electric field (123) and the feed (45) into:

$$Z_{in}^f = -\frac{1}{I_o^2}\iiint_V E_\rho(\vec{J}_f)\, J_{f1} dv \quad (125)$$

which results:

$$Z_{in}^{f} = -\frac{j\omega\mu_d h}{\pi d L_{zc}} \sum_{m=-\infty}^{\infty} \sum_{n=1}^{\infty} \frac{1}{k_d^2 - k_{mn}^2} \sin^2\left[\frac{n\pi}{L_{zc}}(z_f - z_{1c})\right] \mathrm{sinc}^2\left[\frac{m\Delta_{\phi_f}}{2\pi}\right] \left(\sum_{i=1}^{N_p} e^{-jm(i-1)2\pi/N_p}\right) \tag{126}$$

From (116), (119) and (126) the input impedance at each probe will be equal to:

$$Z_{in} = Z_{in}^{M} + Z_{in}^{f} \tag{127}$$

8. Results

Initially the field distribution is presented as a function of the number of feeds. The cylinder has radius of 26 mm, and a cavity of thickness equal to 1 mm is filled with a dielectric of relative permitivitty of 2.1. A wraparound antenna of 34.5 mm (from $z_{1a} = -17.25$ mm to $z_{2a} = 17.25$ mm) is printed onto the dielectric surface. Figure 7 shows the magnitude of the electric field at xy plane, at $f = 3$ GHz when the antenna is fed by 1, 2, 3 or 4 equally spaced probes at $z_f = 8.625$ mm, modeled as strips of 2 mm of width. In this case the circumference of the antenna equals 2.4 times the dielectric wavelength. It can be observed from Fig. 7 that the pattern becomes close to omnidirectional when at least 3 probes are used. In general the spacing between the feed points of a cavity-backed wraparound antenna should be smaller than one wavelength in the dielectric substrate in order to obtain an omnidirectional pattern. But the exact number of probes will depend on the accuracy required. This is similar to that observed by (Munson,1974) for a microstrip wraparound antenna. Figure 8 shows a comparison of the magnitude of the electric field at xz plane for the cavity-backed wraparound antenna obtained with this formulation, and that of a microstrip wraparound antenna obtained with cavity method. Four probes were used, which is enough to radiate an omnidirectional pattern. Despite the differences in the geometry and methods of analysis, the radiation patterns show similar behavior.

Figure 9 shows the input impedance of a cavity-backed wraparound antenna of length $L_{zc} =$ 20 mm. The perfect cylindrical conductor has radius $b = 21$ mm, and the thickness of the cavity values $(b - a) = 1$ mm. It is filled with a dielectric of relative permitivitty equal to 9.6. No dielectric losses were considered. Four probes of coaxial cables were used to reach an omnidirectional pattern. They were placed at $z_f = 5$ mm, and modeled by a strip of width equal to 1.45 mm. The results were plotted around the frequency 2.27 GHz, which corresponds to the first omnidirectional mode. Figure 9 also shows results obtained from Ansoft HFSS, and it is observed that the input impedances from both methods are very close. The cavity effect can be observed in Fig. 10, where input impedances of the same antenna is shown for cavities of 40 mm, 50 mm, and 60 mm. As can be observed the cavity size has little influence on the input impedance, unless the perfect electric walls of the cavity at $z = z_{1c}$ and $z = z_{2c}$ becomes too close to the antenna. The effect of the dielectric thickness for the same antenna is also shown in Fig. 11. As the thickness increases the resonance frequency diminishes, and a significant bandwidth increase is observed. We can also observe a reactance shift, including the self-contribution of the probe. These properties of the cavity-backed wraparound antenna are also observed for the microstrip wraparound antenna with infinite dielectric layer.

9. Conclusion

This chapter has described an accurate model for cavity-backed wraparound antenna using method of moments and equivalence principle. It addresses the numerical challenges of

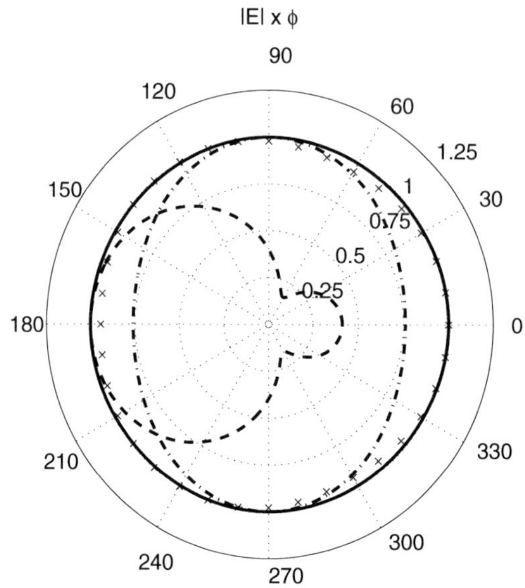

Fig. 7. $|E| \times \phi$ at xy plane. - - - -: 1 feed, $-.-.-$.: 2 feeds, x x x: 3 feeds, ____: 4 feeds. f=3 GHz, a=25 mm, h=1 mm, z_{1c}=-25 mm, z_{2c}=25 mm, z_{1a}=-17.25 mm, z_{2a}=17.25 mm, z_f=8.625 mm, W_f=2 mm, ε_r=2.1.

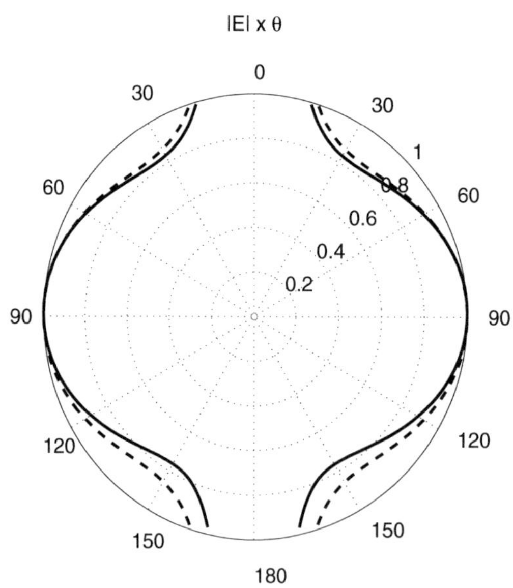

Fig. 8. $|E| \times \theta$ at xz plane. ____: this formulation for a cavity-backed wraparound antenna, - - -: cavity method for a microstrip wraparound antenna. f=3 GHz, a=25 mm, h=1 mm, z_{1c}=-25 mm, z_{2c}=25 mm, z_{1a}=-17.25 mm, z_{2a}=17.25 mm, z_f=8.625 mm, W_f=2 mm, ε_r=2.1, 4 probes at $\phi = 0^o, 90^o, 180^o, 270^o$.

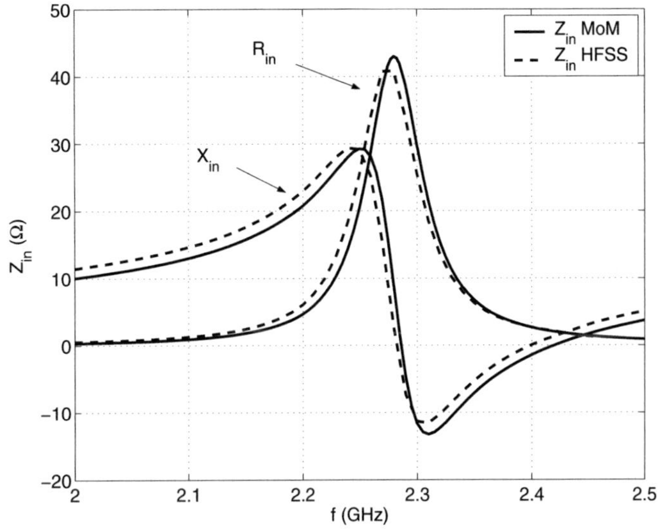

Fig. 9. Input impedance for a cavity-backed wraparound antenna. b=21 mm, a=20 mm, z_{1c}=-25 mm, z_{2c}=25 mm, z_{1a}=-10 mm, z_{2a}=10 mm, z_f=5 mm, W_f=1.45 mm, ε_r=9.6, 4 probes at $\phi = 0^o, 90^o, 180^o, 270^o$

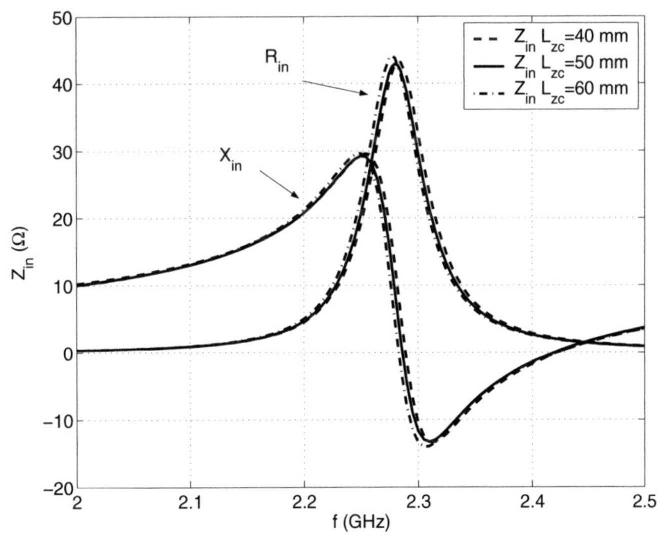

Fig. 10. Input impedance for a cavity-backed wraparound antenna: - - -: L_{zc}=40 mm, ___: L_{zc}=50 mm, -.-.-.: L_{zc}=60 mm. b=21 mm, a=20 mm, z_{1a}=-10 mm, z_{2a}=10 mm, z_f=5 mm, W_f=1.45 mm, ε_r=9.6, 4 probes at $\phi = 0^o, 90^o, 180^o, 270^o$

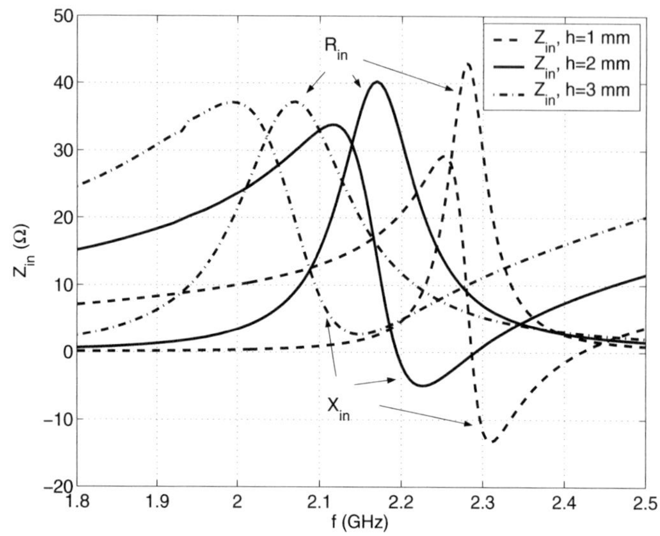

Fig. 11. Input impedance for a cavity-backed wraparound antenna. - - -: h=1 mm, ___: h=2 mm, -.-.-.: h=3 mm. b=21 mm, z_{1c}=-25 mm, z_{2c}=25 mm, z_{1a}=-10 mm, z_{2a}=10 mm, z_f=5 mm, W_f=1.45 mm, ε_r=9.6, 4 probes at $\phi = 0^o, 90^o, 180^o, 270^o$

computational implementation of such solution, and proposes a procedure that allows the analysis for larger cylinders. An approximation for the self-impedance of the probe has also been introduced. The results are validated by comparing to commercial software results. In general it was observed a similar behavior for cavity-backed antenna as for microstrip one. In order to reach an omnidirectional pattern the spacing between probes should be no greater than the dielectric wavelength. The effect of the cavity side wall are limited, and the thickness of the dielectric can be used for obtaining a larger bandwidth.

10. References

[Ali et al.,1989] Ali, S.M.; Habashy, T. M.; Kiang, J. F. & Kong, J. A. (1989). Resonance in cylindrical-rectangular and wraparound structures. *IEEE Trans. on Microwave Theory and Techniques*, Vol. 37, No. 11, pp. 1773-1783

[Ashkenazy et al.,1985] Ashkenazy, J.; Shrikman, S. & Treeves, D. (1985). Electric surface current model for the analysis of microstrip antennas on cylindrical bodies, *IEEE Trans. on Antennas and Propagation*, Vol. 33, No. 3, pp. 295-300

[Erturk & Rojas,2002] Erturk, V. B. & Rojas, R. G. (2002). Paraxial space-domain formulation for surface fields on a large dielectric coated circular cylinder. *IEEE Trans. on Antennas and Propagation*, Vol. 50, No. 11, pp. 1577-87

[Erturk & Rojas,2003] Erturk, V. B. & Rojas, R. G. (2003). Efficient analysis of input impedance and mutual coupling of microstrip antennas mounted on large coated cylinders. *IEEE Trans. on Antennas and Propagation*, Vol. 51, No. 4, pp. 739-749

[Fonseca & Giarola,1983] Fonseca, S. B. A. & Giarola A. J. (1983). Analysis of microstrip wraparound antennas using dyadic Greent's function, *IEEE Trans. on Antennas and Propagation*, Vol. AP-31, No. 2, pp. 248-253

[Habashy et al.,1990] Habashy, T. M.; Ali, S. M. & Kong, J. A. (1990). Input impedance and radiation pattern of cylindrical-rectangular and wraparound microstrip antennas, *IEEE Trans. on Antennas and Propagation*, Vol. 38, No. 5, pp. 722-731

[Harrington,1961] Harrington, R. F. (1961). *Time-harmonic Electromagnetic Fields*, McGraw Hill, New York

[Harrington,1968] Harrington, R. F. (1968). *Field Computation by Moment Method*, IEEE Press, New Jersey

[Heckler et al.,2003] Heckler, M.V.T.; Bonadiman, M., Schildberg, R.; Cividanes, L. & Lacava, J. C. S. (2003). CAD package to design rectangular probe-fed microstrip antennas conformed on cylindrical structures. *Proceedings SBMO/IEEE MTT-S IMOC 2003*, pp. 747-752

[Ke & Wong,1994] Ke, S. Y. & Wong, K. L. (1994). Input impedance of a probe-fed superstrate-loaded cylindrical-rectangular microstrip antenna. *Microwave and Optical Technology Letters*, Vol, 7, No. 5, pp. 232-36

[Kempel & Volakis,1995] Kempel L. C. & Volakis, J. L. (1995). Radiation by cavity-backed antennas on a circular cylinder, *IEE Proceedings on Microwaves, Antennas & Propagation*, Vol. 142, No. 3, pp. 233-239

[Krowne,1983] Krowne, C.C. (1983). Cylindrical-rectangular microstrip antennas. *IEEE Trans. on Antennas and Propagation*, Vol. 31, No. 1, pp. 194-199

[Luk et al.,1989] Luk, K. M.; Uk, K.M.; Lee K. F. & Dahele J. S. (1989). Analysis of the cylindrical-rectangular microstrip patch antenna. *IEEE Trans. on Antennas and Propagation*, Vol. 37, No. 2, pp. 143-147

[Lumini,1991] Lumini, F. (1991). *Analysis and design of rectangular microstrip antennas over cylindrical surfaces*. MsC. Thesis, Technological Institute of Aeronautics, (in portuguese)

[Mang & Xiaowen,2004] Mang, H. & Xiaowen, X. (2004). Full-wave analysis and wide-band design of probe-fed multilayered cylindrical-rectangular microstrip antennas. *IEEE Trans. on Antennas and Propagation*, Vol. 52, No. 7, pp. 1749-57

[Munson,1974] Munson, R. E. (1974). Conformal microstrip antennas and microstrip phased arrays. *IEEE Trans. on Antennas and Propagation*, Vol. 22, No. 1, pp. 74-78

[Pereira Filho,2009] Pereira Filho, O. M. C. (2009) Flush-mounted cylindrical-rectangular microstrip antennas. *IET Microwaves, Antennas & Propagation*, Vol. 3, No. 1, pp. 1-13

[Raffaelli et al.,2005] Raffaelli, S.; Sipus, Z. & Kildal, P. S. (2005). Analysis and measurements of conformal patch array antennas on multilayer circular cylinder. *IEEE Trans. on Antennas and Propagation*, Vol. 53, No. 3, pp. 1105-13

[Silva et al.,1991] Silva, C. M.; Lumini, F.; Lacava, J. C. S. & Richards, F. P. (1991). Analysis of cylindrical arrays of microstrip rectangular patches. *Electronic Letters*, Vol. 27, No. 9, pp. 778-780

[Silva & Lacava,1995] Silva, C. M. & Lacava, J. C. S. (1995). Mutual impedance of conformal cylindrical microstrip antenna arrays with a protection layer. *Proceedings 1995 SBMO/IEEE MTT-S International Conference*, pp. 314-319

[Silva et al.,1991b] Silva, F. C.; Fonseca, S. B. A.; Soares, A. J. M. & Giarola A. J. (1991). Analysis of microstrip antennas on circular-cylindrical substrates with a dielectric overlay. *IEEE Trans. on Antennas and Propagation*, Vol. 39, No. 9, pp. 1398-1403

[Tam et al.,1995] Tam W. Y.; Lai, A. K. Y. & Luk, K. M. (1995). Cylindrical rectangular microstrip antennas with coplanar parasitic patches, *IEE Proc. on Microwave, Antennas and Propagation*, Vol. 142, No. 4, pp. 300-306

[Tulintseff et. al.,1991] Tulintseff, A. N.; Al, S. M. & Kong, J. A. (1991). Input impedance of a probe-fed stacked circular microstrip antenna. *IEEE Trans. on Antennas and Propagation*, Vol. 39, No. 3, pp. 381-390

[Wong & Ke,1993] Wong, K. L. & Ke, S. Y. (1993). Characteristics of the cylindrical wraparound microstrip patch antenna, *roc. Natl. Sci. Counc. ROC(A)*, Vol. 17, No. 6, pp. 438-42

[Wong et al.,1993b] Wong, K. L.; Cheng, Y. T. & Row, J. S. (1993). Resonance in a superstrate-loaded cylindrical-rectangular microstrip structure. *IEEE Trans. on Microwave Theory and Techniques*, Vol. 41, No. 5, pp. 814-819

[Wong,1999] Wong, K. L. (1999). *Design of Nonplanar Microstrip Antennas and Transmission Lines*, Wiley-Interscience

[Yang & Ruan,1993] Yang, C. & Ruan, Y.Z. (1993). Radiation characteristics of wraparound microstrip antenna on cylindrical body. *Electronic Letters*, Vol. 29, No. 6, pp. 512-14

[Zheng & Chang,1991] Zheng, J. X. & Chang, D. C. (1991). End-correction network of a coaxial probe for microstrip patch antennas. *IEEE Trans. on Antennas and Propagation*, Vol. 39, No. 1, pp. 115-118

8

Analysis into Proximity-Coupled Microstrip Antenna on Dielectric Lens

Lawrence Mall
Mall Science and Engineering Research
Australia

1. Introduction

Analysis into proximity-coupled microstrip antenna on dielectric lens for characteristics with high gain and pencil beam is presented. Earlier to this work, the University of Michigan developed double-slot antennas on a silicon material lens (Filipovic et al., 1993) and the Swiss Federal Institute of Technology developed aperture-coupled patch antenna on a substrate lens (Eleftheriades et al., 1997). In order to achieve the stringent requirements for efficiency and low cost necessary for a practical radio communication system, we propose our lens antenna for which lens is made of polyethylene plastic for commercial applications. Since the objective is to result in an efficient dielectric lens antenna system, substrates for the proximity-coupled microstrip antenna and the material for the dielectric lens are chosen that they have the same dielectric constant so as to eliminate surface waves and thereof no surface-wave power losses attributed to these selections. We have shown theoretical investigation for how the lens' geometry comes up with dielectric ellipsoid lens and subsequently modified to extended hemispherical dielectric lens as synthesized ellipsoid. Determination of antenna return loss is carried out using method of moments (MoM) which involves spectral domain Green's functions for a proximity-coupled microstrip antenna residing in homogeneous dielectric half space. The Sommerfeld type double integrals are numerically solved and obtained MoM impedance and excitation voltage matrix elements. After solving MoM matrix equation, current column matrix is resulted in and one of which elements is the equivalent current reflection coefficient at the open-end of the microstrip feedline. The return loss is calculated from this factor over a frequency band of 38 GHz and compared with measurement and found very good agreement. Radiation pattern theory is derived in such a way that the proximity-coupled microstrip antenna mounted on the back of the lens radiates into it and illuminates the interior side of the hemispherical surface. We ensure that the fields arrive at this surface are far fields so that we can apply ray tracing method in order to calculate the fields. We then calculate the fields on the exterior side of the hemisphere using appropriate transmission coefficients (parallel and perpendicular polarizations) for the interface between lens' dielectric and free space media. Equivalent surface electric and magnetic current densities on the hemispherical surface are obtained from these fields. These two current densities are incorporated into radiation integral equations and far field patterns in free space are determined. This method for determination of far field patterns is called Schelkunoff Principle or Huygens' Principle or Field Equivalence Principle. Once again we achieved good agreement between theoretical and measured radiation patterns.

This chapter is organized as follows. Section 2 presents determination of dielectric lens geometry for which exiting rays from the lens travel in parallel with optical axis of the lens. The determined geometry is true ellipsoid lens and verified that the Snell's Law of Refraction is satisfied. Modification of true ellipsoid lens to extended hemispherical lens is presented in Section 3. Section 4 presents formulating and computing for radiation patterns of extended hemispherical dielectric lens. It comprises analysis of radiation of patch into lens by cavity model and analysis of radiation of lens into free space. Design of proximity-coupled microstrip antenna and the lens, and comparison between theoretical and measured return loss and radiation patterns are presented in Section 5. Results and discussion, and conclusion are followed there after.

2. Determination of dielectric lens geometry

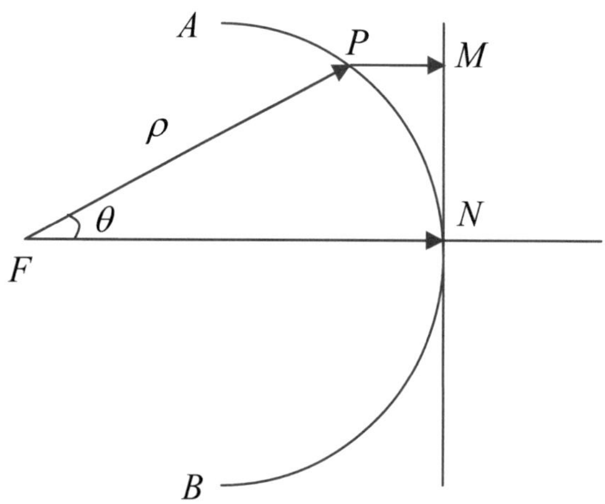

Fig. 2-1 Unknown Geometry of the dielectric lens

Since the dielectric lens' size is chosen large compared with wavelength, the far fields can be determined using ray tracing method. The rays are perpendicular to wavefronts. We now consider a dielectric lens with unknown geometry and having dielectric constant ε_r, refractive index η and holds $\eta = \sqrt{\varepsilon_r}$. In Figure 2-1, ANB is unknown surface curvature of the lens under determination. F is focal point and FN is taken as optical axis. We place an optical source at focal point, F and use spherical coordinate system (ρ,θ,ϕ) for determination of required geometry of the lens while taking symmetry about ϕ. P is a point on the lens' surface and MN is a plane perpendicular to the FN axis. For a wavefront travelled a distance x, its phase change is βx, where wave number $\beta = \frac{2\pi}{\lambda}$. If λ_d is the wavelength in the lens' dielectric medium and λ_0 is the corresponding wavelength in free space, then $\lambda_d = \frac{\lambda_0}{\eta}$. We want that the optical source at F emits rays which are incident on lens' surface and refract into free space as parallel rays and they are also parallel to the optical axis of the lens. These parallel rays will form a wavefront on the plane MN which perpendicular to the optical axis. For a plane wave, there is equal phase on each and every point on wavefront. Consider (1) a ray travels from point F to point P in the dielectric lens, then refracts at point P and travels in free space to point M; (2) another ray travels from point F to point N along the optical axis in the dielectric lens. On arrival of, ray in (1) at point M, and ray in (2) at point N, change of their phases must be the same in order to form a plane wave on the plane MN. A phase change equation can be written as follow for ray in (1) and ray in (2).

$$\beta_d[FP] + \beta_0[PM] = \beta_d[FN] \tag{2-1}$$

The appropriate quantities being substituted in (2-1) will yield

$$\frac{2\pi\eta}{\lambda_0}\rho + \frac{2\pi}{\lambda_0}[FN - \rho.\cos(\theta)] = \frac{2\pi\eta}{\lambda_0}[FN] \tag{2-2}$$

This equation can be further simplified as

$$\rho = \frac{k}{1 - e.\cos(\theta)} \tag{2-3}$$

Where

$$k = \frac{\eta - 1}{\eta}[FN] \tag{2-4}$$

and, the eccentricity

$$e = \frac{1}{\eta} \tag{2-5}$$

It is obvious that $e < 1$ since the refractive index $\eta > 1$, and Equation (2-3) represents an ellipsoid. This finding characterizes the dielectric ellipsoid lens with *F* as one of the two focal points and other is at the right side. The rays diagram is shown in Figure 2-2.

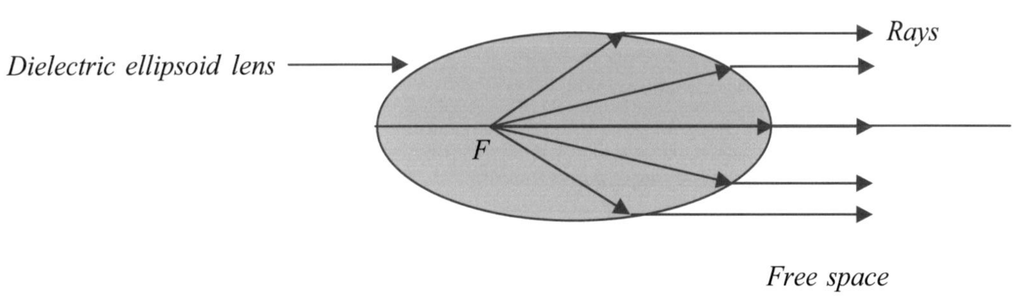

Fig. 2-2 Ray diagram of dielectric ellipsoid lens and an optical source placed at a focal point F.

Figure 2-2 shows an optical source placed at a focal point, F of dielectric ellipsoid lens, emanating rays to its right hand side directions. They are incident on the lens' surface, refract and exit into free space as parallel rays. We will investigate further whether this dielectric ellipsoid lens satisfies the well known Snell's Law of Refraction which states as follows.

$$\frac{\sin(\psi_i)}{\sin(\psi_r)} = \frac{1}{\eta} \tag{2-6}$$

ψ_i and ψ_r are incident and refracted angles respectively, and η is the refractive index of the dielectric ellipsoid lens. Drawing for this investigation is shown in figure (2-3).

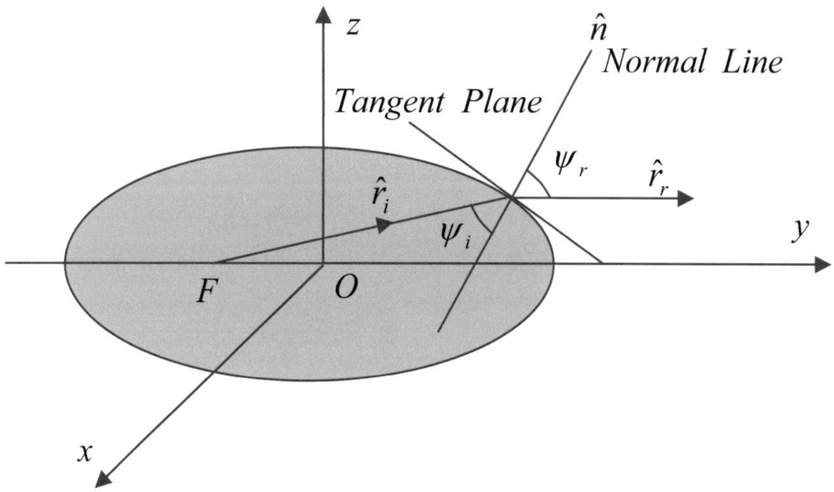

Fig. 2-3 Incident ray emitted from focal point F and its refracted ray of dielectric ellipsoid lens

Equation for this dielectric ellipsoid in Cartesian coordinate system and its associated can be written as,

$$\frac{x^2+z^2}{b^2}+\frac{y^2}{a^2}=1 \tag{2-7}$$

$$b^2=a^2(1-e^2) \tag{2-8}$$

a and b are semi-major and semi-minor axes respectively. We will rewrite (2-7) as,

$$\phi=\frac{x^2+z^2}{b^2}+\frac{y^2}{a^2} \tag{2-9}$$

Then $\phi=1$ and is a constant. Using Vector Calculus, we can determine a unit vector ($\hat{n}$) along normal line on the lens surface as,

$$\hat{n}=\frac{\vec{\nabla}\phi}{|\vec{\nabla}\phi|}=\frac{\frac{2}{b^2}(x.\hat{i}+z.\hat{k})+\frac{2}{a^2}y.\hat{j}}{\sqrt{(\frac{2}{b^2})^2(x^2+z^2)+(\frac{2}{a^2})^2y^2}} \tag{2-10}$$

Making use of (2-7) and (2-8) above equation can be simplified as,

$$\hat{n}=\frac{\frac{x.\hat{i}+z.\hat{k}}{b^2}+\frac{y}{a^2}.\hat{j}}{\sqrt{a^2-e^2.y^2}}ab \tag{2-11}$$

A unit vector ($\hat{r}_i$), parallel to the incident ray emanating from focal point F can be determined as,

$$\hat{r}_i = \frac{x.\hat{i} + (y + a.e).\hat{j} + z.\hat{k}}{\sqrt{x^2 + (y + a.e)^2 + z^2}} \tag{2-12}$$

Note that distance between focal point F and origin O is $a.e$. Simplifying above equation using (2-7) and (2-8) yields,

$$\hat{r}_i = \frac{x.\hat{i} + (y + a.e).\hat{j} + z.\hat{k}}{a + e.y} \tag{2-13}$$

On arrival of the incident ray at a point on the tangent plane, it refracts, exits into free space as refracted ray and travels in parallel with the optical axis or y-axis. A unit vector ($\hat{r}_r$) parallel to this refracted ray is,

$$\hat{r}_r = \hat{j} \tag{2-14}$$

$$\cos(\psi_i) = \hat{n}.\hat{r}_i = \frac{\dfrac{x^2 + z^2}{b^2} + \dfrac{y(y + a.e)}{a^2}}{(a + e.y).\sqrt{a^2 - e^2.y^2}} ab \tag{2-15}$$

Simplifying above equation with (2-7) and (2-8) yields

$$\cos(\psi_i) = \frac{b}{\sqrt{a^2 - e^2.y^2}}$$

$$\sin(\psi_i) = \sqrt{1 - \cos^2(\psi_i)} = e\sqrt{\frac{a^2 - y^2}{a^2 - e^2.y^2}} \tag{2-16}$$

$$\cos(\psi_r) = \hat{n}.\hat{j} = \frac{\dfrac{y}{a^2}}{\sqrt{a^2 - e^2.y^2}} ab$$

$$\sin(\psi_r) = \sqrt{1 - \cos^2(\psi_r)} = \sqrt{\frac{a^2 - y^2}{a^2 - e^2.y^2}} \tag{2-17}$$

Then, from (2-16), (2-17) and (2-5),

$$\frac{\sin(\psi_i)}{\sin(\psi_r)} = e = \frac{1}{\eta} \tag{2-18}$$

We now have verified that the Snell's Law of Refraction is satisfied with (2-18) for our dielectric ellipsoid lens.

3. Modification of true ellipsoid lens to extended hemispherical lens

For ease of manufacture, true ellipsoid dielectric lens is modified to extended hemispherical dielectric lens as synthesized ellipsoid. The reason for using an extended hemispherical lens is based on what can be made practically and can be synthesized to give the properties of a true ellipsoid. Producing of extended hemispherical dielectric lens, Figure 3-1, is in such a way that the surface-curvature of the hemispherical surface has the close match to that of the true ellipsoid lens. Comparison is shown in Figure 3-2, which is for two dimensions since the geometry is rotationally symmetric.

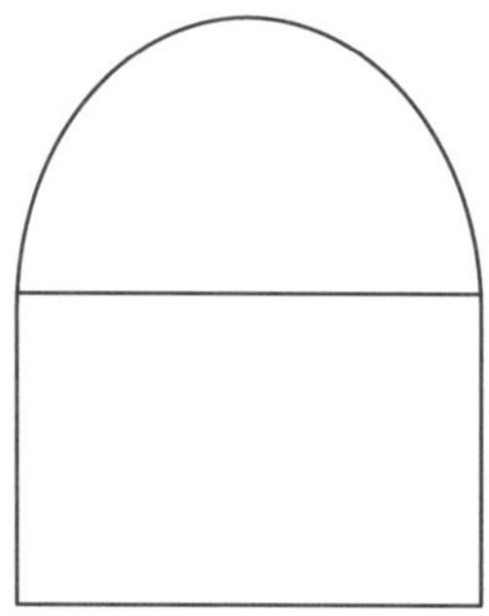

Fig. 3-1 Extended hemisphere

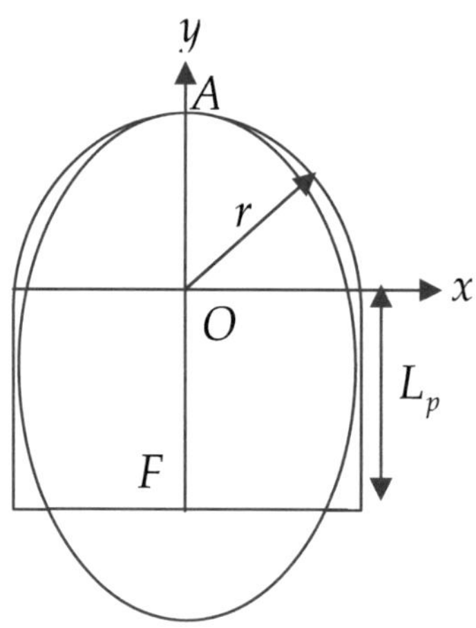

Fig. 3-2 Matching of extended hemisphere to ellipse

In Figure 3-2, F is the more distant focus of the ellipse from the vertex point A. The equation for an ellipse is,

$$\frac{x^2}{b^2}+\frac{y^2}{a^2}=1 \tag{3-1}$$

The foci are at $\pm c$ and

$$c=\sqrt{a^2-b^2} \tag{3-2}$$

The eccentricity e is given by,

$$e = \frac{\sqrt{a^2 - b^2}}{a} = \frac{1}{\eta} \tag{3-3}$$

From Equation (3-3), following can be obtained,

$$b = a\sqrt{1 - \frac{1}{\eta^2}} \tag{3-4}$$

From Equation (3-2) and (3-3), following can be found,

$$c = \frac{a}{\eta} \tag{3-5}$$

In Figure 3-2, the equation for semicircle of unit radius is (for two dimensions),

$$x^2 + y^2 = 1 \tag{3-6}$$

For the ellipse in Figure 3-2,

$$\mathrm{AF} = a + c \tag{3-7}$$

In Figure 3-2, with reference to the extended hemispherical lens,

$$\mathrm{AF} = 1 + L_p \tag{3-8}$$

Since Equation (3-7) and (3-8) are the same in magnitude,

$$L_p = a + c - 1 \tag{3-9}$$

In Figure 3-2, the ellipse has been shifted down by a value $y_0 = L_p - c$ so that the focus of the ellipse has the same coordinates as that of the extended hemisphere. For $a > 1$, the vertex of ellipse is $(a-1)$ unit above that of unit radius semi-circle. In order for the two vertexes overlap, the ellipse must be shifted down by $(a-1)$ unit and (3-1) will change as

$$\frac{x^2}{b^2} + \frac{(y + a - 1)^2}{a^2} = 1 \tag{3-10}$$

By (3-6) a semi-circle ($y > 0$) with unit radius can be plotted. For a given refractive index, using (3-10), in combination with (3-4) for elimination of b, various ellipses ($y > 0$) can be traced with different parameters for a. We have seen that $a = 1$ does not give close match between semi-circle and ellipse. For a given refractive index, a only was varied until a close match was achieved between the curvature of the ellipse and that of the semi-circle with unit radius. Our selected material for the extended hemispherical dielectric lens is ultra high density polyethylene plastic with $\varepsilon_r = 2.35$ which yields $\eta = \sqrt{\varepsilon_r} = 1.53$. From our comparison, we are satisfied with $a = 1.6$. Using (3-5) and (3-9), we received extension length of the hemisphere, with unit radius, as 1.65. This yields extension length, L, of hemisphere with radius R using ultra high density polyethylene plastic is,

$$L = 1.65R \tag{3-11}$$

4. Formulating and computing for radiation patterns

This section documents the analysis into far field radiation patterns of extended hemispherical dielectric lens. Since the proximity-coupled microstrip patch antenna is placed at the back of the lens the radiation will illuminate the whole interior side of the lens' hemispherical surface.

4.1 Analysis of radiation of patch into lens by cavity model

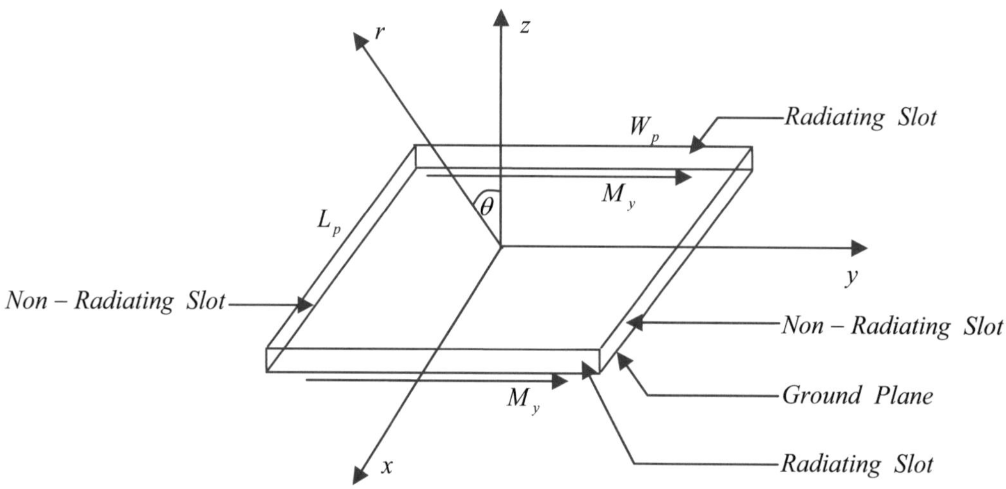

Fig. 4-1 Generation of magnetic current, M_y, on radiating slots

Figure 4-1 shows the microstrip patch in the x-y plane at z = 0 above infinitely large conducting ground plane which is located at z = -d. The length and width of the patch are L_p and W_p respectively. The centre of the patch is placed at the origin of the Cartesian coordinate system. It is supposed that the electric current is flowing along positive-x direction. By using the cavity model technique as shown in Section 14.2.2 of (Balanis, 1997), radiation from the patch can be derived from the two same phase and equal magnitude magnetic current sources on the radiating slots. The two magnetic current sources, M_y, are y-directed and uniformly distributed along the slots. The process of derivation is started according to the equation (12-12c) and (12-15a) of the reference (Balanis, 1997) as follows:

$$L_\theta = \int_{-d}^{0} \int_{-\frac{W_p}{2}}^{+\frac{W_p}{2}} M_y \cos\theta \sin\phi \; e^{jk(y'\sin\theta\sin\phi + z'\cos\theta)} dy'dz' \tag{4-1}$$

this yields

$$L_\theta = \cos\theta\sin\phi \frac{\sin\{\frac{kW_p \sin\theta \sin\phi}{2}\}}{\frac{k\sin\theta \sin\phi}{2}} \frac{\sin\{\frac{k\,d\cos\theta}{2}\}}{\frac{k\cos\theta}{2}} e^{-j\frac{k\,d\cos\theta}{2}} \tag{4-2}$$

M_y in Equation (4-1) is the y-directed uniform magnetic current and is normalized by setting it to unity in Equation (4-2). According to Equation (12-12d) and (12-15a) of reference (Balanis, 1997) the following can be derived:

$$L_\phi = \int_{-d}^{0} \int_{-\frac{W_p}{2}}^{+\frac{W_p}{2}} M_y \cos\phi \, e^{jk(y'\sin\theta\sin\phi + z'\cos\theta)} dy'dz' \tag{4-3}$$

which results

$$L_\phi = \cos\phi \frac{\sin\{\frac{kW_p \sin\theta \sin\phi}{2}\}}{\frac{k\sin\theta \sin\phi}{2}} \frac{\sin\{\frac{k d\cos\theta}{2}\}}{\frac{k\cos\theta}{2}} e^{-j\frac{kd\cos\theta}{2}} \tag{4-4}$$

Again, M_y in Equation (4-3) is the y-directed uniform magnetic current and normalized by setting it to unity in Equation (4-4). The E-plane array factor (AF) is given below (Balanis, 1997),

$$AF = 2\cos\{\frac{kL_p}{2}\sin\theta \cos\phi\} \tag{4-5}$$

According to Equation (12-10b) and Section 14.2.2 of (Balanis, 1997) the following can be found:

$$E_\theta = -\frac{jke^{-jkr}}{4\pi r} L_\phi (AF) \tag{4-6}$$

$$E_\phi = \frac{jke^{-jkr}}{4\pi r} L_\theta (AF) \tag{4-7}$$

By Equation (4-6),

$$E_\theta = -\frac{jke^{-jkr}}{4\pi r}\cos\phi \frac{\sin\{\frac{kW_p \sin\theta \sin\phi}{2}\}}{\frac{k\sin\theta \sin\phi}{2}} \frac{\sin\{\frac{k d\cos\theta}{2}\}}{\frac{k\cos\theta}{2}} e^{-j\frac{kd\cos\theta}{2}}$$

$$\times 2\cos\{\frac{kL_p}{2}\sin\theta \cos\phi\} \tag{4-8}$$

and by Equation (4-7),

$$E_\phi = \frac{jke^{-jkr}}{4\pi r}\cos\theta \sin\phi \frac{\sin\{\frac{kW_p \sin\theta \sin\phi}{2}\}}{\frac{k\sin\theta \sin\phi}{2}} \frac{\sin\{\frac{k d\cos\theta}{2}\}}{\frac{k\cos\theta}{2}} e^{-j\frac{kd\cos\theta}{2}}$$

$$\times 2\cos\{\frac{kL_p}{2}\sin\theta \cos\phi\} \tag{4-9}$$

4.2 Analysis of radiation of lens into free space

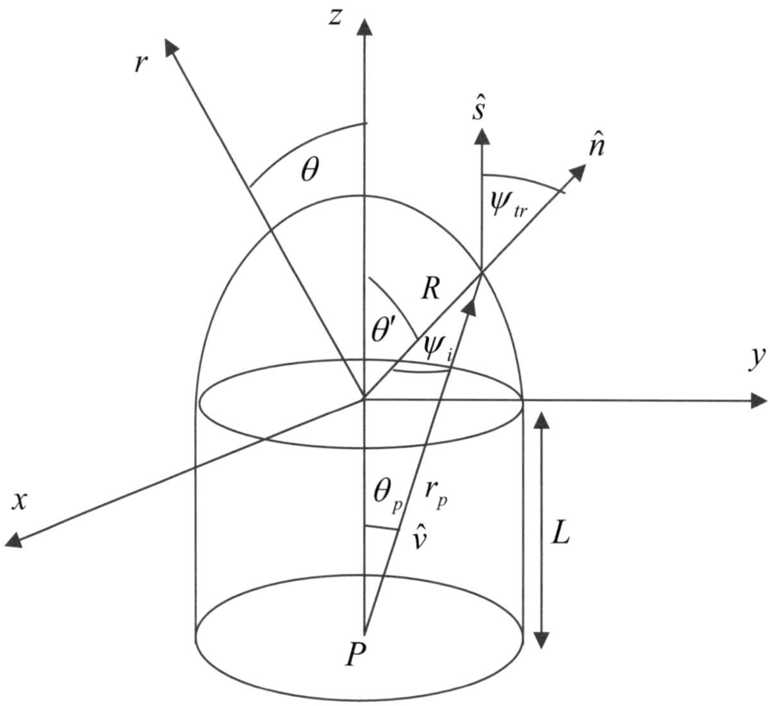

Fig. 4-2 Lens geometry and rays diagram

In Figure 4-2, the intersection point of the x, y and z-axis is the origin (0,0,0) of the Cartesian coordinate system. R is the radius of the hemisphere and L is the extension length. The point P is the centre of the patch which is lying in the x-y plane at z = -L. The patch orientation is the same as in Figure 4-1 and the electric current flowing on the patch is along positive-x direction. For transformations of Equation (4-8) and (4-9) to Figure 4-2, r , θ and ϕ will change to r_p, θ_p and ϕ_p respectively. The (r_p,θ_p,ϕ_p) coordinates have their origin at the point P. Next transformation is performed from (r_p,θ_p,ϕ_p) to (R,θ',ϕ'), whose origin is at the intersections of x,y,z axes and R is constant.

R and L are known constant parameters of the lens, and so for a given θ' , θ_p can be found. Equation (4-8) and (4-9) can be expressed in terms of (R,θ',ϕ') as follows:

$$E_{\theta p} = E_{\theta p}(R,\theta',\phi') \tag{4-10}$$

$$E_{\phi p} = E_{\phi p}(R,\theta',\phi') \tag{4-11}$$

Next the unit vectors shown on Figure 4-2 are to be determined. The surface normal unit vector is expressed as follows:

$$\hat{n} = \sin\theta'\cos\phi'\,\hat{x} + \sin\theta'\sin\phi'\,\hat{y} + \cos\theta'\,\hat{z} \tag{4-12}$$

The ray path inside the lens is,

$$\vec{V} = R\sin\theta'\cos\phi'\hat{x} + R\sin\theta'\sin\phi'\hat{y} + (R\cos\theta' + L)\hat{z} \tag{4-13}$$

Its unit vector is,

$$\hat{v} = \frac{\vec{V}}{|\vec{V}|} \tag{4-14}$$

The vector that defines the component of electric field in the perpendicular plane of incident ray can be found as:

$$\vec{P}_{\perp} = \hat{n} \times \hat{v} \tag{4-15}$$

where its unit vector is:

$$\hat{p}_{\perp} = \frac{\vec{P}_{\perp}}{|\vec{P}_{\perp}|} \tag{4-16}$$

The unit vector that defines the component of the electric field in the parallel plane of incident ray can be found as:

$$\hat{p}_{||} = \hat{p}_{\perp} \times \hat{v} \tag{4-17}$$

The parallel polarization transmission coefficient and perpendicular polarization transmission coefficient can be found respectively based on Section 2-8 of (Collin, 1966):

$$T_{||} = \frac{2\eta\cos\psi_i}{\cos\psi_i + \eta\sqrt{1-\eta^2\sin^2\psi_i}} \tag{4-18}$$

$$T_{\perp} = \frac{2\eta\cos\psi_i}{\eta\cos\psi_i + \sqrt{1-\eta^2\sin^2\psi_i}} \tag{4-19}$$

Where η is the refractive index of the material of lens and ψ_i is the ray incident angle as shown in the Figure 4-2. Equation (4-18) and (4-19) are valid for $0 \le \psi_i \le \psi_{ic}$ where ψ_{ic} holds as:

$$1 - \eta^2\sin^2\psi_{ic} = 0$$

or

$$\psi_{ic} = \arcsin\left(\frac{1}{\eta}\right) \tag{4-20}$$

ψ_{ic} is called the critical angle. When $\psi_i > \psi_{ic}$, the right hand sides of Equation (4-18) and (4-19) are not valid. All the incident rays reflect back into the lens and so, this phenomenon is called the total internal reflection. In such a situation, the left hand side of Equation (4-18) and (4-19) are zero, $T_{||} = T_{\perp} = 0$.
From Figure 4-2, it can be seen that

$$\psi_i = \theta' - \theta_p \tag{4-21}$$

θ_p can be found, using elementary trigonometry, and substituted in (4-21)

$$\psi_i = \theta' - \arcsin[\frac{R\sin\theta'}{\sqrt{R^2 + L^2 + 2RL\cos\theta'}}] \quad (4\text{-}22)$$

From Equation (4-22), it is noticed that the variable ψ_i is only a function of the variable θ'. The refracted angle, ψ_{tr}, can be found by Snell's Law as,

$$\frac{\sin(\psi_i)}{\sin(\psi_{tr})} = \frac{1}{\eta}$$

or

$$\psi_{tr} = \arcsin[\eta\sin(\psi_i)] \quad (4\text{-}23)$$

Equation (4-22) is substituted into Equation (4-23) to express the variable ψ_{tr} in term of the variable θ':

$$\psi_{tr} = \arcsin[\eta\sin(\ \theta' - \arcsin\{\frac{R\sin\theta'}{\sqrt{R^2 + L^2 + 2RL\cos\theta'}}\})] \quad (4\text{-}24)$$

The components of the transmitted electric field at the exterior side of the hemispherical surface of the lens have been presented in (Filipovic et al., 1997) and they are given as follows:

$$E_x = (-E_{\phi p})T_\perp(\hat{x}\cdot\hat{p}_\perp) + (-E_{\theta p})T_{||}[(\hat{x}\cdot\hat{p}_{||})\cos(\psi_{tr} - \psi_i) - (\hat{x}\cdot\hat{v})\sin(\psi_{tr} - \psi_i)] \quad (4\text{-}25)$$

$$E_y = (-E_{\phi p})T_\perp(\hat{y}\cdot\hat{p}_\perp) + (-E_{\theta p})T_{||}[(\hat{y}\cdot\hat{p}_{||})\cos(\psi_{tr} - \psi_i) - (\hat{y}\cdot\hat{v})\sin(\psi_{tr} - \psi_i)] \quad (4\text{-}26)$$

$$E_z = (-E_{\phi p})T_\perp(\hat{z}\cdot\hat{p}_\perp) + (-E_{\theta p})T_{||}[(\hat{z}\cdot\hat{p}_{||})\cos(\psi_{tr} - \psi_i) - (\hat{z}\cdot\hat{v})\sin(\psi_{tr} - \psi_i)] \quad (4\text{-}27)$$

Where $E_{\theta p}$ and $E_{\phi p}$ are given in Equation (4-10) and (4-11) respectively, based on Equation (4-8) and (4-9). The resultant transmitted electric field at the exterior side of the hemispherical surface of the lens is:

$$\vec{E} = E_x\hat{x} + E_y\hat{y} + E_z\hat{z} \quad (4\text{-}28)$$

The ray path unit vector outside the lens can be written as follows:

$$\hat{s} = s_x\hat{x} + s_y\hat{y} + s_z\hat{z} \quad (4\text{-}29)$$

The components have already been given in (Filipovic et al., 1997) and they are shown as follows:

$$s_x = v_x\cos(\psi_{tr} - \psi_i) + (\hat{x}\cdot\hat{p}_{||})\sin(\psi_{tr} - \psi_i) \quad (4\text{-}30)$$

$$s_y = v_y\cos(\psi_{tr} - \psi_i) + (\hat{y}\cdot\hat{p}_{||})\sin(\psi_{tr} - \psi_i) \quad (4\text{-}31)$$

$$s_z = v_z \cos(\psi_{tr} - \psi_i) + (\hat{z} \cdot \hat{p}_{||}) \sin(\psi_{tr} - \psi_i) \quad (4\text{-}32)$$

The normalized resultant magnetic field at the exterior side of the hemispherical surface of the lens is:

$$\vec{H} = \hat{s} \times \vec{E} \quad (4\text{-}33)$$

The equivalent surface electric current density and surface magnetic current density on the exterior side of the hemispherical surface of the lens are:

$$\vec{J}_s = \hat{n} \times \vec{H} \quad (4\text{-}34)$$

$$\vec{M}_s = -\hat{n} \times \vec{E} \quad (4\text{-}35)$$

By Schelkunoff's principle (Balanis, 1997), Equation (4-34) and (4-35) are incorporated into the radiation integral equations and far fields of the lens are determined.

$$N_\theta = \iint_S [J_x \cos\theta \ \cos\phi + J_y \cos\theta \sin\phi - J_z \sin\theta \,] e^{+jkR\cos\psi} ds' \quad (4\text{-}36)$$

$$N_\phi = \iint_S [-J_x \sin\phi + J_y \cos\phi \,] e^{+jkR\cos\psi} ds' \quad (4\text{-}37)$$

$$L_\theta = \iint_S [M_x \cos\theta \ \cos\phi + M_y \cos\theta \sin\phi - M_z \sin\theta \,] e^{+jkR\cos\psi} ds' \quad (4\text{-}38)$$

$$L_\phi = \iint_S [-M_x \sin\phi + M_y \cos\phi \,] e^{+jkR\cos\psi} ds' \quad (4\text{-}39)$$

$\cos\psi$ in Equation (4-36), (4-37), (4-38) and (4-39) is

$$\cos\psi = \hat{a}_r \cdot \hat{n} \quad (4\text{-}40)$$

$\hat{a}_r$ in Equation (4-40) is expressed as,

$$\hat{a}_r = \sin\theta \cos\phi \ \hat{x} + \sin\theta \sin\phi \ \hat{y} + \cos\theta \ \hat{z} \quad (4\text{-}41)$$

$\hat{n}$ was defined in Equation (4-12). Using Equation (4-41) and (4-12) in (4-40), $\cos\psi$ can be determined. ds' is the differential area element on the hemispherical surface of the lens and can be expressed in the polar coordinate form as,

$$ds' = R^2 \sin\theta' \, d\theta' \, d\phi' \quad (4\text{-}42)$$

Substituting Equation (4-40) and (4-42) into Equation (4-36), (4-37), (4-38) and (4-39) allows the solution to be determined by means of solving the double integrals numerically since all the integrands are functions of the variables, θ' and ϕ'. The integration limits are from 0 to 2π for ϕ' and 0 to $\frac{\pi}{2}$ for θ'. The normalized far field can be expressed as:

$$E_\theta = -(L_\phi + Z_0 N_\theta) \quad (4\text{-}43)$$

$$E_\phi = (L_\theta - Z_0 N_\phi) \quad (4\text{-}44)$$

where Z_0 is intrinsic impedance of free space. Since the patch current is x-directed, varying θ while ϕ is kept constant at 0 degree or 90 degrees will yield E-plane or H-plane pattern respectively.

5. Design of proximity-coupled microstrip antenna and the lens

The configuration of cross-sectional view of proximity-coupled microstrip antenna on extended hemispherical dielectric lens is shown in figure 5-1. h and $(d-h)$ are the thickness of the feedline-substrate and patch-substrate. L_{OS} is the offset of the feedline from below the center of the patch. R_L and L_L are radius of hemisphere and extension length respectively. Using C++ programming language on Linux operating system, we developed a design software tool in moment method for a proximity-coupled microstrip antenna residing in dielectric half space. The derived Green's functions in the code are for electric current sources residing in homogeneous dielectric medium. Since the details were given in (Pozar and Voda, 1987) and (Mall, 2008) we will not repeat here for the sake of brevity. We simulated our code for design of antenna, operating at 38 GHz, and obtained the parameters. Two 0.127 mm thick RT/Duroid 5870 substrates ($\varepsilon_r = 2.33$) were used for both patch and feedline layers. The patch dimensions were 2.175 mm x 2.000 mm and the open circuit termination was 0.1 mm from below the centre of the patch. To couple power to and from the antenna, a K-type connector was mounted through the ground-plane and soldered to the microstrip feedline 15 mm below from the centre of the microstrip patch. The track with of microstripline for 50Ω was 0.33 mm. As ultra high density polyethylene plastic material ($\varepsilon_r = 2.35$) was used and manufactured extended hemispherical dielectric lens, the

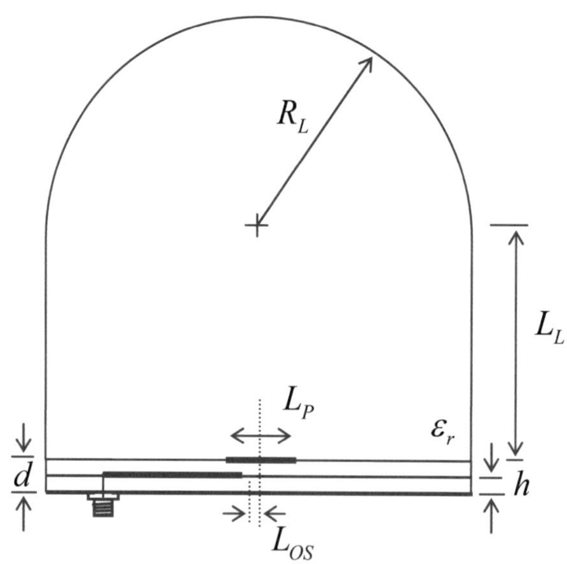

Fig. 5-1 Cross-sectional view of proximity-coupled microstrip antenna on extended hemispherical dielectric lens

wavelength in the lens at 38 GHz is 5.16 mm. Our selection of hemisphere's radius as 50 mm is quite good for the requirement that the lens' radius is much larger than the wavelength. According to (3-11), extension length has to be 82.5 mm.

Figure 5-2 shows the theoretical and measured return loss of the proximity-coupled lens antenna. The measurements were performed using a *Wiltron 360B* 40 GHz Vector Network Analyzer. As can be seen from these results, good agreement between theory and experiment was achieved. The predicted and measured 10 dB return loss bandwidths are 8.0% and 8.2% respectively.

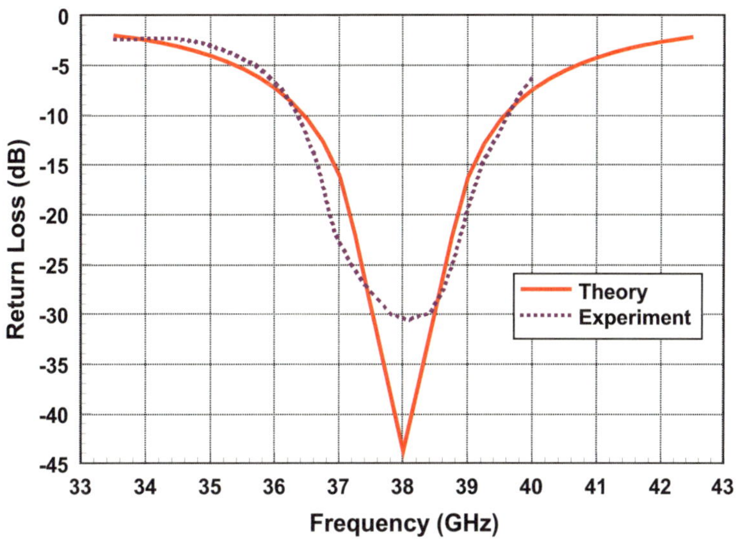

Fig. 5-2 Predicted and measured return loss curves

The predicted and measured *E*-plane and *H*-plane copolar radiation patterns at 38 GHz are shown in Figure 5-3 and 5-4 respectively. Once again, good agreement between theory and experiment is evident from these plots, particularly in the main lobes.

5.1 Results and discussion

Figure 5-2 shows the predicted and measured return loss of the proximity-coupled lens antenna. The measurements were done using a *Wiltron 360B* 40 GHz Vector Network Analyzer. As can be seen from these results, good agreement between theory and experiment was achieved. A high degree of precision is required in the manufacture to achieve such a good match between measured and simulated results. The predicted and measured 10 dB return loss bandwidths are 8.0 % and 8.2 %, respectively. The slight shift in resonant frequency of 0.3 % can be attributed to uncertainties in the dielectric properties of the materials ($\Delta\varepsilon_r \pm 0.02$) or even slight misalignment of the layers.

The predicted and measured E-plane and H-plane co-polar radiation patterns at 38 GHz are shown in Figure 5-3 and 5-4, respectively. Once again good agreement between theory and experiment is evident from these plots, particularly in the main lobes. The slight discrepancy in the sidelobe level is due to total internal reflection in the dielectric lens. These reflected rays leaving opposite wall and enforce sidelobe levels. It is interesting to note that the measured first sidelobes in both planes are less in magnitude than the predicted values,

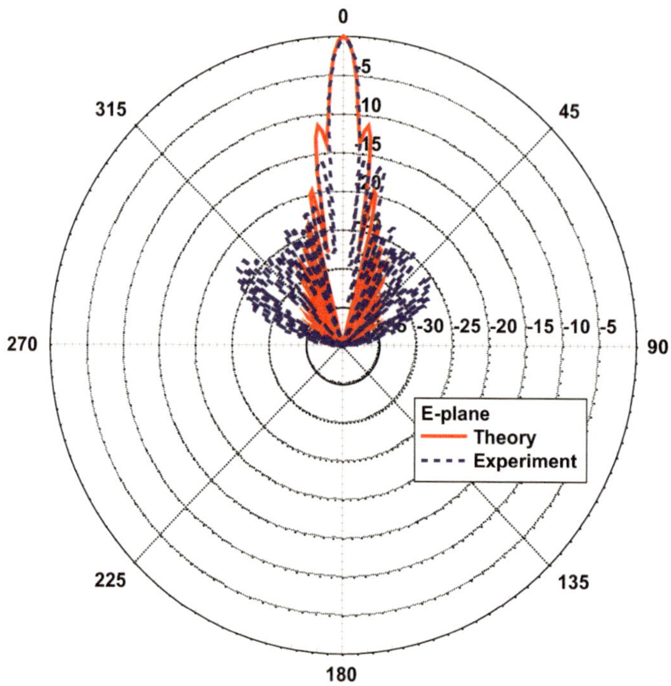

Fig. 5-3 Predicted and measured E-plane pattern at 38 GHz

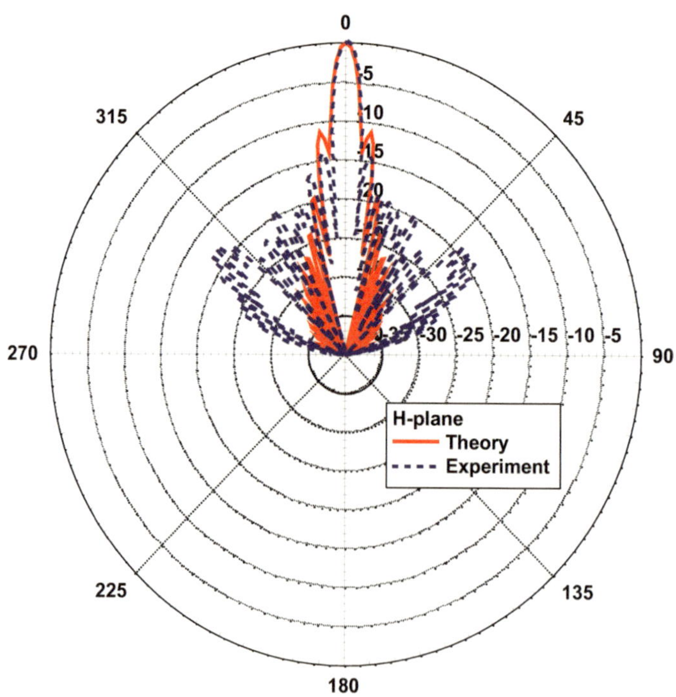

Fig. 5-4 Predicted and measured H-plane pattern at 38 GHz

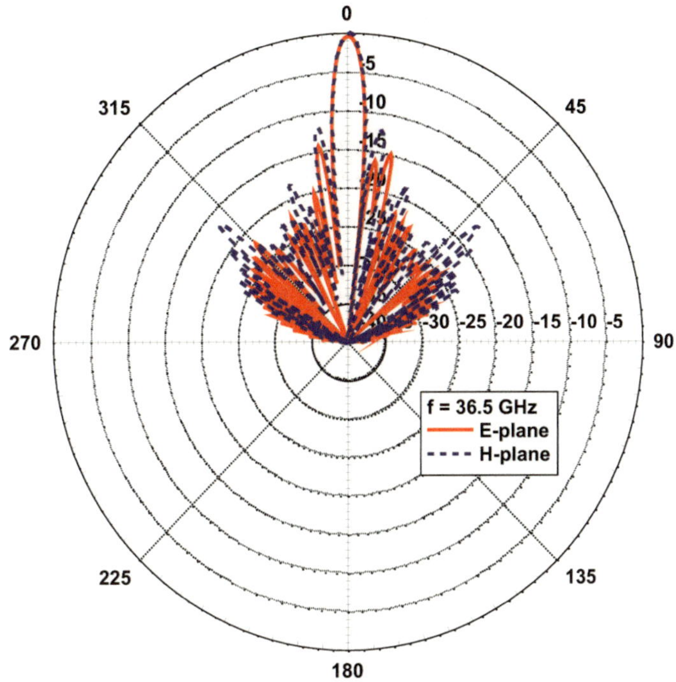

Fig. 5-5 Measured E-plane and H-plane pattern at 36.5 GHz

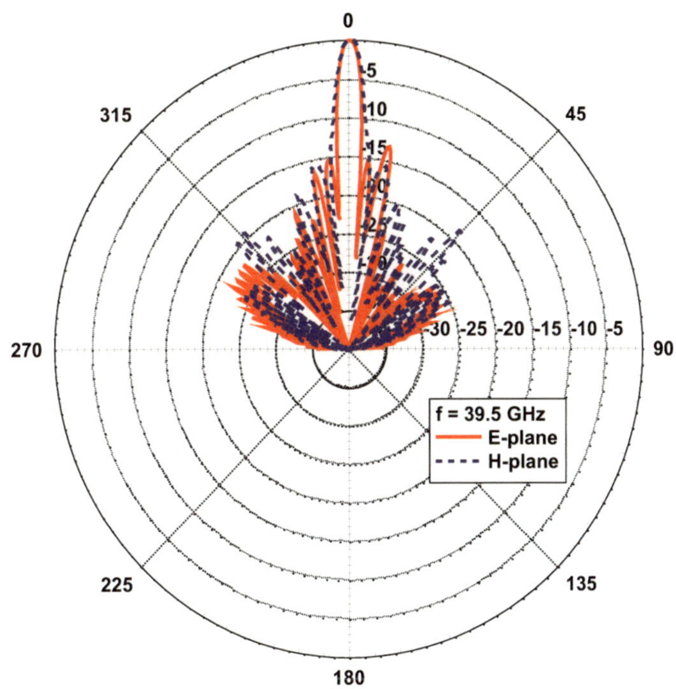

Fig. 5-6 Measured E-plane and H-plane patterns at 39.5 GHz

at the expense of higher order sidelobes. The 3 dB beamwidth is 4.5° in the E-plane and 4.5° in the H-plane. Directivity is computed using numerical integration method and found to be 30.2 dB. The measured cross-polarization levels in both the E- and H-planes were 20 dB below the co-polar levels.

Figure 5-5 and 5-6 show the measured radiation performance of the proximity-coupled lens antenna at the band-edges of the operating frequency range. As can be seen from these plots, the radiation patterns remain relatively constant over this bandwidth. The 3 dB beamwidths in both the E- and H-planes are 4.8° at 36.5 GHz and 4.2° at 39.5 GHz. For the proximity-coupled patch lens-antenna, a slight asymmetry in the E-plane patterns due to spurious radiation from the feed-line, which is not taken into consideration in the numerical model, could be expected. However, as is evident from Figure 5-3, 5-4, 5-5 and 5-6 the measured patterns appear to be quite symmetrical. This indicates that the spurious radiation from these sources is minimal. Once again the cross-polarized levels were greater than 20 dB below the co-polarized fields for both frequencies in both planes. The front-to-back ratio of the antenna, as measured, is greater than 60 dB over this frequency range.

6. Conclusion

This chapter presented analysis into proximity-coupled microstrip antenna on dielectric lens for the characteristics with high gain and pencil beam. In order to achieve this type of dielectric lens we showed theoretical determination of lens geometry for which exiting rays from the lens travel in parallel with optical axis of the lens. The determined geometry is found true ellipsoid lens and verified that the Snell's Law of Refraction is satisfied. Modification of true ellipsoid lens to extended hemispherical lens is presented. We later presented formulation and computation for radiation patterns of extended hemispherical dielectric lens. It comprises the analysis of radiation of patch into lens by cavity model and the analysis of radiation of lens into free space. After design of the proximity-coupled microstrip antenna and the lens, we presented our achievement of good agreements between theoretical and measured results over return loss and radiation patterns.

7. References

Filipovic, D.F.; Gearhart, S.S. and Rebeiz, G.M. (1993). Double-Slot Antennas on Extended Hemispherical and Elliptical Silicon Dielectric Lenses. *IEEE Trans. on Microwave Theory & Techniques,* Vol. 41, No. 10, pp. 1738-1749.

Eleftheriades, G.V.; Brand, Y.; Zurcher, J-F and Mosig, J.R. (1997). ALPSS: A millimeter-wave aperture-coupled patch antenna on a substrate lens. *Electronics Letters,* Vol. 33, No. 3, pp. 169-170, 30th January.

Balanis, C.A. (1997). *Antenna Theory: Analysis and Design,* John Wiley & Sons, Inc.

Collin, R.E. (1966). *Foundations for Microwave Engineering,* McGraw-Hill Inc.

Filipovic, D.F.; Gauthier, G.P., Raman, S. and Rebeiz, G.M. (1997). Off-Axis Properties of Silicon and Quartz Dielectric Lens Antennas. *IEEE Trans. on Antennas and Propagation,* Vol. 45, No. 5, pp. 760-766.

Pozar, D. M. and Voda, S.M. (1987). A Rigorous Analysis of Microstripline Fed Patch Antenna. *IEEE Trans. on Antennas and Propagation,* Vol. 35, No. 12, pp. 1343-1349.

Mall, L. (2008). Improved Formulation for Investigation of Proximity-Coupled Microstrip Antenna in Homogeneous Dielectric Half Space. *IEEE Trans. on Antennas and Propagation,* Vol. 56, No. 4 pp. 927-932

9

Methods to Design Microstrip Antennas for Modern Applications

K. Siakavara
Aristotle University of Thessaloniki
Dept. of Physics, Radiocommunications Laboratory,
54124 Thessalonik,
Greece

1. Introduction

The evolution of modern wireless communications systems has increased dramatically the demand for antennas, capable to be embedded in portable, or not, devices which serve a wireless land mobile or terrestrial-satellite network. With time and requirements, these devices become smaller in size and hence the antennas required for transmit and receive signals have also to be smaller and lightweight. As a matter of fact, microstrip antennas can meet these requirements. As they are lightweight and have low profile it is feasible them to be structured conformally to the mounting hosts. Moreover, they are easy fabricated, have low cost and are easy integrated into arrays or into microwave printed circuits. So, they are attractive choices for the above mentioned type of applications.

For all that, the design of a microstrip antenna is not always an easy problem and the antenna designer is faced with difficulties coming from a) the inherent disadvantages of a printed resonant antenna element, for example the narrow impedance bandwidth, and b) the various requirements of the specific applications, which concern the operation of the radiating element, and can not be satisfied by a printed scheme with an ordinary configuration. For example, it would be demanded, the microstrip element to have gain characteristics that potentially incommensurate to its size or/and frequency bandwidth greater than the element could give, taking into account that it operates as a resonant cavity. Moreover, the rapid development in the field of Land Mobile Telephony as well as in the field of Wireless Local Area Networks(WLANs) demands devices capable to operate in more than one frequency bands. So the design of a printed antenna with intend to conform to multiple communications protocols, for example the IEEE 802.11b/g, in the band of 2.4GHz, and the IEEE 802.11a at 5.3GHz and 5.8GHz, would be a difficult task but at the same time a challenge for the designer. Counting in the above the possibility the device, and so the antenna, to serve terrestrial and also satellite navigation systems the problem of the antenna design is even more complicated.

In this chapter techniques will be analysed, to design microstrip antennas that combine the attributes mentioned above which make them suitable for modern communications applications. Specific examples will be also presented for every case.

2. Bandwidth enhancement and multiband operation

2.1 Bandwidth

Conventional microstrip antennas have a conducting patch printed on a grounded dielectric substrate and operate as resonant cavity elements. This operation leads inherently to narrow impedance bandwidth which is a barrier for microstrip antennas applications in wireless communications. Moreover in many of these applications, as further requirement would be a multi-frequency operation. So, the enhancement of the bandwidth and the achievement of multifrequency operation are major challenges for the antenna designer and many techniques have been proposed for this purpose.

The fundamental definition of the bandwidth of an antenna is the difference between the upper and lower frequencies of operation(f_H and f_L respectively)

$$bw = f_H - f_L \tag{1}$$

For all that, the spectrum managers often use a variety of different bandwidth definitions, including fractional or percent bandwidth. These measures of relative bandwidth require the calculation of a central frequency, which is either the arithmetic or geometric average of the upper and lower frequencies. The center frequency is defined as the arithmetic average of the upper and lower frequencies

$$f_C = \frac{f_H + f_L}{2} \tag{2}$$

An arithmetic average yields the central frequency when frequency is considered on a linear scale. An alternate definition of center frequency involves the geometric average

$$f_C = \sqrt{f_H f_L} \tag{3}$$

The geometric average yields the center frequency when frequency is considered on a logarithmic scale and is less commonly used. So the arithmetic average should be assumed unless otherwise is specified.

The fractional bandwidth of a system is the ratio of the bandwidth to the center frequency(either the geometric or the arithmetic definition is used)

$$BW = \frac{bw}{f_C} \tag{4}$$

Alternatively, fractional bandwidth may be defined on a percentage basis

$$BW\% = \frac{bw}{f_C} 100\% \tag{5}$$

Since the geometric definition of center frequency always yields a frequency smaller than the arithmetic average, fractional bandwidths calculated using the geometric definition are always larger than the arithmetic ones. So, the designer of the antenna has to pay attention on which definition is used. Although specific limit values of BW in order an antenna to be considered as broadband do not exist , criteria to rank an antenna as ultra-wideband have been enacted. In accordance with these criterions an antenna is characterized as UWB if its fractional bandwidth (based on the arithmetic central frequency) exceeds the value of 0.25(Defense Advanced Research Projects Agency_DRPA) or the value of 0.2(Federal Communications Commission_FCC).

2.2 Bandwidth enhancement

The impedance frequency bandwidth of a microstrip antenna depends primarily on both the thickness and the dielectric permittivity of the substrate. A thick substrate with a low dielectric permittivity can increase the bandwidth of the printed patch. Both these selections could be a solution of the problem of bandwidth enhancement if the thickness of the substrate did not a)pose difficulties in integration of the antenna with other microwave circuits, and b)cause some other problems such as the surface wave propagation and the large inductive image part of the input impedance of the antenna, which makes its resonance unfeasible. Thus, a reasonable thickness should be considered in the selection of substrate and the bandwidth would be enhanced using additional techniques. The most common and effective of them, are: a) the loading of the surface of the printed element with slots of appropriate shape b)the texturing of narrow or wide slits at the boundary of the microstrip patch . Other effective techniques used for the enhancement of the bandwidth is the utilization of a)stacked, shorted or not patches, [1]-[3] and b) extra microstrip resonators [4]. The technique of stacked patches is based on the fact that the bandwidth is in general proportional to the antenna volume measured in wavelengths but at the same time a relatively large volume is a disadvantage for many applications. The utilization of additional parasitic patches of different size directly- or gap-coupled to the main patch is an effective method but results to an increased antenna size which would also be undesired. Superior to these methods are the techniques of slot loading or texturing the patches by slits because they ensure the small size and the low profile of the antennas.

2.2.1 Slot loaded patches

The slot loading is made by two ways. By one of them, slots are embedded in the printed patch. Their dimensions and positions are properly selected in order to the first two broadside-radiation modes of the patch be perturbed such that their resonance frequencies get close to each other to form a wide impedance bandwidth. The slots would be of various shapes as, toothbrush(Fig. 1a)[5], double bend (Fig. 1b)[6] , cross (Fig. 1c) [7], or U-shape.

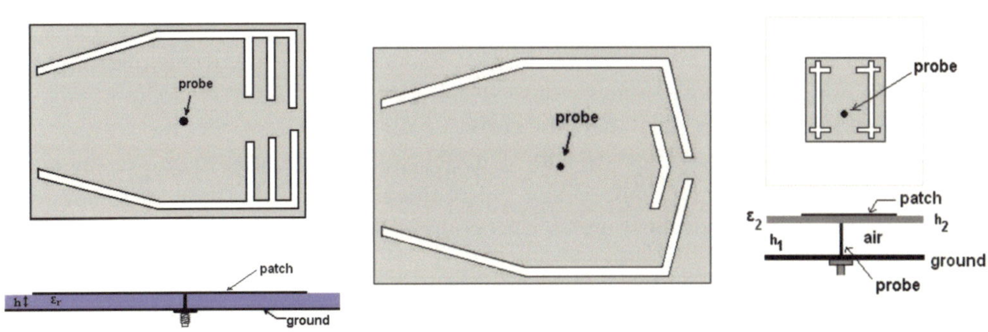

Fig. 1. Microstrip antennas loaded with slots of various shapes : a) 'toothbrush' b) 'double bend c)cross

The microstrip elements etched with a U-slot could be rectangular or triangle patches (Fig. 2)and are perhaps the most popular among the antennas of this type[8]-[15], as they have been proved to be versatile radiating elements: they can be designed not only for wideband applications but also for dual- and triple-band as well as for circular polarization operation. It has been found that the U-slot loaded patch can provide impedance bandwidth in excess of 30% for an air substrate thickness of 0.08λo and in excess of 20% for material substrate of similar thickness [16].

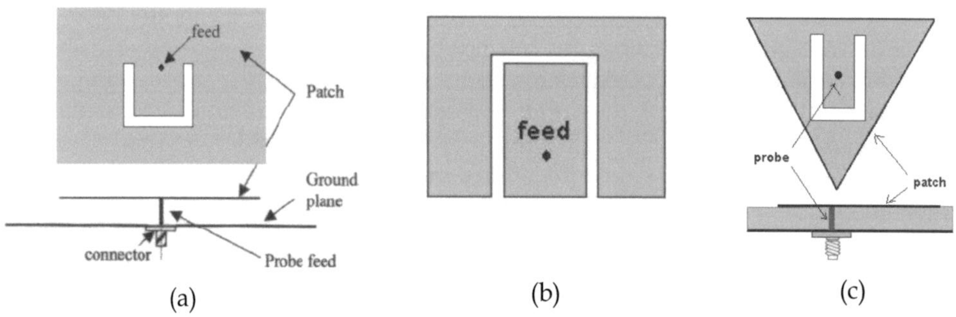

Fig. 2. The geometry of the U-slot patch antenna.

The U-slot is consisted of two parallel vertical rectangular slots and an horizontal rectangular one. The parameters that affect the broadband performance of the patch antenna are the slot length and width and the position of the slot. It is noted that the slot width should be small relative to the slot length and the higher resonant mode is sensitive to the length variation of the horizontal slot, whereas the lower resonant mode strongly depends on the perimeter of the U-slot. Unfortunately mathematical formulas, even empirical, by which the accurate geometry of the structure patch-U-slot could be find in order the antenna to have a pre-specified operation, are not available yet. Empirical formulas introduced in the past[10] are limited to electrically thin substrates which lead to small bandwidth. Thus the antenna designers should adjust the dimensions and the position of the slot by iterative trials or, potentially, by employment of an evolutionary technique of prediction and optimization as in §4 is discussed.

An alternative version of the U-slot patch is the half U-slot patch Fig3a, which maintain an impedance bandwidth similar to that of a full U-slot patch having the further advantage of the smaller size[17]-[18]. This is due to the fact that the electrical current distribution is symmetrical along the line of symmetry of the full U-slot printed element. So, removing half of the patch does not appreciably affect the current paths and , hence, the resonant behavior of the structure. Moreover, a shorting wall (Fig. 3b) can be integrated to reduce the size of the full U-slot [19]-[20] or the half -U-slot patch antenna[17]- [21]. These small-size wide bandwidth designs are particularly suitable for handset devices.

2.2.2 Slit loaded patches

By slit cutting at the boundaries, instead of slot cutting on the surface, of the patch, similar broadband operation can be obtained. The resulting configuration is E-shaped or suchlike and the design can be applied to the antennas with rectangular, circular or triangular patches(Fig. 4)[22]- [24].

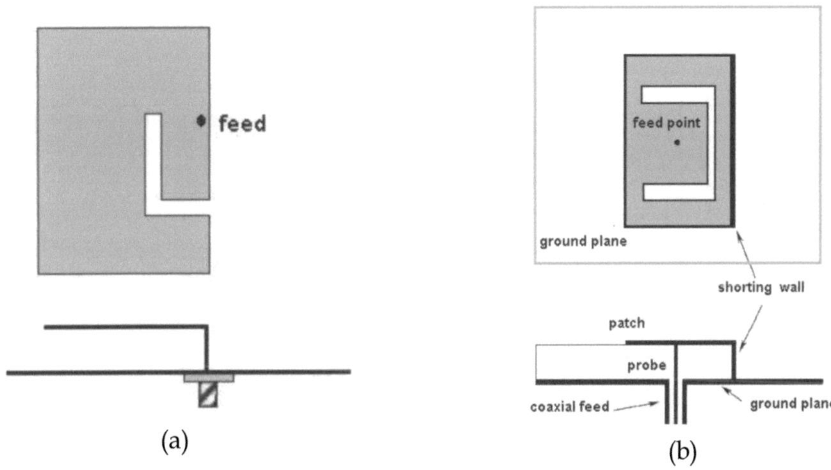

Fig. 3. Patterns of a)Half U-slot loaded patch b) U-slot loaded patch with shorting wall

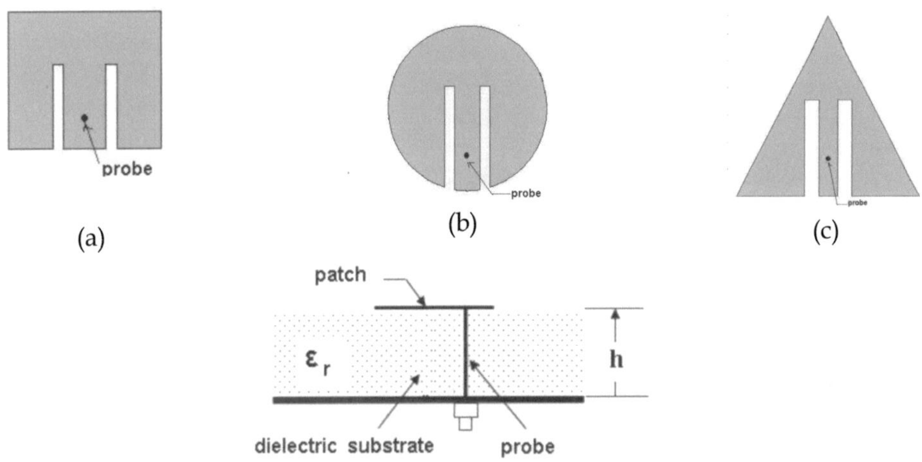

Fig. 4. The geometry of slit loaded microstrip antennas

The wideband performance of the slit loaded patch is based, similarly to the method of slot-loading, on the excitation of more than one adjacent resonant modes. Moreover the presence of the slits in the vicinity of the feeding probe could add a capacitive load at the input impedance of the patch. This capacitive load could effectively contribute to the resonance of the patch because can counteract the inductive part of the probe's input impedance. It is noticed that this inductive part would inevitably be large if a thick substrate is chosen for wideband operation. So, the insertion of slits enhances by two ways the width of the operation band, and it has been reported that bandwidth greater than 25% can be achieved.

The width of the frequency band of the antenna can be controlled by the slits' length and width and the slits' position. The slits divide the patch in three or more parts and at each one corresponds an equivalent circuit of resonance(Fig. 5)[25]-[26]. In order the three regions of resonance to overlap, forming a unified wideband range of operation, the slits could potentially have unequal size.

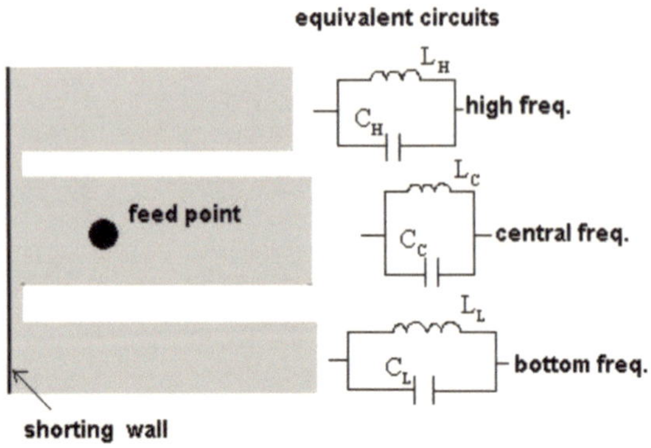

Fig. 5. The structure and the equivalent circuits of E-shaped printed antenna

A lightly different patch with as much bandwidth as the E-patch can be realized by configuring the slits in a way that results in a printed element with meander shape (fig 6) [27].

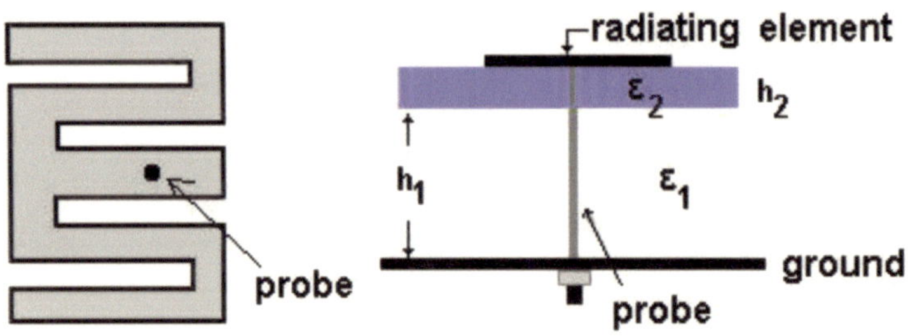

Fig. 6. Top and side view of the meander shaped printed antenna

Another slit loading process to obtain broad band operation is to create several slits at the boundary of the patch. This technique provides the patch with the additional advantage of a size smaller than that of the previous configurations. As an example a patch with five-couple staggered slits is presented in figure 7a [28]. A hybrid feed, inset feed combined with probe, is used for easy impedance matching. A quantitative analysis can prove the possibility to expand the operation bandwidth and reduce the antenna size. When the structure parameters of the antenna are selected properly, two TM_{10} modes can be excited simultaneously. The simple electrical current sketches of the two modes on the patch are shown in Fig. 7b. It can be found that the current paths of the modes overlap at the top portion of the patch whereas they occupy a different structure branch at the bottom portion of the patch.

In addition, the current path lengths are determined by the patch length and the slit depths. The asymmetric slit loading technique leads to different resonance lengths of the two TM_{10} modes. The two modes can be matched by adjusting the depth of the feed slits and the patch length because their input impedances vary with the inset feed position. Moreover due to the slits, the currents of the modes flow on meandering paths that ensure the required length for resonance at the modes' frequencies, whereas the entire length of the patch is smaller compared with that of ordinary rectangular microstrip antennas. Furthermore, the current paths of the TM_{10} modes are along the same direction, thus the radiation characteristics of the modes are similar, in accordance to the radiation theory. Due to the above mechanism of operation, the bandwidth and the size of the antenna could be controlled by the slit number and the geometry of the whole configuration.

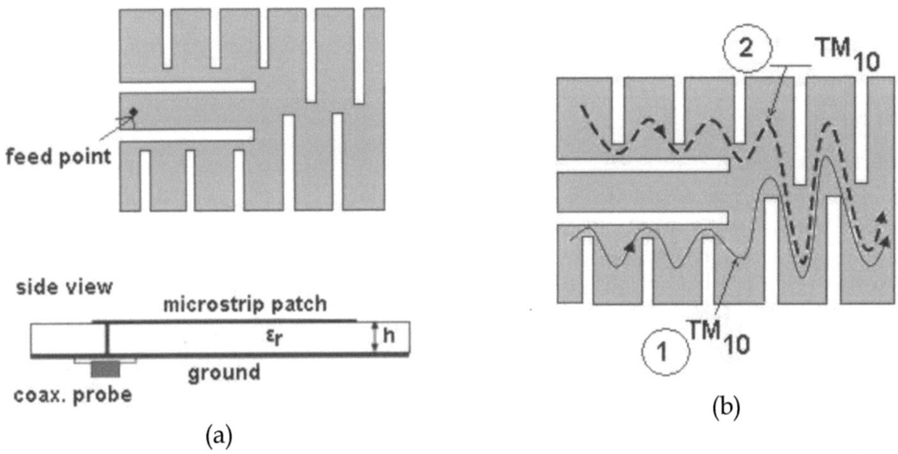

Fig. 7. a)Configuration of multi-slit loaded low profile microstrip antenna b)Current sketches of the two modes on the patch

In all cases of the slit loading, the slits play an important role to control the wideband behavior of the antenna. However to find the proper geometry of the textured patch is usually a difficult process because there are no mathematical formulas, thus, no prediction could be made. The whole process is iterative and is applied by simulations using a high frequency electromagnetic field simulation software. The set up of the process is decided by the designer. Usually the first step of the procedure is to select the dielectric substrate and design the initial unmodified patch. Then slots are etched and their position and geometry are iteratively adjusted to tune the required frequency and the bandwidth. The whole strategy potentially includes steps of simultaneous small variations of the patch dimensions and of the substrate with intend to control the frequency tuning and the bandwidth[24], [27]. Besides this strategy, which is substantially a trial and error one, Artificial Network Algorithms or Optimization techniques as Genetic Algorithms, Particle Swarm Optimization(PSO) or Differential Evolution could be applied(See §4). By these methods all the material and geometrical parameters of the antenna structure are simultaneously taken into account in every step of the design process.

2.2.3 Ultra-Wideband microstrip antennas

There are two criteria available for identifying when an antenna may be considered ultra-wideband(UWB). One definition(by Defense Advanced Research Projects Agency report) requires a UWB to have fractional bandwidth greater than 0.25. An alternate and more recent definition, by Federal Communications Commission(FCC) places the limit at 0.2. Using fractional bandwidth (eq. 4), in mathematical form it is

$$BW = 2\frac{f_H - f_L}{f_H + f_L} = \begin{cases} 0.25 & \text{DARPA} \\ 0.2 & \text{FCC} \end{cases} \tag{6}$$

In the past, different types of small antennas meeting the above requirement have been investigated: resistively loaded pyramidal horn antennas, stacked patch antennas and tapered slot antennas or spiral-shaped antennas. Among the existing designs of printed structures, the ones that provide the required operational bandwidth while maintain sufficient radiation efficiency, simple design structure and low fabrication cost are: a) the spiral shaped antenna b)the tapered slot, especially the antipodal tapered, as the Vivaldi antennas and c) the bowtie shaped antennas.

a. Spiral antennas

Spiral antennas have been the subject of research for several decades and have emerged as leading candidates for applications requiring circularly polarized broadband antennas. Spirals can have a single arm or multiple arms and have been realized in microstrip form as well as in slot form. But spiral antennas have shortcomings as well. They require deeper lossy cavities and special attention to impedance transformer network design. Recently several designs of planar spiral antennas without complicated feeding networks were proposed. They are planar, easy to fabricate and do not need a matching network. The spiral could have annular(fig. 8a) or rectangular(fig. 8b) shape. The microstrip Archimedean annular spiral antenna can be achieved by the traditional Archimedean spiral concept with a dielectric-substrate and a metal ground plane. The spiral configuration could increase the gain of the antenna. However the main lobe splits at the higher frequency band.
The radiation principle of this kind of microstrip configuration is based on the mirror concept. To reach the property of ultra-wide band, the height of the dielectric -slab and the value of the dielectric constant is very important. At the lower part of frequency band, the maximum radiation of the antenna can be found along the axis. By the rising of the frequency the front to back ratio increases and one-sided radiation can be achieved. The direction of the maximum radiation starts to deflect at a specific frequency, depending on the structural characteristics of the spiral. As the frequency rises further the direction deflects to both sides of the axis which is perpendicular to the antenna's surface. The decreasing of the radiation along the axis is explained as follows: The height of the dielectric-slab is related to the central frequency of the antenna, being about a quarter of the wavelength at this frequency, as it is defined inside the dielectric. So, at the higher frequency band the height tends to be one-half of the wavelength at the respective frequencies. In this case, the phase of the original signal, compared with that of the wave reflected by the the ground plane is just the opposite when it reaches the spiral plane . After the superposition of the signals, the radiation at the broadside decreases and the main lobe splits. A modification of the substrate is proposed in this case. A cylinder with a certain radius should be subtracted out from the center of the dielectric so as to improve the gain along the axial direction at higher frequency band. The physical thickness of the substrate in the hollow part corresponds to an electrical length smaller than that in the rest of the dielectric slab. The parameters of the hollow have proper values in order this length to be equal to $\lambda/4$

(a)

(b)

Fig. 8. a) Structural pattern of the microstrip circular spiral printed antenna b)Sketch of the rectangular spiral antenna instead of $\lambda/2$. So, the condition for maximum radiation toward the direction perpendicular to the surface of the antenna is fulfilled. At the operation inside the wide frequency band, the effective region of the antenna at each frequency includes the annuluses of the Archimedean spiral whose perimeters are approximately equal to the respective wavelength. By this modification, the proper radius of the hollow cylinder and the proper dielectric constant of the substrate, we can lead the spiral configuration to a wide band operation, broadside radiation and a gain greater than that of the ordinary spiral structure.

b. Vivaldi antenna

Vivaldi antenna was invented by Gibson in 1979, and has been widely investigated and used in various applications due to its broad bandwidth, low cross polarization and high directivity[30]-[31]. It is a kind of traveling-wave planar antenna, guiding the wave from a narrower slotline to a wider slotline, which varies by exponential rule, where the wave is radiated out from the horn opening. Due to the non-uniform shape of the printed area, different parts are activated at different frequencies. In virtue of this performance the entire scheme can operate in a wide band of frequencies. However it is difficult to select the best parameters and variances to lead the antenna up to the best radiation performance.

The standard model of the Vivaldi antenna is shown in Fig. 9. There are more than ten parameters to be designed. Furthermore, one important part of the antenna is the microstrip-slotline transition feeding structure. This type of feeding is the most commonly used. Microstrip and slotline are deposited in the different sides of the substrate. The incident wave power can be coupled from the microstrip to slotline and then the power can be transmitted to the air through the exponential tapered parts. The impedance match at the coupling point is usually difficult. Three are the parameters that affect the radiation performance of the antenna: the microstrip-slotline transition magnitudes of strip stub, the slotline circle cavity and the gradual rates of the antenna. So, the design of the slotline structure, and the calculation of the antenna' s exponential tapered parts are the basic steps of the design.

The configuration of the exponential tapered part is governed by three parameters: a) the slotline hatch(H) b)the antenna tapered length(L) and c) the exponential gradual (α) , as shown in Fig. 9. According to experience, the higher (F_H) and lower F_L cut-off frequencies determine the slotline hatch. Generally, the widest hatch(H) is equal to $1.3\lambda_{cL}$, where λ_{cL} corresponds to the wavelength of the lower cut-off frequency; the narrowest hatch is equal to $0.2\lambda_{cH}$, where λ_{cH} corresponds to the wavelength of the higher cut-off frequency. The antenna tapered length(L) is equal to 3~5 times the wavelength of the central operation frequency.

The exponential gradual ratio can be determined by the exponential gradual curve

$$y = C_1 e^{\alpha x} + C_2 \qquad (7)$$

Where C_1 and C_2 can be calculated using $C_1 = \dfrac{y_2 - y_1}{e^{\alpha x_2} - e^{\alpha x_1}}$ and $C_2 = \dfrac{y_1 e^{\alpha x_2} - y_2 e^{\alpha x_1}}{e^{\alpha x_2} - e^{\alpha x_1}}$

And (x_1, y_1) and (x_2, y_2) are the begin point and the end point of the exponential gradual curve.

A modified form of the traditional Vivaldi antenna is the dual exponentially tapered antipodal antenna(DETASA)[31]. It is realized by exponentially tapering both the inner and

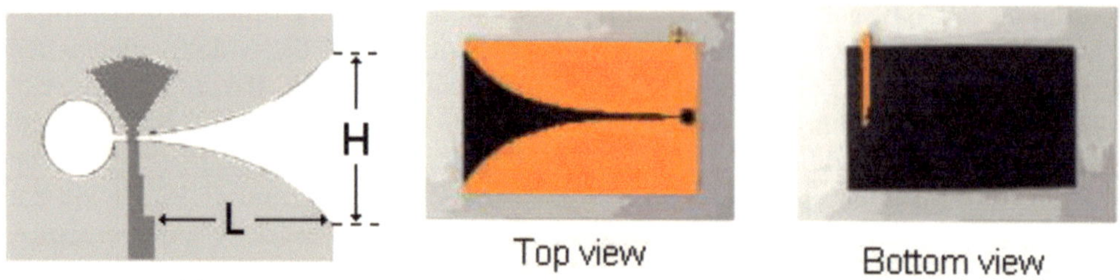

(a) (b)

Fig. 9. Printed Vivaldi antenna a) the model b) the manufactured structure

the outer edges of the radiating flare and placing the mirrored flares on both sides of the substrate(Fig. 10a). It is a slow leaky end-fire traveling wave antenna. The electromagnetic wave travels down the gradually curved paths of the flares. As the separation between the flares increases, the wave becomes progressively weaker and radiates away. A wideband performance can be achieved using DETASA, in virtue of its inherently simple wideband transition from microstrip line to tapered slot flare through parallel strips.
The rule for the design of the radiating flare is described by equations 8a and 8b.

$$x_{inner} = \pm\left[-(c_s + c_w)/2 + c_s \cdot \exp\left(k_s y_{inner}{}^{sft}\right)\right] \tag{8a}$$

$$x_{outer} = \pm\left[-(c_s + c_w)/2 + c_w \cdot \exp\left(k_w y_{outer}{}^{sfo}\right)\right] \tag{8b}$$

Where x_{inner} and x_{outer} are the horizontal distances from the inner and outer exponential edges to the center line of the antenna. y_{inner} and y_{outer} are the vertical distances from the edges to the bottom line of the antenna. The other parameters are obtained by the trial-and error optimization procedure using full-wave electromagnetic simulation or applying optimization techniques as Genetic Algorithms, Particle Swarm Optimization, etc.
In order to further extend the operational bandwidth and he radiation performance, a modification of DETASA is introduced by merging the exponential flare with a circular tapering termination before the edges reach the end of the substrate(Fig. 10b). It creates a longer path for the current flow making the antenna bandwidth greater and smoother. In order to avoid sharp gradient change between the exponential edges and the circular

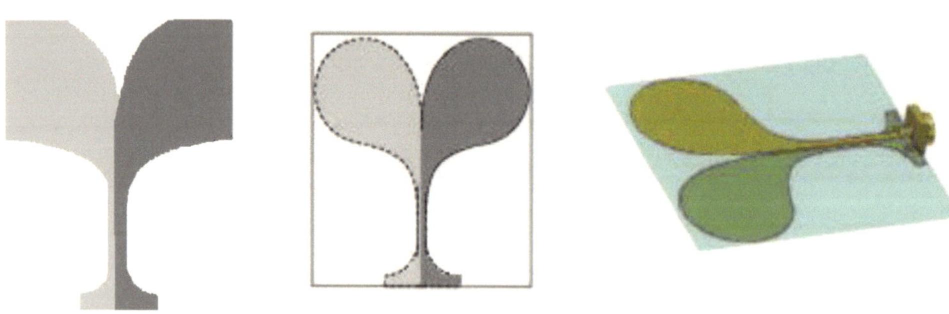

(a) (b)

Fig. 10. Two Configurations of the Dual Exponential Antipodal Vivaldi Antenna

tapering, optimization is performed to search for appropriate tapering parameters along the pre-fixed inner and outer exponential edges and to provide defining parameters for the antenna configuration. Extra spacing is kept between the flares and the edge of the substrate. This additional space helps to reduce ringing of the time-domain impulse response.
With respect to the feeding method of this type of UWB antennas, the transition from the feeding source to the radiation flares is designed by the high frequency transmission line theory. The exact theoretical analysis in detail, and the corresponding realization depend on the Vivaldi antenna configurations(traditional or modified).

c) Bowtie printed antennas

Another patch configuration that drives the antenna system to multifrequency operation is the bowtie shape. Bowtie microstrip antennas have become attractive candidates in the present day communication systems due to their size, that is smaller than the size of a conventional rectangular patch although they have similar characteristics and operation at the same frequency. The fundamental shape of a bowtie microstrip antenna is shown in fig. 11.

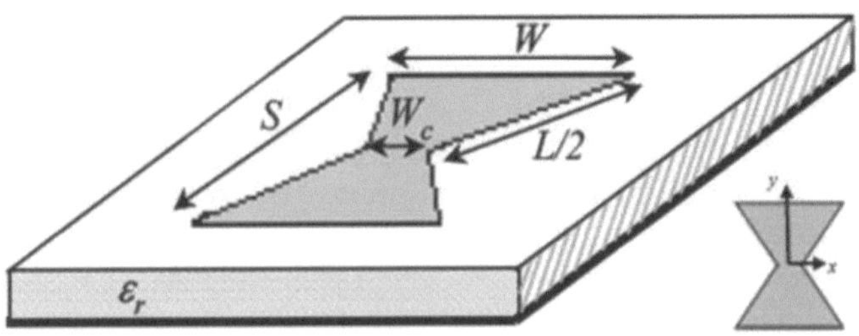

Fig. 11. Geometry of the bow-tie microstrip antenna

The bowtie printed object comes, substantially, from a rectangular patch via modification and the equations for the approximate calculation of the resonance frequency f_r in terms of the geometrical and material parameter values are,[32]

$$f_r = 1.152 \frac{c}{2\sqrt{\varepsilon_{eff}}L} \frac{L}{2} \frac{\left[(W+2\Delta L)+(W_c+2\Delta L)\right]}{(W+2\Delta L)(S+2\Delta L)} \tag{9}$$

Where

$$\Delta L = \frac{0.412h(\varepsilon_{eff}+0.3)\left(\frac{W+W_c}{2h}+0.262\right)}{\left[(\varepsilon_{eff}-0.258)\left(\frac{W+W_c}{2h}+0.813\right)\right]} \tag{10}$$

and

$$\varepsilon_{eff} = \left(\frac{\varepsilon_r+1}{2}\right)+\left(\frac{\varepsilon_r-1}{2}\right)\left(\frac{24h}{W+W_c}+1\right)^{-1/2} \tag{11}$$

By altering the dimensions W_c and L and keeping W constant it is possible to change the fundamental frequency. The equations given above are a good starting point for the antenna design. However for the best impedance matching, the coordinates of the feed point must be found through iterative simulations.
Modified bowtie type antennas have also been proposed fig.12, [33]-[34]. By the appropriate values of the geometrical parameters a bandwidth value that exceeds 90% can be obtained whereas the antenna exhibits small size.

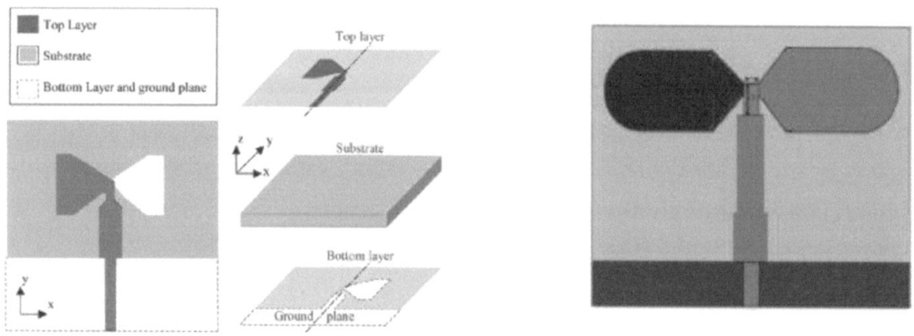

(a) (b)

Fig. 12. Two configurations of the double sided wideband bow-tie antenna

The antenna of fig 12a consists of two identical printed bows, one on the top and one on the bottom of the substrate material. The top and the bottom bows are connected to the microstrip feedline and the ground plane through a stub and mitered transition to match the bow-tie with the selected characteristic impedance of the feedline.
By further modification of the bowtie antenna, namely by rounding of the bow patches, as shown in fig. 12b, wider bandwidth, higher co-polarization and lower cross-polarization for the UWB range can be achieved[34].

2.3 Multiband microstrip antennas

The rapid advances in the wireless communication industry demand novel antenna designs that could be used in more than one frequency bands and that will allow size reduction. For example mobile telephony's services require portable devices compatible with GSM900/DCS1800/UMTS2000 technology and the same equipment should also connect the users to WLAN networks based on 802.11 standards (2.5GH/5GHz). So, the design of small antennas suitable for these devices is of great interest. Many techniques have been proposed for the design of radiating elements of this type, the great majority of which are microstrip antennas. The common characteristic, of almost the total, of the multiband printed elements is that they usually come from an initial patch of ordinary shape which in the following is perturbed. On the basis of the way of the shape perturbation, the multiband microstrip antennas would be classified in categories a) printed elements with incorporated slits or slots. [35]-[40] b) patches with more than one radiating elements conductively connected or/and inductively coupled[41] c) patches of specific shape as conductively connected cross dipoles[42], the bowtie[46], and the spiral [47]-[48] d) more than one stacked patches[49]. A separate category includes the fractal microstrip antennas which come from an initial simple printed element and are developed by a recursive process(§3.5).

At all the aforementioned categories the patches are produced starting from basic theoretical concepts and the procedures of development focus to similar targets. However, it has to be noticed that in no one of them a rigorous design process exists, namely a process which would lead the designer to begin from the same initial concept and, aiming at specific attributes of operation, to end in all cases in the same shape of patch. So, the design of a multi-frequency patch is substantially an art and the ways of texturing the patch' s surface as well as the techniques, used for optimization of the configuration, are just useful tools.
The designing of a multi-band printed antenna requires that the efficient operational features namely, gain greater than 0dB, almost uniform space distribution of the radiated power, if necessary, circular polarization and mainly small reflection coefficient at the feeding port must be ensured in all the frequency bands of operation. Furthermore all these properties must not be obtained at the cost of a complex feeding network, a non compact fabrication or an antenna arrangement of large size. Therefore the design of this type of antennas is a difficult task and beyond the used standard techniques any novel concept or modification of the existing methods would be useful and would lead to interesting antenna schemes.

2.3.1 Slot loaded multiband microstrip antennas

The technique of slot cutting the surface of the printed antenna, besides the broadening of the bandwidth has been proved effective in driving the patch to multi-frequency operation [35]-[40]. Various slot shapes have been proposed for the texturing of the patch; some indicative results are presented in figures 13 to 16. Cutting a rectangular patch with L-shaped(Fig. 13a) or folded slit(fig. 13b) the antenna is considered to consist of two connected resonators of different sizes. The shorting pins at the corner of the patch permit to reduce its size, making this type of compact dual-band antenna suitable for applications in handset mobile communication units. The dimensions of the larger and the smaller sub-patches(Fig. 13a) can be designed to roughly resonate as quarter-wavelength structures at the pre-specified frequencies. In the alternative configuration(Fig. 13b), the smaller sub-patch, that resonates at the higher frequency, starts from the feed point and extends into the center portion of the rectangular patch. In this way it is encircled by the slit and is surrounded by the outer larger sub-patch which resonates at the lower frequency.
Instead of using an L-shaped slit or a folded slit to obtain two separate sub-patches, an embedded U-slot can be used(Fig. 13c). In this configuration the smaller rectangular, of dimensions $L_2 \times W_2$, resonates at the higher of the pre-specified frequencies and occupies the central portion of the original rectangular patch of dimensions $L_1 \times W_1$ which resonates at the lower frequency. The lower f_L and the upper f_H operating frequencies of this design can be approximately determined from

$$f_L \cong \frac{c}{4(L_1 + W_1)} \tag{12}$$

$$f_H \cong \frac{c}{4(L_2 + W_2)} \tag{13}$$

Where c is the speed of light in free space

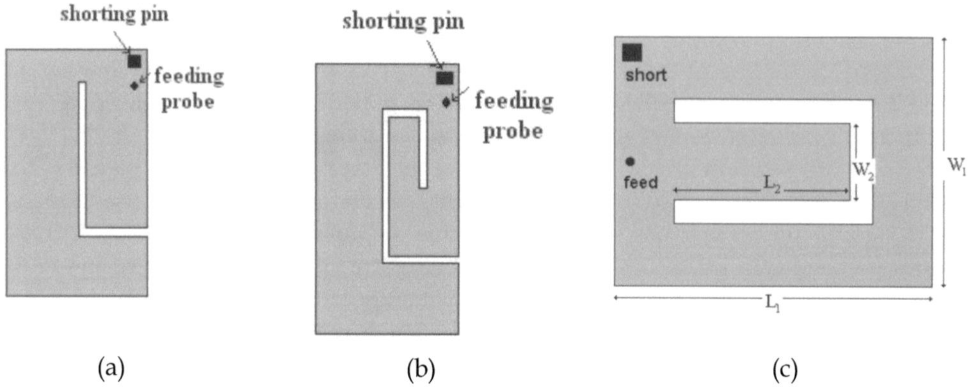

Fig. 13. Geometries of shorted rectangular patch antenna with a) an L-shaped slit b) a folded slit and c) U-slot, for dual frequency operation

In figures 14a and 14b alternate types of perturbation of the patch's surface are shown[38]. The non radiating edges (with respect to the basic TM_{01} mode)of the initial rectangular patch are modified by T-shaped notches in such a way that the current distribution of TM_{01} mode is similar to the current of the higher order TM_{21} mode (fig. 14b) driving the patch to similar operational characteristics at both frequencies. A further modification of the patch by two pairs of smaller T-notches with proper size results in triple frequency band operation(Fig.14c).

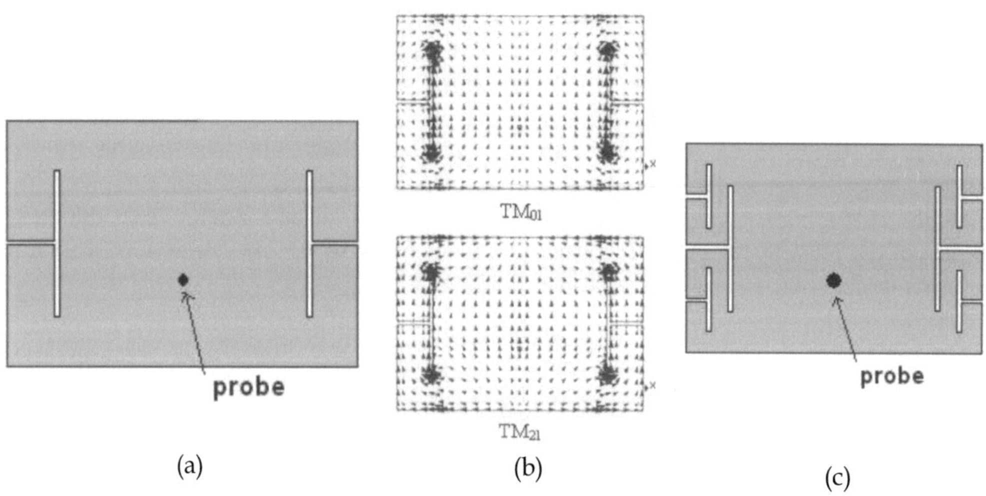

Fig. 14. (a), (c)Microstrip elements textured with single or multiple T-notches. (b) surface current density at TM_{01} and TM_{21} modes.

The concept of texturing the surface of the patch by U-slots that can effectively lead to a broadband operation can also be efficient in driving the element in multiband operation. Two U-slots [15],of different size (fig. 15a) or unequal slits combined with tiny circular slots (Fig15b)can ensure triple band operation[39].

Another configuration of microstrip antenna, loaded with slits is proposed in[40]. The patch (fig. 16a) has annular ring shape, is etched on a two layered dielectric substrate and is

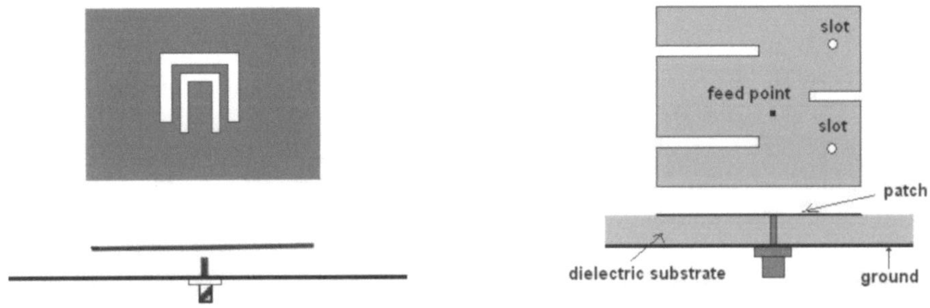

(a) (b)

Fig. 15. Configurations of tri-band microstrip antennas a) dually U-slot loaded patch b)patch loaded with unequal slits and tiny circular slots

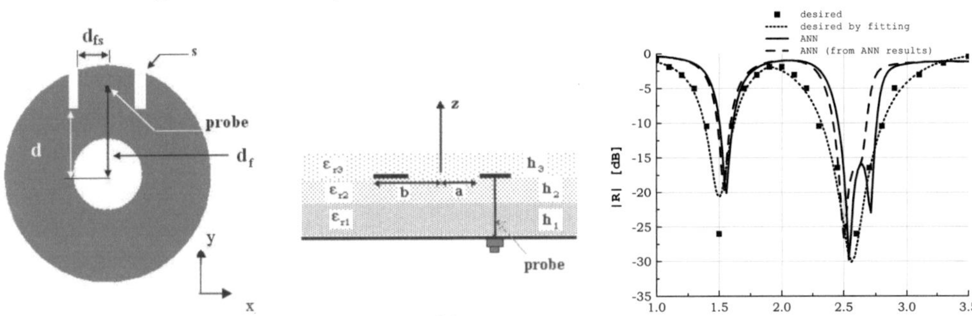

(a) (b) (c)

Fig. 16. (a), (b) Top and side view of a dual-band microstrip structure with multilayered substrate. (c) Reflection coefficient of the signal at the feeding point for a specific set of geometrical and material parameter values.

covered by a dielectric superstrate (fig. 16b). The inherent advantage of a ring antenna is, the property to resonate for a diameter less than $\lambda g/2$ (λg is the guiding wavelength of the equivalent linear microstrip line having width equal to that of the ring). This attribute means a physical size smaller than the size of the respective circular disc resonating at the same frequency. In the proposed configuration the total height of the substrate is high and ensures the broadness of frequency bands(fig. 16c). Additionally the slits in the vicinity of the position of the probe compensate efficiently the large inductive input impedance that inevitably comes from the thickness of the substrate. Due to the large number of the structural parameters of such an antenna an ordinary procedure, for example iterations with gradual variations of the values of the parameters, could be laborious and perhaps ineffective or unfeasible. So, a stochastic technique, which following an objective process , would yield a suitable antenna configuration.

The antenna of fig. 16a was designed using an Artificial Neural Network(ANN) algorithm(see §4). The values of the structural parameters, as shown in figures 16a and 16b, were calculated by a properly designed and trained ANN at the input of which, the plot (sampled) of the scattering parameter of the signal at the feeding point was given. By this plot an antenna capable to resonate at 1.572GHz(GPS)and at 2.45GHz(WLAN based on protocol 802.11) was demanded from the ANN. The antenna implemented by the results of the output of the ANN, exhibits operation that meet these requirements, as shown in fig. 16c.

2.3.2 Multiple patches

An alternative approach to the design of a multiband printed antenna is to use more than one elements with different size that resonate at different frequencies. An indicative structure [41] is shown in fig 17a. In detail, the antenna is an array of concentric microstrip annular rings printed on a double dielectric layer. The multi-frequency performance comes from the resonances of the individual rings and from the further modification made to the rings. The entire configuration has two inherent advantages a) each annular ring resonates having diameter less than $\lambda_g/2$, as explained for the antenna of fig 16a, b) a ring shaped printed element allows other rings of smaller radius , resonating at different frequencies, to be printed in the internal area, ensuring multi-frequency operation and at the same time compactness.

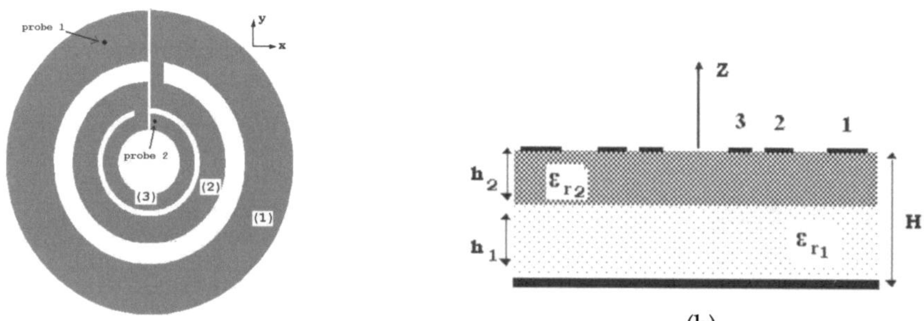

Fig. 17. Top and side-view of a multi-frequency microstrip antenna with three conductively connected rings

To find the frequencies of resonance of each ring it is necessary to solve the electromagnetic problem of printed annular ring antennas fed by probes. The basic and rigorous method for this solution, in spatial domain, is to determine the respective Green function [43]-[44]. The final form of this Green function, expanded in series of orthonormal eigenfunctions, is:

$$
\begin{aligned}
G(\rho,\varphi\,|\,\rho',\varphi') = \sum_n\sum_m \Big[\sqrt{\frac{1}{\pi C_{Jn}}}J_n(\sqrt{\lambda_{mn}}\rho)+\sqrt{\frac{1}{\pi C_{Yn}}}Y_n(\sqrt{\lambda_{mn}}\rho)\Big]\Big[\sqrt{\frac{1}{\pi C_{Jn}}}J_n(\sqrt{\lambda_{mn}}\rho)+ \\
+\sqrt{\frac{1}{\pi C_{Yn}}}Y_n(\sqrt{\lambda_{mn}}\rho')\Big]\cdot\frac{\cos(n\varphi)\cos(n\varphi')}{\lambda-\lambda_{mn}}
\end{aligned}
\qquad (14)
$$

where ρ',φ' are the coordinates of the feeding probe, a and b are the inner and outer radius of the ring, λ_{mn} is the mn^{th} eigenvalue , J_n, Y_n are Bessel functions of the first and second kind respectively and C_{Jn} and C_{Yn} are the constants produced via the ortho-normalization process of the eigenfunctions

$$
\begin{Bmatrix} C_{Jn} \\ C_{Yn} \end{Bmatrix} = \frac{b^2}{2}\Big[\begin{Bmatrix} J_n' \\ Y_n' \end{Bmatrix}(\sqrt{\lambda_{mn}}b)\Big]^2 - \frac{a^2}{2}\Big[\begin{Bmatrix} J_n' \\ Y_n' \end{Bmatrix}(\sqrt{\lambda_{mn}}a)\Big]^2 \qquad (15)
$$

In order the orthonormal eigenfunctions to satisfy the electromagnetic boundary conditions of the structure, the following equation must be satisfied

$$J_n'(\sqrt{\lambda_{mn}}a)Y_n'(\sqrt{\lambda_{mn}}b) + Y_n'(\sqrt{\lambda_{mn}}a)J_n'(\sqrt{\lambda_{mn}}b) = 0 \qquad (16)$$

Equation 16 is the characteristic equation of the problem and its solution gives the frequencies at which the ring resonates, that is

$$f_{mn} = \frac{\sqrt{\lambda_{mn}}}{2\pi\sqrt{\mu\varepsilon}} \qquad (17)$$

The aforementioned formulation can be used to solve the inverse problem, namely to find the geometrical and material parameters of the ring antenna in order to resonate at the desired frequencies.

On the basis of the above theory, an antenna configuration with three rings was implemented in [41]. The rings were printed on the top side of a dielectric layer of 3.17mm height(h_2) and dielectric constant 2.15(RT/Duroid 5880). The layer was suspended at a distance 4mm over the metallic ground plane, namely $\varepsilon_{r1} = 1$ and $h_1 = 4\text{mm}$. The frequencies of resonance of each one of the rings were calculated applying Eq. 17 and using as ε_r the effective dielectric constant of the corresponding suspended microstrip line[45]

$$\sqrt{\varepsilon_{re}} = [1 + \frac{h_2}{h_1}(c_1 - c_2 \ln\frac{w}{h_1})(\frac{1}{\sqrt{\varepsilon_{r2}}} - 1)]^{-1} \qquad (18)$$

where $c_1 = (0.8621 - 0.125\ln\frac{h_2}{h_1})^4$, $c_2 = (0.4986 - 0.1397\ln\frac{h_2}{h_1})^4$

The theoretically calculated frequencies of resonance are the frequencies at which each one of the rings would resonate without the presence of any other element. Therefore, they are just an approximate assessment of the expected resonances of the entire structure, due to the mutual coupling between the rings and the additional modification made to the construction. The modification was necessary because the input impedance of each ring was large and caused strong reflection of the feeding signal. In order to reduce the input impedance, narrow slits were created and the rings were conductively connected (Fig. 17a). As a consequence the entire printed structure has the shape of a spiral. The width of the slits and those of the printed interconnections of the rings, were found via repeated simulations, the object of which was the achievement of better feed matching. This type of antenna would also be designed using the techniques either of ANN or GA and PSO.

With respect to the feeding, two feeding probes were used and their positions were found with the criterion of feed matching. The employment of two probes was necessary because a single one could not excite all the theoretically predicted modes. Therefore at some frequency ranges (e.g. at 1.39-1.5GHz and at 2.82-2.83GHz) both probes should be activated while at the other frequencies the antenna should operate via one of them.

In Fig.18 the measured and simulated scattering coefficients at the input of both probes are depicted. Due to the aforementioned reasons a difference between the theoretically predicted resonate frequencies and the measured ones, appears (see Table 1). Moreover, frequencies of operation smaller 1GHz exist. Their appearance is due to the fact that the path of the current along the spiral is larger than the perimeter of even the larger of the rings. These frequencies would not be predicted by the theoretical analysis, not even

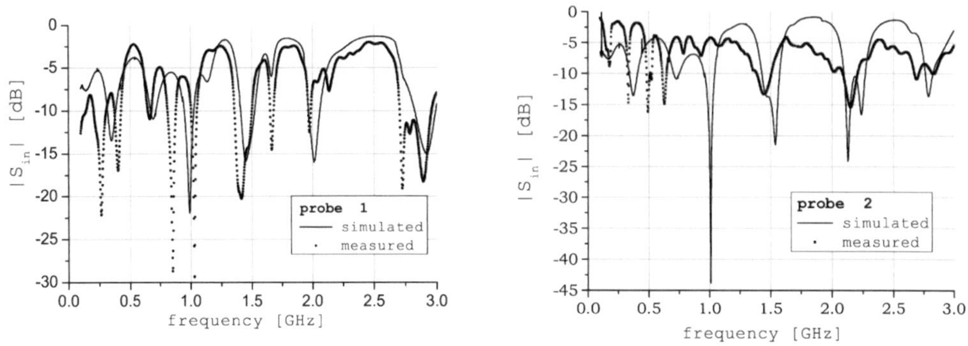

Fig. 18. Reflection coefficient of the signal at the feeding probes of the antenna shown in fig. 17

approximately, because in accordance to equations 16 and 17 the lower resonant frequency of all three rings is the basic resonant frequency of the larger one and it was found equal to 1.26GHz, as shown in Table 1. The results obtained via simulation approximate the respective measurements, in most of the frequency bands.

Table 1 includes the theoretical results for the frequencies of resonance and the respective ones found experimentally.

Theoretical results		Measured values [GHz]
Frequency[GHz]	mode	
1.26	n=0 m=1 (ring 1)	1.005-1.05 (probe1) (probe1) 1.389-1.506 (probe2)
1.7	n=0 m=1(ring 2)	1.653-1.682 (probe2)
1.785	n=0 m=2(ring 1)	1.967-1.987 (probe1)
2.18	n=0 m=1(ring 3) n=0 m=3(ring 1)	2.087-2.208 (probe2)
2.83	n=0 m=5(ring 1)	(probe2)
2.94	n=0 m=3(ring 2)	(probe2)
3.06	n=0 m=2(ring 3)	2.7-2.975 (probe1)

Table 1.

Due to the annular ring shape of the elements of the antenna, the components of the radiated field do not differ by more than 5dB on both main planes, in all frequency bands. Moreover the utilization of two probes offer the potential to increase the level of the gain applying a proper phase shift between the probes. Indicative results of the field pattern at 2GHz are presented in fig. 19.

2.3.3 Bowtie patches loaded with slots

The basic principle of creating slots on the surface of a printed antenna in order to drive it in multi-frequency operation can be applied to bowtie patches, as well. An indicative configuration(see Fig. 20) is proposed in [46]. The first step of this design is to choose the appropriate dimensions of the arms of the bowtie in order to ensure the first band of frequency. The calculations can be made either using equations 9 to 11 or equations for the design a triangular patch antenna.

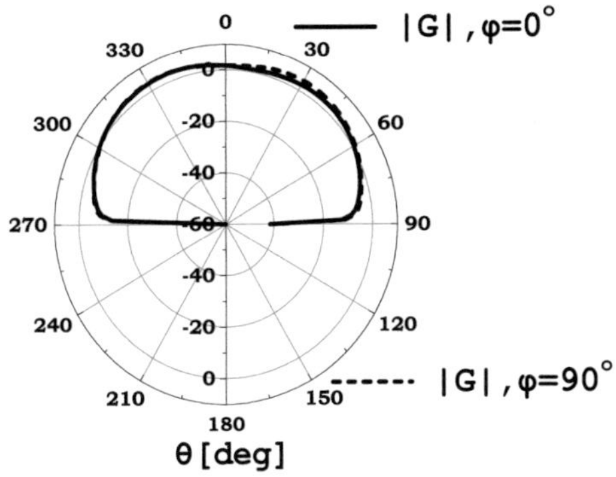

(a)

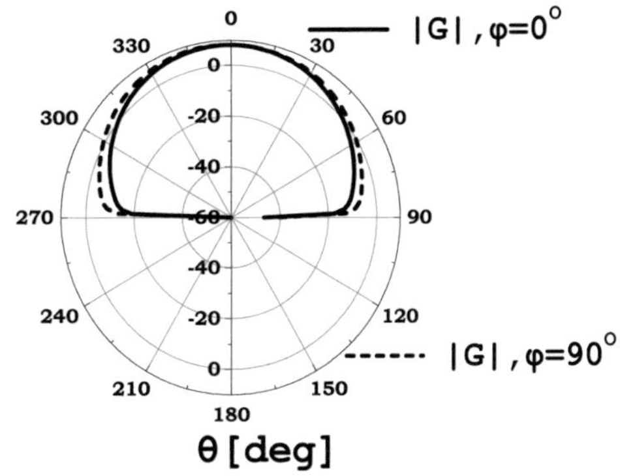

(b)

Fig. 19. Gain pattern at 2.0 GHz of the antenna of fig. 17: a) probe 1 is fed, and b) probes are fed by 180° out of phase.

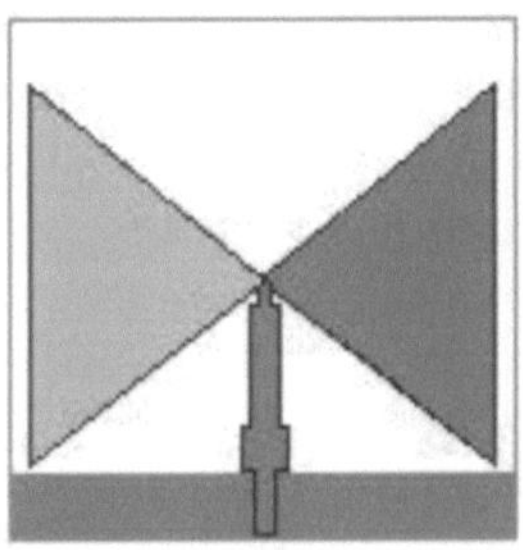

(a)

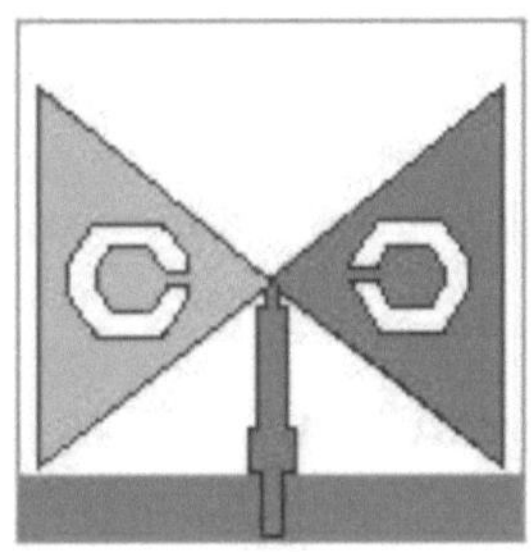

(b)

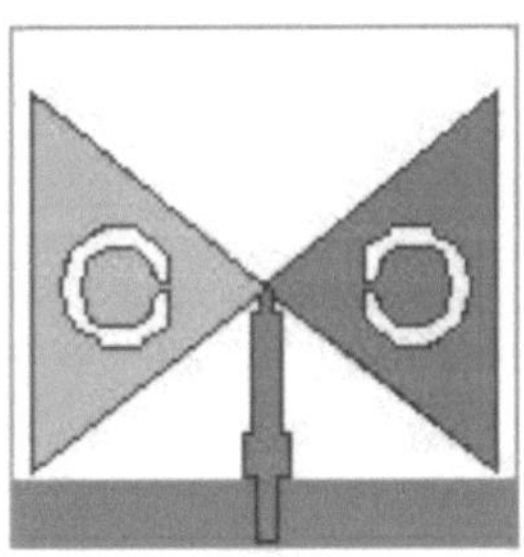

(c)

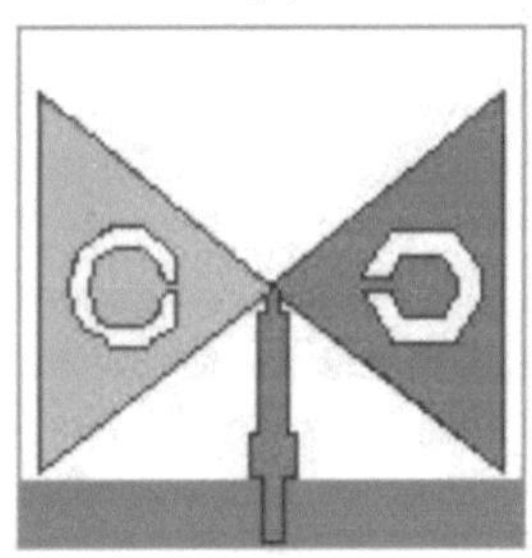

(d)

Fig. 20. Bowtie microstrip antennas: a) the ordinary configuration, (b),(c),(d)modified patches, by loading with slots of various shapes, for multiband operation.

The resonant frequency corresponding to the various modes TM_{mn} of the equilateral triangular microstrip antenna is [45]

$$f_r = \frac{2c}{3\alpha\sqrt{\varepsilon_r}}\sqrt{m^2 + mn + n^2} \tag{19}$$

Where c is the velocity of light in free space, α is the side length of the bow-tie patch and ε_r is the dielectric constant of the substrate.

The above equation is valid when the triangular resonator is enclosed by a perfect magnetic wall. In the case it is not valid the replacement of side length α by an effective value α_e has been suggested. An approximate expression for α_e produced by curve fitting experimental and theoretical results for the resonant frequency for TM_{10} mode is given [45], by

$$\alpha_e = \alpha[1 + 2.199\frac{h}{\alpha} - 12.853\frac{h}{\alpha\sqrt{\varepsilon_r}} + 16.436\frac{h}{\alpha\varepsilon_r} + 6.182(\frac{h}{\alpha})^2 - 9.802\frac{1}{\sqrt{\varepsilon_r}}(\frac{h}{\alpha})^2] \quad (20)$$

where h is the thickness of the substrate.

In the second step of the design process circular or six sided polygon slots are incorporated to the bow patches. This texturing drives the antenna to dual-frequency operation: the initial frequency, obtained by the first step of design, and a second desired one that can be obtained by properly choosing the size of the slots (Figs. 20b and 20c). A tri-band performance can be achieved cutting, in the two parts of the antenna, slots either of different size or of different shape, as shown in Fig. 20d.

In all configurations implemented in [46], the printed elements are fabricated one on the bottom and the other on the top side of the dielectric substrate. The feeding method depends on the selected bow-tie configuration and the parameter values of the feeding network are calculated by the high frequency transmission microstrip line theory

2.3.4 Spiral antennas

Multi-frequency operation can also be achieved using spiral printed antennas. A suitable modification of the ordinary shapes of the printed spirals, especially of the rectangularly shaped, has been proved to be effective to multi-band performance. Three indicative proposed structures are shown in figure 21. The design of all three is based on a modified spiral line - technique model and on the use of shorting -pin method[47]-[48]. An advantage of these antennas is that they have the ability to resonate in frequencies much smaller than the other multiband anennas can, having at the same time very small size. The key for this performance is the increased length of the current path along the conductor area due to the spiral shape of this conductor. For the structures of figures 21a and 21b, two ways of folding the conductor in order to maximize its length were used. One is the standard rectangular folding and the other is the folding the conductor back onto itself in the two dimensions. Each configuration has three dependent variables for a constant antenna area, namely the number of bends in the conductor, the conductor width and the gap between the conductors. Attention has to be drawn to the thickness of the conductor because it has a lower limit, set by the diameter of the probe and the shorting pins. Using these configurations, in [47], dual band operation was obtained at frequencies being both smaller than 1GHz.

A multi band operation can also be obtained by the configuration of fig. 21c. In this case a small number of bends for the spiral shape are used and the corners are truncated to allow smooth current flow. The frequencies of resonance and the corresponding bandwidths as well as the matching, are controlled by the line's width and length, the gaps between the lines, the position of the shorting pin and the feed point. The utilization of a dual slab substrate, the lower slab being of air or foam, may enhance the bandwidth. This spiral configuration due to the relatively small length of the conductor is recommended for high frequency applications. In [48] by proper selection of the parameter values a dual band operation was achieved. One from 2.37GHz to 2.47GHz and another, of large width, from 5GHz to 6.3 GHz.

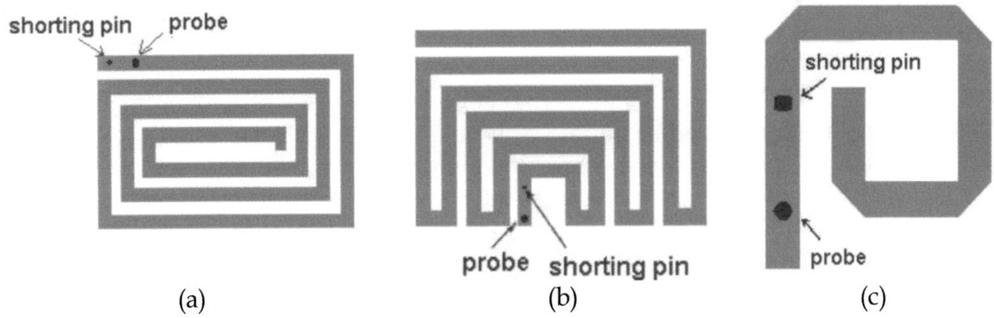

Fig. 21. Microstrip spiral configurations a) the ordinary shape b)the conductor is folded back onto itself c) small number of bends and truncated corners.

2.3.5 Fractal antennas

Fractal technique has been applied to many areas of science including fractal electrodynamics in which fractal concepts are combined with electromagnetic theory for the purpose of investigating a new class of radiation, propagation and scattering problems. Thus, fractal antennas are an innovative type of radiating elements coming from the inspired combination of these two theories.

The fractal antenna engineering focuses in two promising research areas : the first deals with the analysis and design of fractal radiating elements and the second, concerns to the application of the fractal theory to the design of antenna arrays[50]-[54]. Both antenna types have attributes, highly desirable in military as well as in commercial sectors. The majority of fractal antenna elements, are printed configurations, have compact size, low profile and cost, multi-band operation, easy feeding and, potentially, their operation could be optimized by suitable modification of their shape. On the other hand the application of the fractal technique to the design of an antenna array can produce radiating systems of large size, and as a consequence of high gain, with frequency-independent or multi-band operation and radiation patterns with low side-lobe level. Moreover, the elements of the array can be fed one by one, thus operating as Direct Radiating Antennas(DRAs) and can function as phased arrays[55]-[57].

Fractal objects can be classified in two categories: deterministic and random. Deterministic are those that are generated of several scaled-and rotated copies of themselves using a recursive algorithm. Random fractals also contain elements of randomness that allow simulation of natural phenomena. Procedures and algorithms for generating fractals both deterministic and random have been found. However most of the proposed fractal antennas have been designed with a deterministic fractal process.

The fractal technique is based on the idea of realization the operational characteristics of the antenna by repeating an initial radiating structure in arbitrary or regular scales. The self similarity properties of the produced antenna configurations are translated into their electromagnetic behavior and since the initial antenna was introduced, the higher stages of fractaly developed geometries would have, for example, the attribute of multi-frequency operation. The basic scheme of a fractaly designed antenna is a generating, termed also initiating, radiating element. In particular, the entire antenna can be formed recursively through repetitive application of the generating element under a specified scaling factor which is one of the parameters of the problem. This process is realized following potentially two different strategies: By one of them, the repetition of the generating antenna is made in such a way that its entire size gets larger from stage to stage of fractal building. By the second strategy, the entire area, which the final antenna is permitted to occupy, is defined a priori. Then, by the process of the proper repetition of the generator, the available area is filled by scaled replicas of the generator.

This space filling leads to antenna configurations with electrically large lengths although they have small size thus, constituting substantially a miniaturization technique producing elements efficiently packed into small areas, therefore suitable for installation in portable telecommunication devices.

The theoretical analysis as well as the interpretation of the operation of a microstrip fractal antenna is based on the underlying relation between a fractal resonator and a microstrip antenna. The physical problem of the operation of fractal resonators is often referred to, in the literature, as the determination of the 'fractal drum' vibration modes. The simplest example of a surface fractal resonator is the fractal drum, where a vibrating membrane is bounded by a fractal curve, its vibration modes named 'fractions'. The vibration modes of fractal drums exhibit some interesting properties such as the existence of localized modes. In these modes the vibration is strongly localized in certain parts of the membrane. These vibrational states are obtained after the solution of the Helmholtz equation with the appropriate boundary condition. For surface fractals, a distinction is made between Neumann and Dirichlet fractions according to the boundary condition that is applied.

On the other hand microstrip antennas can be modeled, in a first approximation, as a cavity. For a microstrip patch antenna of electrically small height of substrate, the field distribution can be found with very good accuracy from the eigenfuntions of the Helmholtz equation subject to the Neumann boundary condition. Therefore, it is expected that fractal boundary microstrip antennas will exhibit vibration modes similar to those ones of the fractal drum.

The most popular fractal geometries, referred in the literature, which were proved efficient in antenna design are the Koch fractal, the Sierpinski gasket or carpet fractal, the Hilbert fractal, the Minkowski and the Square Curve fractals. All of them were proved to yield printed antennas, planar or conformal, with multi-frequency performance as well as easy feeding, satisfactory polarization properties and gain, whereas they have small size. All these attributes make them an attractive choice for portable telecommunication equipment.

2.3.5 a) Koch: direct and inverse fractal islands

The Koch fractal microstrip patches are commonly used in virtue of their attractive properties: they have small size, a single feeding port is enough and their higher order modes result in directive radiation patterns[58]-[60]. The basic geometry of the Koch island is obtained by replacing the sides of an equilateral triangle by a Koch curve. The triangle is the generator and the higher stages of fractal development follow the respective stages of the Koch fractal curve, as shown in fig. 22.

The Koch fractal curve is constructed as follows: A straight line of length α, is the generator and is first divided into three equal segments. The middle segment is removed and replaced by two segments having the same length to generate an equilateral triangle. This 4-sides line is the first fractal stage, its direct length from one end to the other is equal the length of the generator but the summation of the lengths of all four segments is greater and is calculated by the general expression $L_n = \alpha\left(\frac{4}{3}\right)^n$, where n is the number of iteration. It is noticed that to use the Koch curve in order to build the Koch island, the length of the generator should be equal to the side of the triangle. In the next stage the process is repeated for the 4 segments generated at the first iteration, leading to the drawing of fig. 22c for the curve, and the 22g for the island. The process in the next steps is evident.

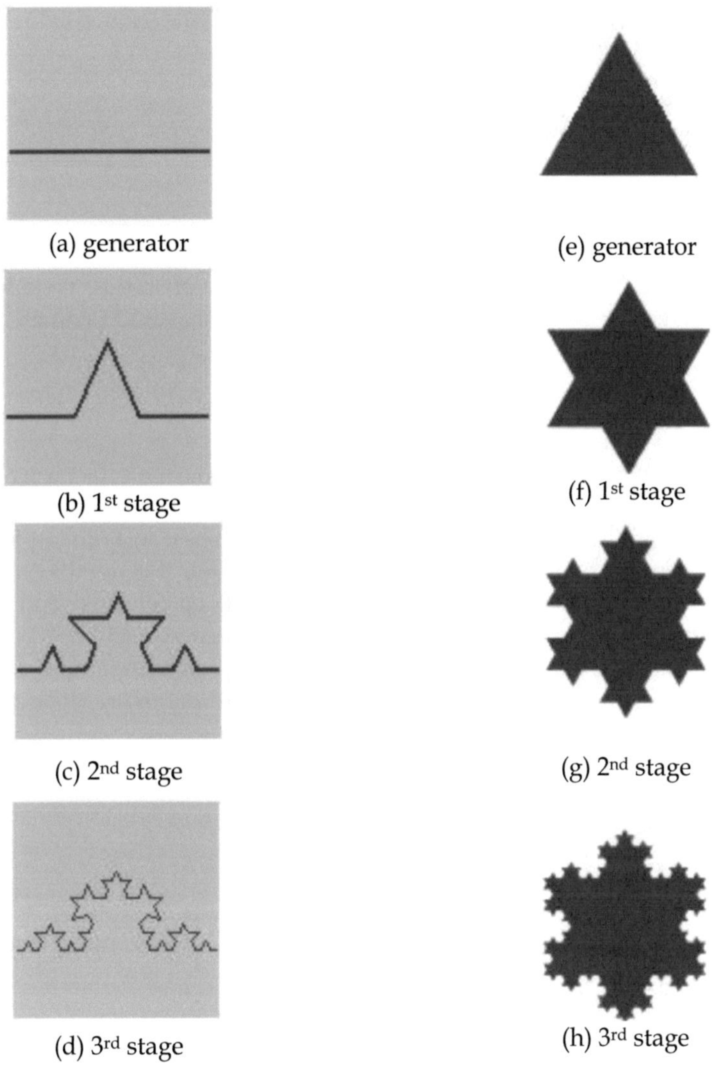

Fig. 22. (a)-(d) The generator and the first three stages of the Koch fractal curve. (e)-(h) The generator and the first three stages of the Koch fractal island and potentially printed antenna.

A basic advantage of the Koch patch, as mentioned above, is its attribute to resonate efficiently at frequencies lower than those of a regular patch of equal size. It has been verified via electromagnetic simulation as well as experimentally. To estimate the superiority of the Koch fractal to a conventional triangular patch a comparison between their sizes is necessary. At each new iteration n the area of the Koch island increases. Let A_n be the area at n^{th} iteration , then the area of the next iteration can be computed as

$$A_{n+1} = A_n + \frac{\sqrt{3}}{12}\left(\frac{4}{9}\right)^{n-1} \alpha^2 \quad (21)$$

Where α is the side of the initial triangle that has an area

$$A_0 = \frac{\sqrt{3}}{4}\alpha^2 \quad (22)$$

The geometry series given by (21) converges to

$$A = \frac{2}{5}\sqrt{3}\alpha^2 \quad (23)$$

All the iterations are circumscribed inside a circumference of radius $r = \sqrt{3}\alpha / 3$. On the other hand the perimeter increases at each new iteration. The overall perimeter for iteration k is given by $l_n = 3\alpha\left(\frac{4}{3}\right)^n$.

This evolution process leads to a fractal object with an infinite perimeter bounding a finite area. Despite of the increasing irregularity of the boundary, the manufacturing process does not become more complex at each new iteration. The patch can be manufactured by standard photo-etching techniques. The fundamental limitation in building the antenna is given by the resolution of the photo-etching process. When the number of iterations is increased the new added details in the structure cannot be resolved and they are not reproduced in the manufactured element.

In fig 23 the frequency dependence of the input impedance of microstrip Koch atennas for first five iterations are presented. All cases concern to the fundamental mode and the results were received by numerical analysis with the Method of Moments(MoM)[58].

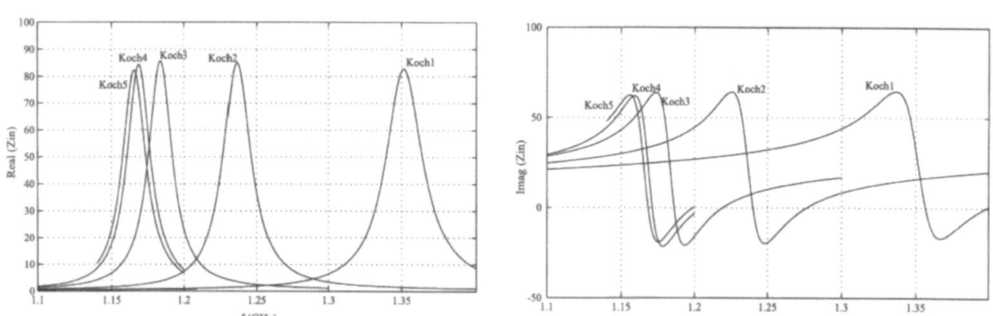

Fig. 23. Indicative results of the input resistance and reactance of a Koch fractal microstrip patch for several fractal iterations

The Koch patches were generated by an equilateral triangle whose side was118.2mm. They were printed on dielectric substrate with dielectric constant ε_r=4.17 and thickness h= 1.6mm. The printed metallic patch was placed 3.4 mm from the ground plane. So, a dual layer substrate was created. The patches were fed by a coaxial probe and the feed point was at the same location for all the patches and was placed at 15.5mm from the center of the patch. An interesting conclusion can be derived from the input impedance plot. The fundamental resonant frequency decreases when the number of iterations increases. Nevertheless this reduction tends to limit by the increment of iterations. So, the difference between the resonant frequencies of stage 4 and stage 5 is very small. This performance proves the great attribute of this type of antenna to resonate at small frequencies having at the same time small size. For example the generator, namely the ordinary equilateral triangle, of the above structure resonates approximately at 2.2 GHz. It can be verified if we use eq. (19) and set as ε_r for the dual layer substrate an equivalent value resulting from the equation $\varepsilon_r \approx \sum_{i=1}^{2} h_i \Big/ \sum_{i=1}^{2} \frac{h_i}{\varepsilon_{ri}}$. So, the ratio of 2.2GHz over the ~1.15GHz at which resonates the Koch island of the 5th stage is about 1.9. At the same time, in accordance to equations (21) and (22) the ratio $\frac{\text{Area of the } 5^{\text{th}} \text{ stage Koch island}}{\text{Area of the triangle}} \simeq 1.6$. If , instead of using the 5th stage Koch island, an ordinary equilateral triangular patch was used, the length of its side had to be 1.9 times the side of the initial triangle in order to resonate at ~1.15GHz . In this case the ratio of the surfaces of the two triangles had to be equal to $1.9^2 = 3.61$ instead of 1.6.

It is worthwhile to note that the concept Koch fractal can be used to build monopole antennas, enlongated or loop formed, which exhibit multi-frequency operation. In [59] a Koch loop monopole is proposed(fig. 24). It was designed to resonate at the bands of 2.45GHz and 5.25GHz therefore capable to serve a 802.11b/g WLAN system.

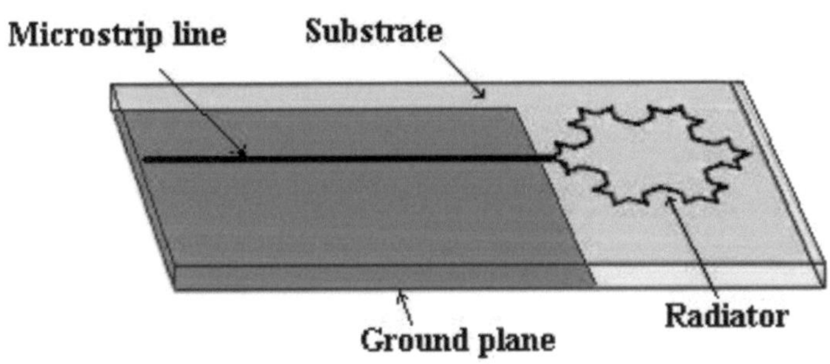

Fig. 24. The geometry of a planar monopole antenna with Koch fractal loop.

An alternative design of a fractal planar object, based on the Koch concept is shown in figure 25a. In this case, the generator is a square patch. The first stage of the fractal building can be produced by dividing each side of the initial square element into three equal segments. Then the middle segment of each size was removed and replaced by two segments of equal length and generating an equilateral triangle. Up to this point the process is similar to that of the Koch curve. In the following this triangle was directed inward to the initial square patch and was subtracted from it.

This configuration has been proposed in [60] and was successfully used to build a tri-band microstrip antenna configuration for GPS and terrestrial radio services. The initial square patch had side length equal to 78.25mm and was printed on a dielectric substrate having ε_r=1.046 and height 6mm. The patch of the first stage fed solely by one probe had small gain and single band operation. The target was the antenna to operate in three frequency ranges

(GPS, DCS1800 and 2.7GHz). The desired performance was obtained by the configuration shown in Fig. 25b. It came from the combination of four patches of the first stage, lightly overlapped. Two probes were used and the operation was enhanced by two pins. The suitable position of the probes and the pins was found by a number of simulations whereas the relatively large size of the entire printed scheme, due to the four fractal patch , ensured the satisfactory high gain. For best performance, the presence of the probes and pins was exploited. A phase difference between the feeding probes was inserted and on-off switching of the pins, depending on frequency area, was imposed. These situations are shown in Table 2.

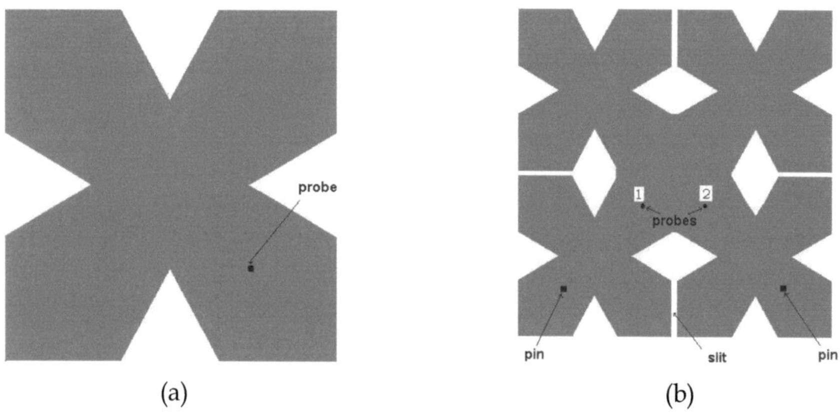

Fig. 25. Microstrip patch antenna for GPS and terrestrial radio services: a) A square patch modified via the first stage inverse van Koch fractal curve b) The compact radiating structure produced by the union of four fractal elements.

	GPS (1.575GHz)	DCS 1.78GHz -1.9 GHz	2.67GHz - 2.78GHz
Current of probe 1	1	1	1
Current of probe 2	$1e^{j\pi/2}$	1	$1e^{j\pi/2}$
pins	off	switched to ground	switched to ground

Table 2.

The results for the scattering coefficient of the signal at the feed input are depicted in figure 26 and show the tri-band operation.

The gain patterns of the antenna are illustrated in figures 27 and 28. At 1.575GHz(fig. 27), the system appears a broadside radiation suitable for a GPS link when the antenna is horizontally positioned and the total maximum gain is 6.8 dB.

The distribution of the radiated power in the DCS frequency range is shown in fig. 28. In this case the field is linearly polarized on yz-plane having a maximum gain value 4.9dB, approximately constant in a range $\pm 40^{\circ}$ about the broadside direction. In xz-plane it is also linearly polarized and the G_{θ} component has maximum value(4dB) at the broadside and minimum(-23dB) at $\pm 36^{\circ}$ apart from the broadside whereas the G_{φ} component has maximum (4dB) at a direction 30° apart from the broadside).

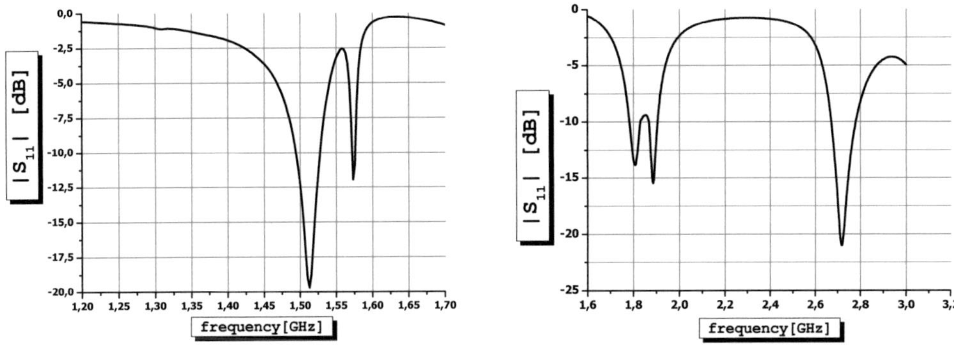

Fig. 26. Operation of the antenna of fig. 25b a) Scattering coefficient at the input of each probe in the range1.2GHz to1.7GHz and b) Scattering coefficient at the input of each probe in the range1.6GHz to 3GHz

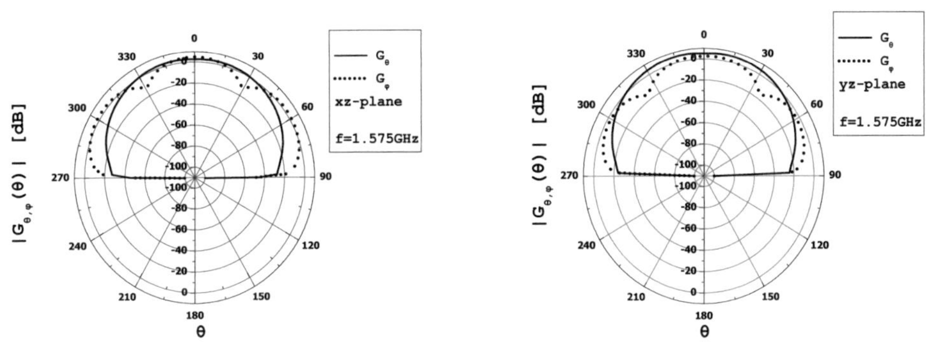

(a) (b)

Fig. 27. Radiation patterns of the antenna of fig. 25 : the power gain components at 1.575GHz on xz- and yz-plane

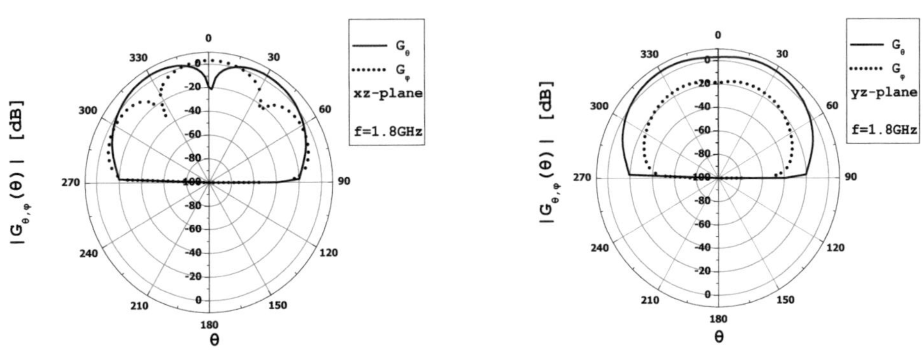

Fig. 28. Radiation patterns of the antenna of fig. 25 : the power gain components at 1.8 GHz on xz- and yz-plane

2.3.5 b) Sierpinski fractals

Another fractal concept widely used for the design microstrip antennas is the Sierpinski fractal[61]-[69]. Various Sierpinski fractal objects have been proposed: The Sierpinski Gasket(or Triangle), the Sierpinski Carpet (or rectangle), the Sierpinski Pentagon and the Sierpinski Hexagon. Judging from the literature the most efficient shapes for antenna applications are the carpet and especially the gasket. Monopole or dipole gasket fractal microstrip schemes have been proposed as multifrequency antennas.

Although the Sierpinski objects are based on different geometrical basis, they share the same construction principle. The geometrical construction of the popular Sierpinski gasket begins with an equilateral triangle which is considered as generator(fig. 29a). The next step in the construction process is to remove the central triangle, namely the one with vertices that are located at the midpoints of the sides of the original triangle. After the substruction, three equal triangles remain on the structure, each one being half of the size of the original one(fig. 29b). This process is then repeated for the three remaining triangles etc(figures 29c, 29d). If the iteration is carried out an infinite number of times the ideal fractal Sierpinski gasket is obtained. In each stage of the fractal building each one of the three main parts of the produced structure is exactly similar to the whole object, but scaled by a factor. Thus the Sierpinski gasket, as well as the other Sierpinski objects, are characteristic examples of self similar schemes.

It has to be pointed out that from an antenna engineering point of view the black triangular areas represent a metallic conductor whereas the white triangular represent regions where metal has been removed.

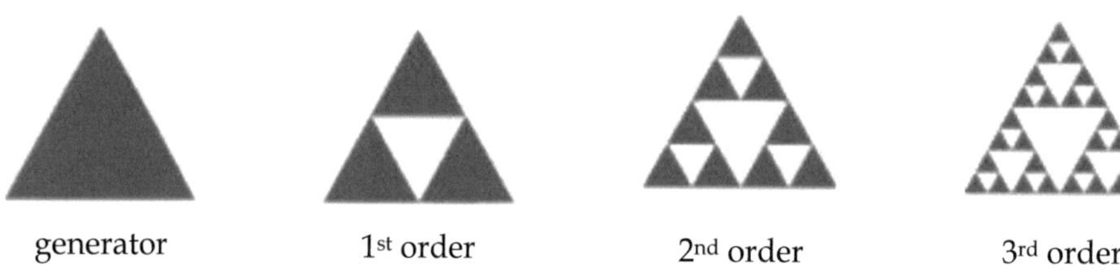

Fig. 29. The generator and the first three stages of the Sierpinski fractal gasket.

Figure 30 shows a Sierpinski gasket monopole printed antenna. Typically such antennas exhibit a log-periodic spacing of resonant frequencies as well as an increase in the impedance bandwidth at higher bands. It is interesting to note that the band number n and the iteration k are interchangeable. For example the band zero and the 0th iteration correspond to the fundamental resonance of the antenna. The first band and the first fractal iteration correspond to the first log-periodic resonant frequency. Therefore after the first fractal iteration two resonant frequencies are available : the fundamental and the first log periodic frequency. This is valid for other higher fractal iterations.

The specific positions of the frequency bands depend on the geometry of the generator and the parameter values of the dielectric substrate. It has to be noticed that the generator would potentially be not an equilateral triangle, namely the angle(flare) that corresponds to the vertex at which the feeding is applied would be not equal to 60°. Many such configurations have been proposed. The potential to select another value for this angle is an advantage because there are two geometrical parameters to control the frequencies of resonance. The height of the triangle and the flare angle. Indicative configurations are shown in fig.31a, and the respective input impedance diagrams are depicted in fig. 31b [64].

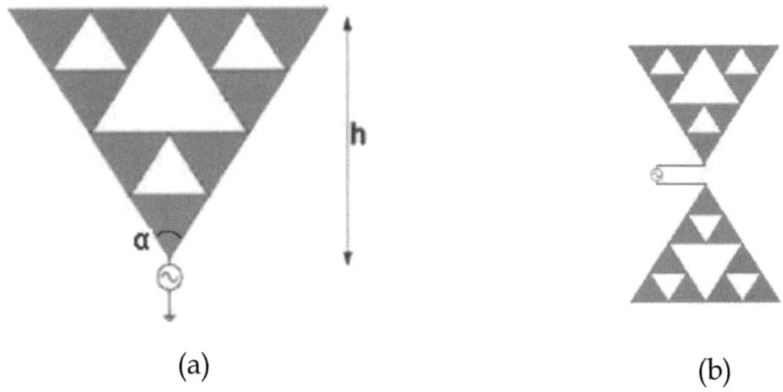

Fig. 30. The Sierpinski monopole(a) and dipole(b) printed antenna

From the design point of view, mathematical expressions for the calculation of the frequencies of resonance of the Sierpinki gasket, are necessary. The most recent available formula[65] in terms of the structural parameters and the order of iteration, for a monopole(fig. 30a) with flare angle equal to 60º is that of eq. (23). This expression includes the parameters of the geometry of the gasket as well as the thickness and the dielectric constant of the substrate.

$$f_r \cong \begin{cases} (0.15345+0.34\rho x)\dfrac{c}{h_e}\left(\xi^{-1}\right)^n & \text{for } n=0 \\ 0.26\dfrac{c}{h_e}\delta^n & \text{for } n>0 \end{cases} \tag{23}$$

Where $\xi = \dfrac{h_n}{h_{n+1}}$ is the ratio of the height of the gasket in the nth iteration to that in the $(n+1)^{th}$ iteration, $\delta=1/\xi$ is the scale factor $\rho = \xi - 0.230735$ and $x = \begin{cases} 0, & n=0 \\ 1, & n>0 \end{cases}$.

Moreover

$$h_e = \frac{s_e\sqrt{3}}{2} \tag{24a}$$

$$s_e = s + t(\varepsilon_r)^{-0.5} \tag{24b}$$

where s is the length of the side of the gasket and t and ε_r are the thickness and the dielectric constant of the substrate. The above equation is valid even in cases where the geometry is perturbed to get different scale factors

In practice the given parameter value is the frequency of resonance and the values of t and ε_r are selected by the designer. So, for a specific value of n, the required parameters are those of the geometry of the gasket. For these calculations the side length of the generating triangle of the gasket is given by the expression

$$s \cong \begin{cases} \dfrac{1}{\sqrt{3}}(0.3069+0.68\rho x)\dfrac{c}{f_r}\left(\xi^{-1}\right)^n - \dfrac{t}{\sqrt{\varepsilon_r}} & \text{for } n=0 \\ \dfrac{0.52}{\sqrt{3}}\dfrac{c}{f_r}\delta^n - \dfrac{t}{\sqrt{\varepsilon_r}} & \text{for } n>0 \end{cases} \tag{25}$$

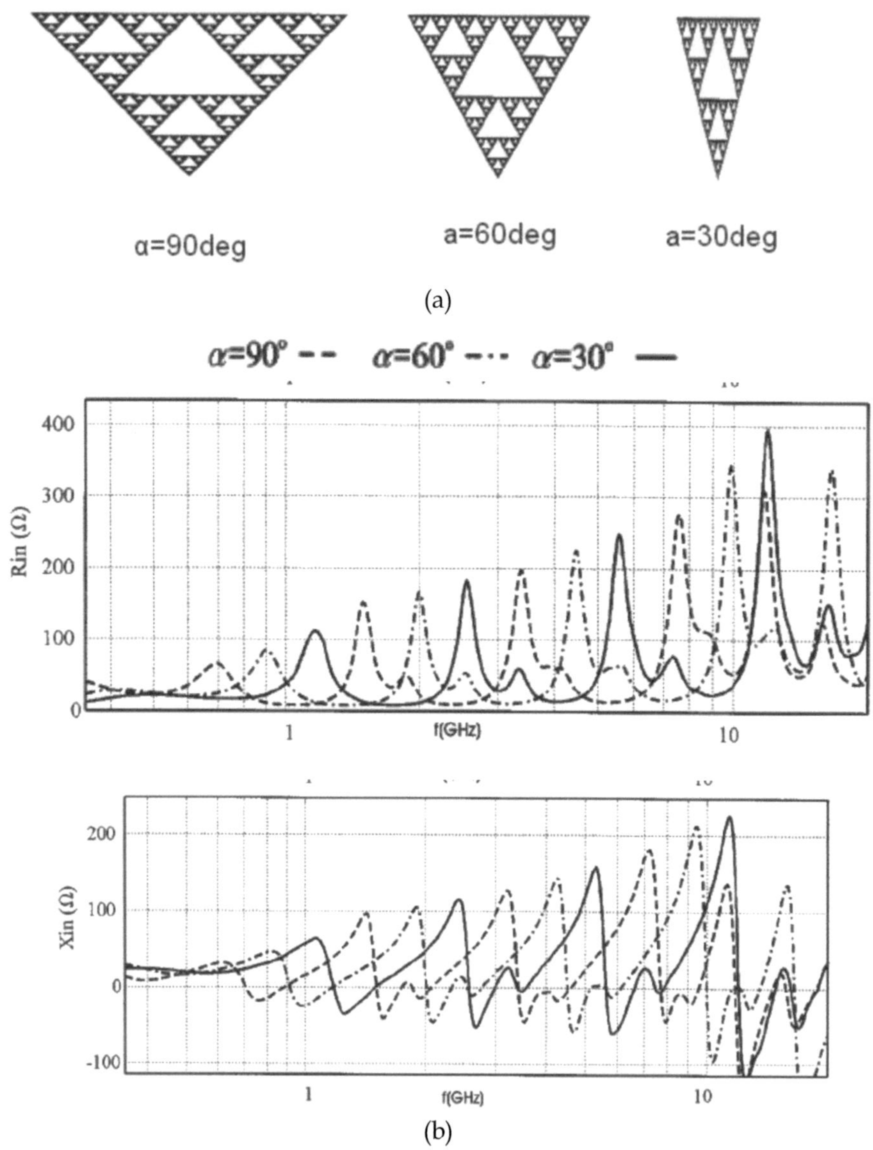

Fig. 31. a)Sierpinski gasket antennas with different flare angle b) Indicative results of their performance: Real and imaginary part of the input impedance for specific geometrical and material parameter values

It is worthwhile to mention that by additional modification of the Sierpinski gasket as proposed in[62] or in [69] (see fig. 32), the bands of resonance could be further controlled in order to meet the technical requirements of the applications for which the antenna is designed.

The Sierpinski carpet is another Sierpinski fractal configuration reported in antenna applications. Sierpinski carpet dipole antennas are shown in figures 33 and 34. The study of these configurations guide to the conclusion that no multiband performance can be obtained. It is due to the fact that the fractal iterations do not perturb the active current carrying region . So, their performance is similar to that of a simple square patch.

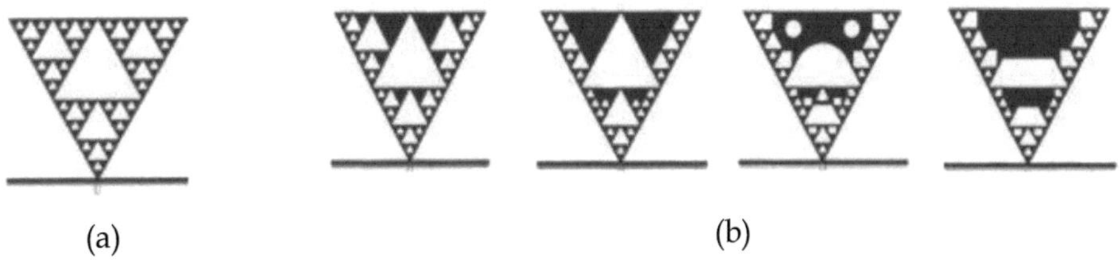

Fig. 32. Printed multiband antenna monopoles a)The self similar ordinary Sierpinski gasket. b) Modified Sierpinski gaskets.

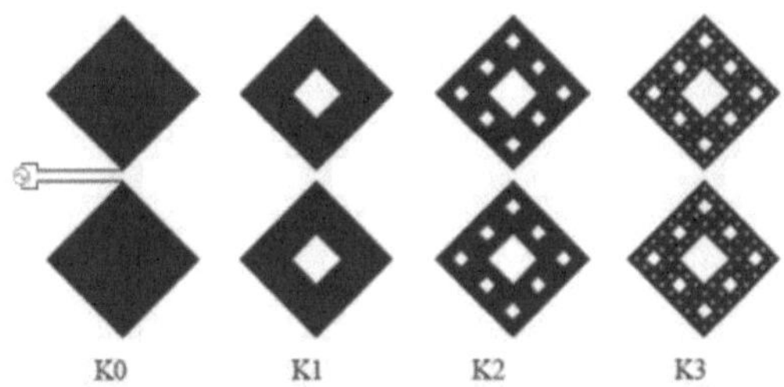

Fig. 33. Sierpinski carpet dipole antennas: the generator and the first three fractal orders

The negative version of the above fractal scheme is shown in Fig. 34. The geometrical method to design this carpet is the following: The generator is that of fig. 34a. In the first iteration the area around the central patch is divided in nine sub-areas of equal size and at the center of each sub-area a rectangular patch with side length three times smaller than the initial central patch is located. The same process is applied in the next iteration. In this antenna only the central element is driven and the energy of the other smaller patches is coupled parasitically from the driven patch.

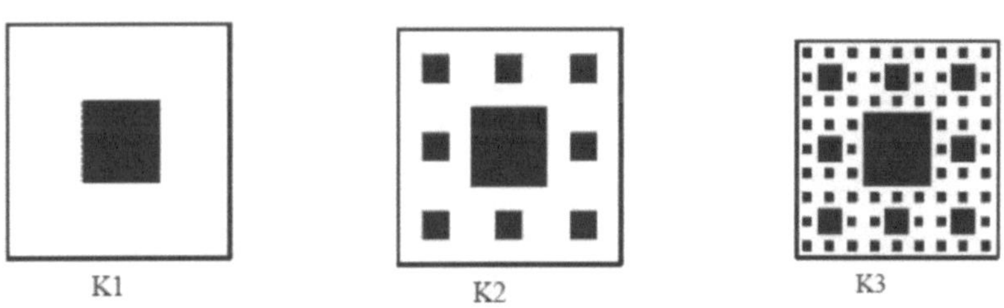

Fig. 34. Sierpinski carpet fractal antennas: the generator and the first two orders.

This fractal microstip configuration exhibits multifrequency performance, Fig. 35, but it was found[63] that the results come from the driven element, not from the parasitic ones.

2.3.5 c)Hilbert fractals

The properties of the Hilbert curve make them attractive candidates for use in the design of fractal antennas. These curves apart from being self similar have the additional property of approximately filling a plane and this attribute is exploited in realizing a 'small' resonant antenna. Hilbert fractal antennas with size smaller than $\lambda/10$ are capable to resonate, with performance comparable to that of a dipole whose resonant length is close to $\lambda/2$.

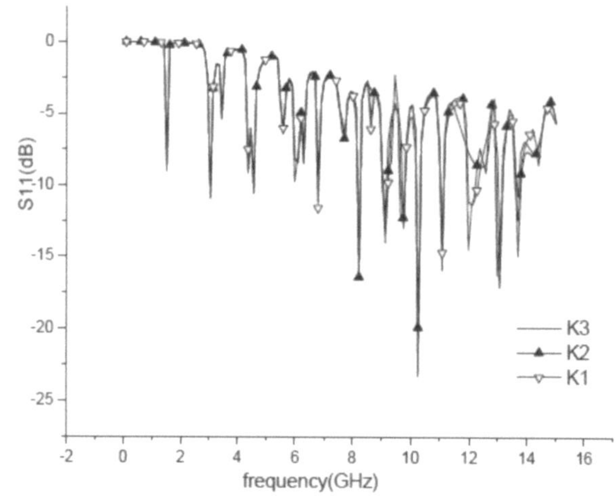

Fig. 35. The reflection coefficients of Sierpinski carpet microstrip antenna(fig. 34) in different iterations

The generator of the Hilbert curve has the form of a rectangular U as shown in fig. 36a. The Hilbert curves for the first several iterations are shown in figures 36b-36d. The construction at a stage is obtained by putting together four copies of the previous iteration connected by additional line segments.

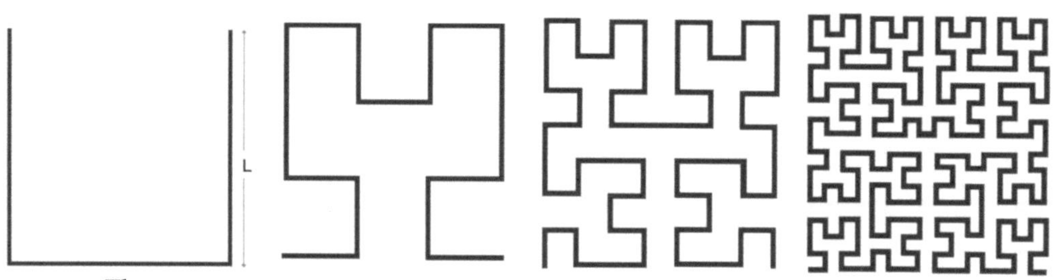

The generator(order=0) | 1st order | 2nd order | 3rd order

Fig. 36. The Hilbert fractal printed antennas of various stages.

It would be interesting to identify the fractal properties of this geometry. The space-filling nature is evident by comparing the first few iterations shown in figure 36. It may however be mentioned that this geometry is not strictly self similar since additional connection segments are required when an extra iteration order is added to an existing one. But the contribution of this additional length is small compared to the overall length of the geometry, especially when the order of the iteration is large. Hence, this small length can be disregarded which makes the geometry self similar. Moreover the curve is almost filling a plane. In other words the total length, if sum the line segments, tends to be extremely large. This could lead to a significant advantage, since the resonant frequency can be reduced considerably for a given area by increasing the fractal iteration order. Thus, this approach strives to overcome one of the fundamental limitations of antenna engineering with regard to small antennas.

For an accurate study of the operational features of a Hilbert fractal printed antenna information about its geometric parameters are necessary. It is obvious that as the iteration order increases, the total length of the line segments is increased in almost geometric progression if the outer dimension is kept fixed. Thus, for a Hilbert curve antenna with side dimension L and order n, the total segment length S(n) is calculated by the formula

$$S(n) = \frac{2^{2n} - 1}{2^n - 1} L \tag{26}$$

and the length of each line segment is given by

$$d = \frac{L}{2^n - 1} \tag{27}$$

A theoretical approach for the calculation of the resonant frequencies of the antenna considers the turns of the Hilbert curve as short circuited parallel-two-wire lines and begins with the calculation of the inductance of these lines[70], [71]. This approach is illustrated in figure 37. The self inductance of a straight line connecting all these turns is then added to the above, inductance multiplied by the number of shorted lines, to get the total inductance. To find the frequencies of resonance, the total inductance is compared with the inductance of a regular half wavelength dipole.

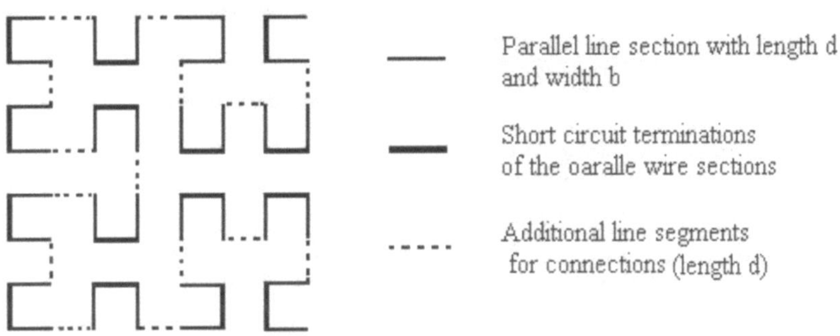

Fig. 37. The 2nd order of fractal building. The segments used to connect the geometry of the previous iteration are shown in dashed lines

In detail for a Hilbert curve fractal antenna with outer dimension of L and order of fractal iteration n, there are $m = 4^{n-1}$ short circuited parallel wire connections each of length d. Moreover the segments not forming the parallel wire sections amount to a total length of

$$s = d\left(2^{2n-1} - 1\right) = \frac{L}{2^n - 1}\left(2^{2n-1} - 1\right) \tag{28}$$

The characteristic impedance of a parallel wire transmission line consisting of wires with diameter b , spacing d , are given by

$$Z_o = 120 \log\left(\frac{2d}{b}\right) \tag{29}$$

The above expression can be used to calculate the input impedance at the end of the each line section , which is purely inductive

$$Z_{in} = L_{in} = \frac{Z_o}{\omega}\tan(\beta d) = \frac{Z_o}{\omega}\tan\left(\beta\frac{L}{2^n - 1}\right) \tag{30}$$

It is noticed that at the nth stage of fractal building there are $m = 4^{n-1}$ such sections.
The self inductance due to a straight line of length s is

$$L_s = \frac{\mu_o}{\pi} s\left(\log\frac{8s}{b} - 1\right) \tag{31}$$

So, the total inductance is

$$L_T = L_s + mL_{in} = \frac{\mu_o}{\pi} s\left(\log\frac{8s}{b} - 1\right) + m\frac{Z_o}{\omega}\tan\left(\beta\frac{L}{2^n - 1}\right) \tag{32}$$

To find the resonant frequency of the antenna, this total inductance is equated with that of a resonant half-wave dipole antenna with approximate length equal to $\lambda/2$. Taking into account that regular dipole antennas also resonate when the arm length is a multiple of quarter wavelength we can obtain the resonant frequencies of the multi-band Hilbert curve fractal antenna by the expression

$$\frac{\mu_o}{\pi} k\frac{\lambda}{4}\left[\log\left(\frac{8}{b}\frac{k\lambda}{4}\right) - 1\right] = m\frac{120}{\omega}\log\frac{2d}{b}\tan(\beta d) + \frac{\mu_o}{\pi} s\left[\log\left(\frac{8s}{b}\right) - 1\right] \tag{33}$$

where k is an odd integer. It is noticed that this expression does not account for higher order effects and hence may not be accurate at higher resonant modes.
At these antennas the feeding point is located at a place of symmetry or at one end of the curve, thus driving the structure to operate as a monopole antenna. It is noticed that the bandwidth at resonances is generally small, whereas the positions of resonant frequencies can be controlled by perturbing the fractal geometry.
In the basis of the above theory, several applications of this type of fractal antenna have been reported. Antennas that can efficiently operate in the range of UHF, as well as in multiple bands, at 2.43GHz and 5.35GHz, serving Wireless Local Area Networks [71]-[73].

2.3.5 d)Square Curve fractals

The design of microstrip antennas by the square curve fractal algorithm can yield radiating structures with multiband operation. The generator of this type of fractal objects is a rectangular ring and as a consequence the curves of the various stages are closed curves. The square curve fractals do not belong to the category of the space filling curves. However the increment of their total length from stage to stage is not significant, thus permitting the antennas to meet the requirement of the small size and at the same time to exhibit an increasing gain in virtue of their increasing length.
The staring point of the construction process is the selection of the size of the generator which is a rectangular ring with side length L(Fig. 38a). At the next step of the recursive process, the four corners of the square ring are used as the center of four smaller squares

each having side, half that of the main square. Overlapping areas are eliminated. The curve produced by this first iteration is shown in Fig.38b. Following the same algorithm the second stage of the fractal antenna can be derived(Fig. 38c). The building of the higher stages is evident.

The total length of the curve is calculated as follow.

a. generator ring: the perimeter is $4 \cdot L$
b. 1st stage : each side of the generator is divided in four segments of equal length. Two segments are removed, at each corner and they are replace by smaller squares with side length equal to L/2. So, the length of the curve is equal to the sum of segments, common between generator square and the first recursion, plus the length of the newly added segments. The total length of the two segments removed at each corner, is $L/2$, so the total removed is $(L/2) \cdot 4$. Looking only at the added segments the length increase of the curve is $L_{st1}^{inc} = (L/2) \cdot 8$.
c. 2nd stage: On the second iteration, the corners of the four small squares added at the first iteration are replaced by four even smaller squares with side length L/4. Here the length of the segments removed at each square corner is equal to $L/4$ and the length of the smaller squares added is equal to $(L/4) \cdot 3$. Taking into account only the added segments the length increase is $L_{st1}^{inc} = (L/4) \cdot 24$.

The general formula for the length increase is $L^{inc} = (L/2^n) \cdot 8 \cdot 3^{(n-1)}$ where n is the iteration number .

The total curve length is $L_1 \cdot ((r^n \cdot 8) - 4)$, where r is the ratio of the length increase between two sequential iterations.

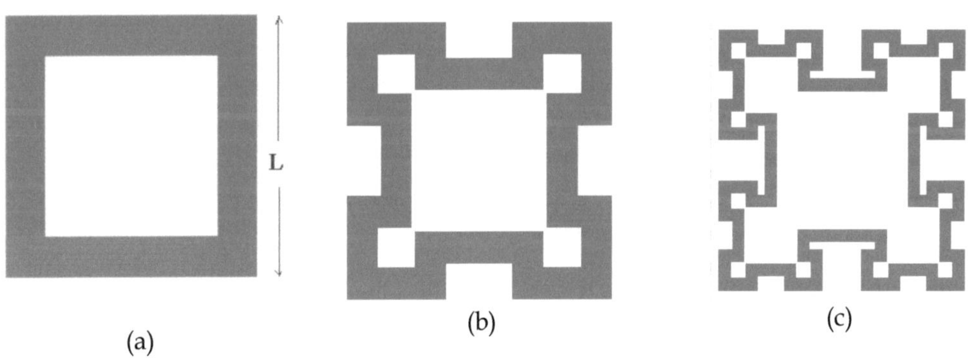

Fig. 38. The Square Curve fractal a) the generator and (b)-(c) the lower two stages

In[75] a microstrip fractal structure designed with the aforementioned algorithm is proposed. It was printed on a dielectric substrate with $\varepsilon_r = 1.046$ (Rohacell 51HF, Northern Fiber Glass Service, Inc.) and height $h = 1mm$. A value for r equal to 1.5 and the second stage of development gave an object with outer dimensions 8.4cm x 8.4cm. A fundamental parameter of the structure is the width of the printed strip which forms the curve. Attention must be paid to the proper selection of the values of the strip's width because there is a trade off between this value and the input impedance of the antenna. A narrow strip guides to high input impedance and inserts difficulties to the matching of the antenna. On the other side a wider strip would yield input impedance suitable for direct matching but could produce difficulties related with the space filling during the process of the fractal expanding. More over, the keys to drive this antenna in multi-band operation are the proper number and positions of the feeding points. The incorporation of a pin can also enhance the performance of the antenna.

In figure 39, results received using three different feedings are depicted. Figure 39a shows the variation of the scattering coefficient at the feeding input using one probe, positioned at a point on an axis of symmetry. This choice is common at many fractal antennas. It is observed that only two frequency bands give scattering coefficient lower than -10dB. It is due the high input impedance of the antenna, as shown in figure 39b. A better performance with seven frequency bands is obtained with two probes(Fig. 39b); and an even satisfactory operation is achieved when a shorting pin is installed between the probes. The pin

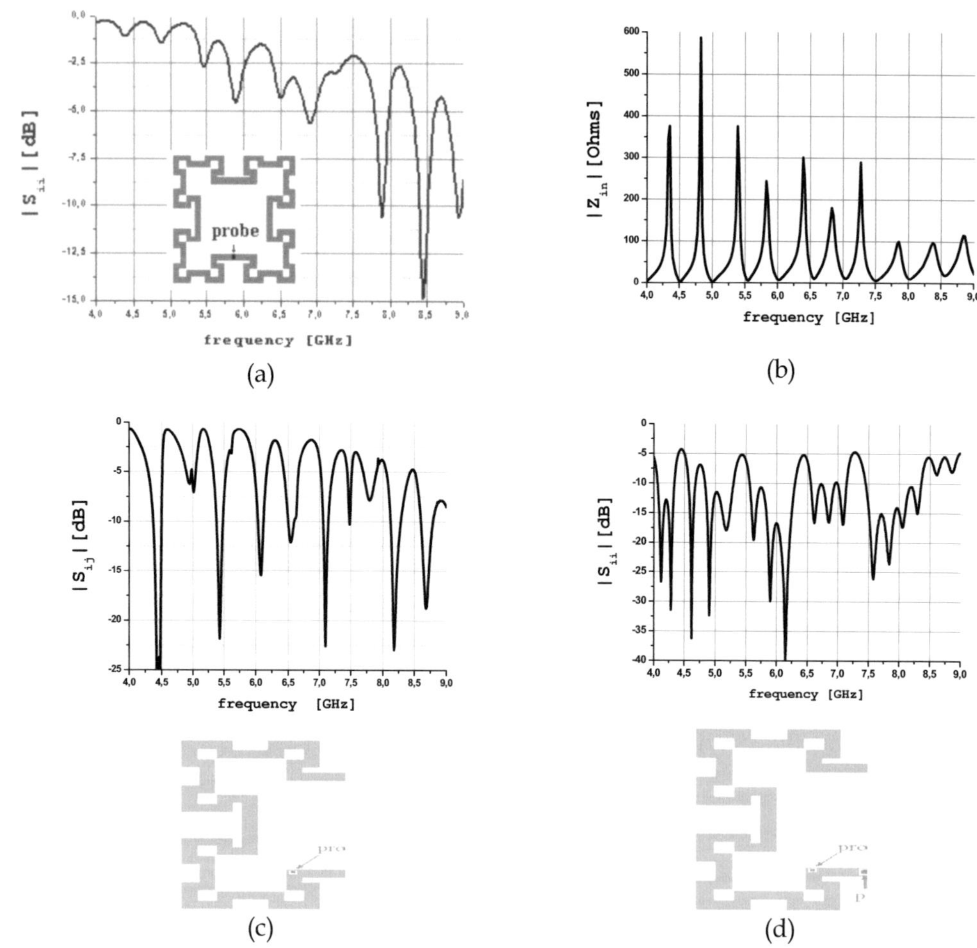

Fig. 39. a) Scattering coefficient at the input of the second stage fractal antenna fed with one probe and b) the respective input impedance. c) Scattering coefficient when fed with two probes and d)fed with two probes and loaded with one pin.

acts as a short circuit between the trace and the ground plane, reflecting the wave produced by the probe. So, two complex impedances combined in parallel appear at the point of the feed. The one is due to the line between the probe and the pin and the other to the remaining trace of the structure. These two parallel impedances involve a lower total impedance which would be suitable for direct match to an 75Ohm probe. Moreover, this matching is attainable in wide ranges around the frequencies of resonance. The results of fig 39c, for the scattering coefficient, show a multiband - and at the same time wideband operation.

3. Electromagnetic Bandgap Structures (EBG) in antenna applications

3.1 The EBG structure and properties

Electromagnetic Band Gap (EBG) structures constitute a specific class of recently discovered microwave objects that, due to their special electromagnetic behavior, reveal promising solutions to several microwave problems, especially in the area of communications[76],[77]. The EBG structures are generally defined as 'artificial periodic or non periodic objects that prevent the propagation of electromagnetic waves in a specified band of frequency for all incident angles and all polarization states'. They can be categorized into three groups according to their geometric configuration: Three dimensional volumetric structures, two-dimensional planar surfaces and one dimensional transmission lines. Among these three categories the planar EBG objects are the most commonly used in antenna systems. They consist of a two dimensional lattice of metal plates conductively connected to a ground plane by metal-plated vias, as shown in figure 40, and are easily fabricated using printed - circuit board technology.

The effective application of EBG surfaces to the antenna design is based on the exploitation of their distinctive electromagnetic properties with respect to the type of the incident electromagnetic waves:

i. When a plane wave impinges on an EBG surface it is reflected with a phase that varies with frequency as shown in figure 41. At a certain frequency the reflection phase is zero degrees. The value of this frequency depends on the structural parameters of the EBG object. This performance resembles a perfect magnetic conductor that does not exist in the nature.
ii. When the incident wave is a surface wave the EBG structures show a frequency band gap through which the surface wave cannot propagate for any incident angles and polarization states.

Both the above attributes contribute to the enhancement of the performance of printed or not printed antenna elements or arrays of elements. For example in the case of a microstrip antenna integrated with EBG structures, the suppression of the surface waves could reduce the mutual coupling between the antenna elements, if the antenna is an array, and also prevent the wave to reach the boundaries of the configuration and be diffracted. This prevention involves lower radiation towards the back space of the antenna, an attribute that would ensure low interference with adjacent microwave elements or low radiation towards the user of the equipment that hosts the antenna. Moreover, due to the property of an EBG cell to work as a resonator, enhancement of the antenna gain could be obtained and also an easier and effective matching of the system to the feeding probe. Furthermore an EBG object could drive the microstrip antenna to a dual frequency operation modifying the higher order radiation patterns, thus making them similar to those of the basic mode. On the other hand an EBG surface is a unique object to obtain low profile antennas if the radiating element is not a microstrip antenna but a wire dipole , thus providing simple and effective radiation systems.

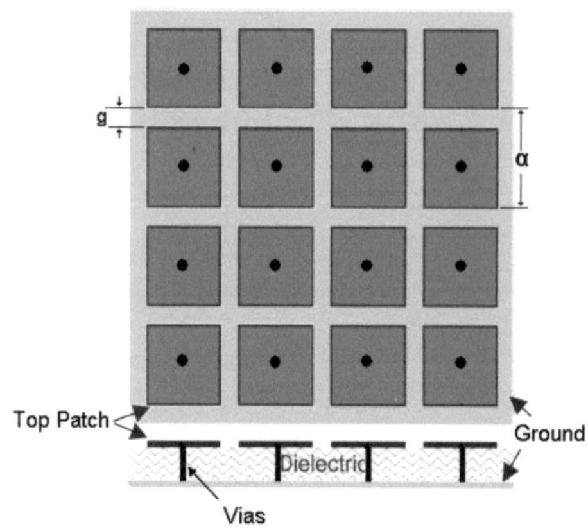

Fig. 40. Geometry of the mushroom-like EBG structure

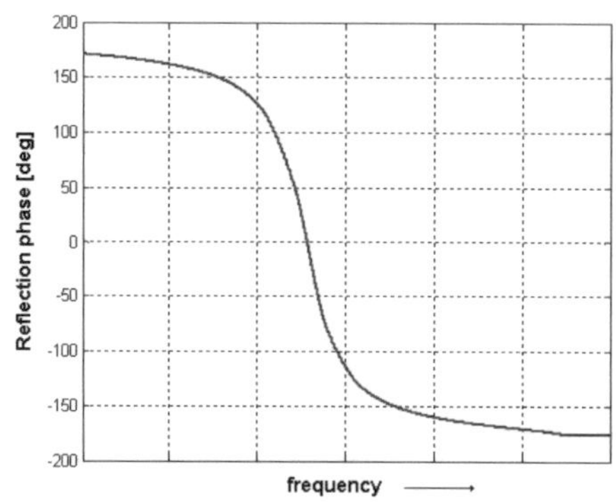

Fig. 41. The phase of a plane wave reflected by the EBG surface

A theoretical analysis of an EBG structure, via various models, gets an insight into the way by which they perform and can yield mathematical expressions for their operational parameters. The analysis would explain the mechanisms by which the EBG objects enhance the performance of the antennas and would give to the antenna designer the ability to properly exploit all the EBG properties, potentially useful in a specific antenna application.

3.2 Theoretical analysis

3.2.1 Low and high impedance surfaces

Flat metal sheets have low surface impedance and are used in many antennas as a reflector or a ground plane. The boundary conditions on these surfaces impose that the tangential component of the electric field intensity has to be equal to zero and this requirement involves that the metal sheet reflects an impinging wave, shifting the phase of its intensity by an amount of π. Moreover the metal sheet redirects one-half of the radiation into the opposite direction improving the antenna gain by 3dB and partially shielding objects on the other side. However, if the antenna is too close to the conductive surface, the out of phase image currents 'cancel' the currents in the antenna, resulting in poor radiation efficiency. This problem is often addressed by positioning the radiating element at a quarter-wavelength distance from the ground plane but this arrangement requires minimum thickness of $\lambda/4$.

By incorporating a special texture on a conducting surface it is possible to alter its radio-frequency electromagnetic properties. A proper modification (see for example fig. 40)would yield a specific surface with high surface impedance. On these textured surfaces the tangential component of the magnetic field intensity tends to zero and this condition means that the surface reflects an incident wave with an almost zero phase shift. This minimization of the magnetic field is due to the minimization of the surface currents, that inevitably comes from the cutting up of the metallic surface, into small patches. It is noticed that although the magnetic field intensity is very small, the electric field may have a large value due to the high voltage induced between the edges of the adjacent patches of the modified sheet. The edges and the narrow gaps between them realize capacitors. The ratio of the high electric field intensity, which is high, over the low magnetic field intensity defines the impedance of the surface which, in this case, is obviously very high. The almost zero tangential magnetic field permit us to term the surface as an artificial magnetic conductor. This unusual boundary condition involves that the image currents are in phase rather than out of phase, allowing radiating elements to lie in very close proximity to the surface while still radiating efficiently.

As a consequence of the performance described above, this type of high impedance surface can function as a new type of ground plane for low profile antennas. For example a dipole positioned in parallel to a high impedance ground plane is not shorted out as it would be on an ordinary metal sheet. In addition to their unusual reflection-phase properties, these textured structures have a surface wave bandgap, within which they do not support bound surface waves of either transverse magnetic (TM) or transverse electric(TE) polarization. They may be considered as a kind of electromagnetic bandgap structures or photonic crystals for surface waves. It is noticed that although bound surface waves are not supported, leaky TE waves can propagate within the bandgap, and they are useful for certain applications.

The theoretical basis to explain the electromagnetic behavior of an EBG structure is the general theory of surface waves and the useful tool for the description of its performance is the surface impedance of the EBG object along with equivalent electric circuit of this impedance.

3.2.2 Surface waves

Surface waves can occur on the interface between two dissimilar materials, for example metal and free space[76]-[78]. They are bound to the interface and decay exponentially into the surrounding materials although at radio frequencies the fields associated with these waves can extend thousands of wavelengths into the surrounding space. To describe theoretically and by a simple manner their physical entity, let us suppose an interface parallel to yz plane as shown in figure 42.

Assume a wave that is bound to the surface in the +x direction with decay constant α , and in -x direction with a decay constant γ. The wave propagates in the z-direction with propagation constant k. For a TM polarized surface wave $E_y = 0$. The electric field in the upper half-space has the following form

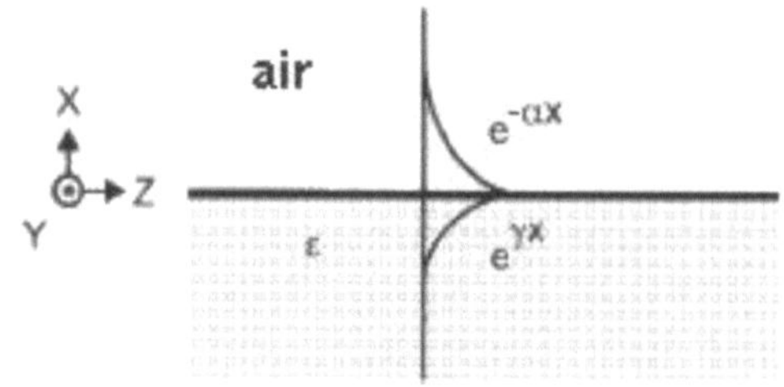

Fig. 42. A surface wave is bounded around the interface of two media and decays exponentially into the surrounding space

$$\vec{E}_1 = \left(\hat{x}E_{1x} + \hat{z}E_{1z}\right)e^{j\omega t - jkz - \alpha x} \tag{34}$$

In the lower half-space the electric field is

$$\vec{E}_2 = \left(\hat{x}E_{2x} + \hat{z}E_{2z}\right)e^{j\omega t - jkz - \gamma x} \tag{35}$$

The solution of Maxwell equations yields the following results for the propagation constants

$$k = \frac{\omega}{c}\sqrt{\frac{\varepsilon}{1+\varepsilon}} \tag{36a}$$

$$\alpha = \frac{\omega}{c}\sqrt{\frac{-1}{1+\varepsilon}} \tag{36b}$$

$$\gamma = \frac{\omega}{c}\sqrt{\frac{-\varepsilon^2}{1+\varepsilon}} \tag{36c}$$

If ε is real and positive, then α and γ are imaginary and the waves do not decay with distance from the surface and propagate through the dielectric interface. Thus TM surface waves do not exist on nonconductive dielectric materials. On the other hand if ε is less than -1, or if it is imaginary or complex, the solution describes a wave that is bound to the surface. These TM surface waves can occur on metals or other materials with non-positive dielectric constants.

The complex permittivity of a metal can be expressed in the following form

$$\dot{\varepsilon} = \varepsilon_o \varepsilon_r \left(1 - j\frac{\sigma}{\omega \varepsilon_o \varepsilon_r}\right) \tag{37}$$

where σ is the conductivity of the metal, ε_0 is the permittivity of the free space and ε_r is the dielectric constant.

For relatively low frequencies, including the microwave spectrum, the conductivity is primarily real and much greater than unity, thus, the permittivity is a large imaginary number which, supposing that $\varepsilon_r = 1$, is calculated by $\dot{\varepsilon} \approx j\frac{\sigma}{\omega\varepsilon_0}$ and in accordance to (36a) the dispersion relation for surface waves at radio frequencies is approximately $k \approx \frac{\omega}{c}$.

Thus, surface waves propagate at nearly the speed of light in the vacuum and they travel for many wavelengths along the metal surface with little attenuation.

The decay constant of the fields into the surrounding space is derived by inserting (37) into (36b). It is easy to ascertain, by an arithmetic application at microwave frequencies, that the surface waves extend a great distance into the surrounding space.

Constant γ ,the inverse of which is related to the wave penetration depth into the metal, is approximately

$$\gamma \approx (1+j)\sqrt{\frac{\omega\mu_0\sigma}{2}} = \frac{(1+j)}{\delta} \tag{38}$$

where δ is the skin depth. The surface currents penetrate only a very small distance into the metal. They can be expressed in terms of the skin depth, the conductivity and the electric field intensity and can be used for the determination of the magnetic field intensity. Then, the surface impedance is derived by the electric and magnetic field intensities as follows

$$Z_s = \frac{E_z}{H_y} = \frac{1+j}{\sigma\delta} \tag{39}$$

The above results show that the surface impedance has equal positive real and positive imaginary parts, so the small surface resistance of the metal surface is accompanied by an equal amount of surface inductance.

By texturing the metal surface we can alter its surface impedance and thereby change its surface-wave properties. Thus the surface impedance would appear inductive or capacitive imaginary part, depending on the frequency. The derivation of the impedance is made considering that a wave decays exponentially away from the boundary with a decay constant α , whereas the boundary is taken into account by its surface impedance. In this case it has been proved that TM waves occur on an inductive surface, in which the surface impedance is given by the following expression

$$Z_S^{TM} = \frac{j\alpha}{\omega\varepsilon} \tag{40}$$

Correspondingly, TE waves can be excited on a capacitive surface. In this case the surface impedance is

$$Z_S^{TE} = \frac{-j\omega\mu}{\alpha} \tag{41}$$

The wave vector k, in terms of the frequency and the decay constant α, helps to get an insight into the behavior of the surface

$$k^2 = \mu_0\varepsilon_0\omega^2 + \alpha^2 \tag{42}$$

Combining equation (42) with the equations (40) and (41) we can derive the dispersion relation for TM waves

$$k^{TM} = \frac{\omega}{c}\sqrt{1-\frac{\left(Z_S^{TM}\right)^2}{\eta^2}} \tag{43}$$

And for TE waves

$$k^{TE} = \frac{\omega}{c}\sqrt{1-\frac{\eta^2}{\left(Z_S^{TE}\right)^2}} \tag{44}$$

where $\eta = \sqrt{\frac{\mu_0}{\varepsilon_0}}$ is the impedance of the free space.

The complete dispersion diagrams of k versus frequency (Fig. 43), show that for a specific textured surface there is a certain frequency, termed resonance frequency, below of which TM surface waves are supported. At frequencies much lower than the resonance frequency the dispersion curve is very near to the light line, revealing the propagation of the wave, thus the field extends many wavelengths along the surface, as does on a metal sheet and the surface is considered of low impedance. Near the resonance the surface waves are tightly bound to the surface and, as the slope of the curve is very small, the group velocity is low. All these mean the surface impedance of the structure becomes high. Above the resonance frequency, the surface appears a capacitive behavior and TE waves are supported. Near the resonance frequency the TE waves are weakly bound to the surface extending far into the surrounding space. As the frequency increases the curve bends away from the light line and it means that the waves are more tightly bound to the surface.

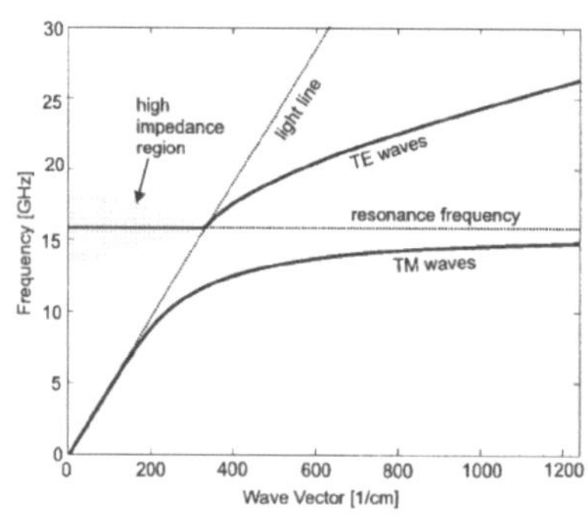

Fig. 43. The predicted surface wave dispersion diagram

3.2.3 Equivalent circuit model

They are exactly the above features of the propagating waves that permit us to approach the electromagnetic behavior of the EBG surface using an equivalent parallel resonant circuit which can be tuned to exhibit high impedance over a pre-specified frequency band. From the physical side of view this equivalence can be explained as follows: as the EBG interacts with electromagnetic waves, currents are induced in the top metal plates. A voltage applied parallel to the top surface causes charges to concentrate around and on the ends of the plåtes which can be considered as a capacitance. As the charges move back and forth, they flow around a long path through the vias and the bottom plate(Fig. 44). Associated with these currents is a magnetic field and, thus, an inductance.

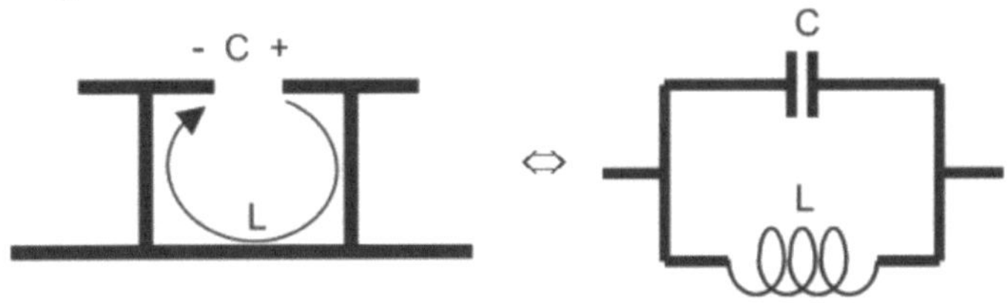

Fig. 44.

By this circuit model, the surface impedance, assigned to the sheet, corresponds to the impedance of a parallel resonant circuit, consisting of the sheet capacitance and the sheet inductance

$$Z = \frac{j\omega L}{1 - \omega^2 LC} \tag{45}$$

The surface is inductive at low frequencies, capacitive at high frequencies and the impedance is very high near the resonance frequency ω_0,

$$\omega_0 = \frac{1}{\sqrt{LC}} \tag{46}$$

This high impedance is associated with a forbidden frequency band. In the 2D geometry of the lattice of the EBG patches the capacitors are formed by the fringing electric fields between adjacent metal patches and the inductance is fixed by the thickness of the structure. An approximate expression for the capacity [79], in the case of the geometry shown in fig. 40, is equation (47), [79]

$$C_{fringe} \approx \frac{w(\varepsilon_1 + \varepsilon_2)}{\pi} \cosh^{-1}\left(\frac{\alpha}{g}\right) \tag{47}$$

In the above expression g is the gap between the plates, w is the width of the plates, α is the lattice constant namely $\alpha = g + w$, and ε_1 and ε_2 are , respectively, the dielectric constants of the substrate of the EBG and the material surrounding the surface which may be free space. Equation (47), although approximate, is adequate for first order designs. The inductance of a high -impedance surface is determined entirely by its thickness. This can be understood by considering a solenoid of current that includes two rows of plates and their associated vias. Current flows up one row of vias across the capacitors and down the next set of vias to return through the ground plane. The length and width of the solenoid are canceled to obtain the sheet inductance, L

$$L = \mu h \tag{48}$$

For the effective surface impedance approximation to be valid the lattice constant should be small compared to the wavelength.
Another, also approximate, equation for C is proposed in [80]. It was produced considering normally-incident plane waves and the vias conductors, connecting patch to the ground, not excited. In this case also square patches, small compared with the wavelength, and gaps between the patches, very narrow compared with the length side of the patches were considered. In the used model the higher order Floquet modes generated by the periodical mesh were neglected. Assuming only the fundamental mode plane waves between the lattice of patches and the ground, the equivalent surface impedance was considered as the parallel connection of two impedances: the impedance of the lattice and the input impedance of a TEM line section of length h. The expression of the surface impedance, normalized to the free space impedance was found

$$Z_{sn} = \frac{j\frac{1}{\sqrt{\varepsilon_r}}\tan(\omega h\sqrt{\varepsilon_r \varepsilon_0 \mu_0})}{1-\frac{(\varepsilon_r+1)\left(\omega\alpha\sqrt{\varepsilon_0\mu_0}\right)}{\pi\sqrt{\varepsilon_r}}\log\left(\frac{2\alpha}{\pi g}\right)\tan(\omega h\sqrt{\varepsilon_r\varepsilon_0\mu_0})} \tag{49}$$

Where α and g were defined previously. The equation for the capacitance C is

$$C = \frac{D\varepsilon_0\left(\varepsilon_r+1\right)}{\pi}\log(\frac{2\alpha}{\pi g}) \tag{50}$$

3.2.4 Reflection phase and Bandwidth

Besides their unusual surface wave features, the high impedance surfaces also have unusual reflection phase properties. Using the effective surface impedance model, described previously, we can determine the reflection phase for the resonant textured surface. For a normally incident wave, the reflection phase of the surface is given as

$$\Phi = \mathrm{Im}\left[\ln\left(\frac{Z_S-\eta}{Z_S-\eta}\right)\right] \tag{51}$$

where Z_S is the surface impedance and $\eta = 120\pi$ is the free space impedance. At very low frequencies the reflection phase is π (see fig. 41) and the structure behaves like a smooth metal surface. At higher frequencies the reflection phase slopes downward and crosses through zero at the resonance frequency. This situation corresponds to the artificial magnetic conductor behavior. Above the resonance frequency the phase converges to -π. The phase varies between π/2 and -π/2 when the magnitude of the surface impedance exceeds the impedance of the free space. It is worth noting that for a wide range of geometries the bounds of the surface' s wave band gap, occur at the same frequencies where the reflection phase crosses through π/2 and -π/2.
To determine the bandwidth of an antenna positioned in parallel to an EBG surface , its radiation is modeled as a resistor, valued the impedance of free space, in parallel to the LC equivalent circuit of the EBG lattice. The amount of power dissipated in the resistor is a

measure of the radiation efficiency of the antenna. The maximum radiation efficiency occurs at the LC resonance frequency of the EBG surface, where the surface reactance is infinite. At very low or at very high frequencies, the radiated power is reduced. It can be shown that the frequencies where the radiation become equal to half of its maximum value, occur when the magnitude of the surface impedance is equal to the impedance of free space. It is easily proved that the frequencies at the bounds of the operating band are approximately given by the expressions

$$\omega \approx \left(1 \pm \frac{1}{Z_0}\sqrt{\frac{L}{C}}\right) \tag{52}$$

The two frequencies designated by the plus and minus signs delimit the frequency range over which an antenna would radiate efficiently on such a surface. The relative bandwidth, defined as the ratio of $\Delta\omega$ over the frequency of resonance, is approximately equal to

$$BW = \frac{\Delta\omega}{\omega_0} \approx \sqrt{\frac{L\varepsilon_0}{C\mu_0}} \tag{53}$$

It is the bandwidth, over which the phase of the reflection coefficient falls between $\pi/2$ and -$\pi/2$ and image currents are more in phase than out of phase. As noted in the previous section, this range often coincides with the surface wave band-gap.

The geometry of the EBG and the parameters of the materials can be adjusted to provide the desired resonance frequency and the bandwidth and it is usually done via simulations. However it is useful to have an initial solution to more rapidly converge on the correct design. So, to design an EBG surface for pre-specified frequency of resonance ω_0 and bandwidth BW, in accordance to the aforementioned analysis the thickness of the substrate is determined by the equation

$$h = \frac{cBW}{\omega_0} \tag{54}$$

and the equation for the required sheet capacitance is

$$C = \frac{1}{\omega_0 \eta BW} \tag{55}$$

It is worth noting the there is a relation between the bandwidth and the variation of the phase reflection. The rate by which the reflection phase varies, conditions the frequency bandwidth of the mushroom type EBG and depends on the length of the side of the patch, the width of the gap between the patches, the dielectric constant and the thickness of the substrate[81].

Diagrams illustrating this performance are depicted in figure 45. In fig. 45a is shown that the increment of the side length reduces the frequency of resonance, as expected, but also reduces the bandwidth. From fig. 45b it is concluded that when the gap width is increased, both the frequency band position and its bandwidth increase. Fig. 45c shows that when the substrate permittivity ε_r is increased the frequency band position and the bandwidth decrease. The results of fig. 45d prove that when the substrate thickness is increased the frequency band position decreases while its bandwidth increases.

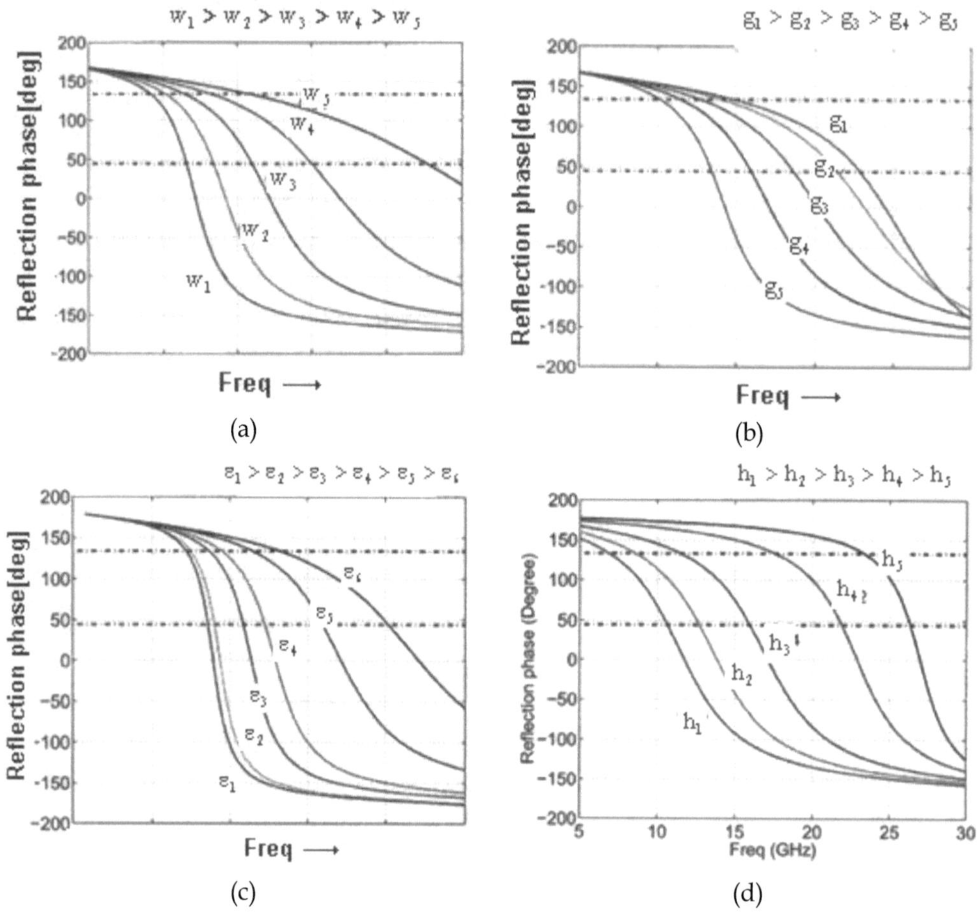

Fig. 45. EBG parameters analysis. Reflection phase variation versus frequency. Effects of a) patch width b)gap between the patches of the EBG lattice c) substrate permittivity d)substrate thickness.

3.3 EBG structures in antenna applications

Although the mushroom type cell is the most commonly used, various EBG types of cells (fig. 46) have been proposed for the enhancement of the antennas they host.

Whatever the type of the used EBG is, the determination of its proper configuration for the improvement of the antenna performance is not an easy task. The band-gap of an EBG, designed in advance, cannot guarantee the effective operation of the antenna that will be incorporated in the following, because complicated interactions occur between the antenna and the EBG lattice. Due to these interactions the improvement of the operation of the system antenna-EBG, as well as the satisfactory feed matching of the antenna, are problems that must be simultaneously solved. An efficient strategy for this purpose is the parametric study of the entire configuration, which is supported by simulations via a high frequency electromagnetic software, before the manufacturing the antenna system. Alternative ways to solve the problem is the employment of deterministic or stochastic algorithms, as will be discussed in the next paragraph.

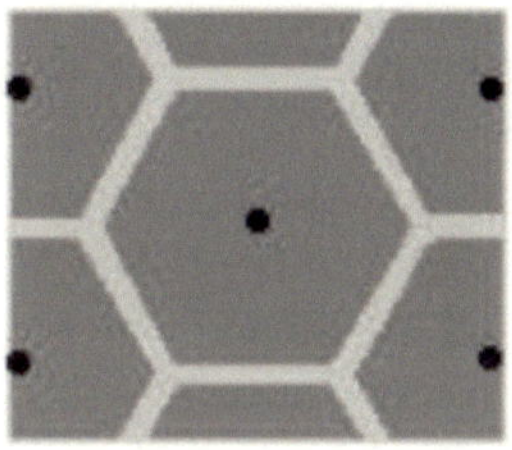

Honeycomb

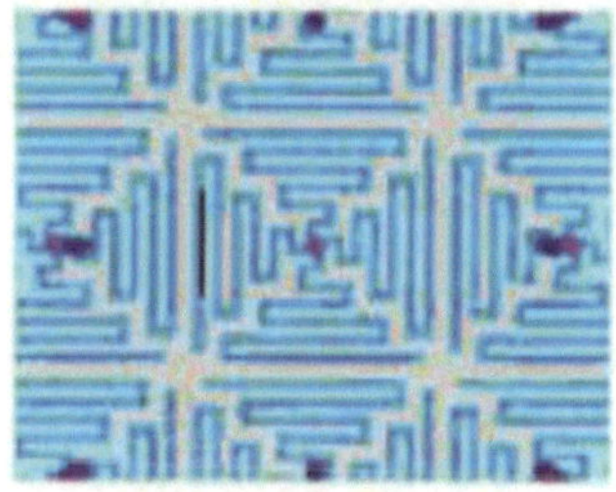

Winding Cross

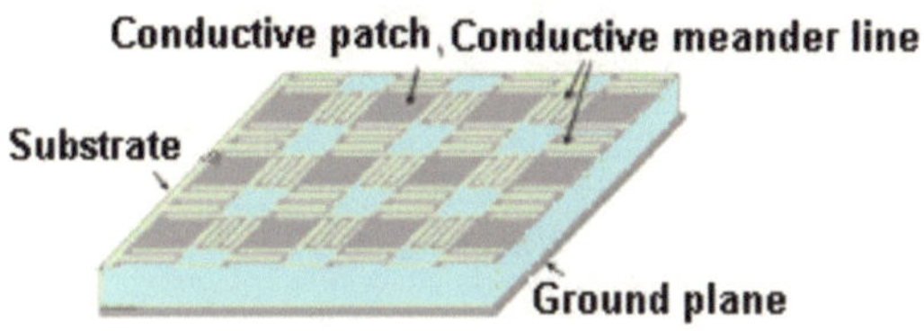

Rectangular patches conductively connected with meander strip lines

Spiral slotted rectangular patches

Fig. 46. Various printed schemes suitable for EBG implementation

Many articles have reported the contribution of the EBG surfaces to the enhancement of the operation of antennas incorporated in their structure. Some typical applications are referred in the following.

In [82] a structure consisted of a 'diamond dipole' of two square patches over a mushroom type EBG was studied(Fig. 42a). The suitable parameter values of the scheme yield the results shown in fig. 47b. The presence of the EBG drives the dipole to operate effectively within a bandwidth much wider than that of the ordinary metallic ground sheet.

The configurations shown in figures 48 and 49 concern in antenna array arrangements, in which reduction of mutual coupling between their elements was obtained, via EBG schemes.

Fig. 48b depicts the scattering parameter S_{12} between the printed elements in the arrangement shown in fig. 48a. The significant reduction of S_{12} in the band of EBG resonance, proves the effectiveness of the EBG lattice, embedded between them, in reducing their mutual coupling[83].

In [84] an antenna array of microstrip elements with high directivity was designed. A large directivity can be obtained by an array with large size and it means that the distance between the elements has to be large. However the large distance would involve undesirably high side lobes. So, to obtain high directivity the designers chose to keep the inter-element distance small and to increase the size of the patch. Small inter-element distance causes strong mutual coupling between the elements of the array. This problem was confronted inserting an EBG configuration between adjacent elements as shown in

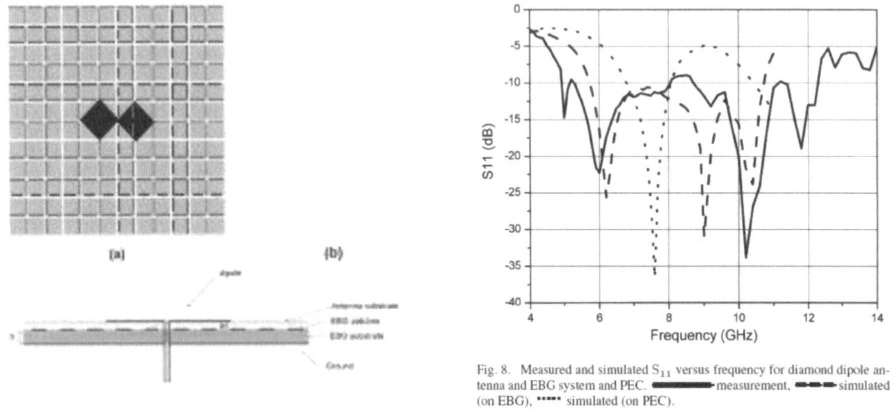

(a) (b)

Fig. 47. a)Top and side view of a diamond planar dipole antenna over a mushroom type EBG b) the broadband performance of the structure: scattering coefficient of the signal at the feeding probe.

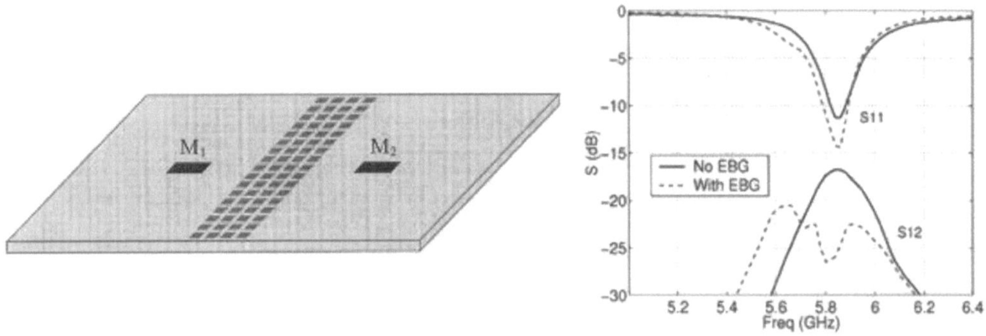

(a) (b)

Fig. 48. a)Microstrip antenna separated by the mushroom-like EBG b) Mutual coupling between the patches with and without EBG

Fig.49a. The increasing of the patch's size was achieved using a dual layer dielectric substrate. A large permittivity was selected for the lower layer while the upper layer was the air. The additional advantage of this configuration is that a substrate with large thickness and low effective permittivity was created and as a result the bandwidth was enhanced. Moreover the patches of the EBG printed on the top of the lower layer with high permittivity had size enough small, to be placed between the antenna radiating patches. In the proposed array, by proper selection of the structural parameter values, a reduction of mutual coupling about 10dB or 15dB, depending on the frequency bandwidth, was obtained.

Another representative example of mutual coupling reduction using EBG in a microstrip antenna array, was recently proposed in [85]. In this configuration (fig. 49b) the thickness of the substrate was large and this fact permitted to embed inside it an elongated mushroom EBG lattice(fig. 49c). By this novel configuration, isolation exceeding 12dB, was obtained between the antenna array's elements.

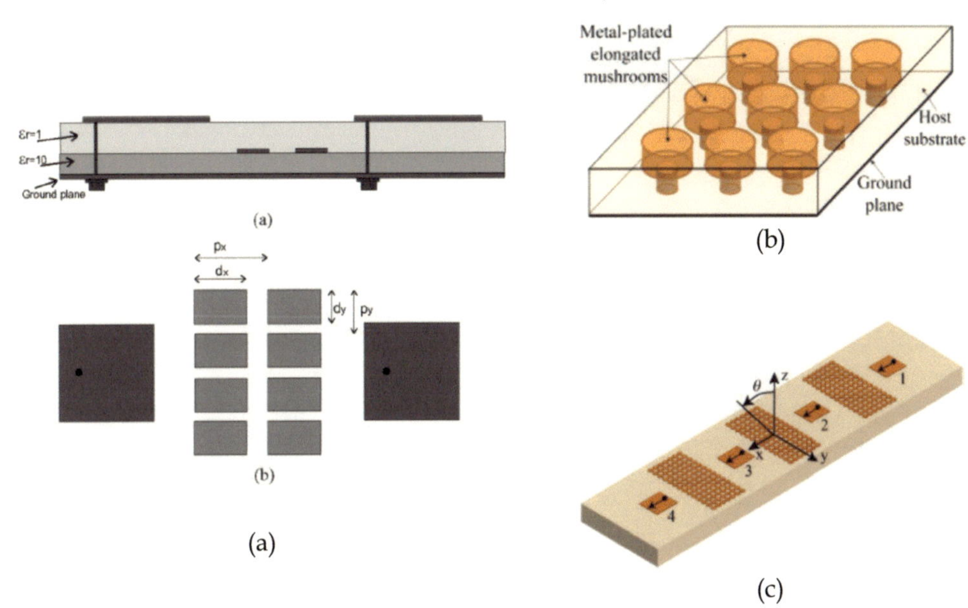

Fig. 49. Printed antennas integrated with EBG Structurs for mutual coupling reduction:
a) Array of two patches structured in a dual-layer substrate. The EBG embedded between the patches reduces the mutual coupling b) a novel type of compact elongated mushroom EBG (EM-EBG)structure c) Linear array of four microstrip antenna elements with isolating (EM-EBG) objects

EBG structures are also effective to the design of dual frequency microstrip antennas. It has been proved, in [86], that a properly designed EBG lattice can drive a printed antenna to dual frequency operation, exploiting the higher order modes of the printed element. Generally speaking, a microstrip antenna would be used as dual frequency band radiating system when operates at the fundamental and at one of the higher order resonance frequencies. However, the distributions of the radiated fields, at these two frequencies, are not similar. So, the two bands can not be used for the same type of services. In [86], this problem was confronted integrating the antenna with an EBG surface and dual frequency operation with similar radiation patterns was obtained. The key of this performance is that the antenna works at the two bands with different dielectric layers.

As an application example in[86], results for a 'diamond' antenna consisted of a pair triangular patches were presented(fig. 50). The EBG structure was formed of equilateral triangular microstrip patches.

In figure 51a the variation of the input impedance of the twin triangles without the EBG is illustrated. The triangle patches of the diamond were designed for the TM_{10} mode at 3.17GHz. At the figure is shown that the second order mode, namely the TM_{11} ,appears at 5.6GHz. In figures 51b and 51c the radiation patterns of both modes are depicted. Obviously the patterns are different, whereas the disadvantage of the pattern of TM_{11} , is the deep minimum at the broadside.

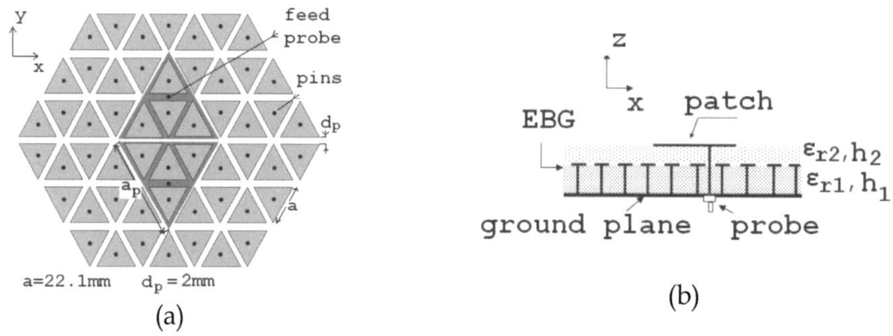

Fig. 50. Top and side view of a triangular probe fed dual patch antenna over an EBG lattice of triangular printed elements.

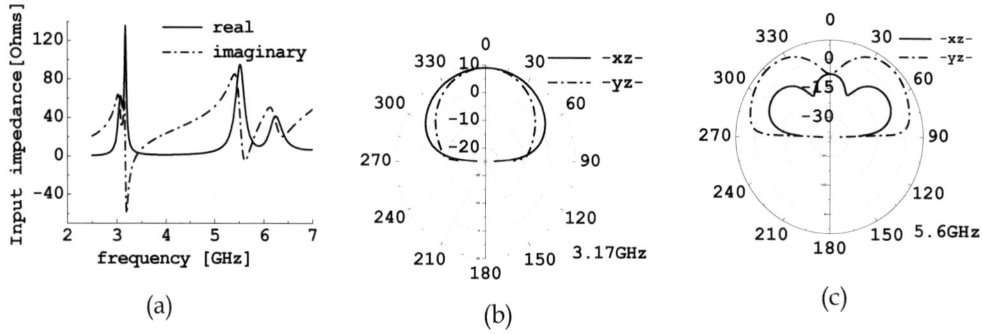

Fig. 51. (a) Input impedance of the patch antenna of fig 50, without EBG. (b), (c) the xz-and yz-plane directivity patterns of the TM_{10} (3.17GHz) and TM_{11} (5.6GHz) modes.

In figure 52 the respective results for the diamond antenna integrated with the EBG are presented. The EBG was designed to resonate in a band around 5.6GHz, namely in the band of the second order mode of the diamond. In fig 52a a new resonance appears at 4.3GHz. It is due to the fact that this frequency is out of the EBG band , thereby the EBG performs as an electric conductor, namely as an ordinary metallic grounded sheet. So, the entire structure operates as a microstrip antenna with a single dielectric layer, which is the upper dielectric layer, operating at TM_{10}. It can be also verified using the expressions for the frequency of resonance of a triangular microstrip patch.
The radiation pattern at 4.3GHz is depicted in fig. 52c and is a pattern similar to respective ones of microstrip antennas when operate at the fundamental mode. The radiation pattern at 5.6GHz is depicted in figure 52d. Around this frequency the EBG is in resonance. Comparing the figures 51c and 52d it is ascertained that the diamond exhibits a different performance in the presence of the EBG. The pattern of 52d is similar to that of the fundamental frequency rather, than to that of the ordinary second order mode. It is noted that at both frequency ranges the diamond is driven only by its feeding probes and the matching is satisfactory, as shown in figure 52b. So, the hybrid structure, microstrip diamond-EBG, operates at two frequencies with similar radiation features.
The electric current distribution gives an explanation of the structure's performance. In Fig. 53a the current at 3.17GHz, namely the TM_{10} mode of the antenna without the EBG, is shown. This distribution is similar to that of the antenna with the EBG at 4.3GHz (Fig.53c).

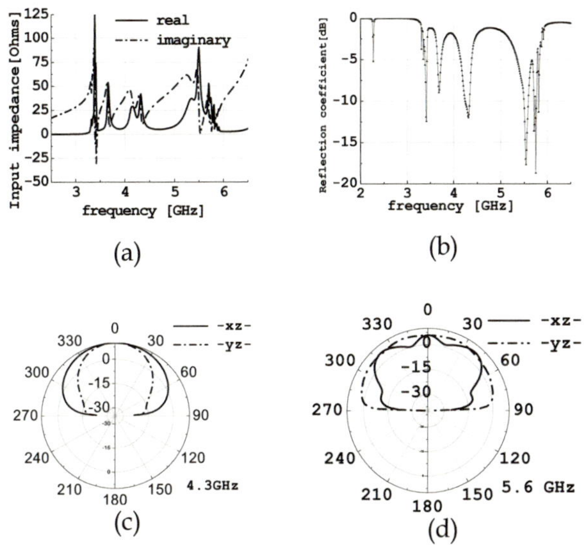

Fig. 52. (a) Input impedance of the dual triangular patch antenna with EBG (b) The reflection coefficient at the feed ports. (c), (d) the xz-and yz-plane directivity patterns of the TM_{10} (4.3GHz) and TM_{11}(5.6GHz) modes.

In this case, current does not exist on the EBG surface patches. That's why the patterns of directivity shown in Figures 51b and 52c, are similar. Comparing the current patterns of the TM_{11} mode without and with EBG(Figs. 53b, 53d), we see that the current distribution is modified. Intense currents flow on the EBG which resonates at this frequency. This performance leads to the field depicted in Fig. 52d.

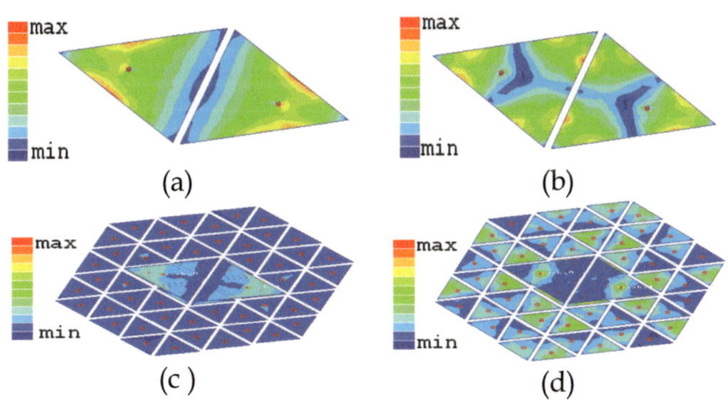

Fig. 53. (a),(b) Current distribution, at 3.17 GHz(TM_{10}) and 5.6GHz(TM_{11}), on the surface of the triangular dual patch antenna without EBG. (c) and (d) The current distribution, at 4.3GHz and 5.6GHz when the EBG was embedded in the structure. In all cases the phase difference between the probes is 180°.

It is worth to mention that the above described performance would lead to a design procedure for various pairs of operation frequencies and various shapes of antenna and EBG patches. The pair of frequencies have been given, the steps of the design are summarized as follows:

a. The smaller frequency of the pair is defined as the f_{10} of the radiating patch and the larger one as the frequency f_{mn} of one of the higher modes as well as of the resonance of the EBG.
b. Taking into account that at f_{10} the EBG will perform as a perfect electric conductor and the patch will work solely with the upper dielectric layer, we select the values of the structural parameters of this layer and calculate the size of the patch.
c. At the selected f_{mn} the EBG resonates, so the patch works with both dielectric layers and we determine the characteristics of the lower layer in order the patch, with the already calculated size, to resonate at f_{mn} with the double dielectric substrate.
d. We calculate the size of the EBG patches to resonate at f_{mn} in the determined dielectric environment.

For example, suppose that 900MHz(GSM) and ~2GHz(UMTS) are the parts of the frequency pair. We set f_{10}=900MHz and f_{11}=2GHz. Then selecting for the upper layer, $\varepsilon_{r2} = 4.4$ and h_2=0.5mm, the length of the side of the antenna has to be a_p=105mm , in order to resonate at 900MHz. The calculations are made by equations 19 and 20. In order the patch, with side length a_p=105mm, to resonate at 2GHz with the TM_{11} mode, in the presence of the double dielectric layer, the lower layer must have $\varepsilon_{r1} = 2.2$ and h_1=1.5mm. The calculations are made by equations 19, 20 and the expression $\varepsilon_r = \sum_{i=1}^{2} h_i \Big/ \sum_{i=1}^{2} \left(h_i / \varepsilon_{ri} \right)$. The side length of the EBG patch has to be a=65mm in order to resonate at 2GHz, if structured between the determined dielectric layers. These values are of first order approximation. The simulation of the configuration, in virtue of the interaction between the antenna and the EBG, is expected to show that a small perturbation of the size of the elements is necessary to resonate at the pre-specified frequencies.

4. Stochastic algorithms for antenna design and optimization

As in the previous sections has been discussed, the design of a printed antenna with advanced characteristics of operation for multi-frequency applications is often a laborious and difficult procedure. To obtain such a performance, the techniques of texturing the surface of the printed patches, the utilization of fractal design procedure or the incorporation of EBG objects are efficient tools for the design, but they are not always enough. Moreover they produce complicated antenna schemes inserting additional parameters that have to be taken into account during the design process. Besides all these, there is a trade off between the antenna's properties which are imposed by the technical requirements. For example, the small size opposes the high gain; the small thickness of the substrate reduces the surface waves and, in this way, the mutual coupling, but also reduces the bandwidth; multi-frequency operation is, in general, not difficult to be achieved but it is not always easy to obtain the frequencies specified by a particular application. On the top of all, there is the difficulty that in most of the cases there are not mathematical expressions in closed form that could help the designer of the antenna. Solution to the above problems would be given by evolutionary techniques, most of which are inspired from the nature and the biology. The Neural Network(NN) method, the Genetic Algorithms(GA), The Particle Swarm Optimization(PSO) are some of the most commonly used methods which were proved efficient in antenna design and more generally in many electromagnetic problems.
In the following a brief discussion and a very simple analysis of the NNs and GAs and will be done. The detailed study is beyond the scope of this chapter and would be the content of a separate self-existent book.

4.1 Artificial Neural Networks

A Neural Network is a network of many simple processors connected by communication channels that carry numerical data[87],[88]. The NNs' structure resemble that of the human brain and their operation also imitates the operation of the brain. Two are the basic similarities between them: a)Knowledge is acquired by both through the learning process and b)Interneuron connection strengths, known as synaptic weights, have a small amount, of local memory and are used to store the knowledge. It is exactly the property of the NNs to work as the human brain that make them effective in antenna design. As mentioned above, in an antenna design problem the given data are the technical requirements, for example the gain, the frequency bands of operation etc and what is demanded is the configuration of the antenna. The difficulty in finding the solution of this problem is that the relation between the given and the demanded quantities can not be expressed in closed form and moreover is usually complicated and non linear.

The NN algorithms are ideal for solving problems of the above type, because, working as the human brain, can be trained to find the underlying relation between the values of the antenna parameters, given to them, and the operational features at which these parameters would drive the antenna.

Two are the basic steps when decide to construct and employ a NN: to select the NN's architecture and the strategy of its learning. For all that, it has to be pointed out that it is difficult to know a priori if a specific NN architecture or learning strategy is suited for a given antenna design problem. So, these selections are substantially a part of the whole problem.

Figure 54 depicts the architecture of a fundamental fully feed-forward NN with Multiple Layer Perceptron Architecture which is commonly used in antenna design.

In this fundamental form the NN consists of three layers. The input layer that includes a M nodes. One (or more) 'hidden' layers with K nodes and the output layer that includes J nodes. From the mathematical and computer code point of view, the M nodes of the first layer are the positions at which the data x_i , $i = 1, 2 \ldots M$ of the problem will be given. The synapses, or connecting links, between these nodes and those of the hidden layer will transfer the signals, namely the input data, from one layer to the other weighting them by an amount, depending on the link. Each node of the hidden layer gather the accepted signals, adds them and uses this summation as argument of a selected function called 'activation' function. The results are led to the output nodes. The arithmetic values y_j that appear in the output nodes are substantially the response of the NN and are the results of the problem under solution. The value of M is equal to the number of data and J is equal the required parameters. The number of the hidden layers and their nodes, the type of the activation function, the interconnection, or not, of all the nodes are part of the study of the NN.

In the case of the design of a rectangular printed antenna, for example, if the technical requirements are a specific value of gain and the resonance frequency, two input nodes are necessary. The output nodes would potentially be four, giving the values of the lengths of the sides of the patch the dielectric constant and the height of the substrate.

In order to receive from the NN the correct response it must be trained at first. The training is the fundamental procedure of a NN and the first step of this procedure is the construction of a training set. It has to contain a number N_t of pairs of vectors. If , for example we suppose that x_i , $i = 1, 2 \ldots M$ describe the parameters of operation(gain, bandwidth, etc) and y_i , $i = 1, 2 \ldots J$ stand for the geometry and material parameters of the antenna, the first part

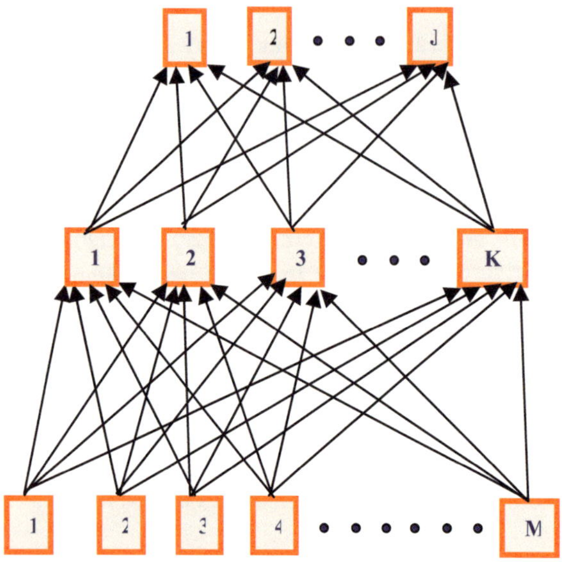

Fig. 54. The basic fully feed forward NN with Multiple Layer Perceptron Architecture

of the pair is the input vector containing the values of the input parameters x_i, whereas the second part of the pair is a vector with the y_i values, that result from the x_i and the functions that relate all x_i and y_i. Exactly these functions, which are unknown, are the cause we turn to the employment of NNs techniques. A well trained NN would successfully make the mapping from the x_i parameters to the y_i results, thus finding the relation between them without be necessary to find the mathematical function which connects them. In the cases of antenna design the training set can be created simulating the performance of N_t antennas with N_t different combinations of parameter values y_i and recording the corresponding x_i values.
The training of the NN is made by inserting at its input repetitively and in random order, the input vectors of the training collection and simultaneously presenting at its output nodes, the respective output vector. If the actual NN output does not match the respective output then the NN' S interconnection weights are modified to move the actual output closer to the desired output. The criterion of this matching is an error function defined as

$$E_p(w) = \frac{1}{2}\sum_{i=1}^{J}[d_j(p) - y_j^o(p)]^2 \tag{56}$$

where $d_j(p)$ is the value at the jth output node, expected in accordance with the jth input element of the pth training set, and $y_j^o(p)$ is the value of the node calculated by the the NN with a specific set w of weights.
The above process is repeated for all the input -output pairs of the train collection until we reach a point, that is a set of proper weights, where the actual output of the NN for every input of the collection matches to our satisfaction the desired output, namely the minimization of the function $E_p(w)$. Then, it is considered that the learning process is over and that the trained NN has successfully approximated the desired mapping between the inputs and the corresponding outputs of the collection and in this way has found out the

non linear relation between the configuration parameters y_i of the antenna and its operational features. To test the effectiveness of the NN, it is exposed in a collection of input-output vectors which were not included in the training set. If the results are satisfactory the NN is ready for generalization. In other words we can give to its input a set of values of desired properties of operation and the well trained NN will give us the parameter values to construct the antenna that will have these properties.

The described learning strategy is characterized as 'supervised'. The main feature of the process is that the NN 'knows' externally the expected output vector for each pair of vectors it accepts during the training. So, it construct an error quantity defined as the difference between the actual response of the network and the desired response of the network and the NN is trained trying to minimize the error quantity. Alternate categories of learning are the unsupervised and the hybrid learning. In unsupervised or self-organized learning, we can not provide the NN with a desired or target response for each training vector it accepts at its input, because it is not known. Instead, a provision is made to identify a measure of the quality of the representation that the network is required to learn and the free parameters of the network are optimized with respect to that measure. On the other hand, there are problems at which a purely supervised learning procedure is not very efficient or many input-output training pairs have to be used. In the last case the difficulty is that as the training data set increases in size the training time of the multilayer perceptron increases exponentially. A way of dealing with this problem is to incorporate unsupervised and supervised learning. This is what we call hybrid learning. In this approach we first group the data into a number of categories which are a lot fewer than the number of input-output pairs in the original training set. It is accomplished by using an appropriate unsupervised learning algorithm. Having reduced the training data set significantly, we can subsequently utilize a supervised learning NN architecture.

In many applications of NN in microstrip antenna design, the training vectors include the parameters of the geometry of the antenna and those of the substrate and the required results at the output of the NN are the indices of operation, that is the gain, the resonant frequency, the bandwidth, the polarization etc,[89], or the radiation pattern[90]. In [40], however, the inverse problem is solved. Vectors of the sampled radiation pattern are used for the training of the NN, whereas the structural parameters of the antenna are provided by the NN at the output nodes. By the constructed NN, the synthesis of a dual band, multilayered printed antenna , textured with slits was obtained. The NN methodology has also been used successfully in the synthesis of EBG surfaces with specific reflection phase characteristics[91], suitable for low profile polarization diversity antenna systems.

It is worth noting that the solution given by the NN is not necessarily the optimum. This fact is due to the nature of the NN algorithms. A NN gives an answer that fits to the imposed requirements which has accepted, in the form of data, and if it is well trained the results fulfils the requirements. In many applications it is enough. What would be done in the case in which the prescriptions are imposed in the form of constraints? For example suppose to design an antenna with gain value greater, as possible, than a specific value and Side Lobe Level lower, as possible, than a pre-defined value. This type of specifications have the meaning of optimization and techniques with a different nature would be used. An ordinary efficient method for this purpose is a Genetic Algorithm.

4.2 Genetic Algorithms(GA)

Optimization is generally the process of making something better. The process consists in making variations on an initial concept and using the gained information to improve the ideas. From the application side of view, optimization is the process of adjusting the inputs or the characteristics of a device or generally a creation, using a mathematical or experimental process in order to find the minimum or maximum output results. In antenna applications the scope of an optimization procedure is to find out not just a solution but the best one for a specific application. The terminology of 'best' solution implies that a)there is more than one solutions and b) the solutions are not of equal value. The definition of 'best' is relative to the application, namely the technical specifications imposed, and the tolerance allowed.

One of the most commonly techniques for antenna optimization is the employment of Genetic Algorithms[92]-[94]. GAs model natural selection and evolution using biological processes to optimize highly complex functions. The method was developed by John Holland, in 1975, and finally popularized by one of his students David Goldberg, in 1989.

Genetic Algorithms have been widely used in antenna synthesis and optimization problems. They have been proved efficient in finding the proper excitations or the suitable division of the arrays in sub-arrays in order the antenna to produce a desired radiation pattern with respect to the SLL and the directions of the nulls, to optimize the antenna by broadening the frequency bandwidth or in thinning an array the target being, in all cases, the achievement of special performance features. GAs are also a frequent choice for the design and optimization of microstrip antennas, especially broadband, slit textured, fractal printed, or not, as well as optimized EBG antennas with enhanced gain [95]-[98].

Several antenna optimization problems are characterized multi-dimensional. It means that there are more than one parameters that are taken into account for the evaluation of their performance. These parameters are ordered in a vector $\overline{p} = [p_1, p_2, ...p_N]$, where N is the number of the parameters, termed chromosome. At the same time the problem would be multi-objective[97], if the indices of performance are more than one; for example , the gain, the SLL, bandwidth etc. These indices are functions, defined as $g(\cdot)$, of the parameters of the problem, termed objective functions. At each chromosome $\overline{p}$, a set of values of these functions corresponds, making a vector $g_C(\overline{p}) = [g_1(\overline{p}) - c_1, g_2(\overline{p}) - c_2, ...g_M(\overline{p}) - c_M]$ where M is the number of the functions-indices and c_i is a constrained value for the respective function. Vector $g_C(\overline{p})$ is termed, cost or fitness vector and these constrained values are imposed by the requirements of the problem. The aim of the GA is to minimize the cost vector.

The general steps of a GA are the following:

Step 1. Generation of an initial population of N_{pop}^{in} chromosomes $\overline{p}$ applying a procedure of random selection. This population, termed generation, does not remain constant but changes gradually in the following steps.

Step 2. Calculation of the cost vectors for all chromosomes, $g_{ci}(\overline{p}^i)$, $i = 1,2,...N_{pop}^1$

Step 3. From this population only a number of members is kept for the next generation whereas the rest are discarded. The selection of the members which will survive is made by pre-specified constraints expressed by qualities or inequalities. The chromosomes which survive undergo the process of mating: Pairs of chromosomes are selected and then are used as parents for the generation, via pre-specified rules, of a new pair of offspring. In these rules, steps of randomness are incorporated. The total of the N^i offspring produced in this way constitute a new generation.

Step 4. The cost vectors $g_{ci}(\overline{p}^i)$ for all i=1,2....N^i chromosomes are calculated.

Step 5. Using, potentially, a threshold value for each component of each cost vector, or for the total of them, we select a subtotal of the N^i chromosomes which will be the population of the next generation. The thresholds may be not constant and the new generation includes the individuals that correspond to all the $g_{ci}(\overline{p}^i)$ which passed the test of thresholding and at the same time ensure the feasibility of the antenna. In this way the number chromosomes in the populations gradually decrease and better performers are included from generation to generation.

Step 6. The process goes back to step 3 for the mating of the members of the new generation. This procedure is repeated iteratively, either a pre-specified number of times or until all the chromosomes of the population fulfill a pre-defined criterion.

Additional steps as the mutation, which is a separate process to change the chromosomes or/and elitism strategies can be incorporated in the GA depending on the problem. Elitism strategies give solution to the following problem: during the process of a simple genetic algorithm it is possible for the next generation to have a best individual with a lower fitness, with concern to the cost function, than the best individual encountered in a preceding generation. This loss of the best individual occurs due to the probabilistic nature of the GA selection and mutation. A simple test can be added to verify that the best individual in the new generation is at least as good as the one from the preceding generation. Saving and inserting the best individuals from the last generation is known as an elitist action. Elitism can be used to ensure that there is a monotonic increase in the best fitness in the population as a function of time during the GA process.

The basic characteristics and the advantages of the GAs are summarized as follows:

1. The GAs can be used in problems with one or more objective functions. In these cases of more than one optimum solutions would be produced.
2. Constraints, which ensure the feasibility of the antenna solutions and the their compliance the technical specifications can be incorporated.
3. In GAs the population from one generation to the next changes through mutation and perhaps additional constrains imposed by the problem under solution. In this way, the space of possible solutions is explored more efficiently and the GA, looking simultaneously for more than one minima, can give a final solution, which in the case of one objective function, would be not simply a local optimum but a global optimum.
4. The GAs start with a population of feasible solutions instead of a single one(as in more traditional approaches)
5. Gas use probabilistic transition and selection rules, not deterministic rules.
6. GAs optimize with continuous or discrete parameters
7. A GA doesn' t require derivative information
8. GAs deal with a large number of parameters
9. GAs optimize parameters with extremely complex cost surfaces because they can jump out of a local minimum
10. Works with numerically generated data, experimental data or analytical functions

It has to be pointed out that , as in the case of NNs, it is difficult to know a priori if a specific GA is the proper for a given antenna design problem or how to improve the GA when it fails. So, an extended study is often necessary and the study concerns not only the specific antenna problem but the GA itself.

5. References

[1] R. Chair, K.M. Luk, and K.F. Lee, ' Miniature multilayer shorted patch antenna,' *Electron. Lett.*, Vol. 36, pp. 3-4, Jan. 6, 2000

[2] L. Zaid, G. Kossiavas, J. Dauvignac, J. Cazajous, and A. Papiernik, 'Dual-frequency and broad-band antennas with stacked quarter wavelength elements,' *IEEE Trans. Antennas Propagat.*,Vol. 47, pp. 654-660, April 1999

[3] Kin-Lu Wong, *'Compact and broadband microstrip antennas'*, John Wiley & Sons, New York, 2002.

[4] C.K. Wu and K.L. Wong, 'Broadband microstrip antenna with directly coupled and gap coupled parasitic patches', *Microwave Opt. Technol. Lett.*, Vol. 22, pp. 348-349, Sept. 5, 1999

[5] J.Y. Sze and K.L. Wong, 'Broadband rectangular microstrip antenna with a pair of toothbrush-shaped slots', *Electron Lett.*, Vol. 34, pp. 2186-2187, Nov. 12, 1998.

[6] J.Y. Sze and K.L. Wong, 'Single-layer single -patch broadband rectangular microstrip antenna,' *Microwave Opt. Technol. Lett.* , Vol. 22, pp. 234-236, Aug. 20, 1999.

[7] C.-L. Li, H-,H. Wang, H.-J.Lin, X-.E. Shi, W.-T. Li and L. Xu, 'Analysis and design of broadband microstrip patch antenna with a pair of double cross-shaped slots', *Proc. Of International conference on microwave and millimeter wave technology(ICMMT)*, 2010.

[8] T. Huynh and K. F. Lee , 'Single-layer single-patch wideband microstrip antenna,' *Electron. Lett.* , Vol. 31, pp. 1310-1311, Aug. 3, 1995

[9] K.L. Wong and W.H. Hsu, 'Broadband triangular microstrip antenna with U-shaped slot', *Electron. Lett.* , vol. 33, pp. 2085-2087, Dec. 4, 1997

[10] S. Weigand, G.H. Huff, K.H. Pan, and J.T. Bernhard, 'Analysis and design of broadband single-layer rectangular U-Slot microstrip patch antenna,' *IEEE Transactions on Antennas and Propagation,* AP-51, 3, pp. 457-468, 2003.

[11] K.F. Tong and T.P. Wong, 'Circularly polarized U-slot antenna', *IEEE Transactions on Antennas and Propagation*, AP-55, 8, pp. 2382-2385, 2007

[12] K.F. Lee, S.L. Steven Yang, A. A. Kishk, 'Dual and multiband U-slot patch antennas', *IEEE Antennas and Wireless Propagation Letters*, Vol. 7, pp. 645-647, 2008

[13] H. Wang, X.B. Huang and D.G. Fang, 'A single layer wideband U-slot microstrip patch antenna array,' *IEEE Antennas and Wireless Propagation Letters,* Vol. 7, pp. 9-12, 2008

[14] Lingjian Li and Zhirum Hu, 'A wideband linear U-slot microstrip patch antenna array for wireless applications', *International Journal of Electronics*, Vol. 96, 7, pp. 755-765, 2009

[15] Kai Fong Lee, Shing Lung , Steven Yang, Ahmed A. Kishk and Kwai Man Luk, 'The versatile U-slot patch antennas,' *IEEE Antennas and Propagation Magazine,* Vol. 52, 1, pp. 71-88, 2010.

[16] K.F. Tong, K.M. Luk, K.F. Lee and R.Q. Lee, 'A broadband U-slot rectangular patch antenna on a microwave substrate', *IEEE Transactions on Antennas and Propagation,* AP-48, 6, 954-969, 2000.

[17] R. Chair, K.F. Lee, C.L. Mak, K.M. Lukand A.A. Kishk, 'Miniature wideband half U-slot and half E-Shaped patch', *IEEE Transactions on Antennas and Propagation,* AP-53, vol. 8, pp. 2645-2652, 2005.

[18] A.A. Deshmukh and K.P. Ray, 'Compact broadband slotted rectangular microstrip antenna', *IEEE Antennas and Wireless Propagation Letters* ,Vol. 8, pp. 1410-1413, 2009.
[19] Y.X. Guo, A. Shackelford, K. F. Lee and K.M. Luk, 'Broadband quarter-patch antenna with a U-shaped slot', *Microwave Opt. Technol. Lett.* , Vol. 28, pp. 328-330, 2001
[20] A. Shackelford, K.F. Lee and K.M. Luk, 'Design of small-size wid-bandwidth microstrip patch antennas', *IEEE Antennas and Propagation Magazine*, Vol. 45, 1, pp. 75-83, 2003.
[21] C.L. Mak, R. Chair, K.F. lee, K.M. Luk AND a.a. Kishk, 'Half U-slot patch antenna with shorting wall', *Electron. Lett.* , Vol. 39, 1779-1780, 2003.
[22] K.L. Wong and W.H. Hsu, 'A broadband rectangular patch antenna with a pair of wide slits", *IEEE Trans. Antennas and Propagation,* Vol. 49, 9, pp. 1345-1347, 2001.
[23] W.H. Hsu and K.L. Wong, 'A wideband circular patch antenna', *Microwave Opt. Technol. Lett.,* Vol. 25, pp. 327-328, June 5, 2000
[24] K.M. Pramod, R. Jyoti, S. S. Kumar, V.S.K. Reddy, 'Simplified and efficient technique for designing of broadband patch antenna,' *Proc. of Applied Electromagnetics Conference (AEMC)*, 2009
[25] L. Peng, C. Ruan, Y. Zhang, 'A novel compact broadband microstrip antenna,' *Proc. Asia-Pasific Microwave Conference(APMC2007)*, 2007
[26] Y. Chen, S. Yang and Z. Nie, ' Bandwidth enhancement method for low profile E-shaped microstrip patch antennas,' *IEEE Trans. Antennas and Propagation,* Vol. 58, 7, pp. 2442-2447, 2010.
[27] K.M. Pramod, S. S. Kumar, R. Jyoti, V.S.K. Reddy, and P.N.S. Rao, 'Novel structural design for compact and broadband patch antenna,' *International Workshop on Antenna Technology (IWAT)*, 2010.
[28] S. Xiao, Z. Shao, B.Z. Wang, M.T. Zhou and M. Fujise, 'Design of low profile microstrip antenna with enhanced bandwidth and reduced size,' *IEEE Trans. Antennas and Propagation,* Vol. 54, 5, pp. 1594-1599, 2006.
[29] S. Zhaohui, L. Meijia and D. Zhiyong, ' An improved design of microstrip Archimedeam spiral antenna', *Proc. of International Conference on Microwave and millimeterWave Technology(ICMMT)*, 2007
[30] L. Bian and X.-Q. Che, 'Application of the equivalent Circuit Method in Designing the Vivaldi UWB antenna', 4th *International Conference on Wireless Communications, Networking and Mobile Computing(WiCOM '08)*, 2008.
[31] X. Zhuge and A. Yarovoy, ' Design of low profile antipodal Vivaldi antenna for ultra-wideband near field imaging', *Proc. of the Fourth European Conference on Antennas and Propagation(EuCAP)*, 2010.
[32] J. George, M. Depukumar, C.K. Aanandan, P.Mohanan and K.G. Nair, 'New compact microstrip antenna', *Electron Let.* , Vol. 32, pp.508-509, 1996.
[33] A.A. Eldek, A.Z. Elsherbeni and C. E. Smith, 'Wide-band modified printed bow-tie antenna with single and dual polarization for C- and X-Band applications', *IEEE Transactions on Antennas and Propagation,* Vol. 53, no. 9, pp. 3067-3072, 2005
[34] T. Karacolak and E. Topsakal, 'A double-sided rounded bow-tie antenna (DSRBA) for UWB communication', IEEE Antennas and Wireless Proagation Letters, Vol. 5, pp. 446-449, 2006

[35] Z. D. Liu, P.S. Hall and D. Wake 'Dual-frequency planar inverted F antenna', *IEEE Trans. Antennas and Propagation,* Vol. 45, 10, pp.1451-1458, 1997.

[36] P. Salonen, M. Keskilammi and M. Kivikoski, 'New slot configurations for dual-band planar inverted -F antenna', *Microwave Opt. Technol. Lett.,* Vol. 28, No. 5, pp. 293-298, 2001

[37] M. Martinez-Vasquez, M. Geissler, D. Heberling, A. Martinez-Gonzalez and D. Sanchez-Hernandez, 'Compact dual-band antenna for mobile handsets', *Microwave Opt. Technol. Lett.* , Vol. 32, No. 2, pp.87-88, 2002

[38] M. Polívka, 'Multiband behavior of the rectangular microstrip patch antenna modified by T-notch perturbation elements', *Proc. Of 18th International Conference on Applied Electromagnetics and Communications(ICECom),* 2005

[39] I. Sarkar, P.P Sarkar and S.K. Chowdhury, 'A novel compact, microstrip antenna with multifrequency operation', International Seminar/Workshop on Direct and Inverse Problems of Electromagnetic and Acoustic Wave Theory(DIPED 2009), 2009

[40] K. Siakavara , 'Artificial Neural Network Based Design of a Three-Layered Microstrip Circular Ring Antenna with Specified Multi-Frequency Operation', *Neural Computing and Applications,* Vol. 18, pp. 57-64, 2009

[41] A. Manassas, T. Kaifas, K. Siakavara, 'Multiband printed antenna for low frequenies WLAN applications', *International Journal of Microwave and Optical Technology* , Vol 2, No 3, pp. 182-186, 2007

[42] K. Siakavara and T. Ganatsos, 'A compact quadruple crossed dipole microstrip antenna design for dual frequency and beam scanning operation', *Electrical Engineering,* Vol. 86, pp. 117-119, 2004

[43] C.A. Balanis, *Advanced Engineering Electromagnetics,* John Wiley & Sons, New York , 1989

[44] D. M. Kokotoff, J.T. Aberle and R.B. Waterhouse, 'Rigorous analysis of probe fed printed annular ring antennas', *IEEE Trans. on Antennas and Propagation,* Vol. 47, No. 2, pp.384-388, Feb. 1999.

[45] R. Garg, P. Bhartia, I. Bahl,, A. Ittipiboon, *Microstrip Antenna Design Handbook,* Artech House, Boston-London, 2001

[46] Y. Tawk, K.Y. Kabalan, A. El-Hajj, C.G. Christodoulou and J. Constantine, 'A simple multiband printed bowtie antenna', *IEEE Antennas and Wireless Propagation Letters,* Vol. 7, pp. 557-560, 2008

[47] H. K. Kan and R.B. Waterhouse, 'Shorted spiral-like printed antennas', *IEEE Trans. on Antennas and Propagation,* Vol. 50, No. 3, pp. 396-397, 2002.

[48] JoongHan Yoon, 'Fabrication and measurement of modified spiral-patch antenna for use as a triple-band antenna', *Microwave Opt. Technol. Lett.,* Vol. 48, No 7, pp.1275-1279, 2006

[49] J. Anguera, G. Font, C. Puente, C. Borjia and J. Soler, 'Multifrequency microstrip patch antenna using multiple stacked elements', *IEEE Microwave and Wireless Components Letters,* Vol. 13, no 3, pp.123-124, 2003

[50] D. H. Werner and S. Ganguly, 'An Overview of Fractal Antenna Engineering Research', *IEEE Antennas and Propagation Magazine,* Vol. 45, No 1, pp. 38-57, 2003

[51] J. P. Gianvittorio and Y. Rahmat-Samii, 'Fractal antennas: A novel antenna miniaturization technique and applications', *IEEE Antennas and Propagation Magazine*, Vol. 44, No. 1, pp. 20-36, 2002.

[52] C. A. Balanis, *Antenna Theory_Analysis and Design.* John Wiley & Sons, New Jersey, third edition, 2005.

[53] E. -O Peitgen and D. Saupe, The science of fractal images, Springer-Verlag, Berlin/ New York, 1988

[54] D.H. Werner, R.L. Haupt, and P.L. Werner, "Fractal Antenna Engineering: The Theory and Design of Fractal Antenna Arrays", *IEEE Antennas and Propagation Magazine*, Vol. 41, No 5, 37-59, 1999.

[55] K. Siakavara, E. Vafiadis, and J.N. Sahalos, "On the Design of a Direct Radiating Array by Using the Fractal Technique," *Proc. of the 3rd European Conference on Antennas and Propagation*, pp. 1242-1246, Berlin, Germany 23-27 , March 2009

[56] K. Siakavara, 'Novel fractal antenna arrays for Satellite Networks: Circular ring Sierpinski carpet arrays optimized by Genetic Algorithms', *Progress in Electromagnetic Research_PIER*, Vol. 103, pp. 115-138, 2010.

[57] K. Siakavara, 'Hybrid-Fractal Direct Radiating Antenna Arrays with Small Number of Elements for Satellite Communications', *IEEE Trans. on Antennas and Propagation*, Vol. 58, No 6, pp. 2102-2106, June 2010.

[58] C. Borja and J. Romeu, ' On the behavior of Koch island fractal boundary microstrip antenna', *IEEE Transactions on Antennas and Propagation*, Vol. 51, pp. 1281-1291, 2003.

[59] M. Bledowski and M. Kitlinski, 'Compact-planar monopole loop antennas for 802.11b/g WLAN systems', *Microwave and Optical Technology Letters*, Vol. 49, No 9, pp. 2299-2303, 2007.

[60] K. Siakavara, 'A compact fractal microstrip antenna for GPS and terrestrial radio services,' *International Journal of Electronics*, Vol. 94, No. 3, pp. 277-283, March 2007

[61] C. Puente-Baliarda, J. Romeue, R. Pous and A. Cardana, 'On the behavior of the Sierpinski multiband fractal antenna', *IEEE Transactions on Antennas and Propagation*, Vol. 46, No. 4, p. 518-523, 1998.

[62] S. R. Best, ' On the significance of self-similar fractal geometry in determining the multiband behavior of the Sierpinski gasket antenna', *IEEE Antennas and Wireless Propagation Letters*, Vol. 1, p. 22-25, 2002.

[63] Y. Na and S. Xiao-Wei, 'Analysis of the multiband behavior on Sierpinski carpet fractal antennas', *Proc. of Asia Pasific Conference-Microwave Conference*, 2005

[64] C. Puente Baliarda, C. Borja Borau, M. Navarro Rodero and J. Romeu, 'An iterative model for fractal antennas : application to the Sierpinski gasket antenna', *IEEE Transactions on Antennas and Propagation*, Vol. 48, No. 5, pp. 713-719, 2000.

[65] R. K. Mishra, R. Ghatak and D.R. Poddar, 'Design formula for Sierpinski Gasket, pre-fractal planar -monopole antennas', *IEEE Antennas and Propagation Magazine*, Vol. 50, No. 3, pp.104-107, 2008

[66] G.F. Tsachtsiris, C.F. Soras, and M.P. Karaboikis, and V.T. Makios, 'Analysis of a modified Sierpinski gasket monopole antenna printed on dual band wireless devices', *IEEE Transactions on Antennas and Propagation*, Vol. 52, pp. 2571-2579, 2004

[67] M.P. Salmasi, F. Hodjatkashani, and M. N. Azarmanesh, "A Novel Broadband Fractal Sierpinski Shaped, Microstrip Antenna" , *Progress In Electromagnetics Research,* PIER C 4, pp. 179-190, 2008.

[68] K. Zhang, Ch. Chen, Ch. Guo, J. Xu, ' On the behavior of conformal Sierpinski fractal microstrip antenna', *Proc. of International conference on Microwave and Millileter Wave Technology(ICMMT),* pp. 1073-1076, 2008.

[69] J. Yeo, and R. Mittra, 'Design of conformal multiband antennas based on fractal concepts', *Microwave and Optical Technology Letters,* Vol. 36, pp. 333-338, 2003.

[70] K. J. Vinoy, K.A. Jose, V.K. Varadan and V.V. Varadan, ' Resonant frequency of Hilbert curve fractal antennas', *Proc. of IEEE Antennas and Propagation International symposium,* pp. 648-651, 2001.

[71] K. Wang, Z. Wang, X. Lei, M. Zhou, ,Resonant frequency and sensitivity analysis of Hilbert curve fractal antennas for DRM communications', *Proc. of 8th International Symposium on Antennas, Propagation and EM Theory(ISAPE 2008),* pp. 137-139, 2008.

[72] D. Gala, J. Soler, C. Puente, C. Borja and J. Anguera, 'Miniature microstrip patch antenna loaded with a space-filling transmission line based on the fractal Hilbert curve', *Microwave and Optical Technology Letters,* Vol. 38, No. 4, pp. 311-312, 2003.

[73] G.S.A. Shaker and S. Safavi-Naeini, 'Highly miniaturized ractal antennas', *Proc. of IEEE Radio and Wireless Symposium,* pp. 125-128, 2007.

[74] J.-C. Liu, B.-H. Zeng, H.L. Chen, S.-S. Bore And D.-C. Chang, 'Compact fractal antenna with self-complimentary Hilbert-curves for WLAN dual-band and circular polarization applications', *Microwave and Optical Technology Letters,* Vol. 5, No. 11, pp. 2535-2539, 2010.

[75] K. Siakavara and F. Tsaldaris , "A Multi-wideband Microstrip Antenna Designed by the Square Curve Fractal Technique," *Microwave and Optical Technology Letters,* Vol. 41, No 3, pp180-185, 2004.

[76] D. Sievenpiper, l. Zhang , R.F. Jimenez Broas , N.G. Alexopoulos , E. Yablonovitch , 'High impedance electromagnetic surfaces with a forbidden frequency band,' *IEEE Trans. Microwave Theory and Techniques,* Vol. 47, No 11, pp. 2059-2074, Nov. 1999.

[77] Fan Yang and Yahya Rahmat-Samii, *Electromagnetic Band Gap Structures in Antenna Engineering,* Cambridge University Press, 2009.

[78] R. Collin, *Field theory of guided waves* , 2nd ed. New York , IEEE PRESS 1991

[79] N. Engheta, R.W. Ziolkowski, *Metamaterials-Physics and Engineering Extrapolations,* IEEE Press and John Wiley & Sons, 2006.

[80] S. A. Tretyakov and C.R. Simovski, 'Dynamic model for artificial reactive impedance surfaces', *J. of Electromagnetic waves and applications,* Vol. 17, No. 1, pp. 131-145, 2003

[81] F. Yang and Y. Rahmat-Samii, 'Reflection phase characterizations of the EBG ground plane for low profile wire antenna applications', *IEEE Transactions on Antennas and Propagation,* Vol. 52, No. 10, pp. 2691-2703, 2003.

[82] L. Akhooondzadeh-Asl, D. J. Kern, P. S. Hall and D. H. Werner, 'Wideband dipoles on Electromagnetic bandgap ground planes', *IEEE Transactions on Antennas and Propagation,* Vol. 55, No. 9, pp. 2426-2434, 2007

[83] F. Yang and Y. Rahmat-Samii, 'Microstrip antennas integrated with electromagnetic band-gap(EBG) structures: a low mutual coupling design for array applications', *IEEE Transactions on Antennas and Propagation*, Vol. 51, No. 10, pp. 2936-2946, 2003.

[84] E. Rajo-Iglesias, O. Quevedo-Teruel and L. Inclan-Sanchez, 'Mutual coupling reduction in patch antenna arrays by using a planar EBG structure and a multilayer dielectric substrate', *IEEE Transactions on Antennas and Propagation*, Vol. 56, No. 6, pp. 1648-1655, 2008.

[85] M. Coulombe, S. F. Koodiani and Ch. Caloz, 'Compact enlogated mushroom (EM)-EBG structure for enhancement of patch antenna array performance', *IEEE Transactions on Antennas and Propagation*, Vol. 58, No. 4, pp. 1076-1086, 2010.

[86] K. Siakavara , T. Ganatsos, 'Modification of the Radiation Patterns of Higher Order Modes of Triangular Printed Antennas by EBG Ground Planes,' *IEEE Antennas and Wireless Propagation Letters*, Vol. 8 , pp. 124-128 , 2009

[87] S. Haykin, *Neural networks - A comprehensive foundation,* (Second edition),Prentice Hall Inc. , Boston, 1999.

[88] C. Christodoulou and M. Georgiopoulos, *Applications of neural networks in electromagnetics* , Artech House, Boston, 2001.

[89] S. Lebbar, Z. Guennoun, M. Drissi and F. Riouch, 'A compact and broadband microstrip antenna design using a geometrical-methodology based artificial neural network', *IEEE Antennas and Propagation Magazine*, Vol. 48, No 2, pp.146-154, 2006

[90] D.K. Neog, S.S. Pattnaik, D. C. Panda, S. Devi, B. Khuntia and M. Dutta, 'Design of a wideband microstrip antenna and the use of artificial neural networks in parameters calculation', *IEEE Antennas and Propagation Magazine*, Vol. 47, No 3, pp. 60-65, 2005

[91] T. Ganatsos, K. Siakavara and J.N. Sahalos, 'Neural Network - Based Design of EBG Surfaces for Effective Polarization Diversity of Wireless Communications Antenna Systems', *PIERS On-line*, Vol. 3, No 8, pp. 1165-1169, 2007

[92] Haupt, R. and S.E. Haupt, *Practical Genetic Algorithms,* John Wiley & Sons, 1998.

[93] D.E. Goldberg, Genetic Algorithms in Search Optimization &Machine Learning, Addison Wesley Longman,1989

[94] Y. Rahmat-Samii and E. Michielssen, *Electromagnetic Optimization by Genetic Algorithms,* Wiley , New York, 1999

[95] R. K. Shaw and D.H. Werner, 'Design of optimal broadband microstrip antenna elements in the array environment using genetic algorithms', Proc. of *IEEE Antennas and Propagation Society International Symposium* , pp. 3727-3730, 2006

[96] R. Ghatak, D.R. Poddar, R.K. Mishra, 'Design of Sierpinski gasket fractal microstrip antenna using real coded genetic algorithm', *Microwaves, Antenna & Propagation, IET,* Vol. 3, Issue 7, pp. 1133-1140, 2009

[97] K. Siakavara, 'Novel Fractal Antenna Arrays for Satellite Networks: Circular Ring Sierpinski Carpet Arrays Optimized by Genetic Algorithms', *Progress In Electromagnetics Research,* Vol. 103, pp. 115-138, 2010.

[98] Y. Ge, K.,P. Esselle, Y. Hao, 'Design of low-profile high-gain EBG resonator antennas using genetic algorithm' , *IEEE Antennas and Wireless Propagation Letters*, Vo. 6, pp. 480-483, 2007

10

Fractal-Shaped Reconfigurable Antennas

Ali Ramadan, Mohammed Al-Husseini, Karim Y. Kabalan and Ali El-Hajj
American University of Beirut
Lebanon

1. Introduction

With the tremendous advancements in wireless communications, there is an increasing demand for miniature, low-cost, easy-to-fabricate, multiband and wideband antennas for use in commercial communications systems. As a part of an effort to further enhance modern communications systems technology, researchers have been studying different approaches for creating novel and innovative antennas. The fact that different wireless standards, such as UMTS, WLAN and WiMAX, use different operation bands, pushes the need for terminal antennas that are multiband and/or wideband. The antennas should also be well-suited in terms of cost, size, radiation patterns, gain and ease of integration in the circuit boards of communication devices.

Microstrip antennas have received increasing attention in satellite and communications applications because of their low profile, small size, light weight, low cost and ease of fabrication. Their simple feed methods, especially microstrip-line and coplanar waveguide (CPW) feeds, make them compatible with wireless communication integrated circuitry.

In this chapter, fractal-shaped and reconfigurable microstrip antennas are discussed. The space-filling and self-similarity properties of fractal geometries, from an antenna engineering perspective, are presented. Moreover, the recent techniques used in microstrip antennas with frequency-, polarization- and pattern-reconfigurability are surveyed. A separate section will focus on hybrid antenna design approaches, which combine fractal shapes and electronic reconfigurability.

2. Fractal antenna engineering

Fractal antenna engineering is a swiftly evolving field that aims at developing a new class of antennas that are multiband, wideband and/or compact in size (Werner & Ganguly, 2003). A fractal is a self-repetitive geometry which is generated using an iterative process and whose parts have the same shape as the whole geometry but at different scales, as shown in Fig. 1. Accordingly, fractal-based radiators are expected to operate similarly at multiple wavelengths and keep similar radiation parameters over several bands (El-Khamy, 2004).

Another property of fractal geometries, which makes them attractive candidates for use in the design of fractal antennas, is their space-filling property (Werner & Ganguly, 2003). Fig. 2 demonstrates the first four stages in the construction of a space-filling fractal curve known as Hilbert curve. This feature can be exploited to miniaturize classical antenna elements, such as dipoles and loops, and overcome some of the limitations of small antennas. The line that is used to represent the fractal geometry can meander in such a way that effectively fills the available space, leading to curves that are electrically long but compacted in a small physical

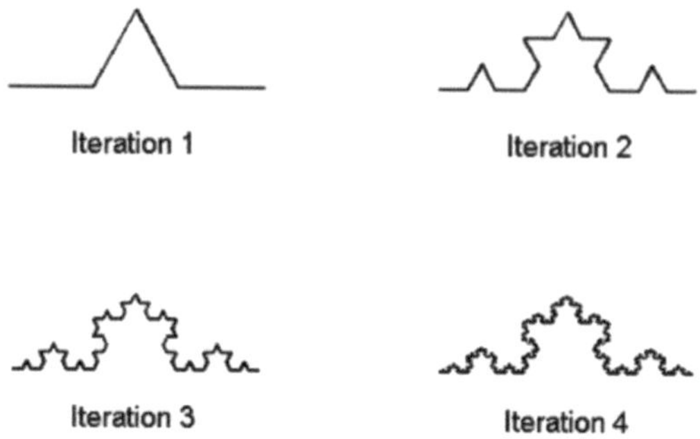

Fig. 1. The first four iterations in the construction of the standard Koch curve

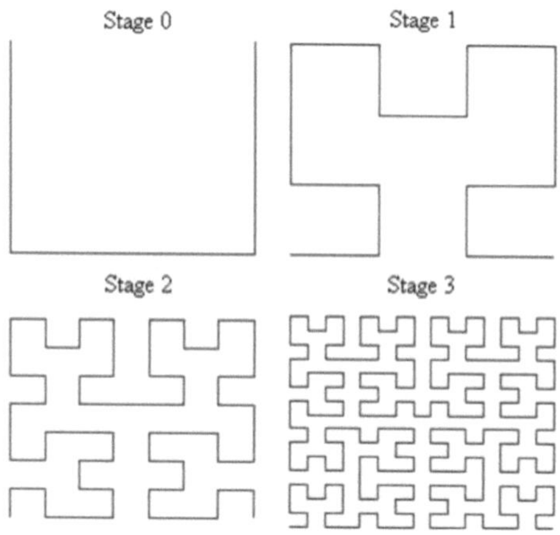

Fig. 2. The first four stages in the construction of a Hilbert Curve

space (Balanis, 2005). Fractal geometries have also been used to design antenna arrays. Fractal arrays have shown to possess desirable attributes, including multiband performance, low side-lobe levels and the ability to develop rapid beamforming algorithms based on the recursive nature of fractals. Fractal elements and arrays have been also recognized as perfect candidates for use in reconfigurable systems (Werner & Ganguly, 2003).

3. Fractal-shaped antennas

Koch curves are a good example of self-similar space-filling fractals which have been used to develop wideband/multiband and/or miniaturized antennas. In the paper by (Krupenin, 2006), it was shown that self-similar fractals affect the electromagnetic properties of antennas created on the basis of these geometries, and that Koch fractal antennas are multiband structures. (Vinoy et al., 2003) related multiple resonant frequencies of Koch fractal antennas to their fractal dimension. (Krishna et al., 2008) proposed a dual wide-band CPW-fed modified Koch fractal printed slot antenna for WLAN and WiMAX operations. In the paper by (Anagnostou et al., 2008), Koch fractal dipoles were introduced as the basic structural elements of a planar Log-Periodic Koch-Dipole Antenna (LPKDA) array, thus replacing the full-sized Euclidean monopoles. Compared to the Euclidean LPDA, the proposed design revealed very similar characteristics, while achieving 12% less space. The geometrical structure of the proposed antenna along with its VSWR and radiation pattern plots, are shown in Fig. 3.

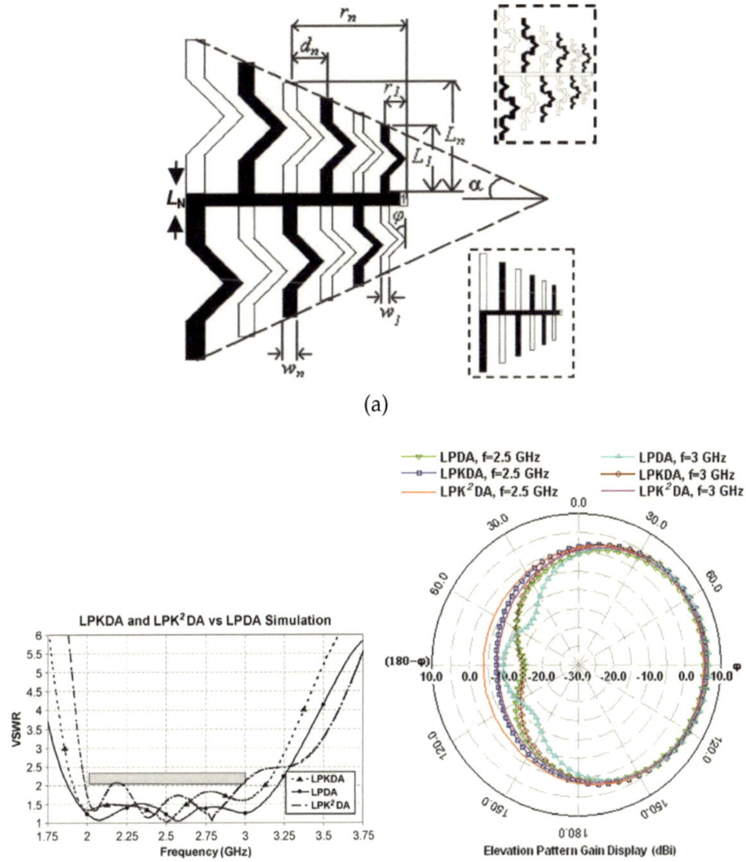

(a) (b) (c)

Fig. 3. (a) Design parameters of the LPKDA, Euclidean LPDA (bottom right), and LPK^{2}DA antennas (b) VSWR and (c) radiation pattern plots (Anagnostou et al., 2008)

Another Koch-based antenna-size-compacting scenario was proposed by (Rao et al., 2008). Here, the authors introduced a second-iterated Koch fractal, with an indentation angle of 20° as depicted in Fig. 4, along the sides of a regular Euclidean shaped patch to increase the overall electrical length of the patch. With this proposed method, a 3 dB axial ratio bandwidth of 1.2% with a minimum value of 0.31 dB is obtained at 2.245 GHz. Moreover, antennas with smaller sizes can be designed if further iterations are considered.

Another fractal geometry, which has been used to design ultra-wideband antennas, is the Sierpinski carpet. In the paper by (Ramadan et al., 2009a), rotated square slots forming a 45°-rotated 2nd-iterated-Sierpinski-carpet are integrated in the patch. Herein, the computed VSWR showed a 3.5–11 GHz impedance bandwidth for a compact 2 cm × 2.5 cm PCB antenna. Fig. 5 illustrates the geometrical structure of a compact Sierpinski-carpet-based patch antenna for ultra-wideband applications, discussed by (Ramadan et al., 2009b). The return loss of the proposed antenna is given in Fig. 6. It is worth mentioning that satisfactory omnidirectional

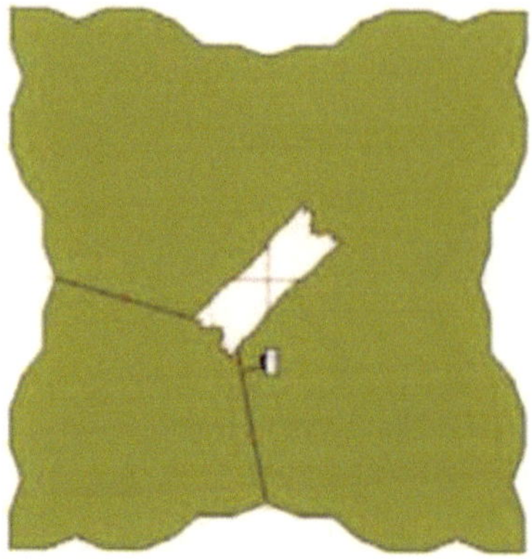

Fig. 4. Geometry of the circularly polarized antenna proposed by (Rao et al., 2008)

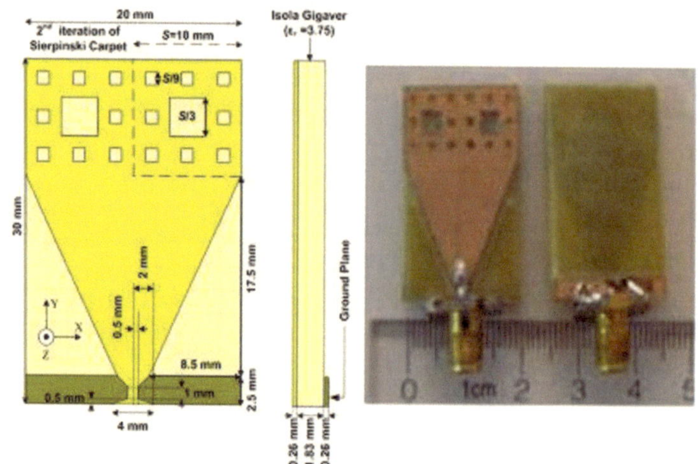

Fig. 5. Geometry and photo of the fabricated antenna presented by (Ramadan et al., 2009b)

radiation patterns, due to the self-similarity property, were attained over the 3.1–10.6 GHz frequency range.

4. Reconfigurability

Reconfigurability in antenna systems is a desired feature that has recently received significant attention in developing novel and pioneering multifunctional antenna designs. Compared to conventional antennas, reconfigurable antennas provide the ability to dynamically adjust various antenna parameters. The active tuning of such antenna parameters is typically achieved by manipulating a certain switching behavior. Reconfigurable antennas reduce any unfavorable effects resulting from co-site interference and jamming (Peroulis et al., 2005). In addition, they have a remarkable characteristic of achieving diversity in operation, meaning that one or multiple parameters, including operating frequency, radiation pattern, gain and/or polarization, can be reconfigured with a single antenna. The use of reconfigurability in coordination with a self-similar antenna leads to a considerable improvement in antenna performance. This is because not only a wider selection of frequencies is achieved, but also similar radiation properties for all designed frequency bands are obtained. Electronic,

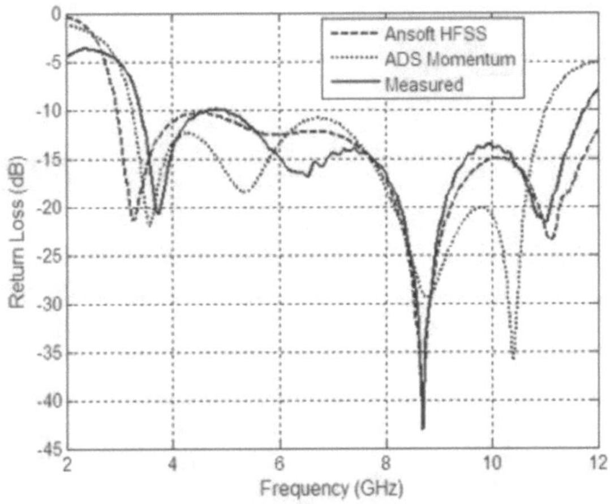

Fig. 6. Measured and simulated return loss of the antenna discussed by (Ramadan et al., 2009b)

mechanical or optical switching may be employed with reconfigurable antennas (Zhang et al., 2008). Nonetheless, electronic tunability is more frequently used because of its efficiency and reliability especially in dynamic bandwidth allocation. Electronic reconfigurability is often attained using lumped components such as PIN diodes, FET transistors or RF MEMS switches (Liu et al., 2007; Xiao et al., 2007). Compared to PIN diodes and FET transistors, RF MEMS switches have better performance in terms of isolation, insertion loss, power consumption and linearity (Xiao et al., 2007).

5. Reconfigurable antennas

Frequency-reconfigurable antennas allow either for smooth transitions within or between operating bands without jumps or for distinct switching mechanism to operate at separate frequency bands. In the paper by (Ramadan et al., 2009c), a small-sized reconfigurable antenna, which employs electronic reconfigurability, for ultra-wideband, C-band and X-band Operation was presented. The geometrical structure of the proposed antenna is given in Fig. 7. Four switching conditions were specifically selected to achieve multiband/wideband behavior, as shown in Fig. 8. For each switching condition, the antenna's computed peak gain and radiation efficiency show acceptable values, as illustrated in Fig. 9 and Fig. 10, over the frequency span of interest. The computed radiation patterns, for each case, of the proposed antenna reveal satisfactorily omnidirectional patterns over the desired frequency bands.

Pattern reconfigurability without significant changes in the operating frequency band is a desired feature in antenna systems. However, the relationship between the source currents and the resulting radiation makes this process difficult, but not impossible, to achieve. A novel frequency/pattern-reconfigurable microstrip antenna for WLAN applications was proposed by (Ramadan et al., 2010). Herein, the presented antenna features a circular patch fed using a microstrip line, a shape-optimized partial ground plane, and two PIN diodes mounted over two slots in the ground plane, as depicted in Fig. 11. Three switching cases were considered. The first resulted in a single-band operation at 5.2 GHz, whereas the other two cases offered a dual-band operation, at 2.4 GHz and 5.2 GHz. In all three cases, an omnidirectional radiation pattern was obtained in the 5.2 GHz band. However, in the two cases where operation at 2.4 GHz is possible, an equal-gain E-plane pattern and 180°-switchable H-plane patterns were

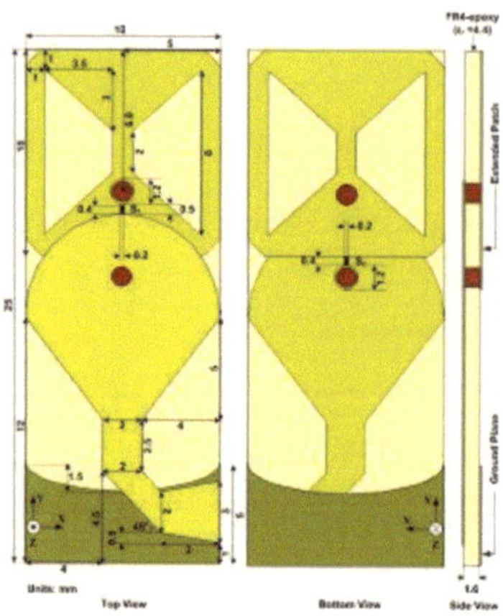

Fig. 7. Geometry of the antenna presented by (Ramadan et al., 2009c)

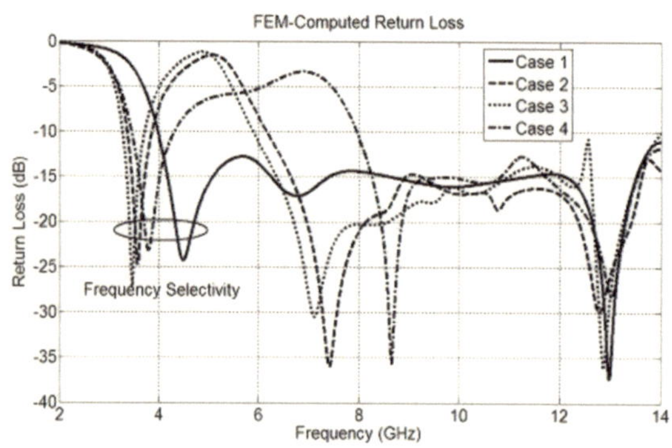

Fig. 8. Superimposed FEM-Computed S11 plots of the antenna for four switching conditions (Ramadan et al., 2009c)

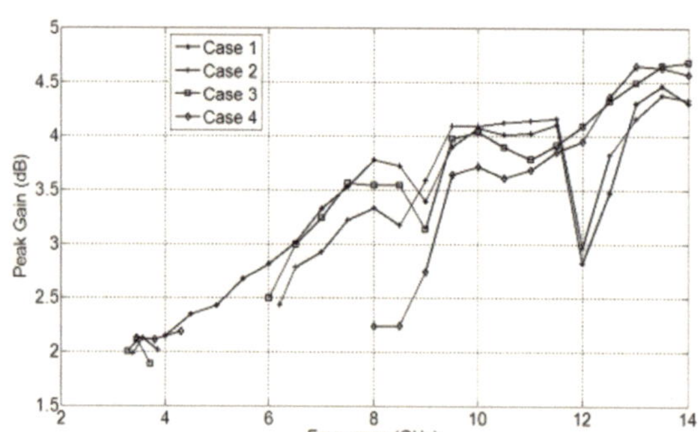

Fig. 9. Peak gain of the antenna (Ramadan et al., 2009c)

offered depending on the switching condition. The configuration, measured return loss and normalized gain patterns of the antenna are demonstrated in Fig. 11, Fig. 12 and Fig. 13, respectively.

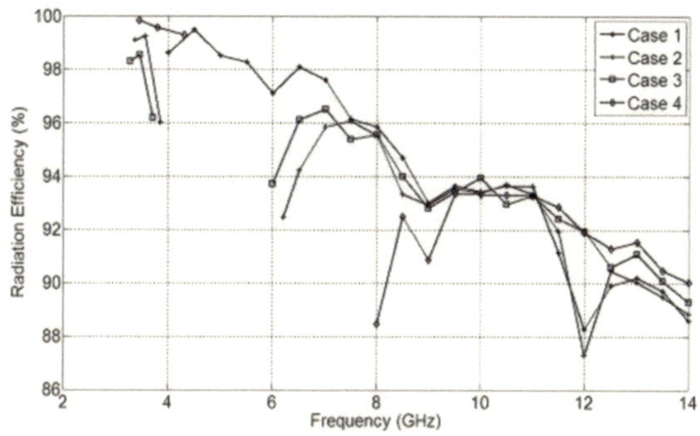

Fig. 10. Radiation efficiency of the antenna (Ramadan et al., 2009c)

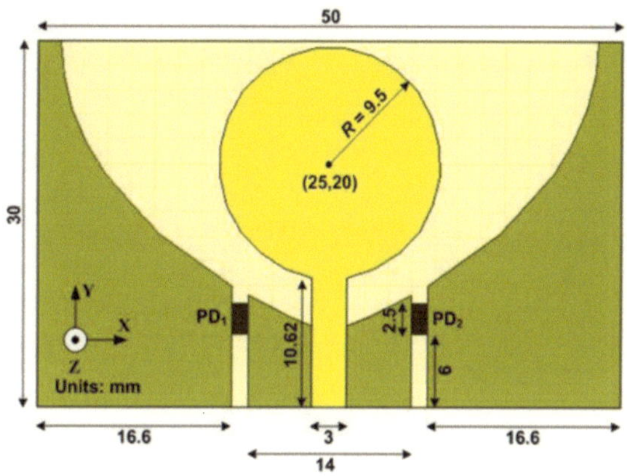

Fig. 11. Antenna configuration (Ramadan et al., 2010)

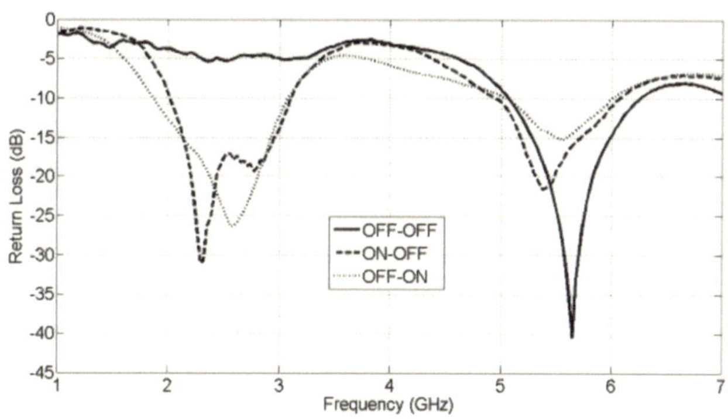

Fig. 12. Measured return loss of the antenna for the three cases (Ramadan et al., 2010)

Another type of printed antennas, which deals with both frequency and polarization reconfigurability, was reported by (Monti et al., 2009). The proposed design approach, illustrated in Fig. 14, is based on the use of two pairs of switches in order to obtain both types of reconfigurability. Specifically, three different polarization states were achieved. These are right-hand and left-hand circular polarization in one frequency band, and linear polarization in other bands, as given in Fig. 15 and Fig. 16, respectively.

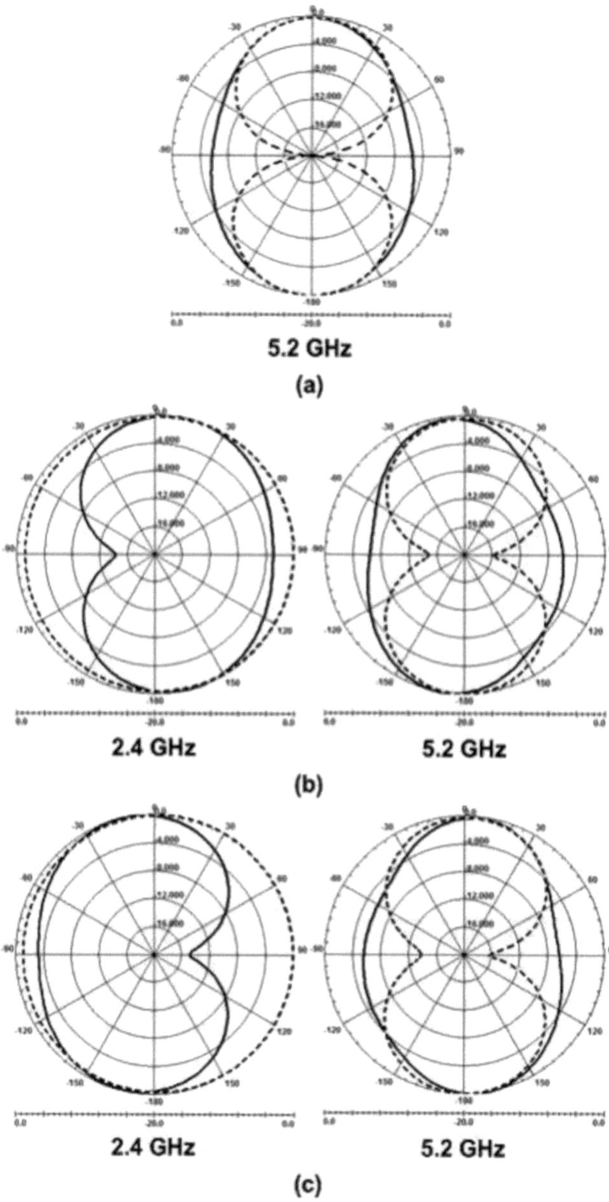

Fig. 13. Normalized gain pattern of the antenna in the X–Z (solid line) and Y–Z (dotted line) planes for (a) Case 1, (b) Case 2 and (c) Case 3 (Ramadan et al., 2010)

6. Fractal-shaped reconfigurable antennas

Hybrid antenna design approaches, which combine fractal shapes and electronic reconfigurability, are presented in this section. A low-cost multiband printed-circuit-board (PCB) antenna that employs Koch fractal geometry and tunability was demonstrated by (Ramadan et al., 2009). In their work, the authors combined the space-filling property of Koch fractal with reconfigurability in order to a multiband/wideband operation with the antenna

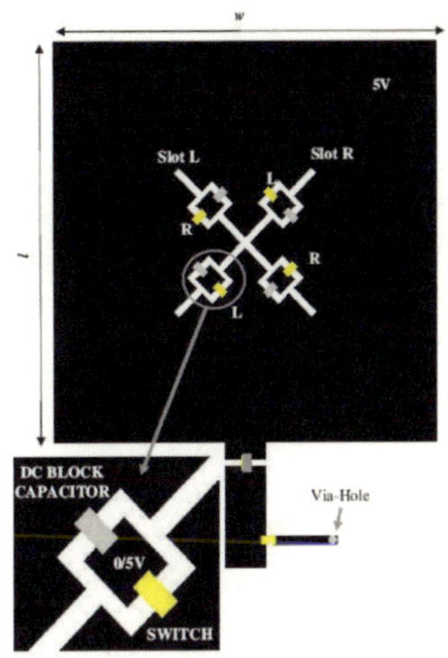

Fig. 14. A patch antenna with diagonal rectangular slots along both diagonals. Two couples (*R* and *L*) of switches are used to select the active slot (*L* or *R*) (Monti et al., 2009)

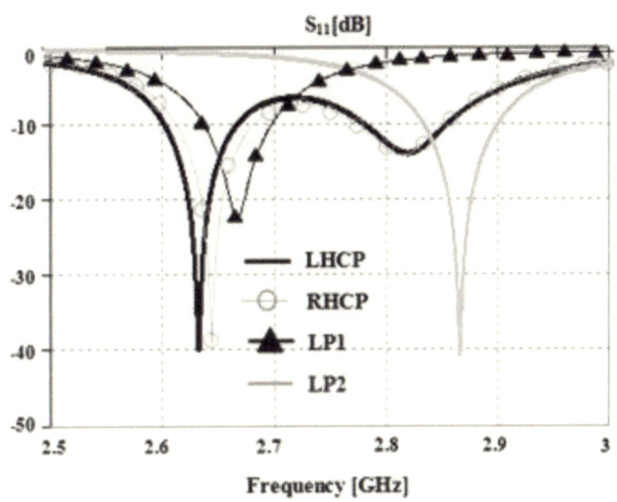

Fig. 15. Measured return loss of the antenna for several switching conditions (Monti et al., 2009)

still compact in size. A prototype of the fabricated design along with its return loss plot, for three switching conditions, is shown in Fig. 18. Fig. 19 illustrates the design of a Sierpinski Gasket-shaped reconfigurable antenna (Anagnostou et al., 2006). Herein, the authors got the switching components, mainly RF-MEMS, integrated with the proposed self-similar structure. The return loss plots of the Sierpinski Gasket-shaped reconfigurable antenna when all switches are off or on are shown in Fig. 20. Accordingly and due to Sierpinski fractal's self-similarity feature, the antenna was found to operate over three frequency bands with similar radiation patterns, as depicted in Fig. 21.

A pattern-reconfigurable fractal-shaped microstrip antenna, operable at 10 GHz, was presented by (Zhang et al., 2005). The geometry of this antenna, which is based on a Hilbert curve, is shown in Fig. 22. Eight slots were etched in the patch, and two RF MEMS switches

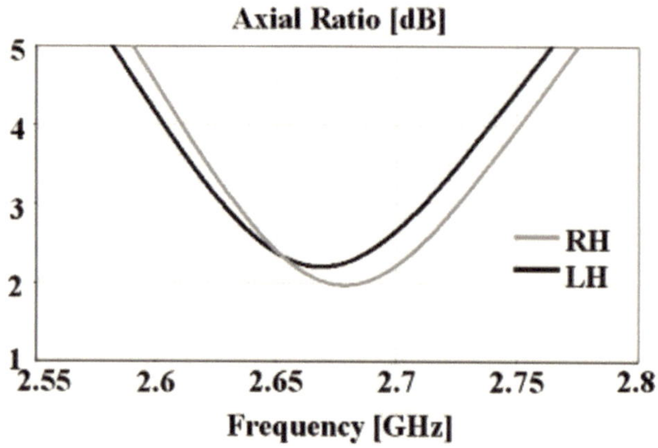

Fig. 16. Computed axial ratio of the antenna for the circular polarization cases (Monti et al., 2009)

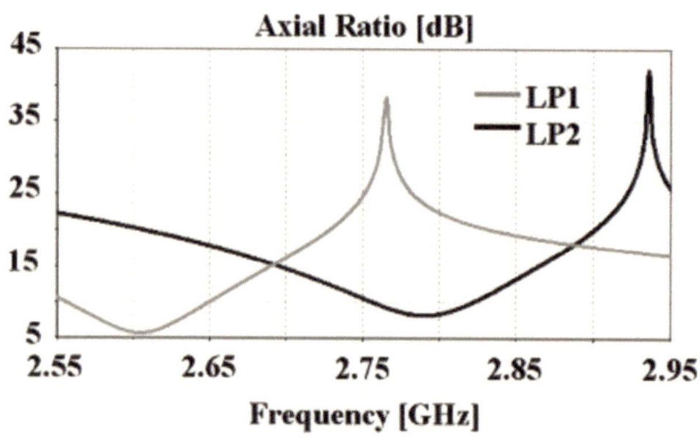

Fig. 17. Computed axial ratio of the antenna for the linear polarization cases (Monti et al., 2009)

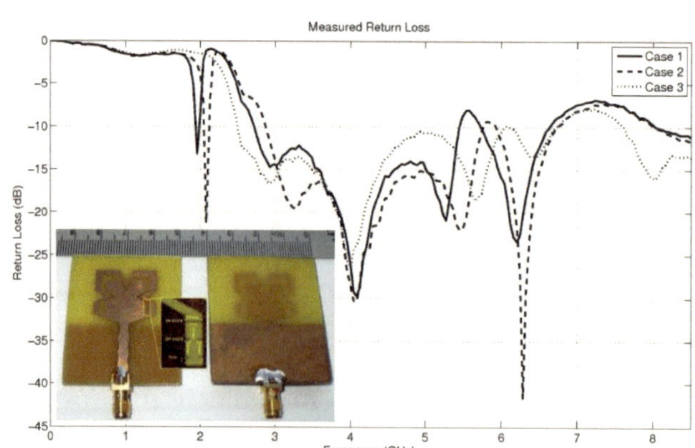

Fig. 18. Measured S11 of the proposed antenna in the 0.3–8.5 GHz frequency range (Ramadan et al., 2009)

were mounted across each slot. These switches control the direction of the current flow, thus leading to radiation pattern reconfigurability. Two switching scenarios were adopted. The resulting radiation patterns, at 10 GHz, are depicted in Fig. 23. The E-plane beam is directed along $\theta = -32°$ for the first switching case, and along $\theta = +32°$ for the other switching case. The H-plane pattern is almost identical for both switching scenarios.

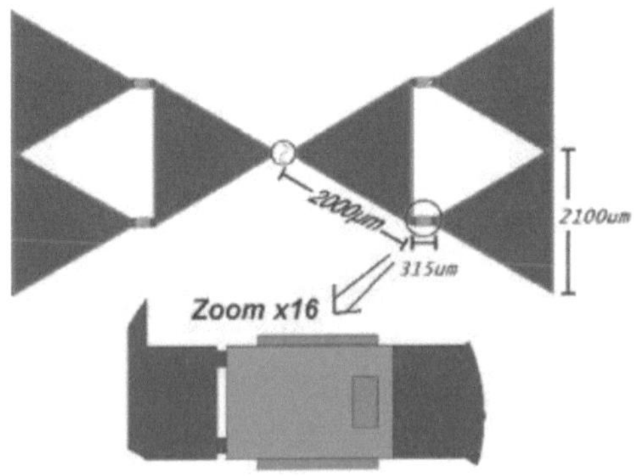

Fig. 19. Initial antenna design and RF-MEMS switch connections (Anagnostou et al., 2006)

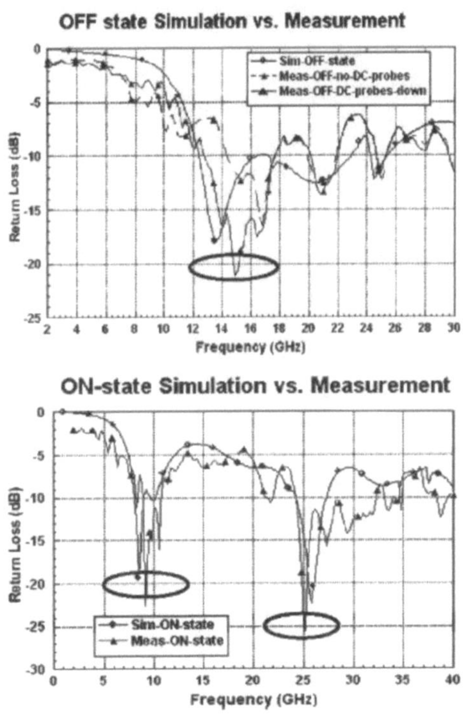

Fig. 20. Return loss of the antenna when switches are all OFF or ON (Anagnostou et al., 2006)

7. Conclusion

Fractal-shaped reconfigurable microstrip antennas are discussed in this chapter. The space filling and self-similarity properties of fractal geometries are presented from an antenna

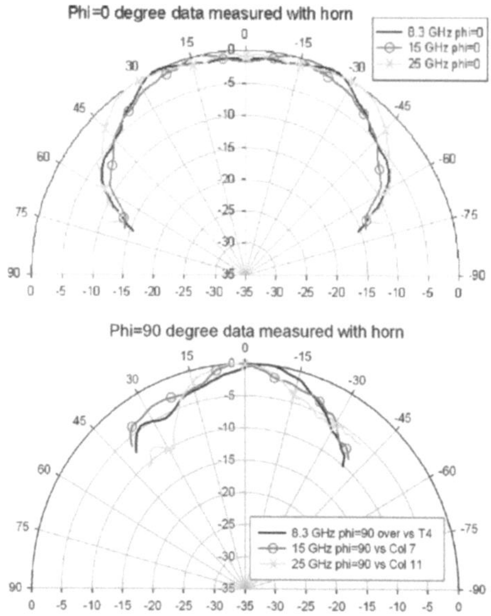

Fig. 21. Measured radiation patterns at different frequencies in the $\phi = 0°$ and $\phi = 90°$ cut-planes (Anagnostou et al., 2006)

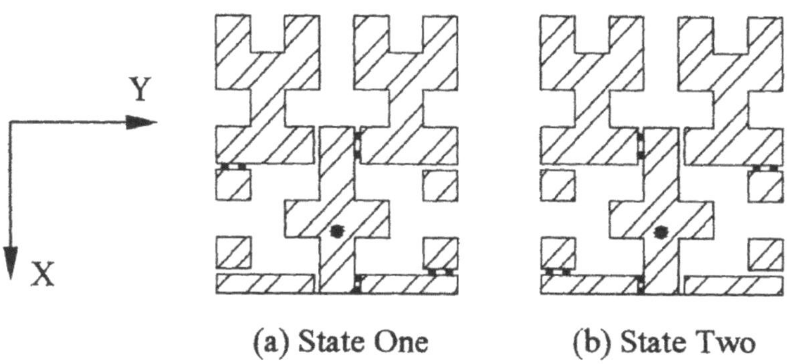

Fig. 22. Configuration of the pattern-reconfigurable Hilbert-shaped antenna in its two switching states (Zhang et al., 2005)

engineering perspective. Moreover, a survey on recent fractal-shaped microstrip antennas that are compact in size or multi-band/wideband in operation has been included. Some recent frequency-, pattern- and polarization-electronically reconfigurable microstrip antennas were also reviewed in this chapter. It was shown that reconfigurability is a demanding antenna design traits in building multiple-in-1 antennas. Three hybrid antenna design approaches, which combine fractal shapes and electronic reconfigurability, were also presented. The first employs Koch fractal geometry and tunability to cover several frequency bands while keeping the antenna compact in size. The second is based on the design of a Sierpinski Gasket-shaped reconfigurable antenna and achieves similar radiation patterns at several frequency bands. The latter employs Hilbert curves in the design of a compact pattern-reconfigurable antenna.

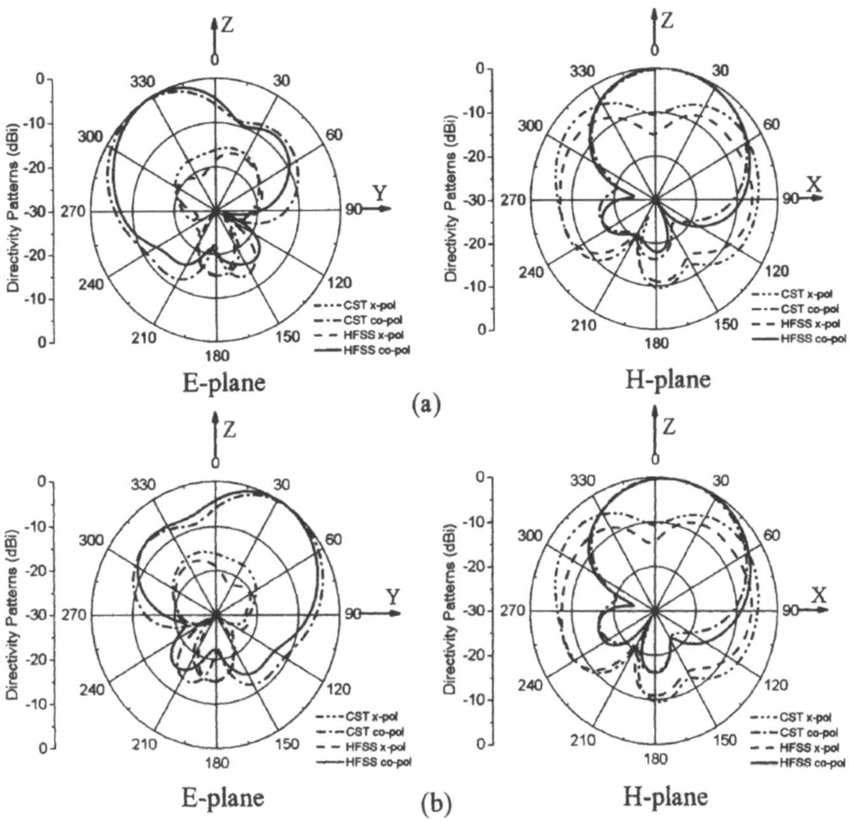

Fig. 23. Radiation patterns at 10 GHz of the antenna in (Zhang et al., 2005) for the two switching states

8. References

Anagnostou, D.E.; Zheng G.; Chryssomallis, M.T.; Lyke, J.C.; Ponchak, G.E.; Papapolymerou, J. & Christodoulou, C.G. (2006). Design, fabrication, and measurements of an RF-MEMS-based self-similar reconfigurable antenna. IEEE Transactions on Antennas and Propagation. Vol. 54, No. 2, February 2006, pp. 422–432, 0018-926X

Anagnostou, D.E.; Papapolymerou, J.; Tentzeris, M.M. & Christodoulou, C.G. (2008). A printed log-periodic Koch-dipole array (LPKDA). *IEEE Antennas and Wireless Propagation Letters*, Vol. 7, June 2008, pp. 456–460, 1536-1225

Balanis, C.A. (2005). *Antenna Theory, Analysis and Design*, John Wiley and Sons, 0-471-66782-X, Hoboken, New Jersey, USA

El-Khamy, S.E. (2004). New trends in wireless multimedia communications based on chaos and fractals, *Proceedings of the 21st National Radio Science Conference*, pp. 1–25, 977-5031-77-X, March 2004, Cairo, Egypt

Krishna, D.D.; Gopikrishna, M.; Anandan, C.K.; Mohanan, P. & Vasudevan, K. (2008). CPW-fed Koch fractal slot antenna for WLAN/WiMAX applications. *IEEE Antennas and Wireless Propagation Letters*, Vol. 7, May 2008, pp. 389–392, 1536-1225

Krupenin, S.V. (2006). Modeling of fractal antennas. *Journal of Communications Technology and Electronics*, Vol. 51, No. 5, May 2006, pp. 561–568

Liu, J.; Zhang J.; Wang, W & Wang, D (2007). Compact reconfigurable microstrip antenna for multi-band wireless applications. *IEEE International Symposium on Microwave, Antenna, Propagation and EMC Technologies for Wireless Communications*, pp. 608–611, 9781424410446, August 2007, Hangzhou, China

Monti, G.; Corchia, L & Tarricone, L. (2009). Patch antenna with reconfigurable polarization. *Progress In Electromagnetic Research C*, Vol. 9, 2009, pp. 12–23

Peroulis, D.; Sarabandi, K. & Katehi, L.P.B. (2005). Design of reconfigurable slot antennas. *IEEE Transactions on Antennas and Propagation*, Vol. 53, No. 2, February 2005, pp. 645–654, 0018-926X

Ramadan, A.; Kabalan, K.Y.; El-Hajj, A.; Khoury, S. & Al-Husseini, M. (2009). A reconfigurable U-Koch microstrip antenna for wireless applications. *Progress In Electromagnetic Research*, PIER 93, 2009, pp. 355–367

Ramadan, A.; Al-Husseini, M.; El-Hajj, A. & Kabalan, K.Y. (2009). A 2×2.5 cm^2 low-cost PCB antenna for UWB applications,âĂİ *Proceedings of The 5th IEEE-GCC Conference*, March 2009, Kuwait City, Kuwait

Ramadan, A.; Al-Husseini, M.; Kabalan, K.Y.; El-Hajj, A. & Costantine, J. (2009). A compact Sierpinski-carpet-based patch antenna for UWB applications. *Proceedings of IEEE Antennas and Propagation Society International Symposium (APS-URSI 2009)*, pp. 1–4, 1522-3965, June 2009, Charleston, SC, USA

Ramadan, A.; Al-Husseini, M.; Kabalan, K.Y. & El-Hajj, A. (2009). A small-sized reconfigurable antenna for ultra-wideband, C-band and X-band operation. *Proceedings of International Conference on Electrical and Electronics Engineering (ELECO 2009)*, pp. II-157–II-160, 978-1-4244-5106-7, November 2009, Bursa, Turkey

Ramadan, A.; Al-Husseini, M.; Tawk, Y.; Kabalan, K.Y. & El-Hajj, A. (2010). A novel frequency/pattern-reconfigurable microstrip antenna for WLAN applications. *Proceedings of the Fourth European Conference on Antennas and Propagation (EuCAP 2010)*, pp.1–3, 978-84-7653-472-4, April 2010, Barcelona, Spain

Rao, P.N. & Sarma, N.V.S.N. (2008). A single feed circularly polarized fractal shaped microstrip antenna with fractal slot, *Progress In Electromagnetics Research Symposium Proceedings*, pp. 195–197, China, March 2008, Hangzhou

Vinoy, K.J.; Abraham, J.K. & Varadan, V.K. (2003). On the relationship between fractal dimension and the performance of multi-resonant dipole antennas using Koch curves. *IEEE Transactions on Antennas and Propagation*, Vol. 51, No. 9, September 2003, pp. 2296–2303, 0018-926X

Werner, D.H. & Ganguly, S. (2003). An overview of fractal antenna engineering research. *IEEE Antennas and Propagation Magazine*, Vol. 45, No. 1, February 2003, pp. 38–57, 1045-9243

Xiao, G. & Zhu, S. (2007). Novel fractal and MEMS fractal antennas. *International Conference on Microwave and Millimeter Wave Technology*, pp. 1–4, 1-4244-1049-5, April 2007, Gulin, China

Zhang, Y.; Wang, B.-Z.; Yang, X.-S. & Wu, W. (2005). A fractal Hilbert microstrip antenna with reconfigurable radiation patterns, *Proceedings of IEEE Antennas and Propagation Society International Symposium (APS-URSI 2005)*, pp. 254–257, 0-7803-8883-6 , July 2005, Washington DC, USA

Zhang, J.; Wang, A. & Wang, P. (2008). A survey on reconfigurable antennas. *Proceedings of the International Conference on Microwave and Millimeter Wave Technology*, pp. 1156–1159, 978-1-4244-1879-4, April 2008, Nanjing, China

11

A Microstrip Antenna Shape Grammar

Adrian Muscat and Joseph A. Zammit
University of Malta
Malta

1. Introduction

In this chapter a microstrip antenna shape grammar is described and its use is demonstrated as a design tool for compact microstrip antennas and in the real-time control of reconfigurable pixel microstrip structures. The microstrip shape grammar formalizes the cut and try methodology used widely by microstrip antenna engineers, who make use of rules of thumb, simple formula models and intuition to combine and manipulate various patch shapes that yield the required characteristics. Simple formula models linked to shape attributes are developed to provide immediate feedback to the shape evolution process and microstrip patch shapes whose electrical characteristics closely match the required specifications are generated. The machine can therefore explore a wide variety of configurations in a relatively short time.

2. Compact microstrip antennas

Compact microstrip antennas are devices that are small in size, but at the same time exhibit electrical characteristics that are similar to those enjoyed by antennas of standard size. More specifically the electrical length of compact antennas is similar to that of a standard antenna. Fig.1(a & b) are two typical examples of compact multi-band antennas, where the non-standard shapes of the microstrip patches are mainly responsible for the electrical characteristics exhibited by these structures. The structure of fig.1(b) is an evolution of the design in fig.1(a), which is modified to accommodate a third element in the same space. The design process for these two prototypes is not obvious and difficult to formalize. Additionally, the positions of feeds are often fixed by the overall system considerations and the antenna designer has to work out the shape around these fixed components. Compact antennas are also characterized by a strong coupling between the electrical characteristics and the physical composition of the structure and a small change in the topology usually results in an invalid design. A significant amount of design time and effort is therefore invested in developing these antennas.

An antenna that can be switched between wide-band and narrow-band operation is shown in fig.1(c). The synthesis task for this case is equally, if not more, difficult to formalize and the designer has to develop patch shapes that when connected together via switches yield the required electrical properties.

Fig.1(d) shows a pixel reconfigurable antenna. This structure consists of an $M \times N$ matrix of small metal patches, or *pixels*, interconnected with electronic switches. When all switches are switched *ON* the structure behaves like a rectangular patch. When some of the switches are turned *OFF*, the electrical current path is modified and therefore the electrical characteristics are altered. A single device can therefore, configure itself to operate over a wide frequency

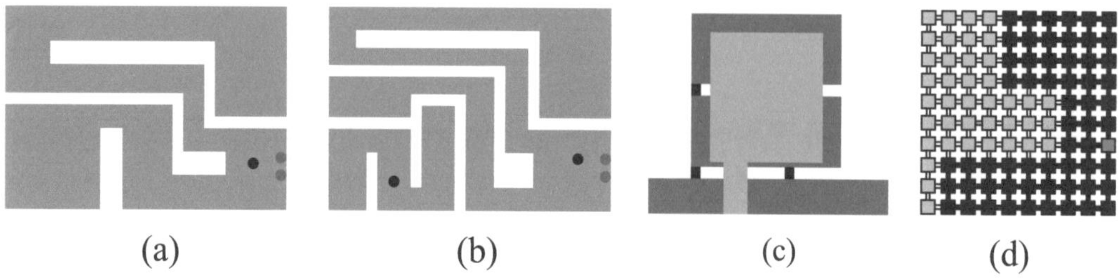

Fig. 1. Compact microstrip antenna designs (a) Dual-Band single feed, (b) Tri-band with two separate feeds, (c) a structure that can be switched from wide-band to narrow-band operation, from (J. R. Kelly et al.,2008), and (d) a pixel microstrip reconfigurable structure.

range, optimize the signal strength and change polarization in a short period of time as demanded by the system. The delivery of such devices certainly means a change in the way antennas are designed and deployed. Most of the research effort to-date on reconfigurable antennas has been targeted towards the hardware issues, ((B. A. Cetiner et al.,2004) and (A. Grau and Lee Ming-Jer et al.,2007)) and the development of efficient control algorithms still needs to be adequately addressed. The switching operation is equivalent to altering the shape of the patch and the skills required to carry out this task are similar to the expert skills exhibited by compact antenna designers. It is therefore expected that algorithms proposed for the control of reconfigurable pixel antennas closely follow those studied in the automation of the synthesis task in compact antenna design.

3. Microstrip antenna CAD

The design of compact antennas is a process that iterates between two phases. The first phase is essentially a manual task, where the antenna engineer selects, combines and fits in a constrained volumetric and electromagnetic space, a number of geometric shapes that together yield the desired electrical characteristics. During the second phase the engineer makes extensive use of numerical Computer Aided Design (CAD) tools to refine and tune the initial design. Much research effort has focused on CAD for the second phase and very little has been done in developing antenna CAD for the first phase. These designs are therefore carried out by expert designers and modifying a structure to suit some additional specifications requires a significant time investment and effort. A CAD tool that formalizes some of the design processes would therefore help not only in the timely delivery of the product, but also in the exploration of novel structures.
Most of the research work carried out in the area of patch antenna CAD related to the first phase or the synthesis task considers meta-heuristic algorithms, such as the Genetic Algorithm (GA), Simulated Annealing (SA) and Tabu Search (TS). Systems or methodologies based on the GA are by far the most studied. GAs are search methods based on the *survival of the fittest* technique found in nature. Initially a large population of randomly generated designs (in this case antennas) is instantiated and each design is encoded as a *chromosome*. These designs, called *individuals*, are simulated and ranked according to how well they perform, or how *fit* they are. The next generation is constructed by first selecting two parents and then combining the designs or chromosomes together, in the same way as offsprings in nature are characterized by a chromosome that is made up from parts of the parents' chromosomes. Superior individuals are given a greater chance in the reproduction process, ensuring that the fitness in a population improves as the number of generations increases.

The evolutionary process continues until a design that closely matches the requirements is found. The performance of GAs depends mainly on two parameters, the *crossover* operator and the *encoding* of the chromosome. The *crossover* operator defines how two chromosomes are combined together to produce the offsprings, whereas the encoding techniques defines how a *real-world* design is represented as a string of symbols.
(J. M. Johnson et al.,1997) pioneered the use of the GA for machine evolved planar microstrip antenna shapes that exhibit wide-band characteristics. In this work a rectangular design space is defined in terms of a matrix of rectangular or square pixels. The entire patch shape is defined by the presence or absence of a pixel. The shape is therefore represented or encoded as a binary bit string of length $M \times N$, where each element of the string, or *gene*, defines the presence or absence of a pixel. The crossover operator is a straightforward single point crossover. In (N. Herscovici et al.,2002) the same encoding and crossover techniques are used in the search of compact microstrip antenna and a 30% reduction in size is obtained. Not surprisingly the GA has been applied to the control of reconfigurable antennas. In (D. S. Linden,2002) and (Zhang Min et al.,2004), the GA is applied to tune the antenna for maximum signal strength on-situ and in (Coleman,C.M. et al.,2000) and (L. N. Pringle et al.,2004) the GA is used to tune the antenna over a large range of frequencies. In (Adrian Muscat et al.,2010) it is shown experimentally that the encoding scheme described in (J. M. Johnson et al.,1997) fails to evolve compact structures of the types in Fig.1(a& b) and concludes that a different encoding scheme and corresponding crossover operator is required. Another drawback with a GA based search algorithm is the large number of iterations required to converge to a result. (E. A. Jones et al.,2000) proposed an encoding scheme for wire antenna arrays based on a language, defined by a set of rules that combine a set of components to generate wire antenna array configurations. A GA is deployed to produce designs in this language. The search space is therefore narrowed, rendering a more efficient search. In (Adrian Muscat,2002), a Knowledge-Based Genetic Algorithm (KBGA) is developed to reduce the time required to evolve novel microstrip antennas. The KBGA makes use of the abstract shape representation described in (J. M. Johnson et al.,1997) and selects antenna design heuristics (rules of thumb) to influence the genetic operators. The KBGA is similar to a set of design techniques, but does not deal directly with shape.
Furthermore, the design paradigms described above are required to call an accurate numerical model for each configuration developed and more importantly they do not link the topology of the shape to the electrical properties. They cannot therefore carry out a modification in the shape or topology in a logical way and thus cannot mimic reasoning and intuition, two important characteristics in design.

4. Shape grammars

Shape grammars consist of a finite set of elements and a finite set of rules that define how the elements are connected together to generate designs in the specified language. The rules applied can be either chosen by the human designer or by the machine through an algorithm. Shape grammars where originally developed in architecture to generate designs in a particular style, (G. Stiny et al.,1978) and (U. Flemming,1987). The functional requirements are first specified and then the generating engine generates physical configurations in a particular style. Shape grammars where later applied in civil and mechanical engineering such as in the design of optimal truss structures (K. Shea et al.,1997), and the design of coffee makers (Manish Agarwal et al.,1998).

(Manish Agarwal et al.,2000) then proposed shape grammars as a framework for the design of geometry-based engineering systems. The functional component of most engineering systems is strongly coupled with its physical and geometrical composition and a small change in the topology can result in a significant change in function and performance. This characteristic limits human designers to very few configurations. Therefore a formal design tool that is automated on a computational machine should be able to explore a much wider variety of configurations. (Manish Agarwal et al.,2000) demonstrated the concept by developing a coupled form-function shape grammar for the design of micro-electromechanical resonators. The grammar enforces the inclusion of all the necessary components for the device to function as a resonator. However, this does not guarantee that the device is close to meeting the specifications and the only way to know this, is through a numerical simulation, which is computationally expensive.

(E. A. Jones et al.,2000) developed a wire antenna array language that defines structure in the style of Yagi and log-periodic arrays and combines this with a genetic programming (GP) algorithm . The *antenna language* defines how wire dipoles, gaps, sources and transmission lines are connected to form antennas. A genetic algorithm, coupled with a numerical wire antenna model, uses this language to search for functional designs. As in (Manish Agarwal et al.,2000), many of the designs generated are invalid, and a time-consuming numerical model is required. This is however the first time a grammar based technique is applied to an antenna problem.

The shape grammar developed in (Adrian Muscat,2010) and described in this chapter uses immediate feedback to guide the shape evolution and avoids the problem of generating and numerically simulating invalid configurations. Feedback is obtained by analyzing the geometry of the 2D shape and linking the shape attributes to simple computationally inexpensive formula models. Shape analyzes is carried out by decomposing the shape into a chain of labeled rectangular shapes. The labels and the sub-shape attributes are used for the analysis. This paradigm is suitable to generate the first phase or initial design, which is then passed through an optimization process. Additionally, the microstrip shape grammar solves the problem of encoding compact microstrip shapes.

5. The microstrip antenna shape grammar

The microstrip antenna shape grammar generates shapes that fit in a 2D design space divided into square pixels. Two examples of design spaces are shown in fig.2(a), a rectangular design space, and fig.2(b), an L-shaped design space. Fig.2(c, d, e, & f) are examples of microstrip shapes generated on a 12×10 matrix of square pixels. The grammar rules described in this chapter evolve a shape as two labeled branches or paths coming out of the probe feed or a shorting post. The paths are composed of cascaded rectangles. The grammar ensures that no branch overlaps or touches another branch and there are no loops along one path. This grammar is suitable to model structures resonating at the fundamental mode, including shorted patch structures, that are the most common building blocks in the design of compact microstrip antennas. Figs.2(c) and (f) are shapes that can be evolved by this grammar, while figs.2(d) and (e) cannot be generated by this grammar.

The rest of this section is organized as follows: The grammar elements and rules are described first, followed by an explanation of the feedback mechanism. The grammar is then demonstrated in the design of a multi-band antenna and in the control of the reconfigurable microtrip antenna.

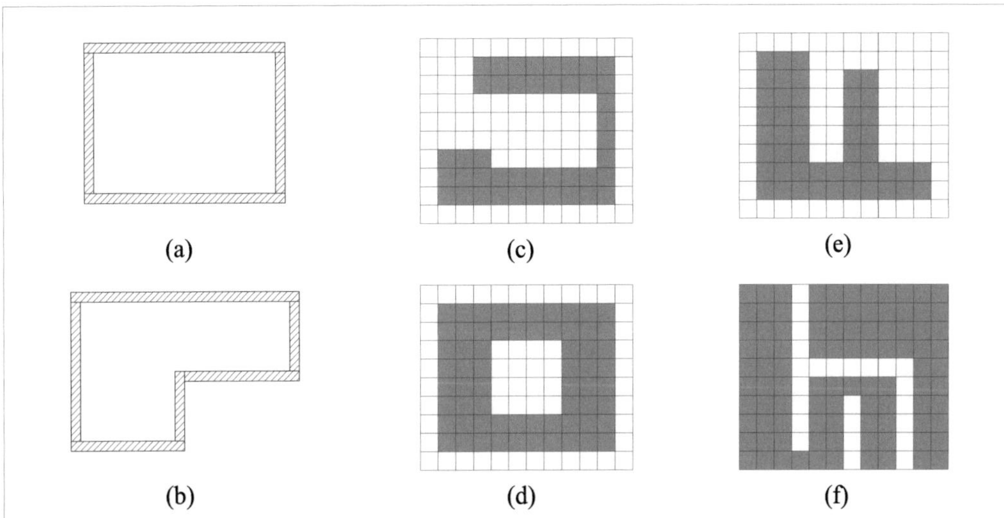

Fig. 2. (a) rectangular shape 2D design space, (b) L-shape design space, (c) to (f) Shapes generated over a matrix of square pixels. Examples (c) and (f) conform to shapes generated by the grammar described in this chapter.

5.1 The grammar elements and rules

Fig.3 shows the elements and the initial rules that define the initial patch shape, that consists of a probe feed and two narrow branches. The basic shape is the square pixel. Rectangles emerge from the union of adjacent pixels and more complex shapes emerge from the union of rectangles. Therefore as the shape evolves rectangles combine into new rectangles and the overall shape topology changes. Furthermore the pixels and rectangles are labeled. An integer number, label 'r', is given to an emergent rectangle. The arrow and diamond symbols define how the pixels are connected to form a rectangle. The arrows define the general direction of the electrical current and the diamonds define transverse paths. The labels are used during shape evolution as well as during analysis for feedback. As an example, fig.4(b) shows rectangles *1, 3, and 4* connected in cascade. The *dot* symbol defines the interface points in between rectangles. The *width* and *length* of each rectangle is used by the feedback mechanism. Finally the grammar rules ensure that the necessary components are not overwritten during the evolution of the shape. The microstrip shape grammar is therefore a 2D parametric, labeled and weighted grammar.

Fig.3 gives the first set of grammar rules. *Rules 1* to *4* are termed the *initial rules*. A configuration starts with a feed pixel that is defined by *rule 1*. *Rule 2* defines and labels the starting points for the two branches *a* and *b*. *Rule 3* attaches a pixel to one of the chosen edges and the first rectangle is defined. *Rule 4* attaches a pixel to the other chosen edge and the second rectangle is so defined. The latter rectangle is composed of two pixels. The initial shape thus comprises two branches, labeled *a* and *b*, and the labeled feed pixel. The radiating edges are defined at the end of each branch. The *line generation rules 5* to *7*, fig.3, are applied to evolve the starting shape. The rules define both the shape and label algebra, and prior to applying a grammar rule, the machine algorithm checks that the application of the rule does not violate a constraint.

To ensure a concise grammar two rule operators, the *ROTATE* and the *FLIP* are defined in fig.6. These operators transform rules to suit different orientations, and two examples on *Rule*

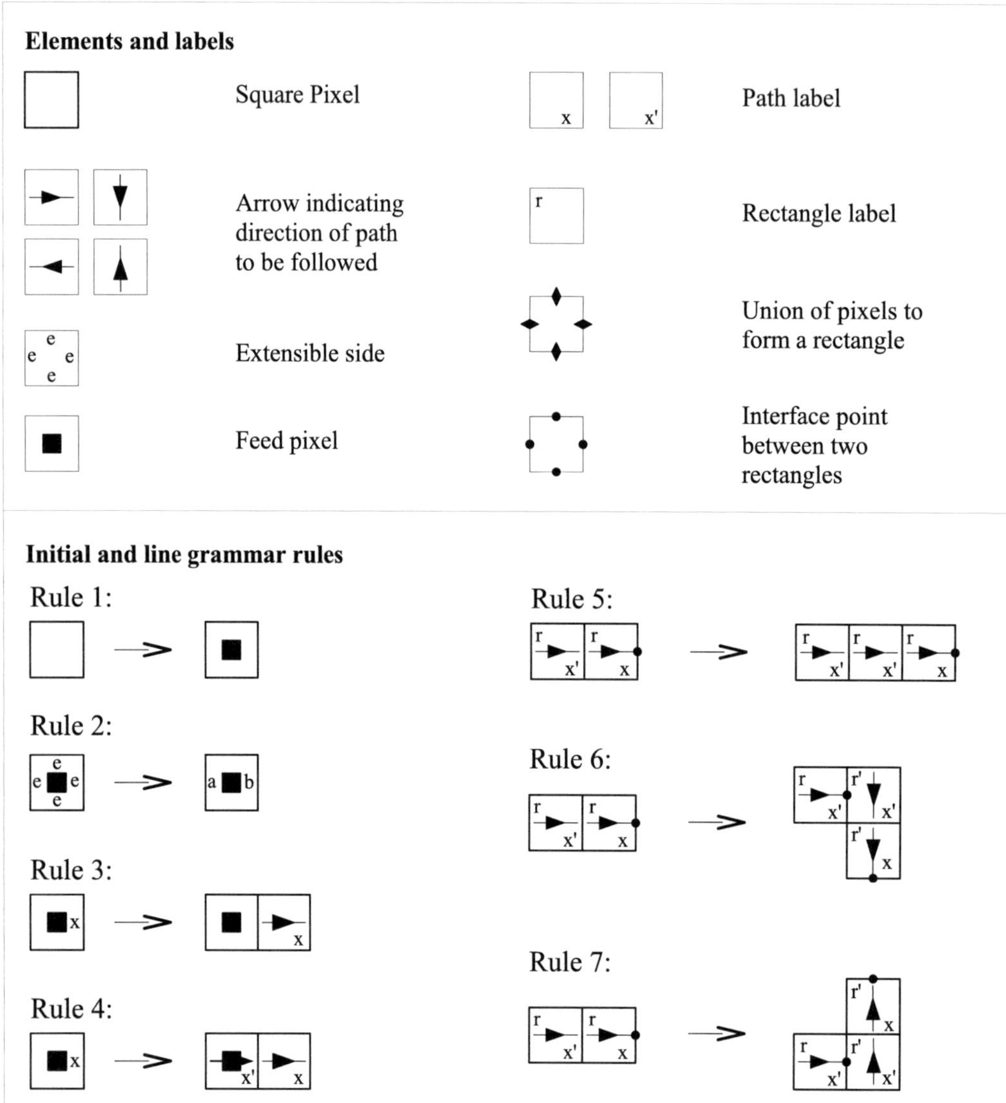

Fig. 3. The grammar elements and labels, initial and line grammar rules.

7 and *Rule 6* are given. The second example shows how in fact rules *6* and *7* are related with the *FLIP* operator.

In summary, the line grammar ensures that one feed is included, a maximum number of two branches are present, the width of the branches is equal to one pixel and the branches do not form loops. An example of a shape generated by the *line grammar* is given in Fig.5. The arrow label is very important as this indicates the current flow and subsequently the path. The line grammar generates one-pixel wide lines. To widen the rectangles or sub-shapes, the *extended grammar* is applied. The initial shape for the extended grammar is the final shape evolved with the line grammar rules. Another important objective of the *line grammar* is to explore the design space, taking into consideration other concurrent shapes as in multi-band designs. The

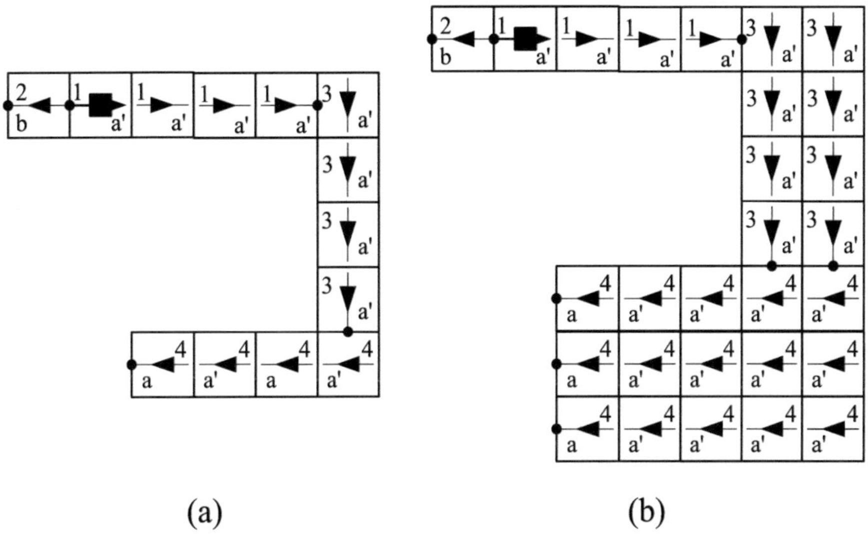

Fig. 4. Two patch shapes decomposed into rectangles, ready for analysis.

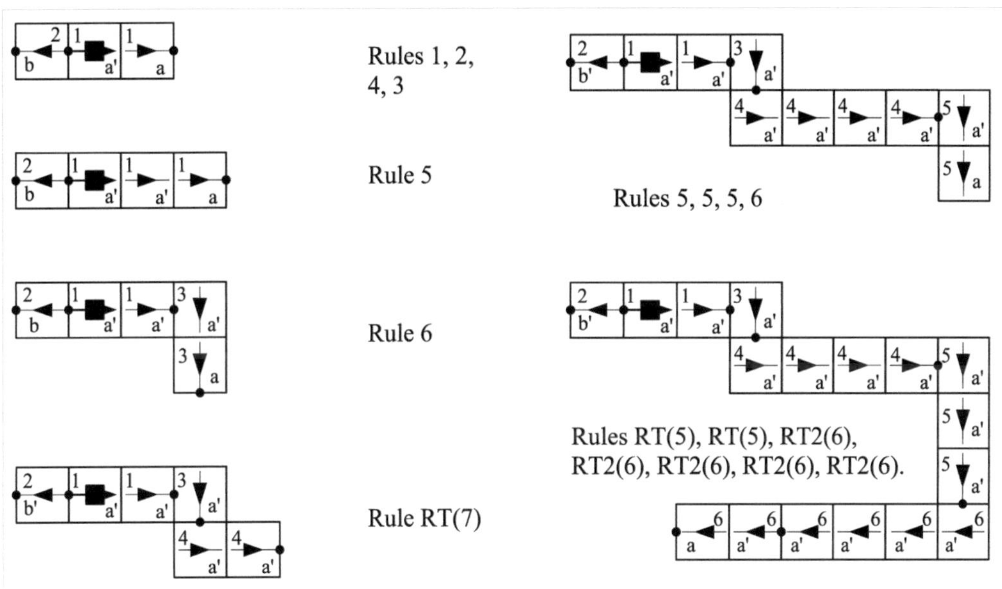

Fig. 5. A Shape generated by the grammar defined in fig.3.

final shape for the extended grammar is considered as a microstrip line defined by a chain of rectangles.

Fig.7 and fig.8 define the extended grammar *rules 8* to *18*. The rules are applied to widen arbitrary line sections and extend the line ends. *Rule 8, 9 and 10* in fig.7 define how the line in the vicinity and including the probe feed can be evolved. The application of the *FLIP* operator defined in fig.6 is useful to apply these rules to other orientations. *Rule 11* extends the branch ends, thus contributing to a longer current path, while *rule 12* widens the rectangle at the end of the branch. *Rule 13* is a special rule that does not change the shape, but changes the

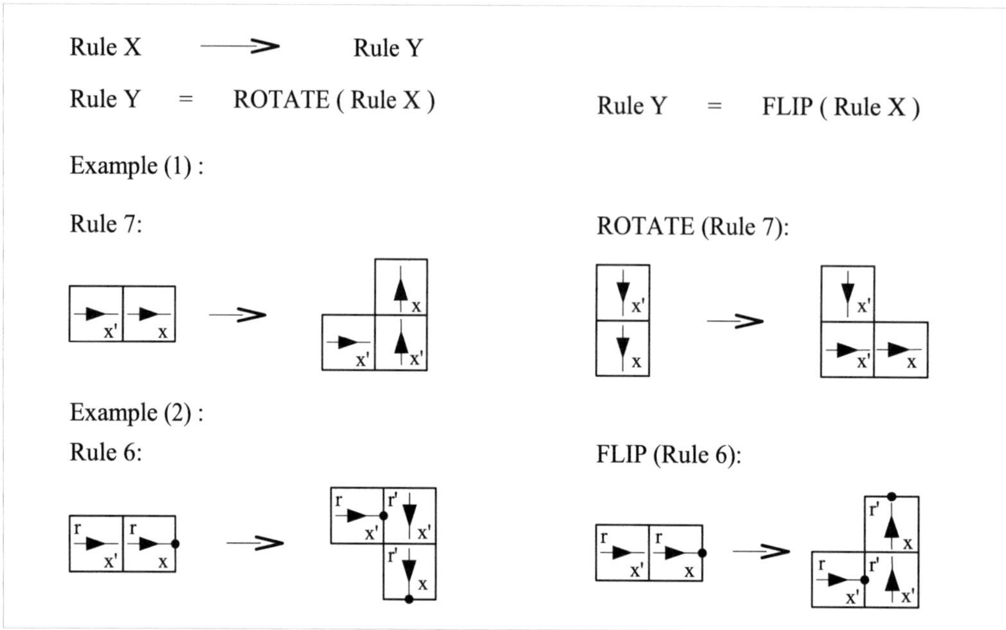

Fig. 6. Rule operators.

current direction along the rectangle at the end of the branch and therefore this rule is useful during analysis described in section 5.2. Rule 14 in fig.8 evolves a δ-bend into an L-shape. *Rule 15* widens the outgoing inside corner in an L-shape. *Rule 16* widens a straight branch from start to end, or a branch that consists of one rectangle. This rule is instrumental in evolving rectangular shapes. *Rule 17* widens the inside of a U-shape, while *rule 18* widens the outside of an S-shape. For the grammar to be complete some more rules that extend for example the inside of an S-shape and the outside of a U-shape need to be defined.

Fig.9 shows how the final shape generated by the line grammar in fig.5 is evolved by the application of the *extended grammar rules*. The rules update the symbols and labels accordingly and the chain of rectangles emerge as a product of the application of the rules. Fig.10 gives some examples of shapes generated by the shape grammar. The grammar therefore has the capability of generating microstrip shapes and decomposing these shapes into a chain or rectangles, ready for analysis.

5.2 The feedback mechanism

An antenna design can be considered valid if it satisfies the electrical specifications and fits in the designated physical design space. The grammar described in section 5.1 generates the shape of the radiating patch and includes the probe feed. This however does not guarantee that the antenna electrical characteristics are close to the required ones. To generate valid shapes the grammar requires immediate feedback on the values of the electrical properties, throughout the generation process. Needless to say, the model used to compute the approximate values should be very fast to compute.

An antenna is usually characterized by the frequency of operation, input impedance, efficiency, radiation pattern and bandwidth. During the initial phase compact microstrip antenna designers concentrate mainly on the frequency of operation and input impedance. The other electrical properties are either ignored or taken into consideration at a higher level,

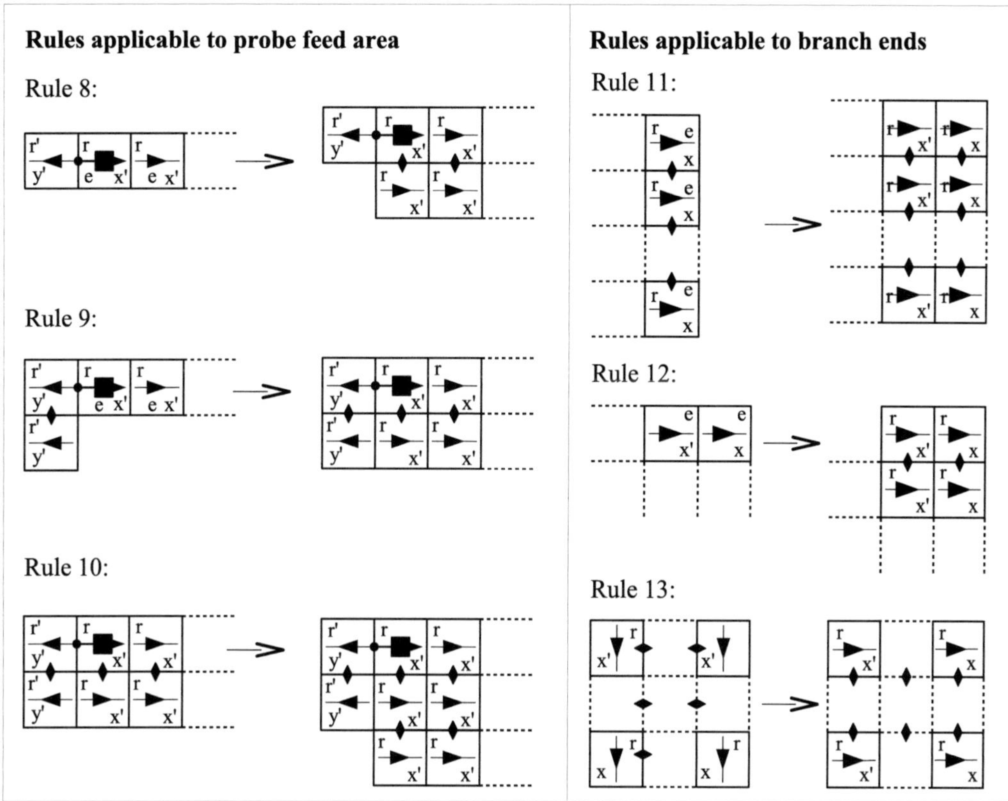

Fig. 7. Extended grammar rules applicable to probe feed area and branch ends.

(Brian S. Collins,2007). For example, it is practically a certainty that the radiation pattern of a compact antenna is essentially an omni-directional pattern with relatively high cross-polar levels and the designer has very little control over this. Efficiency depends on the materials used and bandwidth is taken care of during the selection of topologies and design guidlines which are taken into consideration when selecting the grammar rules and the sequence in which these are applied. Additionally it is common practice to add a final matching circuit to the antenna port once the antenna design is finalized, (Kingsley,S.P.et al.,2008). The frequency of operation and the input impedance are therefore the two properties that need to be considered explicitly in the analysis task and this section describes how these are estimated.

The frequency of resonance and the input impedance are a function of the physical composition of the patch antenna, which consists mainly of the ground-plane size, size and topology of the radiating patch, height of the supporting substrate, the relative permittivity of the substrate, shorting posts or planes and capacitive or inductive components that load the main patch. The microstrip cavity supports an infinite number of modes. However compact antennas are usually operated in the fundamental mode, which is the most useful. For a rectangular patch this mode is excited when the length of the patch is approximately half a wavelength long, or quarter of a wavelength for shorted patches. The rectangular patch can be considered as a wide open-ended transmission line and a standing wave analysis of the $\lambda/2$ line yields good estimates of the fundamental mode frequency as well as the impedance along the line. The width of the rectangular patch determines the impedance bandwidth and the

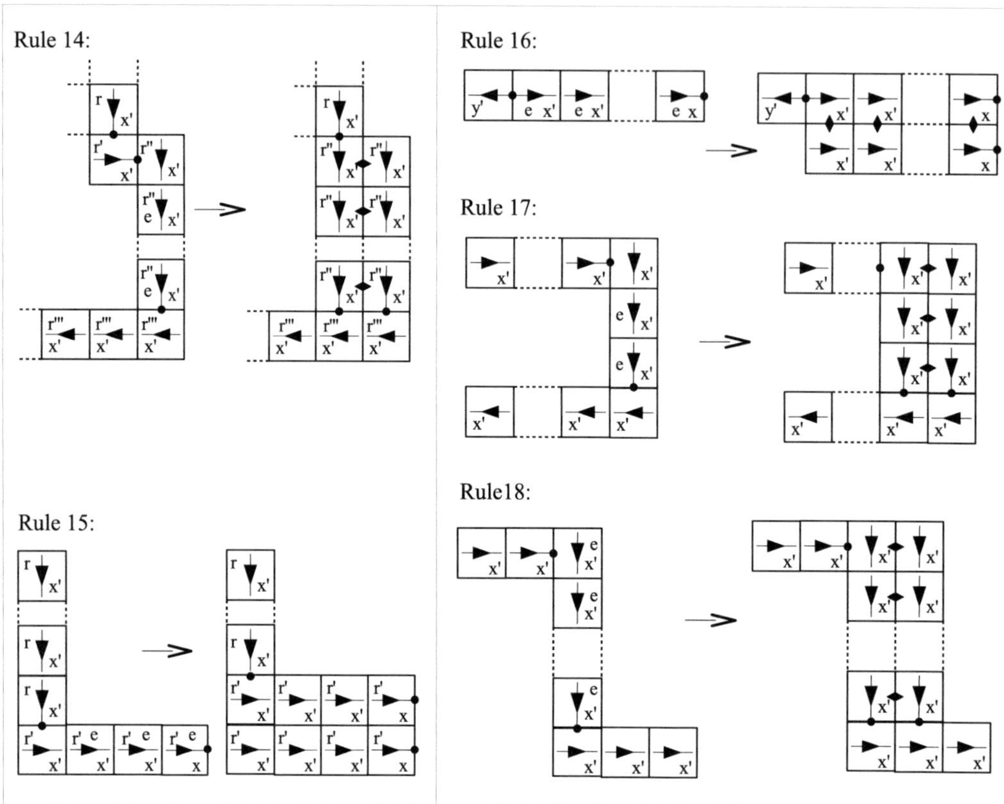

Fig. 8. Extended grammar rules applicable along a branch in between and excluding the start and branch ends.

height of the substrate has an effect on the inductive part of the input impedance. Furthermore the substrate height extends the fringing fields and therefore has a small effect on the value of the resonant frequency.

During the first phase of design the transmission line model and the *design guidelines* are used extensively by compact antenna designers to explore and evaluate a number of shape topologies. During such work high accuracy is not important and the designer uses a combination of qualitative modeling and a crude form of the transmission line model to relate the shape of the radiating patch to the electrical properties. The designer iterates through this process until he is satisfied with the prototype. This iterative process is modelled in the shape grammar with feedback, which can therefore be thought of as a formalization of the design process itself.

The main task for the feedback algorithm is therefore to extract important *geometrical attributes and dimensions*, that are used as inputs to the transmission line model. The resonant frequency and the input impedance depend mostly on the effective current path length and the relative position of the probe feed along this path. Fig.11 is used to explain the process of calculating the effective path length. The first task, which is carried out by the grammar rules during the shape evolution process, is to decompose the shape into rectangles and trace the current paths. The radiating edges are labeled as *a* and *b* and the rectangle interfaces by the *dot* label. The current path direction is labeled by the *arrow* label and the transverse direction by the

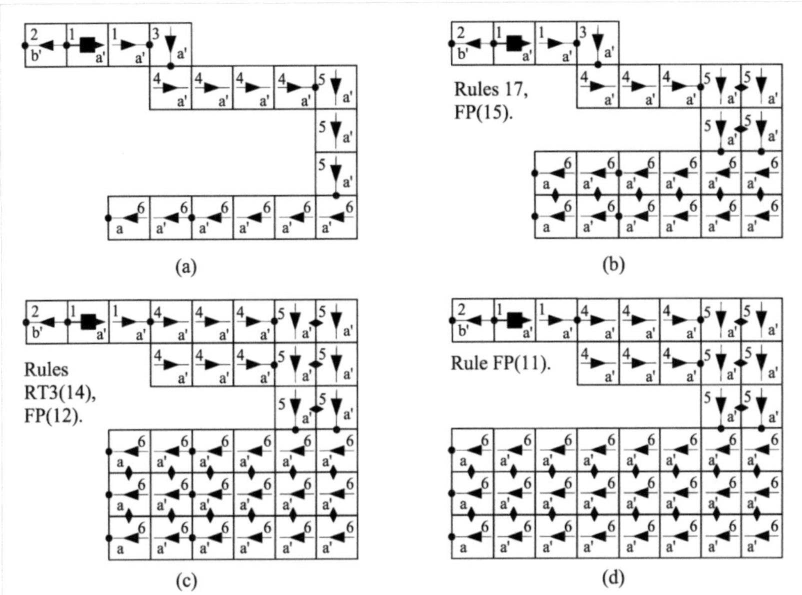

Fig. 9. The evolution of the final shape in fig.5 by the application of the extended rules.

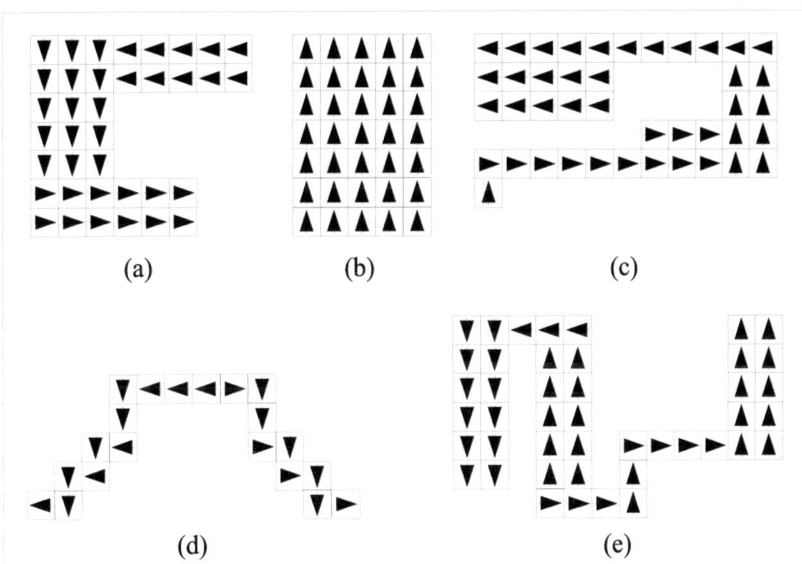

Fig. 10. Examples of shapes generated by the shape grammar.

diamond label. The midpoints of the rectangle interfaces are marked and linked together with straight lines starting from the probe feed. The length of these lines are labeled as L_a for branch a and L_b for branch b. The number of corners (when the path direction changes) are counted for each branch and labeled as NC_a and NC_b. The L_a, L_b, NC_a and NC_b variables are used to obtain an approximation for the resonant frequency and the input impedance.

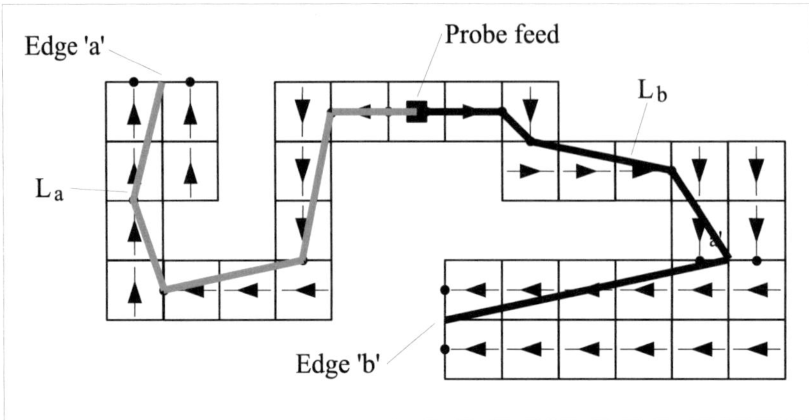

Fig. 11. Evaluation of the effective length.

Intuitively the resonant frequency, f_0 is dependent mostly on L_a, L_b and to a lesser extent on NC_a and NC_b. This intuition is confirmed using scatter plots in (Adrian Muscat,2010). The relationship based on the simple transmission line model for f_0, the frequency of resonance, is derived in (Adrian Muscat,2010) and repeated here,

$$f_0 = 3 \times 10^4/((L_a + L_b + (2a_0 - (NC_a + NC_b) * a_1) * L_p) * 2) \tag{1}$$

where, L_p is the width of the square pixel in millimeters. The coefficient a_0 accounts for the field edge extension effect and generally depends on the substrate height and relative permittivity. The coefficient a_1 weights the number of corners. The values for these coefficients are obtained by minimizing the error for a set of prototypes, that is representative of the range of configurations that can be generated.

The input impedance is mainly a function of its position along the length of the patch and on the width of the patch. Compact antennas are characterized by relatively narrow widths and so the width parameter is ignored in the shape grammar model. The input impedance is estimated on the position of the feed only. Furthermore the model does not give a numerical value, but gives a qualitative indication of how far away it is from the system impedance, assumed to be 50Ω. The ratio of the effective length for branch a to that of branch b is used to judge on this deviation from the system impedance. Experiments reported in (Adrian Muscat,2010) show that when the ratio is in between 0.75 and 0.90 the input impedance is within range of 50Ω. When the ratio is greater than 0.90 the input impedance is significantly smaller than the system impedance and when it is smaller than 0.75 the input impedance is significantly larger.

The coefficients, a_0 and a_1, in eqn.1 are fitted over a set of fifty prototypes that operate over a frequency range of $1.0 - 5.0GHz$. These prototypes are generated randomly and the shapes cover rectangular, L, C and U-shapes, as well as meander lines. The designs are accurately analyzed with a Finite-Difference-Time-Domain (FDTD) model, (Adrian Muscat,2002), and the FDTD results used to tune the coefficients. The average error in estimating the resonant frequency is in the region of 5% with a standard deviation of 3. The errors are smaller for the shapes characterized by narrow rectangles and greatest for the lines characterized by wider rectangles. However for the conceptual or first phase design the accuracy of the model is adequate. Nevertheless, the error can be reduced by fitting the model over a smaller frequency range and a more specific topology. In the next two sections examples are used to demonstrate the use of the shape grammar with feedback.

5.3 Example in multi-band design

In this section the shape grammar is deployed in the conceptual design of a mobile terminal antenna consisting of a single feed dual-band antenna operating at $0.925GHz$ and $1.8GHz$ and a separately fed antenna for $2.45GHz$. These frequencies correspond to cellular licensed mobile communications bands and the unlicensed Industry, Scientific and Medical (ISM) band. The prototype is projected on a rectangular design space. The single feed cellular antenna consists of two combined shorted patch elements. The two patch elements are first evolved separately and then joined together at a later stage. The *line grammar rules* are applied to evolve one-pixel wide elements as well as to explore the design space. Most of the designs evolved at this stage are discarded and some are stored as candidates to be further evolved by the *extended grammar rules* that widen the rectangles, which make up the initial shape, starting from the one at the end of the line. During the second stage the process is allowed to remove any one of the other elements to make space for the current element. This however necessitates the re-application of the *line grammar rules*. An extracted sequence of interim designs during the evolution of the antenna is shown in fig.12. The initial shape generated with the *line grammar rules* is shown in fig.12(a), where the *rules* are applied simultaneously to the three elements. The *extended grammar rules* are then applied to the $1.8GHz$ element on the left-hand-side and extends the last rectangle. This results in a shape that does not satisfy the specifications and there is no more space to correct the error, fig.12(b). Therefore the conflicting element is removed and the first element is allowed to evolve. The *line grammar rules* are applied again which in turn conflict with the third element and these two elements are re-designed simultaneously, fig.12(c). The *extended grammar rules* are then applied to the central element starting from the rectangle at the end of the line with no success,fig.12(d). So the third element is removed and the rectangle is widened. The *line grammar rules* are then applied to the third element and the initial design is complete, fig.12(e).

The estimated frequencies of resonance are $1.86GHz$, $1.05GHz$, and $2.47GHz$, and the respective deviations from the target values are 3.5%, 13.2% and 2.1%. The $1.8GHz$ and the $0.925GHz$ elements are combined to create a single feed dual-band structure and shorting planes are added to the dual-band patch as well as to the ISM patch. The structure shown in fig.12(f), is analyzed with an FDTD model. The three bands resonate at $0.85GHz$, $1.69GHz$, and $2.52GHz$ and the deviations from the grammar model are 18%, 9% and 2%. The smaller resonant frequencies are due to the increase in length when the two elements are combined as well as due to the shorting plane which is narrow than the line width. Additionally both feeds are sufficiently closely matched to the system impedance.

At this point in time the antenna designer proceeds to the second phase - the detailed design, where the structure is optimized using a numerical model. Fig.12(f) indicates some variables for optimization. It should be noted that the optimization process will not change the topology of the shape itself, but only the dimensions of the sub-shapes or rectangles.

5.4 Example in the control of a reconfigurable antenna

The pixel reconfigurable structure described in section 2 requires algorithms that search in real-time for configurations that yield the required electrical specifications. The transient performance of such algorithms is therefore important. In this example the shape grammar is used as part of a control algorithm that can efficiently tune the reconfigurable pixel microstrip

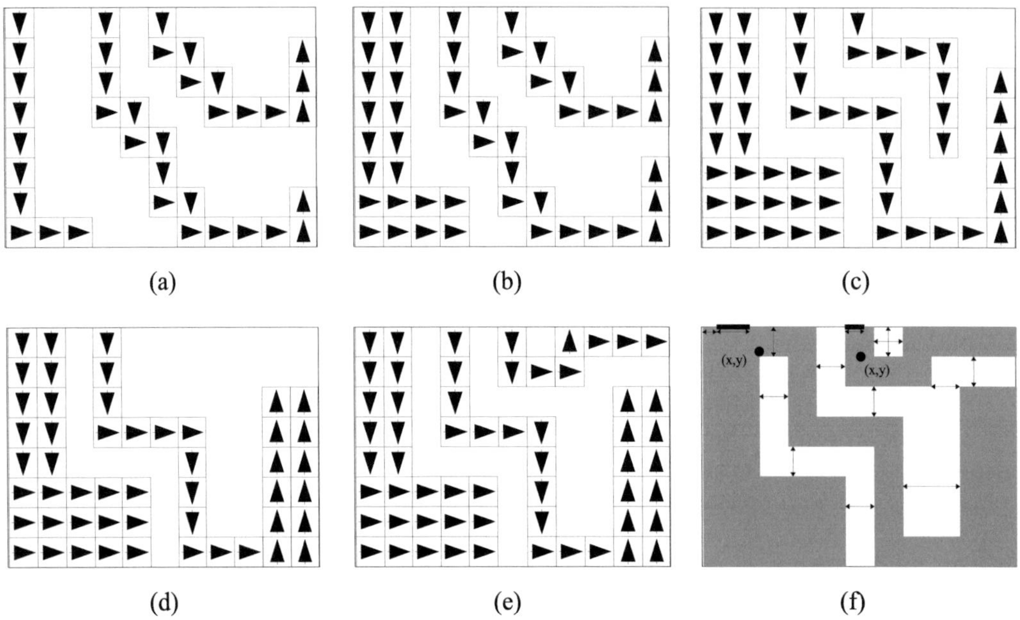

Fig. 12. (a) to (e) stages during the design process of a tri-band separately fed structure, and (f) the numerical model ready for optimisation. The arrows and positions for the probe feeds are suggested variables for the optimisation process.

antenna over the range of mobile frequencies that span from a few hundred *MHz* to a few *GHz*. The problem is formulated as a search for a patch shape that yields the required frequency of operation, while minimizing the amount of hardware switching taking place.

A system diagram for the algorithm is shown in fig.13. The search algorithm instructs the *shape grammar model* to suggest a valid shape that is likely to satisfy the specifications received from the transceiver. The search algorithm accepts or rejects the suggestion, depending on whether the estimated electrical characteristics fall within a specified range. If accepted the shape is hardware switched and measured feedback is used to terminate or proceed with the search. This process continues until an acceptable solution is found. The measurements can also be used to tune the model coefficients. This algorithm works on the premise that the *designs* exhibit characteristics that are close to the intended targets. As used here the shape grammar model reduces a global search problem to a local random search. Furthermore for this application the shape grammar details needs to be modified since the shape is synthesized by switching the interconnections rather than the pixel itself. The modifications are described in detail in (Adrian Muscat et al.,2010).

The control algorithm is demonstrated on two cases (a) $\lambda/2$ patch shape operating at $1.8GHz$, and (b) $\lambda/4$ shorted patch shape operating at $0.9GHz$. For these two examples, the candidate shapes are generated with the algorithms given in fig.14 and fig.15. *Algorithm A* generates the one-pixel wide shape, while *Algorithm B* evolves the rectangles that make up the initial shape. For case (a) the antenna is a 12×12 pixel structure and the total size of the square antenna patch is $41mm \times 41mm$ with a pixel size of $2.9mm \times 2.9mm$. The substrate height is $3.0mm$ and its relative permittivity $\varepsilon_r = 1.0$. In this example the coefficients are tweaked to $a_0 = 0.6$ and $a_1 = -0.1$. The candidates are then simulated with the FDTD model, which is used as a

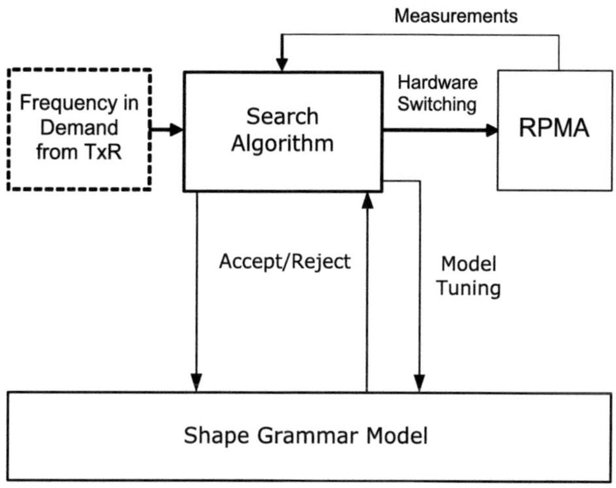

Fig. 13. Block diagram for the control algorithm based on a random search method and a shape grammar model, modified from (Adrian Muscat et al.,2010).

benchmark and replaces measurements. Table 1 list the first 30 candidates in the run. The best candidate is off the frequency mark by 0.556% and for this case occurs at the 25th iteration. The error for the second best candidate is 0.833% and occurs at the 14th iteration. The topologies for these two candidates are shown in fig.16.

In case (b) the antenna is a 10×16 pixel structure and the total size of the square antenna patch is $26mm \times 41mm$ with a pixel size of $2.05mm \times 2.05mm$. The substrate height is $3.0mm$ and its relative permittivity $\varepsilon_r = 1.0$. The candidate shapes are generated with algorithms A and B and in this case a shorting post is added. Equation 1 is therefore adjusted to,

$$f_0 = 300/4(L_e + (2a_0 + a_1 NC)L_p) \tag{2}$$

Table 1 lists the first 30 candidates in the run. The error for best candidate is 0.889% and occurs at the 19th iteration. The second best candidate is off the frequency mark by 1.778% and occurs at the 28th iteration.These two designs are shown in fig.17.

The above cases show that within $20 - 30$ iterations a solution is usually found and demonstrate how effective a grammar based qualitative model can be in reducing the number of switching iterations required. This result is a major gain over systems that rely solely on a GA.

6. Conclusions and future work

This chapter described a shape grammar that generates compact microstrip antenna patch shapes in a constrained 2D space. A feedback loop based on an approximate transmission-line model is used during the shape generation process such that the shapes suggested are valid and closely satisfy the specifications in hand.

The shape grammar with feedback tool formalizes and mimics the informal and intuitively based *cut and try* process that compact microstrip antenna designers follow. The shape grammar generates shapes, decomposes these shapes into a chain of rectangles and positions the feed to match the structure to the system impedance. Labels are used to derive the topology of the shape and to extract shape attributes and parameters that are

```
01 Set frequency of resonance and desired input
   impedance;
02 Start Synthesis
03 Define feed position, rule 1;
04 Define branch 'x', rule 2;
05 Define branch 'y', rule 3;
06 Obtain an estimate for the input impedance;
07 If the input impedance estimate is greater than the
   target, extend the shortest branch from the set x,y,
   rules 4 or 5; if branch is not extensible goto line 10;
08 If the input impedance estimate is less than the target,
   extend the longest branch from the set x,y, rules 4 or 5;
   if branch is not extensible goto line 10;
08 Obtain an estimate of the resonant frequency;
09 If frequency estimate is greater than the target value
   goto line 06;
10 End Synthesis
11 Return estimate of resonant frequency and input
   impedance;
```

Fig. 14. Algorithm Synthesize A: Generates meander-line elements whose width is equal to one pixel, from (Adrian Muscat et al.,2010)

utilized estimating the frequency of resonance and the input impedance. These electrical characteristics are exploited to guide the selection of rules and therefore influence the shape evolution process.

When deployed as a tool in design, the shape grammar can generate a wide variety of potentially useful structures and can form the basis of an Intelligent Computer Aided Engineering (ICAE) software, that acts as a junior partner as described by (Kenneth D. Forbus, 1988). Such a tool can therefore reduce costly design time and can also be used to capture and re-use antenna design knowledge. This concept is demonstrated in the synthesis of a multi-band compact antenna.

The shape grammar is also illustrated in the real-time control of reconfigurable antennas, where a fast and efficient control algorithm is desired. A random search algorithm considers a candidate solution by the grammar and based on measured results accepts or rejects the candidate. This process continues till an acceptable solution is found. Due to the relatively good accuracy of the model, the algorithm converges much faster than a Genetic Algorithm.

The approximate transmission line model performs very well for narrow element designs and when fitted over a narrow range of shapes and frequencies. However the accuracy degrades as more variables are introduced. Nevertheless, the accuracy of the model is still good enough for its intended purpose, the initial design phase. On the other hand it is always desired to have a single model applicable to a wide range of topologies and frequencies. Neural Network architectures (NN) have been proposed as a replacement to the CAD formula for microwave devices, (K. C. Gupta,1998), where physical attributes are assumed as inputs to the NN which in turn yields the frequency of operation or wide-band input impedance. This approach has been shown to work for microstrip antennas of standard shapes,(Kerim Guney et al.,2002) and

```
01 Set frequency of resonance and desired input
   impedance;
02 Start Synthesis;
03 Call Algorithm Synthesize A to generate an initial
   shape;
04 Define and reset subsetFlag to FALSE;
05 For Each branch from the set x,y do
   {
07 For Each rectangle along a branch (starting from the
   end) do
   {
09 Build a subset of applicable rules from the set 6 ... 13;
10 If a subset is not NULL set subsetFLAG to TRUE;
11 Choose a rule from the subset and apply it with a
   probability of Pa = 0.8;
   }
   }
14 If subsetFLAG == FALSE goto line 16;
15 Goto Line 04 with a probability of Pr = 0.7
16 End Synthesis
17 Obtain an estimate for the input impedance;
18 Obtain an estimate of the resonant frequency;
19 Return estimate of resonant frequency and
   recomputed input impedance
```

Fig. 15. Algorithm Synthesize B: Generates meander-line elements whose width is greater than one pixel, from (Adrian Muscat et al.,2010)

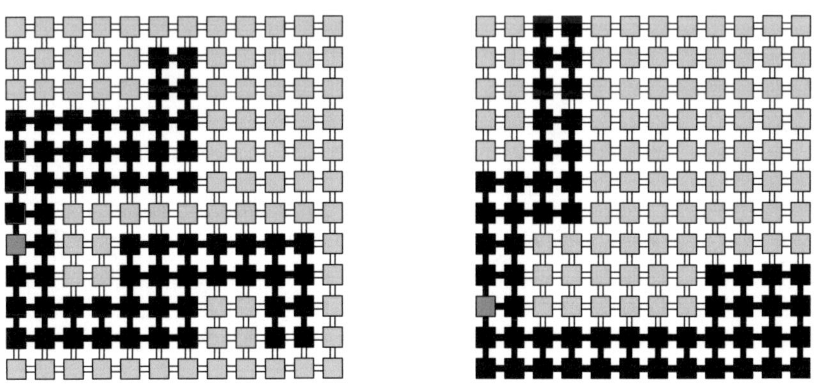

(a) (b)

Fig. 16. The best two candidates resonating at 1.8GHz, from (Adrian Muscat et al.,2010)

(Heriberto Jose Delgado et al.,1998). It is therefore of interest to research on whether NNs can improve the accuracy of the shape grammar in analysis.

Table 1. Candidates for the 1.8GHz set and the 0.9GHz set, from (Adrian Muscat et al.,2010).

	1.8GHZ Set			0.9 GHz Set		
#	Freq Model	Freq FDTD	Target Error	Freq Model	Freq FDTD	Target Error
0	1.774	1.755	2.481	0.912	0.928	3.089
1	1.817	1.858	3.228	0.889	0.873	2.997
2	1.866	1.871	3.962	0.914	0.950	5.559
3	1.856	1.914	6.327	0.936	0.957	6.300
4	1.774	1.751	2.716	0.989	1.023	13.619
5	1.764	1.692	6.025	0.855	0.848	5.767
6	1.787	1.828	1.542	0.937	0.881	2.109
7	1.908	2.088	15.982	0.937	0.999	11.016
8	1.742	1.637	9.030	0.918	0.982	9.074
9	1.831	1.890	5.006	0.851	0.878	2.446
10	1.805	1.827	1.527	0.833	0.936	3.953
11	1.789	1.730	3.911	0.936	0.850	5.527
12	1.705	1.688	6.244	0.894	0.918	1.963
13	1.815	1.775	1.388	0.823	0.861	4.389
14	1.892	1.785	0.833	0.967	0.925	2.747
15	1.723	1.605	10.807	0.959	1.033	14.752
16	1.750	1.638	8.976	0.827	0.748	16.943
17	1.875	1.723	4.271	0.884	0.879	2.385
18	1.806	1.864	3.558	0.884	0.863	4.157
19	1.856	1.757	2.381	0.831	0.892	0.889
20	1.914	2.019	12.152	0.894	0.878	2.456
21	1.786	1.819	1.040	0.892	0.877	2.508
22	1.735	1.523	15.372	0.912	0.952	5.830
23	1.951	1.763	2.029	0.862	0.824	8.445
24	1.781	1.560	13.345	0.864	0.853	5.273
25	1.975	1.810	0.556	0.877	0.923	2.512
26	1.930	1.936	7.562	0.982	0.951	5.680
27	1.806	1.896	5.340	0.835	0.810	10.014
28	1.822	1.778	1.196	0.928	0.884	1.778
29	1.772	1.994	10.766	0.876	0.766	14.853

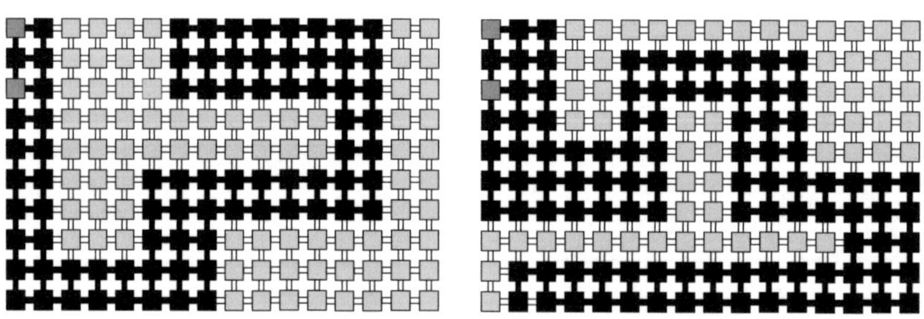

(a) (b)

Fig. 17. The best two candidates resonating at 0.9GHz, from (Adrian Muscat et al.,2010)

7. References

[Seref Sagiroglu et al.,1999] Seref Sagiroglu and Kerim Guney and Mehmet Erler, Calculation of Bandwidth for Electrically Thin and Thick Rectangular Microstrip Antennas with the Use of Multilayered Perceptrons, *International Journal of RF and Microwave Computer Aided Engineering*, Vol. 9, p. 277–286, (1999)

[Adrian Muscat,2010] Adrian Muscat, A Shape Grammar with Feedback Generative Model for the Design of Compact Microstrip Antennas, *International Journal on Advances in Systems and Measurements*, `http://www.iariajournals.org/systems_and_measurements/`, Vol. 3, No. 1 & 2, p. 57–70 (2010)

[Adrian Muscat et al.,2010] Adrian Muscat and Joseph A. Zammit, A Coupled Random Search-Shape Grammar Algorithm for the Control of Reconfigurable Pixel Microstrip Antennas, *submitted for publication to International Journal on Advances in Systems and Measurements*

[Kerim Guney et al.,2002] Kerim Guney and Seref Sagiroglu and Mehmet Erler, Generalized Neural Method to Determine Resonant Frequencies of Various Microstrip Antennas", *International Journal of RF and Microwave Computer Aided Engineering*, Vol. 12, p. 131–139, (2002)

[Kenneth D. Forbus,1988] Kenneth D. Forbus, Intelligent Computer-Aided Engineering, *AI Magazine*, Vol. 9, No. 3, p. 23–36, (1988)

[P. Burrascano et al.1999] P. Burrascano and S. Fiori and M. Mongiardo, A Review of Artificial Neural Networks Applications in Microwave Computer-Aided Design, *International Journal of RF and Microwave Computer Aided Engineering*, Vol. 9, p. 158–174 (1999)

[Coleman,C.M. et al.,2000] Coleman, C.M. and Rothwell, E.J. and Ross, J.E., Antennas and Propagation Society International Symposium, 2000. IEEE, title: Self-structuring antennas, Vol. 3, p. 1256 –1259, doi 10.1109/APS.2000.874431, (2000)

[Kingsley,S.P.et al.,2008] Kingsley, S.P. and Ireland, D.J. and O'Keefe, S.G. and Langley, R.J. and Luyi Liu, Antennas and Propagation Conference, 2008. LAPC 2008. Loughborough, In search of the perfect handset antenna, month 17-18,p. 62–65, keywords:FM radio band;mobile handset antenna;mobile antennas;mobile handsets;, doi 10.1109/LAPC.2008.4516866, (2008)

[Heriberto Jose Delgado et al.,1998] Heriberto Jose Delgado and Michael H. Thursby, A Novel Neural Network Combined With FDTD for the Synthesis of a Printed Dipole Antenna, *IEEE Transactions On Microwave Theory And Techniques*, Vol. 53, No. 7, month July, p. 747–755, (1998)

[Manish Agarwal et al.,1998] Manish Agarwal and Jonathan Cagan, A blend of different tastes: the language of coffemakers, *Environment and Planning B: Planning and Design*, Vol. 25, p. 205–226, (1998)

[G. Stiny et al.,1978] G. Stiny and W. J. Mitchell, The Palladian grammar, *Environment and Planning B*, Vol. 5, p. 5–18, (1978)

[U. Flemming,1987] U. Flemming, More than the sum of parts: the grammar of Queen Anne houses, *Environment and Planning B: Planning and Design*, Vol. 14, p. 323–350, (1987)

[P. A. Fitzhorn,1990] P. A. Fitzhorn Formal Graph Languages of Shape, *Artificial Intelligence for Engineering Design, Analysis and Manufacturing*, Vol. 4, No. 3, p. 151–164, (1990)

[K. Shea et al.1997] K. Shea and J. Cagan, Innovative dome design: applying geodisic patterns with shape annealing, *Artificial Intelligence for Engineering Design, Analysis and Manufacturing*, Vol. 11, No. 3, p. 379–394, (1997)

[K. Shea et al.,1997] K. Shea and J. Cagan, The design of novel roof trusses with shape annealing: assessing the ability of a computational method in aiding structural designers with varying design intent, *Design Studies*, Vol. 20, p. 3–23, (1997)

[Patrick Henry Winston,1984] Patrick Henry Winston, Artificial Intelligence, PUBLISHER: Addison-Wesley, file F, (1984)

[Timothy Masters,1993] Timothy Masters, Practical Nueral Network Recipes in C++, PUBLISHER: Academic Press, file F, (1993)

[Benny Raphael et al.,2003] Benny Raphael and Ian F. C. smith, Fundamentals of Computer-Aided Engineering, PUBLISHER: Wiley, file: F, (2003)

[Adam Drozdek,2005] Adam Drozdek, Data Structures and Algorithms in C++, PUBLISHER: Thomson, file: F, (2005)

[J. S. Gero et al.,1994] J. S. Gero and S. J. Louis and S. Kundu, Evolutionary Learning of Novel grammars for Design Improvement, *Artificial Intelligence for Engineering Design, Analysis and Manufacturing*, Vol. 8, No. 3, p. 83–94, (1994)

[Manish Agarwal et al.,2000] Manish Agarwal and Jonathan Cagan, A micro language: generating MEMS resonators by using a coupled form-function shape grammar, *Environment and Planning B: Planning and Design 2000*, Vol. 27, p. 615–626, (2000)

[Manish Agarwal et al.,2000] Manish Agarwal and Jonathan Cagan On the use of shape grammars as expert systems for geometry-based engineering design, *Artificial Intelligence for Engineering Design, Analysis and Manufacturing*, Vol. 14, p. 431–439, (2000)

[C. G. Christodoulou et al.,2007] C. G. Christodoulou and D. A. Anagnostou and L. M. Feldner, Re-configurable Antennas, IEEE International Workshop on Anti-counterfeiting, Security, Identification., p. 0–12, month: April, (2007)

[D. S. Linden,2002] D. S. Linden, Optimizing Signal strength In-Situ Using an Evolvable Antenna System, Proceedings of the 2002 NASA/DOD Conference on Evolvable Hardware, (2002)

[A. Grau and Lee Ming-Jer et al.,2007] A. Grau and Lee Ming-Jer and J. Romeu and H. Jafarkhani and L. Jofre and F. De Flaviis, A multifunctional mems-reconfigurable pixel antenna for narrowband MIMO communications, Antennas and Propagation International Symposium, p. 489–492, month: June, organization IEEE, (2007)

[Brian S. Collins,2007] Brian S. Collins, editor: Zhi Ning Chen, Antennas for Portable Devices, CHAPTER: Handset Antennas, p. 9–57, PUBLISHER: John Wiley & Sons, (2007)

[B. A. Cetiner et al.,2004] B. A. Cetiner and H. Jafarkhani and Jiang-Yuan Qian and Hui Jae Yoo and A.Grau and F. De Flaviis, Multifunctional reconfigurable MEMS integrated antennas for adaptive MIMO systems, *IEEE Communications Magazine*, Vol. 42, No. 12, p. 62–70, month: December, (2004)

[A. Patnaik et al.2005] A. Patnaik and D. Anagnostou and C. G. Christodoulou and J. C. Lyke, A frequency reconfigurable antenna design using neural networks, *Antennas and Propagation Society International Symposium*, p. 409–412, month: July, (2005)

[Zhang Min et al.,2004] Zhang Min and Luo Xiao-Wu and Wang Guang-Hui, Preliminary research of the reconfigurable antenna based on genetic algorithms, BOOKTITLE:

3rd International Conference on Computational Electromagnetics and Its Applications, p. 137–140, month: November, (2004)

[Muscat,A.,2009] Muscat, A., Advanced Engineering Computing and Applications in Sciences, 2009. ADVCOMP '09. Third International Conference on, title: A Shape-Function Grammar Approach for the Synthesis and Modelling of Pixel-Microstrip-Antennas, month: 11–16, p. 23–28, doi 10.1109/ADVCOMP.2009.12, (2009)

[Adrian Muscat.,2009] Adrian Muscat, Advanced Engineering Computing and Applications in Sciences, 2009. ADVCOMP '09. Third International Conference on, A Shape-Function Grammar Approach for the Synthesis and Modelling of Pixel-Microstrip-Antennas, month: 11-16, (2009)

[Sushil J. et al.,1995] Sushil J. Louis and Fang Zhao, Domain Knowledge for Genetic Algorithms, *International Journal of Expert Systems Research*, Vol. 8, No. 3, p. 195–212, file: F, (1995)

[Adrian Muscat,2002] Adrian Muscat, The Design of Low Gain, Wideband and Multi-band Antennas Employing Optimisation Techniques, CHAPTER: FDTD Model for The Patch Antenna, p. 68–118, PUBLISHER: Queen Mary University of London, month: January, (2002)

[Adrian Muscat,2002] Adrian Muscat, The Design of Low Gain, Wideband and Multi-band Antennas Employing Optimisation Techniques, CHAPTER: Optimisation Based Design, p. 120–174, PUBLISHER: Queen Mary University of London, month: January, (2002)

[A. K. Goel,1997] A. K. Goel, Design, Analogy and Creativity, *IEEE Expert*, p. 62-70, month: May/June, file: F, (1997)

[J. R. Kelly et al.,2008] J. R. Kelly and E. Ebrahimi and P. S. Hall and P. Gardner and F. Ghanem, Combined Wideband and Narrowband Antennas for Cognitive Radio Applications, Cognitive Radio and Software Defined Radio: Technologies and Techniques, month: 18th September, organization: IET, file: F, (2008)

[E. Walton et al.,2000] E. Walton and E. Lee and Y. Bayram and A. Duly and B. Salisbury and G. Bruce and B. Montgomery, Reconfigurable Antenna Arrays Using Pixel Elements, file: F, (2000)

[L. N. Pringle et al.,2004] L. N. Pringle and P. H. Harms and S. P. Blalock and G. N. Kiesel and E. J. Kuster and P. G. Friederich and R. J. Prado and J. M. Morris and G. S. Smith, A Reconfigurable Aperture Antenna Based on Switched Links Between Electrically Small Metallic Patches, *IEEE Transactions on Antennas and Propagation*, Vol. 52, No. 6, p. 1434–1445, month: June, file: F, (2004)

[N. Herscovici et al.,2002] N. Herscovici and M. F. Osorio and C. Peixeiro, Miniaturization of Rectangular Microstrip Patches Using genetic Algorithms, *IEE Antennas and Wireless Propagation letters*, Vol. 1, p. 94–97 (2002)

[J. M. Johnson et al.,1997] J. M. Johnson and Y. Rahmat-Samii, month: August, Genetic Algorithms in Engineering Electromagnetics, *Antennas and Propagation Magazine*, Vol. 39, No. 4, p. 7–25 (1997)

[P. E. Frandsen et al.,2007] P. E. Frandsen and M. Ghilardi and F. Mioc and M. Sabbadini and F. Silvestri, The Electromagnetic Data Exchange: Much More Than A Common Data Format, booktitle: Proc. EuCAP, (2007)

[J. R. Koza et al.,2007] J. R. Koza and S. H. Al-Sakran and L. W. Jones and G. Manassero, Automated Synthesis of a Fixed-Length Loaded Symmetric Dipole Antenna Whose Gain Exceeds That of A commercial Antenna and Matches The Theoretical Maximum, BOOKTITLE: GECCO, p. 2074–2081, address: London, UK, month: July, organization: ACM, file: F, (2007)

[E. A. Jones et al.,2000] E. A. Jones and W. T. Jones, Genetic Design of Linear Antenna Arrays, *IEEE Antennas and Propagation Magazine*, Vol. 42, No. 3, p. 92–100, month: June, file: F, (2000)

[Q. Vo et al.,2000] Q. Vo and D. A. Lowther, A Paradigm for the Non-Routine Design of Electromagnetic Devices Using a Case Based Reasoning Approach, *IEEE Transactions on Magnetics*, Vol. 36, No. 4, p. 1669–1672, month: July, file: F, (2000)

[A. Esposito et al.,2008] A. Esposito and L. Tarricone and L.Vallone and M. Zappatore, A Proposal for an Electromagnetic Ontolgy Framework, BOOKTITLE: International Conference on Complex, Intelligent and Software Intensive Systems, organization: IEEE Computer Society, file: F, (2008)

[A. Esposito et al.,2006] A. Esposito and L. Tarricone and L.Vallone, Semantic-Driven Grid-Enabled Computer-Aided Engineering of Aperture-Antenna Arrays, *IEEE Antennas and Propagation Magazine*, Vol. 48, No. 2, p. 106–117, month: April, file: F, (2006)

[A. Esposito et al.,2007] A. Esposito and L. Tarricone and L.Vallone, Knowledge Modelling in Electromagnetic Applications, *IEEE Antennas and Propagation Magazine*, Vol. 49, No. 5, p. 130–137, month: October, file: F, (2007)

[M. Green,1992] M. Green, Knowledge Aided Design, PUBLISHER: New York:Academic, file F (1992)

[M. Mantyla,1996] M. Mantyla, editor: T. Tomiyama, M. Mantyla, S. Finger, Knowledge Intensive CAD, CHAPTER: Knowledge Intensive CAD: introduction and a reserach agenda, PUBLISHER: Springer, Vol. 1, file: F, (1996)

[D. C. Brown,1998] D. C. Brown, Revision of 1993 Article on Intelligent Computer-Aided Design, *Encyclopdeia of Computer Science and Technology*, month: September, file: F, (1998)

[G. E. Dieter et al.,2009] G. E. Dieter and L. C. Schmidt, Engineering Design, PUBLISHER: McGraw Hill, edition 4th, isbn 978-0-07-283703-9, file F (2009)

[Tony Holden,1987] Tony Holden, Knowledge Based CAD and Microelectronics, PUBLISHER: Elsevier Science Publishers B. V., isbn 0444701508, file: F, (1987)

[Heng Li,1994] Heng Li, Machine Learning of Design Concepts, PUBLISHER: Computational Mechanics Publications, isbn 1853123587, file: F, (1994)

[Michael D. Rychener,1998] Michael D. Rychener, Expert Systems for Engineering Design, PUBLISHER: Academic Press, Inc., isbn: 0126051100, (1988)

[R. C. Booton,1999] R. C. Booton, Microwave CAD in the Year 2010 - A Panel Discussion, *International Journal of RF and Microwave Computer-Aided-Engineering*, Vol. 9, p. 439–448, file: F, (1999)

[K. C. Gupta,1998] K. C. Gupta, Emerging Trends in Millmeter-Wave CAD, *IEEE Transactions on Microwave Theory and Techniques*, Vol. 46, No. 6, p. 747–755, month: June, file: F, (1998)

12

Electrically Small Microstrip Antennas Targeting Miniaturized Satellites: the CubeSat Paradigm

Constantine Kakoyiannis and Philip Constantinou
Mobile Radio Communications Laboratory
School of Electrical and Computer Engineering
National Technical University of Athens
Greece

1. Introduction

A CubeSat is a type of *miniaturized satellite* used primarily by universities for space exploration and research, typically in low Earth orbits (e.g. sun-synchronous). The design protocol specifies maximum outer dimensions equal to $10 \times 10 \times 10$ cm^3, i.e., a CubeSat occupies a volume up to 1 litre (CubeSat programme, 2010). CubeSats weigh no more than 1.0 kg, whereas their electronic equipment is made of Commercial Off-The-Shelf (COTS) components. Several companies have built CubeSats, including large-satellite maker Boeing. However, the majority of development comes from academia, with a mixed record of successfully orbited Cubesats and failed missions (Wikipedia, 2010a).
Miniaturized satellites, or small satellites, are artificial orbiters of unusually low weights and small sizes, usually under 500 kg. While all such satellites can be referred to as *small satellites*, different classifications are used to categorize them based on mass (Gao et al., 2009):

1. Mini-satellite (100–500 kg)
2. Micro-satellite (10–100 kg)
3. Nano-satellite (1–10 kg)
4. Pico-satellite (0.1–1 kg)
5. Femto-satellite (0.01–0.1 kg)

CubeSats belong to the genre of *pico-satellites*; their maximum weight lies on the borderline between pico- and nano-satellites. The main reason for miniaturizing satellites is to reduce the cost of deployment: heavier satellites require larger rockets of greater cost to finance; smaller and lighter satellites require smaller and cheaper launch vehicles, and are often suitable for launch in multiples. They can also be launched "piggyback", using the excess capacity of larger launch vehicles (Wikipedia, 2010b). But small satellites are not short of technical challenges; they usually require innovative propulsion, attitude control, communication and computation systems. For instance, micro-/nano-satellites have to use electric propulsion, compressed gas, vaporizable liquids, such as butane or carbon dioxide, or other innovative propulsion systems that are simple, cheap and scalable. Micro-satellites can use radio-communication systems in the VHF, UHF, L-, S-, C- and X-band. On-board communication systems must be much smaller, and thus more up-to-date than what is used in conventional satellites, due to space constraints. Furthermore, miniature satellites usually lack the power supply and size required for conventional bulky radio transponders. Various compact innovative communication solutions have been proposed for small satellites, such as optical (laser) transceivers, antenna arrays and satellite-to-satellite data relay. Electronics need to be rigorously tested and modified to be "space hardened", that is, resistant to the outer space environment (vacuum, microgravity, thermal extremes and radiation exposure) (Wikipedia, 2010b).

The CubeSat programme was developed through the joint efforts of research laboratories from California Polytechnic State University (Cal Poly) and Stanford University, beginning in 1999. The concept was introduced to the scientific community as an opportunity for all universities to enter the field of space science and exploration. A large group of universities, along with certain companies and organizations, participate actively in the CubeSat programme; it is estimated that 40 to 50 universities were developing CubeSats in 2004. Featuring both small size and weight, a CubeSat can be built *and* launched for an estimated total of $65,000–80,000 (per fiscal year 2004 values). The standard $10 \times 10 \times 10$ cm^3 basic CubeSat is often called a "1U" CubeSat, meaning *one unit*. CubeSats are roughly scalable in 1U increments and larger. The four basic sizes are 0.5U, 1U, 2U and 3U. The number corresponds to the length of the CubeSat in decimetres; width and depth are always 10 cm, or 1 dm. Orbiters such as a "2U" CubeSat ($20 \times 10 \times 10$ cm^3) and a "3U" CubeSat ($30 \times 10 \times 10$ cm^3) have been both built and launched. Since CubeSats are all 10×10 cm^2 (regardless of length) they can all be launched and deployed using a common deployment system. CubeSats are typically launched and deployed from a mechanism called a Poly-PicoSatellite Orbital Deployer (P-POD), developed and built by Cal Poly. The low cost of the CubeSat programme, compared to standard satellite missions, has formed a cost-effective independent means of getting a payload into orbit for many institutions around the world. Most CubeSats carry one or two scientific (measuring) instruments as their primary mission payload (Wikipedia, 2010a). Only few of them are equipped with a propulsion system that enables orbit correction or attitude control. One such example is the CubeSat built by the University of Illinois, which was loaded with an array of small ion thrusters.

CubeSats enable a vast array of research possibilities and applications. One of the key areas of study is *Earth remote sensing*: efforts there focus on earthquake detection through the detection of magnetic signals, study of the air-glow phenomenon in the Earth's atmosphere, cosmic dust detection and the possibility of terrestrial gamma-ray bursts originating from lightning. Another field of study, and a rather expensive one, is *biology*. For instance, the GeneSat-1 project by NASA was not cheap by CubeSat standards: total expenditure on the satellite and its experiments reached $6 million before GeneSat was launched on a Minotaur rocket. Its mission is to establish methods for studying the genetic changes in bacteria exposed to a space environment (Wikipedia, 2010a). Modern small satellites are also useful for other applications, such as telecommunications, space science, mitigation and management of disasters (floods, fire, earthquake, etc.), in-orbit technology verification, military applications, education, and training (Gao et al., 2008; 2009).

CubeSat missions started in 2003; that year 5 academic and 1 commercial CubeSat were carried into orbit. Successful launches continued in 2005 with 3 more CubeSats built by universities; 17 more satellites were carried into orbit between 2007 and 2010. But, in 2006, July saw the greatest disaster in the short history of CubeSats: with a payload of 14 satellites from 11 universities and a private company, a DNEPR-1 rocket was launched from Baikonur Cosmodrome, Kazakhstan; it was the largest planned deployment of CubeSats to date. The

rocket failed and crashed into the ground, obliterating the CubeSats and 4 other satellites aboard. The launch was lost after the engine responsible for the first stage of lift off stopped working prematurely. Thrust termination occurred at 74 seconds after lift off (Wikipedia, 2010a). An extensive (but incomplete) list of CubeSat missions can be found in (Wikipedia, 2010c). Out of the 52 past missions documented, 24 are currently active; 1 was successful but has been de-activated; 23 satellites failed; and the status of 4 others is unclear. Out of the 23 missions that failed, 17 cases were launch failures; the other 6 were successfully carried into orbit, but a malfunctioning system prevented them from becoming operational.
Since 2001, there has been a growing number of European universities that build and contribute their own pico-satellite(s) to the programme. The efforts of research groups from universities in Denmark, Germany and the UK have been particularly instrumental to the development of CubeSat technology.

2. Scope of the chapter

Modern small satellites require antennas to realize the following four fundamental functions, and CubeSats are no exception to the norm:

1. Telemetry, tracking and command (TTC),[1] which includes both uplink and downlink, at different frequencies
2. High-speed downlink for payload data, e.g., in Earth-observation missions
3. GPS/GNSS signal reception
4. Inter-satellite cross links

These functions often require several different antennas. Basic radiator configurations used are normally helices, monopoles, patches, and patch-excited cups, depending on frequency range, coverage requirements, and application (Gao et al., 2008; 2009).
Before moving on to the objectives of the Chapter, a few comments on frequency allocations are in order. Since CubeSats are mostly developed in academic research centres, it is easy to deduce that they are almost solely intended for educational and research (i.e., *non-commercial*) purposes. Therefore, there is a frequency allocation problem, because CubeSats could not occupy commercial spectrum. In Europe, certain frequency bands have been allocated for amateur satellite communication purposes. Examples of these frequencies are the 434.8–438, 1260–1270 and 2400–2450 MHz bands. In the US, although not strictly termed "amateur satellite", nine frequency bands have been allocated for space research. These bands fall in the 400–2700 MHz range. The most commonly used ones are 2025–2110 and 2200–2290 MHz.
This Chapter addresses the problem of building nano-/pico-satellites from the antenna designer's point-of-view. Our objective is to describe the implementation of a *planar*, *low-cost* antenna solution for the TTC subsystem of a 1U CubeSat orbiter, operating in the 434.8–438 MHz band; this band, also known as the "70-cm" band, is often chosen by system designers due to favourable path loss characteristics. A radiator backed by a ground plane is required, so as to obtain "single-sided" directivity. These three initial specifications can be met with proper design of microstrip "patch" antennas. Patch antennas are manufactured using standard Printed Circuit Board (PCB) techniques; space-graded substrate materials must be used to satisfy the tighter mechanical and thermal constraints of space applications. A planar antenna is essentially a two-layer PCB, and thus it costs much less than a standard 4-layer or 6-layer PCB. What is more, if the designer implements a feeding network without

[1] Often abbreviated also as "TT&C".

any metallized holes (Vias), manufacturing costs drop even further. Nonetheless, it is a rather challenging design task to integrate a planar radiator that resonates at a wavelength $\lambda_0 = 687.3$ mm on a 100×100 mm^2 surface. Herein, λ_0 denotes the *free-space* wavelength.

The implementation of the CubeSat antenna is based on the following key assumptions. The satellite is built on a cubic conductive chassis having 10-cm sides. One of the six faces is totally used up by the antenna. Thus, the surface on which the antenna will be printed is a 100-mm square that will be occupied by the substrate in its full extent. The patch antenna must have both sides shorter than 100 mm. Moreover, it has been also assumed that the feed network of the antenna will reside either on a different face of the cube, or on the inside with the rest of the electronics, but definitely not on the same face as the antenna. In the latter case, antenna excitation is done with a protruding coaxial probe, i.e., a coaxial transmission line whose center conductor runs through the substrate. Last but not least, since this application is oriented towards *satellite communications*, it has been assumed that the microstrip antenna should have the ability to produce circularly polarized waves. A robust technique for the generation of circular polarization from patch antennas is the dual feed with signals in phase-quadrature. For this scheme to work, the shape of the microstrip patch, which determines the shape of the Surface Current Distribution (SCD), must display two perpendicular symmetry axes; otherwise, polarization will come out elliptical at best.

At the centre frequency $f_0 = 436.5$ MHz we get a free-space wavelength $\lambda_0 = 687.3$ mm. The size of a patch antenna should be at least equal to $\frac{\lambda_g}{2} \times \frac{\lambda_g}{2}$, where λ_g is the *guided* wavelength inside the cavity formed by the substrate and the two copper layers (Top and Bottom). If we had used a low-permittivity substrate with $\varepsilon_r \approx 1.0$ (e.g. Arlon FoamClad™), which is desirable for microstrip antennas, then the dimensions of the patch would be $\frac{\lambda_0}{2} \times \frac{\lambda_0}{2} = 343.6 \times 343.6$ mm^2. In that case, the antenna would require 12 times the area that is available on the satellite.

From the above introductory design notes it becomes obvious that the designer is forced to use extensive Dielectric Loading (DL) to obtain an initial degree of miniaturization. As it will be demonstrated in the following sections, DL alone is not sufficient in order to achieve the required total miniaturization. In any case, the high–ε_r material is bound to have a detrimental effect on bandwidth. To lessen this side-effect, a thick substrate has been chosen. To avoid further degradation of the radiation efficiency, the material should display low loss tangent ($\tan \delta_e$). The above issues preclude the use of a low-cost material, such as FR-4; the choices left to the designer are high-quality space-graded materials, such as PTFE, ceramic and alumina.

This Chapter is intended for serving the antenna engineering community as a concise *design guide* to a specific class of microstrip antennas, particularly *inductive-slit-loaded microstrip antennas*. However, the design approach and the electromagnetic modelling are applicable to any sort of microstrip antenna. This design guide will be useful for senior undergraduate and graduate students, research engineers, and practising antenna engineers in the field of printed/planar antennas. A basic understanding of electromagnetic theory and antennas is required.

The Chapter is organized as follows. Section 3 attempts an exhaustive review of existing literature on small-satellite-oriented antennas. Section 4 documents the design of the hybrid coupler which feeds the antenna with two equal-amplitude signals in phase-quadrature. The simulation environment is also described, along with details of the simulation setup used throughout the Chapter. Section 5 describes in detail the design procedure of the antenna and the miniaturization techniques employed. A large array of numerical results is given for this two-step design procedure. Section 6 discusses the simulation results and examines the electrical performance of the antenna in terms of its electrical size. Finally, Section 7 concludes the Chapter with a summary of key findings and suggestions for further research.

3. Antennas for modern small satellites: Literature survey

The development of the antenna described herein can commence directly from system specifications. However, good engineering practice dictates that the relevant literature be surveyed first. After the state-of-the-art of the field has been determined, the antenna engineer can make a more educated guess on the course of action. Non-planar designs used are normally helices, monopoles, and open waveguides. Planar structures are usually patches and patch-excited cups. The choice depends on frequency range, coverage requirements, and application.

This Section aims to serve as an introduction to what is undoubtedly a fascinating and important part of the future of satellite antennas.

3.1 Wire Antennas and other non-planar structures

The literature review first revealed a number of studies oriented towards small satellites operating in the 70-cm band that are equipped with *linear* monopole or dipole antennas built with measuring tape. The tape remains folded while the satellite is not yet in orbit, and is held in place with Nylon fibres, which are secured using a short length of Nichrome wire inside the satellite. Once the satellite is in orbit, current passes through the Nichrome wire, the fibres melt and the linear antenna is released. Small satellites using this linear-antenna technique are studied in (Dabrowski, 2005; Galysh et al., 2000; Heidt et al., 2000; Hunyadi et al., 2002; LaBerteaux et al., 2007; Mizuno et al., 2005; Puig-Suari et al., 2001; Schaffner & Puig-Suari, 2002).

But this is not the only available option in terms of wire antenna elements. (Moghaddam et al., 2004) used a separated turnstile antenna (STA) to obtain saddle-shaped and hemispherical patterns for small low-earth-orbit (LEO) satellites at VHF and UHF bands. This STA is an array of four monopoles that are mounted symmetrically on the satellite and are electrically driven in phase-quadrature. The antenna is built with 55 cm–long wire elements and it resonates at 130 MHz. Dual-band operation in the UHF band is possible by exploiting the next natural odd (third-harmonic) resonance of the monopoles at 390 MHz. Gain in the UHF band was specified at 5 dBi.

Helices, despite being protruding radiators like monopoles, are also popular solutions. Quadrifilar helical antennas (QHAs) are suitable for small LEO satellites not only for their gain pattern, but also due to their low weight, size and cost. The QHA is made up of four coaxial identical elements, which are fed in phase-quadrature to produce circular polarization. (Rezaei, 2004) designed an S-band QHA for the TTC subsystem of the Small Multi-Mission Satellite that covers the Asia-Pacific region. The antenna operates at 2.26 GHz providing a 2% fractional bandwidth at VSWR $= 1.3 : 1$. If we denote the VSWR level by S, then fractional bandwidth scales by the factor $(S-1)/(2\sqrt{S})$ (Yaghjian & Best, 2005). Thus, at a level VSWR $= 2 : 1$ the bandwidth is estimated at 5.4%. Based on the author's description, the electrical size of this radiator, omitting the ground plane, was estimated at $ka = 2.11$ rad.[2] The result was anticipated, since QHAs are invariably electrically large antennas.

The G-shaped wire monopoles designed by (Yousuf et al., 2008) specifically target CubeSat missions. This rhombic structure is based on the concept of the loaded monopole, where a

[2] See Section 5.6 and Fig. 32 for the definition and importance of electrical size.

short monopole is loaded with two rectangular rings. Three versions were designed; two for the VHF (150 and 180 MHz) and one for the UHF band (370 MHz). Initial designs were done on infinite ground planes. Mounted this way, the achieved numerical bandwidths ranged between 32–42%. Bandwidth definition is ambiguous, since the authors used multiple system impedances. In any case, once the G-shaped monopoles where mounted on a finite wire-grid model of a CubeSat, their numerical electrical performance was severely affected, particularly that of the VHF versions. This is also to be expected: monopole antennas work fine as long as their ground plane is large enough. Below a certain limit, chassis-coupling-and-excitation considerations must be included in the design cycle. The electrical size of the VHF antennas was $(ka)_{\mathrm{VHF}} = 1.26$ rad, whereas that of the UHF antenna was $(ka)_{\mathrm{UHF}} = 1.41$ rad, i.e., slightly smaller than the size of the half-wavelength dipole. The ground plane was ignored in these estimations.

The size of helices can be reduced either by modifying the helical structure or by introducing dielectric loading (DL). Little design freedom is obtained by varying the pitch angle and diameter of the helix to reduce its size without destroying its performance. Moreover, the ground plane of helical antennas needs to remain large enough for good performance. However, DL strengthens the near field, increases the quality factor, and reduces the operational bandwidth of the helix. (Niow et al., 2009) proposed a well-balanced combination of a modified helical antenna and DL to reduce antenna size. The modified structure was a backfire bifilar helical antenna (BBHA), which, unlike conventional helices, does not require a ground plane. This structure is still relatively large, thus a dielectric rod was introduced around the feeding coaxial cable to reduce the size further. The dielectric rod was bound to affect antenna gain and bandwidth. Therefore, trade-offs were made between size reduction and antenna performance. All developed antennas operated at 2.6 GHz. The chosen performance metrics were broadband gain and axial ratio (AR) bandwidth. The initial (unloaded) BBHA featured a gain of 4 dBi and an AR bandwidth equal to 0.8 GHz. Its electrical size is calculated at $(ka)_{\mathrm{ini}} = 2.22$ rad. The second BBHA was loaded with a Teflon rod. It featured a peak gain of 4 dBi and an AR bandwidth equal to 0.4 GHz. Its electrical size was reduced by 19.5% to a new value of $(ka)_{\mathrm{Teflon}} = 1.78$ rad. The third BBHA was loaded with a Macor rod. It featured a peak gain of 3.7 dBi and an AR bandwidth equal to 0.4 GHz. Its electrical size was reduced by 24% to a new value of $(ka)_{\mathrm{Macor}} = 1.72$ rad. Note that the electrical size of a half-wavelength dipole equals $(ka)_{\lambda/2} = 1.57$ rad.

High-frequency bands, like the X-band, provide for physically small antennas regardless of structure. (Galván & Colantonio, 2009) implemented a waveguide-based antenna for the data-downlink subsystem of the SAC-D/AQUARIUS mission, which is a LEO earth observation satellite. The radiating part is a compact choke ring antenna with reduced back-radiation. The structure consists of a segment of circular waveguide surrounded by equally spaced concentric rings ("corrugations"). Flaring of the waveguide was not allowed, in order to avoid narrowing the beam; a hemi-spherical pattern was thus obtained, according to specifications. The antenna operates at 8.2 GHz with a 5% fractional bandwidth and a gain of 7.4 dBi. The electrical size of the radiating element alone was calculated at $ka = 2.99$ rad.

This part of the literature survey is concluded with the recent study by (Nohmi et al., 2010), in which the authors implemented a solar paddle antenna for the "KUKAI" pico-satellite mission of the Kagawa University, Takamatsu, Japan. The solar paddle antenna is a bent wire radiator installed around the circumference of the solar paddle. This is a low-cost implementation, since it requires only a segment of wire and a few ferrite beads to separate the power supply line from high-frequency signals. This type of antenna simplifies the structure of the satellite, reduces its weight and provides the necessary directivity. From the manuscript it is not particularly clear whether a single antenna or two antennas have been integrated into the periphery of the solar paddle. The radiating system operates at the VHF (145 MHz) and the UHF band (435 MHz); these two frequencies are harmonically related. Furthermore, it is clear that at the lower frequency the radiator is a $\lambda/4$–long monopole, whereas at the upper frequency the radiator is a $3\lambda/4$–long monopole.

3.2 Planar antennas for modern small satellites

One of the earliest studies on planar antennas for small satellites appeared in (Tanaka et al., 1994). To preserve area for both antenna and solar cells on a micro-satellite, the authors designed the radiating system with the solar panels attached on top of the patch antenna. The concept behind this stacked configuration is that a patch antenna, being a lossy cavity, radiates through the fringing fields appearing at its open-circuited edges. Therefore, it is irrelevant whether the solar panel is transparent to radio waves or not; the solar panel needs only to be placed in such a manner that its own power generation function as well as the radiation performance of the patch beneath it are both maintained. To this end, solar cells can occupy the whole surface of the patch and of the surrounding substrate, *except* for the region where the radiating fringing fields appear, that is, the area immediately surrounding the patch. The prototype antenna operated at 2.225 GHz with a 1.5% fractional bandwidth.

Microstrip antennas have three main feed methods: transmission line feed, optionally inset; coaxial probe feed; and aperture coupling. In (He & Arichandran, 2001), where the focus is again on micro-satellites, the authors designed a physically small, aperture-coupled patch antenna at 10.74 GHz (X-band). The antenna displays a 5.6% fractional bandwidth and a gain of 6.5 dBi (both numerical). Radiation efficiency was estimated at $n_{\text{rad}} = 0.85 = -0.7$ dB. The electrical size of the radiating element was calculated at $ka = 1.28$ rad.

In (Mathur et al., 2001) the authors describe the design of two patch antennas for the USUsat nano-satellite which is part of the ION–F constellation. The uplink antenna works at 450 MHz, and it is printed on a substrate with $\varepsilon_r = 10.2$. It also achieves a bandwidth of 7 MHz at VSWR $= 2:1$ (1.6%), whereas the square patch has a side length equal to 106.7 mm. To achieve circular polarization with a single feed, a coaxial probe feeds the patch along its diagonal. Its electrical size was calculated from the authors' description to be equal to $(ka)_{\text{UHF}} = 0.71$ rad. The downlink antenna resonates at 2.26 GHz and displays 4.9 dBi gain and a 17 MHz bandwidth (0.8%). It was designed on the same substrate, and it is a 20-mm square patch, thus it is estimated that $(ka)_{\text{S-band}} = 0.67$ rad. Size estimations are somewhat optimistic, since they take into account only the size of the patch, leaving out the spread of the current distribution on the ground plane.

The next study, which deals specifically with a CubeSat planar antenna, was conducted by (Fujishige et al., 2002) and (Tamamoto & Shiroma, 2002) in the framework of the CubeSat programme of the University of Hawaii. It addresses the design of a type of active antenna better known as a "grid oscillator", which is essentially a C-band active antenna array. An array of active semiconductor devices (transistors) are embedded in a grid of copper traces printed on a substrate, which serves as a DC bias distribution circuit, RF-embedding circuit and radiator. The authors in (Fujishige et al., 2002) used pHEMT devices and built a 6×6 transistor matrix on a Rogers Duroid™ substrate ($\varepsilon_r = 10.2$) at 5.85 GHz. The structure is backed by the conductive chassis of the CubeSat, which serves as a mirror to provide the necessary excess feedback for oscillation. The horizontal traces of the matrix function as DC bias lines. The vertical traces are the radiating elements, producing a vertically polarized array that does not interact with the bias lines. The grid oscillator is actually the microwave equivalent of a laser:

1. A lossy cavity is formed by the structure and the mirror.
2. As soon as DC power is applied, oscillation is triggered by transients and/or noise; each transistor oscillates at a different frequency.
3. A non-coherent wave originates from the grid, bounces off the mirror and injection-locks the active devices; the cavity starts to reverberate.
4. Different eigenmodes compete inside the cavity, just as in a laser. Higher-order modes lose most of their power to diffraction. Single-frequency, self-locked, coherent oscillation is what remains from this process.
5. The output power from each device is combined in the far-field, making this power-combining scheme very efficient.

Unlike regular phased arrays, the spacing between the vertical traces is on the order of $\lambda_0/10$, making the grid oscillator very compact at microwave frequencies. Indeed, the outer dimensions of the grid oscillator in (Fujishige et al., 2002) were a mere 50×70 mm^2.

Several planar antennas were built to address the communication needs of European small satellite missions, such as the ESEO and SSETI-Express student missions. The size of the mini-satellite described in (Wincza et al., 2004) is $60 \times 60 \times 60$ cm^3 and it communicates at 2.025 GHz and 8.45 GHz. However, very little information is given on the topology and the electrical performance of the antenna, except an operational bandwidth of 50 MHz. Besides, the design procedure is unclear, since the microstrip patch is printed on a complex sandwiched structure to increase the bandwidth. The ESEO satellite, studied in (Idzkowski et al., 2004), communicates at 2.080 GHz and 2.260 GHz, and bears a total of six microstrip antennas for communications and telemetry. The authors cite a 7-dBi gain for this antenna system, without clarifying whether this gain holds for each antenna separately or if the antennas were grouped in two 3-element linear arrays. Their study focuses mainly on the details of the link budget of the ESEO mission.

(Muchalski et al., 2004) studied low-gain TTC antennas for the ESEO and SSETI-Express missions. The main objective of their research was the optimization of antenna placement. The authors designed a 61.4 mm square patch that was suspended 10 mm above the ground plane (air dielectric). The ground plane measured 60 cm $\times$ 70 cm and corresponded to the wall size of the spacecraft. Input impedances and radiation patterns were numerically calculated for five different simulation scenarios, which included centre-mounted, edge-mounted and corner-mounted antennas. The model corresponding to each scenario included a single antenna. The results confirmed that both $Z_{\text{in}}(j\omega)$ and the radiation pattern were severely affected by antenna placement. The centre-mounted antenna was initially designed, and it operated at 2.45 GHz. It featured a 8.2% fractional bandwidth and an electrical size $(ka)_{\text{centre}} = 2.24$ rad. When this patch was moved to either of the edges of the ground plane, the centre frequency shifted to 2.35 GHz. The operational bandwidth changed to 8.5% and the new electrical size was $(ka)_{\text{edge}} = 2.15$ rad. Finally, when the patch was moved to one of the corners of the ground plane, the centre frequency shifted to 2.275 GHz. The operational bandwidth changed to 7.9% and the new electrical size was $(ka)_{\text{corner}} = 2.08$ rad.

The antenna system of the ESEO orbiter has drawn considerable attention. An innovative, light weight, high gain antenna for the high-speed downlink of payload data of the ESEO was described by (Arnieri et al., 2004). The design is based on the shorted annular patch (SAP), which was integrated with a stacked parasitic element. The mechanical attributes

of SAPs provide for an all-metal antenna (very desirable for space applications) realized in suspended technology, leading to a compact and robust radiator. The stacked configuration provides the necessary design freedom to adjust gain and beam aperture simply by changing the distance between the two radiating elements. The nominal half-wavelength distance turns the structure essentially into a two-element end-fire linear array. The authors excited their antenna through an aperture-coupled feed, which provided circular polarization. The antenna operated at 2.425 GHz with an 8% fractional bandwidth. Peak measured gain was 12.2 dBi. The specified dimensions of the radiating parts lead to an estimated electrical size $ka = 2.50$ rad. This structure was later refined in (Arnieri et al., 2007). The updated dimensions produce a value $ka = 2.73$ rad. The stacked SAP antenna is also described in (Gao et al., 2009).
Most antenna implementations target either the TTC or the payload data downlink subsystem. One of the very few studies on inter-satellite cross-link antennas picks up a topic of tremendous interest to array designers: modified Van Atta retrodirective arrays were proposed for pico-satellites operating at 10.5 GHz (Mizuno et al., 2005). The first obvious choice for cross-linking satellites is to use omnidirectional antennas. However, this choice is energy-inefficient, wasting valuable satellite resources, while at the same time it creates a satellite network that is vulnerable to eavesdropping. The second obvious alternative is to design dynamically steerable antenna arrays ("smart beamformers"). Such an implementation would tax a significant amount of resources in terms of processing power, and would introduce a complexity level that would cancel out the simple, low-cost nature of small satellites. For pico-satellite applications, a suitable alternative to dynamic beam steering is a *self-steering* retrodirective array (Mizuno et al., 2005). Retrodirective antennas are able to sense the direction of an incoming radio signal and send a reply back in the same direction. A fascinating property of retrodirective arrays is that this ability results purely from analog/RF signal processing; no digital signal processing algorithms are required at the digital baseband part of the transceiver. The topic of modified Van Atta arrays is also discussed in (Gao et al., 2009).
The S-band is also used for communication by the commercial SSTL micro-satellite built by Surrey Satellite Technology Ltd. Out of the three antennas described in (Hadj Abderrahmane et al., 2006), one is a circular patch citing a 4.9-dBi gain and main lobe beamwidth equal to 80°; this one is used for command uplink. Apart from a gain variation of 3 dB, no other details are given. Moreover, the provided radiation pattern (see Fig. 8 in (Hadj Abderrahmane et al., 2006)) corresponds to 400 MHz, not 2 GHz. Finally, the antenna depicted in Fig. 9 of (Hadj Abderrahmane et al., 2006) does not agree with the cited main-lobe beamwidth, since the circular patch seems to be equally wide with its ground plane.
(Maleszka et al., 2007) describe briefly the design of a low-profile, low-gain planar antenna for a SSETI-Express mini-satellite; the size of the satellite is $60 \times 60 \times 70$ cm^3. The antenna is mounted on a 350×350 mm^2 ground plane and operates at 2.4 GHz. Particular emphasis is placed on circular polarization generation and on maintaining an acceptable axial ratio, so that controlled degeneration to elliptical polarization is achieved.
During the initial acquisition period following the separation of the satellite from the launch vehicle, satellite stabilization has not been achieved yet; thus an omnidirectional antenna is required for communication between space and ground segment (the first established TTC link). Various low-gain antennas have been developed for TTC of small satellites at VHF, UHF and the S-band. These antennas are simple, cheap, easily fabricated, and have nearly omnidirectional or broad-beam radiation patterns, thus the satellite does not need accurate control of its attitude. One such antenna is the microstrip patch described in (Gao et al.,

2008). It uses a circular patch fed by a coaxial probe at the bottom. It is circularly polarized and operates within a tunable frequency range of 2.0–2.5 GHz, that is, inside an aggregate fractional bandwidth equal to 22%. This patch achieves a gain of 6.5 dBi at an electrical size equal to $ka = 1.99$ rad at 2.25 GHz.

Design ideas for small satellite antennas can be borrowed just as well from other segments of the antenna field, as long as the structure undergoes the necessary "space" modifications. In this context, the compact, dual-band, circularly polarized microstrip antenna (CPMA) by (Lee & Woo, 2008) is reported here. This CPMA was designed for satellite communication handsets by combining a folded patch and a plate into a stacked 3-D structure. However, the design was carried out on an electrically large ground plane, thus chassis-coupling-and-excitation considerations were not included in the design iterations. The lower band, which corresponds to the downlink, is centred around 1.61 GHz. The CPMA achieves a fractional bandwidth equal to 4.8% and a gain of 2.4 dBi. The electrical size was calculated at $(ka)_{\text{lower}} = 0.90$ rad (the ground plane was ignored in this estimation). The upper band, which corresponds to the uplink, is centred around 2.4865 GHz. The CPMA achieves a fractional bandwidth equal to 6.6% and a gain of 5.3 dBi. The electrical size was calculated at $(ka)_{\text{upper}} = 1.39$ rad.

The next study deals specifically with planar antennas targetting a CubeSat programme. (Hamrouni et al., 2009) designed and prototyped two microstrip antennas on a 1.6 mm–tall substrate. The antennas were intended for use on the first Tunisian pico-satelite ERPSat-1 and operated at 2.4 GHz (S-band). Very few details were given on the design strategy. The first prototype achieved a 2.9% fractional bandwidth with an electrical size equal to $(ka)_1 = 1.62$ rad. The second prototype achieved a 3.6% fractional bandwidth and, even though dimensions differed, its electrical size was again $(ka)_2 = 1.62$ rad.

The little space available on small satellites brings about limited capabilities for the analog/RF section. (Marrocco et al., 2010) suggested that these limitations can be overcome by exploiting the idea of "structural radiators", which has already been implemented in avionics, naval communications and hand-held terminals. This concept diverges from the use of independent sets of self-consistent radiators, and instead relies on deliberate exploitation of antenna-chassis coupling. Accordingly, radio functions are no longer fulfilled by stand-alone antennas (whips, patches, helices, etc.), while, at the same time, the satellite structure is exploited as part of the radiation mechanism. Thus, the problem of antenna integration onto small satellites can be addressed through a distributed approach, where multiple "exciter" antennas are placed all over the structure and stimulate it to produce a controllable radiation pattern. Strong coupling is expected among the exciters of the satellite, hence they must be treated as a multi-port network, i.e., designed as a whole. In their effort to demonstrate that such a system can be optimized to produce variable patterns and polarizations, the authors initially developed a tunable vertical inverted–F antenna mounted on a finite horizontal ground plane. The antenna operated at 2.3 GHz and achieved a 5.7% instantaneous fractional bandwidth. When multiple such exciters were used to design an eight-element circular array on a finite satellite body, the active reflection coefficients revealed that bandwidth was increased to 13.9% because of the configuration and coupling between the elements.

In a very recent study, (Maqsood et al., 2010) presented dual-band, circularly polarized, planar antennas for GNSS-based remote sensing applications. The authors developed a zenith and a nadir antenna that can be body mounted on-board a small UK-DMC satellite. Both antennas cover both L1 and L2 bands, centred at 1.575 GHz and 1.227 GHz respectively. In terms of compact antenna design, attention is drawn to the zenith antenna, which was initially designed as a slot-coupled stacked patch radiator. Three layers were stacked together, with the square patch antenna on top of the upper two layers. The initial feed network comprised of three Wilkinson power dividers, which fed the bottom patch through aperture coupling. The upper patch was electromagnetically coupled to the lower patch. However, since the numerical results revealed that the radiation pattern had a strong back-lobe, the feed network was changed. Instead of aperture coupling with the bottom patch, the top patch was eventually directly connected to the feed network using vias; the bottom patch was electromagnetically coupled to the upper patch. Wilkinson dividers were replaced by a broadband three-branch coupler. The measured prototype showed a 6.3 dBi gain at the L1 and a 4.0 dBi gain at the L2 band. Calculated electrical sizes based on author data are $(ka)_{\mathrm{L1}} = 1.82$ rad and $(ka)_{\mathrm{L2}} = 1.42$ rad, respectively.

The literature survey is concluded with three recent review papers of great educational value, particularly for young engineers in the field (Gao et al., 2009; Wettergren et al., 2009; Zackrisson, 2007). A distinct feature of these papers is that, through an abundance of photographs depicting commercial antennas and arrays, they present state-of-the-art antennas for modern small satellites *from the perspective of the industry*. Industrial perspective is often very different from that of academia, so it could be the "rude awakening" that reveals types of antennas that have proven viable over time.

3.2.1 Antenna development at Saab Space

The first two contributions were made by Saab Space, now RUAG Aeropace Sweden. In (Wettergren et al., 2009; Zackrisson, 2007), the authors present wide coverage antennas for small satellites. Depending on frequency range, coverage requirements and application, the proposed solutions are

- Helical antennas, which are suitable for L-, S- and X-band applications
- Toroidal antennas, which are suitable for S-, Ku- and Ka-band applications
- Horns, which are suitable for Ku- and Ka-band applications
- Waveguide radiators, which are suitable for C-, X-, Ku- and Ka-band applications, and
- Patch-excited cups, which are suitable for L-, S- and X-band applications

Focusing on a few specific applications and frequency bands, it is first noted that helices and patch-excited cups are the preferred solutions for GPS/GNSS applications. The latter antenna structure is also the main choice for S-band applications, since it can implement both compact/low-gain and large/medium-gain antennas. At the X-band front, helices are used for high-speed data downlink. TTC antennas with hemispherical coverage come in two "flavours": waveguide radiators for dual-frequency operation, and patch-excited cups for single-frequency operation.

Patch-excited cups are truly versatile performers. They are light weight, robust, all-metal radiators, able to produce gains as high as 15 dBi. It can be roughly stated that their radiation pattern is designed separately from their input matching. The pattern is mainly influenced by cup diameter, rim height, and radius and height of the top-most patch. Input matching is mostly handled by adjusting the geometry of the cavity formed by the lower two patches and by changing the radial position of the feed probes. The cross-influence among these parameters is weak enough (Wettergren et al., 2009). The GPS cup described in (Zackrisson, 2007) covers both L1 and L2 bands, achieving peak gains equal to 8.5 dBi and 7.4 dBi, respectively. With a cup diameter of 160 mm, and given that rim height is about a quarter-wavelength, it is estimated that the corresponding electrical sizes are $(ka)_{\mathrm{L1}} = 2.82$ rad and $(ka)_{\mathrm{L2}} = 2.20$ rad. On the other hand, the S-band cup that was designed for the LCROSS mission is a medium-gain antenna (MGA). It scored a peak gain of 12.5 dBi at 2.2 GHz, combined with an 18.2% fractional bandwidth. Nonetheless, higher gains demand larger apertures (Harrington, 1960; Skrivervik et al., 2001), thus it comes as no surprise that its electrical size is estimated in the range $4.10 < ka < 4.35$ rad.[3] Lastly, two patch-excited cups were implemented for X-band TTC applications, using different receive (RX) and transmit (TX) frequencies. The RX antenna achieves 8.9 dBi of gain, whereas the estimated electrical size of the radiating parts is $(ka)_{\mathrm{RX}} = 2.63$ rad at 8 GHz; the choke ring has been excluded from this calculation. The TX antenna achieves 7.5 dBi of gain, whereas the estimated electrical size of the radiating parts is $(ka)_{\mathrm{TX}} = 2.24$ rad at 8 GHz; the mounting flange has been excluded from this calculation.

3.2.2 Work described by Gao et al., IEEE Antennas Propag. Mag., 2009

The third and final contribution is by (Gao et al., 2009). The authors presented an excellent overview of the status of antennas for small satellites until the end of 2007. Work from many groups was included, albeit the focus was on the work done in this area by the University of Surrey, Surrey Satellite Technology Ltd., and the Surrey Space Centre.

The article takes off with an introduction to small satellites, describing their *modular* structure and the modules they have in common. Examples of several small satellite projects are then given, including remote-sensing micro-satellites; a student-built multiple–pico-satellite system, with a "mother-ship"–"daughter-ship" arrangement, used for communications and as a scientific test-bed; multi-university student-built nano-satellites used for a variety of purposes; and pico-satellite systems being built and used by a wide-ranging international collaboration (the CubeSat programme).

The functions and commonly used types of antennas for such small satellite systems are discussed, followed by a discussion of antenna design challenges. Due to the special environment in space and the requirements of modern satellites, there should be careful consideration of electrical, mechanical and thermal performance constraints. Yet another major consideration for antenna design is the interaction between antennas and modern small-satellite structures. The chassis of the spacecraft is a finite 3-D ground plane, which couples strongly to the radiating elements, causing electromagnetic scattering and radiation pattern blockage. These problems intensify the importance of electromagnetic simulations and measurements in an effort to obtain antenna placement for optimum performance and coverage.

The article then presents a detailed examination of the antennas used for each of the major functions associated with small satellites. Antennas for TTC applications include monopoles, PIFAs, patches, QHAs and patch-excited cups. The range of frequencies covers the VHF, UHF, S-, C- and X-bands. Antennas for high-speed payload data downlinking include S- and X-band QHAs; compact MGAs based on the SAP principle; X-band, mechanically-steered, high-gain horn antennas; deployable parabolic reflectors; S-band patch-excited cups; and UHF through S-band active antennas for CubeSat missions based on the grid oscillator concept. Antennas for satellite navigation and positioning include medium-gain patch arrays, patch-excited cups and ceramic-loaded QHAs. Inter-satellite cross links can be facilitated by high-gain patch arrays and deployable reflectors, however the most attractive option seems to be the use of low-complexity, self-steering, retrodirective arrays. In each of these areas, the requirements for the antennas are explained, followed by many examples of antennas that have been used to meet these requirements.

[3] The ambiguity stems from the fact that the top-most patch stands much higher than the edge of the rim.

The conclusions of the article display that this research area is flourishing, while numerous challenges remain wide open to further investigation. Innovative concepts such as "satellite-on-PCB" (PCB-Sat), where the whole satellite is built on a single PCB, as well as "satellite-on-chip" (Chip-Sat), where the satellite is built in a single chip, are paving the way for cooperative small-satellite networks with sophisticated functionalities. To satisfy the requirements of next-generation satellite communication systems, much more work is clearly pending at the antenna front.

3.3 Overview of techniques pertinent to the chapter

One of the techniques used in this Chapter is an extension of the method introduced by (Notis et al., 2004). The authors etched 40 equally long slits along the periphery of a square microstrip antenna, thus producing a meandering (ragged) outline. The ten slits on each side of the patch covered one-third of its length. (Notis et al., 2004) showed that the disturbance of the current distribution could easily bring about a 30% size reduction for a given frequency; this reduction could even reach levels up to 44%. Alternatively, a reduction in operating frequency up to 25% can be obtained, given the size of the patch.

In (Kakoyiannis & Constantinou, 2008), the authors extended this slit-loading technique by tapering the length of the slits, and thus modulating their length in the spatial domain. It was briefly shown therein that the manipulation of the spatial distribution of the slits can result in greater reduction in size and is also a way to control the input impedance $Z_{\mathrm{in}}(j\omega)$ of the antenna. This Chapter is a detailed record of that modified miniaturization technique, which modulates slit length in order to spread the slit distribution further along the periphery of the patch.

One of the earliest studies on the use of slits in microstrip antennas was presented by (Zhang & Yang, 1998), who studied the effect of transverse slots in patches, i.e., straight slots that are perpendicular to the SCD and parallel to the phase front of the current. The authors proposed an equivalent circuit that characterizes the electrical behaviour of the slot; it is a multi-port network of inductances appearing in series with current flow. Experimental data displayed a 23% reduction in resonant frequency. This reduction can climb up to 40% if the straight slot is replaced by an "H"-shaped slot. Optimal slot positioning was also investigated: since the slot is inductive, it should be etched where the current is *maximum*, that is, at the very centreline of the patch. On the other hand, the authors did not consider the effect of slot *width*, which, in our opinion, provides a significant extra degree of freedom to further increase the *electrical length* of the patch.

A slightly more complex technique was presented in (Row et al., 2000), where four "Γ"-shaped slots are arranged in a cross order. By proper choice of slot length and inter-slot distances, a 25% reduction in f_0 can be achieved. Instead, the two wide longitudinal slits used by (Wong & Hsu, 2001) do not increase the electrical length of the patch; their placement does little to alter the SCD. However, it was experimentally proven that they augment bandwidth. The same holds for the application of "U"-shaped slots; U-slotted antennas have been investigated extensively in the open literature, and a complete design methodology is presented in (Weigand et al., 2003).

Particularly useful are techniques that achieve simultaneous size reduction (an increase in electrical length) and bandwidth enhancement. One such method unfolds in (Xiao et al., 2005), where an inset-fed patch antenna is loaded by asymmetric (alternate) peripheral slits. The slits were etched along the non-radiating edges of the patch, and achieved a 60% reduction in antenna size. At the same time, with a little help from the inset feed, two different Transverse-Magnetic (TM) modes are excited: TM_{10} and TM_{01}. These modes are frequency-adjacent, and thus manage to double the Voltage Standing Wave Ratio (VSWR) bandwidth: these compact patch antennas displayed a fractional bandwidth close to 4%. The same research group capitalized once more on asymmetric slit loading in (Xiao et al., 2006). The new slit arrangement excites two frequency-adjacent TM_{10} modes, and manages to produce even better results in terms of size reduction and bandwidth augmentation. However, none of the above techniques is suitable for circularly polarized antennas, since they fail to preserve the required double symmetry. Therefore, in order to implement the CubeSat antenna, we decided to extend the technique presented in (Notis et al., 2004).

4. Hybrid Feed Network Design and Simulation Setup

4.1 A hybrid feed network for circular polarization

The communications antenna is used by the orbiter in both transmit (Tx) and receive (Rx) modes. Therefore, it is essential that identical radiation characteristics are maintained between modes. According to the Reciprocity Theorem (see §1.3 and §9.4 in (Stutzman & Thiele, 1998)), an antenna will maintain its properties as long as it is made out of bidirectional elements. This requirement holds not only for the body of the radiator, but for its feeding network as well; no diodes or transistors are allowed in the circuit for reciprocity to hold exactly. On the other hand, a purely passive antenna-plus-feeding-network combination will be reciprocal by definition. The hybrid coupler presented in this Section is a purely passive, bidirectional four-port microwave circuit (Bahl & Bhartia, 2003; Gustrau & Manteuffel, 2006). The model of the coupler used in simulations is shown in Fig. 1, which also indicates port numbering; ports **1** and **4** lie on the transceiver side, whereas ports **2** and **3** lie on the microstrip antenna side.

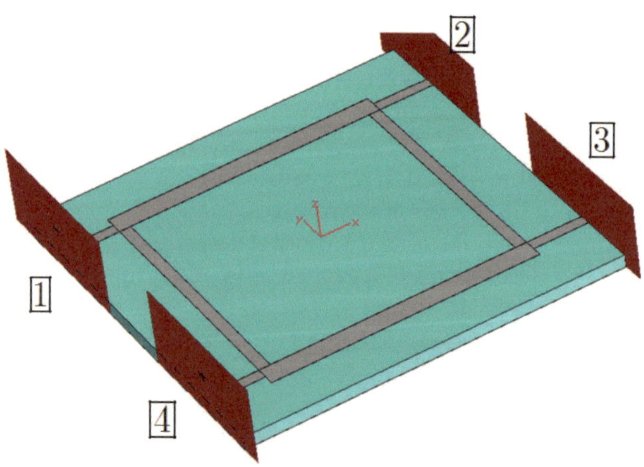

Fig. 1. Perspective view of the hybrid coupler model.

When in Rx mode, the coupler combines the signals arriving at ports **2** & **3** and produces a single output at port **4**. Port **1** is (ideally) completely decoupled from port **4**, that is, in theory we get $S_{14} \triangleq 0$ and no signal power appears at port **1**. When in Tx mode, the output signal coming from the High-Power Amplifier (HPA) drives port **1**. This signal is divided into two signals in phase-quadrature ($\Delta\phi = 90°$), with each signal carrying 3 dB less power than the input signal. Thus, this circuit is essentially a power divider that introduces an excess phase

delay to one of the two branches. Port **4** is isolated from port **1**, that is, theoretically $S_{41} \triangleq 0$ (Gustrau & Manteuffel, 2006).
The hybrid coupler is very well documented in the literature pertaining to microwave solid-state circuits (Bahl & Bhartia, 2003). As a result, initial predictions of the dimensions of the transmission lines that make up the coupler are readily available from theory. Before calculating any estimated values, it is important to reach a decision regarding the microwave substrate on which the coupler will be printed. The coupler is essentially composed of four $\lambda_g/4$-long microstrip lines. The designer must make sure that the guided wavelength λ_g is short enough for the whole coupler to fit on a 100×100 mm^2 area. After surveying the substrate market, the military/space-graded Rogers TMM 10i™ ceramic substrate was chosen ($\varepsilon_r = 9.80$, $\tan\delta_e = 0.0020$). Other substrates suitable for the coupler are Rogers RT/Duroid 6010LM; Rogers TMM 10; Rogers RO3010; Rogers RO3210; and Arlon AD1000. The chosen substrate height was $H = 3.2$ mm, and this is one of the advantages of TMM 10i. The thickness of the copper cladding was assumed to be 35 µm (1 oz Cu).
The design of a high-frequency circuit is an iterative procedure, even with today's advanced field solvers. The solvers decrease the number of iterations significantly, but a few iterations are always required for first-pass success on the test bench. In this context, the initial simulations revealed that the theoretical estimations for the dimensions of circuit elements needed some improvement. On the one hand, the coupler did not resonate at the desired frequency, whereas on the other hand, power was not divided equally between the two branches. To improve the design further, a parametric study was undertaken, which converged quickly to the proper circuit dimensioning that satisfies coupler specifications. The converged values are shown in Fig. 2.

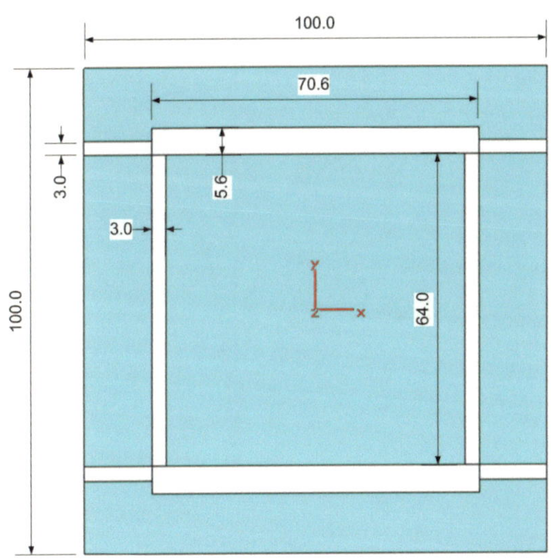

Fig. 2. Converged values of the lengths and widths of the microstrip lines of the hybrid coupler model. All dimensions are in millimetres.

4.2 Numerical results

The scattering (S-) parameters of the 90° hybrid are depicted in Fig. 3. The excitation was applied to port **1**, which is properly matched to the 50-Ω system impedance: $S_{11} < -40$ dB.

Port **4** is strongly decoupled from port **1**: $S_{41} < -30$ dB. Power impinging on port **1** is equally divided between ports **2** and **3**, since $S_{21} \approx S_{31} \approx -3$ dB. The corresponding phase diagrams (not shown here) indicate that the phase lead/lag between the signals at ports **2** & **3** is $\Delta\phi_{23} = 90.1°$.

Fig. 4 and Fig. 5 illustrate the amplitude and phase balance of the coupler, respectively. The operational bandwidth of this circuit can be defined in various ways, exactly like an antenna. So, let us choose four proper definitions:

1. The available bandwidth for a 2:1 VSWR, that is $S_{11} < -10$ dB, equals $\mathrm{BW}_{\mathrm{V}}^{\mathrm{hybrid}} = 149$ MHz or, stated in fractional terms, $\mathrm{FBW}_{\mathrm{V}}^{\mathrm{hybrid}} = 34\%$.
2. Instead, the frequency range where the isolation between ports **1** & **4** is better than 20 dB is much narrower: $\mathrm{BW}_{\mathrm{iso20dB}}^{\mathrm{hybrid}} = 46$ MHz.
3. The operational bandwidth for a $\pm5°$ phase balance is $\mathrm{BW}_{\pm5°}^{\mathrm{hybrid}} = 148$ MHz.
4. The operational bandwidth for a ±1 dB amplitude balance is $\mathrm{BW}_{\pm1\mathrm{dB}}^{\mathrm{hybrid}} = 140$ MHz.

Fig. 6 is an illustration of the surface current distribution on the conductive parts of the coupler; the dielectrics have been made invisible to get a clear view of the currents. Current amplitude is shown as concurrent maximum values at every point along the circuit. Of course, concurrent maxima are physically impossible because of phase shifting across the circuit; this is just a graphical tool that helps the designer identify "hot" and "cold" areas. From this rough thermograph it is obvious that power entering the circuit at port **1** is equally divided between ports **2** & **3**, whereas almost no power exits from port **4**.

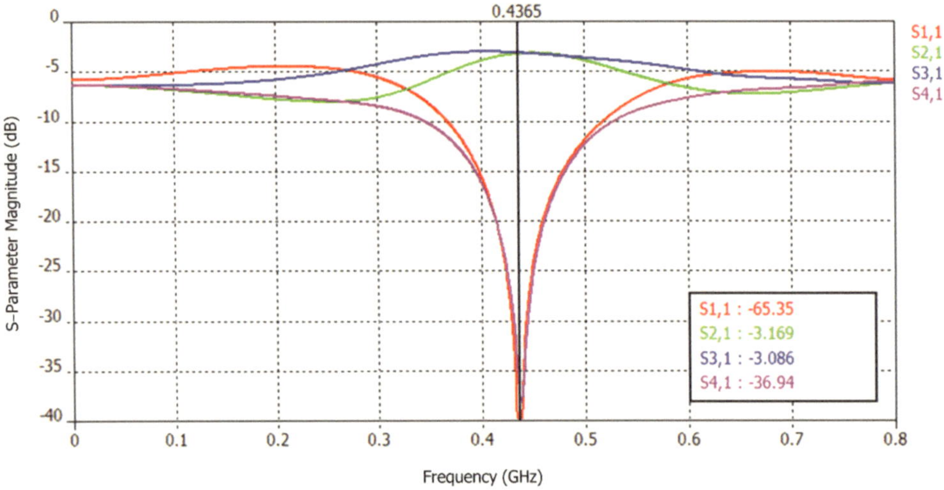

Fig. 3. Magnitudes of the S-parameters of the converged hybrid model. These parameters were obtained by exciting port **1** and terminating all other ports at the system impedance.

4.3 Field solver and generic simulation setup

All microwave systems were designed and simulated in a Transient Solver (TS) that is part of the CST Microwave Studio™ full-wave electromagnetic suite. The TS applies the Finite Integration Technique (FIT) in the time domain to reformulate Maxwell's integral equations

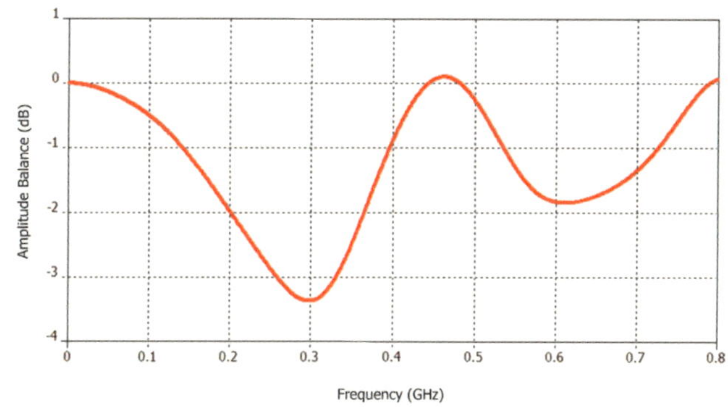

Fig. 4. Amplitude balance variation (in dB) between ports **2** & **3**. Amplitude imbalance less than ±1 dB is maintained for a 140-MHz bandwidth.

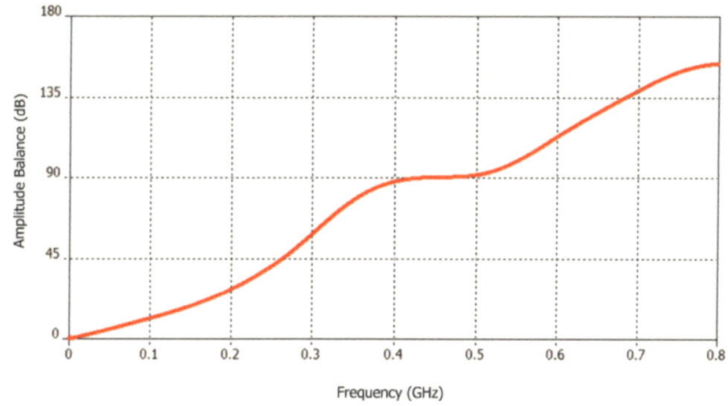

Fig. 5. Phase balance variation (in degrees) between ports **2** & **3**. Phase imbalance less than ±5° is maintained for a 148-MHz bandwidth.

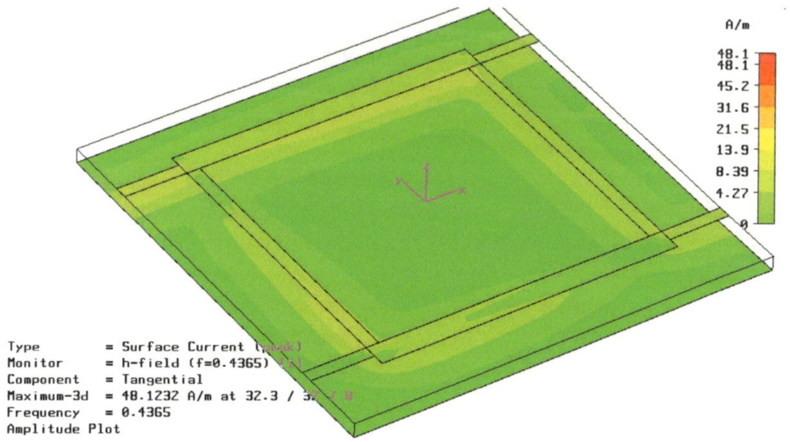

Fig. 6. Concurrent maximum values of the surface current distributed along the coupler. Note the absence of current flowing out of port **4**.

into the so-called "Maxwell Grid Equations" (Gustrau & Manteuffel, 2006; Munteanu et al., 2010; Vasylchenko et al., 2007a; Weiland et al., 2008). By applying Yee's spatial discretization scheme and Courant's maximum stable time-step, FIT results in the same set of equations as the Finite-Difference Time-Domain (FDTD) technique (Gustrau & Manteuffel, 2006). The TS calculates the broadband behaviour of electromagnetic (EM) devices in a single simulation run, with an arbitrarily fine frequency resolution, thus without missing any resonance peaks (Vasylchenko et al., 2007a).

Time-domain solvers are particularly suitable for designing wideband antennas and passive microwave systems such as waveguide components, filters, couplers and connectors. For *active* microwave design, co-simulation is required between the EM solver and a non-linear circuit simulator, such as Agilent Advanced Design System™ or APLAC™. TSs like CST MWS can easily handle exotic materials, such as frequency-dependent (dispersive) and ferri/ferro-magnetic materials. The ability to naturally include such difficult materials in models is one of the main strengths of TSs over FEM- and MoM-based solvers, although the two latter have recently improved their material-handling capabilities (Vasylchenko et al., 2007b).

A spatially non-uniform (adaptive) hexahedral mesh discretized the objects and the solvable space in between. The mesh was refined four-fold near the edges of Perfect Electric Conductor (PEC) objects and inside the substrate to capture the large gradients of the E-field. Tetrahedral meshing is possible through Floquet modes only when FIT is applied in the frequency domain, in which case the technique results in the Finite Element Method (FEM). Nevertheless, the FIT engine used here employs the Perfect Boundary Approximation (PBA) technique (Munteanu et al., 2010; Weiland et al., 2008), and therefore the hexahedral mesh did not result in any object staircasing whatsoever.

A wideband Gaussian pulse excited the structures; its spectral content ranged from DC to 0.8 GHz. The simulator stopped when the initial system energy decayed by 50 dB. This was a good trade-off between simulation speed and truncation error in the FFT engine that translates the results from the time- to the frequency-domain. It is also a good trade-off for the near-to-far-field transformation that produces the far-field pattern of an antenna out of the fields calculated in the near-field. The maximum cell size at the maximum frequency f_{max} (smallest wavelength λ_{min}) was set to a small fraction of λ_{min}. The solvable space was terminated at an adequate number of Bérenger Perfectly Matched Layers (PML) (Bérenger, 1994), which had a normal (broadside) reflectivity of -80 dB. The distance of every object from the boundary of the solvable space was set equal to $\lambda_c/8 = 300/(8 \times 0.4)$ mm $= 94$ mm, unless otherwise noted.

Whenever a model featured topological symmetry and satisfied the appropriate boundary conditions for the electric/magnetic tangential components and the magnetic/electric flow, an electric/magnetic wall was placed across the plane of symmetry. This boundary condition reduced the computational burden significantly without loss of accuracy, because only a fraction of the structure needed solving. Complexity depends upon the level of detail exhibited by the objects comprising the model and the electrical size of the solvable space.

All structures that were modelled as part of this Chapter were fully parametrised. The key concept here is that, if the objects in a model are defined with parameters instead of numbers, then the designer benefits from parametric studies and optimization. In a sense, parametrisation creates "inflatable" models—like an accordion—instead of fixed, "frozen" models. Parametric sweeping and optimization jobs can be distributed across many "worker" computers through the corporate LAN and run in parallel. This basic form of laboratory/company distributed computing power exploitation brings about significant time savings for the design team.

4.4 Solver settings applied to the hybrid coupler

Fig. 7 depicts the spatial discretization (better known as *grid formation* or *meshing*) of the model used to design the 90° hybrid. The structure was excited in the time domain by a Gaussian pulse having spectral content in the range DC–0.8 GHz. The excitation signal along with the four output signals are shown in Fig. 8.

For efficient simulations, that is, simulations that strike a good balance between speed of execution and result accuracy, a spatially non-uniform (adaptive) grid was designed; maximum allowed grid step was equal to $\lambda_g/50$ at 0.8 GHz. No form of packaging was adopted, thus the rectangular solvable space surrounding the PCB of the coupler was terminated at a 4-layer Bérenger PML structure (Bérenger, 1994); these are *open-space* boundary conditions. This circuit is non-radiating, thus a 4-layer boundary absorber is more than adequate. However, to increase the speed of iterations, the solvable space was trimmed to half by terminating the area below the substrate at a PEC condition, i.e., $\mathbf{E}_t \equiv 0$. This approximation is valid because, when the coupler is studied in solitude, it is a non-radiating system (at least intentionally). Therefore, it is safe to assume that the ground plane of the PCB extends to infinity—this is precisely the computational effect of the electric boundary condition. The complexity of the model was $60 \times 79 \times 14 = 66,360$ Yee cells.

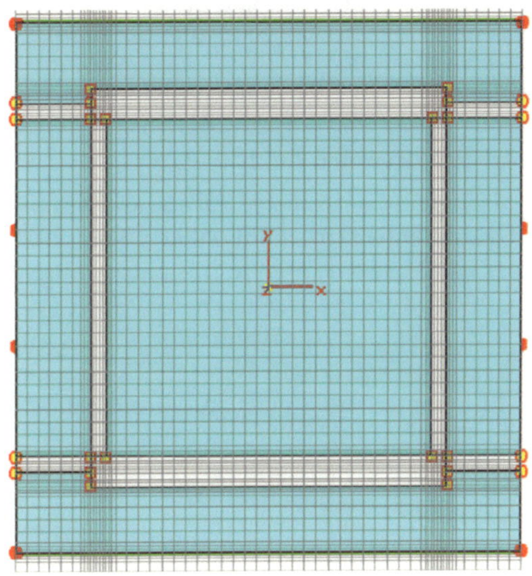

Fig. 7. The grid on which the electromagnetic problem was solved displayed a variable step ranging from 0.6 mm to 2.4 mm. Maximum grid step corresponds to $\lambda_g/50$ (or $\lambda_0/156$) at the maximum frequency $f_{\max} = 0.8$ GHz.

5. Inductive-slit-loaded Microstrip Antenna design

5.1 Antenna design considerations

Examples of microwave substrates suitable for the antenna are Rogers RT/Duroid™ 6006 and Rogers RO3006™. Both present a dielectric strength $\varepsilon_r = 6.15$, and approximately the same loss tangent; the former displays $\tan\delta_e = 0.0019$, whereas $\tan\delta_e = 0.0020$ for the latter. However, production heights differ: Duroid 6006 can be purchased laminated with a maximum height $H_{\max}^{6006} = 2.54$ mm= 100 mil, whereas RO3006 is sold at a maximum height

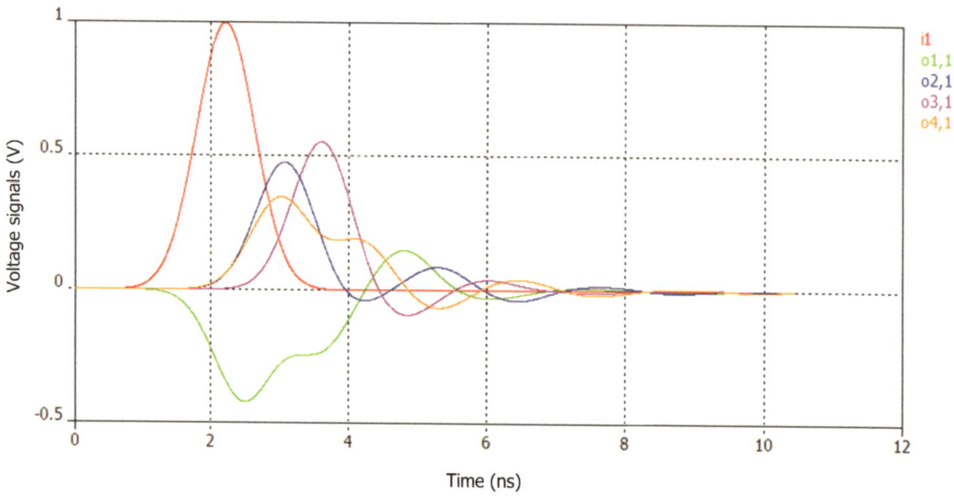

Fig. 8. Voltage signals, in the time domain, being input ("**i1**") and output ("**ox,1**", where $x \in \{1, 2, 3, 4\}$) from the ports of the hybrid coupler. System stored energy decayed to the point of terminating the execution after 10.5 ns of simulation time.

$H_{\text{max}}^{3006} = 1.27$ mm$=$ 50 mil. In any case, the height chosen for the CubeSat antenna substrate was equal to

$$H_{\text{patch}} = 6.4 \text{ mm} = 252 \text{ mil} \tag{1}$$

and thus bonding of several single-side-laminated substrates with prepregs is required.
The initial design stages of the antenna started out on the assumption that the same dielectric material used for the coupler would also be used for the patch antenna (Rogers TMM 10i™, $\varepsilon_r = 9.80$, $H = 3.2$ mm). On the TMM 10i the antenna resonated at 440 MHz for a square patch length $L_{\text{ini}} = 108.3$ mm. With a rough frequency scaling, it was estimated that the antenna would resonate at $f_0 = 436.5$ MHz for a length $L'_{\text{ini}} = 109.2$ mm. After scaling the dielectric constants, it was estimated that an antenna built on Duroid 6006 or RO3006 would have a resonant length

$$L''_{\text{ini}} = L'_{\text{ini}} \sqrt{\frac{9.80}{6.15}} = 137.8 \text{ mm}. \tag{2}$$

In theory, the resonant length of a microstrip patch antenna that corresponds to the considered parameters equals

$$L_{\text{res}}^{\text{theory}} = 0.49 \frac{\lambda_0}{\sqrt{\varepsilon_r}} = 135.8 \text{ mm}. \tag{3}$$

The deviation between the results in (2) and (3) is a mere 1.5%.
From the handy analysis unfolding in Chapter 5 of (Stutzman & Thiele, 1998) the following estimation on the real part of the input impedance of the patch can be extracted,

$$R_{\text{in}} \triangleq Z_{\text{A}} = 90 \left(\frac{\varepsilon_r^2}{\varepsilon_r - 1} \right) \left(\frac{L_{\text{patch}}}{W_{\text{patch}}} \right)^2. \tag{4}$$

Substituting $\varepsilon_r = 9.80$ and $L_{\text{patch}} = W_{\text{patch}}$, we obtain

$$R_{\text{in}} \triangleq Z_{\text{A}} = 90 \frac{(9.80)^2}{9.80 - 1} \ \Omega = 982 \ \Omega. \tag{5}$$

This uselessly high resistance is a side-effect of the high ε_r . This value can be reduced down to 50 Ω by setting

$$\frac{L'_{\text{patch}}}{W'_{\text{patch}}} = \frac{1}{4.43}. \tag{6}$$

Because of the severe space constraints on the spacecraft, we cannot afford to design a rectangular patch; the initial study indicated clearly that only a square patch can fit in the allocated area. All of the above, combined with the fact that the high ε_r threatens to eliminate the minimal bandwidth of the patch antenna, led us to the choice of the lower $\varepsilon_r = 6.15$. Thus, the theoretical estimation for the input resistance of the antenna now becomes

$$R_{\text{in}} = 90\left(\frac{\varepsilon_r^2}{\varepsilon_r - 1}\right)\left(\frac{L_{\text{patch}}}{W_{\text{patch}}}\right)^2 = 90\frac{(6.15)^2}{6.15-1}\,\Omega = 661\,\Omega. \tag{7}$$

By extending the microstrip feed line inside the patch by a proper length Δx_i the *inset feed* technique is employed; the modified input resistance becomes (Stutzman & Thiele, 1998)

$$R_{\text{inset}} = R_{\text{in}}\cos^2\left(\pi\frac{\Delta x_i}{L}\right). \tag{8}$$

Solving for $\Delta x_i/L$, which is the *fractional insertion depth*, we get

$$\frac{\Delta x_i}{L} = \frac{1}{\pi}\cos^{-1}\left(\sqrt{\frac{R_{\text{inset}}}{R_{\text{in}}}}\right). \tag{9}$$

Substituting $R_{\text{in}} = 661\,\Omega$ and $R_{\text{inset}} = 50\,\Omega$ we obtain the following insertion depth

$$\frac{\Delta x_i}{L} = \frac{1}{\pi}\cos^{-1}\left(\sqrt{\frac{50}{661}}\right) = 0.411. \tag{10}$$

The result in (10) means that the inset feed has to penetrate half-way along the surface of the patch; to avoid this, we used a quarter-wavelength transformer ($\lambda_g/4$–Xformer). By trading off *transformer impedance* for a mechanically robust *copper trace width*, we chose the width $W_{\text{quarter}} = 1.0$ mm, which gives a characteristic impedance $Z_{\text{quarter}} = 117.5\,\Omega$. The length of the transformer equals $L_{\text{quarter}} = 87.0$ mm, whereas the resistance that can be matched to 50 Ω is

$$Z_x = \frac{Z^2_{\text{quarter}}}{50}\,\Omega = 276\,\Omega. \tag{11}$$

Now the initial estimation for the depth of the inset feed can be derived,

$$\left(\frac{\Delta x_i}{L}\right)_{\text{quarter}} = \frac{1}{\pi}\cos^{-1}\left(\sqrt{\frac{50}{276}}\right) = 0.360. \tag{12}$$

However, this effort did not produce any significant reduction in inset depth; the reason is the slope of the curve shown in Fig. 9.

The 2:1 VSWR bandwidth is approximately estimated by equation (13) for $H \ll \lambda_0$ (Stutzman & Thiele, 1998)

$$\text{FBW}_{\text{V}} = 3.77\left(\frac{\varepsilon_r - 1}{\varepsilon_r^2}\right)\left(\frac{W_{\text{patch}}}{L_{\text{patch}}}\right)\left(\frac{H}{\lambda_0}\right). \tag{13}$$

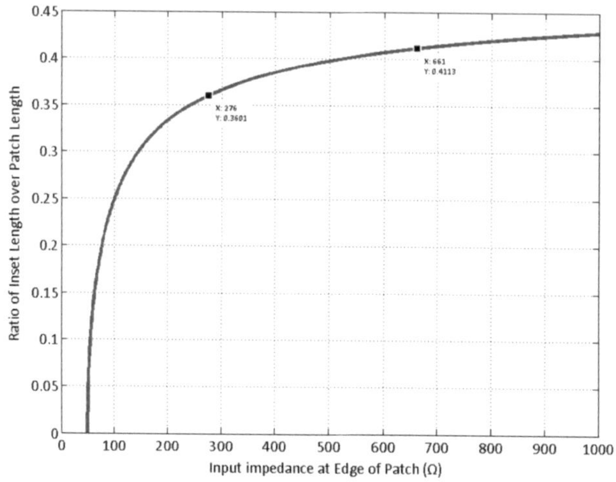

Fig. 9. Variation in required fractional inset depth as a function of the impedance seen at the edge of the patch.

Applying (13) for the parameters of this design gives the following fractional VSWR bandwidth,

$$\mathrm{FBW_V} = 3.77 \left[\frac{6.15 - 1}{(6.15)^2} \right] \left(\frac{6.4}{687.3} \right) = 0.0048. \tag{14}$$

The value corresponds to just 2.1 MHz of BW at the center frequency $f_0 = 436.5$ MHz. Thus, it is expected that the total application bandwidth BW $= 3.2$ MHz will be covered with a VSWR value higher than 2:1.

5.2 Antenna geometry evolution

The number and arrangement of the peripheral slits along the edges of the CubeSat patch antenna have been influenced by the design strategy of (Notis et al., 2004). The authors used slits of maximum length equalling 380 mil (9.5 mm) for a patch length 1620 mil (40.5 mm); slit depth was 23% of patch length. The slits were 20-mil (0.5-mm) wide, whereas inter-slit distance was 40 mil (1 mm). The starting centre frequency of that study was $f_0^{\mathrm{Notis}} = 2.36$ GHz. The frequency ratio between the two studies equals 5.4, therefore frequency scaling leads to roughly 2.5-mm wide slits with inter-slit spacing equal to 5.0 mm. The total edge length occupied by the 10 slits in (Notis et al., 2004) was

$$W_{\mathrm{total}}^{10} = 10 \cdot 20 + 40\,(10 - 1)\ \mathrm{mil} = 560\ \mathrm{mil}. \tag{15}$$

Consequently, the portion of the patch edge occupied by the slits is

$$\frac{560}{1620} = \frac{28}{81} \simeq \frac{1}{3},$$

which is a reasonable design choice, since provisions for circular polarization were made in (Notis et al., 2004) as well.

The CubeSat antenna was designed by etching 10 slits on each of the 4 sides of the square patch. The slits have variable width (but equal for all), and also variable length that follows a certain set of values $\{a_1, a_2, a_3, a_4, a_5\}$. To preserve the potential for circular polarization, slit configuration has been chosen in a way that maintains the two perpendicular symmetry axes (see Fig. 10). This is the reason why there are only 5 length variables in the previous set, instead of 10.

Since the slits have been etched on the periphery of the patch and not, for example, on the ground plane, the most natural way of spatially modulating (tapering) their lengths is the *triangular distribution*. Theoretically, this tapering would force the current to go through the center of the patch, and thus produce an effective physical length

$$L_{\text{eff}} = L_{\text{patch}}\sqrt{2}. \tag{16}$$

Simulations of antenna models using the triangular tapering started out with a 33% total edge coverage. The parametric sweeps indicated that the estimation of (12) was quite correct: the optimal fractional inset depth lies between 0.36 and 0.38. It is obvious from Fig. 10 that the shape of the slits enables us to increase their width, and therefore occupy a larger part on each side of the square; this leads to a greater miniaturization degree. The study indicated that good results are obtained when the slits take over 70–80% of every side. Furthermore, it was discovered that a good compromise between miniaturization and bandwidth is obtained when the ratio of slit width to slit gap is set around unity. Simulations showed beyond any doubt that this ratio affects both f_0 and $Z_{\text{in}}(j\omega)$. Changes in input impedance are critical and must occur in a controlled manner: the CubeSat antenna is electrically small, thus it is rather challenging to tune it ($X_{\text{in}}(j\omega) \neq 0$) and match it ($R_{\text{in}}(j\omega) \ll 50\ \Omega$).

Fig. 10 illustrates in perspective the final antenna geometry; the tall substrate is evident. This particular model, which is just a 2-layer PCB, represents the first completed design stage; it is designated as the *CubeSat Patch Prototype version 1* and abbreviated herein as "CSPP–1". In Section 5.5 and Section 5.6 we will present the second completed design stage, abbreviated herein as "CSPP–2".

The 40 transverse slits along the periphery of the patch increase the distance that current must travel to reach the opposite edge, and thus increase the effective electrical length of the radiator. The increased electrical path, in turn, reduces the physical size of the patch below 100 mm. If the antenna were designed on a foam substrate ($\varepsilon_r \simeq 1.0$), then the nominal patch size would be 344×344 mm^2. This nominal area was initially reduced by 84% due to dielectric loading; the area of the resulting patch was further reduced by 55% due to the slit distribution. Patch side length was reduced by 60% and 33%, respectively. The final antenna converged to dimensions 93×93 mm^2. Its area was reduced 13.7 times; side length was reduced 3.7 times. Fig. 11 depicts all important dimensions of both the microstrip patch and the substrate. The size of the substrate has been increased beyond 100×100 mm^2 to facilitate the incorporation of the feed network in the same model (i.e., the $\frac{\lambda_g}{4}$ transformer and a small segment of 50-Ω microstrip).

The length of the slits was modulated according to the triangular distribution. Other distributions can also be used, such as binomial, uniform, geometric and cosine-on-pedestal. In fact, (Notis et al., 2004) used the *uniform* distribution. A first estimation on the electrical size of the radiator is $ka = 2\pi\frac{93\sqrt{2}/2}{687.3}$ rad $= 0.60$ rad < 1 rad, and thus the CubeSat antenna is indeed *electrically small*. It remains to be seen how well can such a small cavity-like antenna perform in terms of *gain*, *radiation efficiency*, *quality factor*, *bandwidth* and *half-power beamwidth*.

5.3 CSPP–1 simulation setup

Fig. 12 depicts the spatial discretization (*meshing*) of the model used to design the CSPP–1 antenna. The structure was excited in the time domain by a Gaussian pulse having spectral

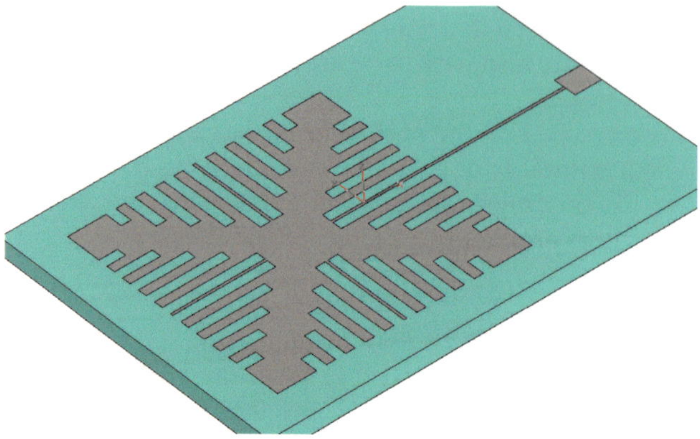

Fig. 10. The square printed CSPP–1 antenna loaded with inductive peripheral slits. The $\lambda_g/4$ transformer and a small segment of 50-Ω microstrip line are also shown.

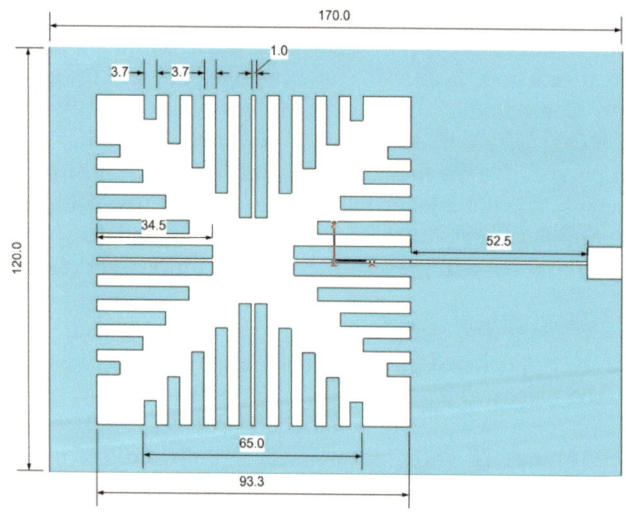

Fig. 11. Converged dimensions of the final model, as they resulted from the optimization process. The circuit is printed on a Rogers RO3006 substrate ($\varepsilon_r = 9.80$, $\tan \delta_e = 0.0020$, $H = 6.4$ mm). All dimensions are in millimetres.

content in the range DC–0.8 GHz. For efficient simulations, a spatially non-uniform (adaptive) grid was designed; maximum allowed grid step was equal to $\lambda_g/36$ at $f_{\max} = 0.8$ GHz. The rectangular solvable space surrounding the PCB of the antenna was terminated at a 6-layer Bérenger PML structure (Bérenger, 1994); an antenna cannot be properly simulated unless it is terminated at open-space ("radiating") boundary conditions. Based on our prior experience with the T-Solver, the 6-layer PML provides an excellent accuracy/speed trade-off, in the sense that antenna radiation characteristics converge while simulation time does not increase noticeably (Kakoyiannis et al., 2010; Kakoyiannis & Constantinou, 2010a;b).
Setting the Boundary Conditions (BCs) properly in an antenna simulation is always a major issue. During the CSPP–1 design stages, where high simulation speed was preferred over extreme accuracy, the solvable space was trimmed to half by terminating the area below the substrate at a PEC boundary condition, i.e., $\mathbf{E}_t \equiv 0$. This causes the ground plane of the PCB

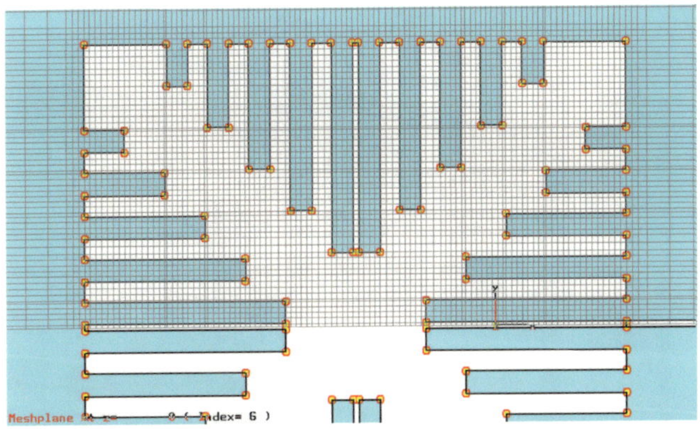

Fig. 12. The grid on which the electromagnetic problem was solved displayed a variable step ranging from 0.3 mm to 4.2 mm. Maximum grid step corresponds to $\lambda_g/36$ (or $\lambda_0/90$) at the maximum frequency $f_{\max} = 0.8$ GHz.

to extend to infinity; now, the mirror backing the patch becomes electrically huge. This is only a practical approximation, since in reality the patch occupies approximately the same area as the ground plane does.

Yet another important BC-related issue is raised by the geometry of the antenna, i.e., the two perpendicular symmetry planes: could electric/magnetic BCs be applied to the model so as to reduce the solvable space to a fraction ($\frac{1}{2}$, $\frac{1}{4}$, $\frac{1}{8}$) of the original? Indeed, after studying the volume field distributions inside the solvable space, one can notice right away that at the boundary of xz-plane the magnetic field $\mathbf{H}(x, y, z, t)$ is normal to the plane ($\mathbf{H}_t \equiv 0$), whereas the electric field $\mathbf{E}(x, y, z, t)$ is tangential. This means that through the xz-plane there is only magnetic flux, and no electric. Therefore, at the xz-plane a magnetic BC (or *magnetic wall*) is applicable; this BC reduces the computational burden to one-half without any loss of accuracy. Fig. 13 illustrates the application of the magnetic wall. Magnetic symmetry was maintained throughout the design stages of the CubeSat antenna. The complexity of the CSPP–1 model was $127 \times 57 \times 39 = 282{,}321$ cells.

5.4 Numerical electrical performance of the CSPP–1 antenna

After the geometry of the antenna was established, the next (and most important) step towards design closure was to resonate the antenna. Tuning ($X_{\text{in}}(j\omega_0) \approx 0\ \Omega$) and matching to 50 Ω ($R_{\text{in}}(j\omega_0) \approx 50\ \Omega$) must be accomplished at the desired frequency without violating any of the other specifications. The result of this procedure is documented through S-parameters and the Smith chart in Fig. 14 and Fig. 15 respectively. The matching/resonance depth is satisfactory ($|S_{11}| \approx -18$ dB), albeit resulting in a narrowband antenna, as expected. It achieves a 2:1 VSWR bandwidth $\text{BW}_{-10\text{dB}} = 2$ MHz, and a 3:1 VSWR bandwidth equal to $\text{BW}_{-6\text{dB}} = 4$ MHz.

The Smith chart shows that the antenna behaves like a capacitor or inductor for most of the frequencies. This is an anticipated result; the antenna is electrically small (this is proven in Section 5.6). Inside its operational bandwidth, a single resonance exists ($dX_{\text{in}}/d\omega > 0$). The swift crossing of the curve through the central area of the chart recounts the small achievable bandwidth.

According to the *cavity model*, which provides an adequate theoretical treatment of microstrip antennas, the dominant component of the electric field is $\mathbf{E}_z = E_z \mathbf{i}_z$. Fig. 16 illustrates the

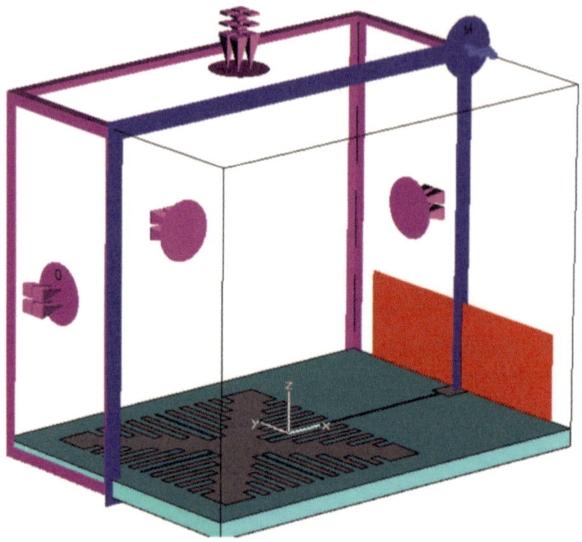

Fig. 13. By applying the magnetic BC $\mathbf{H}_t \equiv 0$ across the xz-plane (blue frame), we get to simulate only the left half of the solvable space, while obtaining equally accurate results. The combination of the magnetic symmetry with the electric BC below the substrate have reduced the computational burden to 1/4.

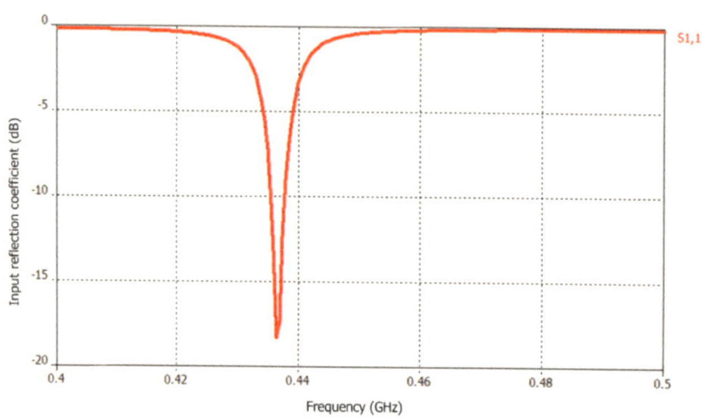

Fig. 14. Resonance and matching at the input of the CSPP–1 antenna given by the magnitude of the input reflection coefficient.

magnitude of the total electric field $\mathbf{E}(x, y, z) = E_x\mathbf{i}_x + E_y\mathbf{i}_y + E_z\mathbf{i}_z$ taken at a snapshot when the field is maximum. Notice that the electric field is strongest not only at the two radiating edges, but also along other vertices along the patch; this is due to the presence of the 40 slits. However, as is well-known from theory, the broadside radiation of microstrip antennas does not result from the z-component of the $\mathbf{E}$-field, but from the two tangential components producing the *fringing field* at the radiating edges of the patch. Fig. 17 illustrates a snapshot of the peak magnitude of the tangential electric field $\mathbf{E}_t = \mathbf{E}_x + \mathbf{E}_y$. Notice how the high-$\varepsilon_r$ keeps the fringing fields too close to the patch. This an ominous conjecture in terms of radiation efficiency; the antenna will tend to behave as a resonant cavity with a small radiating leakage.

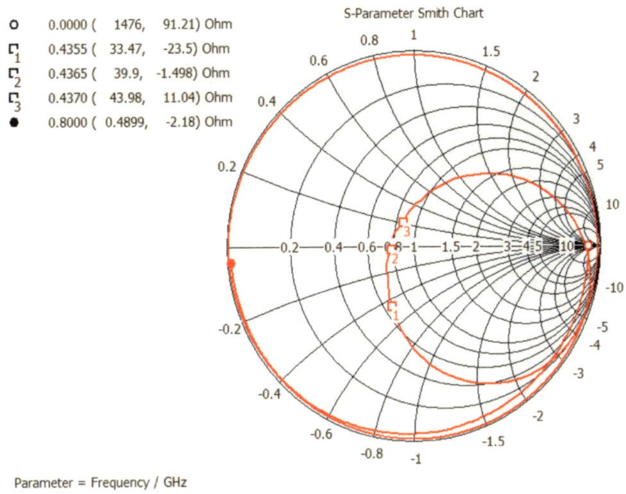

Fig. 15. The CSPP–1 antenna resonates near $f_0 = 436.5$ MHz showing an input resistance $R_{\text{in}} \approx 40\ \Omega$.

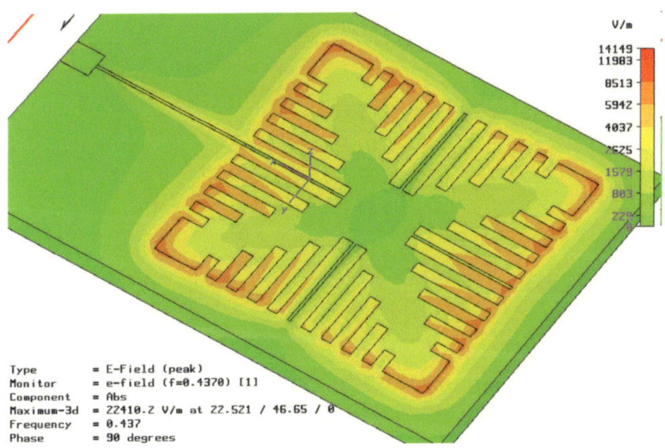

Fig. 16. Maximum magnitude of total **E**-field in the vicinity of the patch. The colour code was set to a logarithmic scale for better resolution.

During all design stages it was very important to check whether the slit distribution would force antenna miniaturization. Visualization of the surface distribution of the current is the way to go; this extremely useful design tool is illustrated by the vector field in Fig. 18. It is evident that the slits do not allow current to travel on a straight line from one radiating edge to the other. Rather, the current moves on a broken line, crossing the center of the patch. Thus, the electrical length of the antenna indeed increases by a factor dependent on maximum slit length.

The study of the CSPP–1 antenna is concluded with the presentation of numerical results pertaining to far-field radiation characteristics. Fig. 19 depicts the three-dimensional (3-D) far-field pattern of the antenna embedded into the model. This pattern was calculated at $f = 437$ MHz and its key aspects are low gain and average radiation efficiency, as it was

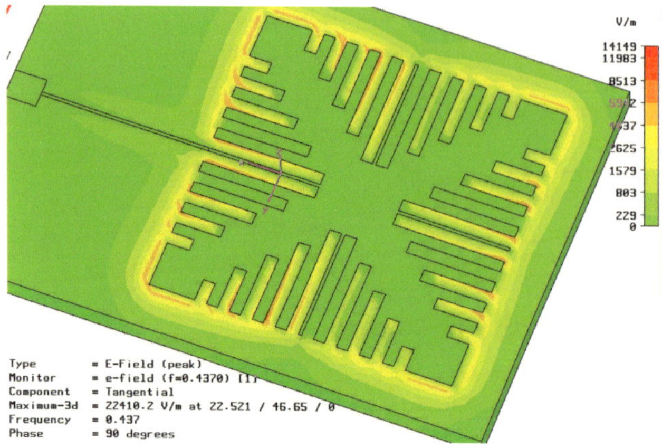

Fig. 17. Surface distribution of the peak magnitude of the tangential **E**-field around the patch. The colour code was set to a logarithmic scale for better resolution.

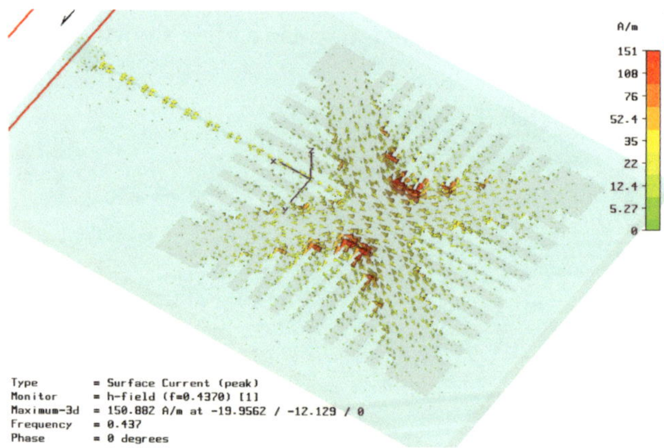

Fig. 18. Visualization of the current distribution on the CSPP–1 patch as a vector field. The color code has been set to logarithmic scale for greater detail.

already predicted:

$$\left.\begin{array}{r} G_{\max} = -1.2 \text{ dBi} \\ n_{\text{rad}} = 0.53 = -2.8 \text{ dB} \end{array}\right\} \tag{17}$$

As a direct consequence of the results in (17), the input resistance of the CSPP–1 antenna $R_{\text{in}}\left(j\omega\right) \triangleq \Re\left\{Z_{\text{in}}\left(j\omega\right)\right\}$ is comprised at resonance by the following two parts:

$$\left.\begin{array}{r} R_{\text{rad}} = 21.2\ \Omega \\ R_{\text{loss}} = 18.8\ \Omega \end{array}\right\} \tag{18}$$

The shape of the far-field pattern is anticipated from theory. The abrupt cut is due to the electric boundary condition applied below the substrate; this BC prohibits radiation in the lower semi-space. Therefore, at this design stage there is no information available on the back-lobe. Significant backward radiation is anticipated because of the comparable dimensions of the patch and the ground plane.

A CubeSat is intended to communicate with its ground control within a given frequency range, which the antenna designer must match to the operating bandwidth of the antenna.

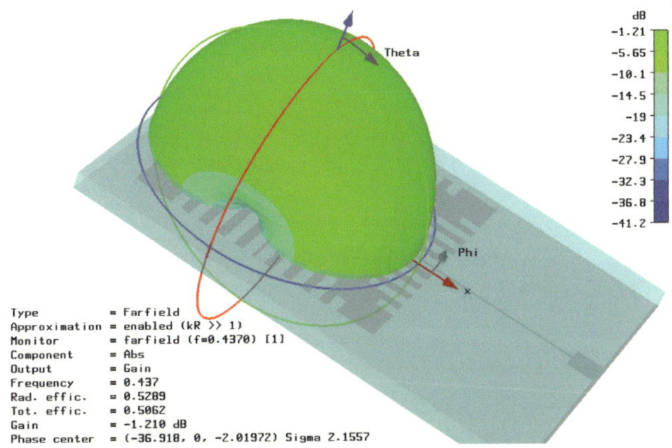

Fig. 19. The gain values of CSPP–1 remain below 0 dBi for all directions in the upper semi-space owing to low radiation efficiency. Radical changes are expected in the more realistic CSPP–2, where the back-lobe will enter the visible region of the pattern.

It would be very useful for the satellite link designer to have information on antenna gain not just for one frequency, but for the whole range of frequencies. To this end, broadband far-field monitoring and recording was facilitated by 17 field monitors uniformly spaced across the 3.2-MHz bandwidth. This recording provided the broadband variation in radiation efficiency n_{rad} (see Fig. 20), total efficiency n_{total} (see Fig. 21) and gain (see Fig. 22).

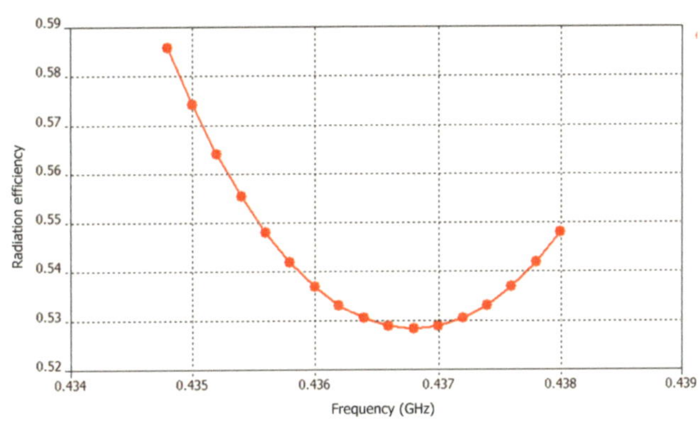

Fig. 20. Broadband variation in the radiation efficiency of the CSPP–1 antenna.

5.5 CSPP–2 simulation setup

The CSPP–2 antenna is a far more realistic model compared to CSPP–1. Two important changes were made to this second prototype: first, the electric boundary condition below the substrate was removed. Thus, all 6 BCs were set to open-space (radiating). The solvable space was terminated at a 6-layer PML, which was separated from every side of the model by $\lambda_c/8$ at $f_c = 0.4$ GHz. Second, a solid PEC rectangular object was mounted right below the substrate, pressing against the ground plane. This PEC object is 100 mm tall, whereas its transverse dimensions are equal to those of the ground plane. It models the chassis of the satellite; it is useful for studying the behaviour of the antenna in what looks like its true operating environment. The realistic CSPP–2 model is depicted in Fig. 23.

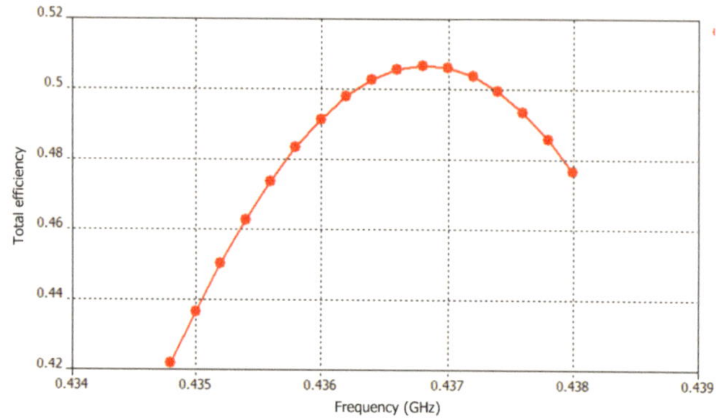

Fig. 21. Broadband variation in the total efficiency of the CSPP–1 antenna.

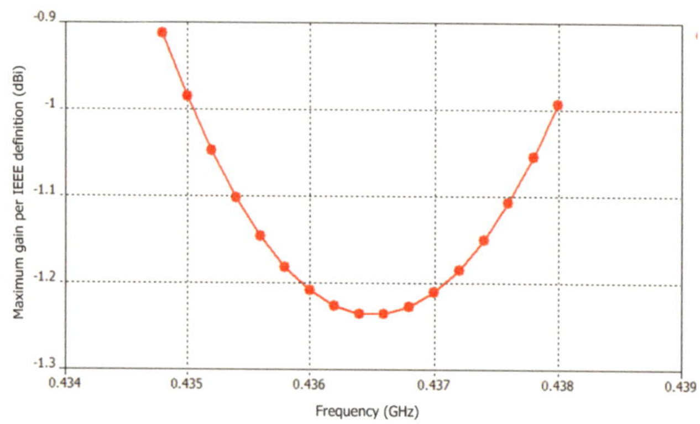

Fig. 22. Broadband variation in the IEEE gain of the CSPP–1 antenna. Peak-to-peak gain flatness equals 0.3 dB.

This was the most complex model that was solved in the framework of this Chapter; the complexity of the solvable space was $143 \times 68 \times 44 = 427,856$ hexahedral cells. Note that the magnetic symmetry across the xz-plane is still applicable. Fig. 24 shows the full solvable space annotated by the 6 boundary conditions.

5.6 Numerical electrical performance of the CSPP–2 antenna

With the planar antenna mounted on the PEC chassis, the first thing that was noticed about its behaviour was that the resonant frequency remained almost constant; it shifted slightly downwards from $f_0 = 436.5$ MHz to $f_0' = 436.0$ MHz. However, there was a significant change in input *resistance*, which dropped from $R_{\text{in}}(j\omega_0) = 40\ \Omega$ to $R_{\text{in}}'(j\omega'_0) = 23\ \Omega$. The drop can be compensated by reducing the depth of the inset feed, but this is hardly the issue here: this 42% reduction epitomizes the fact that antenna design iterations end only *after* the antenna has been embedded into its working environment, be it a satellite, a cell-phone held by a hand next to a head, a vessel at sea, and so on. If the working environment is not taken into account, even as a very crude model, the system designer will have to settle for an antenna of sub-optimal performance.

The shape of the $|S_{11}|$ curve is essentially preserved, but minimum reflection coefficient increases from -18 dB to -9 dB. The operational bandwidth for a 3:1 VSWR ($S_{11} \leq -6$ dB) remains at $\text{BW}_{-6\text{dB}} = 4$ MHz. Fig. 25 illustrates the variation in $Z_{\text{in}}(j\omega)$ on the Smith chart.

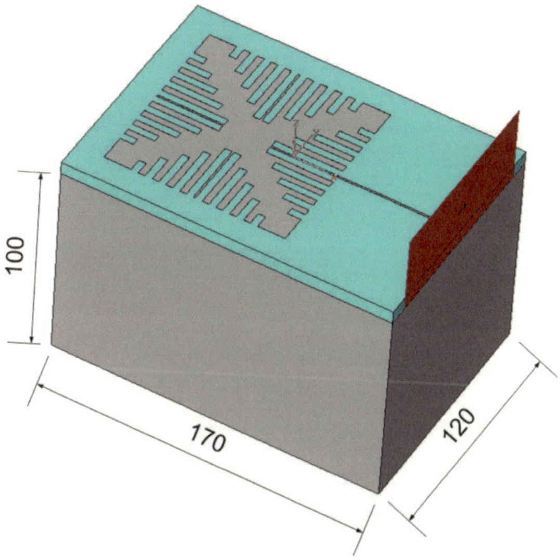

Fig. 23. A 100-mm tall conductive chassis was embedded into the realistic model. The other two sides are longer to fit the size of the substrate. All dimensions are in millimetres.

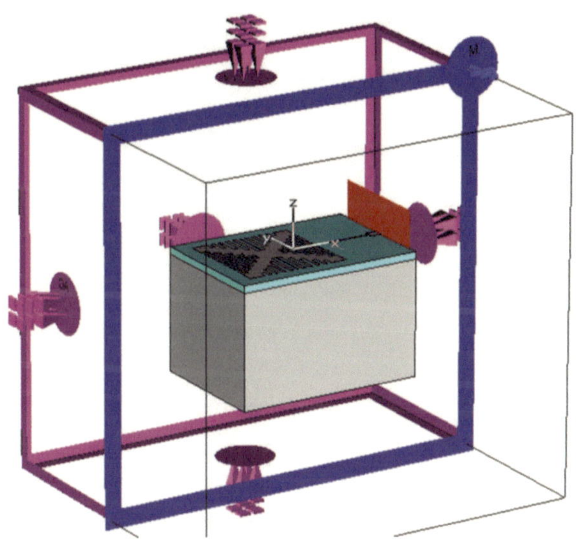

Fig. 24. By applying the magnetic BC $\mathbf{H}_t \equiv 0$ across the xz-plane (blue frame), we get to simulate only the left half of the solvable space, while obtaining equally accurate results. This time, however, there was no electric BC below the substrate.

The far-field pattern of the CSPP–2 model is certainly more intriguing than the textbook-looking one of CSPP–1. The 3-D pattern shown in Fig. 26 confirms the existence of a strong back-lobe, which is created by near-field diffraction on the edges of the ground plane and the PEC chassis. Maximum directivity is $D_{\max} = 4.9$ dBi, but the low efficiency results in a maximum gain $G_{\max} = 0.7$ dBi at $f_0' = 436$ MHz.
The pattern cuts at the principal planes $\phi = 0°$ and $\phi = 90°$ are illustrated in Fig. 27. They too indicate that there is a strong back lobe caused by the patch and the ground plane having

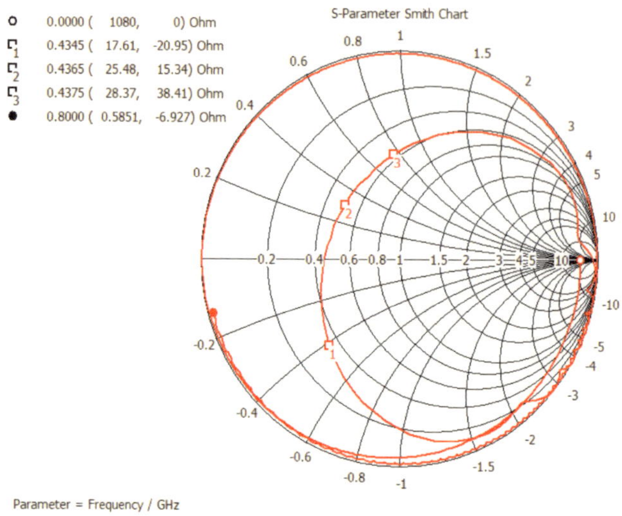

Fig. 25. The CSPP–2 antenna resonates at 436 MHz showing an input resistance $R_{\text{in}} \approx 23\ \Omega$.

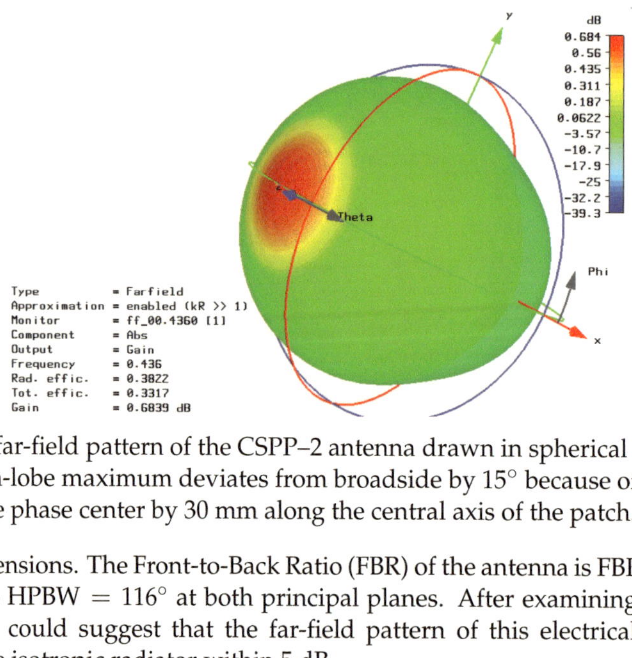

Fig. 26. The 3-D far-field pattern of the CSPP–2 antenna drawn in spherical coordinates. The direction of main-lobe maximum deviates from broadside by 15° because of the inset feed, which shifted the phase center by 30 mm along the central axis of the patch.

comparable dimensions. The Front-to-Back Ratio (FBR) of the antenna is FBR $= 5$ dB, whereas its beamwidth is HPBW $= 116°$ at both principal planes. After examining the three patters as a whole, one could suggest that the far-field pattern of this electrically small antenna approximates the isotropic radiator within 5 dB.

Broadband far-field monitoring and recording was also performed for the CSPP–2 model. Recordings of the broadband variation in radiation efficiency n_{rad}, total efficiency n_{total} and gain are depicted in Fig. 28, Fig. 29 and Fig. 30 respectively.

This section is concluded with the calculation of the *electrical sizes* of the two antennas. The electrical size is required for the calculations tabulated in Table 1. The vector SCD in Fig. 31 indicates that the slit distribution works equally well for the CSPP–2 antenna. Therefore, the following calculations apply to both antenna models.

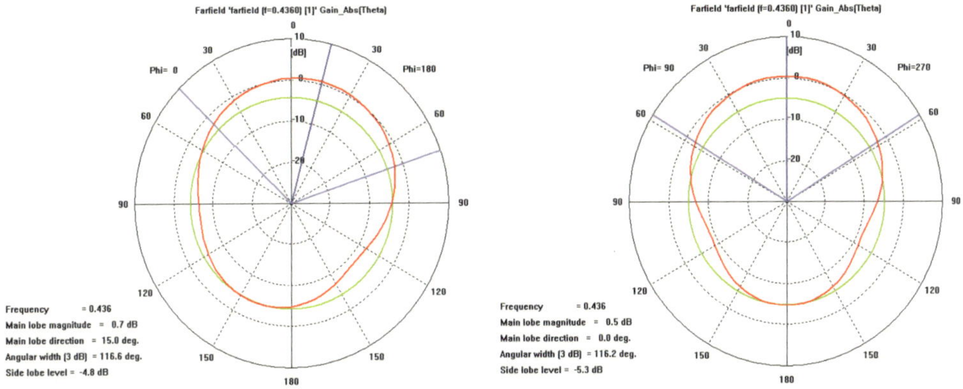

(a) $\phi = 0^\circ$ (b) $\phi = 90^\circ$

Fig. 27. Pattern cuts at the $\phi = 0^\circ$ and $\phi = 90^\circ$ principal planes. On the latter plane the front-to-back ratio is 0.5 dB higher (FBR = 5.3 dB).

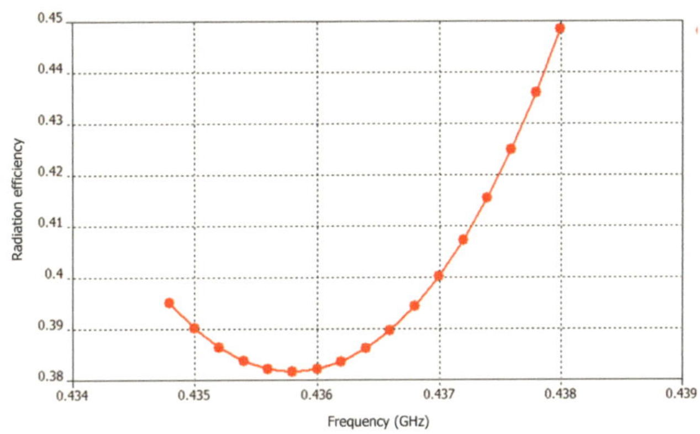

Fig. 28. The radiation efficiency of the CSPP–2 antenna varies between 38–45% throughout the service band of the CubeSat.

First of all, we need to calculate the radius of the *circumscribing sphere*, that is, the thought spherical surface that circumscribes all *radiating parts* of the antenna. To this end, we have used the surface current distribution shown in Fig. 32. According to Sten's theoretical treatment, the radius of the circumscribing sphere should be calculated by placing the centre at the feed point of the planar antenna (Bancroft, 2004; Sten et al., 2001). This holds for printed monopole antennas, which, according to Image Theory, use a parallel ground plane to "mirror" the other half of the antenna that is missing, thus becoming asymmetric dipoles. Obviously, microstrip patch antennas do not form an image, since they are nominally half-wavelength antennas. Therefore, the centre of the sphere was placed at the centre of the square patch. Apart from all the above, the radius must be wide enough to accommodate the whole of the surface current distribution, i.e., the radiating parts of the antenna. Including just the patch may not suffice; currents exist also on the upper face of the ground plane below, and these currents contribute to radiation, too. This SCD is free to extend beyond the limits of the patch (Bancroft, 2004). This phenomenon occurs in CSPP–1 and CSPP–2: the lower SCD extends to the edges of

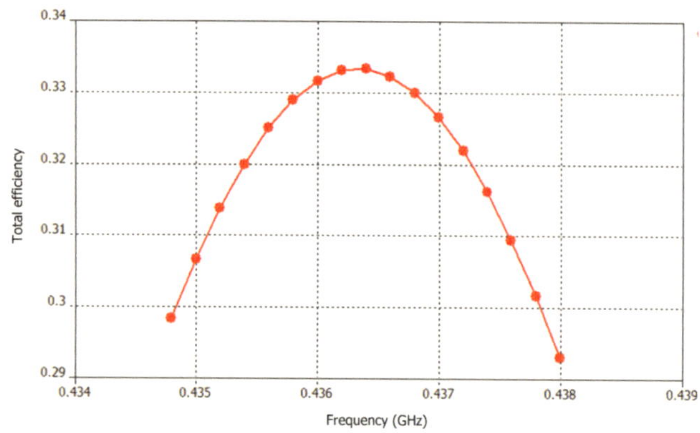

Fig. 29. The total efficiency of the CSPP–2 antenna varies between 29–33% throughout the service band of the CubeSat.

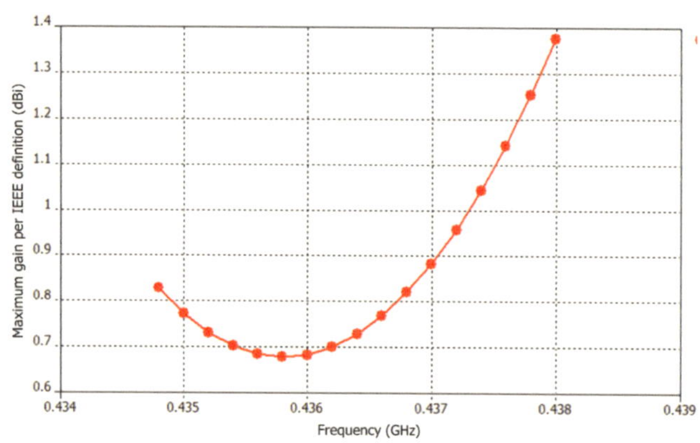

Fig. 30. The maximum gain of the CSPP–2 antenna varies between 0.7–1.4 dB throughout the service band of the CubeSat.

the ground plane. Finally, the radius of the circumscribing sphere is $a = 64.2$ mm, and the resulting electrical size of the antenna is

$$(ka)_{\text{CubeSat}} = 2\pi\frac{64.2}{687.3}\ \text{rad} = 0.587\ \text{rad} < 1\ \text{rad}. \tag{19}$$

Therefore, the radiator proposed in this Chapter is indeed an *Electrically Small* Antenna (ESA) (Hansen, 1981).

6. Discussing the results

Table 1 summarizes the electrical performance of the CubeSat antenna. Based on the tabulated data, it is easy to derive the Quality Factor of the antenna from the relation (Yaghjian & Best, 2005)

$$Q = \left(\frac{2}{B}\right)\cdot\left(\frac{S-1}{2\sqrt{S}}\right) \tag{20}$$

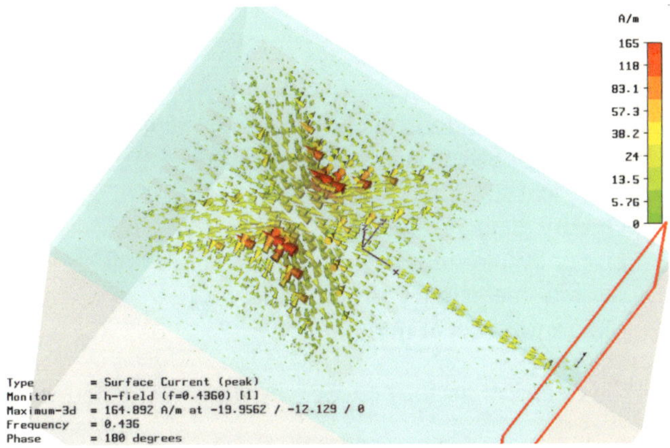

Fig. 31. Visualization of the current distribution on the CSPP–2 patch as a vector field. The color code has been set to logarithmic scale for greater detail.

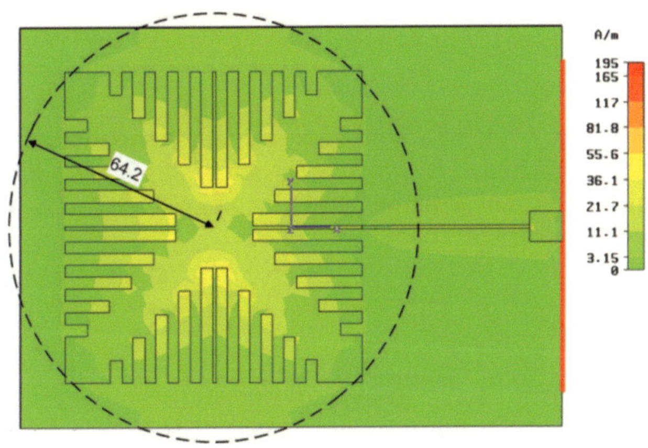

Fig. 32. Definition of the radius of the circumscribing sphere for the CubeSat antenna.

where B is the fractional VSWR bandwidth and S is the corresponding VSWR. Setting $S = 3$, we get $B = 0.0092$, and (20) leads to a loaded Quality Factor

$$Q^{\text{Cube}} = \frac{1}{0.0092} \cdot \frac{3-1}{\sqrt{3}} = 125.5. \tag{21}$$

Another useful expression for calculating the loaded Q of antennas comes from (Pues & Van de Capelle, 1989)

$$Q_{ST} = \frac{1}{B}\sqrt{\frac{(TS-1)(S-T)}{S}} \tag{22}$$

where T is the VSWR level that corresponds to the minimum reflection coefficient. In our case, $\min\{|\Gamma_{\text{in}}(j\omega)|\} = -18$ dB, so the corresponding standing-wave ratio is $S = 1.29$. Substituting these values in (22) we obtain

$$Q_{ST}^{\text{Cube}} = \frac{1}{0.0092}\sqrt{\frac{(1.29 \times 3 - 1)(3 - 1.29)}{3}} = 139 \tag{23}$$

which is 10% higher than the estimation of loaded Q in (21).

Antenna Attribute	Value
Centre frequency, f_0	436.5 MHz
Square patch size, L_{patch}	93.3 mm
3:1 VSWR bandwidth, $\text{BW}_{-6\text{dB}}$	4 MHz
Fractional VSWR bandwidth, FBW_{V}	0.92%
Electrical size, ka	0.587 rad
Quality factor, Q	125.5
Lower bound on Q, $Q_{\ell b}$	2.53
Radiation efficiency at mid-band, n_{rad}	38%
Maximum gain at mid-band, $G_{\text{max}}^{\text{Cube}}$	0.7 dBi

Table 1. Overview of the electrical performance of the proposed antenna

The lower bound on Q according to Chu and McLean is calculated by the following equation after substituting the values $n_{\text{rad}} = 0.38$ and $ka = 0.587$ rad (Lopez, 2006; Skrivervik et al., 2001; Yaghjian & Best, 2005)

$$Q_{\ell b}^{\text{Cube}} = n_{\text{rad}} \left[\left(\frac{1}{ka} \right)^3 + \frac{1}{ka} \right] = 2.53. \tag{24}$$

It has been shown that, as a rule-of-thumb, it is safe to assume that a feasible antenna can achieve a quality factor 75% higher than the lower bound $Q_{\ell b}$ (Lopez, 2006). Assuming that the antenna maintains a constant radiation efficiency and electrical size, then the bandwidth corresponding to a 75% higher Q is equal to

$$\text{BW}_{-6\text{dB}}^{\text{max}} = 113.5 \text{ MHz} \tag{25}$$

or equivalently,

$$\text{BW}_{-10\text{dB}}^{\text{max}} = 56.8 \text{ MHz}. \tag{26}$$

From the above we gather that a generic antenna of electrical size $ka = 0.587$ rad could achieve 28 times as much bandwidth as the cavity-like CubeSat patch antenna with its inherently narrow bandwidth.

Similarly, the *upper bound* on *gain* for an antenna of given electrical size ka that also achieves some meaningful operational BW is given by Harrington's equation (Harrington, 1960; Skrivervik et al., 2001)

$$G_{ub}^{\text{Cube}} = 10 \log \left[(ka)^2 + 2\,(ka) \right] = 2.3 \text{ dBi}. \tag{27}$$

The upper bound on gain was approached by the CubeSat antenna within a margin

$$G_{ub}^{\text{Cube}} - G_{\text{max}}^{\text{Cube}} = 1.6 \text{ dB} \tag{28}$$

which is very satisfactory. However, in the conclusions of his seminal paper, Harrington notes that none of the concepts described therein holds for *small antennas*; thus the calculation in (27) is not valid, the reason being that small antennas are actually capable of significantly higher gains than the values predicted by the upper bound. But, Harrington's equation can be useful in the ESA case if we mirror the question: given the gain of the small antenna, what would its electrical size be if it were *electrically large*? This introduces the notion of *effective electrical size*, $(ka)_{\text{eff}}$. In our case, the effective size turns out to be $(ka)_{\text{eff}} = 0.48$ rad, that is, even smaller than the actual electrical size of the CubeSat antenna due to the low radiation efficiency. As a further example, let us assume that radiation efficiency could climb to 100%. This would produce a new gain value $G_{\text{peak}}^{\text{Cube}} = 4.9$ dBi, and the effective size would then become $(ka)_{\text{eff}} = 1.02$ rad.

6.1 Comparison with existing studies

The first goal of a comparison is to be meaningful; only then will it produce some useful conclusions. The literature review in Section 3 surveyed over fourty different antenna topologies, which spanned four different functions (see Section 2) and six different frequency bands. The statistics of the examined set reveal that 7% of antennas operated in the VHF band; 29% operated in the UHF band; 10% operated in the L-band; 37% operated in the S-band; 2% operated in the C-band; and 15% operated in the X-band. Different frequencies do not pose a problem: fractional bandwidth and electrical size are frequency–normalized quantities, whereas gain is related to directivity, which is determined by the electrical aperture. Hence, an inter-band comparison is still valid. Radiation efficiency is weakly dependent on frequency, so this attribute could also be included in the comparison. However, only (He & Arichandran, 2001) have included details on n_{rad} in their report.

Different functions, on the other hand, can certainly be a problem. GPS/GNSS and TTC subsystems rely on low-gain, broad-beam antennas. Data downlinking and satellite cross-linking is done through medium-/high-gain, directive antenna systems. Increased gain requires a large electrical size, period.[4] Therefore, the electrical performance of the electrically small TTC antenna described herein is compared with the performance of previously reported antennas aimed at GPS/GNSS and TTC subsystems. The comparison is shown in Table 2.[5] The upper part of the Table is reserved for non-planar structures, whereas the lower part tabulates planar ones.

Commenting on the tabulated results, it is first noted that quadrifilar helices (Niow et al., 2009; Rezaei, 2004) are wideband MGAs with clean circular polarization but they are also electrically large structures ($ka > 2$ rad). With proper structure modification and extensive dielectric loading they can be shrunk down to resonant size, that is, the electrical size of the half-wavelength monopole. Bent-wire monopoles (Yousuf et al., 2008) are wideband antennas with reasonable size. Their main drawback is that their performance degrades when mounted on a small satellite.

Turning to the planar antenna regime, the antennas presented by (Mathur et al., 2001) are directly comparable with the CSPP antennas, particularly the UHF antenna. These are electrically small, narrowband antennas as well. Their size exceeds that of the CSPP by 21% and 14% in the UHF and S-band case, respectively. At the other end of the design spectrum, the work by (Muchalski et al., 2004) is a typical example of how suspended, electrically large patches can exhibit quite significant bandwidths. Furthermore, resonant-size patches can produce gains in the range 4–6.5 dBi and bandwidths of a few percent (Gao et al., 2008; Hamrouni et al., 2009; Maqsood et al., 2010). On the other hand, 3-D structures consisting of folded and stacked radiating parts offer moderate bandwidths and gains combined, and can even become electrically small (Lee & Woo, 2008). And last but not least, patch-excited cups are electrically large MGAs offering significant gain values with respect to their size (Zackrisson, 2007). However, in their current form, their size is prohibitive for UHF CubeSat

[4] Only compound field antennas threaten to alter this otherwise fundamental relationship, however that topic is outside the scope of and space available in this Chapter.

[5] The FBW cited for (Gao et al., 2008) is aggregate, not instantaneous.

Antenna Configuration	f_0 (GHz)	FBW (%)	G_{max} (dBi)	ka (rad)
(Rezaei, 2004)	2.260	5.4	n/a	2.11
(Yousuf et al., 2008) I	0.370	37.8	n/a	1.41
(Yousuf et al., 2008) II	0.180	32.8	n/a	1.26
(Yousuf et al., 2008) III	0.150	41.8	n/a	1.26
(Niow et al., 2009) initial	2.600	30.8	4.0	2.22
(Niow et al., 2009) Teflon	2.600	15.4	4.0	1.78
(Niow et al., 2009) Macor	2.600	15.4	3.7	1.72
(Mathur et al., 2001) UHF	0.450	1.6	n/a	0.71
(Mathur et al., 2001) S-band	2.260	0.8	4.9	0.67
(Muchalski et al., 2004) centre	2.450	8.2	n/a	2.24
(Muchalski et al., 2004) edge	2.350	8.5	n/a	2.15
(Muchalski et al., 2004) corner	2.275	7.9	n/a	2.08
(Zackrisson, 2007) GPS L2	1.227	n/a	7.4	2.20
(Zackrisson, 2007) GPS L1	1.575	n/a	8.5	2.82
(Zackrisson, 2007) X-band RX	8.000	n/a	8.9	2.63
(Zackrisson, 2007) X-band TX	8.000	n/a	7.5	2.24
(Gao et al., 2008)	2.250	22.2	6.5	1.99
(Lee & Woo, 2008) TX	2.487	6.6	5.3	1.39
(Lee & Woo, 2008) RX	1.610	4.8	2.4	0.90
(Hamrouni et al., 2009) 1	2.400	2.9	n/a	1.62
(Hamrouni et al., 2009) 2	2.400	3.6	n/a	1.62
(Maqsood et al., 2010) L1	1.575	1.0	6.3	1.82
(Maqsood et al., 2010) L2	1.227	0.3	4.0	1.42
This work, *CSPP–2*	0.437	0.5	0.7	0.59

Table 2. Comparison of electrical performance among GPS/GNSS/TTC antennas

missions since it is 3.7–4.8 times larger than the size of the CSPP, which fits marginally on the wall of the spacecraft.

Judging by the sizes listed in the right-most column of Table 2, it becomes immediately clear that antenna miniaturization techniques are utterly relevant to modern small satellites.

7. Conclusions and further research

Modern small satellites are designed and developed at costs and timescales that are revolutionizing the satellite industry. These satellites range in size from 10–100 kg down to less than 100 g, take a year or less to develop and launch, and have individual costs ranging from a few million dollars to under $1,000. They achieve sophisticated functionality by utilizing state-of-the-art commercial off-the-shelf components, which are initially developed for terrestrial applications and later adapted to space applications. Advances in low-power microelectronics and digital signal processing are turning satellites smaller, smarter, faster and cheaper (Gao et al., 2009). The history and framework of small satellites were described in Section 1. Within this framework, Section 2 defined the objectives of this Chapter. Section 3 presented an overview of the status of antennas for such small satellites. Work from many research groups around the world has been included. Although the focus was on planar antenna structures, linear and 3-D antennas were also described.

From the analysis, design procedure and results presented in sections 4, 5 and 6, it is clear that designing a 1U CubeSat antenna at the lower end of the UHF spectrum is a challenging task. There are too many constraints, and the designer cannot enhance one aspect of the antenna without seriously compromising another. The specifications lead inevitably to the implementation of an *electrically small* antenna. The antenna must be made so small that its performance suffers in every aspect. From the designer's point-of-view, the feasibility of implementing an electrically large antenna ($ka > 2$ rad), or even a resonant-size antenna ($1 < ka < 2$ rad), would be highly desirable. Such a radiator would have significantly better performance, and it would aid the link budget with its higher gain. The need for higher gain becomes evident if we consider that the path loss for a CubeSat is about 140 dB.

However, since the outer dimensions of CubeSats are fixed, there is really no point in suggesting modifications; the system would fall outside protocol specifications. The designer could suggest using more than one faces of the cube. This scenario also comes to no avail, since there can be only one *Earth Facing Facet*. Therefore, it seems that the only way to get a better antenna would be to switch to higher frequencies and, more specifically, at least double the frequency (0.9 GHz). The higher frequency offers many design advantages and one very important drawback: path loss increases by 6 dB. Will the new antenna system be able to compensate this loss, so that it becomes worthy of the time required for the re-design?

The answer is that there is indeed potential to cover the extra loss. Starting with 40×40 mm^2 square patches over a 100×100 mm^2 ground plane, we get the choice of increasing the patch width and switch to rectangular patches, which show better input matching, higher total efficiency and broader bandwidth. Moreover, two rectangular patches forming a two-element linear broadside array can be fed in-phase with a corporate feeding network. Nonetheless, even if the orthogonal patches maintain the double orthogonal symmetry, axial ratio will shift from unity and the purity of circular polarization will degrade.

Another interesting topic for further research is to study different length distributions (tapering) for the peripheral slits. The triangular tapering seemed like a natural choice, but there are other options, e.g. binomial, cosine-on-pedestal, etc. It would be interesting to investigate how the other distributions perform in terms of the size-bandwidth trade-off, because each distribution will affect L_{eff} and $Z_{\text{in}}(j\omega)$ differently.

Finally, regardless of the frequency of operation of the CubeSat antenna, there is great interest in studying applicable techniques for broadband microstrip antennas. One such technique is the etching of slots on the surface of the patch that are shaped after the letters "U" and "E". This technique is presented in the most elegant way by (Weigand et al., 2003) and (Shafai, 2007). Again, such broadbanding techniques destroy the double orthogonal symmetry of the patch, and hence do not provide for clean circular polarization.

8. Acknowledgements

C. Kakoyiannis thanks Ms Calliope Raptis for proofreading and improving the manuscript, and Dr.-Ing. Dimitris Komnakos for the fruitful discussions on structure and content. The assistance of both colleagues is gratefully acknowledged.

The authors contribute this Chapter in loving memory of their colleague Dimitris T. Notis. Dimitris was a bright PhD student with the Department of Electrical Engineering, Aristotle University of Thessalonike, Greece. His sudden and untimely passing caused us great grief.

9. References

Arnieri, E.; Boccia, L.; Amendola, G. & Di Massa, G. (2004). A high gain antenna for small satellite missions, *IEEE Antennas and Propagation Society Int'l Symposium, 2004*, Vol. 2, pp. 1587–1590, Jun. 2004

Arnieri, E.; Boccia, L.; Amendola, G. & Di Massa, G. (2007). A compact high gain antenna for small satellite applications. *IEEE Transactions on Antennas and Propagation*, Vol. 55, No. 2, pp. 277–282, Feb. 2007

Bahl, I. & Bhartia, P. (2003). *Microwave Solid State Circuit Design*, Wiley-Interscience, ISBN 0471207551, Hoboken, New Jersey

Bancroft, R. (2004). *Fundamental dimension limits of antennas: Ensuring proper antenna dimensions in mobile device designs*. Centurion Wireless Technologies, Westminster, Colorado [Online]. Available: http://www.xertex.com/home/pdf/wp_dimension_limits.pdf (accessed September 30, 2010)

Bérenger, J.P. (1994). A perfectly matched layer for the absorption of electromagnetic waves. *Journal of Computational Physics*, Vol. 114, No. 2, pp. 185–200, Feb. 1994

CubeSat programme & community official website [Online]. Available: www.cubesat.org/ (accessed June 11, 2010)

Dabrowski, M.J. (2005). *The design of a software system for a small space satellite*. M.Sc. thesis, Graduate College, Univ. of Illinois at Urbana-Champaign, Urbana, IL, 2005

Fujishige, T.S.; Ohta, A.T.; Tamamoto, M.A.; Goshi, D.S.; Murakami, B.T.; Akagi, J.M. & Shiroma, W.A. (2002). Active antennas for CubeSat applications, *Proceedings of the 16th AIAA/USU Annual Small Satellites Conference*, paper SSC02-V-2, Utah State University, Logan, USA, Aug. 2002

Galysh, I.; Doherty, K.; McGuire, J.; Heidt, H.; Niemi, D. & Dutchover, G. (2000). *CubeSat: Developing a standard bus for picosatellites*. The StenSat Group, 9512 Rockport Rd, Vienna, VA 22180 [Online]. Available: http://cubesat.net/images/Papers/stensat_hist.pdf (accessed September 30, 2010)

Galván, J. & Colantonio, D. (2009). Low back radiation compact antenna for data downlink in LEO satellites, *Proceedings of the 2009 SBMO/IEEE MTT-S Int'l Microwave and Optoelectronics Conference (IMOC'09)*, pp. 816–820, Belem, Brazil, Nov. 2009

Gao, S.; Brenchley, M.; Unwin, M.; Underwood, C.I.; Clark, K.; Maynard, K.; Boland, L. & Sweeting, M.N. (2008). Antennas for small satellites, *Proceedings of the 2008 Loughborough Antennas and Propagation Conference*, pp. 66–69, Loughborough, UK, Mar. 2008

Gao, S.; Clark, K.; Unwin, M.; Zackrisson, J.; Shiroma, W.A.; Akagi, J.M.; Maynard, K.; Garner, P.; Boccia, L.; Amendola, G.; Massa, G.; Underwood, C.; Brenchley, M.; Pointer, M. & Sweeting, M.N. (2009). Antennas for modern small satellites. *IEEE Antennas and Propagation Magazine*, Vol. 51, No. 4, pp. 40–56, Aug. 2009

Gustrau, F. & Manteuffel, D. (2006). *EM Modeling of Antennas and RF Components for Wireless Communications Systems*, Springer-Verlag, ISBN 3540286144, Berlin, Germany

Hadj Abderrahmane, L.; Benyettou, M.; & Sweeting, M.N. (2006). An S band antenna system used for communication on earth observation microsatellite, *Proceedings of the 2006 IEEE Aerospace Conference*, Big Sky, Montana, USA, Mar. 2006, doi: 10.1109/AERO.2006.1655813

Hamrouni, C.; Neji, B.; Alimi, A.M. & Schilling, K. (2009). Design and prototype of a flight microstrip antennas for the Pico satellite ERPSat-1, *Proceedings 4th Int'l Conf. Recent Advances in Space Technologies*, pp. 750–755, Constantinople, Turkey, Jun. 2009

Hansen, R.C. (1981). Fundamental limitations in antennas. *Proceedings of the IEEE*, Vol. 69, No. 2, pp. 170–182, Feb. 1981

Harrington, R.F. (1960). Effect of antenna size on gain, bandwidth, and efficiency. *J. Res. Nat. Bur. Stand*, Vol. 64, No. 1, pp. 1–12, Jan. 1960

He, Y. & Arichandran, K. (2001). The design of X band dual feed aperture coupled patch antenna for microsatellites, *Proceedings of the IEEE 2001 Int'l Geoscience and Remote Sensing Symposium (IGARSS '01)*, vol. 6, pp. 2784–2786, Sydney, Australia, 2001

Heidt, H.; Puig-Suari, J.; Moore, A.S.; Nakasuka, S. & Twiggs, R.J. (2000). CubeSat: A new generation of picosatellite for education and industry low-cost space experimentation, *Proceedings of the 12th AIAA/USU Annual Small Satellites Conference*, paper SSC00-V-5, Utah State University, Logan, USA, Aug. 2000

Hunyadi, G.; Klumpar, D.M.; Jepsen, S.; Larsen, B. & Obland, M. (2002). A commercial off the shelf (COTS) packet communications subsystem for the Montana EaRth-Orbiting Pico-Explorer (MEROPE) CubeSat, *Proceedings of the 2002 IEEE Aerospace Conference*, Vol. 1, pp. 473–478, Big Sky, Montana, USA, Mar. 2002

Idzkowski, B.; Kalka, T.; Linowski, J.; Preisner, M.; Plywacz, G. & Kabacik, P. (2004). The optimization of communication link performance in scientific minisatellites, *Proceedings of the 15th International Conference on Microwaves, Radar and Wireless Communications (MIKON-2004)*, Vol. 3, pp. 997–1000, Warzaw, Poland, May 2004

Kakoyiannis, C.G. & Constantinou, P. (2008). A compact microstrip antenna with tapered peripheral slits for CubeSat RF Payloads at 436MHz: Miniaturization techniques, design & numerical results, *Proceedings of the IEEE International Workshop on Satellite and Space Communications (IWSSC 2008)*, pp. 255–259, Toulouse, France, Oct. 2008

Kakoyiannis, C.G.; Kyrligkitsi, A. & Constantinou, P. (2010). Bandwidth enhancement, radiation properties and ground-dependent response of slotted antennas integrated into wireless sensors. *IET Microwaves, Antennas & Propagation*, Vol. 4, No. 5, pp. 609–628, May 2010

Kakoyiannis, C.G. & Constantinou, P. (2010a). Radiation properties and ground-dependent response of compact printed sinusoidal antennas and arrays. *IET Microwaves, Antennas & Propagation*, Vol. 4, No. 5, pp. 629–642, May 2010

Kakoyiannis, C.G. & Constantinou, P. (2010b). Compact printed arrays with embedded coupling mitigation for energy-efficient wireless sensor networking. *International Journal of Antennas and Propagation – special issue "Mutual Coupling in Antenna Arrays"*, Vol. 2010, Article ID 596291, 18 pages, 2010

LaBerteaux, J.; Moesta, J. & Bernard, B. (2007). Cajun advanced picosatellite experiment, *Proceedings of the IEEE/AIAA 26th Digital Avionics Systems Conference (DASC'07)*, pp. 5.E.2-1 – 5.E.2-7, Dallas, TX, USA, Oct. 2007

Lee, H.-R. & Woo, J.-M. (2008). Miniaturized dual band circularly polarization microstrip antenna for satellite communication, *Proceedings of the 8th Int'l Symposium on Antennas, Propagation and EM Theory (ISAPE 2008)*, pp. 294–297, Kunming, China, Nov. 2008

Lopez, A.R. (2006). Fundamental limitations of small antennas: Validation of Wheeler's formulas. *IEEE Antennas and Propagation Magazine*, Vol. 48, No. 4, pp. 28–36, Aug. 2006

Maleszka, T.; Gorski, P. & Kabacik, P. (2007). On omnidirectional coverage with minimum number of circularly polarized patch antennas placed on minisatellites, *Proceedings of the 2007 IEEE Antennas and Propagation International Symposium*, pp. 3037–3040, Honolulu, Hawaii, USA, June 2007

Maqsood, M.; Bhandari, B.; Gao, S.; De Vos Van Steenwijk, R. & Unwin, M. (2010). Dual-band circularly polarized antennas for GNSS remote sensing onboard small satellites, *Proceedings of the 7th Int'l Symposium on Communication Systems Networks and Digital Signal Processing (CSNDSP'10)*, pp. 86–90, Newcastle upon Tyne, UK, Jul. 2010

Marrocco, G.; Mattioni, L.; Potenza, A.; Milani, F.; Giacomini, A. & Sabbadini, M. (2010). Distributed multi-function antenna system for micro- and nano-satellites, *Proceedings 4th European Conf. Antennas and Propagation (EuCAP'10)*, Barcelona, Spain, Apr. 2010

Mathur, R.; Haupt, R. & Swenson, C. (2001). Student antenna design for a nanosatellite, *Proc. 2001 IEEE Aerospace Conf.*, Vol. 7, pp. 3683–3688, Big Sky, Montana, USA, Mar. 2001

Mizuno, T.J.; Roque, J.D.; Murakami, B.; Yoneshige, L.; Shiroma, G.; Miyamoto, R. & Shiroma, W.A. (2005). Antennas for distributed nanosatellite networks, *Proceedings of the IEEE/ACES Int'l Conference on Wireless Communications and Applied Computational Electromagnetics*, pp. 606–609, Honolulu, Hawaii, USA, Apr. 2005

Moghaddam, E.S.; Aboutorabian, N.; Amiri, S.; Nikmehr, S. & Rezaei, P. (2004). Design and analysis of a dualband antenna for small LEO satellite applications, *Proceedings of the 3rd International Conference on Computational Electromagnetics and Its Applications (ICCEA 2004)*, pp. 228–231, Beijing, China, Nov. 2004

Muchalski, K.; Jagoda, M.; Tomasiak, M.; Gorski, P.; Akonom, A.; Kulig, M.; Barecki, W. & Kabacik, P. (2004). Optimizing TT&C antenna placement on minisatellites, *Proceedings of the 15th International Conference on Microwaves, Radar and Wireless Communications (MIKON-2004)*, Vol. 2, pp. 489–492, Warzaw, Poland, May 2004

Munteanu, I.; Timm, M. & Weiland, T. (2010). It's about time. *IEEE Microwave Magazine*, Vol. 11, No. 2, pp. 60–69, April 2010

Niow, C.H.; Mouthaan, K.; Coetzee, J.C. & Hui, H.T. (2009). Design of a small size dielectric loaded helical antenna for satellite communications, *Proceedings of the Asia Pacific Microwave Conference (APMC 2009)*, pp. 48–51, Singapore, Dec. 2009

Nohmi, M.; Oi, K.; Takuma, S. & Ogawa, M. (2010). Solar paddle antenna mounted on pico-satellite "KUKAI" for amateur radio communication, *Proceedings of the Second International Conference on Advances in Satellite and Space Communications (SPACOMM'10)*, pp. 31–36, Athens, Greece, Jun. 2010

Notis, D.T.; Liakou, P.C. & Chrissoulidis, D.P. (2004). Dual polarized microstrip patch antenna, reduced in size by use of peripheral slits, *Proceedings of the 34th European Microwave Conference*, Vol. 1, pp. 125-128, Amsterdam, Netherlands, Oct. 2004

Pues, H.F. & Van de Capelle, A.R. (1989). An impedance-matching technique for increasing the bandwidth of microstrip antennas. *IEEE Transactions on Antennas and Propagation*, Vol. 37, No. 11, pp. 1345–1354, Nov. 1989

Puig-Suari, J.; Turner, C.; & Ahlgren, W. (2001). Development of the standard CubeSat deployer and a CubeSat class picosatellite, *Proceedings of the 2001 IEEE Aerospace Conference*, Vol. 1, pp. 347–353, Big Sky, Montana, USA, Mar. 2001

Rezaei, P. (2004). Design of quadrifilar helical antenna for use on small satellites, *2004 IEEE Antennas and Propagation Society Int'l Symposium Digest*, Vol. 3, pp. 2895–2898, Monterey, CA, USA, Jun. 2004

Row, J.-S.; Yeh, S.-H. & Wong, K.-L. (2000). Compact dual-polarized microstrip antennas. *Microwave and Optical Technology Letters*, Vol. 27, No. 4, pp. 284–287, Nov. 2000

Schaffner, J.A. & Puig-Suari, J. (2002). The electronic system design, analysis, integration, and construction of the Cal Poly State University CP1 CubeSat, *Proceedings of the 16th AIAA/USU Annual Small Satellites Conference*, Utah State University, Logan, USA, Aug. 2002 [Online]. Available: polysat.calpoly.edu/PublishedPapers/JakeSchaffner_srproj.pdf (accessed September 30, 2010)

Shafai, L. (2007). Wideband microstrip antennas, In: *Antenna Engineering Handbook, 4th ed.*, Volakis, J.L. (Ed.), Chapter 16, pp. 16.27–16.42, McGraw-Hill, New York

Skrivervik, A.K.; Zürcher, J.-F.; Staub, O. & Mosig, J.R. (2001). PCS antenna design: The challenge of miniaturization. *IEEE Antennas and Propagation Magazine*, Vol. 43, No. 4, pp. 12–27, Aug. 2001

Sten, J.C.-E.; Hujanen, A. & Koivisto, P.K. (2001). Quality factor of an electrically small antenna radiating close to a conducting plane. *IEEE Transactions on Antennas and Propagation*, Vol. 49, No. 5, pp. 829–837, May 2001

Stutzman, W.L. & Thiele, G.A. (1998). *Antenna Theory and Design, 2nd ed.*, Wiley, ISBN 0471025909, New York, USA

Tamamoto, M.A. & Shiroma, W.A. (2002). Active antennas and UHF antennas for CubeSat applications, [Online]. Available: www.spacegrant.hawaii.edu/reports/12_FA02-SP03/tamamoto.pdf (accessed September 30, 2010)

Tanaka, M.; Suzuki, R.; Suzuki, Y. & Araki, K. (1994). Microstrip antenna with solar cells for microsatellites, *1994 Antennas and Propagation Society Int'l Symposium Digest*, Vol. 2, pp. 786–789, Seattle, WA, USA, Jun. 1994

Vasylchenko, A.; Schols, Y.; De Raedt, W. & Vandenbosch, G.A.E. (2007a). Challenges in full wave electromagnetic simulation of very compact planar antennas, *Proc. 2nd European Conf. Antennas Propag. (EuCAP 2007)*, pp. 1–6, Edinburgh, UK, Nov. 2007

Vasylchenko, A.; Schols, Y.; De Raedt, W. & Vandenbosch, G.A.E. (2007b). A Benchmarking of six software packages for full-wave analysis of microstrip antennas, *Proc. 2nd European Conf. Antennas Propag. (EuCAP 2007)*, pp. 1–6, Edinburgh, UK, Nov. 2007

Weigand, S.; Huff, G.H.; Pan, K.H.; & Bernhard, J.T. (2003). Analysis and design of broad-band single-layer rectangular U-slot microstrip patch antennas. *IEEE Transactions on Antennas and Propagation*, Vol. 51, No. 3, pp. 457–468, Mar. 2003

Weiland, T.; Timm, M. & Munteanu, I. (2008). A practical guide to 3-D simulation. *IEEE Microwave Magazine*, Vol. 9, No. 6, pp. 62–75, Dec. 2008

Wettergren, J.; Ingvarson, P. & Zackrisson, J. (2009). Data-link antennas for moon-crashing probes, *Proceedings of the Third European Conference on Antennas and Propagation (EuCAP 2009)*, pp. 1891–1895, Berlin, Germany, Mar. 2009

Wikipedia contributors (2010a). CubeSat. *Wikipedia, The Free Encyclopedia*, http://en.wikipedia.org/wiki/Cubesat (accessed June 11, 2010)

Wikipedia contributors (2010b). Miniaturized satellite. *Wikipedia, The Free Encyclopedia*, http://en.wikipedia.org/wiki/Miniaturized_satellite (accessed June 11, 2010)

Wikipedia contributors (2010c). List of CubeSats. *Wikipedia, The Free Encyclopedia*, http://en.wikipedia.org/wiki/List_of_CubeSats (accessed June 11, 2010)

Wincza, K.; Osys, M.; Dudzinski, L. & Kabacik, P. (2004). Lightweight low gain microstrip antennas for use in minisatellites, *Proceedings of the 15th International Conference on*

Microwaves, Radar and Wireless Communications (MIKON-2004), Vol. 1, pp. 314–317, Warzaw, Poland, May 2004

Wong, K.L. & Hsu, W.-H. (2001). A broad-band rectangular patch antenna with a pair of wide slits. *IEEE Transactions on Antennas and Propagation*, Vol. 49, No. 9, pp. 1345–1347, Sept. 2001

Xiao, S.; Wang, B.-Z.; Shao, Z.; Zhou, M.-T. & Fujise, M. (2005). Bandwidth-enhancing ultralow-profile compact patch antenna. *IEEE Transactions on Antennas and Propagation*, Vol. 53, No. 11, pp. 3443–3447, Nov. 2005

Xiao, S.; Shao, Z.; Wang, B.-Z.; Zhou, M.-T. & Fujise, M. (2006). Design of low-profile microstrip antenna with enhanced bandwidth and reduced size. *IEEE Transactions on Antennas and Propagation*, Vol. 54, No. 5, pp. 1594–1599, May 2006

Yaghjian, A.D. & Best, S.R. (2005). Impedance, bandwidth, and Q of antennas. *IEEE Transactions on Antennas and Propagation*, Vol. 53, No. 4, pp. 1298–1324, April 2005

Yousuf, H.J.; Haider, M.M.; Siddique, M.K. & Amin, M. (2008). Analysis of G-shape antennas mounted on a CUBESAT, *Proceedings of the 2nd Int'l Conference on Advances in Space Technologies*, Vol. 2, pp. 28–32, Islamabad, Pakistan, Nov. 2008

Zackrisson, J. (2007). Wide coverage antennas, *Proc. 21st AIAA/USU Annual Conf. Small Satellites*, paper no. SSC07-XIII-7, Utah State University, Logan, USA, Aug. 2007

Zhang, X.-X. & Yang, F. (1998). Study of a slit cut on a microstrip antenna and its applications. *Microwave and Optical Technology Letters*, Vol. 18, No. 4, pp. 297–300, July 1998

13

Circularly Polarized Microstrip Antennas with Proximity Coupled Feed for Circularly Polarized Synthetic Aperture Radar

Merna Baharuddin[1] and Josaphat Tetuko Sri Sumantyo[2]
[1]*Department of Electrical Engineering, Faculty of Engineering, Hasanuddin University*
Jl. Perintis Kemerdekaan, Kampus Tamalanrea
Makassar 90211,
[2]*Center for Environmental Remote Sensing, Chiba University*
1-33, Yayoi, Inage, Chiba 263-8522
[1]*Indonesia*
[2]*Japan*

1. Introduction

Synthetic aperture radar (SAR) is an active sensor that can produce high resolution imagery in microwave bands.

A circularly polarized SAR (CP-SAR) to be launched onboard a micro-satellite is currently developed in the Microwave Remote Sensing Laboratory (MRSL) of the Center for Environmental Remote Sensing (CEReS), Chiba University. As part of the project, an airborne CP-SAR development is also undertaken in order to obtain sufficient knowledge of CP-SAR sensor systems. An L-band CP-SAR system will be designed for operation onboard an unmanned aerial vehicle (UAV).

Conventional SAR systems have been based on linearly polarized (LP) antenna systems (Nemoto et al. 1991, Raney et al. 1991). However, there are limitations due to the propagation phenomena. Especially for space propagation, electromagnetic waves propagating through the ionosphere interact with electrons and magnetic fields. As a result, the polarization vector of the electric field is rotated by the Faraday rotation effect (Rignot, 2000).

The destructive effects mentioned above of a SAR sensor with linear polarization can be solved by the use of CP-SAR. In addition, it has been pointed out (Raney, 2007) that a full characterization of SAR signals backscattered from a random object can only be possible through the use of circular polarization. Hence, compared to a linear SAR sensor, a greater amount of information about scenes and targets being imaged would be provided with a CP-SAR sensor. The operational frequency is chosen to be 1.27 GHz in L-band considering its superior features on particular applications.

As the Microwave Remote Sensing Laboratory, Center for Environmental Technology, Chiba University is developing a CP-SAR sensor onboard an UAV, an antenna for the L-band CP-SAR sensor is required.

Until now, a number of airborne SAR has been developed with LP antennas installed in the SAR sensor onboard. The E-SAR (Horn, 1996), implementing pyramidal horn for its X-band antenna, microstrip patch arrays for its C- and P-band antenna, and passive microstrip phased arrays (L- and S-band). A waveguide phased array antenna is implemented in the C-band airborne Danish SAR system (Madsen et al., 1991). The UAVSAR, a polarimetric UAV SAR system developed by NASA's Jet Propulsion Laboratory implementing a phased array antenna consists of LP microstrip elements.

Therefore, to investigate the new SAR sensor with CP antennas, in this chapter, CP microstrip elements are proposed and developed to be implemented in an array antenna for the CP-SAR sensor onboard an UAV. The numerical simulations and the measurement results are shown and discussed.

The remainder of this chapter is organized as follows. Section 2 explains the methodology used to design and develop the CP microstrip antennas. The numerical simulation and measurement results for the single elements are presented in section 3 along with explanations of the results, whereas for the array antenna is discussed in section 4. Finally, the summary and conclusions of this work are presented in the section 5.

2. Microstrip antennas development methodology

2.1 Specification of the antenna

Specification of the CP-SAR onboard UAV is shown in Table 1.

Parameter	Specification
Frequency f	1.27 GHz (L band)
Chirp bandwidth Δf	10 MHz
Polarization	Transmitter : RHCP
	Receiver : RHCP + LHCP
Gain G	> 20 dBic
Axial ratio AR	< 3 dB (main beam)
Antenna size	2 m (azimuth) × 0.7 m (range)
Beam width	8° (azimuth) × 25° (range)
Resolution	≈ 1.4 m (azimuth) × 15 m (range)
Altitude range	1 - 4 km

Table 1. Specification of CP-SAR Onboard Unmanned Aerial Vehicle.

An array of microstrip antenna elements will be designed to meet the specification listed in the Table 1.

The requirements for the range resolution (15 m) determine the antenna bandwidth of 10 MHz, or less than 1% of the operation frequency of 1.27 GHz.

$$\delta_R = \frac{c}{2B}, \tag{1}$$

where

δ_R = range resolution (m)

c = speed of light (m/sec)

B = bandwidth (Hz)
This bandwidth requirement must be compatible with a low axial ratio (AR) (below 3 dB) for ensuring transmitting/receiving circularly polarized waves. To satisfy the matching of input impedance, the return loss must be smaller than 10 dB in this bandwidth range.
The primary considerations in the design and subsequent fabrication processes are low cost, light weight and ease of manufacturing. One antenna aperture will be used for both transmitting and receiving CP-SAR signals, with a circulator to control the direction of signal flow into/out from the CP-SAR sensor circuit (Chan, 2004). The CP-SAR antenna consists of an array of single antenna elements, each being a microstrip antenna for circular polarization. Even though it is also possible to obtain a CP array comprising of linearly-polarized elements, the electrical performance of a CP-elements array is generally better than that of an LP-elements array (Bhattacharyya, 2008) : namely, (1) bandwidth of a CP-elements array is significantly wider (about twice) than that of LP-elements array; and (2) gain of a CP-elements array is significantly higher than that of LP- elements array for large element spacing. The single element patches which have been optimized are then spatially arranged to form a planar array. A better control of the beam shape and position in space can be achieved by correctly arranging the elements along a rectangular grid to form a planar array. The beam pattern for optimum ground mapping function is cosecant-squared beam in the elevation plane (E-plane) which can correct the range gain variation and pencil beam in the azimuth plane (H-plane) (Vetharatnam et al., 2006). The antenna side lobe levels in the azimuth plane must be suppressed in order to avoid the azimuth ambiguity. To deal with reflection, the antenna side lobes and back lobes also must be suppressed. The antenna gain is mostly determined by the aperture size and inter-element separation.
Feed network is implemented in a separate substrate as the feeding method is proximity coupled. The feeding array is parallel to the antenna array, corresponding to the scheme of proximity-coupled, corporate feeding. This type of feed method allows better optimization of both feeding and antenna array structures individually. Concept of the feed network layout proposed here is the n × n microstrip arrays with a power dividing network, consisting of an element building block of 2 × 2 "H" shaped feed network (Levine et al., 1989). Constructions of a larger array can be achieved by combining the "H" networks.
In addition to the entire system specification, a list of specification for the single element microstrip antenna is also need to be given. The specification is shown in Table 2.

Parameter	**Specification**
Frequency f	1.27 GHz (L band)
Chirp bandwidth Δf	10 MHz
Polarization	Transmitter : RHCP
	Receiver : RHCP + LHCP
Gain G	> 5 dBic
Axial ratio AR	< 3 dB (main beam)
S_{11}	< -10 dB

Table 2. Specification of single-element microstrip antenna for CP-SAR Onboard Unmanned Aerial Vehicle.

2.2 Design procedure

2.2.1 Design and analysis tools: IE3D

A full-wave analysis tool (IE3D Zeland software) based on the Method of Moment (MoM) algorithm is used for design and analyzing by electromagnetically simulate the antenna models. IE3D is an integrated full-wave electromagnetic simulation and optimization package, which is capable of generating high accuracy analysis and design of microwave electronics component including antennas both 3D and planar (Zeland Software Inc., 2006). Results obtained from the IE3D simulation are the S parameter, input impedance, radiation pattern, and current distribution.

2.2.2 Developed antennas

The developed antennas comprise of four types of microstrip antennas and one array configuration. The first model investigated and developed with findings of Axial Ratio disturbance from the presence of holes for installing plastic screws. The other three models are new developed configuration of elliptical microstrip antennas. The array configuration is proposed from the elliptical microstrip antenna. The list of the developed antennas is as follows:

1. Equilateral Triangular Microstrip Antenna
2. Elliptical Microstrip Antenna
3. Elliptical Annular Ring Microstrip Antenna
4. Elliptical Annular Ring Microstrip Antenna Having a Sine Wave Periphery

An array of the elliptical microstrip element is developed and simulated.

2.3 Prototype fabrication

The equilateral triangular microstrip antenna, elliptical microstrip antenna, elliptical annular ring microstrip antenna, and the elliptical annular ring microstrip antenna having a sine wave periphery models have been fabricated to verify the simulation results. Careful and precise fabrication process is required to produce radiating behavior similar to the simulated model. The stages for fabrication is as follows: (1) Microwave Artwork;(2) Etching;(3) Bonding.

After installing the plastic screws, then the antenna is ready for measurement. Pictures of fabricated antennas are shown in Figure 1.

2.4 Measurement

The reflection coefficient and input impedance were measured with the RF Vector Network Analyzer (Agilent, E5062A, ENA-L). Before performing this measurement, a standard calibration process is needed to minimize imperfections which will cause the equipment to yield less than ideal measurements. There are three calibrated reflection standards: a short circuit, an open circuit, and a matched load. This one-port calibration makes it possible to derive the actual reflection S-parameters of the Antenna-under-test (AUT).

The antenna gain, AR, and radiation patterns were measured inside the anechoic chamber of MRSL, having a dimension of $4 \times 8.5 \times 2.4$ m^3. The measurement system is schematically shown in Figure 2 (Wissan et al., 2009). The AR vs. frequency characteristic, AR pattern, gain vs. frequency characteristic and gain pattern were measured using conical log-spiral LHCP/RHCP antennas and a dipole antenna as the standard reference. Precise alignment between AUT and the conical log-spiral antenna is indispensable for obtaining accurate measurement results.

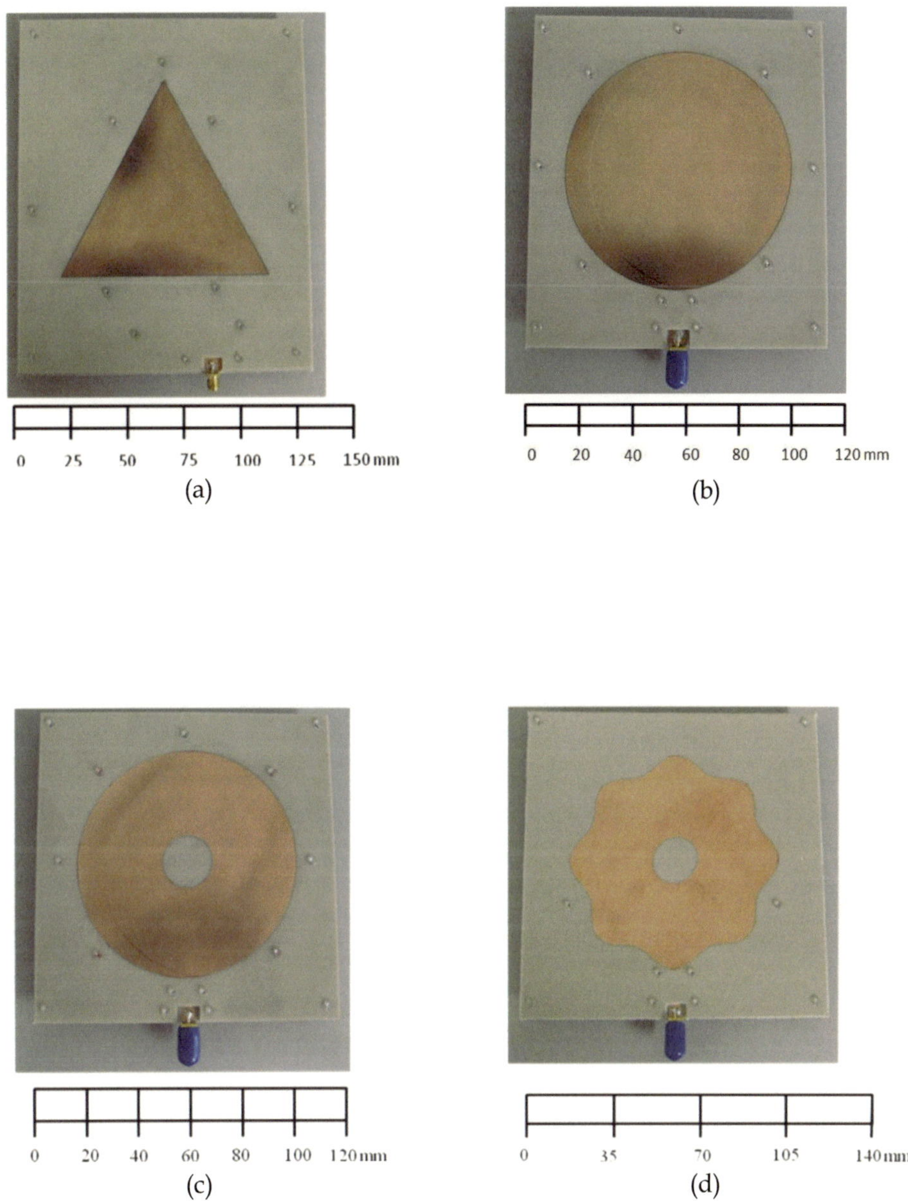

Fig. 1. Fabricated microstrip antennas ; (a) equilateral triangular microstrip antenna, (b) elliptical microstrip antenna, (c) elliptical annular ring microstrip antenna, and (d) elliptical annular ring microstrip antenna having a sine wave periphery.

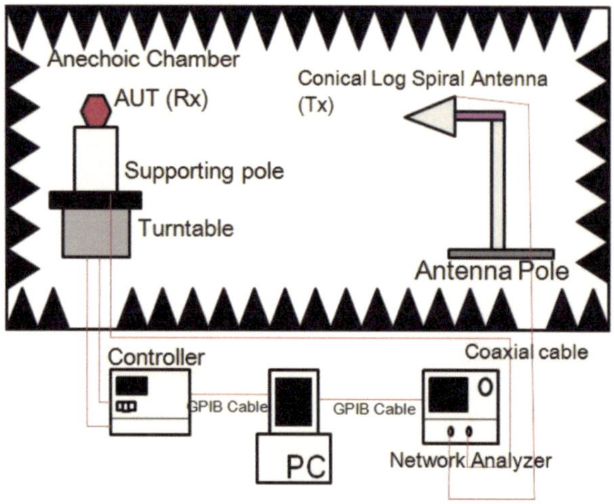

(a)

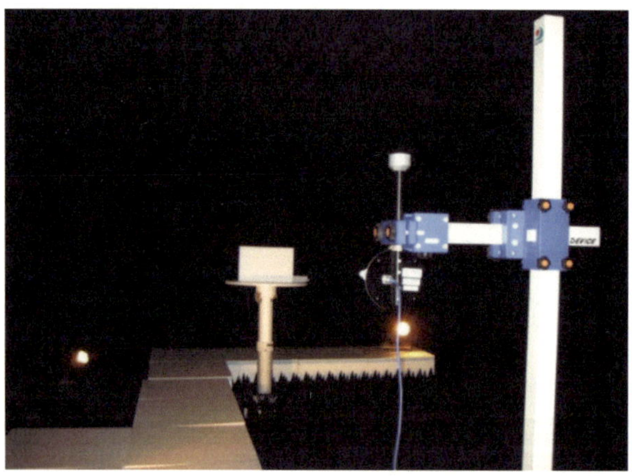

(b)

Fig. 2. (a) Schematic of the measurement system; (b) Anechoic chamber at MRSL, Chiba University.

3. Results and discussion on the simulation and measurement of the microstrip antenna elements

3.1 Equilateral triangular microstrip antenna

Previously, a number of CP triangular microstrip antennas have been developed, some of them are reported by Garg et al. (2001), Suzuki et al. (2007), and Karimabadi et al. (2008). However, almost all the developed models implement single-feed type with coaxial probe feeding method, which possess some problems, namely : (1) the CP radiator (patch) from single feed type antenna will generate an unstable current distribution which will impair the performance of axial ratio in array configuration; (2) single feed type antenna is not preferred type for a multi polarization (RHCP and LHCP) array due to the poorer isolation parameter compared to the dual feed type one (3) probe feed implementation is more complex in fabrication process for a CP antenna. A dual feed equilateral triangular microstrip element antenna has superior properties and would be a good element for the CP-SAR implementation.

The configuration of the radiating element together with the microstrip line feed and ground plane is shown in Figure 3(a), where important parameters are labeled. Side view is depicted in Figure 3(b). The equilateral triangular radiator will generate a left-handed circular polarization (LHCP) by employing the dual feed method as shown in Figure 3(a). In order to generate a 90° phase delay on one of the two modes, the line feed on the left side is approximately $\lambda/4$ longer than the other.

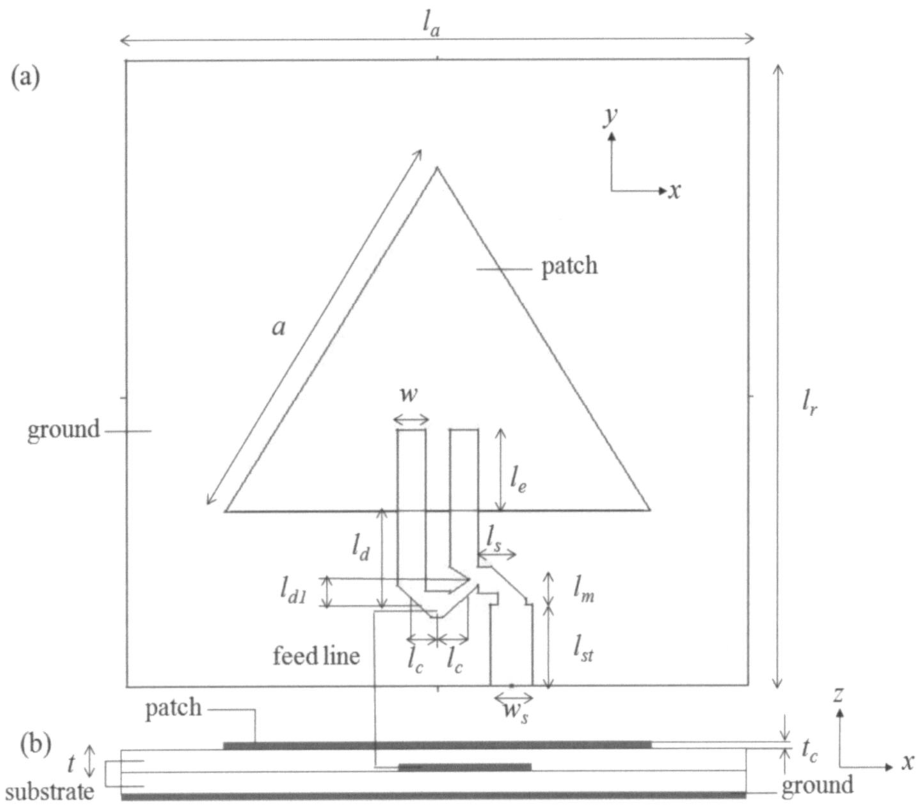

Fig. 3. Configuration of equilateral triangular patch antenna with proximity coupled feed; (a) top view and (b) side view.

Simulations with a finite-ground-plane model have been undertaken to optimize the size parameters using a full-wave analysis tool (IE3D Zeland software) based on the method of moment (MoM) algorithm. The dimensions of the radiator, microstrip feed line and the ground plane for the equilateral triangular patch are tabulated in Table 3 in units of mm. The geometry model is implemented on two substrates, each with thickness t = 1.6 mm, with the conductor thickness $t_c \approx$ 0.035 mm, relative permittivity ε_r = 2.17 and loss tan δ (dissipation factor) 0.0005.

Parameter	Dimension
a	102.75
w	6.8
l_d	21.5
l_e	27
l_{d1}	6.9
l_c	9.2
l_s	10.1
l_m	3.9
l_{st}	21.5
w_s	10.2
l_a	146.1
l_r	163.1

Table 3. Geometry parameters (in units of mm) of the equilateral triangular patch antenna.

3.1.1 Parameter study

Parameter study on the parameter l_c (distance between the two feeds) was conducted since during the optimization process of the microstrip line feed configuration, it was observed that this parameter exerts a strong influence on both the CP frequency and the AR of the antenna. Figure 4 shows the result of the simulation, in which the frequency dependence of the AR is plotted for various values of the parameter l_c while keeping the other parameters unchanged. Thus, the distance must be exact in order to achieve the orthogonality of the two modes fed from the current source to the triangular patch.

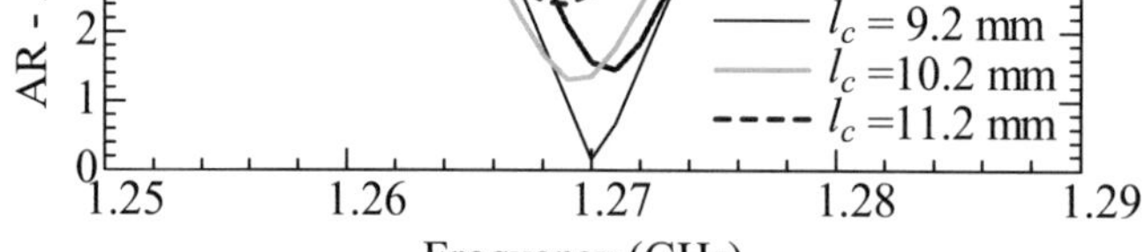

Fig. 4. Simulation results showing the frequency dependence of the axial ratio (AR) of the equilateral triangular microstrip antenna for various values of l_c.

3.1.2 Simulation and measurement results and discussions

The antenna efficiency from the simulation is 86.59%. Simulated input impedance bandwidth is 26.0 MHz whereas the measured one is 21.5 MHz (Figure 5).

Figure 6 shows the gain simulated and measured at $\theta = 0^{\circ}$. While the gain of the antenna has been simulated to be 7.04 dBic at 1.27 GHz, experimental results shows a smaller value by about 0.6 dB. This difference may be ascribed to the fabrication imperfections (such as inaccuracy in the milling and etching processes and connector soldering) and the substrate loss.

Figure 6 also shows the simulated and measured results of AR. From this figure it can be seen that the 3-dB AR bandwidth from the simulation is 7.2 MHz and from the measurement is 7.4 MHz (ranging from 1.2653 GHz to 1.2727 GHz). Even though the measurement result of 3-dB AR bandwidth is slightly better than that of the simulation result, this bandwidth is still narrower than the target specification (10 MHz). To improve this situation, the next work will consider the technique to extend the 3-dB AR bandwidth.

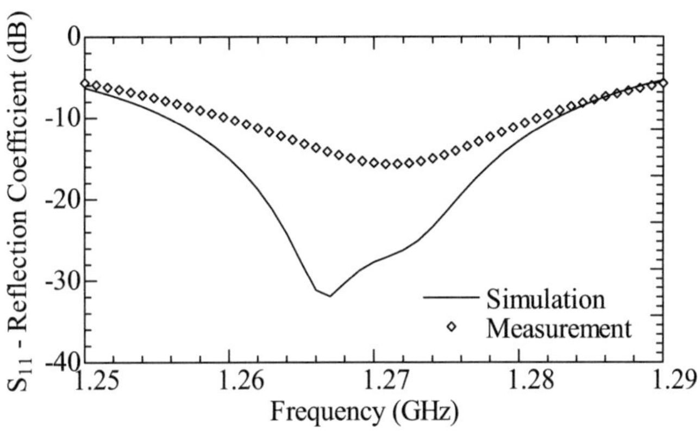

Fig. 5. Simulated and measured reflection coefficient

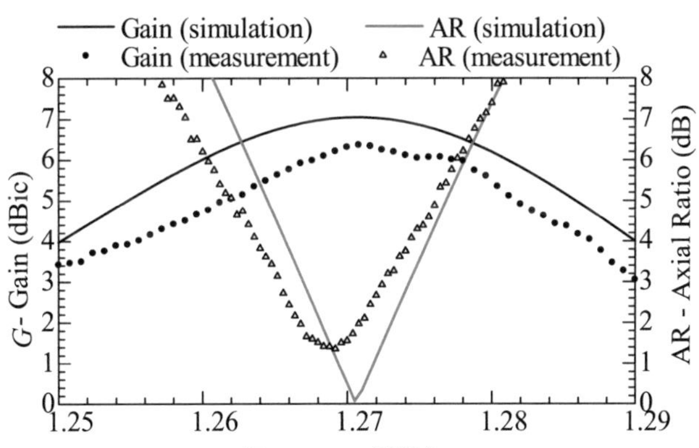

Fig. 6. Simulated and measured gain and AR at $\theta = 0^{\circ}$.

Figures 7 -10 show the radiation pattern in terms of gain and AR at an azimuth angle Az = 0° (and 180°, x-z plane) and 90° (and 270°, y-z) plane and at the frequency of $f = 1.27$ GHz. In Figure 7, a difference of around 0.7 dB is seen between the simulated model and the measured antenna on the gain radiation pattern. Figure 8 shows that the most beam radiated in the direction of Az = 0° (x>0 in Figure 8). Figure 9 shows the 90° azimuth measurement of gain pattern.

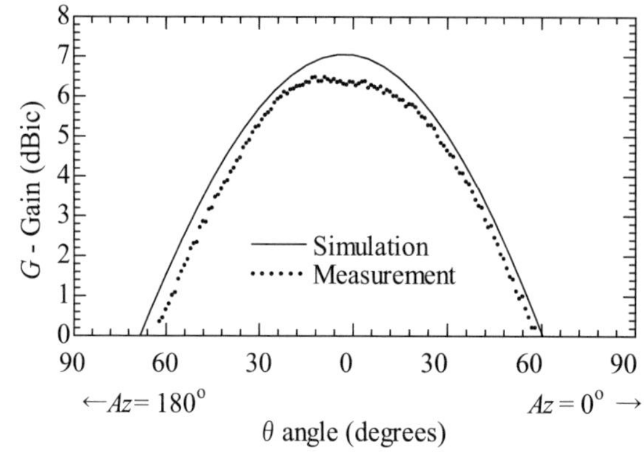

Fig. 7. Gain vs. theta angle (radiation pattern) in the theta plane (Az = 0° and 180°) (x - z plane) at f = 1.27 GHz.

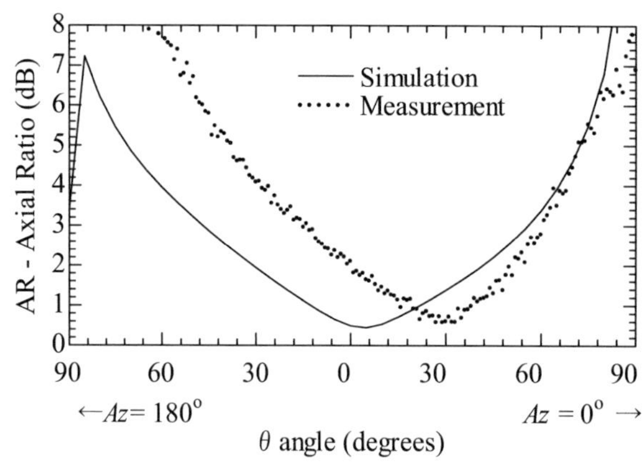

Fig. 8. Axial Ratio vs. theta angle (radiation pattern) in the theta plane (Az = 0° and 180°) (x - z plane) at f = 1.27 GHz.

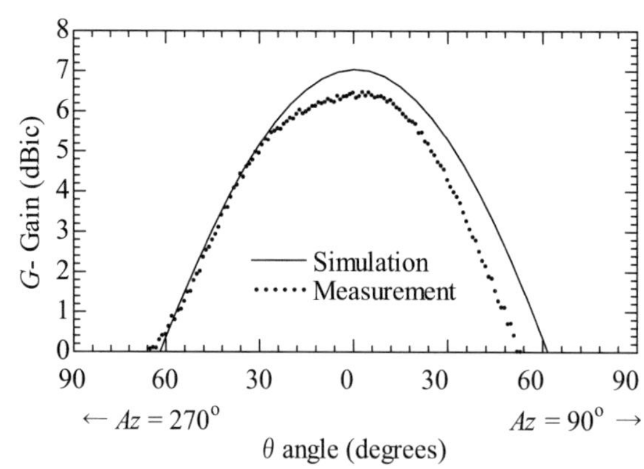

Fig. 9. Gain vs. theta angle (radiation pattern) in the theta plane (Az = 90° and 270°) (y - z plane) at f = 1.27 GHz.

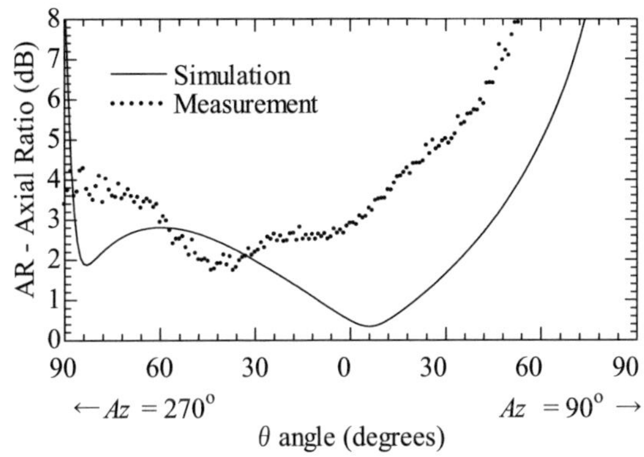

Fig. 10. Axial Ratio vs. theta angle (radiation pattern) in the theta plane (Az = 90° and 270°) (y - z plane) at f = 1.27 GHz.

Most of the beam that has good CP characteristics is radiated in the direction of Az = 270° (y>0 in Figure 10). There are some differences between the simulated and measured pattern of the antenna. This may be due to the slightly altered fabricated model and different measurement environment compared to the simulated model in the IE3D simulation environment. Especially the AR vs. angle results which show a larger difference compared with the gain one. The high sensitivity of AR behaviour to the measurement condition, the infinite lateral substrate extension in IE3D, and a possible of additional radiation from the edges of the substrate in the fabricated model may contribute to the differences.

3.2 Elliptical microstrip antenna

The requirement of a patch element for generating a circularly polarized radiation is that the patch must have orthogonal (in-phase and quadrature) fields of equal amplitude. Slightly elliptical patch can have a circular polarization radiation with a single feeding (Bailey et al. 1985, Shen 1981). In addition, an elliptical antenna element generally has an elliptically polarized radiation, but it has left-handed (or right-handed) circularly polarized (LHCP/RHCP) radiation when the feed point of the antenna element is located on the radial line rotated 45° counterclockwise (or clockwise) to the semi major-axis of the ellipse (Bailey et al., 1985). Also according to Bailey et al. (1985), the best circular polarization radiation may be achieved by limiting the eccentricity of the ellipse to a range of 10 to 20%.

The configuration of the radiating element together with the microstrip line feed and ground plane is shown in Figure 11(a), where important parameters are labeled. Although prior elliptical patch were based on the probe-feed method (Bailey et al. 1985, Shen 1981), in this chapter we adopt the proximity-coupled feeding method (Pozar et al., 1987). This approach has the advantage of easier adjustment in the process of design and fabrication processes, especially in producing good circular polarization with good impedance matching. Also, bandwidth enhancement and reduced parasitic radiation of the feeding network is achieved compared with other direct feeding methods.

The dimensions of the radiator, and the ground plane for the elliptical patch are a = 45.9 mm, b = 44.5 and $l_a \times l_r$ = 120 × 126.65 mm. Side view is depicted in Figure 11(b). The geometry model is implemented on two substrates, each with thickness t = 1.6 mm, conductor thickness $t_c \approx 0.035$ mm, relative permittivity ε_r = 2.17 and dissipation factor

0.0005. The parameters of the microstrip line feed are w = 4.8 mm, d = 10.8 mm, l = 48.7 mm, l_s = 7 mm, and w_s = 7 mm. With the width of the microstrip line of 4.8 mm, the characteristic impedance is approximately 50.9 Ω.
The elliptical radiator will generate LHCP by rotating the patch by -45° around the center of the ellipse. Simulations with a finite-ground-plane model have been undertaken to optimize the size parameters using a full-wave analysis tool (IE3D Zeland software).

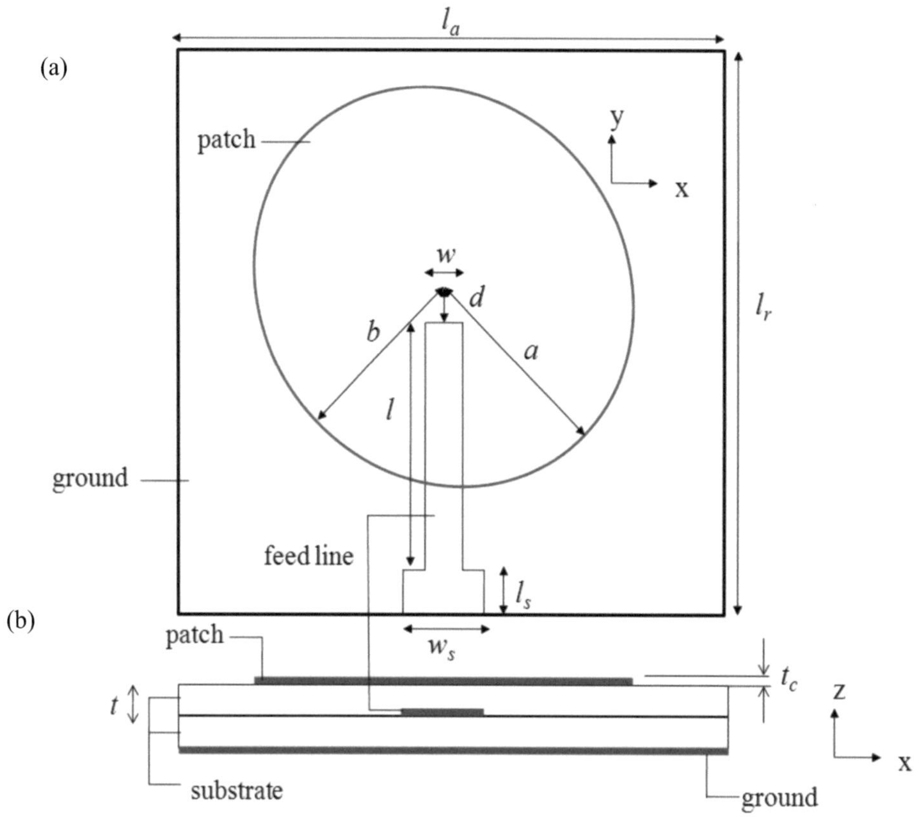

Fig. 11. Configuration of elliptical patch antenna with proximity coupled feed; (a) top view and (b) side view.

3.2.1 Parameter study

During the optimization process of the elliptical patch configuration, it was observed that the parameters a (semi major axis) and b (semi minor axis) have a strong influence on both the CP frequency and the AR of the antenna. Figure 12 shows the result of the simulation, in which the frequency dependence of the axial ratio is plotted for various values of the parameters a and b while keeping the other parameters unchanged (with the optimized values a = 45.9mm and b = 44.5 mm). The best circular polarization radiation is achieved for the eccentricity ranging from 19 to 28%. The difference between the present result (19 to 28%) and that in a previous work (10 to 20 %) (Shen, 1981) is presumably due to the difference in the feeding method applied to the elliptical patch.

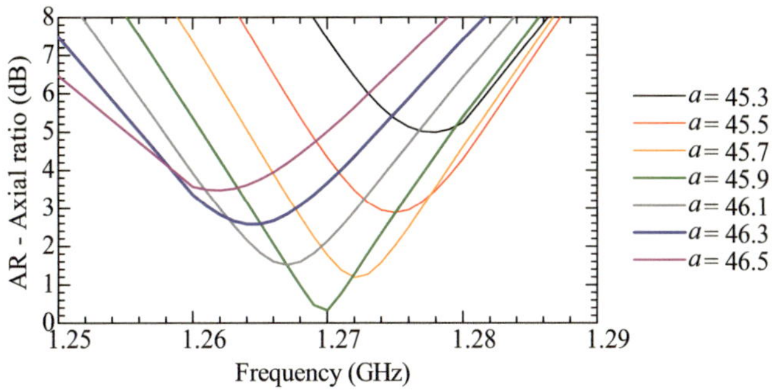

(a)

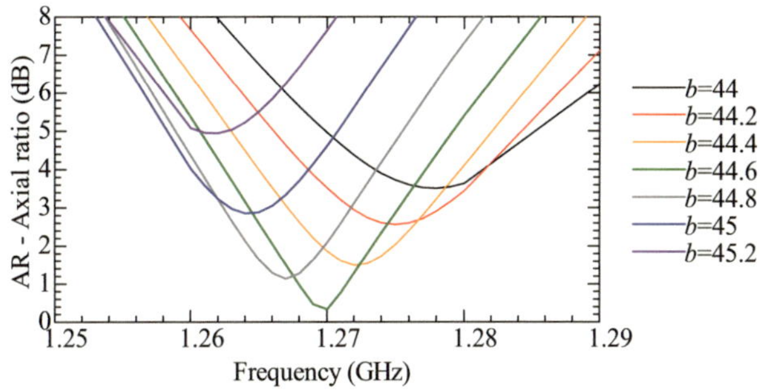

(b)

Fig. 12. Simulation results showing the frequency dependence of the axial ratio (AR) of the elliptical microstrip antenna for various values of (a) the semi-major axis *a* and (b) semi-minor axis *b*.

3.2.2 Input and radiation characteristic

Figure 13 shows the frequency dependence of the S_{11}-parameter (reflection coefficient). Although the measured parameter takes a minimum value at 1.256 GHz, somewhat smaller than the operation frequency of f = 1.27 GHz, an impedance bandwidth (S_{11}<-10 dB) of more than 20 MHz is attained around the operation frequency, in spite of the difference between the measured and simulated curves.

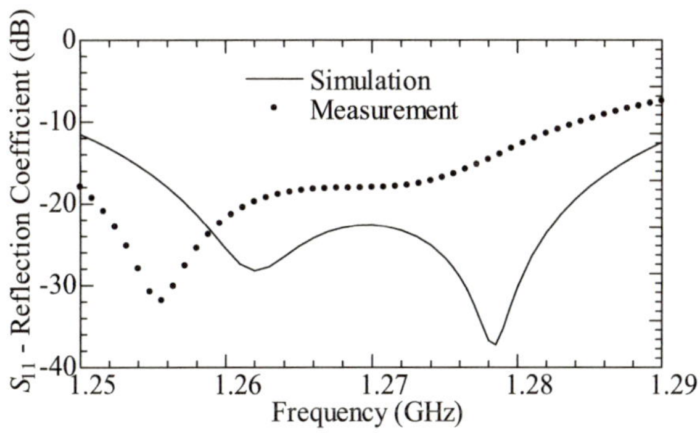

Fig. 13. Reflection coefficient vs. frequency.

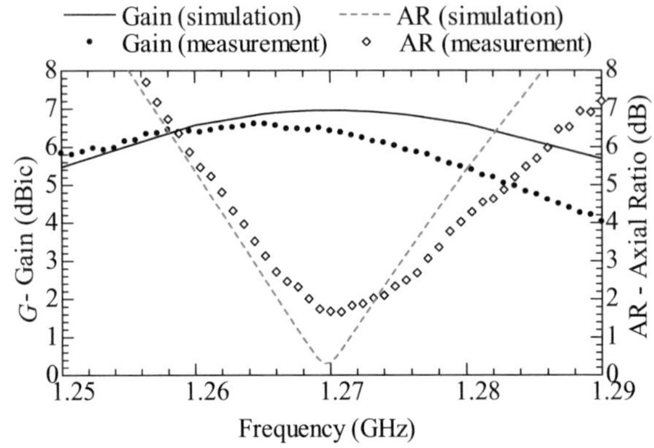

Fig. 14. Gain and AR vs. frequency at θ angle = 0°.

In Figure 14, the antenna gain and AR at $\theta = 0°$ are plotted against the frequency. From this figure, it can be seen that whereas the gain of the antenna is simulated to be 6.96 dBic at 1.27 GHz, the experimental result shows a smaller value by about 0.5 dB. Such a difference between the simulation and experimental results, as also seen in other curves in Figures 15 - 16, can probably be ascribed to the fabrication imperfections (such as inaccuracy in the milling and etching processes, connector soldering and holes with plastic screws), the substrate loss and cable loss and also the infinite lateral extension of the substrate in the IE3D simulation while the fabricated one is a finite substrate with the same size as the ground plane. As for the frequency dependence curves of AR, a crucial parameter for circularly polarized antenna operation, Figure 14 shows that the 3-dB AR bandwidth of the simulation is 10.8 MHz and from observation it is 10.4 MHz, ranging from 1.2658 to 1.2762 GHz. The AR bandwidth of the simulated model has satisfied the target specification (10 MHz) of CP-SAR, but the range is slightly shifted from the ideal range of 1.265 to 1.275 GHz. Again, this shift is possibly caused by fabrication imperfections.

Figures 15 - 16 show the radiation pattern in terms of the gain and AR at an azimuth angle Az = 0° (and 180°, x-z plane) and 90° (and 270°, y-z) plane and at the frequency of f = 1.27 GHz. In Figure 15, a difference of around 0.5 dB is seen between the simulated and

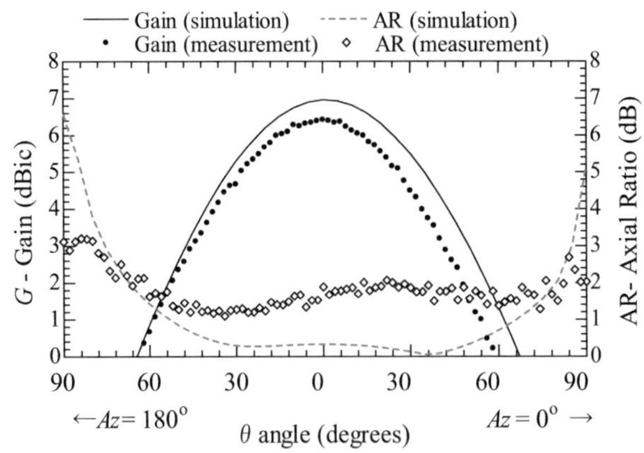

Fig. 15. Gain and AR vs. theta angle (radiation pattern) in the theta plane (Az = 0° and 180°) (x - z plane) at f = 1.27 GHz.

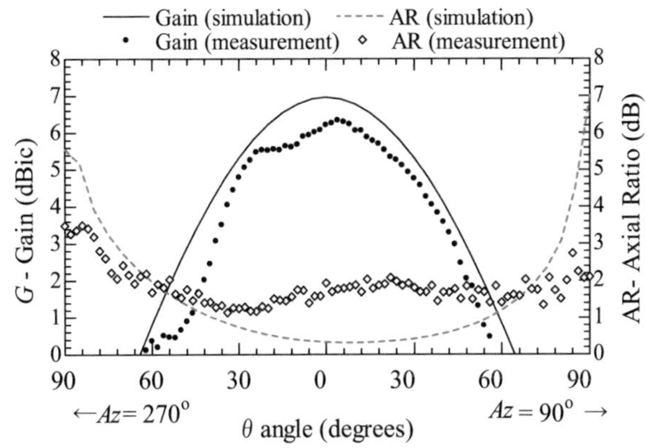

Fig. 16. Gain and AR vs. theta angle (radiation pattern) in the theta plane (Az = 90° and 270°) (y - z plane) at f = 1.27 GHz.

measured results. From Figures 15 and 16, it is apparent that the measured AR performance is better in terms of AR beam width than the simulated one. This may be influenced by the imperfection effects mention above. It can also be noticed that there is a slight shift of gain pattern, and particularly the 90° azimuth gain pattern has a little fluctuation. This is possibly due to the measurement system, i.e. the slight variations in antenna alignment during rotation.

3.3 Elliptical annular ring microstrip antenna

The patch size of the annular ring microstrip antenna is smaller than the other shapes when operated at the TM11 mode due to the longer excited patch surface current path of the TM11 mode (Chen et al., 1999), whereas in this mode the patch will behave as a resonator. One reported work on the CP design of the annular ring microstrip antenna is by inserting a pair of slits at the inner boundary of the annular patch, and for solving the input-impedance problem, a quarter-wavelength impedance transformer is utilized (Chen et al., 1999).

In this work on the elliptical annular ring microstrip antenna, circular polarization is produced by locating the feed point of the antenna element on the radial line rotated 45° counterclockwise (or clockwise) to the semi major-axis of the ellipse for a left-handed (or right-handed) circularly polarized (LHCP/RHCP) radiation. As applied before to the elliptical microstrip antenna, we adopt the proximity-coupled feeding method.

The configuration of the radiating element together with the microstrip line feed and ground plane is shown in Figure 17(a), where important parameters are labeled. The dimensions of the radiator, and the ground plane for the elliptical patch are a = 43.7 mm, b = 42.5, a_1 = 9.8 mm, b_1 = 9.3 and $l_a \times l_r$ = 120 × 120.4 mm. Side view is depicted in Figure 17(b). The geometry model is implemented on two substrates, each with thickness t = 1.6 mm, conductor thickness $t_c \approx 0.035$ mm, relative permittivity ε_r = 2.17 and dissipation factor 0.0005. The parameters of the microstrip line feed are w = 3 mm, d = 13.4 mm, l = 41 mm, l_s = 5 mm, and w_s = 5 mm. With the width of the microstrip line of 3 mm, the characteristic impedance is approximately 68.5 Ω.

The elliptical radiator will generate LHCP by rotating the patch by -45° around the center of the elliptical annular ring. Simulations with a finite-ground-plane model have been undertaken to optimize the size parameters using a full-wave analysis tool (IE3D Zeland software) based on the method of moment (MoM) algorithm.

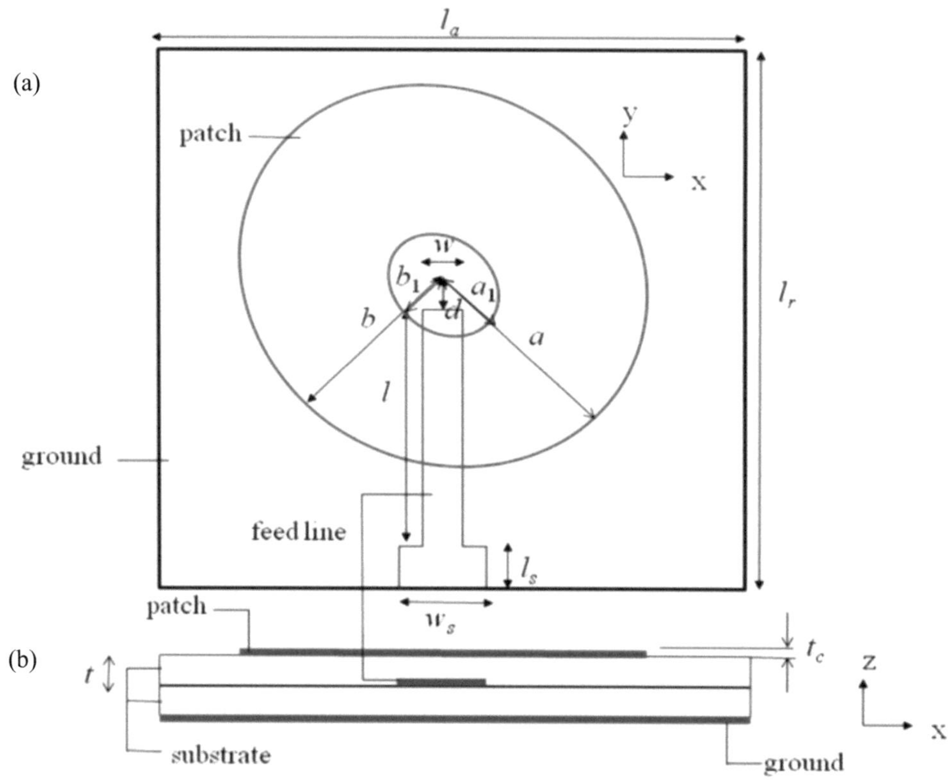

Fig. 17. Configuration of elliptical annular ring microstrip antenna with proximity coupled feed; (a) top view and (b) side view.

Figure 18 shows the frequency dependence of the S_{11}-parameter (reflection coefficient). Although the measured parameter takes a minimum value at 1.276 GHz, somewhat higher than the operation frequency of f = 1.27 GHz, an impedance bandwidth (S_{11}<-10 dB) of more than 25 MHz is attained around the operation frequency, in spite of the difference between the measured and simulated curves.

In Figure 19, the antenna gain and AR at $\theta = 0^{\circ}$ are plotted against the frequency. From this figure, it can be seen that whereas the gain of the antenna is simulated to be 6.87 dBic at 1.27 GHz, the experimental result shows a smaller value by about 0.4 dB. Such a difference between the simulation and experimental results, as also seen in other curves in Figures 20 - 21, can probably be ascribed to the fabrication imperfections (such as inaccuracy in the milling and etching processes, connector soldering and holes with plastic screws), the substrate loss and cable loss. As for the frequency dependence curves of AR, a crucial parameter for circularly polarized antenna operation, Figure 19 shows that the 3-dB AR bandwidth of the simulation is 8 MHz and from observation it is 8.7 MHz. Even though the measurement result of 3-dB AR bandwidth is slightly better than that of the simulation result, its bandwidth is still narrower than the target specification (10 MHz). To improve this situation, the next part will consider the technique to extend the 3-dB AR bandwidth.

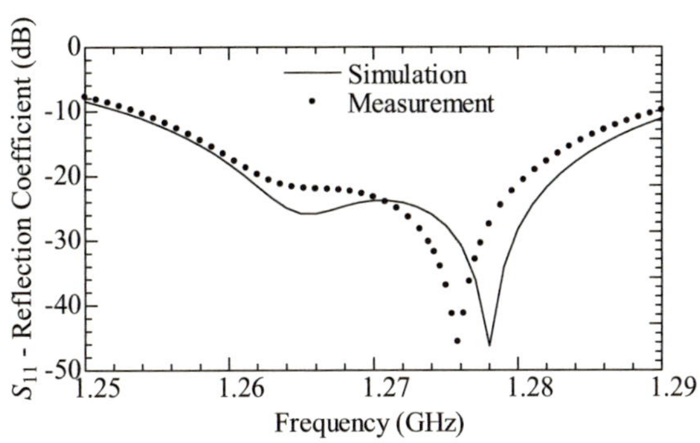

Fig. 18. Reflection coefficient vs. frequency.

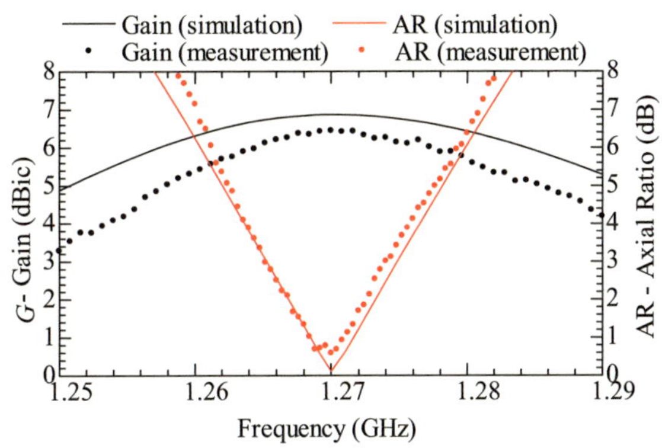

Fig. 19. Gain and AR vs. frequency at θ angle = 0°.

Figures 20 - 21 show the radiation pattern in terms of the gain and AR at an azimuth angle Az = 0° (and 180°, *x-z* plane) and 90° (and 270°, *y-z*) plane and at the frequency of f = 1.27 GHz. In Figure 20, a difference of around 0.4 dB is seen between the simulated and measured results. Figure 21 shows that the beam width simulated for 3-dB AR is 155° and

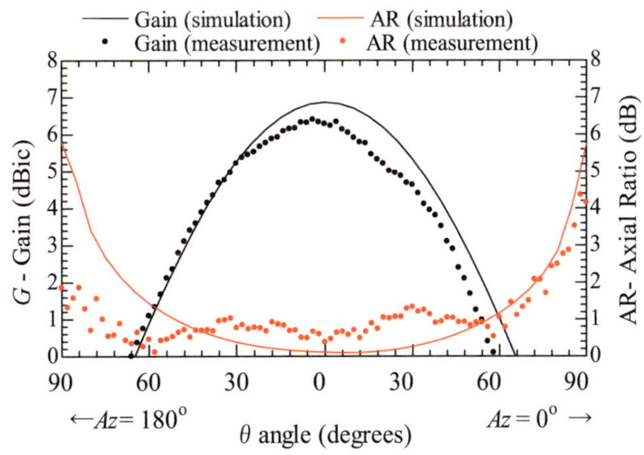

Fig. 20. Gain and AR vs. theta angle (radiation pattern) in the theta plane (Az = 0° and 180°) (x - z plane) at f = 1.27 GHz.

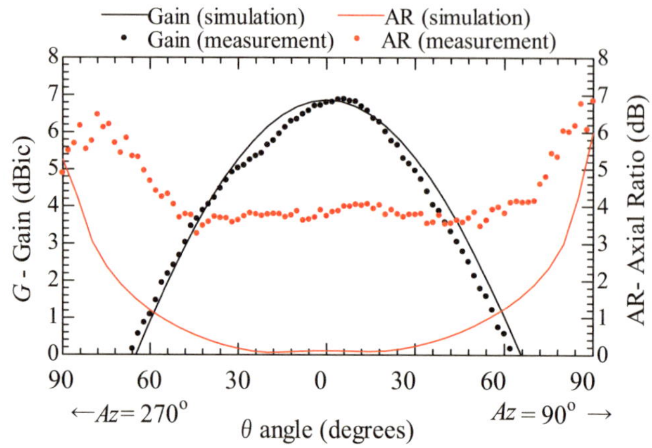

Fig. 21. Gain and AR vs. theta angle (radiation pattern) in the theta plane (Az = 90° and 270°) (y - z plane) at f = 1.27 GHz.

on the other hand, the measurement of AR shows the value of more than 3 dB for all the angle range. This may be influenced by the imperfection effects mention above, most contribution presumably from fabrication imperfection. It can also be noticed that there is a slight shift and fluctuation of gain pattern. This is possibly due to the measurement system, i.e. the slight variations in antenna alignment during rotation.

3.4 Elliptical annular ring microstrip antennas having a sine wave periphery

In this work, we propose a single-feed, circularly polarized, annular-ring microstrip antenna operated in the L-band (1.285 GHz). The antenna periphery is modulated with a sinusoidal wave, giving a flowery appearance. The antenna characteristics are investigated by both simulation and experiment. An annular ring microstrip antenna (ARMA) with a periodically added sector stubs at the outer and inner edge has been reported (Hong et al., 2006) to have a relatively wider bandwidth and smaller size than ordinary ARMAs. Fractal geometry applied to the periphery of microstrip antennas has also led to the reduction of the antenna size and broadening of the bandwidth (Gianvittorio et al. 2002, Krishna et al. 2009, Chen et al. 2009). Therefore, it is expected that by applying sine wave pattern to the periphery of the EARMA, broader bandwidth and reduced antenna size will be obtained. In this work the characteristics of such antennas are investigated by both simulation and experiment. It turns out that favourable characteristics of input parameter and radiation pattern in broadside direction can be achieved, ensuring its application to circularly polarized L-band applications such as global positioning system (GPS), synthetic aperture radar (SAR), television broadcasting, etc.

Circular polarization is produced from the elliptical annular-ring microstrip antenna (EARMA) by locating the feed point of the antenna element on the radial line that is rotated 45° counterclockwise (clockwise) to the semi major-axis of the ellipse for a left-handed (right-handed) circularly polarized (LHCP/RHCP) radiation. We adopt the proximity-coupled feeding method that has the advantage of easier adjustment in the of design and fabrication processes, especially in producing good circular polarization with good impedance matching.

The configuration of the radiating element, together with the microstrip line feed and ground plane, is shown in Figure 22(a), where important parameters are labeled. The dimensions of the radiator, and the ground plane for the elliptical patch are a = 42.4 mm, b = 41.1, a_1 = 9.85 mm, b_1 = 8.9 and $l_a \times l_r$ = 130 × 129.4 mm. The periphery (outer ring) is an ellipse modulated with a sine wave, and in the Cartesian coordinate system, it can be expressed as

$$\frac{x^2}{\left[a+m\sin 8\left(\tan^{-1}\frac{y}{x}\right)\right]^2}+\frac{y^2}{\left[b+m\sin 8\left(\tan^{-1}\frac{y}{x}\right)\right]^2}=1, \tag{2}$$

where m is the amplitude of the sine wave, and a and b are the semi-major and -minor axis of the outer elliptical ring, respectively. The inner ellipse is given by the ordinary ellipse equation:

$$\frac{x^2}{a_1^{\,2}}+\frac{y^2}{b_1^{\,2}}=1, \tag{3}$$

where a_1 and b_1 are the semi-major and -minor axis of the inner elliptical ring, respectively. In this work m parameter is chosen as 3, while the working frequency concerning the axial ratio (AR) is 1.285 GHz. Side view is depicted in Figure 22(b). The geometry model is implemented on two substrates, each with thickness t = 1.6 mm, conductor thickness $t_c \approx 0.035$ mm, relative permittivity ε_r = 2.17 and dissipation factor 0.0005. The parameters of the microstrip line feed are d = 15.3 mm, w = 3 mm, l = 43.1 mm, l_s = 6 mm, and w_s = 7 mm. With the width of the microstrip line of 3 mm, the characteristic impedance is approximately 68.3 Ω.

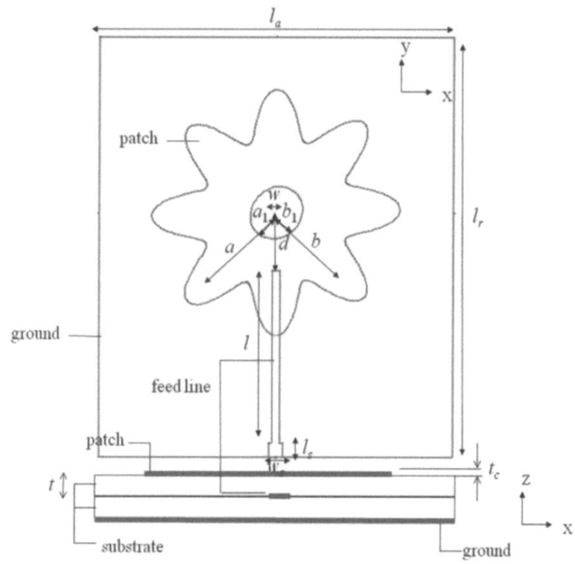

Fig. 22. Configuration of the elliptical annular ring microstrip antenna (EARMA) with proximity coupled feed; (a) top view and (b) side view.

The EARMA radiator will generate LHCP by rotating the patch by -45° around the center of the patch. Simulations with a finite-ground-plane model have been undertaken to optimize the size parameters using a full-wave analysis tool (IE3D Zeland software) based on the method of moment (MoM) algorithm.

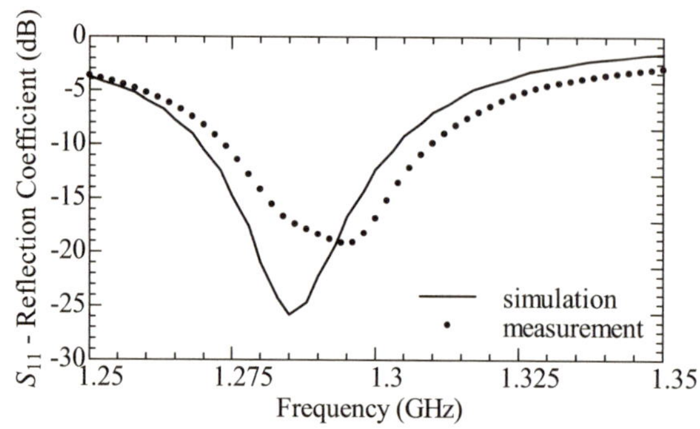

Fig. 23. Frequency dependence of simulated and measured reflection coefficient.

Figure 23 shows the frequency dependence of the S_{11}-parameter (reflection coefficient) obtained from the simulation and measurement. Figure 23 also shows that experimentally, the reflection coefficient takes a minimum value at f_1 = 1.295 GHz. This frequency is somewhat higher than the operation frequency of f_0= 1.285 GHz, though an impedance bandwidth (S_{11}<-10 dB) of more than 30 MHz is attained around at f_1. Such a difference between the simulation and experimental results (seen also in other curves in Figures 25 and 26 shown below) can probably be ascribed to the fabrication imperfections (such as inaccuracy in the milling and etching processes, connector soldering and holes with plastic screws) and/or the substrate loss and cable loss.

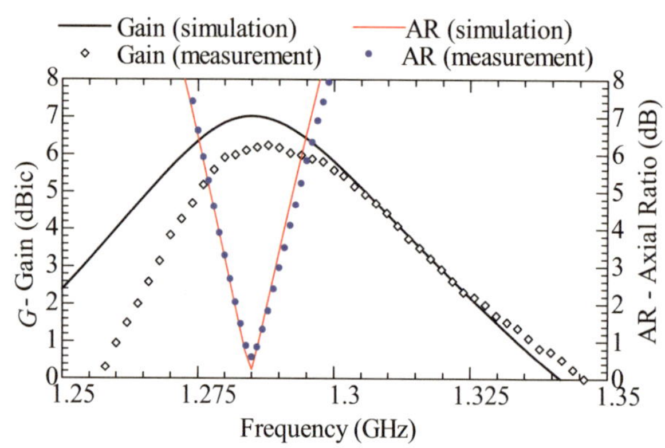

Fig. 24. Frequency dependence of simulated and measured gain and AR at θ angle = 0°.

In Figure 24, the antenna gain and AR at θ angle = 0° are plotted against the frequency. The gain of the antenna is simulated to be around 7.0 dBic at 1.285 GHz. As for the frequency dependence curves of AR, a crucial parameter for circularly polarized antenna operation, the values of 3-dB AR bandwidth simulated is 9.01 MHz (0.70 %) respectively. From the measurement, on the other hand, the gain is obtained to be 6.1 dBic, to some extent smaller than the simulated value of 7.0 dBic at 1.285 GHz. The measured 3-dB AR bandwidth is 9.5 MHz, slightly wider than the simulated result. This model exhibit narrower bandwidth than the ordinary EARMA model simulated at the same working frequency (40.2 MHz or 3.1% of -10 dB S_{11} bandwidth and 9.8 MHz or 0.76% of 3-dB AR bandwidth).

Spatial distributions (radiation pattern) of the gain and AR at an azimuth angle Az = 0° (and 180°, *x-z* plane) and Az = 90° (and 270°, *y-z* plane) are shown in Figure 25 and Figure 26, respectively, both at the frequency of $f = 1.285$ GHz. At Az = 90° the measured AR characteristic shows higher value than 3 dB at the whole 180° range. This particular AR measurement result is probably contributed from the difference in geometry antenna size (between simulated and fabricated model for adjustment) due to the substrate variation. The very sensitive AR characteristic may be improved in the future by further adjustment in fabrication process.

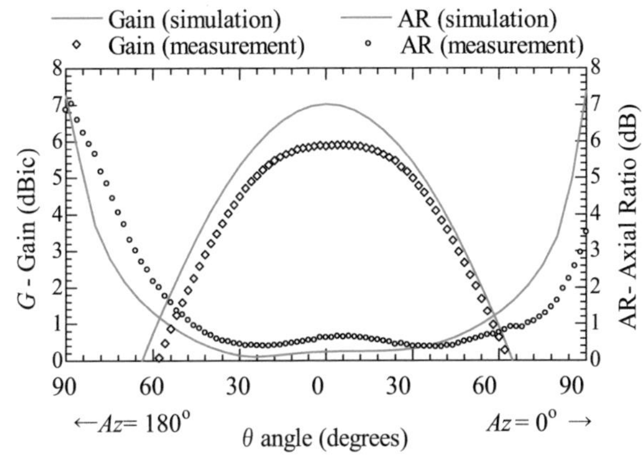

Fig. 25. Simulated and measured gain and AR vs. theta angle (radiation pattern) in the theta plane (Az = 0° and 180°) ($x - z$ plane) at $f = 1.285$ GHz.

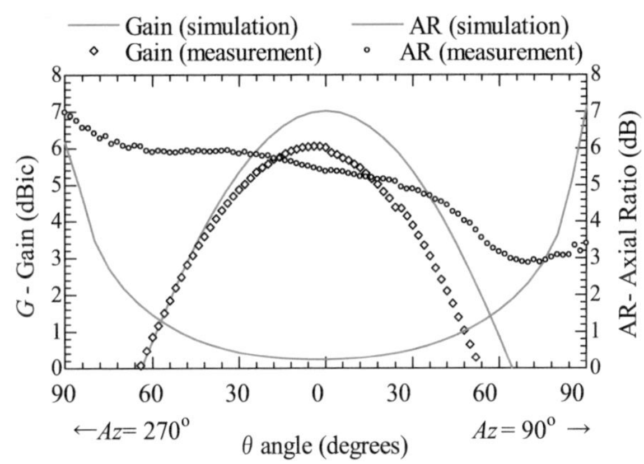

Fig. 26. Simulated and measured gain and AR vs. theta angle (radiation pattern) in the theta plane (Az = 90° and 270°) ($y - z$ plane) at $f = 1.285$ GHz.

4. Elliptical microstrip array antenna

The elliptical microstrip antenna described in the part 3.2 is arranged to form a 5 × 11 elliptical microstrip element array and this array is simulated using Zeland IE3D on an infinite ground, with each elliptical element fed individually. The layer setting is the same as the single element elliptical microstrip antenna, this array has two layers, one is for the feed and another one is for the radiator patch. The thickness of each layer is 1.6 mm, relative permittivity ε_r = 2.17 and dissipation factor 0.0005. Elevation spacing between elements is 125 mm and the azimuth spacing is 140 mm.

The gain obtained in the range frequency 1.25 - 1.29 GHz is 16.0 to 17.6 dBic. The 3-dB AR bandwidth from this array configuration is 11.5 MHz
Figure 27 shows the antenna gain patterns (Az = 0° and 90°). It is the resulting co- and cross-polarization composition patterns from RHCP elements. From this figure, the gain at $\theta = 0°$ is 17.6 dBic, somewhat lower than the target gain of 20 dBic (Table 1). However, mostly satisfactory beam patterns for CP-SAR operation are seen in both the azimuth and elevation planes. The nearest side lobe is suppressed by about 13 dBic. From Figure 28 it can be seen that in the main beam (around $\theta = 0°$) the 3-dB AR beam width in the azimuth plane is 16° and in the elevation plane is 38°.

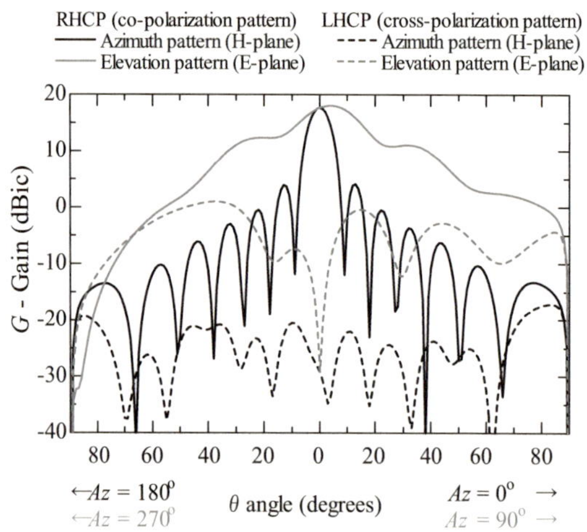

Fig. 27. Simulated radiation pattern of a 5 × 11 elliptical microstrip array at $f = 1.27$ GHz.

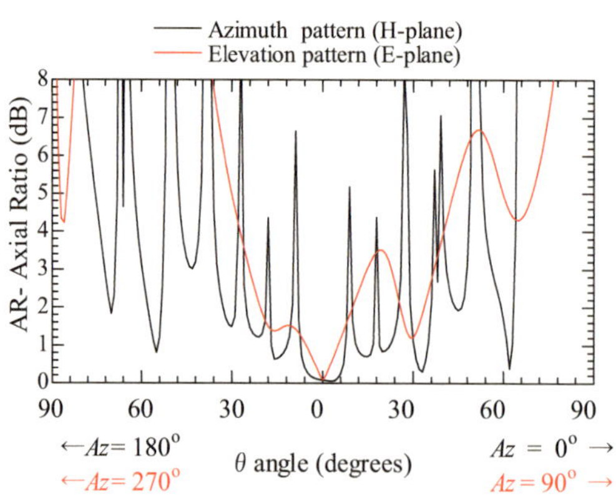

Fig. 28. Simulated AR pattern of a 5 × 11 elliptical microstrip array at $f = 1.27$ GHz.

5. Conclusions

In this chapter, microstrip antennas for circularly polarized synthetic aperture (CP-SAR) radar have been investigated. The CP-SAR antenna will be mounted on an Unmanned Aerial Vehicle (UAV) and operated at L-band frequency. The proposed antenna has been analyzed numerically and developed experimentally.

The MoM analysis results of the antenna models have been presented and discussed. Best effort has been given to achieve an adequate frequency tuning range with the best performance of radiation characteristics of the antenna. Study on geometric parameters is also carried out in order to specify the one that is most influential to the axial ratio characteristic. The measurement results of the equilateral triangular microstrip antenna, elliptical microstrip antenna, elliptical annular ring microstrip antenna and the elliptical annular ring microstrip antenna having a sine wave periphery also have been presented and discussed. From these results the elliptical patch has mostly satisfied the target specification for a CP-SAR antenna element to be used onboard an UAV. The performance of an elliptical microstrip array has also been examined by simulation with good results. As for the equilateral triangular microstrip antenna and the elliptical annular ring microstrip antenna, although the AR bandwidth of these antenna are slightly smaller than the requirement for an airborne CP-SAR system, the present work has indicated that the goals can be met through a precise adjustment in the design and fabrication process in the near future. The elliptical annular ring microstrip antenna having a sine wave periphery is recommended for L-band applications including radar.
Future research will be aimed at considering the dual polarized CP-antenna array antenna design with low cross-polarization and low mutual coupling. Low cross-polarization can be achieved by accomplish a high isolation between orthogonal input ports, lower than -25 dB. Another solution is to avoid uncontrolled radiation, e.g., from the feed network.

6. Acknowledgement

The first author would like to thank to Professor Hiroaki Kuze for his excellent guidance and support.
The authors would like to thank also to Victor Wissan, Basari, Muhammad Fauzan, Prilando Rizki Akbar, Luhur Bayuaji, Bambang Setiadi, and Ilham Alimuddin for their assistance in the antenna fabrication and measurement.
The authors also would like to thank to the Japan Society for the Promotion of Science (JSPS) for Grant-in-Aid for Scientific Research - Young Scientist (A) (No. 19686025); National Institute of Information and Communication Technology (NICT) for International Research Collaboration Research Grant 2008, Chiba University COE Start-up Program, European Space Agency (ESA) Earth Observation Category 1 (No. 6613), the research grant for Mission Research on Sustainable Humanosphere from Research Institute for Sustainable Humanosphere (RISH), Kyoto University and other research grants that have supported our research.

7. References

Bailey, M.C. & Deshpande, M.D.(1985).Analysis of Elliptical and Circular Microstrip Antennas Using Moment Method. *IEEE Transaction on Antennas and Propagation*, 33, 954-959.

Bhattacharyya, A.K. (2008).Comparison Between Arrays of Rotating Linearly Polarized Elements and Circularly Polarized Elements. *IEEE Transaction on Antennas and Propagation*, 56, 2949 - 2954.

Chan, Y. K.; Chung, B. K., & Chuah, H. T. (2004). Transmitter and Receiver Design of an Experimental Airborne Synthetic Aperture Radar Sensor. *Progress In Electromagnetics Research*, 49, 203–218.

Chen, H.-M, & Wong, K.-L. (1999). On the circular polarization operation of annular-ring microstrip antennas. *IEEE Transactions on Antennas and Propagation*, 47, 8, 1289-1292.

Chen, W.-L.; Wang, G. –M., & Zhang, C.-X. (2009).Bandwidth Enhancement of a Microstrip-Line-Fed Printed Wide-Slot Antenna with a Fractal-Shaped Slot. *IEEE Transactions on Antenna and Propagation*, 57, 7, 2176 - 2179.

Garg, R.; Bhartia, P.; Bahl, I. & Ittipiboon, A. (2001). *Microstrip Antenna Design Handbook*, Artech House Inc.

Gianvittorio, J. & Rahmat-Samii, Y. (2002).Fractal antennas: A novel antenna miniaturization technique, and applications. *IEEE Antennas and Propagation Magazine*, 44, 20–36.

Horn, R. (1996).The DLR airborne SAR project E-SAR.*International Geoscience and Remote Sensing Symposium*, 1996. IGARSS '96. 'Remote Sensing for a Sustainable Future.', Volume 3, 1624 - 1628.

Karimabadi, S. S.; Mohsenzadeh, Y.; Attari, A. R. & Moghadasi, S. M. (2008). Bandwidth Enhancement of Single-feed Circularly Polarized Equilateral Triangular Microstrip Antenna. *Progress In Electromagnetics Research Symposium*, 147-150, Hangzou, China, March 24-28.

Krishna, D.D.; Gopikrishna, M.; Aanandan, C.K.; Mohanan, P. & Vasudevan, K. (2009). Compact wideband Koch fractal printed slot antenna. *IET Microwaves, Antennas and Propagation*, 3, 5, 782 - 789.

Levine, E.; Malamud, G.; Shtrikman, S. & Treves, D. (1989). A Study of Microstrip Array Antennas with the Feed Network. *IEEE Transactions on Antennas and Propagation*, 37, 4, 426-434.

Madsen, S.N.; Christensen, E.L.; Skou, N.& Dall, J. (1991).The Danish SAR system: design and initial tests. *IEEE Transactions on Geoscience and Remote Sensing*, 29, 3,417 - 426.

Nemoto Y.; Nishino H.; Ono M.; Mizutamari H.; Nishikawa K. & Tanaka K. (1991). Japanese Earth Resources Satellite-1 synthetic aperture radar.*Proceedings of the IEEE* , 79, 800-809.

Pozar, D.M. & Kaufman, B. (1987). Increasing the bandwidth of a microstrip antenna by proximity coupling. *IEEE Electronics Letters*, 23, 8, 368 - 369.

Raney, R.K.; Luscombe, A.P.; Langham, E.J. & Ahmed S. (1991). RADARSAT. *Proceedings of the IEEE* , 79, pp. 839-49.

Raney, R.K. (2007). Hybrid-Polarity SAR Architecture. *IEEE Transactions on Geoscience Remote Sensing*, 45, 11, 3397 - 3404.

Rignot, J.M. (2000). Effect of Faraday Rotation on L-Band Interferometric and Polarimetric Synthetic-Aperture Radar Data. *IEEE Transactions on Geoscience and Remote Sensing*, 38, 1, 383-390.

Shen C.S. (1981). The Elliptical Microstrip Antenna with Circular Polarization. *IEEE Transactions on Antennas and Propagation*, 29, 1, 90-94.

Suzuki, Y.; Miyano, N. & Chiba, T. (1987). Circularly Polarised Radiation from Singly Fed Equilateral-triangular Microstrip Antenna. *Microwaves, Antennas and Propagation, IEE Proceedings H*, 134, 2, 194 - 198.

Vetharatnam, G.; Kuan, C.B. & Teik, C.H. (2006). Microstrip Antenna for Airbone SAR Applications. Available online at:

http://www.remotesensing.gov.my/images/default/publication_3rdmicrowave/3rdmicrowave_paper5.pdf.

Wissan, V.; Baharuddin, M.; Sri Sumantyo, J.T & Kuze, H., (2009). Development of Measurement System for Circularly Polarized Synthetic Aperture Radar. *The 16th Remote Sensing Forum - SICE*, pp. 11-12, RESTEC, Tokyo, March 2009 (in Japanese).

Zeland Software Inc.(2006). IE3D User's Manual Release 11.2.

14

Circularly Polarized Slotted/Slit-Microstrip Patch Antennas

Nasimuddin, Zhi-Ning Chen and Xianming Qing
Institute for Infocomm Research
Singapore

1. Introduction

In modern wireless communication systems, small circularly polarized microstrip antennas with good performance are desirable mainly at low microwave frequencies. The radio frequency identification (RFID) systems in the ultra-high frequency (UHF) band have gained much interest in many services [Finkenzeller2004]. The reader antenna is one of the important components in the RFID reader system and has been designed with circularly polarized (CP) radiation. The RFID system consists of the reader and the tag. Normally, the UHF tag antennas are linearly polarized. The circularly polarized microstrip antennas (CPMAs) can reduce the loss caused by the multipath effects between a reader and the tag antenna. The circularly polarized reader antenna is useful in particular when tag antenna system is in rotating motion or when its respective orientation cannot be ensured.
Total frequency span of 840 - 960 MHz is used over worldwide for UHF RFID systems. The UHF RFID system operates at the bands of 902 - 928 MHz for North-South America, 865 - 867 MHz for Europe and 840 - 955 MHz for Asia-Pacific region. The compact CPMA design to cover total UHF RFID frequency span for handheld RFID applications is very difficult. Overall sizes of the broadband CPMAs to cover woldwide RFID applications are bulky [Chen2009, Chung2007]. However these broadband antennas are not suitable for handheld or portable reader applications. Design of the compact CPMA is attractive for handheld/portable device applications. Small size of the CPMA can be achieved at the cost of limited gain and narrow 3-dB Axial ratio (AR) bandwidth/10-dB return loss impedance bandwidth. A CPMA with a low-profile, small-size, and light weight is required in a portable/handheld RFID reader. However the compact CP reader antenna should be covered at least one country frequency span used for RFID UHF band [Nasimuddin2010, Nasimuddin2009]. A typical technique for producing CP is to excite two orthogonal linearly polarized modes with a 90° phase difference of the patch antenna. The single-feed CP annular-ring, square and circular microstrip patch antennas with perturbation elements have reported in [Chen1999, Chen2001, Row2004, Huang1998, Hyun2008, Row 2005, Sharma 1983]. Using perturbation or strips or slots on a radiating patch two orthogonal modes can be generated at around resonance frequency with 90° phase-shift for CP radiation requirements. The single-feed circularly polarized microstrip antennas [Haneishi1988, Wong2002] are generally compact when compared with the dual-feed CPMAs [Tragonski1993]. Single-feed CPMA is simple, compact structure, easy manufacture, and low-cost. However, the single-feed conventional microstrip antennas [Sharma1983,

Haneishi1988, Wong2002, Iwasaki1996] usually have very narrow 3-dB axial-ratio bandwidth which is not suitable for many wireless communication applications such as 2.4 - 2.48 GHz. Several kinds of the radiating patch elements such as square, circular, triangular and ring shapes have been used to obtain a CP radiation using single-feed [Wong2002]. Iwasaki has been reported a proximity-coupled fed circular CPMA with a centrally located asymmetric cross-slot on a circular patch radiator. Asymmetric cross-slot provides necessary perturbation to excite two orthogonal modes with 90° phase-shift to generate CP radiation. However, 2-dB axial-ratio bandwidth of the antenna is 0.65%. For improvement of CP bandwidth, a single series feeding cross-aperture coupled microstrip antenna with effect of hybrid-feeding has been presented in [Kim2003]. However, this feeding structure needs more ground-plane size and it is not useful for compact CPMAs design. A proximity-coupled fed ring antenna was presented in [Ramirez2000] for CP radiation. CP radiation is achieved by adding two inner stubs of perturbation method. In [Tong2007], a slotted patch radiator design was proposed; in which CP radiation is achieved by using the asymmetric *U*-shaped slot embedded patch. Aperture-coupled asymmetric cross-slotted microstrip antenna can be used for CP radiation [Huang1999]. Measured 3-dB axial-ratio bandwidth of the antenna is 1.79%. Unfortunately, information on gain, overall antenna size and 3-dB AR beamwidth of the antenna is not available. Aperture-coupled fed antenna structures have attracting much attention because their geometries are suitable for monolithic integration with microwave or millimeter devices. Aperture-coupled structure is also useful for antenna array design. Nasimuddin et al. [Nasimuddin2008, 2009, 2008] have been proposed an aperture-coupled asymmetric *C*-shaped slotted microstrip antenna for CP radiation. The asymmetric *C*-shaped slotted microstrip patch is fabricated on a dielectric substrate and mounted on a thick foam substrate. Aperture-coupled fed asymmetric *S*-shaped slotted microstrip antenna can also be used for dual-band CP radiation [Nasimuddin2010].

First part of this chapter is studied on compact circularly polarized slotted/slit-microstrip patch antennas for UHF RFID reader applications. The slotted/slit-microstrip patch can be used for overall size reduction of the microstrip patch antenna with CP radiation. By slightly changed circumference one of diagonal slot/slit as compared to other diagonal slot/slit, the CP radiation of the antenna can be obtained. Performances of various slotted/slit-microstrip antennas are also compared. The parametric studied of the *V*-shaped slits microstrip antenna is also presented to understand the generation of CP radiation. Second part of chapter is studied and comparison of the aperture-coupled fed circularly polarized slotted-microstrip patch antennas to cover frequency range 2.4 - 2.48 GHz. The antenna consists of slotted square patch and an aperture-coupled feeding structure. The asymmetric *C*-shaped, symmetric *S*-shaped and asymmetric *Z*-shaped slotted patches are fabricated from 0.5 mm copper sheet. Each slotted patch is mounted on a same size thick foam substrate. Slot dimensions are optimized for good CP radiation with fixed antenna and patch sizes. Performance of the slotted microstrip antennas are compared and studied. Design and optimization of proposed antennas were conducted with the help of commercial EM software, IE3D [IE3D2010].

2. Compact CP slotted/slit-microstrip structures and designs for RFID readers

Cross-section of the typical compact circularly polarized slotted/slit-microstrip patch antenna is shown in Figure 1. Various slotted/slit-microstrip patches are shown in Figures

2(a) - 2(e) for CP radiation with compact antenna size. Ground-plane size of 90 mm × 90 mm is fixed for all patch radiators. Square patch length of all patch radiators is also fixed (L = 80.0 mm). The microstrip patch antennas are designed on a RO4003 substrate (H = 4.572 mm, ε_r = 3.38 and tanδ = 0.0027). Coaxial feed location (F) from center of the microstrip patch is y_o.

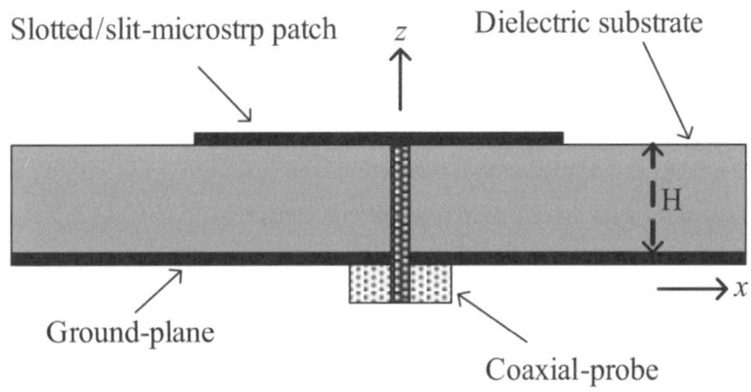

Fig. 1. Cross-section view of the proposed CP slotted/slit-microstrip antenna.

Various slotted/slit-microstrip patch antennas are proposed for CP radiation based on fixed overall antenna and microstrip patch sizes. All microstrip patch antennas are optimized for good CP radiation. Designed and optimized geometrical parameters (in mm) are added on microstrip patch structures. Figure 2(a) shows the conventional truncated corners microstrip patch [Sharma1983] for comparison (Antenna#1). Rectangular slot [Sharma1983] along one of the diagonal axes at the centre of the patch radiator for CP radiation is also designed as illustrated in Figure 2(b) (Antenna#2). Figure 2(c) shows the CP cross-shaped slotted microstrip patch (Antenna#3). Figure 2(d) exhibits an asymmetric- Y-shaped slotted microstrip patch for CP radiation (Antenna#4). V-shaped slits [Nasimuddin2009] are embedded symmetrically along the diagonal directions of the microstrip patch (Antenna#5) as illustrated in Figure 2(e). Locations of four V-shaped slits are located at (P, P) along diagonal directions from centre of the square microstrip patch. A_1, A_2, A_3 and A_4 are areas of the slit along the diagonal directions. Slit areas can be determined by P, d_1, and d_2. A_1 and A_3 are same as well as A_2 and A_4 are also same. For CP radiation of the patch antenna, A_1 / A_3 should be not equal to A_2 / A_4. The antenna is symmetrical along diagonal axes.

2.1 Simulated results of CP slotted/slit-microstrip antennas and comparison

Simulated return loss, axial-ratio at the boresight and gain at the boresight of the CP slotted/slit-microstrip antennas are compared in this section. The square patch without slit/slot is also simulated for comparison. Resonance frequency of the square patch antenna is around 977.5 MHz. Simulated return loss of the slotted/slit CPMAs is plotted in Figure 3(a). Simulated 10-dB return loss bandwidth is 3.13% for truncated corners microstrip antenna, 2.5% for rectangular slotted microstrip antenna, 1.80% for asymmetric-cross shaped slotted microstrip antenna, 1.60% for asymmetric Y-shaped slotted microstrip antenna and 1.63% for V-shaped slits microstrip antenna. Note that the conventional truncated corners microstrip antenna resonance frequency is higher than that of the square patch antenna resonance frequency. So that the truncated corners method is not useful for compact CPMAs designs. The V-shaped slits CP microstrip antenna exhibits the lowest resonance frequency.

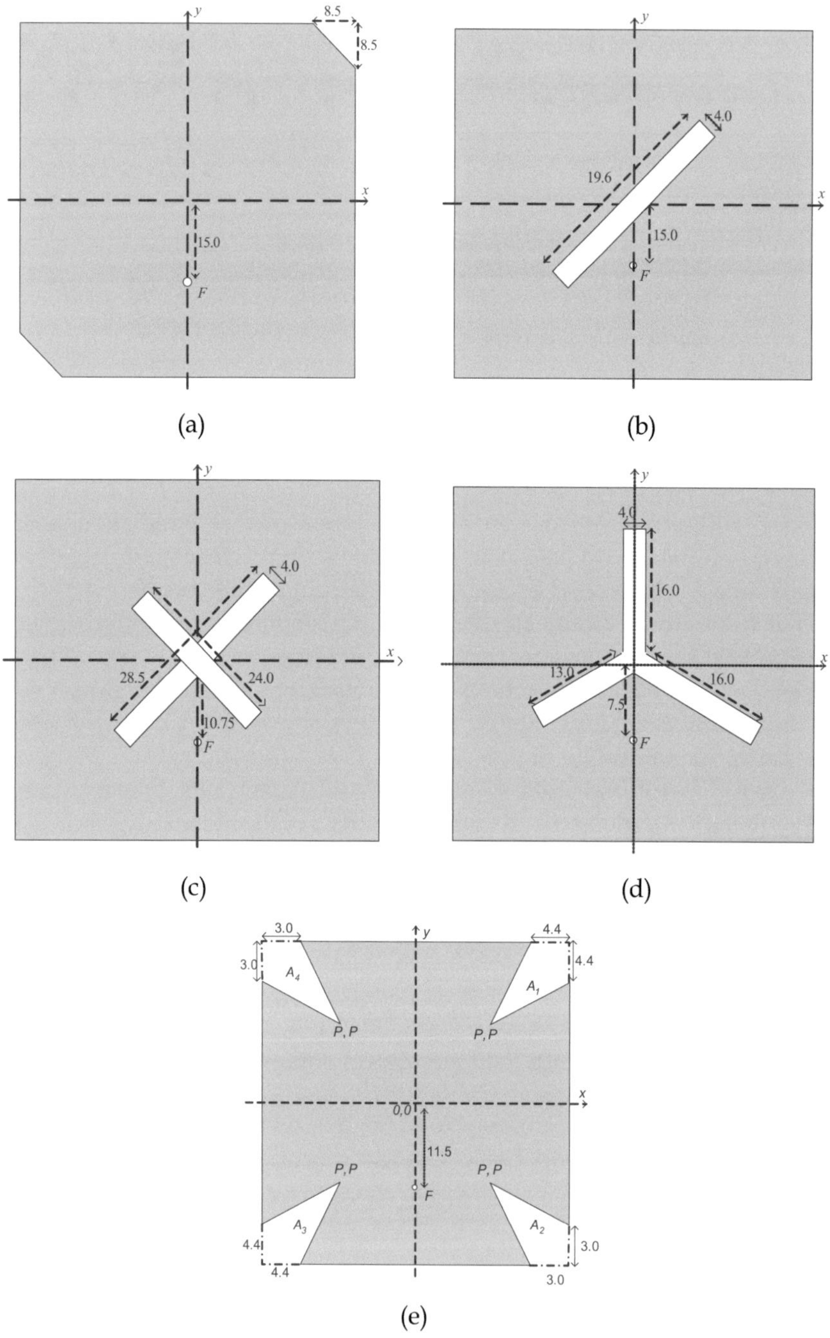

Fig. 2. Slotted/slit-microstrip patch structures for CP; (a) Antenna#1, (b) Antenna#2, (c) Antenna#3, (d) Antenna#4 and (e) Antenna#5.

It means *V*-shaped slits microstrip antenna is electrically small when compared with the other CPMAs. Figure 3(b) shows AR of the slotted/slit CPMAs at the boresight. Simulated 3-dB AR bandwidths of the CP slotted/slit-microstrip antennas are 7.4 MHz (980.0 - 987.4 MHz), 6.3 MHz (960.7 - 967 MHz), 5.0 MHz (934.0 - 939.0 MHz), 4.2 MHz (922.6 - 926.8 MHz) and 5.0 MHz (917.0 - 921.0 MHz), respectively. The 3-dB AR bandwidth of truncated corners patch antenna is larger when compared to the other CPMAs.
Simulated 3-dB AR bandwidth of the *V*-shaped slits and *Y*-shaped slotted microstrip antennas are slightly narrow because these antennas are electrically small. Simulated gain at

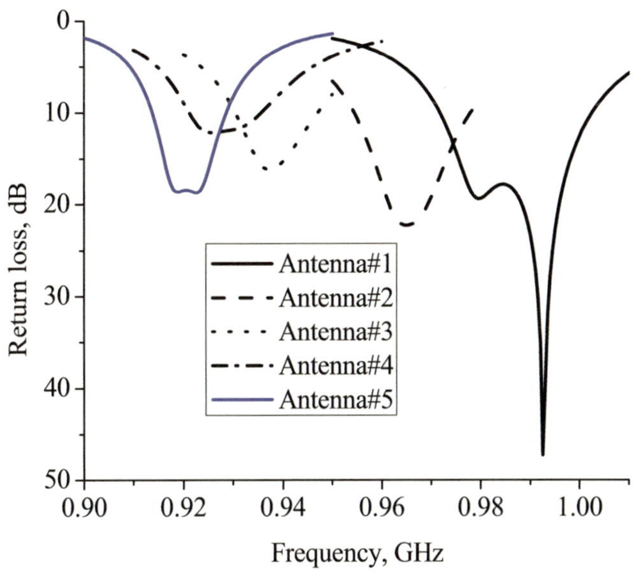

(a)

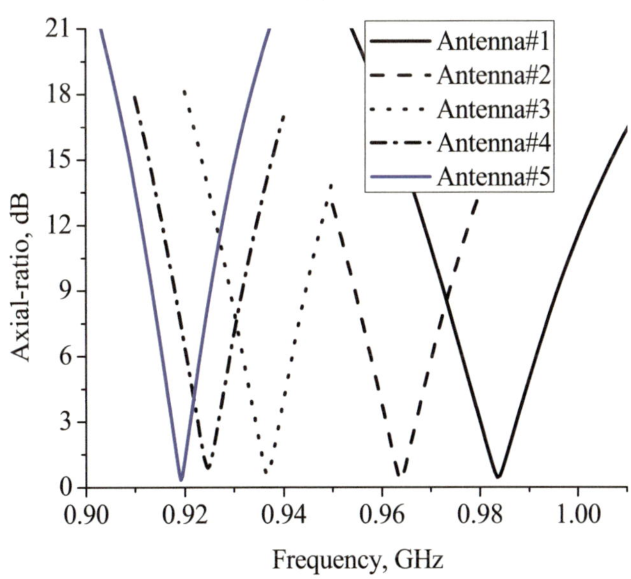

(b)

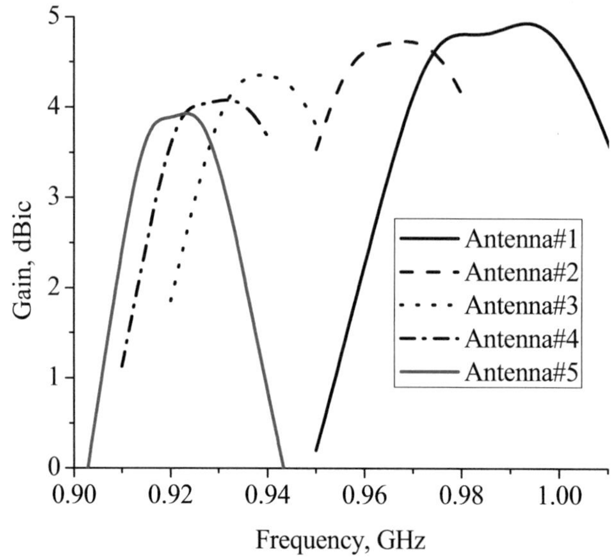

(c)

Fig. 3. Simulated performances of the CP slotted/slit-microstrip antennas; (a) return loss, (b) axial-ratio at the boresight and (c) gain at the boresight.

the boresight of the CPMAs is compared in Figure 3(c). Broadside gain is more than 4.0 dBic for all slotted/slit CPMAs over the 3-dB AR bandwidth except *V*-shaped slits CPMA. Gain of the *V*-shaped slits microstrip antenna is bit lower due to the antenna is electrically smallest. For RFID handheld reader applications, the small antenna size is more important. Broesight gain variation over the 3-dB AR bandwidth is less than 0.2 dB for CP slotted/slit-microstrip antennas. Simulated performances of the CP slotted/slit-microstrip antennas are also summarized in Table 1.

Antenna	3-dB AR frequency range (MHz)	3-dB AR bandwidth (MHz)	10-dB return loss bandwidth (%)	Gain (dBic) (maximum)
1	980.0 - 987.4	7.4 MHz	3.13% (30.0 MHz)	4.8
2	960.7 - 967.0	6.3 MHz	2.5% (24.0 MHz)	4.6
3	934.0 - 939.0	5.0 MHz	1.81 % (16.8 MHz)	4.3
4	922.6 - 926.8	4.3 MHz	1.60% (14.5 MHz)	4.1
5	917.0 - 921.0	5.0 MHz	1.63% (15.0 MHz)	3.9

Table 1. Simulated Performance of the CP Slotted/slit-microstrip Antennas

2.2 Parametric study of slit-microstrip antenna

Parametric analysis of the circularly polarized slit-microstrip antenna (Antenna#5) is conducted to understand effect of slit parameters on the antenna performance. By changing the slit parameters, the two-orthogonal modes with 90° phase shift can be generated for requirements of circularly polarized radiation. For single-feed microstrip antenna, the orthogonal mode resonant frequencies should be much closed to get the good CP radiation. In this study, locations of the *V*-shaped slits and feed-location are fixed (P = 15.5 mm, 11.5 mm). d_2 is also fixed as 3.0 mm. By decreasing or increasing d_1 with respect to d_2, the antenna can be optimized for CP radiation. Simulated return loss, AR and gain at the boresight of the CPMA with different d_1 are plotted in Figures 4(a), (b) and (c), respectively. The d_1 is

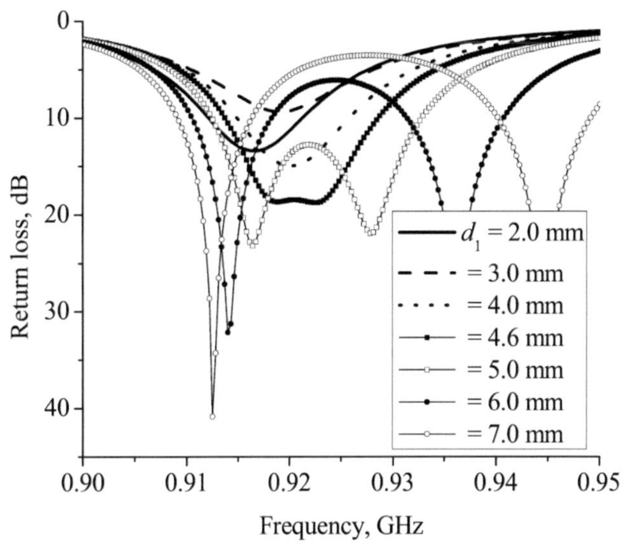

(a)

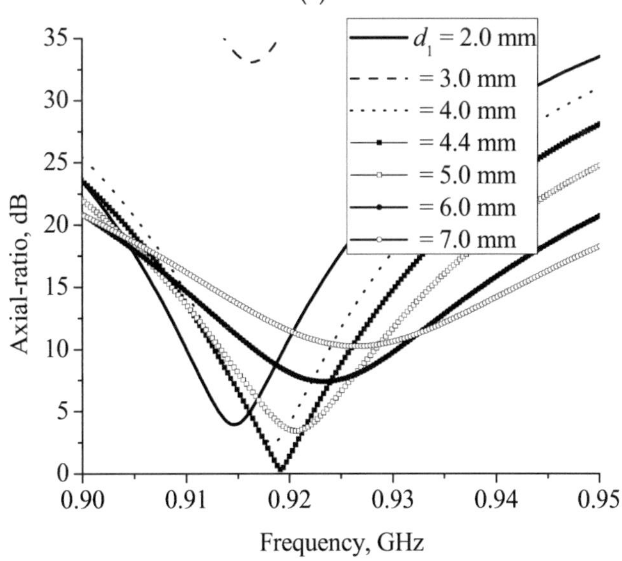

(b)

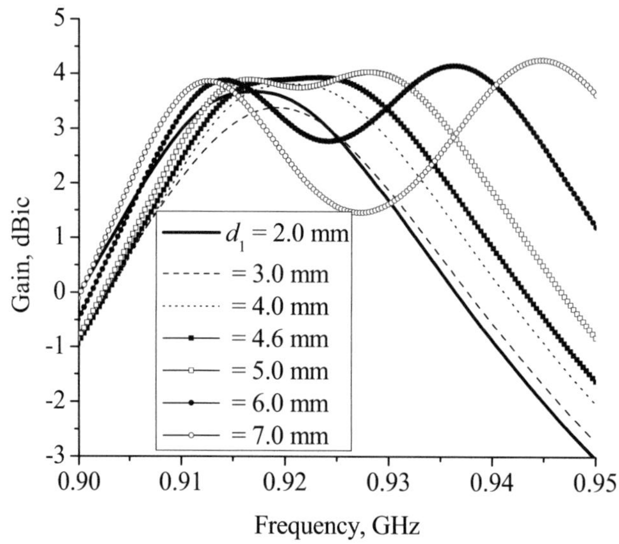

(c)

Fig. 4. Effect of the d_1 on antenna parameters: (a) return loss, (b) axial-ratio at the boresight and (c) gain at the boresight.

changed from 2.0 mm to 7.0 mm. When d_1 increases, the impedance bandwidth is enhanced while the frequency-band is moved down. Two orthogonal modes start to regenerate when d_1 increases and at around d_1 = 4.4 mm two orthogonal modes with their resonance frequencies nearly close to each other. It can be observed that the boresight AR decreases with an increase in d_1 up to 4.4 mm and then start increase AR with increases in d_1. Best AR performance is achieved, when d_1 is around 4.4 mm. When d_1 is 3.0 mm, the patch antenna with symmetric *V*-shaped slits is linear polarized radiation ($d_1 = d_2$). The axial-ratio is more than 30 dB at $d_1 = d_2$ = 3.0 mm. Boresight gain and gain bandwidth increases with increase in d_1. At around d_1 = 4.4 mm, the gain is also more flat.

2.3 Measured results and discussions of the slit-microstrip antenna

The compact circularly polarized *V*-shaped slits microstrip patch antenna was fabricated and tested. Antenna performances were measured by an Agilent vector network analyzer N5230A and MiDas 5 antenna test system. Measured 10-dB return loss bandwidth is 19.0 MHz (914 - 933 MHz). Measured 3-dB AR bandwidth is achieved around 6.0 MHz (920 - 926 MHz) and it is within the measured 10-dB return loss bandwidth. Measured maximum boresight gain is around 3.6 dBic at 921 MHz. Boresight gain remains relatively constant around 3.4 dBic with variation of less than 0.2 dB within the 3-dB AR bandwidth. Measured results on return loss, AR and gain agree well with simulated results.

Radiation patterns were measured using a rotating linear polarized transmitting horn antenna for both (*x-z* and *y-z*) principal planes. Figure 5 shows the measured radiation patterns *x-z* and *y-z* planes at 924 MHz. In both planes (*x-z* and *y-z*), the AR is found to be less than 3-dB across a 100° beamwidth over the 3-dB AR bandwidth frequency range. Proposed antenna is compact and very useful for RFID handheld reader applications. For validation of RFID reader applications of the antenna, the tag reading-range was measured for the antenna. The MP9320 2.8 EPC™ UHF reader and home made UHF tag were used for the measurement of the reading range. The maximum reading range of 60 - 65 cm is achieved at the boresight.

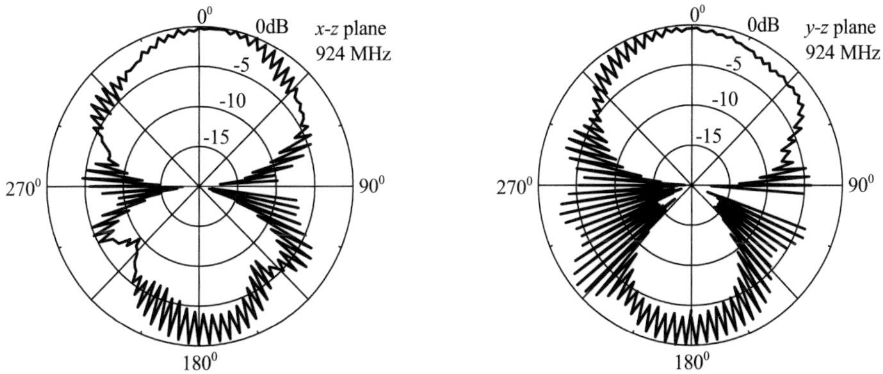

Fig. 5. Measured radiation patterns at 924 MHz.

2.4 Symmetric-slit with slot based compact CP microstrip antennas

In this section, the symmetric-slit microstrip patch based compact circularly polarized microstrip antennas are also designed at frequency around 2.4 GHz. The slit-slotted microstrip patch radiators are printed on the RO4003C substrate (H = 1.524 mm, ε_r = 3.38 and $tan\delta$ = 0.0027). The square patch (L) and ground-plane sizes are 31.8 mm and 36.0 mm × 36.0 mm, respectively. The coaxial feed-location (y_o) is 5.0 mm from centre of the patch radiator. Designed dimensions of the patch radiators are shown in patch sketches.

The V-shaped symmetric-slit along with circular slot at the centre of square patch radiator is shown in Figure 6. The circularly polarized radiation is achieved by changing the area of slits along the diagonal directions ($A_1 > A_2$). Simulated resonance frequency of the square patch without slits and slot is around 2.488 GHz. The measured and simulated return loss is

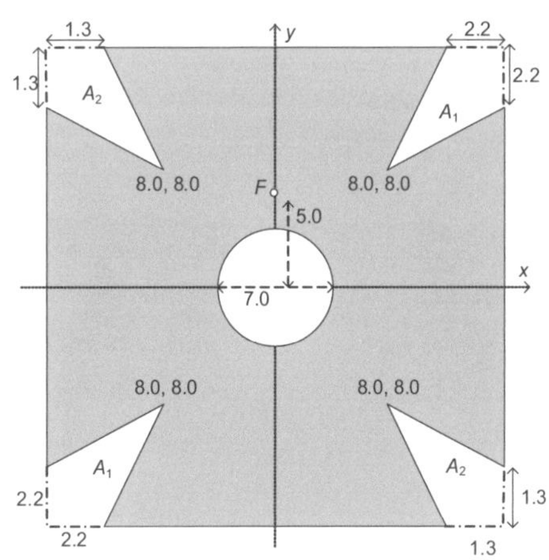

Fig. 6. Symmetric-slit with circular-slot patch radiator.

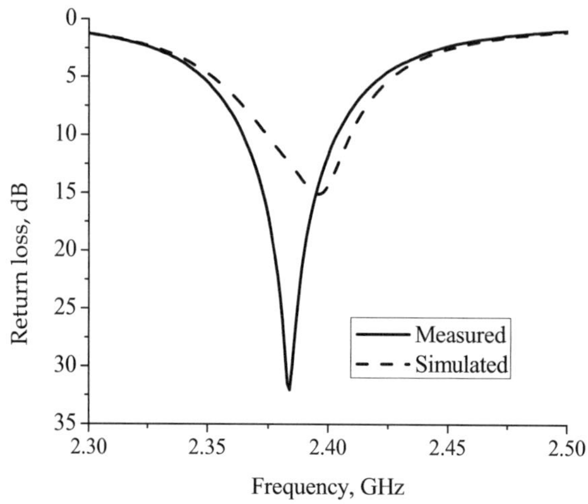

Fig. 7.(a). Simulated and measured return loss of the symmetric-slit patch with circular slot.

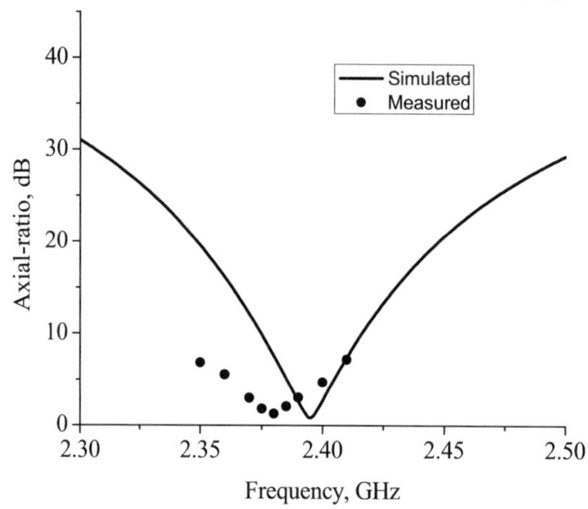

Fig. 7.(b). Simulated and measured axial-ratio at the boresight

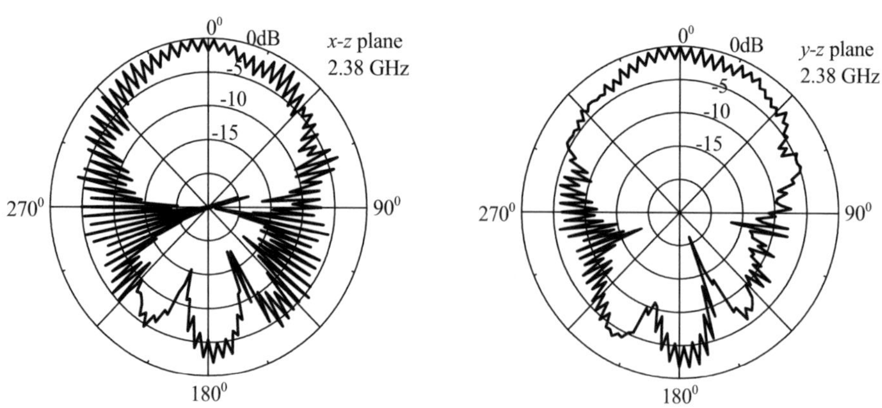

Fig. 8. Measured radiation patterns in *x-z* and *y-z* planes at 2.38 GHz

plotted in Figure 7(a). Measured resonance frequency of the proposed slit-slotted microstrip patch antenna is 2.384 GHz. Boresight minimum axial-ratio is around the 2.38 GHz as shown in Figure 7(b). Simulated and measured results are good agreement. The boresight measured gain is more than 3.5 dBic over the 3-dB AR bandwidth. Measured radiation patterns in *x-z* and *y-z* planes at 2.38 GHz are plotted in Figure 8. Operating frequency of the antenna is lower when compared with the square patch antenna. As a result the slit-slotted microstrip antenna is electrically small.

Other *V*-shaped symmetric-slit along with rectangular slot in diagonal direction at the centre of square patch radiator is shown in Figure 9. The circularly polarized radiation is achieved using rectangular slot along the diagonal direction at centre of the patch. Note that the area of slits along the diagonal directions are same in this antenna (A_1 = A_2). The measured and simulated results on return loss and axial-ratio at the boresight of the antenna are illustrated in Figures 10(a)-(b), respectivily. The measured 10-dB return loss bandwidth is 2.34 - 2.38 GHz. The centre frequency is around 2.36 GHz and minimum axial-ratio is achieved at around 2.355 GHz. The proposed symmetric-slit with slotted microstrip antennas are compact when compared with the conventional square patch antenna. Resonance frequency of the symmetric-slit with rectangular-slot microstrip patch antenna is lowest when compared with the circular-slot and conventional square patch antennas. The measured spining radiation pattrens at x-z and y-z planes are plotted in Figure 10(c).

3. Aperture-coupled slotted-microstrip structures

Cross-section view of the typical aperture-coupled fed CP slotted microstrip antenna is shown in Figure 11(a). The slotted patch radiators (asymmetric *C*-shaped [Nasimuddin2008, Nasimuddin2009], symmetric *S*-shaped and asymmetric *Z*-shaped) are illustrated in Figures 11(b)-11(d). Length of a square microstrip patch is *L*. *H* is the total antenna height and h_2 is the foam thickness. The 50-Ω microstrip feed-line and the coupling aperture are printed on the opposite surfaces of the RO4003C substrate (h_1 = 1.524 mm, ε_{r1} = 3.38 and $tan\delta_1$ = 0.0027). Microstrip feed-line from center of an aperture is S_f as shown in Figure 11(e). W_a is aperture width and L_a is aperture length. Aperture-coupled feed is located at center of the slotted

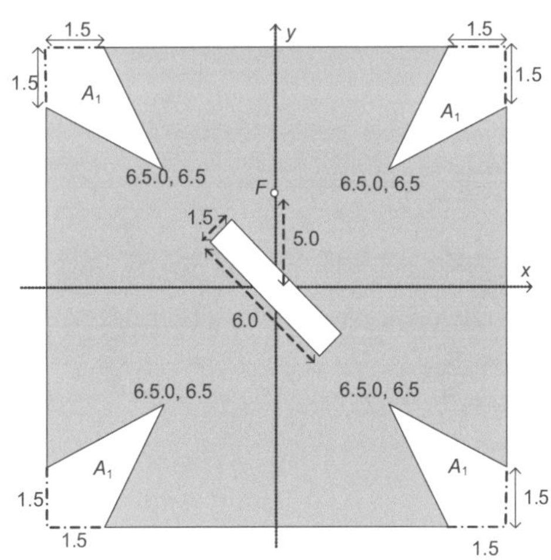

Fig. 9. *V*-shaped symmetric-slit with rectangular-slot patch radiator.

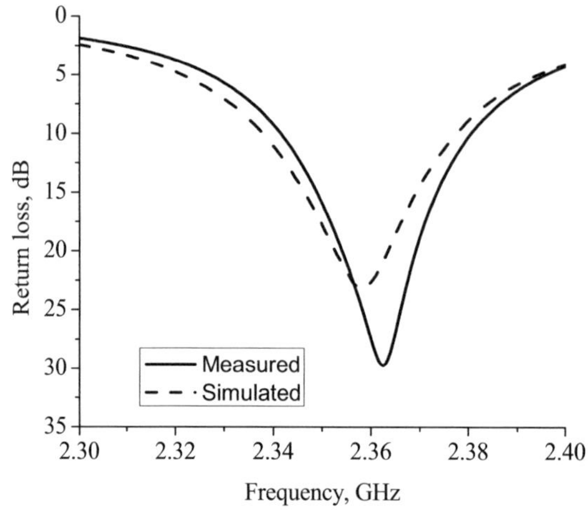

Fig. 10. (a). Simulated and measured return loss of the symmetric-slit with rectangular slot patch antenna.

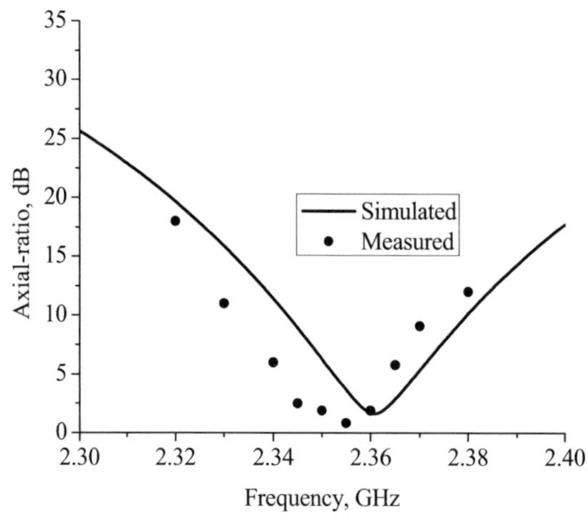

Fig. 10. (b). Simulated and measured axial-ratio at boresight of the symmetric-slit with rectangular slot patch antenna.

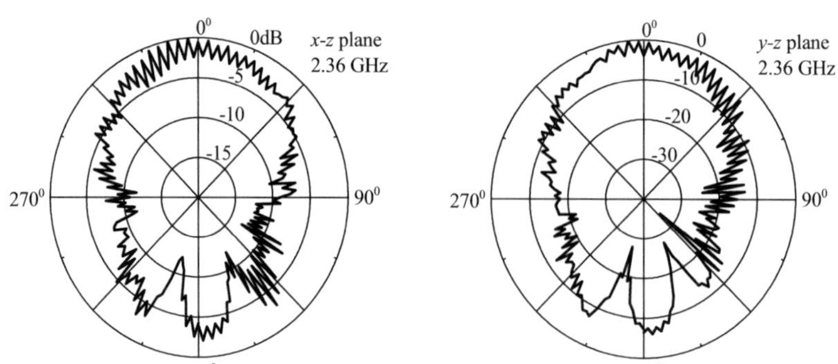

Fig. 10. (c). Measured radiation patterns in x-z and y-z planes at 2.36 GHz of the symmetric-slit with rectangular slot patch antenna

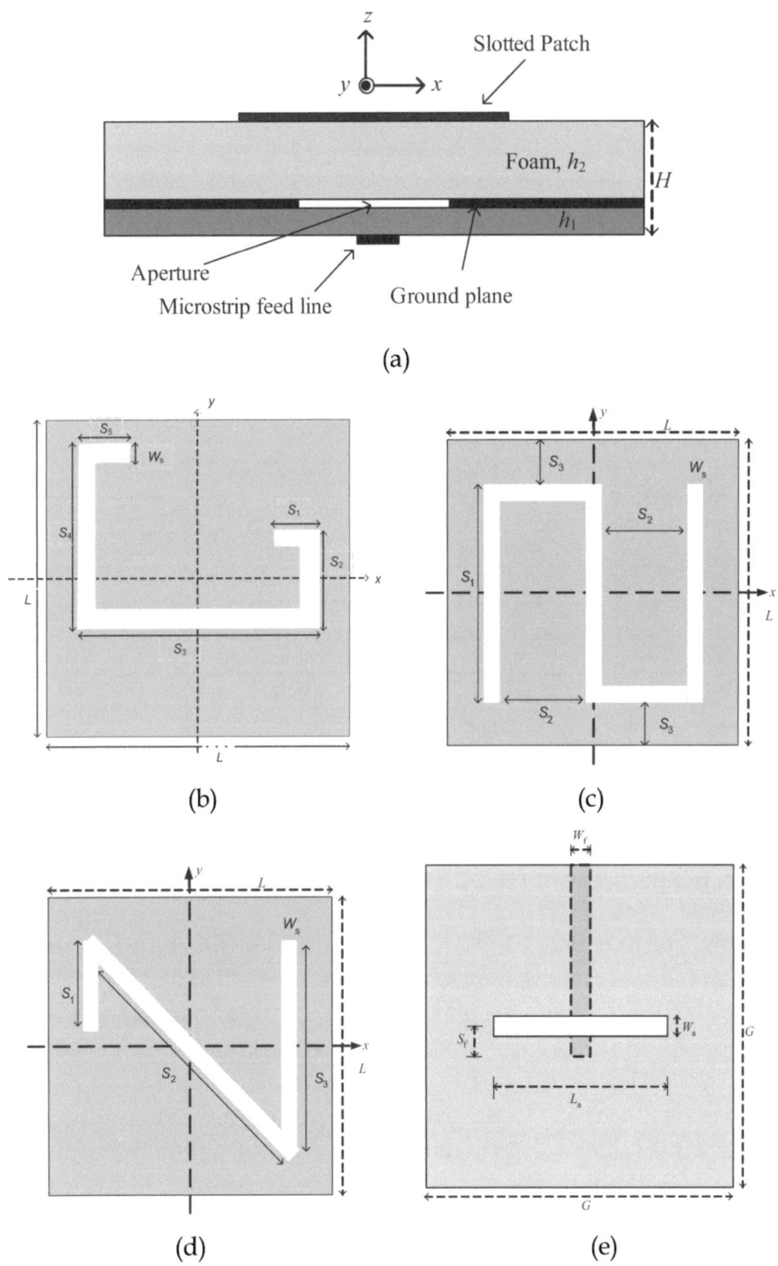

Fig. 11. Aperture-coupled CP slotted microstrip antenna; (a) cross-section view, (b) *C*-shaped slotted patch, (c) *S*-shaped slotted patch, (d) *Z*-shaped slotted patch, and (e) aperture-coupled feed structure.

square radiator. The slotted patch radiator can cause meandering of excited surface current paths and result in lowering resonance frequency. As a result electrical size of the antenna is reduced. By adjusting the slot size, the two near-degenerated resonance modes with 90° phase-shift can be generated for CP radiation of the slotted square patch antenna. Lengths of the slot can be used to optimize the antenna for minimum AR, wide AR bandwidth, wide impedance bandwidth and size reduction. Three sets of the slot shapes are selected and optimized for minimum AR, wide CP bandwidth and compact antenna size with help of [Nasimuddin2008-2010] based on same antenna and patch radiator sizes. Optimized dimensions of the slots are given in Table 2 and all dimensions are in mm.

Parameters	*C*-shaped slot	*S*-shaped slot	*Z*-shaped slot
L	43.0	43.0	43.0
Ground plane ($G \times G$)	60.0 × 60.0	60.0 × 60.0	60.0 × 60.0
W_s	3.0	3.0	3.0
S_1	3.5	30.0	12.5
S_2	8.0	13.5	42.5
S_3	28.0	6.50	33.1
S_4	21.0	-	-
S_5	5.5	-	-
S_f	4.0	4.0	4.0
W_a	33.0	33.0	33.0
L_a	3.0	3.0	3.0
H	11.524	11.524	11.524

Table 2. Designed and optimized dimensions of the slotted microstrip antennas.

3.1 Comparison of aperture-coupled CP slotted microstrip patch antennas

In this section, three types of the slotted patch radiators are studied by simulation for good CP radiation with fixed antennas size. Simulated return loss, AR at the boresight and gain at the boresight of the slotted microstrip patch antennas are studied and compared. Return loss of the slotted microstrip antennas are plotted in Figure 12(a). Simulated 10-dB return loss bandwidth is around 13.5% (2.230 - 2.554 GHz) for asymmetric *C*-shaped slotted patch, 14.0% (2.092 - 2.407 GHz) for symmetric *S*-shaped slotted patch and 14.2% (2.188 - 2.522 GHz) for asymmetric *Z*-shaped slotted patch. Impedance bandwidth of the asymmetric *Z*-shaped slotted antenna is larger. Figure 12(b) shows axial-ratio at the boresight of the slotted antennas. Simulated 3-dB axial-ratio bandwidths of the asymmetric *C*-shaped, symmetric *S*-shaped and asymmetric *Z*-shaped slotted antennas are 3.1% (2.406 - 2.483 GHz), 2.1% (2.250 - 2.310 GHz) and 2.62% (2.409 - 2.473 GHz), respectively. Minimum AR is 0.72 dB at 2.445 GHz for asymmetric *C*-shaped slot, 0.14 dB at 2.267 GHz for symmetric *S*-shaped slot and 0.64 dB at 2.445 GHz for asymmetric *Z*-shaped slot. Simulated gain at the boresight of the slotted antennas is compared in Figure 12(c). Boresight gain is more than 6.0 dBic for all antennas over the impedance bandwidth. Gain variation with frequency over the impedance bandwidth is less than 0.5 dB for all antennas. Lowest resonance frequency of the antenna is

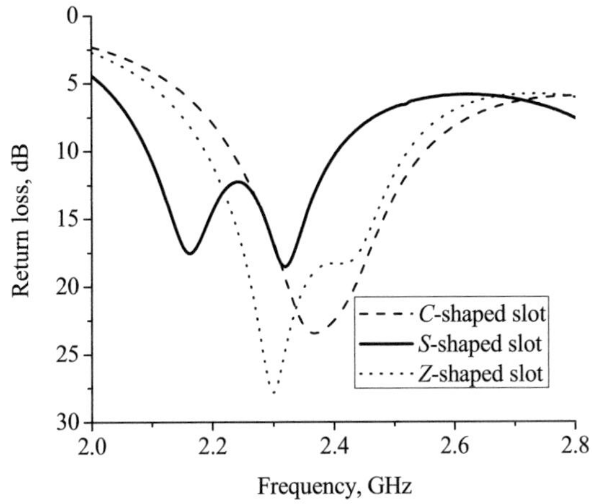

(a)

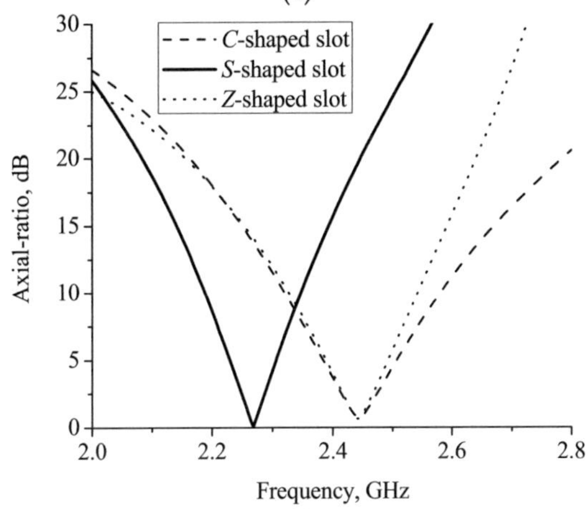

(b)

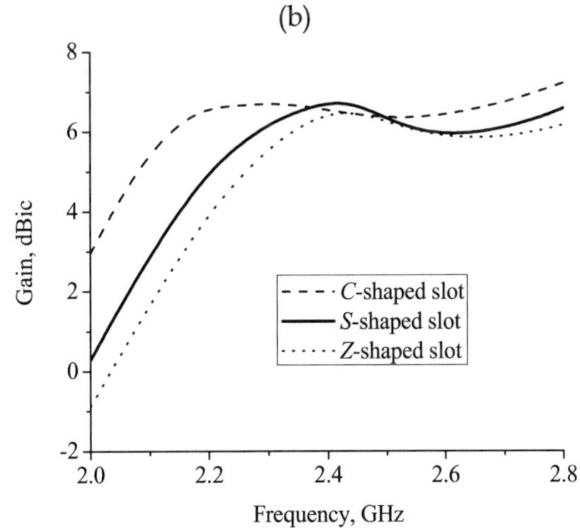

(c)

Fig. 12. Simulated results of the slotted microstrip antennas; (a) return loss, (b) axial-ratio at the boresight and (c) gain at the boresight.

achieved with symmetric *S*-shaped slotted patch. The symmetric *S*-shaped slotted antenna has 7% overall area reduction when compared with the asymmetric *C*-shaped and *Z*-shaped slotted microstrip antennas. However, its the 3-dB axial-ratio bandwidth is narrow when compared with the asymmetric *C*- and *Z*-shaped slotted antennas. Minimum axial-ratio dip frequency for the asymmetric *C*- and *Z*-shaped slotted patch is almost same. The symmetric *S*-shaped slotted patch is useful for compact CP microstrip antenna design. For wide CP bandwidth CPMA design, the asymmetric *C*-shaped slotted patch radiator is better.

3.2 Parametric study of aperture-coupled Z-shaped slotted microstrip antenna

In this section, the parametric studied is conducted of the *Z*-shaped slotted microstrip patch antenna. The S_1 arm of *Z*-shaped slot is varied to study the CP radiation generation for *Z*-shaped slotted patch antenna and variation of antenna performances. The return loss, axial-ratio at the boresight and gain at the boresight with variation of S_1 are plotted in Figures 13(a)-(c) respectively. The 10-dB return loss bandwidth increases with decrease in S_1 of *Z*-shaped slot and starts degenerate two orthogonal modes with 90° phase-shift for CP radiation requirements. At around $S_1 = 12.5$ mm, lowest axial-ratio can be achieved for good CP radiation. The axial-ratio decreases with increase in S_1 as shown in Figure 13(b). The brosight gain is not significant effect with variation of S_1 as plotted in Figure 13(c).
The current distributions of the slotted patch antennas are shown in Figures 14(a) - (c), respectively. Strongest current distribution is at around asymmetric *C*-shaped slot arm S_6 as shown in Figure 14(a). These current distributions are at minimum axial-ratio frequencies of the antennas. The S_6 is very sensitive to optimize the *C*-shaped slotted antenna for best CP (minimum AR) and wide CP bandwidth. Strongest current distribution is at around symmetric *S*-shaped slot. The symmetric *S*-shaped slotted radiating patch can cause meandering of the excited patch surface current paths is more when compared with the asymmetric *C*-shaped and *Z*-shaped slotted patches as shown in Figure 14(b). Lowering resonance frequency of the antenna, which is corresponds to a reduced antenna size for such a fixed overall antenna size. For left hand CP radiation, S_2 should be greater than S_4 (*C*-shaped slotted patch), S_1 should be greater than S_3 (*Z*-shaped slotted patch), and for the right hand CP radiation, S_2 should be less than S_4 (*C*-shaped slotted patch), S_1 should be less than S_3 (*Z*-shaped slotted patch). For changing the polarization sense of the *S*-shaped slot patch antenna, *S*-shaped slot should be rotated by 180° with respect to feed axis.

3.3 Measured results and discussions of the slotted-microstrip antennas

Optimized aperture-coupled fed slotted patch antennas are fabricated and tested. The slotted patch radiators are fabricated from 0.5 mm thick copper sheet and each mounted on a thick foam substrate of height (h_2 = 10 mm). Measured performance of the aperture-coupled slotted patch antennas is summarized in Table 3. The asymmetric *C*-shaped slotted patch antenna has larger CP bandwidth (3-dB axial-ratio bandwidth) when compared with the symmetric *S*-shaped and asymmetric *Z*-shaped slotted antennas. The symmetric *S*-shaped slotted patch antenna is electrically small when compared with the asymmetric *C*- and *Z*-shaped slotted patch antennas. The slotted patch antennas have a wide angle CP of more than 90°. Measured performance of the slotted patch antennas is also compared with some related structures in the published literatures. The slotted patch antennas are compact and have large impedance bandwidth when compared with the other published antennas. All three *C*-, *S*-, and *Z*-shaped slotted patch antennas show overall antenna size is less than

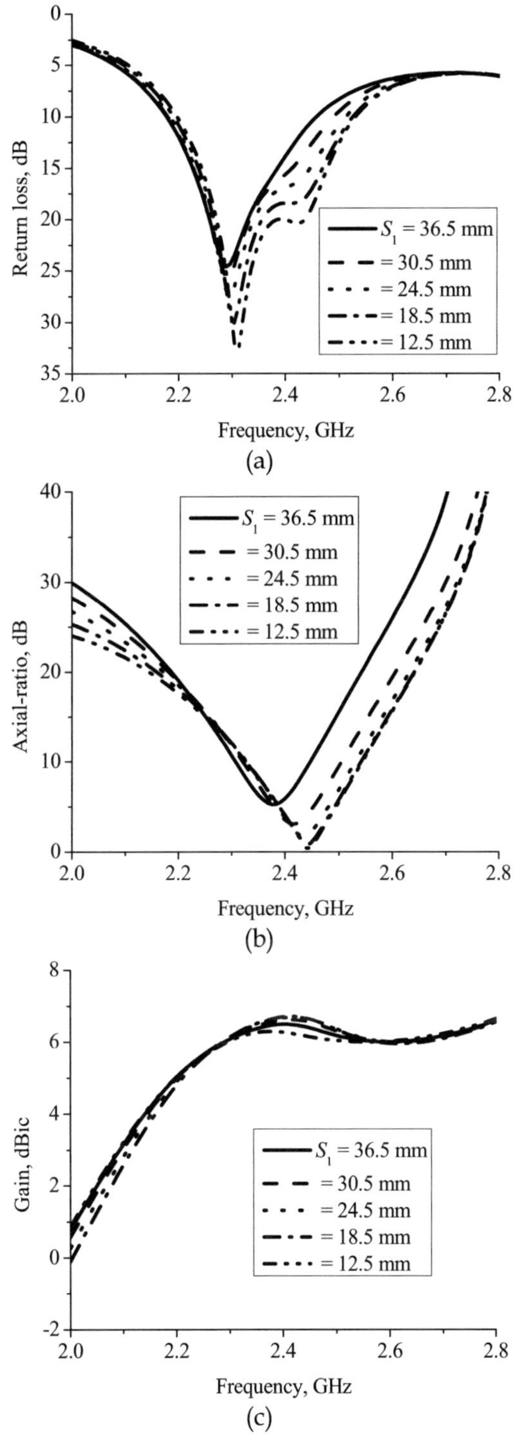

Fig. 13. Effect of the S_1 on antenna parameters of *Z*-shaped slot antenna; (a) return loss, (b) axial-ratio at the boresight and (c) gain at the boresight.

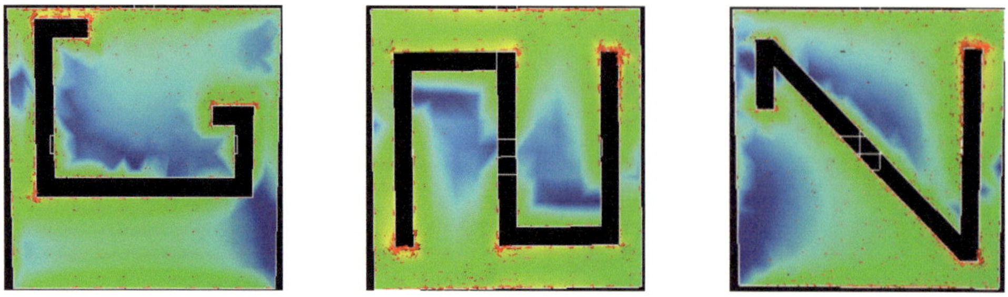

Fig. 14. Current distributions on the slotted patch radiators with slots; (a) *C*-shaped slot, (b) *S*-shaped slot and (c) *Z*-shaped slot.

Antenna structures	Overall antenna size	3-dB Beamwidth	3-dB AR bandwidth (%)	10-dB return loss bandwidth (%)
C-shaped slot	$0.48\lambda_o \times 0.48\lambda_o \times 0.092\lambda_o$ at 2.40 GHz	> 90°	3.31	14.9
S-shaped slot	$0.45\lambda_o \times 0.45\lambda_o \times 0.087\lambda_o$ at 2.25 GHz	> 90°	2.50	13.5
Z-shaped slot	$0.48\lambda_o \times 0.48\lambda_o \times 0.0927\lambda_o$ at 2.40 GHz	> 90°	2.85	14.7
F-shaped slot [Yeh2009]	$0.533\lambda_o \times 0.533\lambda_o \times 0.0815\lambda_o$ at 2.45 GHz	> 90°	4.8	14.4
U-shaped slot [Tong2007]	$0.785\lambda_o \times 0.785\lambda_o \times 0.108\lambda_o$ at 2.30 GHz	58°	4.0	9.0
[Huang1999]	-	-	1.25	3.93

Table 3. Measured performances of the antennas.

half wavelength in free space. The asymmetric *U*-shaped slot antenna [Tong2007] has narrow angle CP when compared with the *C*, *S*, *Z* and *F*-shaped slotted patch antennas. The aperture-coupled asymmetric *S*-shaped antenna has electrically smallest antenna size. However, the *F*-shaped slotted microstrip antenna [Yeh2009] has largest 3-dB AR bandwidth. The 10-dB return loss bandwidth and 3-dB beamwidth of the *C*, *S*, *Z* and *F*-shaped slotted microstrip antennas are almost same. The *U*-shaped slotted patch antenna has lower impedance bandwidth due to this antenna is coaxial feed.

4. Conclusion

The compact CP slit/slotted-microstrip antennas have been presented for UHF RFID handheld reader applications. Various slotted/slit-microstrip patches have also been studied and compared with fixed antenna size. It has been found that the slotted-slit-microstrip patch radiator can be used for small overall size CPMA design. The presented CPMAs have wide angle CP (3-dB AR beamwidth) of more than 100° over the 3-dB AR frequency range. A parametric study of the *V*-shaped slit-microstrip antenna has been conducted. It has also been found that the impedance matching and AR can be optimized easily by adjusting the slit sizes. This chapter has been also presented the aperture-coupled circularly polarized slotted patch antennas. Small overall size of the antenna is achieved with symmetric *S*-shaped slotted microstrip patch. Symmetric *S*-shaped slotted patch radiator is better choice for the compact CPMA design. The *F*-shaped slotted patch antenna has larger 3-dB AR bandwidth. The slotted patch antennas have a wide angle CP radiation (3-dB AR beamwidth) of more than 90° over the 3-dB axial-ratio frequency range. The proposed combined slotted-slit patch technology is useful for compact CPMAs and array designs.

5. Acknowledgment

The authors wish to thank T. M. Chiam for fabrication of the antenna prototypes, C. K. Goh for helping in measurement of reading range of the reader antenna.

6. References

Finkenzeller, K. (2004). *RFID Handbook,* Wiley, 2nd edition, New York, USA

Chen, Z.N., Qing, X., and Chung, H.L. (May2009). A universal UHF RFID reader antenna", *IEEE Trans Microwave Theory and Techniques,* Vol 57, No.5, 2009, pp. 1275-1282

Chung, H.L., Qing, X., Chen, Z. N. (2007). Broadband circularly polarized stacked probe-fed patch antenna for uhf RFID applications, *International Journal of Antennas and Propagation,* Vol. 2007, pp. 1-9

Nasimuddin, Chen, Z.N. and Qing, X. (December2010), Asymmetric-circular shaped slotted microstrip antennas for circular polarization and RFID applications, *IEEE Trans. Antennas and Propagation,* Vol.58, No.12, pp.3821-3828, 2010

Nasimuddin, Qing, X. and Chen, Z.N. (December2009), Compact circularly polarized microstrip antenna for RFID handheld reader applications, *Asia Pacific Microwave Conference,* December 2009, Singapore, pp. 1950-1953.

Chen, H.M., and Wong, K.L. (August1999), On the circular polarization operation of annular-ring microstrip antennas, *IEEE Trans. Antennas and Propagation,* 1999, Vol.47, No.8, pp. 1289–1292

Chen, W.S., Wu, C.K., and Wong, K.L. (March2001), Novel compact circularly polarized square microstrip antenna, *IEEE Trans. Antennas Propagation,* 2001, Vol.49,No.3, pp. 340–342

Row, J.S., and Ai, C.Y.(2004), Compact design of single-feed circularly polarized microstrip antenna, *Electronics Letters,* 2004, Vol.40, No.18, pp. 1093–1094

Huang, C.Y., Wu, J.Y., and Wong, K.L.(1998), Slot-coupled microstrip antenna for broadband circular polarization, *Electronics Letters,* 1998, Vol.34, No.9, pp. 835–836

Hyun, D.H., Baik, J.W., and Kim, Y.S.(2008), Compact reconfigurable circularly polarized microstrip antenna with asymmetric cross slots, *Microwave and Optical Technology Letters,* 2008, Vol.50, No.8, pp. 2217–2219

Row, J.S. (May 2005), Design of aperture-coupled annular-ring microstrip antenna for circular polarization, *IEEE Trans. Antennas and Propagation,* 2005, Vol.53, No.5, pp. 1779–1784

Sharma, P.C., and Gupta, K.C.(June 1983), Analysis and optimized design of single feed circularly polarized microstrip antennas, *IEEE Trans. Antennas and Propagation*, 1983, Vol. 29, No.6, pp. 949-955

Haneishi, M., and Yoshida, S.(1988), A design method of circularly polarized rectangular microstrip antenna by one-point feed, Microstrip Antenna Design, K.C. Gupta and A. Benalla (Eds), Artech House, Norwood, MA, USA, 1988, pp. 313-321

Wong, K.L.(2002), Compact circularly polarized microstrip antennas, Chapter Five, Compact and broad band Microstrip Antenna, John Wiley & Sons, Inc, USA, 2002, pp. 162-220

Targonski, S. D., and Pozar, D.M.(February 1993), Design of wideband circularly polarized aperture-coupled microstrip antennas, *IEEE Transactions on Antennas and Propagation*, February 1993, Vol.41, No.2, pp. 214-219

Iwasaki, H.(October 1996), A circularly polarized small size microstrip antennas with cross slot, *IEEE Trans. Antennas and Propagation*, Vol.44, No.10, 1996, pp. 1399 - 1401

Kim, H., Lee, B.M., and Yoon, Y.J. (2003), A single feeding circularly polarized microstrip antenna with the effect of hybrid feeding, *IEEE Antennas and Wireless Propagation Letters*, Vol.2, 2003, pp. 74-76

Ramirez, R.R., Flaviis, F.D., and Alexopoulos, N.G.(July 2000), Single-feed circularly polarized microstrip ring antenna and arrays, *IEEE Trans. Antennas and Propagation*, Vol.48, No.7, July 2000, pp. 1040-1047

Tong, K.F. and Wong, T.P.(August 2007), Circularly polarized *U*-slot antenna, *IEEE Trans. Antennas and Propagation*, Vol.55, No.8, 2007, pp. 2382-2385

Huang, C.Y.(February 1999), Design for an aperture-coupled compact circularly polarized microstrip antenna, *IEE Proc. Microwaves, Antennas and Propagation*, Vol.146, No.1, Feb. 1999, pp. 13-16

Nasimuddin, Chen, Z.N. and Qing, X.(December 2008), Single fed circularly polarized microstrip antenna with C-slot, *APMC Asia Pacific Microwave Conference* 2008, December 16-20 2008, Hong Kong/Macau, pp. 1-4

Nasimuddin and Chen, Z.N.(April 2009), Aperture-coupled asymmetrical C-shaped slot microstrip antenna for circular polarization, *IET Microwaves, Antennas and Propagation*, Vol.3, No.3, April 2009, pp. 372-378

Nasimuddin, Chen, Z.N. and Qing, X.(December 2008), Aperture-coupled *C*-shape slot cut square microstrip antenna for circular polarization, *Microwave and Optical Technology Letters*, Vol.50, No.12, Dec 2008, pp. 3175-3178

Nasimuddin, Chen, Z.N. and Qing, X.(June 2010), Dual-band circularly polarized *S*-shaped slotted patch antenna with a small frequency-ratio, *IEEE Trans. Antennas and Propagation*, Vol.58, No.6, June 2010, pp. 2112-2115

IE3D EM simulator, version 14.0, 2010.

Yeh,Y., Nasimuddin, Chen, Z.N. and Alphones, A.(April 2009), Aperture-coupled circularly polarized *F*-slot microstrip antenna, *Microwave and Optical Technology Letters*, Vol.51, No.4, April 2009, pp. 1100-1104

15

Microstrip Antenna Arrays

Dr. Albert Sabban
Ort Braude College, Karmiel
Israel

1. Introduction

Microstrip antennas posse's attractive features such as low profile, flexible, light weight, small volume and low production cost. In addition, the benefit of a compact low cost feed network is attained by integrating the RF feed network with the radiating elements on the same substrate. Microstrip antennas are widely presented in books and papers in the last decade (J.R James et al, 1981; A. Sabban, 1981; A. Sabban,1983; A. Sabban & E. Navon, 1983). Microstrip antennas maybe employed in communication links, seekers and in biomedical systems.
In this chapter we present several applications of microstrip antennas. The design of mm wave microstrip antenna arrays with high efficiency is presented in this chapter. Gain limitation in microstrip antenna arrays due to losses in the feed network are presented by (J.R James et. al., 1981). However, this discussion is limited to a 12GHz plane slot array and radiation and dielectric losses are neglected.
The efficiency of microstrip antenna arrays may be improved significantly by reducing losses in the feed network. Losses in the microstrip feed network are due to conductor loss, radiation loss and dielectric loss. Equations to calculate conductor loss and dielectric loss in microstrip lines are given by (J.R James et. al., 1981). In (A. Sabban & K.C. Gupta, 1991; A. Sabban, 1991) a planar multiport network modeling approach has been used to evaluate radiation loss from microstrip discontinuities. Full-wave analysis has been used by (P.B. Kathei & N.G. Alexopoulos, 1985) to calculate radiation conductance of an open-circuited microstrip line. Full-wave analysis methods and software are available for characterization of microstrip discontinuities. These analyses include radiation effects also. The power radiated from a discontinuity may be evaluated from the computed S parameters. However, evaluation of power radiated from the computed S parameters requires a high numerical accuracy of the computed results. Therefore radiation loss values based on full-wave analysis are not widely available.
In this chapter losses in 64 and 256 patch antenna array at Ka band are evaluated. Methods to minimize these losses and to improve the antenna efficiency are presented.
In the literature several applications of mm wave microstrip antenna arrays are described. Gain limitation in microstrip antenna arrays may be solved by employing active microstrip arrays.
However, active microstrip arrays have several disadvantages such as significantly increase in power consumption, weight and dimensions. In some applications reflect arrays are employed. In reflect arrays the received power is transmitted by the same antenna array. Efficiency of microstrip antenna arrays may be improved by using a waveguide feed

network. However, this result in significantly increase in the antenna weight and dimensions. In this case a transition from microstrip to waveguide is required.
Several imaging approaches are presented (M.M. Milkov, 2000; G de Lange et. Al., 1999; A. Rahman et. al., 1996; A. Luukanen et. al., 2001; M. D. Jack et. al., 2001; G. N. Sinclair et. al., 2000). The common approach is based on an array of radiators (antennas) that receives radiation from a specific direction by using a combination of electronic and mechanical scanning. Another approach is based on a steering array of radiation sensors at the focal plane of a lens of reflector. The sensor can be an antenna coupled to a resistor. In this chapter we present the development of millimeter wave radiation detection array. The detection array may employ around 256 to 1024 patch antennas. These patches are coupled to a resistor. Optimization of the antenna structure, feed network dimensions and resistor structure allow us to maximize the power rate dissipated on the resistor. Design considerations of the detection antenna array are given in this chapter.
Microstrip and printed antennas features make them excellent candidates to serve as antennas in biomedical systems. However, the electrical performance of the antenna is altered significantly in vicinity to human body. These facts complicate significantly the antenna design. The electrical performance of a new class of wideband wearable printed antennas for medical applications is presented in this Chapter. RF transmission properties of human tissues have been investigated in several papers (C. Lawrence et. al., 2003; D. Werber et. al., 2006). However, the effect of human body on the electrical performance of the antennas at frequencies that biomedical system operates is not presented. The interaction between microstrip antennas and human body is presented in this Chapter. The antenna bandwidth is around 10% for VSWR better than 2:1. The antenna beam width is around 100°. The antenna gain is around 4dBi. If the air spacing between the sensors and the human body is increased from 0mm to 5mm the antenna resonant frequency is shifted by 5%.

2. Microstrip antenna arrays with high efficiency

One of the major advantages of microstrip antennas is the simplicity of array construction (J.R James et. al., 1981). The radiating elements may be etched jointly with the feed network as an integrated structure leading to a very compact and low cost design. Although the technique for designing feed networks is well established, several difficulties are encountered while implementing it at mm wave frequencies. Microstrip line losses increase considerably at mm wave frequencies. Conductor, dielectric and radiation losses are the major components of loss in mm-wave microstrip antenna arrays. At frequencies ranging from 30 to 40 GHz, conductor losses are around 0.15 to 0.2 dB per wavelength, dielectric losses are around 0.045 dB per wavelength for a 50Ω line on a 10 mil substrate with εr=2.2.
The open nature of microstrip configuration suffers from radiation. In mm-wave microstrip antenna arrays, more bends, T-junctions and other discontinuities are introduced in the feed network and radiation loss increases considerably. The multiport network model is employed to evaluate radiation loss from microstrip discontinuities. Minimization of losses in the microstrip feed network may results in microstrip antenna arrays with high efficiency.

2.1 Evaluation of microstrip feed network losses

Equations to calculate conductor loss and dielectric loss in microstrip lines are given in (J.R James et. al., 1981). Dielectric loss is incorporated in the multiport network analysis by considering a complex dielectric constant. Conductor losses are included in the analysis by defining an equivalent loss tangent δc, given by δc=√ 2/ωμσ/h, where σ is the strip conductivity, h is the substrate height and μ is the free space permeability.

2.1.1 Evaluation of radiation loss

The multiport network model is employed to evaluate radiation loss from microstrip feed networks by adding a number of open ports at the edges of the planar model for the discontinuity structure. The multiport network model is based on the parallel-plate wave-guide model (G. Compa & Mehran, 1975) for microstrip lines. A similar network modeling approach has been used for analysis of microstrip discontinuities (A. Sabban & K.C. Gupta, 1991). The planar waveguide model consists of two parallel conductors bounded by magnetic walls in the transverse directions. In this modeling approach for microstrip structures, fields underneath the microstrip configuration, the external fields (radiated fields, surface waves) are modeled separately in terms of multiport sub networks by adding an equivalent edge admittance network connected to the edges of the microstrip configuration. These sub networks are characterized in terms of Z-matrices that are evaluated by using the Green's function approach. The sub-networks are combined using the segmentation technique to obtain circuit characteristics such as scattering parameters. Equations to compute the sub-networks Z-matrices are given in (A. Sabban, 1991).
The multiport network model is used to evaluate the voltage distributions at the open ports, voltages at the discontinuity edges are represented by equivalent magnetic current sources, as shown in Fig.1. The amplitude M of the magnetic current elements is twice that of the edge voltage at that location and the phase of the magnetic current is equal to the phase of the corresponding voltage. The total radiation is computed using the superposition of the far-field radiated by each section. Referring to the coordinate system, shown in Fig. 2, the far-field pattern F(θ,ϕ) may be written in terms of voltages at the various elements. With the voltage at the ith element as (V(i) $e^{j\alpha(i)}$), we have

$$F(\theta,\phi)=\sum_{i=1}^{N}2V(i)W(i)\exp\{k_0\gamma_0(i)+\alpha(i)\}F_i(\theta,\phi) \tag{1}$$

where

$$F_i(\theta,\phi)=\frac{\sin(\frac{k_0W(i)}{2}\cos\theta)}{\frac{k_0W(i)}{2}\cos\theta}\sin\theta$$

$$\gamma_0(i)=X_0(i)\sin\theta\cos\phi+Y_0(i)\cos\theta$$

and N is the number of ports , $X_0(i)$, $Y_0(i)$ specify the location of the ith magnetic current element, k_0 is the free space wave number, W(i) is the width of the ith element. The factor 2 in (1) accounts for the image of the magnetic current with respect to the ground plane. The radiated power is calculated by the integration of the Poynting vector over the half space and may be written as

$$P_r=\frac{1}{240\pi}\int_{\frac{-\pi}{2}}^{\frac{\pi}{2}}\int_0^{\pi}(\left|E_\theta\right|^2+\left|E_\phi\right|^2)r^2\sin\theta\ d\theta\ d\phi \tag{2}$$

The fields E_θ and E_ϕ are expressed in terms of $F(\theta,\phi)$ as

$$E_\theta = \hat{a}_\theta\left(\frac{-jk_0}{4\pi r}F(\theta,\phi)F_\theta\right) \tag{3}$$

$$E_\phi = \hat{a}_\varphi\left(\frac{-jk_0}{4\pi r}F(\theta,\phi)F_\phi\right) \tag{4}$$

where

$$F_\phi = \sin\phi'\sin\phi + \cos\delta\cos\phi\cos\phi' \tag{5}$$

$$F_\theta = -\sin\phi'\cos\theta\cos\phi + \cos\delta\cos\theta\sin\phi + \sin\delta\cos\phi'\sin\theta \tag{6}$$

$$\cos\theta' = \sin\theta\sin\phi\sin\delta + \cos\theta\cos\delta \tag{7}$$

$$\cos\phi' = \sin\theta\cos\phi / \sqrt{1-\cos^2\theta'} \tag{8}$$

The radiated power may be expressed as a fraction of input power $10\log_{10}(P_r/P_i)$ dB. The radiation loss may be expressed as, Radiation Loss (dB) = $10\log_{10}(1-P_r/P_i)$. Where P_i is the input power at port 1. P_i is calculated from input current and the input impedance of the discontinuity terminated in matched loads at other ports.

2.1.2 Radiation loss from microstrip discontinuities

Computation of microstrip feed network losses are given in (A. Sabban, 1991). As an example radiation loss of a right angle bend in a 50 ohm line on a 10 mil duroid substrate with εr=2.2, is 0.1dB at 30GHz and 0.17 dB at 40GHz. Dielectric loss can be reduced by using substrates with a low dielectric loss. To minimize radiation loss, the number of discontinuities, such as bends and T-Junctions, should be made as small as possible. Radiation from a curved microstrip line is much smaller compared to radiation from right angled bend. Moreover in order to reduce radiation loss in the feed network, the width of the microstrip line is designed to be less than 0.12λ on 0.25mm substrate with εr=2.2.
Conductor loss may be minimized by designing the feed network length per wavelength as short as possible. By using a multilayer feed network design, the feed network length per wavelength is minimized considerably. Gold plating of the microstrip lines decreases conductor losses. Low loss coaxial cables may replace in the microstrip feed network long sections of microstrip lines. As an example the insertion loss of a flexible cable at 30GHz is 0.039 dB per centimeter. However, the insertion loss at 30GHz of a 50 ohm microstrip line on a 10 mil substrate with εr=2.2 is around 0.2dB. By replacing a microstrip line with length of ten centimeters with a coaxial transmission line, the loss decreases by 1.6 dB. The coaxial line may be an integral part of the feed network or it may be embedded in the metallic ground plane. The transition from microstrip line to coaxial line is straightforward. The coaxial line center conductor is soldered to the microstrip line and the outer conductor is soldered or glued to the ground plane, by using conductive glue.

2.2 64 and 256 microstrip antenna arrays with high efficiency

Microstrip antenna arrays with integral feed networks may be broadly divided into arrays fed by parallel feeds and series fed arrays. Usually series fed arrays are more efficient than parallel fed arrays. However, parallel fed arrays have a well controlled aperture distribution.

Two Ka band microstrip antenna arrays which consist of 64 radiating elements have been designed. The first array uses a parallel feed network and the second uses a parallel-series feed network as shown in Fig. 3.a and 3.b. Comparison of the performance of the arrays is given in Table 1. Results given in Table 1 verifies that the parallel series fed array is more efficient than the parallel fed array due to minimization of the number of discontinuities in the parallel series feed network.

The parallel-series fed array has been modified by using a five centimeter coaxial line to replace the same length of microstrip line. Results given in Table 1 indicates that the efficiency of the parallel series fed array that incorporates coaxial line in the feed network is around 67.6% due to minimization of the microstrip line length.

PARAMETER	Corporate feed	Parallel feed	Microstrp & coaxial corporate feed
Number of elements	64	64	64
Beamwidth(deg.)	8.5	8.5	8.5
Computed gain(dBi)	26.3	26.3	26.3
Microstrip line loss(dB)	1.1	1.2	0.5
Radiation loss T-J.(dB)	0.27	0. 45	0.27
Radiation loss bends(dB)	0.4	0.8	0.4
Radiation loss steps(dB)	0.045	-	0.045
Mismatch Loss (dB)	0.5	0.5	0.5
Expected Gain(dBi)	24.0	23.35	24.6
Efficiency (%)	58.9	50.7	67.6

Table 1. Performance of 64 Elements Microstrip Antenna arrays

Two microstrip antenna arrays which consist of 256 radiating elements has been designed. In the first array, Type A as shown in Fig. 4.a, the number of microstrip discontinuities is minimized. The second array, Type B as shown in Fig. 4.b, incorporate more bend discontinuities in the feeding network. Comparison of the performance of the arrays is given in Table 2. The Type (A) array with 256 radiating elements has been modified by using a ten

PARAMETER	Type A	Type B	Type A& Microstrp coaxial feed
Number of elements	256	256	256
Beamwidth(deg.)	4.2	4.2	4.2
Computed gain(dBi)	32	32	32
Microstrip line loss(dB)	3.1	3.1	1.5
Radiation loss T-J.(dB)	0.72	0.72	0.72
Radiation loss bends(dB)	0.13	1.17	0.13
Radiation loss steps(dB)	0.12	-	0.12
Mismatch Loss (dB)	0.5	0.5	0.5
Expected Gain(dBi)	27.43	26.5	29.03
Efficiency (%)	34.9	28.2	50.47

Table 2. Performance of 256 Elements Microstrip Antenna arrays

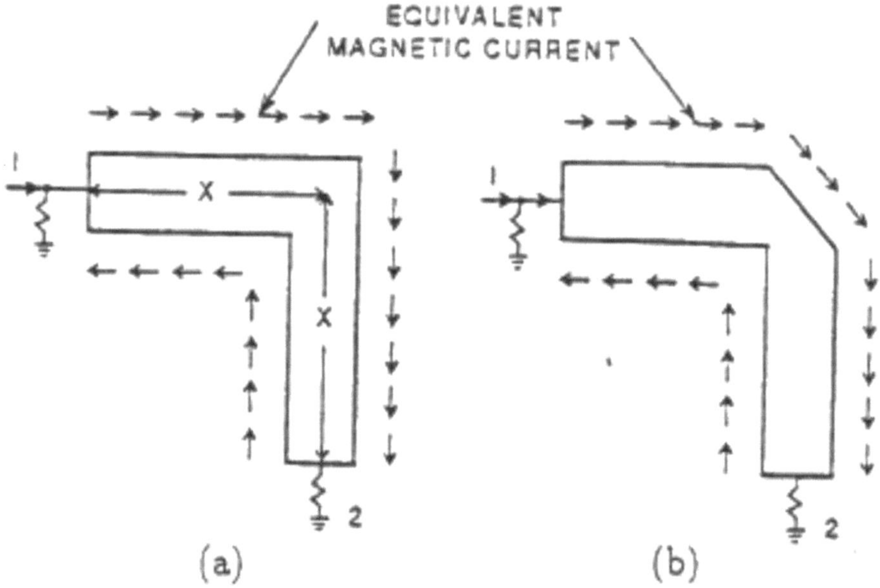

Fig. 1. Equivalent magnetic current distribution at discontinuity edges

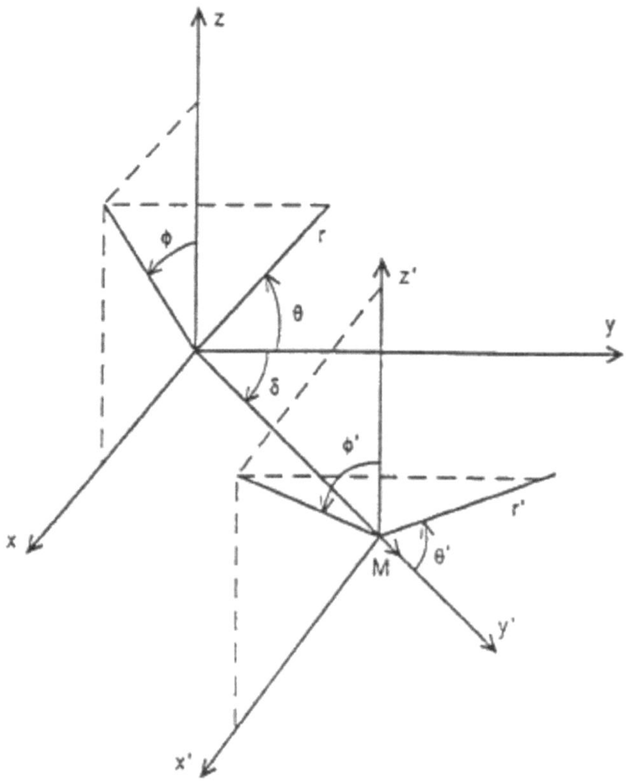

Fig. 2. Coordinate system for external field calculations.

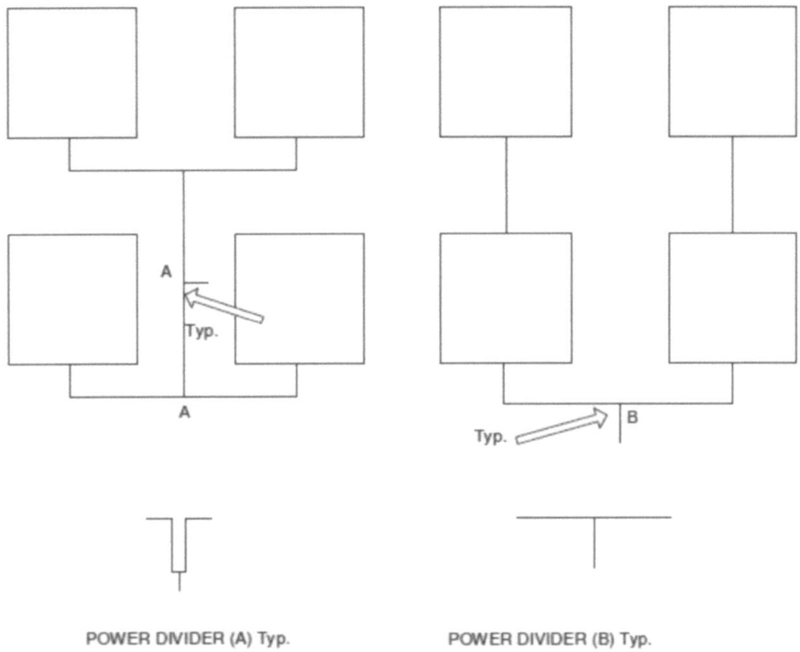

Fig. 3. Configuration of 64 elements microstrip antenna array
a. Parallel feed network. b. Parallel series feed network.

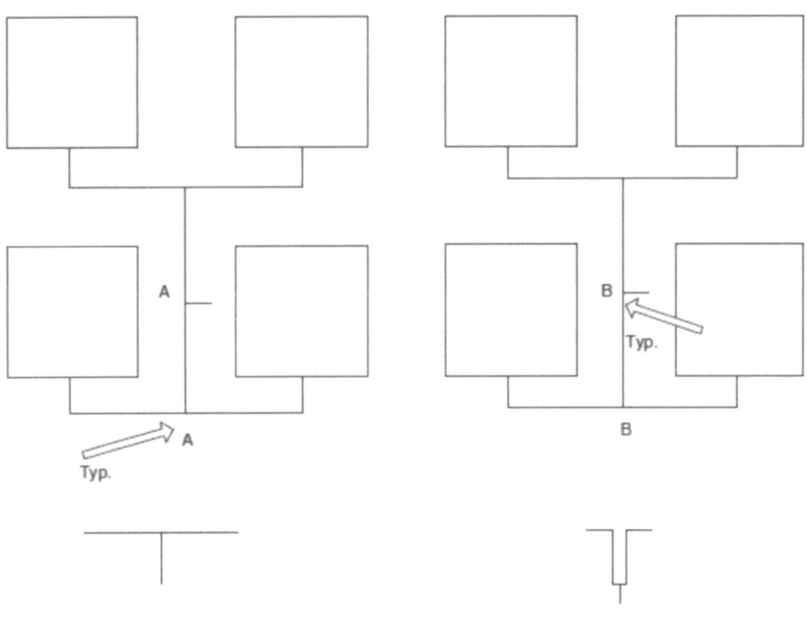

Fig. 4. Configuration of 256 elements microstrip antenna array
a. Array type A. b. Array type B.

centimeter coaxial line to replace the same length of microstrip line. Comparison of the arrays performance is given in Table 2. Table 2 shows that the gain of the modified array has been increased by 1.6 dB.
Results given in Table 2 verifies that the Type A array is more efficient than the Type B array due to minimization of the number of bend discontinuities in the Type A array feed network. The measured gain results are very close to the computed gain results and verify the loss computation presented in this chapter.

3. W band microstrip antenna detection array

Losses in the microstrip feed network are very high in the W band frequency range. In W band frequencies we may design a detection array. The array concept is based on an antenna coupled to a resistor. A direct antenna-coupling surface to a micro machined micro bridge resistor is used for heating and sensing. Analog CMOS readout circuit may be employed as a sensing channel per pixel. Fig. 5 presents a pixel block diagram.

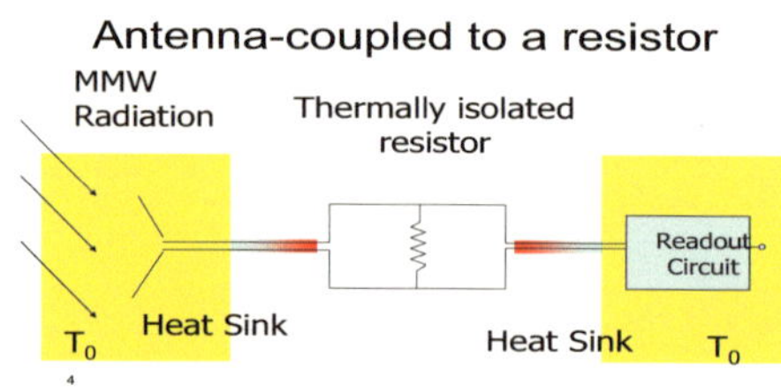

Fig. 5. Antenna coupled to a resistor

3.1 The array principle of operation

The antenna receives effective mm wave radiation. The radiation power is transmitted to a thermally isolated resistor coupled to a Ti resistor. The electrical power raises the structure temperature with a short response time. The same resistor changes its temperature and therefore its electrical resistance. Fig. 6 shows a single array pixel. The pixel consist a patch antenna, a matching network, printed resistor and DC pads. The printed resistor consist Titanium lines, a Titanium resistor coupled to an isolated resistor.

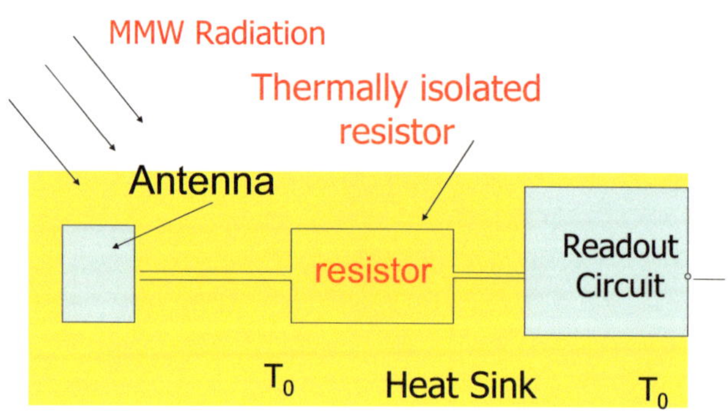

Fig. 6. A single array pixel

The operating frequency range of 92 to 100GHz is the best choice. In the frequency range of 30-150GHz there is a proven contrast between land, sky and high transmittance of clothes.

Size and resolution considerations promote higher frequencies above 100GHz. Typical penetration of clothing at 100GHz is 1dB and 5 to 10dB at 1THz. Characterization and measurement considerations promote lower frequencies. The frequency range of 100 GHz allows sufficient bandwidth when working with illumination. The frequency range of 100 GHz is the best compromise. Fig. 7 presents the array concept. Several types of printed antennas may be employed as the array element such as bowtie dipole, patch antenna and ring resonant slot.

3.2 W band antenna design

The bowtie dipole and a patch antenna have been considered as the array element. Computed results shows that the directivity of the bowtie dipole is around 5.3dBi and the directivity of a patch antenna is around 4.8dBi. However the length of the bowtie dipole is around 1.5mm the size of the patch antenna is around 700x700µm. We used a quartz substrate with thickness of 250µm. The bandwidth of the bowtie dipole is wider than that of a patch antenna. However, the patch antenna bandwidth meets the detection array electrical specifications. We chose the patch antenna as the array element since the patch size is significantly smaller than that of the bowtie dipole. This feature allows us to design an array with a higher number of radiating elements. The resolution of detection array with a higher number of radiating elements is improved. We also realized that the matching network between the antenna and the resistor has smaller size for a patch antenna than that for a bowtie dipole. The matching network between the antenna and the resistor consist of microstrip open stubs. Fig. 8 shows the 3D radiation pattern of the Bowtie dipole. Fig. 9 presents S11 parameter of the patch antenna.
We compared the electrical performance of the antennas.

Fig. 7. Array concept

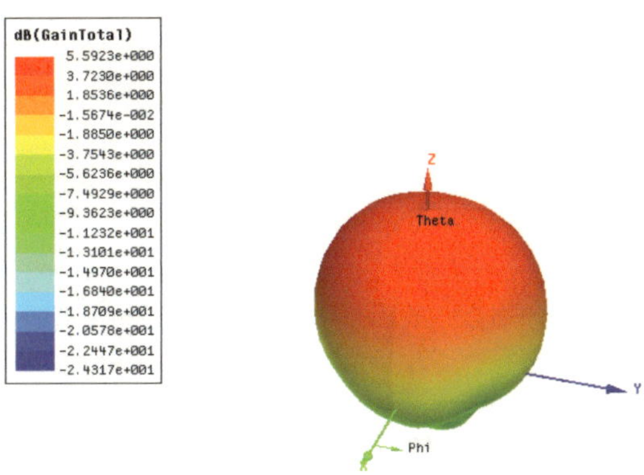

Fig. 8. Dipole 3D radiation pattern

Fig. 7 shows the 3D radiation pattern of the patch antenna.

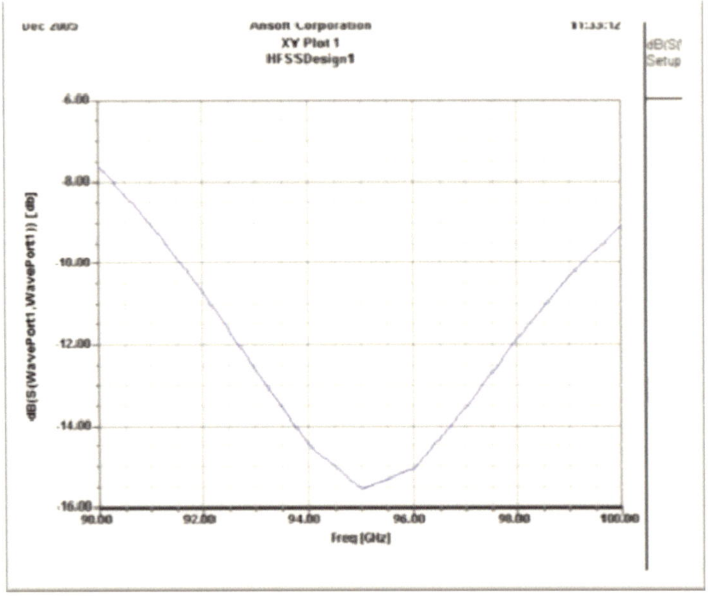

Fig. 9. Patch S11 computed results

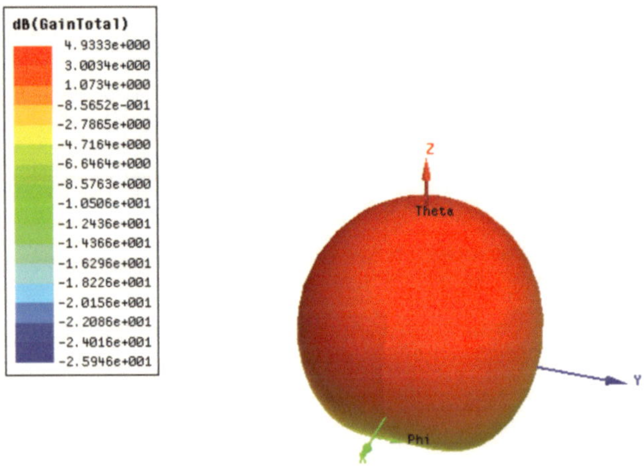

Fig. 10. Patch 3D radiation pattern

3.3 Resistor design

As described by (M.M. Milkov, 2000) the resistor is thermally isolated from the patch antenna by using a sacrificial layer. Optimizations of the resistor structure maximize the power rate dissipated on the resistor. Ansoft HFSS software is employed to optimize the height of the sacrificial layer, the transmission line width and length. Dissipated power on Titanium resistor is higher than the dissipated power on Platinum resistor. The rate of the dissipated power on the resistor is around 25%. Material properties are given in Table 3.

Ti	siNi	Units	Property
7	1.6	W/m/K	Conductivity [K]
520	770	J/Kg/K	Capacity [C]
4.5	2.85	Gr/cm^3	Density [ρ]
90	>1e8	Ω/□	Resistance
0.1	**0.1**	**μm**	**Thickness**

Table 3. Material properties

The sacrificial layer thickness may be 2 to 3μm. Fig. 11 shows the resistor configuration.

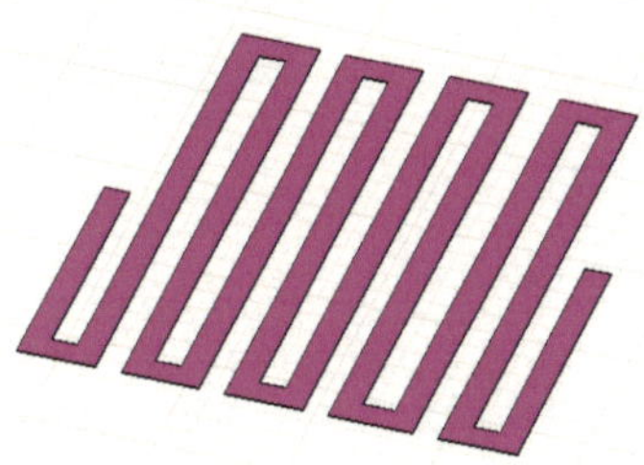

Fig. 11. Resistor Configuration

3.4. 220GHz microstrip patch antenna

Quartz substrate with thickness of 50μm to 100μm has been used to fabricate microstrip antennas at frequencies higher than 200GHz. The size of the patch antenna is around 300x300μm.

Fig. 12 presents S11 parameter of the patch antenna. Fig. 13 shows the 3D radiation pattern of the patch antenna.

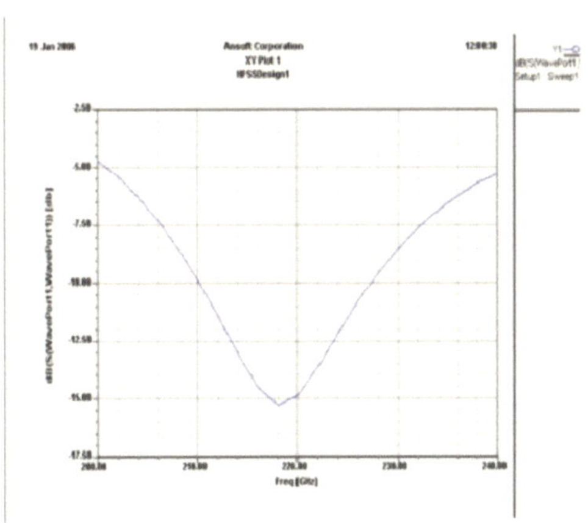

Fig. 12. 220GHz Patch S11 computed results

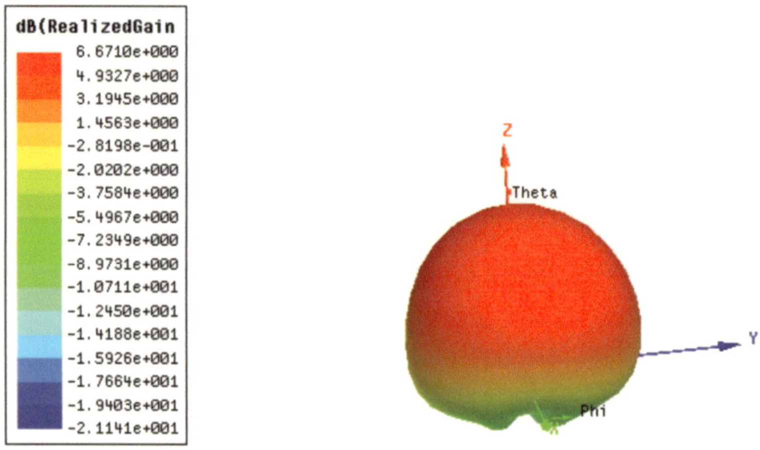

Fig. 13. 220GHz Patch 3D radiation pattern

4. Medical applications of microstrip antennas

Microstrip antennas posse's attractive features such as low profile, flexible, light weight, small volume and low production cost. Microstrip and printed antennas features make them excellent candidates to serve as antennas in biomedical systems. However, the antenna electrical performance is altered significantly in vicinity to human body. These facts complicate significantly the antenna design. The electrical performance of a new class of wideband wearable printed antennas for medical applications is presented in this Chapter. RF transmission properties of human tissues have been investigated in several papers (C. Lawrence et. al., 2003; D. Werber et. al., 2006). However, the effect of human body on the electrical performance of the antennas at frequencies that biomedical system operates is not presented. A new class of wideband compact wearable printed and microstrip antennas for medical applications is presented in this chapter.

4.1 Dual polarized 434MHz printed antenna

A new compact microstrip loaded dipole antennas has been designed to provide horizontal polarization. The antenna consists of two layers. The first layer consists of RO3035 0.8mm dielectric substrate. The second layer consists of RT-Duroid 5880 0.8mm dielectric substrate. The substrate thickness determines the antenna bandwidth. However, with thinner substrate we may achieve better flexibility. We also designed a thicker double layers microstrip loaded dipole antennas with wider bandwidth. A printed slot antenna provides a vertical polarization. The proposed antenna is dual polarized. The printed dipole and the slot antenna provide dual orthogonal polarizations.

The dual polarized antenna is shown in Fig. 14. The antenna dimensions are 26x6x0.16 cm. The antenna may be employed as a wearable antenna on a human body. The antenna may be attached to the patient shirt in the patient stomach zone. Alternatively the antenna may be attached to the patient back. The antenna has been analyzed by using Agilent ADS software. There is a good agreement between measured and computed results. The antenna bandwidth is around 10% for VSWR better than 2:1 as shown in Fig. 15. The antenna beam width is around 100º. The antenna gain is around 4dBi.

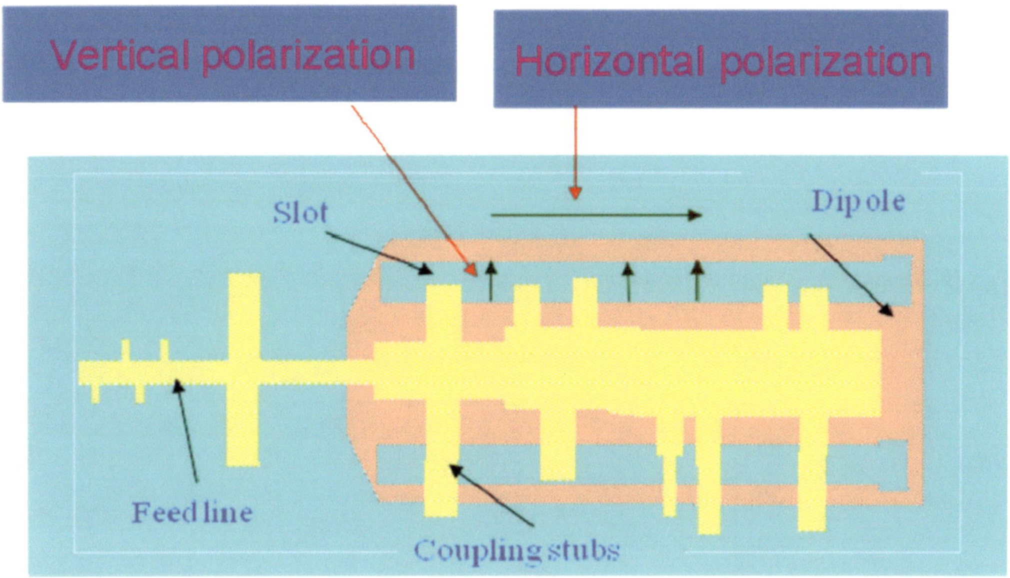

Fig. 14. Printed dual polarized antenna

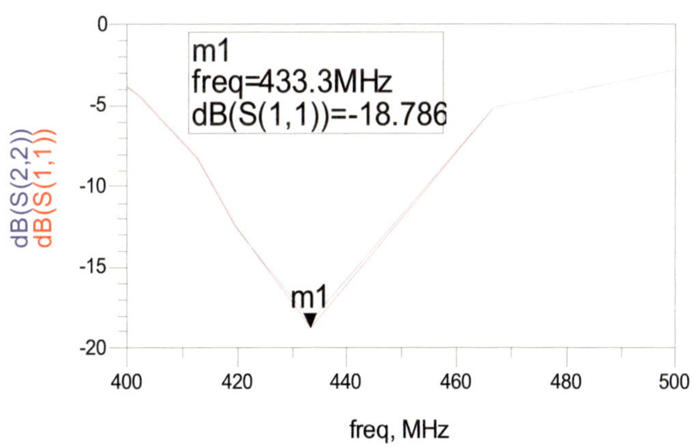

Fig. 15. Computed S11 and S22 results

The computed S11 and S22 parameters are presented in Fig. 15. Fig. 16 presents the antenna measured S11 parameters. The computed radiation pattern is shown in Fig. 17. The antenna cross-polarized field strength may be adjusted by varying the slot feed location.

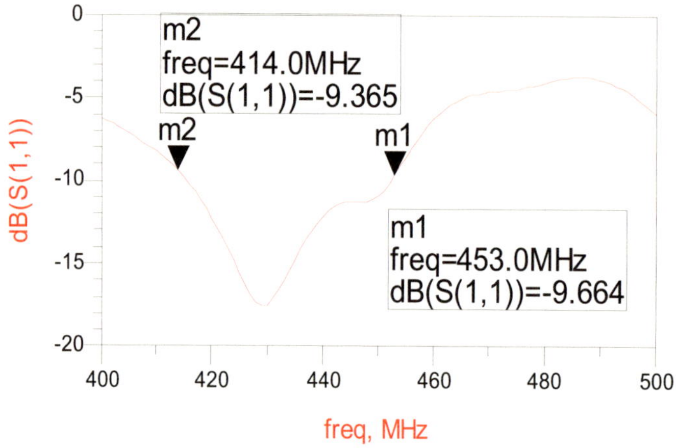

Fig. 16. Measured S11 on human body

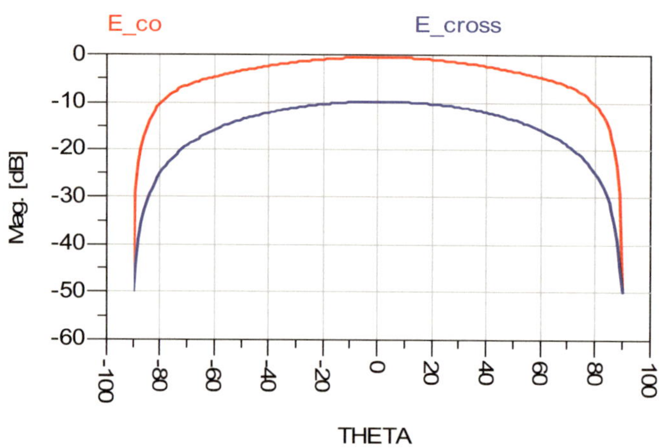

Fig. 17. Antenna Radiation pattern

The antenna dimensions may be reduced to 6X6X0.16cm by employing a folded antenna configuration as shown in Fig. 18.

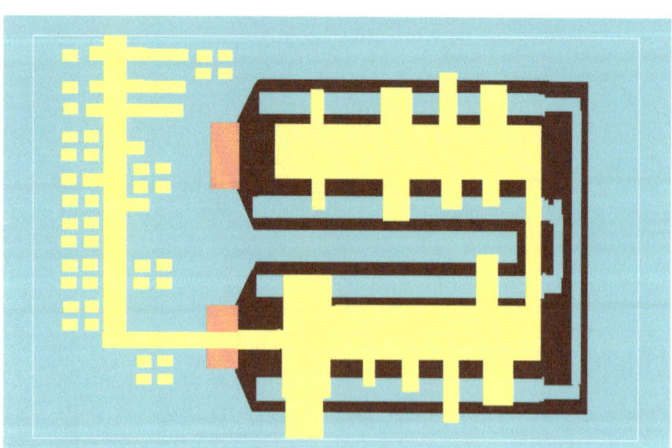

Fig. 18. Folded dual polarized antenna

Fig. 19 presents the antenna computed S11 and S22 parameters. The computed radiation pattern of the folded dipole is shown in Fig. 20.

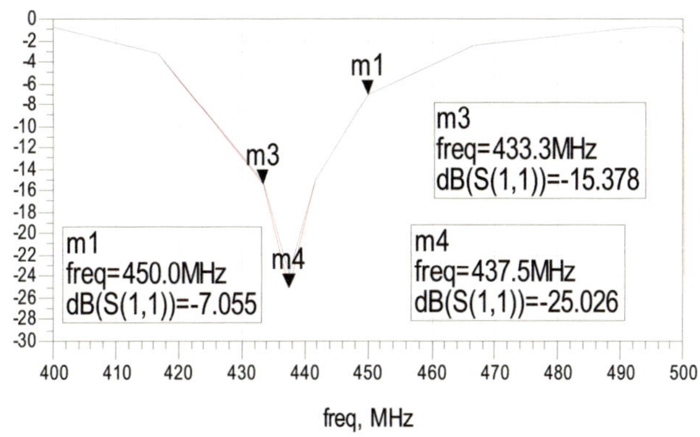

Fig. 19. Folded antenna Computed S11 and S22 results

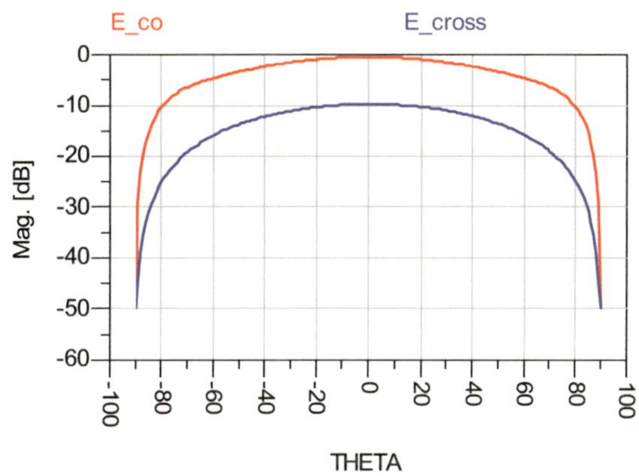

Fig. 20. Folded antenna Radiation pattern

4.2 New loop antenna with ground plane

A new loop antenna with ground plane has been designed on Kapton substrates with thickness of 0.25mm and 0.4mm. The loop antenna diameter is around 50mm. The antenna is shown in Fig. 21. Fig. 22 presents the Loop antenna computed S11 on human body. The computed radiation pattern is shown in Fig 23.

Table 4 compares the electrical performance of a loop antenna with ground plane with a loop antenna without ground plane. There is a good agreement between measured and computed results of antenna parameters on human body.

VSWR	Gain dBi	Beam width 3dB	Antenna
4:1	0	100°	Loop no GND
2:1	0	100°	Loop with GND

Table 4. Comparison of Loop Antennas

Fig. 21. Loop antenna with ground plane

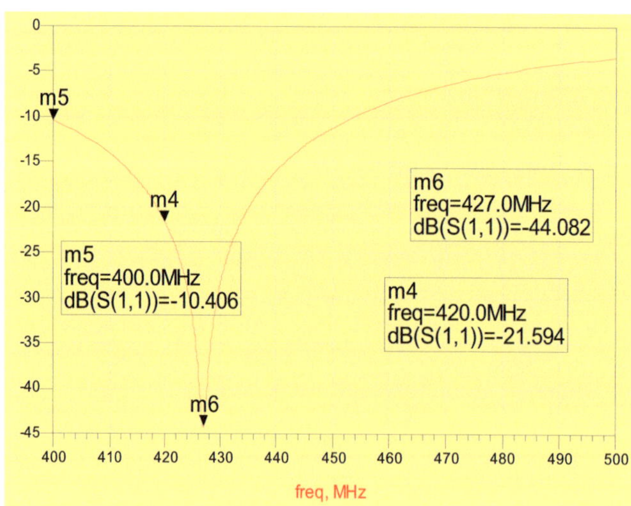

Fig. 22. Computed S11 of Loop Antenna

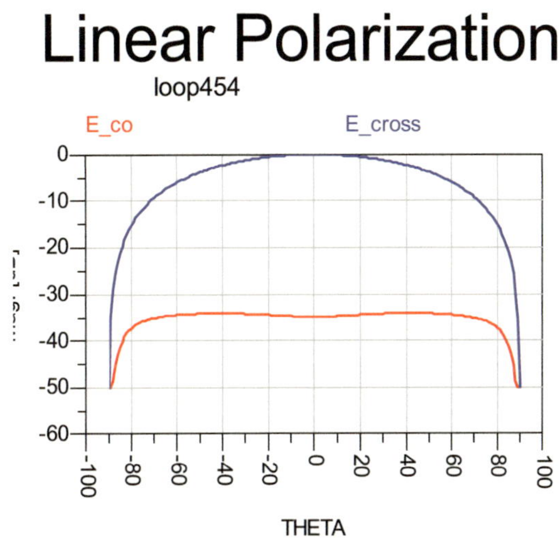

Fig. 23. Loop Antenna Radiation pattern on human body

4.3 Antenna S11 variation as function of distance from body

The Antenna input impedance variation as function of distance from the body has been computed by employing ADS software. The analyzed structure is presented in Fig. 24.

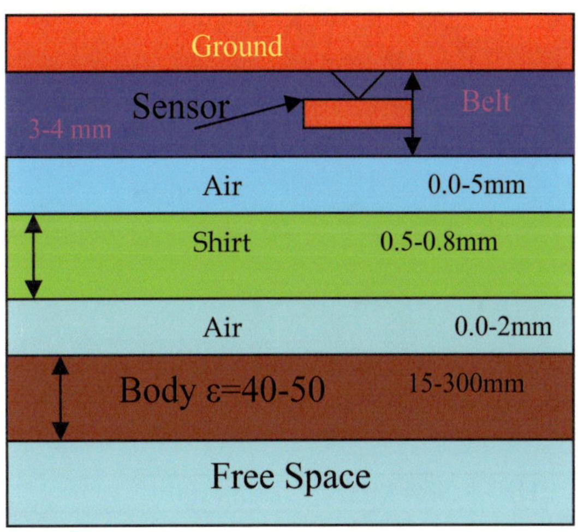

Fig. 24. Analyzed structure for Impedance calculations

Properties of human body tissues are listed in Table 5 see (C. Lawrence et. al., 2003).
These properties were employed in the antenna design. Fig. 25 presents S11 results for different belt thickness, shirt thickness and air spacing between the antennas and human body. One may conclude from results shown in Fig. 26 that the antenna has V.S.W.R better than 2.5:1 for air spacing up to 8mm between the antennas and patient body. For frequencies ranging from 415MHz to 445MHz the antenna is matched when there is no air spacing between the antenna and the patient body.

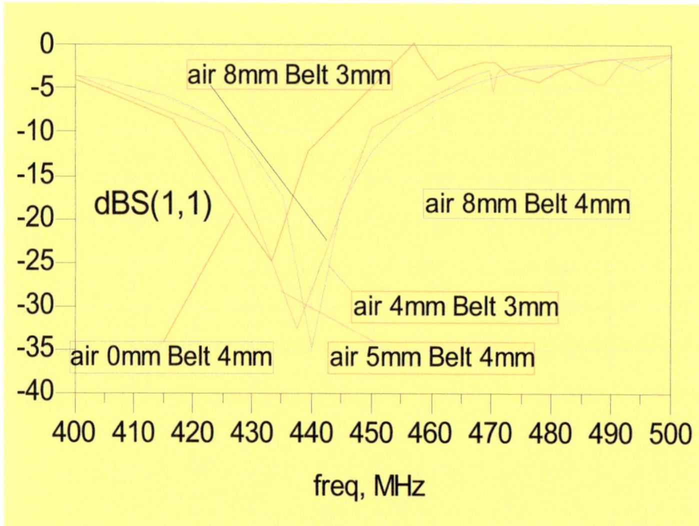

Fig. 25. S11 results for different antenna position relative to the human body

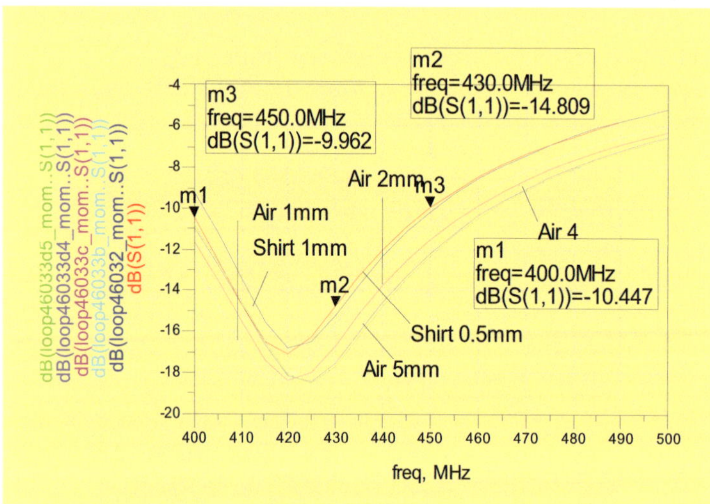

Fig. 26. Folded antenna S11 results for different position relative to the human body

600 MHz	434 MHz	Property	Tissue
0.6	0.57	σ	Skin
40.43	41.6	ε	
0.73	0.67	σ	Stomach
41.41	42.9	ε	
1.06	0.98	σ	Colon, Muscle
61.9	63.6	ε	
0.27	0.27	σ	Lung
38.4	38.4	ε	

Table 5. Properties of human body tissues

Fig. 27 presents S11 results of the folded antenna results for different position relative to the human body. Explanation of Fig. 26 is given in Table 6. From results shown in Fig. 13 we can see that the folded antenna has V.S.W.R better than 2.0:1 for air spacing up to 5mm between the antennas and patient body. If the air spacing between the sensors and the human body is increased from 0mm to 5mm the antenna resonant frequency is shifted by 5%.

Sensor position	Plot colure
Shirt thickness 0.5mm	Red
Shirt thickness 1mm	Blue
Air spacing 2mm	Pink
Air spacing 4mm	Green
Air spacing 1mm	Sky
Air spacing 5mm	Purple

Table 6. Explanation of Fig. 26

Sensor position	Colure
Shirt thickness 0.5mm	Red
Air spacing body to shirt 1mm	Blue
Belt thickness 4mm	Pink
Air spacing shirt to belt 1mm	Sky
Air spacing shirt to belt 4mm	Green

Table 7. Explanation of Fig. 27

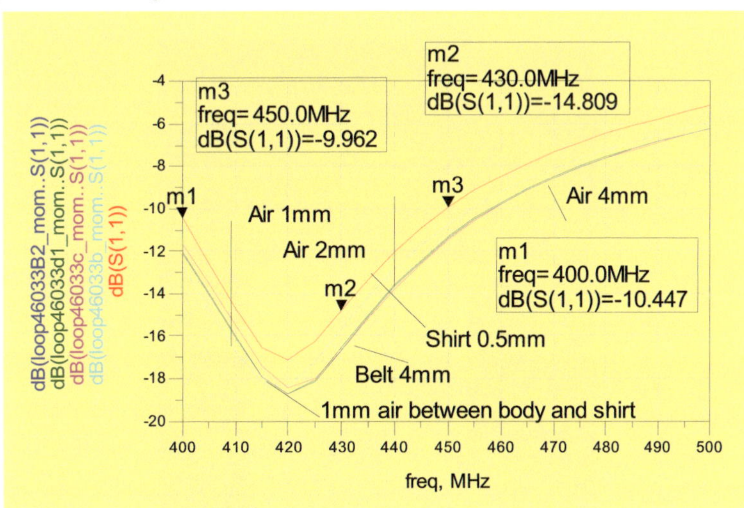

Fig. 27. S11 results for different belt thickness

S11results in Fig. 27 presents the folded antenna matching when the belt thickness has been changed from 2 to 4mm. Explanation of Fig. 27 is given in Table 7. S11results are better than -10dB for belt thickness ranging from to 2 to 4mm. Computed S11 and S22 results was better than-10dB for different body tissues with dielectric constant ranging from 40 to 50. Computed S11 and S22 results were better than -10dB for different shirts and belts with dielectric constant ranging from 2 to 4.

4.4 Medical applications

An application of the proposed antenna is shown in Fig. 28. Three to four folded dipole antennas may be assembled in a belt and attached to the patient stomach. The cable from each antenna is connected to a recorder. The received signal is routed to a switching matrix. The signal with the highest level is selected during the medical test. The antennas receive a signal that is transmitted from various positions in the human body. Folded antenna may be also attached on the patient back in order to improve the level of the received signal from different locations in the human body. Fig. 29 and Fig. 30 show various antenna locations on the back and front of the human body for different medical applications. In several applications the distance separating the transmitting and receiving antennas is less than $2D^2/\lambda$, where D is the largest dimension of the source of the radiation. In these applications the amplitude of the electromagnetic field close to the antenna may be quite powerful, but because of rapid fall-off with distance, they do not radiate energy to infinite distances, but instead their energies remain trapped in the region near the antenna, not drawing power from the transmitter unless they excite a receiver in the area close to the antenna. Thus, the near-fields only transfer energy to close distances from the receivers, and when they do, the result is felt as an extra power-draw in the transmitter. The receiving and transmitting antennas are magnetically coupled. Change in current flow through one wire induces a voltage across the ends of the other wire through electromagnetic induction. The amount of inductive coupling between two conductors is measured by their mutual inductance. In these applications we have to refer to the near field and not to the far field radiation pattern.

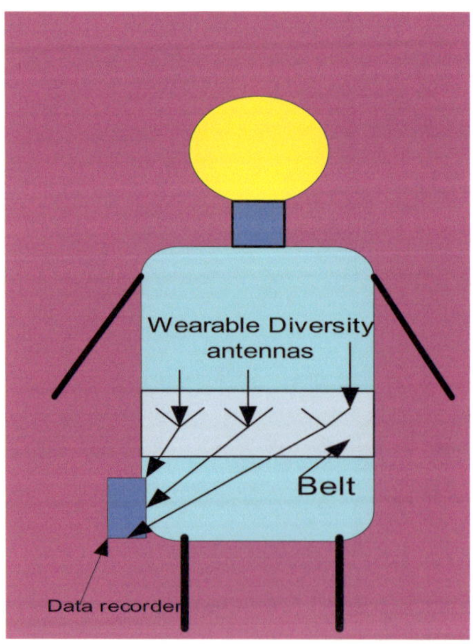

Fig. 28. Wearable antenna

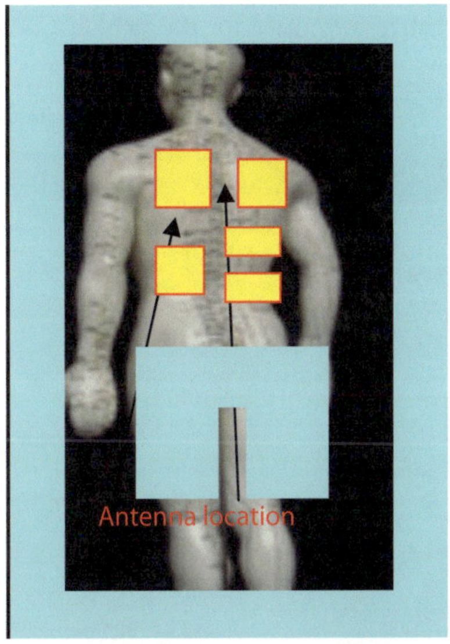

Fig. 29. Printed Antenna locations on the back for medical applications

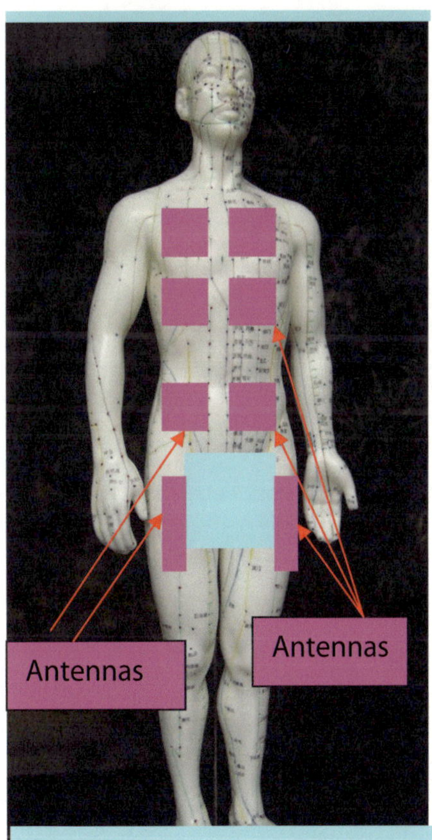

Fig. 30. Printed Patch Antenna locations for medical applications

In Fig. 31and 32 several microstrip antennas for medical applications at 434MHz are shown.

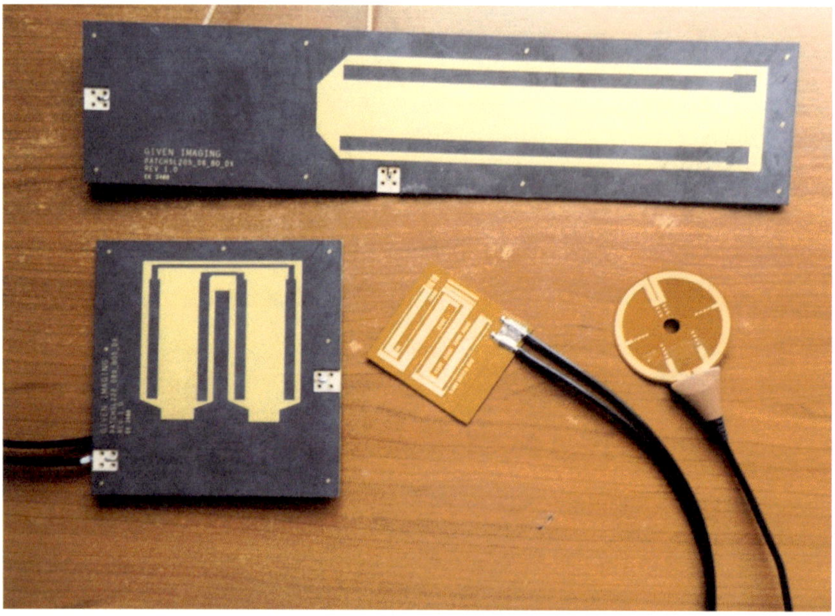

Fig. 31. Microstrip Antennas for medical applications

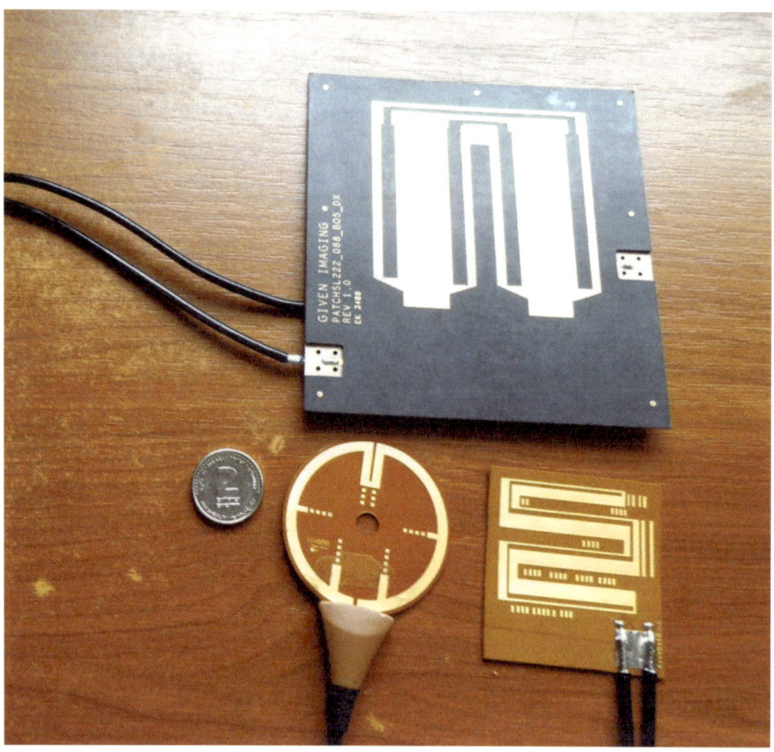

Fig. 32. Microstrip Antennas for medical applications

In Fig. 31 and in Fig. 32 one can see different designs of dual polarized microstrip antennas with 10% bandwidth around 434MHz. The loop antenna is with a ground plane on the antenna back. The loop antenna diameter is around 50mm.

5. Conclusion

A 64 microstrip antenna array with efficiency of 67.6% and a 256 microstrip antenna array with efficiency of 50.47% have been presented in this Chapter. Methods to reduce losses in mm-wave microstrip antenna arrays have been described in this Chapter. The results presented in this chapter point out that radiation losses need to be taken into account for accurate microstrip antenna array design at mm wave frequencies. By minimizing the number of bend discontinuities the gain of the 256 microstrip antenna array has been increased by 1dB.

Several applications of mm wave microstrip antenna arrays have been presented. Losses in the microstrip feed network form a significant limit on the possible applications of microstrip antenna arrays in mm wave frequencies. MM wave microstrip antenna arrays may be employed in communication links, seekers and detection arrays. The array may consist around 256 elements to 1024 elements. Design considerations of the antenna and the feed network are given in this chapter. Optimization of the antenna structure and feed network allows us to design and fabricate microstrip antenna arrays with high efficiency.

This chapter presents wideband microstrip antennas with high efficiency for medical applications. The antenna bandwidth is around 10% for VSWR better than 2:1. The antenna beam width is around 100°. The antenna gain is around 2dBi. The antenna S11 results for different belt thickness, shirt thickness and air spacing between the antennas and human body are given in this chapter. The effect of the antenna location on the human body should be considered in the antenna design process. If the air spacing between the sensors and the human body is increased from 0mm to 5mm the antenna resonant frequency is shifted by 5%.The proposed antenna may be employed in Medicare RF systems.

6. References

J.R. James, P.S Hall & C. Wood, (1981). Microstrip Antenna Theory and Design,1981.

A. Sabban & K.C. Gupta, (1991). Characterization of Radiation Loss from Microstrip Discontinuities Using a Multiport Network Modeling Approach, I.E.E.E Trans. on M.T.T, Vol. 39,No. 4, April 1991,pp. 705-712.

A. Sabban, (1991). PhD Thesis, Multiport Network Model for Evaluating Radiation Loss and Coupling Among Discontinuities in Microstrip Circuits, University of Colorado at Boulder, January 1991.

P.B. Kathei & N.G. Alexopoulos, (1985). Frequency-dependent characteristic of microstrip, MTT-33, discontinuities in millimeter-wave integrated circuits, IEEE Trans. Microwave Theory Tech, vol. pp. 1029-1035, Oct. 1985.

A. Sabban, (1983). A New Wideband Stacked Microstrip Antenna, I.E.E.E Antenna and Propagation Symposium, Houston, Texas, U.S.A, June 1983.

A. Sabban & E. Navon (1983). A MM-Waves Microstrip Antenna Array, I.E.E.E Symposium, Tel-Aviv, March 1983.

A. Sabban, (1981). Wideband Microstrip Antenna Arrays, I.E.E.E Antenna and Propagation Symposium MELCOM, Tel-Aviv,1981.

M. M. Milkov, (2000). Millimeter-Wave Imaging System Based on Antenna-Coupled Bolometer, MSc. Thesis, UCLA UCLA (2000).

G. de Lange et. al., (1999). A 3*3 mm-wave micro machined imaging array with sis mixers, Appl. Phys. Lett. 75 (6), pp. 868-870 (1999).

A. Rahman et. al., (1996). Micromachined room temperature microbolometers for mm-wave detection, Appl. Phys. Lett. 68 (14), pp. 2020-2022 (1996).

A. Luukanen et. al., US Patent 6242740 (2001).

M. D. Jack et. al., (2001). US Patent 6329655 (2001).

G. N. Sinclair et. al., (2000). Passive millimeter wave imaging in security scanning, Proc. SPIE Vol. 4032, pp. 40-45, (2000).

G. Kompa & R. Mehran, (1975). Planar waveguide model for computing microstrip components, Electron Lett., vol. 11, no. 9, pp. 459-460, 1975.

Lawrence C. Chirwa; Paul A. Hammond; Scott Roy & David R. S. Cumming, (2003). Electromagnetic Radiation from Ingested Sources in the Human Intestine between 150 MHz and 1.2 GHz, IEEE Transaction on Biomedical eng., VOL. 50, NO. 4, pp. 484-492, April 2003.

D.Werber; A. Schwentner & E. M. Biebl, (2006). Investigation of RF transmission properties of human tissues, Adv. Radio Sci., 4, pp. 357–360, 2006.

16

Microstrip Antennas for Indoor Wireless Dynamic Environments

Mohamed Elhefnawy and Widad Ismail
Universiti Sains Malaysia (USM), Malaysia

1. Introduction

This chapter is organized in two parts. The first part deals with the design and implementation of a microstrip antenna array with Butler matrix. The planar microstrip antenna array has four beams at four different directions, circular polarization diversity, good axial ratio, high gain, and wide bandwidth by implementing the 4×4 Butler matrix as a feeding network to the 2×2 planar microstrip antenna array. The circular polarization diversity is generated by rotating the linearly polarized identical elements of the planar microstrip antenna array so that the *E*-field in the *x*-direction is equal to the *E*-field in the *y*-direction. Then, by feeding the planar array with Butler matrix, phase delay of $\pm\pi/2$ between those two *E*-fields is provided.

In the second part of this chapter, the analysis, design and implementation of an Aperture Coupled Micro-Strip Antenna (ACMSA) are introduced. A quadrature hybrid is used as a feeder for providing simultaneous circular polarization diversity with a microstrip antenna; but the utilization of the quadrature hybrid as a feeder results in large antenna size. In order to minimize the antenna size, the microstrip antenna is fed by a quadrature hybrid through two orthogonal apertures whose position is determined based on a cavity model theory. The size of the proposed ACMSA is small due to the use of the aperture coupled structure. The cavity model theory is started with Maxwell's equations, followed by the solution of the homogeneous wave equations. Finally, the eigenfunction expansion for the calculation of the input impedance is presented. This chapter is organized as follows. The first part deals with design and implementation of a microstrip antenna array with Butler matrix, which describe the design details of a rectangular microstrip patch antenna and a 4×4 Butler matrix. Further, analysis of planar microstrip antenna array with Butler matrix and the development of the radiation pattern for the planar microstrip antenna array are presented.

In the second part, the design and implementation of an aperture coupled microstrip antenna, the analysis of ACMSA using cavity model, the circular polarization diversity with ACMSA and the geometry of the ACMSA are described.

2. Design and implementation of the microstrip antenna array with Butler matrix

A planar microstrip antenna array with a Butler matrix is implemented to form a microstrip antenna array that has narrow beamwidth, circular polarization and polarization diversity.

This microstrip antenna array improves the system performance in indoor wireless dynamic environments. A circularly polarized microstrip antenna array is designed such that it consists of four identical linearly polarized patches. A 2×2 planar microstrip antenna array and a 4×4 Butler matrix are designed and simulated using advanced design system and Matlab software. The measured results show that a combination of a planar microstrip antenna array and a 4×4 Butler matrix creates four beams two of which have RHCP and the other two have LHCP.

2.1 Design of the rectangular microstrip patch antenna

A rectangular microstrip patch antenna is designed based on the Transmission Line Model (TLM) in which the rectangular microstrip patch antenna is considered as a very wide transmission line terminated by radiation impedance. Figure 1 shows a rectangular microstrip patch antenna of length L and width W. M_s is the magnetic current of each radiating slot of the microstrip patch antenna and s is the width of each radiating slot.

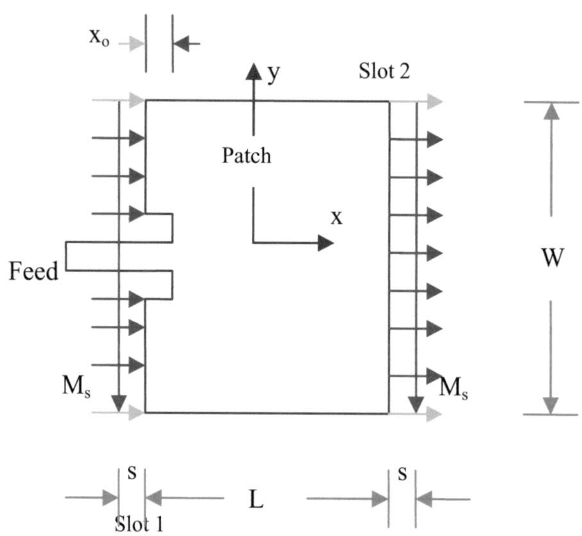

Fig. 1. Inset fed rectangular microstrip patch antenna

Figure 2 shows the transmission line model of the antenna where G_R and C_F represent the radiation losses and fringing effects respectively. The input impedance of an inset fed rectangular microstrip patch antenna is given by the equation (1) [1].

$$Z_{in} = \frac{1}{2\ (G_R + G_{12})} \cos^2\left(\frac{\pi\ x_o}{L}\right) \tag{1}$$

where X_o is the distance into the patch, G_{12} is the coupled conductance between the radiating slots of the antenna [2].

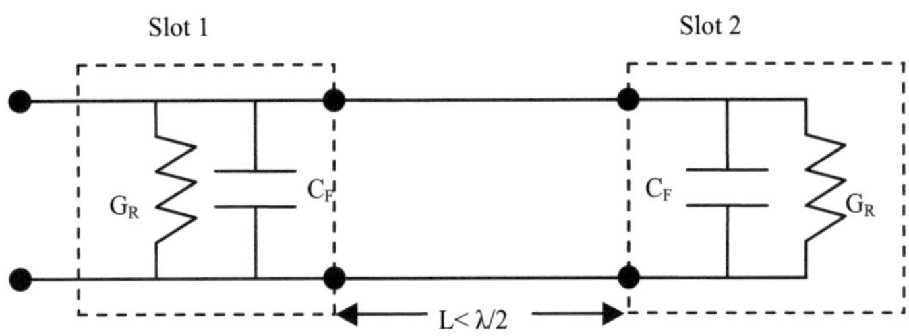

Fig. 2. Transmission line model of the rectangular microstrip patch antenna

The inset fed rectangular microstrip patch antenna is designed using Matlab software based on the expression for the input impedance which is given by equation (1). The input impedance depends on the microstrip line feed position as shown in Figure 3.

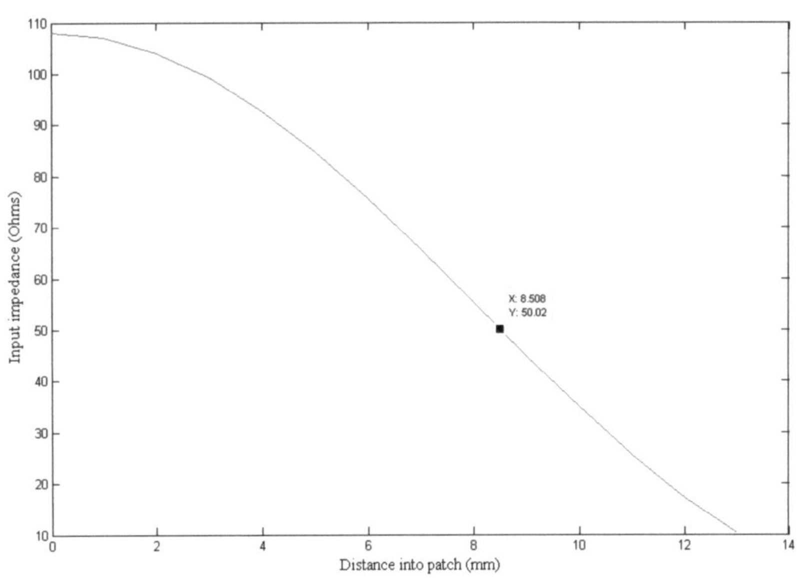

Fig. 3. Dependence of the input impedance on the distance into the patch

2.2 Design of the 4×4 Butler matrix

The Butler matrix is used as a feeding network to the microstrip antenna array and it works equally well in receive and transmit modes. The 4×4 Butler matrix as shown in Figure 4 consists of 4 inputs, 4 outputs, 4 hybrids, 1 crossover to isolate the cross-lines in the planar layout and some phase shifters [3]. Each input of the 4×4 Butler matrix inputs produces a different set of 4 orthogonal phases; each set used as an input for the four element antenna array creates a beam with a different direction. The switching between the four Butler inputs changes the direction of the microstrip antenna array beam.

Advanced design system (ADS) has been used for simulating the 4×4 Butler matrix as shown in Figure 5. Table 1 shows a summary of the simulated and the measured phases that are associated with the selected port of the 4×4 Butler matrix.

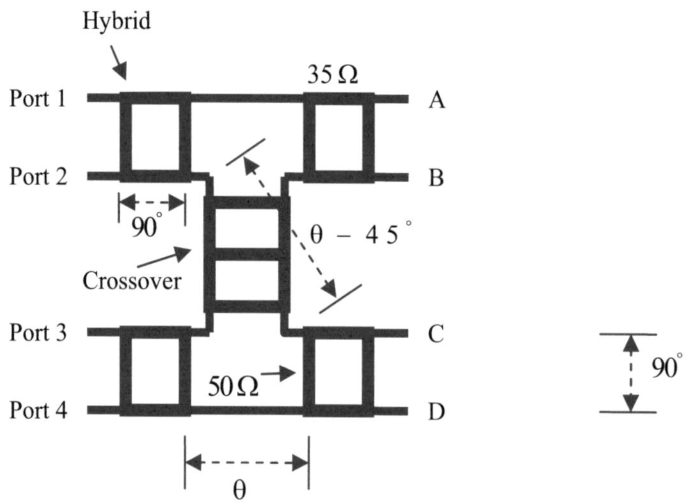

Fig. 4. 4×4 Butler matrix geometry

		Phase A Antenna 1	Phase B Antenna 2	Phase C Antenna 3	Phase D Antenna 4
Port 1 **(set 1)**	**Theoretical**	0	-90	-45	-135
	Simulated	0	-89.813	-45.04	-135.04
	Measured	0	-98	-50	-142.8
Port 2 **(set 2)**	**Theoretical**	0	-90	135	45
	Simulated	0	-90.273	134.142	44.773
	Measured	0	-80.5	140.5	48.5
Port 3 **(set 3)**	**Theoretical**	0	90	-135	-45
	Simulated	0	89.369	-135.046	-44.773
	Measured	0	86.7	-126	-43
Port 4 **(set 4)**	**Theoretical**	0	90	45	135
	Simulated	0	90	45.227	135.04
	Measured	0	89	48	143

Table 1. Phases associated with the selected port of the 4×4 Butler matrix

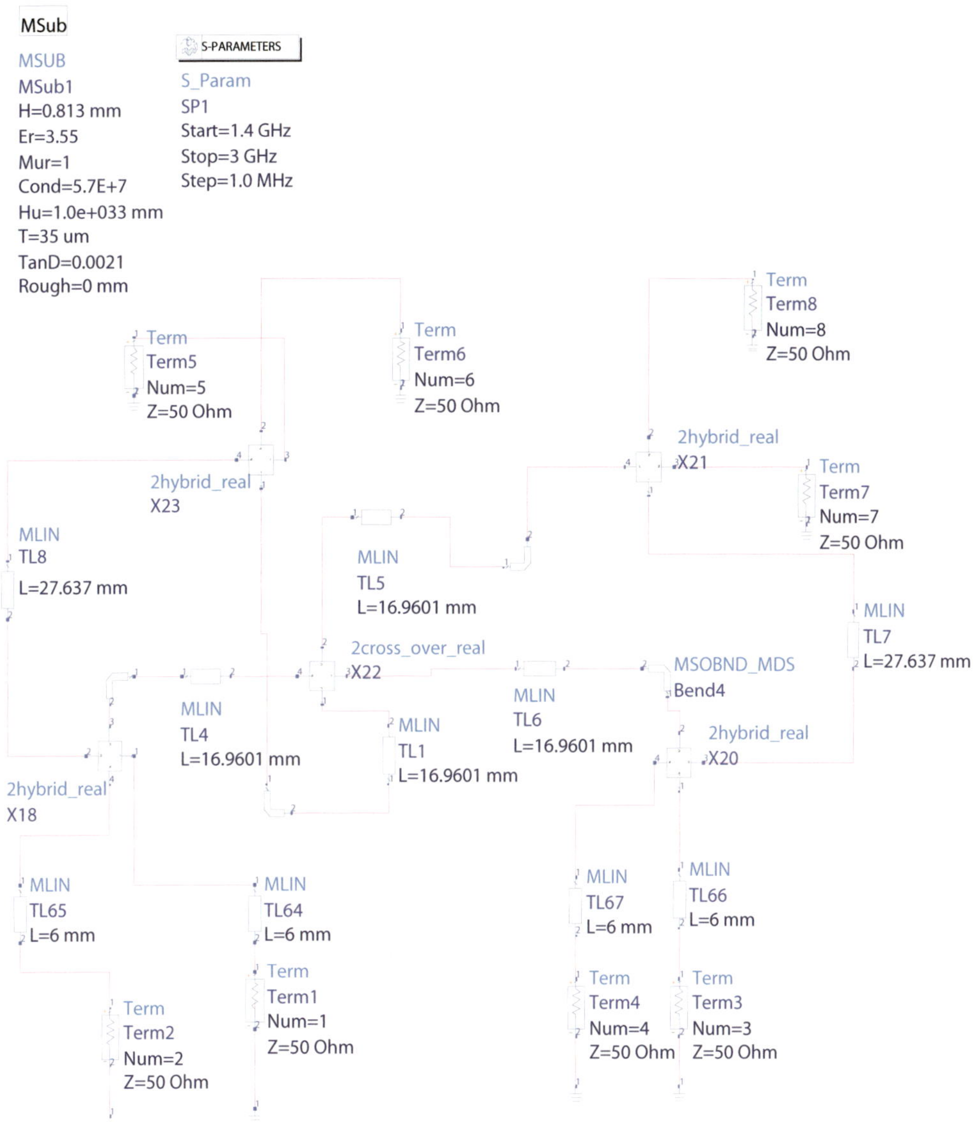

Fig. 5. ADS schematic for 4×4 Butler matrix

2.2.1 Design of quadrature hybrid

The design of a quadrature hybrid is based on the scattering matrix equation which can be obtained as follows [4]:

$$[S]_{Hybrid} = -\frac{1}{\sqrt{2}}\begin{bmatrix} 0 & j & 1 & 0 \\ j & 0 & 0 & 1 \\ 1 & 0 & 0 & j \\ 0 & 1 & j & 0 \end{bmatrix} \quad (2)$$

The design of a quadrature hybrid using ADS is started by creating ADS schematic for the quadrature hybrid using ideal transmission lines as shown in Figure 6 [3].

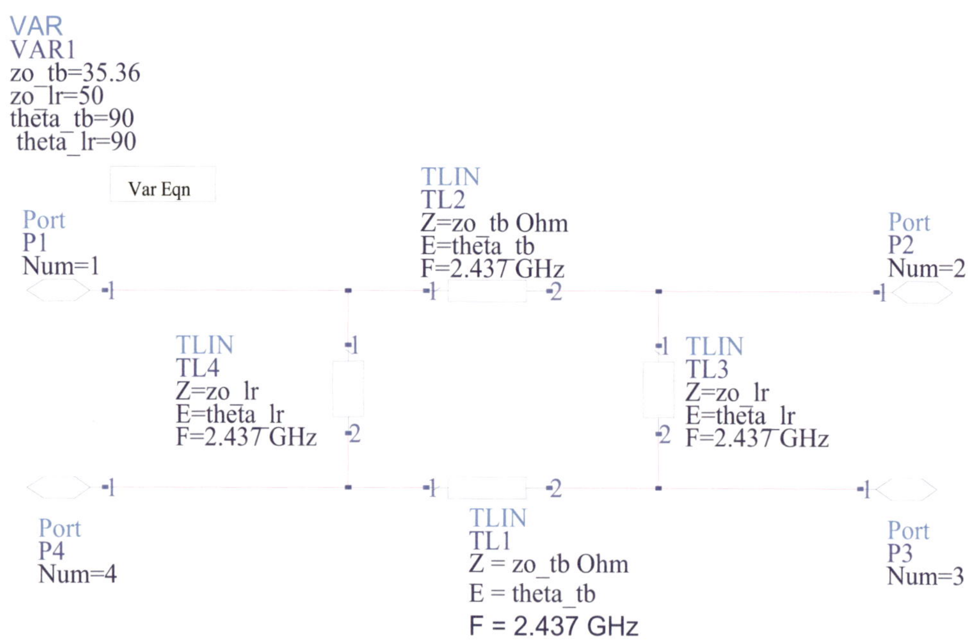

Fig. 6. ADS schematic for an ideal quadrature hybrid

Then a new ADS schematic for the quadrature hybrid is created using a microstrip substrate element (MSUB) and real transmissions lines as shown in Figure 7.

Figure 8 shows the higher level ADS schematic which is used to optimize and modify the dimensions of the real quadrature hybrid to emulate the ideal quadrature hybrid as closely as possible.

The simulated S-parameters of the real quadrature hybrid indicate that the power entering in any port is not reflected, and it is equally divided between two output ports that are existed at the other side of the hybrid, while no power is coupled to the port which is existed at the same side of the input port. This quadrature hybrid can operate over a bandwidth from 1.9 GHz to 2.8 GHz as shown in Figure 9.

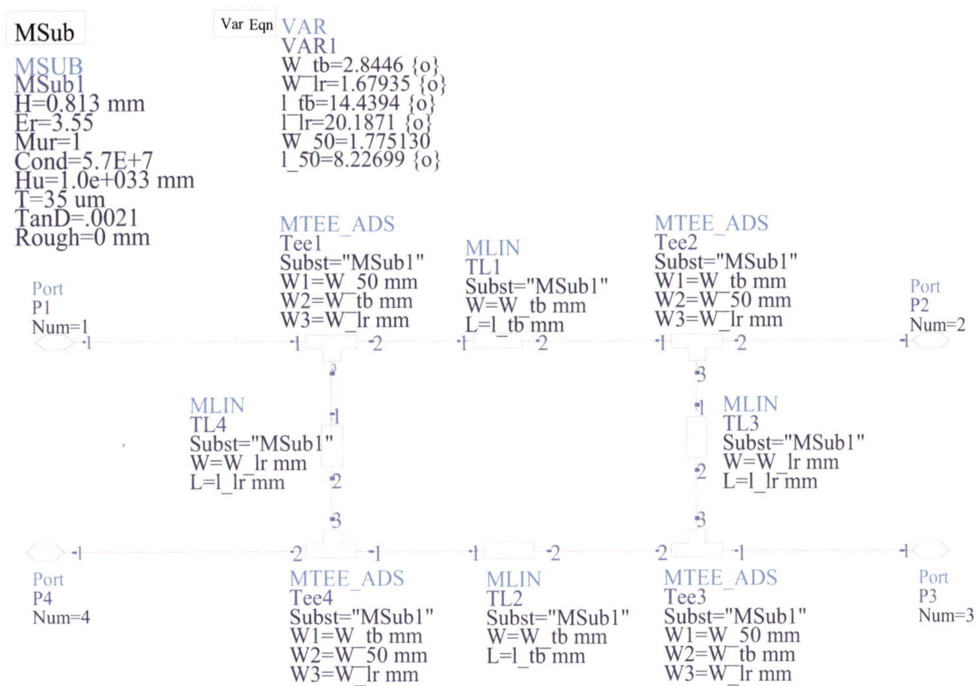

Fig. 7. ADS schematic for a real quadrature hybrid

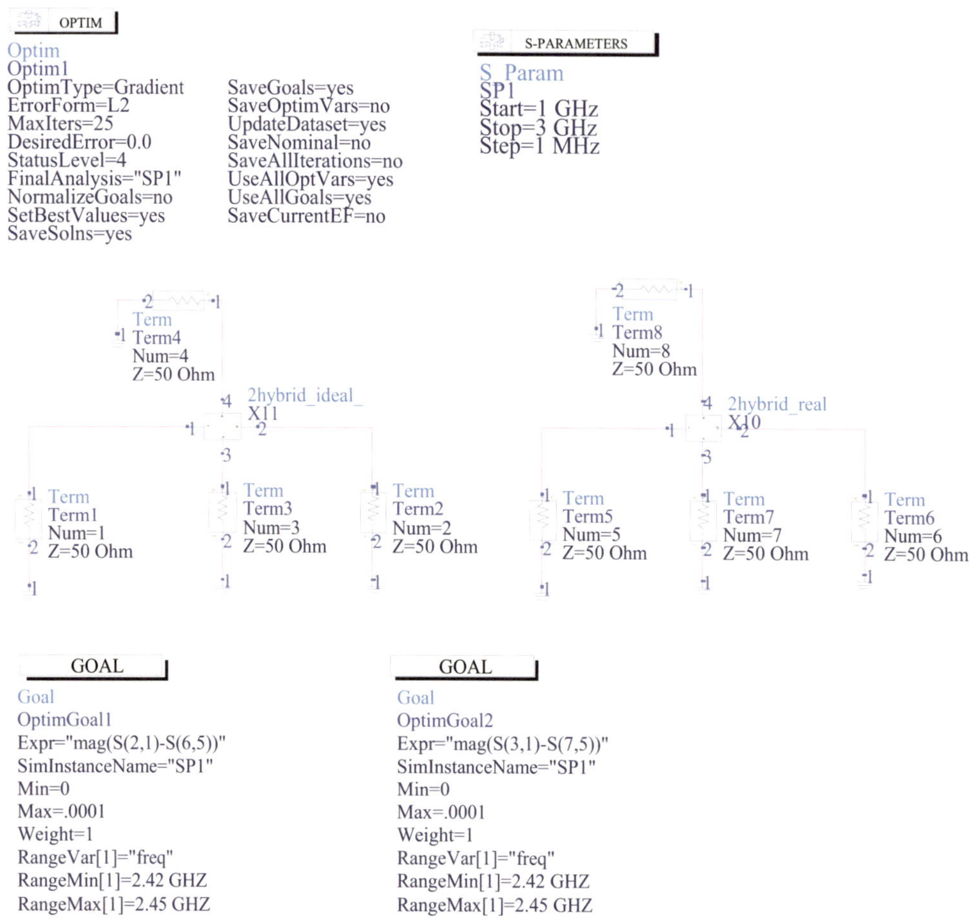

Fig. 8. Higher level ADS schematic to optimize the dimensions of the real quadrature hybrid

2.2.2 Design of crossover

The crossover provides an efficient mean to isolate two crossing transmission lines. The design of the crossover depends on the following scattering matrix equation [5].

$$[S]_{Crossover} = \begin{bmatrix} 0 & 0 & j & 0 \\ 0 & 0 & 0 & j \\ j & 0 & 0 & 0 \\ 0 & j & 0 & 0 \end{bmatrix} \tag{3}$$

A planar crossover can be designed by creating a new ADS schematic for the ideal crossover as shown in Figure 10 [3].

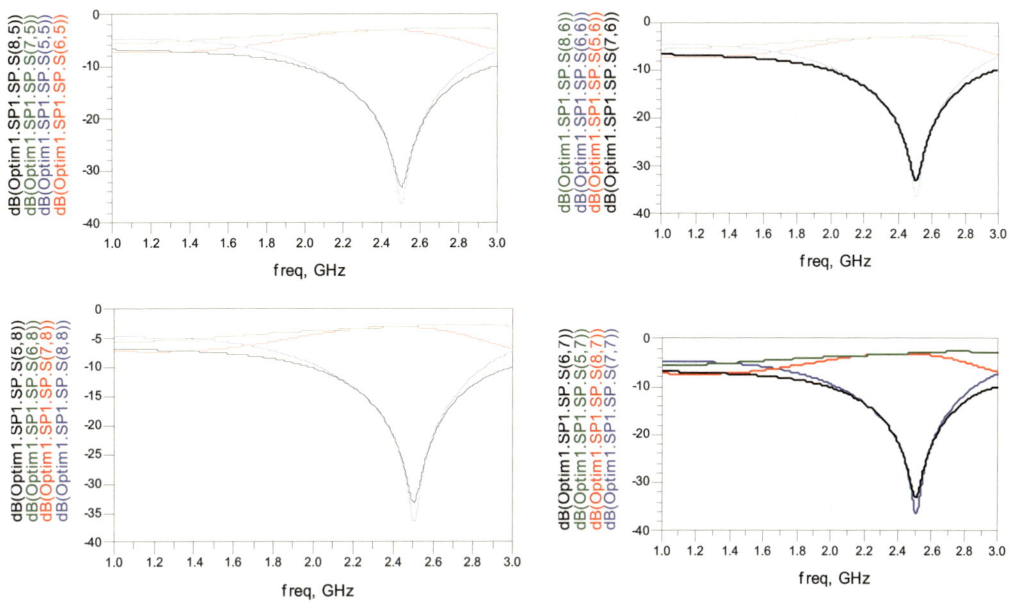

Fig. 9. Simulated S-parameters of the real quadrature hybrid

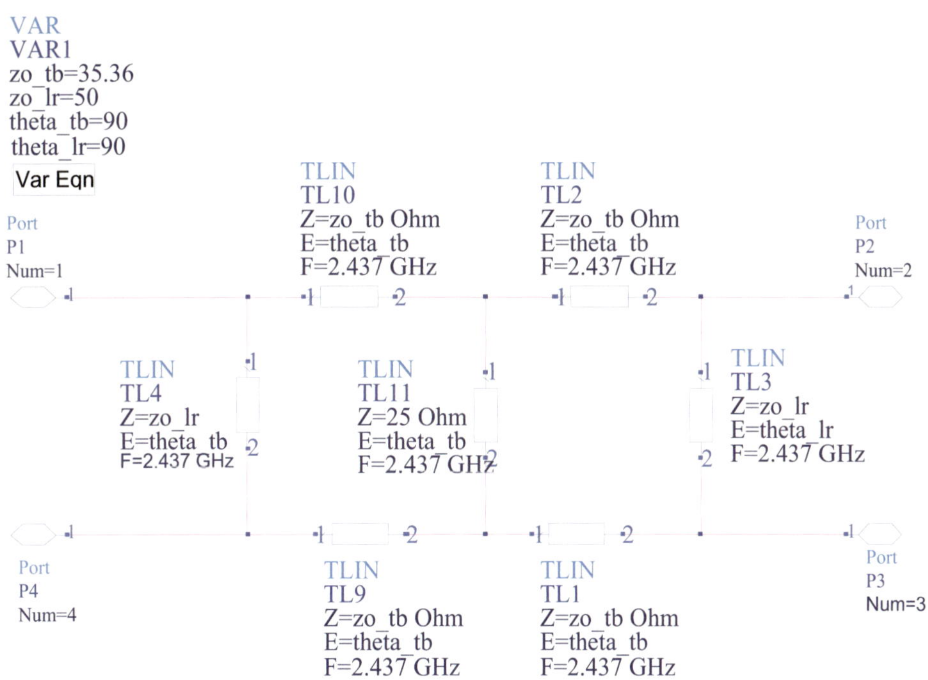

Fig. 10. ADS schematic for the ideal crossover

Figure 11 shows the ADS schematic for the real microstrip crossover.

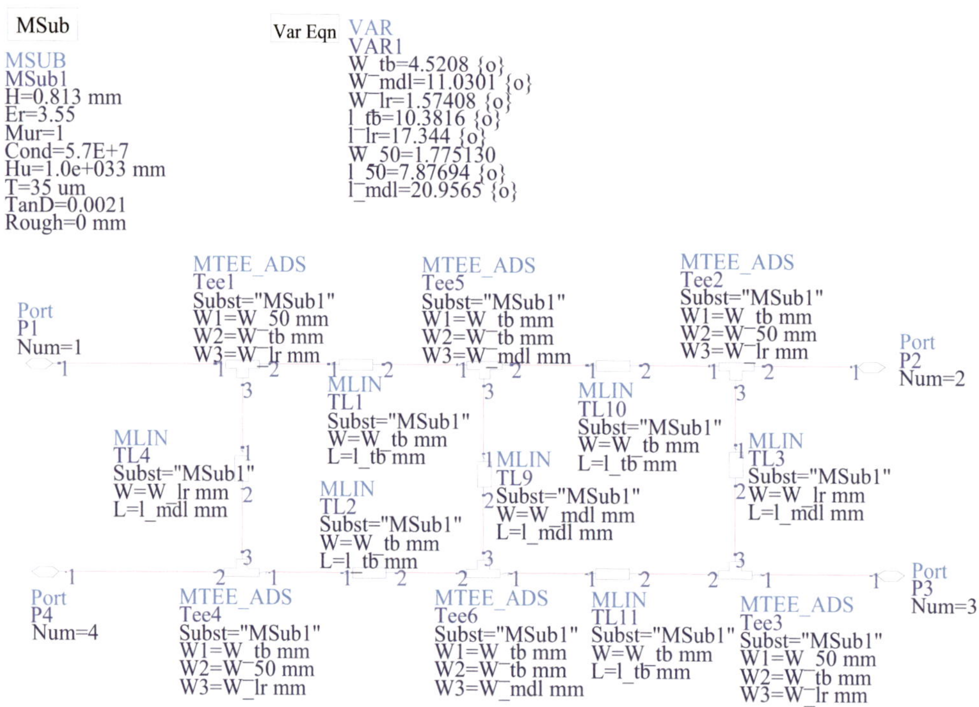

Fig. 11. ADS schematic for the real microstrip crossover

The dimensions of the real crossover are modified and optimised by using the higher level ADS schematic shown in Figure 12.

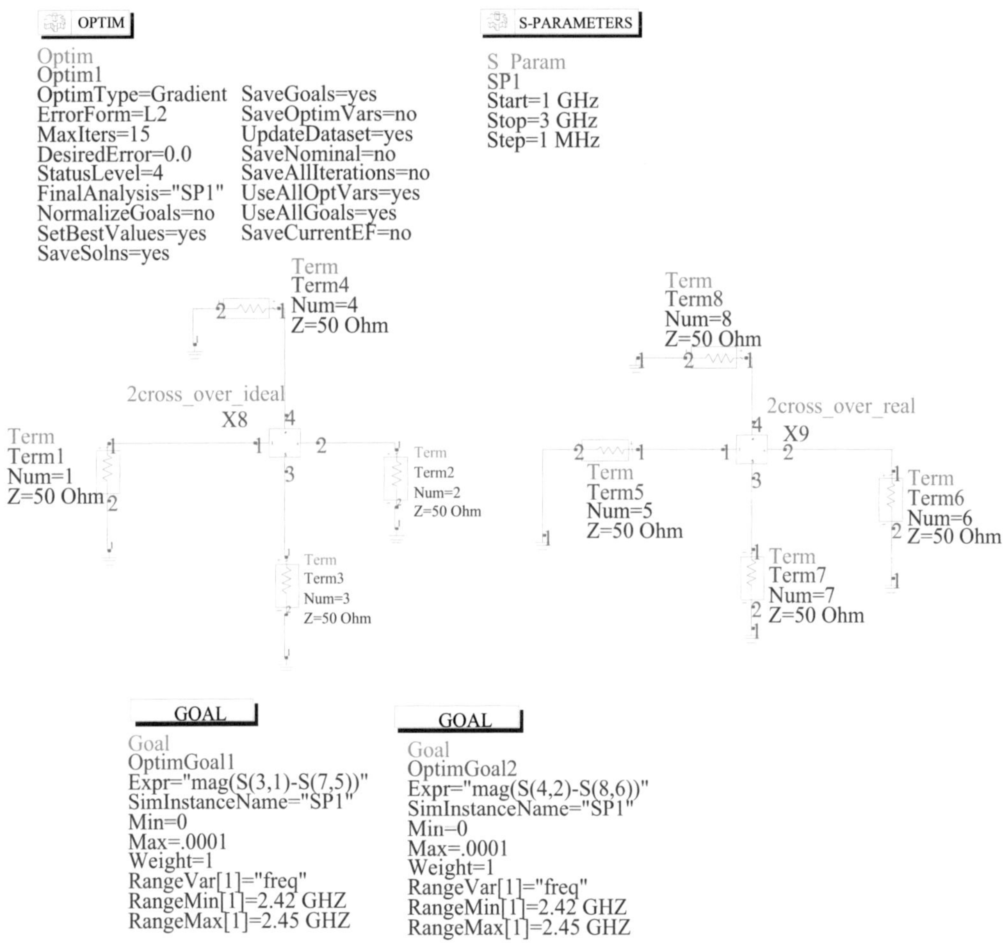

Fig. 12. Higher level ADS schematic to optimize the dimensions of the real crossover

The simulated S-parameters of the real crossover are shown in Figure 13. The bandwidth of the crossover is extended from 2 GHz to 2.8 GHz.

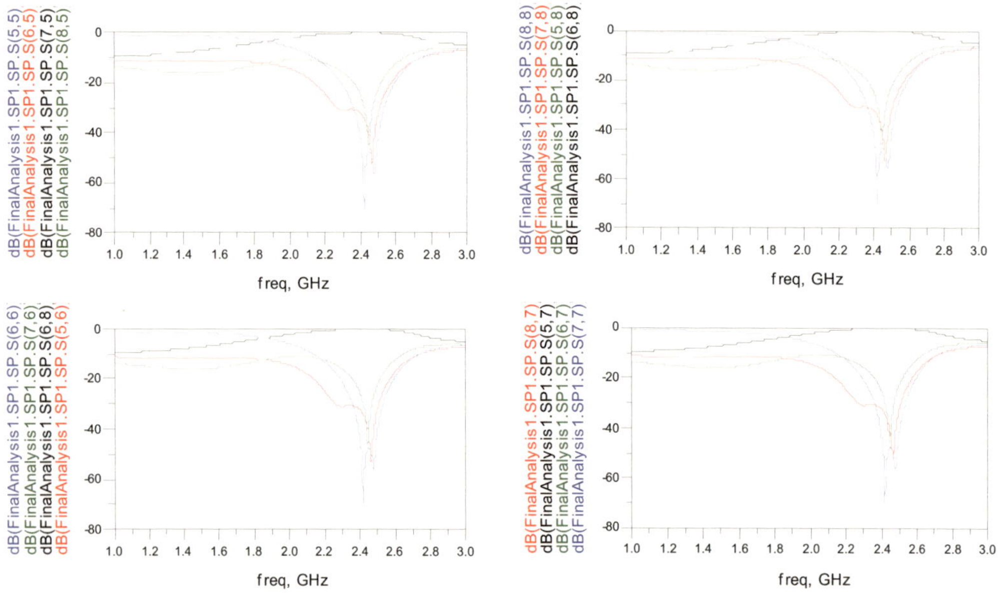

Fig. 13. Simulated S-parameters of the real crossover

2.2.3 Design of phase shifter

The phase shifter introduces a phase shift in the signal and can be implemented by adding a bit of length to a microstrip transmission line. The ADS is used to determine the length of red sections microstrip transmission lines which introduces phase shift (θ°_{shift} – 45°) in the signal that passes through the crossover. θ°_{shift} is the phase shift of the signal passing through the black section microstrip transmission line as shown in Figure 14. θ°_{shift} is determined by finding the length of the black section, then using ADS LineCalc tool to calculate the phase shift. This phase shifter should be replaced by a Schiffman phase shifter for wideband applications [6].

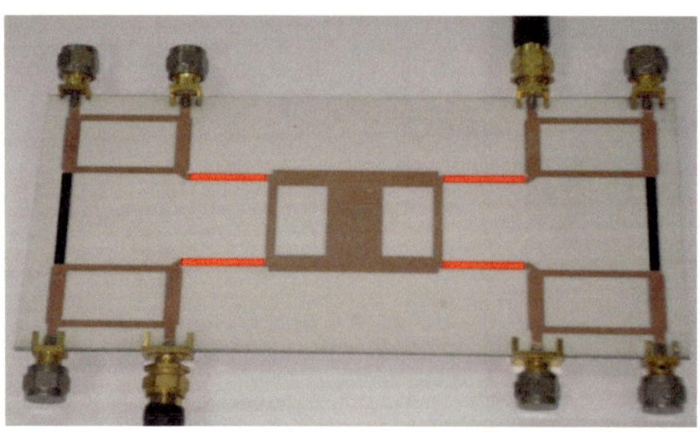

Fig. 14. Fabricated 4×4 Butler matrix

2.2.4 Design of paths between a 4×4 Butler matrix and a planar antenna array

The phases associated with each port of the 4×4 Butler matrix will be changed and the circular polarization cannot be obtained if the paths connecting the 4×4 Butler matrix with the planar antenna array do not have the same phase shift. Hence these paths are designed with the same phase shift. The path connecting antenna 1and Butler matrix is considered as the reference. The ADS is used to modify the length of the other paths to make its phases as close as possible to the phase of the reference path. Figure 15 shows the ADS schematic for path 2 which connects antenna 2 and Butler matrix. The length of this path is modified and optimized using ADS as shown in Figure 16.

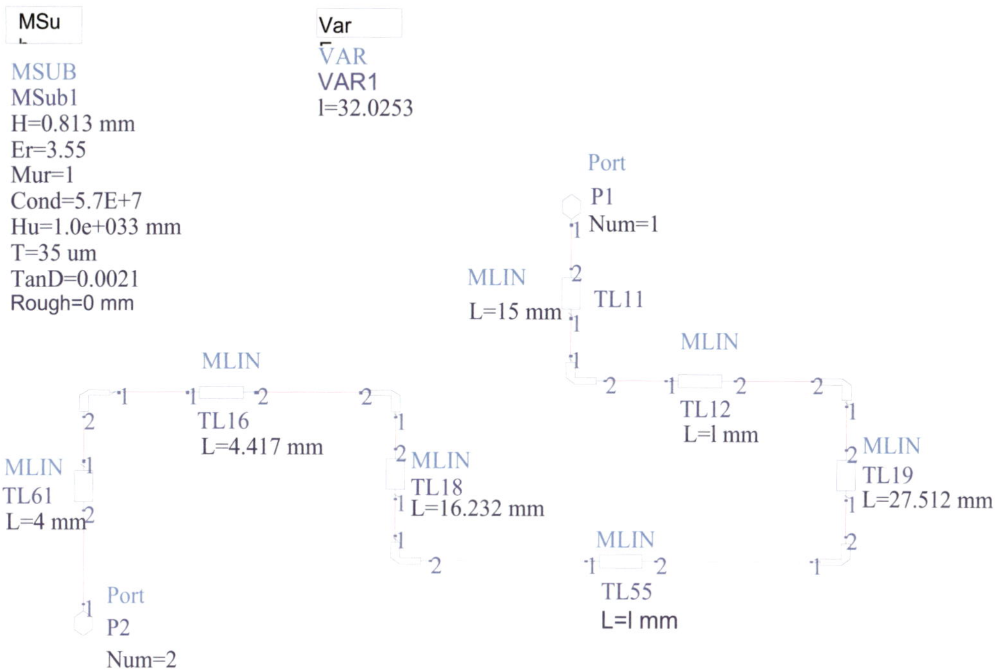

Fig. 15. ADS schematic for the path connecting antenna 2 and Butler matrix

The ADS simulated results for phases of path 2 and reference path are shown in Figure 17. The ADS schematics for designing path 3 and path 4 are given in Appendix A.

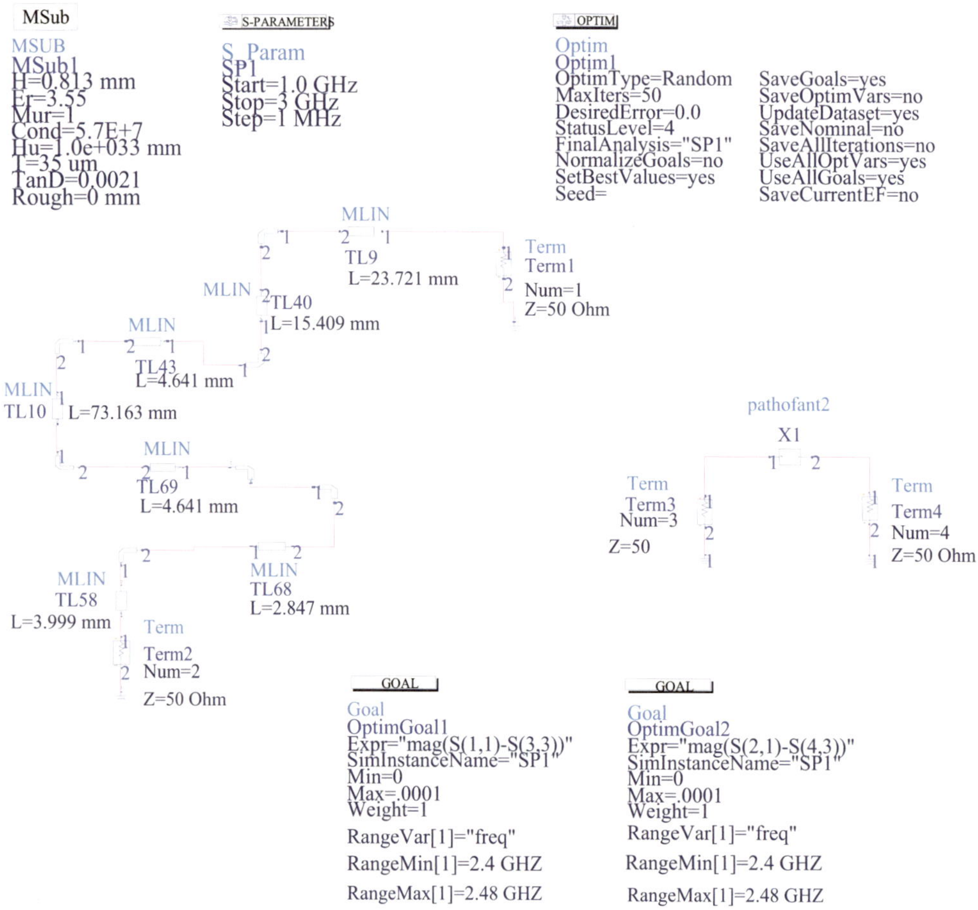

Fig. 16. ADS schematic to modify the length of path 2

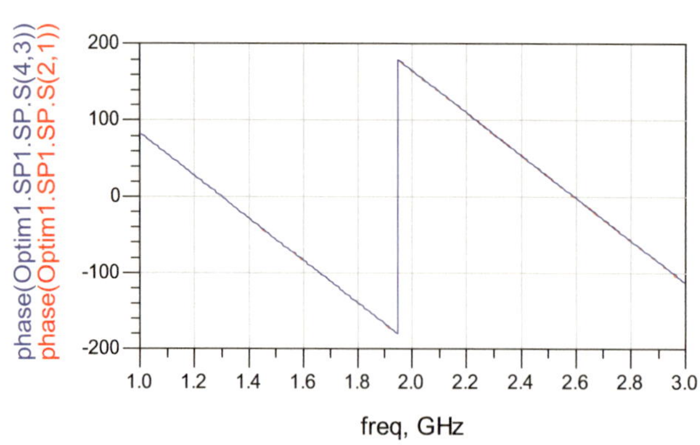

Fig. 17. ADS simulated results for phases of path 2 and reference path

2.3 Analysis of planar microstrip antenna array with Butler matrix

A planar microstrip antenna array consists of four orthogonally oriented inset-fed rectangular patch antennas as shown in Figure 18. The circular polarization can be generated with linearly polarized elements when all the adjacent elements are orthogonally oriented and are fed by a Butler matrix to form two orthogonally polarized *E*-fields from the four linearly polarized *E*-fields of the planar array elements [7,8]. The normalized instantaneous *E*-fields in *x*- and *y*-directions are represented by equations (4) and (5) respectively:

$$E_{x\_planar} = \cos(\omega t - K_o z + A) + \cos(\omega t - K_o z + C) \tag{4}$$

$$E_{y\_planar} = \cos(\omega t - K_o z + B) + \cos(\omega t - K_o z + D) \tag{5}$$

where K_o is the propagation constant in free space and ω is the angular frequency. The values of A, B, C and D phases in the above equations are changed according to the selected port of 4×4 Butler matrix. The instantaneous field of the plane wave traveling in positive *z*-direction is given by

$$E(z,t) = E_{x\_planar}\hat{x} + E_{y\_planar}\hat{y}. \tag{6}$$

For the planar microstrip antenna array that consisted of identical patches, the magnitude of E_{x_planar} is equal to E_{y_planar}.

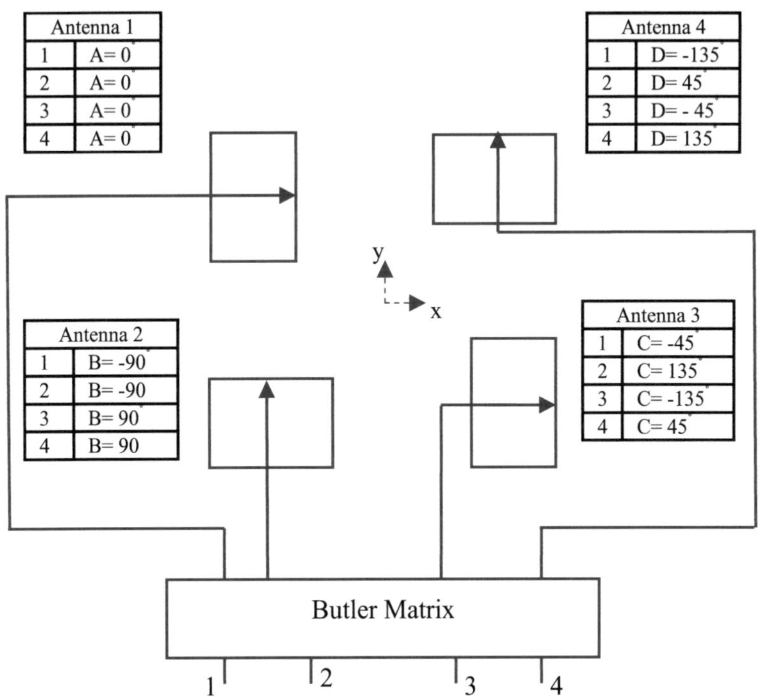

Fig. 18. Phases associated with each port of the planar microstrip antenna array with 4×4 Butler matrix

As shown in Figure 18, when port 1 or port 2 is selected, the phase delay between E_{y_planar} and E_{x_planar} will be $-\frac{\pi}{2}$ and RHCP is generated. The LHCP is obtained when port 3 or port 4 is selected because the phase delay between E_{y_planar} and E_{x_planar} will be $+\frac{\pi}{2}$.

2.4 Radiation pattern for planar microstrip antenna array

The total normalized *E*-plane radiation pattern of the planar microstrip array is obtained by the following equation [9]:

$$E_{T\_\theta} = E_\theta \; |AF_{T\_E}|. \tag{7}$$

where $|AF_{T\_E}|$ is the normalized *E*-plane array factor for the planar microstrip array and can be obtained from equation (8) [10]:

$$|AF_{T\_E}| = \left(1 + e^{j(K_o d_x \sin\theta + \phi_x)}\right) \left(1 + e^{j\phi_y}\right) \tag{8}$$

E_θ is the total normalized *E*-plane radiation pattern of a single microstrip patch antenna and is obtained by equation (9) [11]:

$$E_\theta = \left|1 + e^{jK_o L \sin\theta}\right| \tag{9}$$

where θ is the elevation angle and *L* is the length of the microstrip patch antenna. The total normalized radiation pattern and the normalized array factor for the *H*-plane are obtained from equations (10) and (11) respectively [9]:

$$H_{T\_\theta} = H_\theta \; |AF_{T\_H}| \tag{10}$$

$$|AF_{T\_H}| = \left(1 + e^{j(K_o\, d_y \sin\theta + \phi_y)}\right) \left(1 + e^{j\phi_x}\right) \tag{11}$$

where ϕ_x, ϕ_y are the feeding phases for the antenna 4 and the antenna 2 respectively. d_x, d_y are the spacings between patches in the *x*-direction and *y*-direction respectively. equation (12) is used to determine the total normalized *H*-plane radiation pattern of a single microstrip patch antenna (H_θ) [11];

$$H_\theta = \frac{\sin\left(\frac{K_o W}{2}\sin\theta\right)}{\frac{K_o W}{2}\sin\theta} \cos\theta \; . \tag{12}$$

where *W* is the width of the microstrip patch antenna.

2.5 Simulation and measured results

To design the planar microstrip antenna array, the spacing distance between the patches in *x*-direction (d_x) is determined based on the simulation of equations (7), (8) and (9) using

Matlab. Figure 19 shows the total normalized *E*-plane radiation pattern of the planar microstrip array (E_{T_θ}) versus the elevation angle θ at different values of d_x.

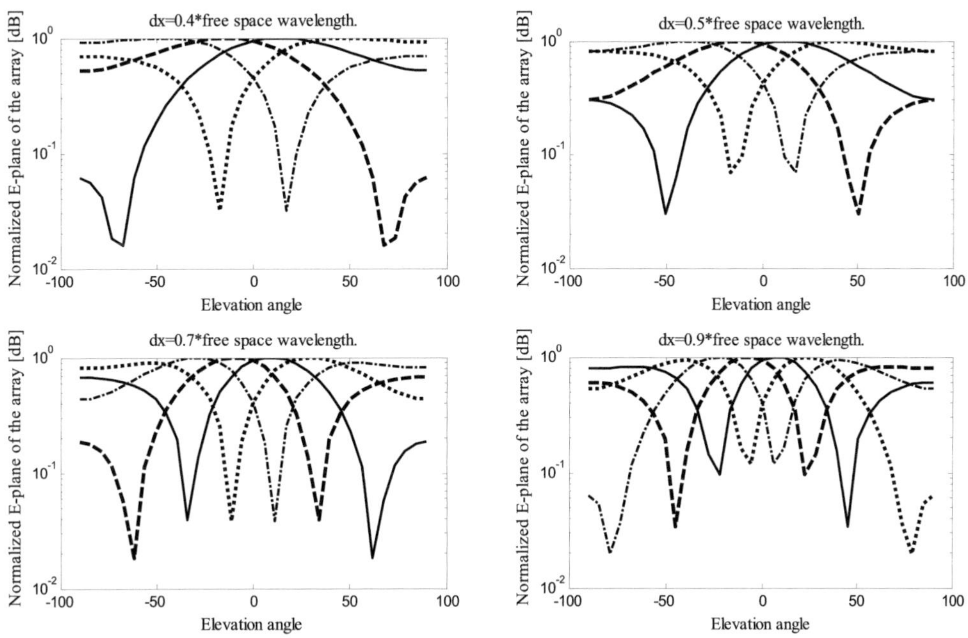

Fig. 19. Simulated normalized *E*-plane radiation pattern of the planar microstrip antenna array versus an elevation angle (— feed at port 1, −· feed at port 2, ... feed at port 3, --- feed at port 4)

The fixed beams get narrowed as d_x is increased, but to maintain balance between the narrowed beams and small-size antenna array, and also to avoid the grating lobes, d_x=0.5*(free space wavelength) is selected. d_y is determined based on the simulation of equations (10), (11) and (12) using Matlab. The total normalized *H*-plane radiation pattern (H_{T_θ}) versus the elevation angle θ is shown in Figure 20.
The value of H_{T_θ} is equal to the value of E_{T_θ} divided by the intrinsic impedance of the free space; so H_{T_θ} is not sensitive to the change in the separation distance d_y because its value is much smaller than that of E_{T_θ}. However, d_y=0.5*(free space wavelength) is selected to maintain a balance between the avoidance of mutual coupling and the small-size antenna array. The set of curves in Figures 19 and 20 represents the normalized *E*-plane and *H*-plane respectively; each curve is generated by selecting a different feed port of the 4×4 Butler matrix.
The planar microstrip antenna array with 4×4 Butler matrix is fabricated using Rogers's substrate of thickness h_t=0.85 mm, loss tangent=0.0021 and dielectric constant $\varepsilon_r = 3.55$. The inset-fed rectangular microstrip patch antenna is designed at a resonant frequency equal to 2.437 GHz using Matlab and then simulated using ADS [12]. The parameters for the substrate layers and metallization layers of the Rogers's substrate are created in the ADS Momentum as shown in Figures 21 and 22 respectively.

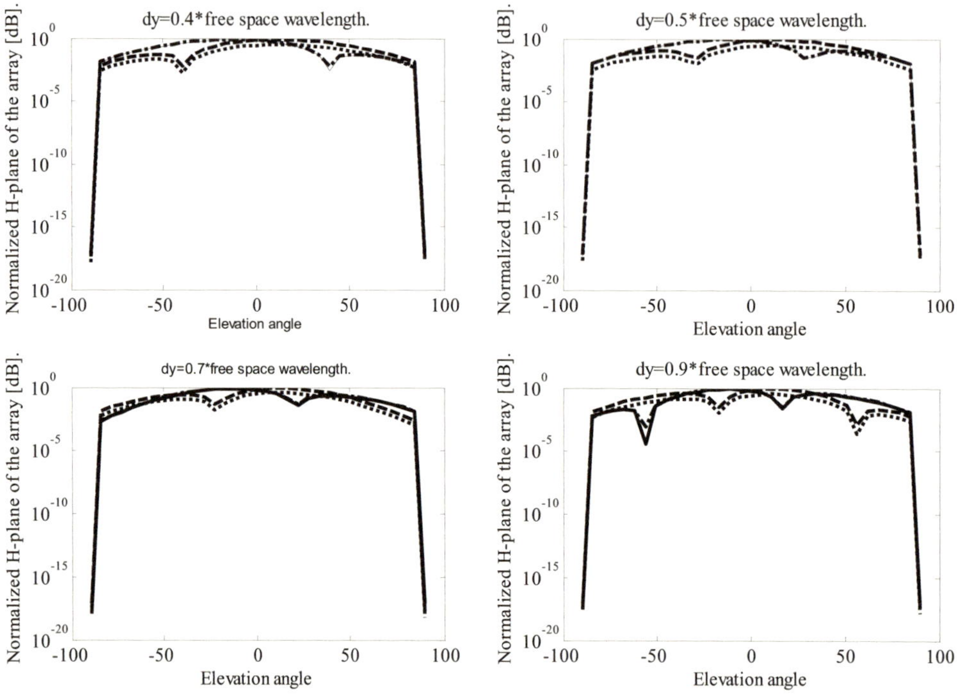

Fig. 20. Simulated normalized *H*-plane radiation pattern of the planar microstrip antenna array versus an elevation angle (— feed at port 1, –· feed at port 2, … feed at port 3, --- feed at port 4)

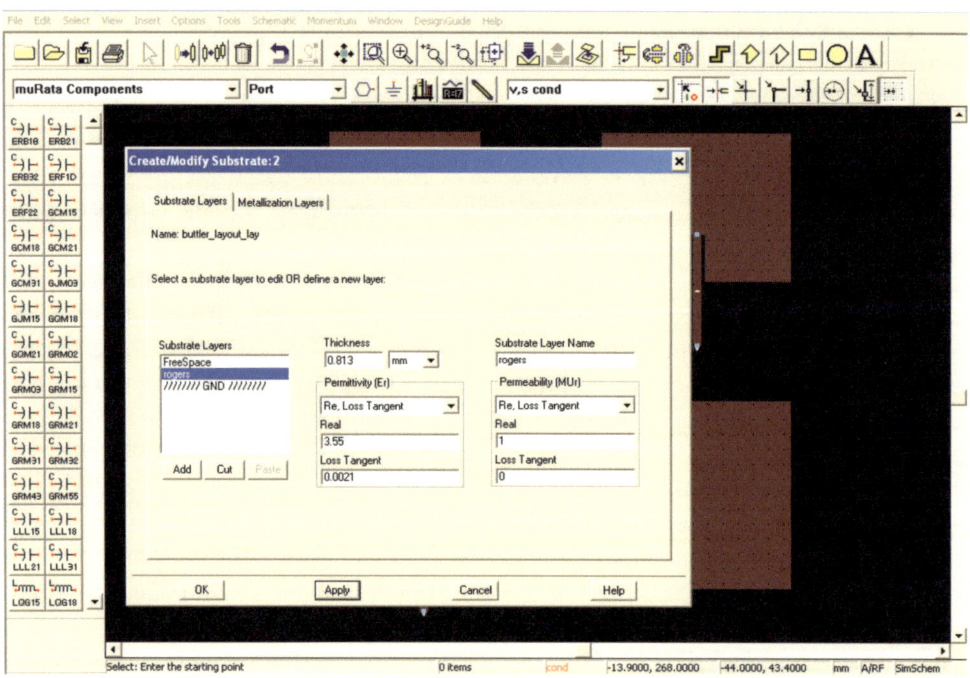

Fig. 21. Parameters setup in ADS Momentum for substrate layers of the Rogers's substrate

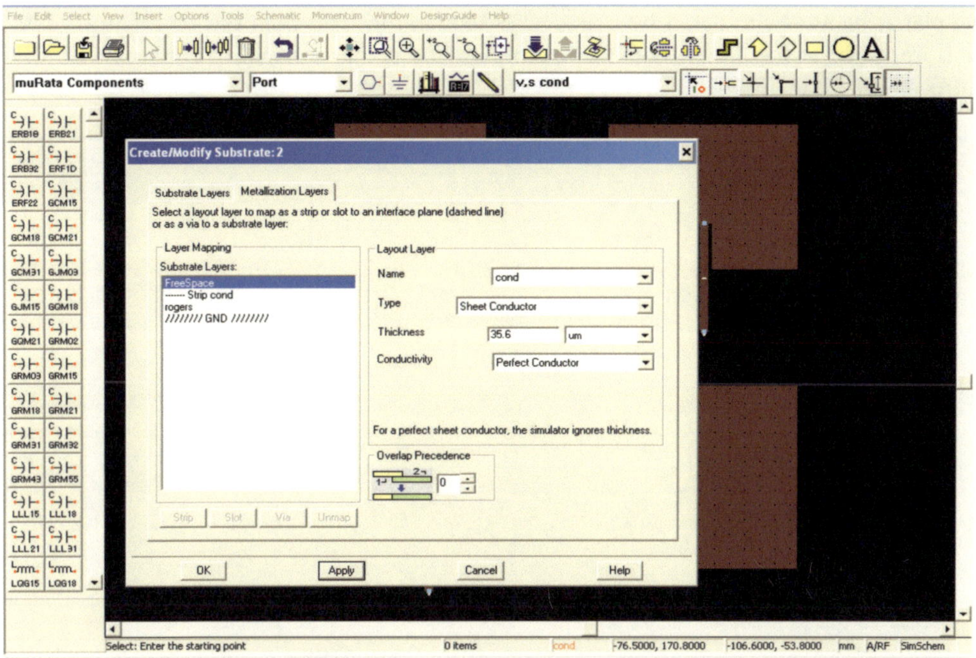

Fig. 22. Parameters setup in ADS Momentum for metallization layers of the Rogers's substrate

Figure 23 shows the ADS Momentum for the planar microstrip antenna array.

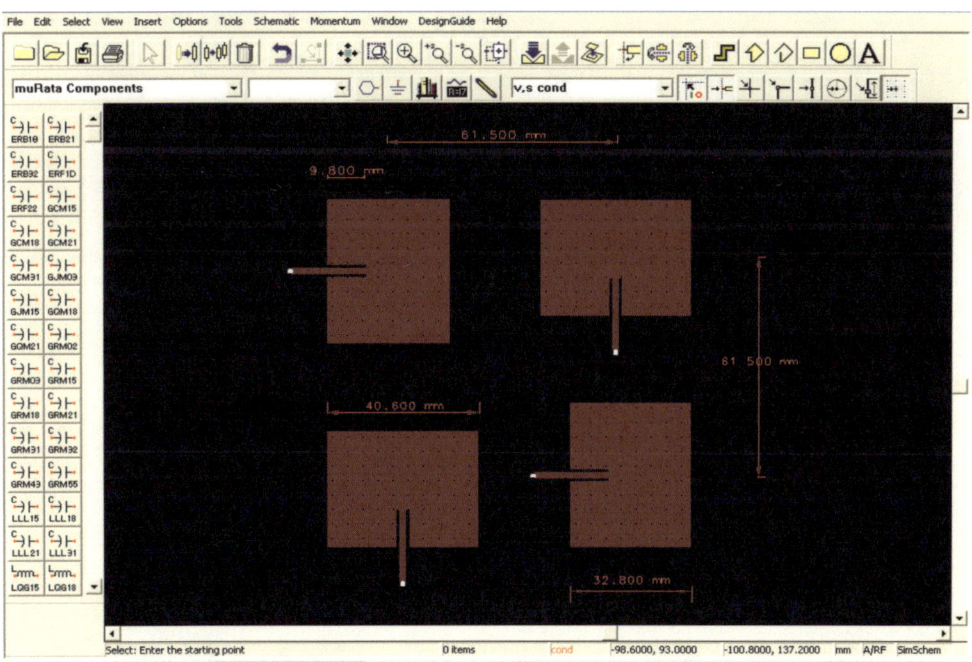

Fig. 23. ADS Momentum for the planar microstrip antenna array

The planar array is simulated by using ADS Momentum and then the Momentum dataset file is imported to the ADS schematic to simulate the planar microstrip antenna with Butler matrix as shown in Figure 24.

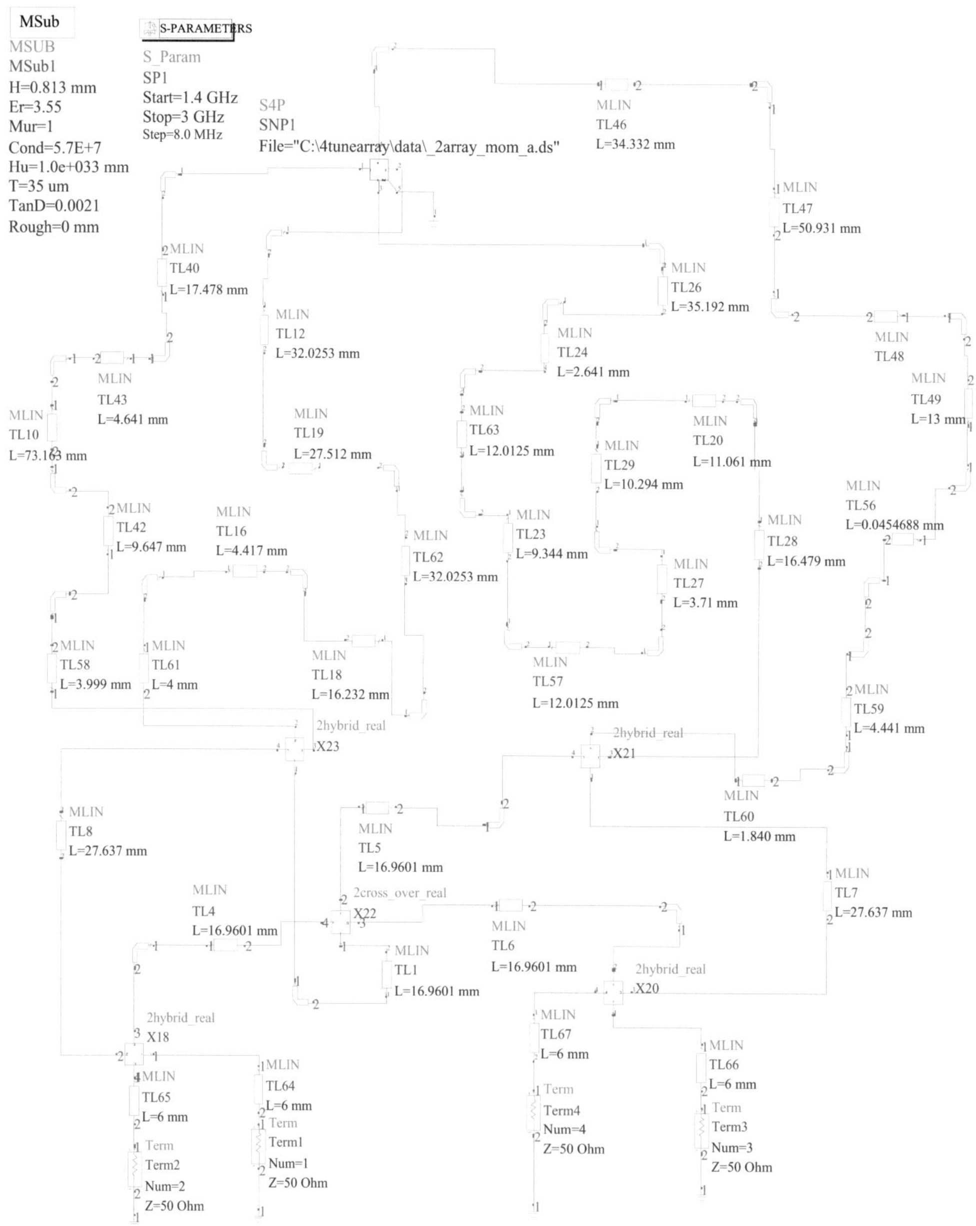

Fig. 24. ADS schematic diagram for simulating the planar microstrip antenna with 4×4 Butler matrix

Figure 25 shows the ADS layout of the planar mirostrip antenna array with 4×4 Butler matrix. The Fabricated planar microstrip antenna array with 4×4 Butler matrix is shown in Figure 26.

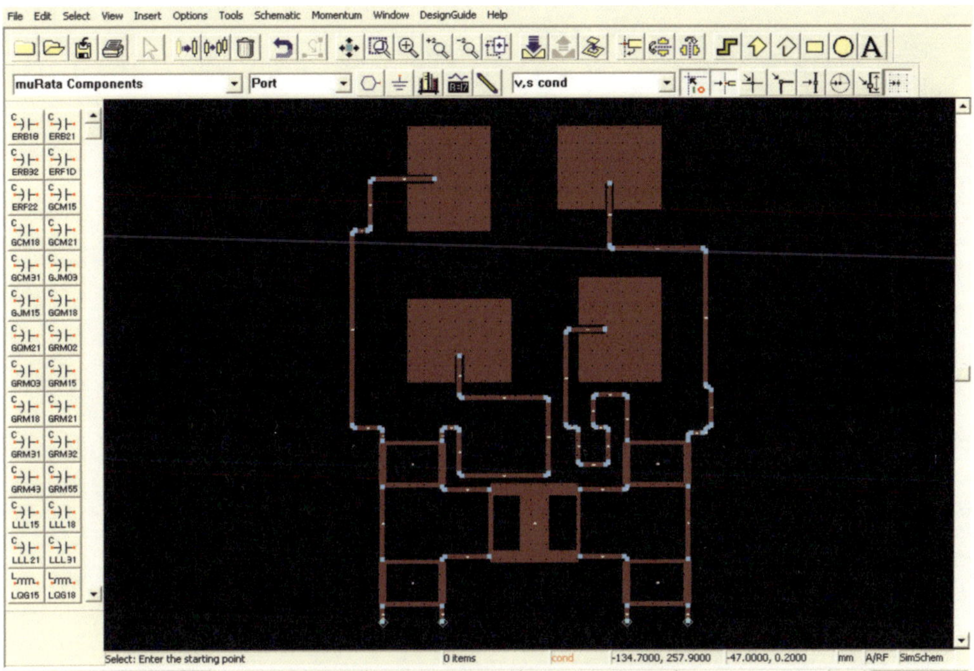

Fig. 25. ADS layout of the planar microstrip antenna array with 4×4 Butler matrix

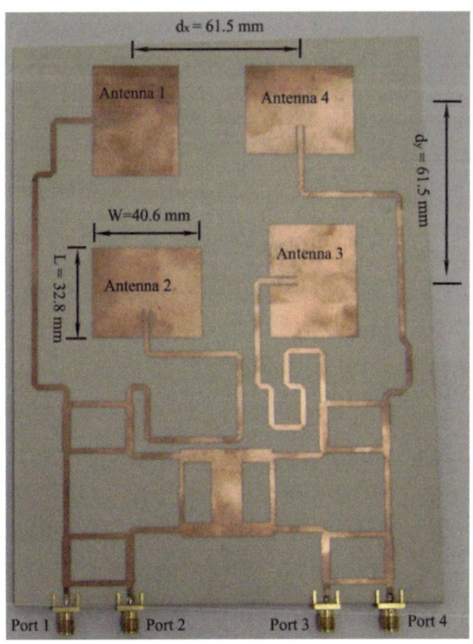

Fig. 26. Fabricated planar microstrip antenna array with 4×4 Butler matrix

The measured and the simulated values of the reflection coefficient at each port of the planar microstrip antenna array with 4×4 Butler matrix versus the frequency band of 1.4–3 GHz are shown in Figure 27.

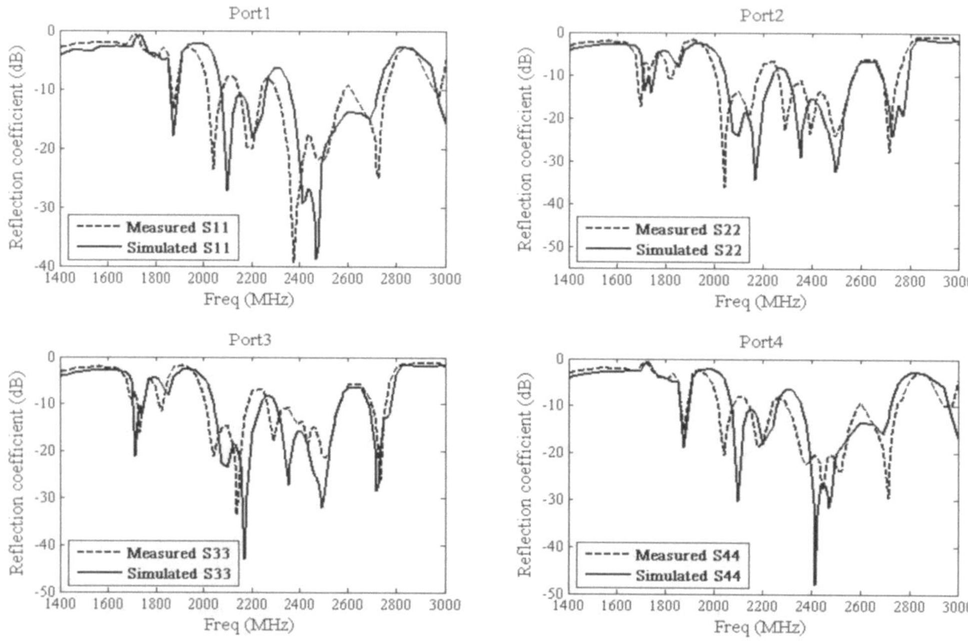

Fig. 27. Reflection coefficients versus frequency for the planar microstrip antenna array with 4×4 Butler matrix

The phases associated with the ports of the Butler matrix result in the existence of different voltages and different input impedances at the ports of the Butler matrix. Hence the reflection coefficients at these ports will not be the same owing to the relationship between the reflection coefficient and input impedance. The mutual coupling between the patches induces frequency modes. These modes are matched with 50 Ω impedance outside the required 2.437 GHz range (i.e. for port 1, the measured reflection coefficient is less than −10 dB at ranges 1864–1885 MHz, 2014–2073 MHz, and 2145–2237 MHz owing to the effect of mutual coupling between the patches). The mutual coupling impedances between the patches are different because of the effect of the phases associated with the ports of the Butler matrix. Due to the dissimilarity of the mutual coupling impedances, each port will have reflection coefficient less than −10 dB over different frequency ranges outside the required bandwidth [13]. Although the Butler matrix reflection coefficients are not the same at all the ports, they have good values over the required bandwidth. The measured impedance bandwidth (for reflection coefficient < −10 dB) at port 1, port 2, port 3, and port 4 are 18.8%, 12.52%, 12.5%, and 18.5 %, respectively. The obtained impedance bandwidth of this planar array is high when compared with a single microstrip patch antenna that achieves 0.7% impedance bandwidth as shown in Figure 28 [14]. The implementation of the Butler matrix gives a wide band due to the absorption of the reflected power in the matched loads connected to the non-selected ports of the butler matrix. Moreover, the mutual coupling between the patches enhances the bandwidth [13].

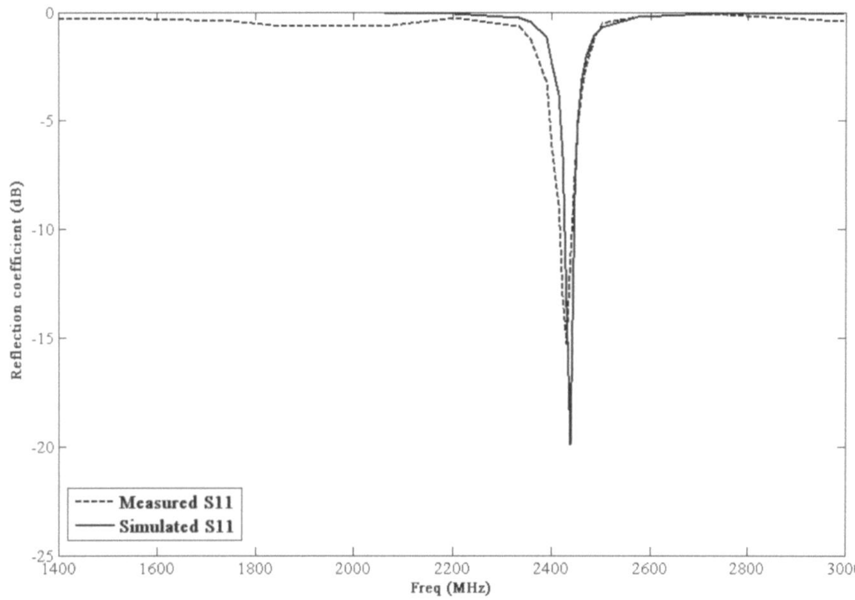

Fig. 28. Reflection coefficient versus frequency for an inset-fed rectangular microstrip patch antenna

The measured normalized radiation pattern of the planar microstrip antenna array with 4×4 Butler matrix is shown in Figure 29. The planar microstrip antenna array that is fed by the 4×4 Butler matrix has four beams at four different directions. These beams have a circular polarization diversity because a beam with RHCP will be obtained when port 1 or 2 is selected, but if port 3 or 4 is selected a beam with LHCP will be generated. The measured gains at 2.437 GHz for ports 1, 2, 3 and 4 are 9.74 dBi, 8.6 dBi, 9 dBi and 10.1 dBi respectively. The gain of port 1 and port 2 is measured using right hand circularly polarized standard antenna (2.4 GHz, 8 dBi, RHCP, Flat Patch Antenna), while the gain for port 3 and port 4 is measured using left hand circularly polarized standard antenna (2.4 GHz, 8dBi, LHCP, Flat Patch Antenna).

The measured axial ratios of the individual ports versus the elevation angle are shown in Figure 30. A good axial ratio (axial ratio < 3 dB) of ports 1, 2, 3 and 4 are achieved over angular ranges -36° to 61°, -55° to -9°, -31° to 81° and -90° to 15°respectively.

Radiation pattern and axial ratio measurements are carried out in a near field Satimo chamber as shown in Figure 31. The measurement system consists of probe antennas mounted with equal spacing on a circular arch. The measurements of the radiation pattern and the axial ratio can be obtained by electronic switching of the probe antennas. The measured data is collected automatically and saved in MS Excel format.

The impedance bandwidth of the planar array with Butler matrix can be extended by using a thicker substrate. The obtained impedance bandwidth is 37% for ports 1 and 4, while ports 2 and 3 have an impedance bandwidth equal to 20% as shown in Figure 32. This is when using FR4 substrate (h_t=1.6 mm, loss tangent=0.035, and dielectric constant $\varepsilon_r = 5.4$). But the FR4 substrate has a poor loss tangent which results in lower efficiency and therefore the gain will be decreased [15].

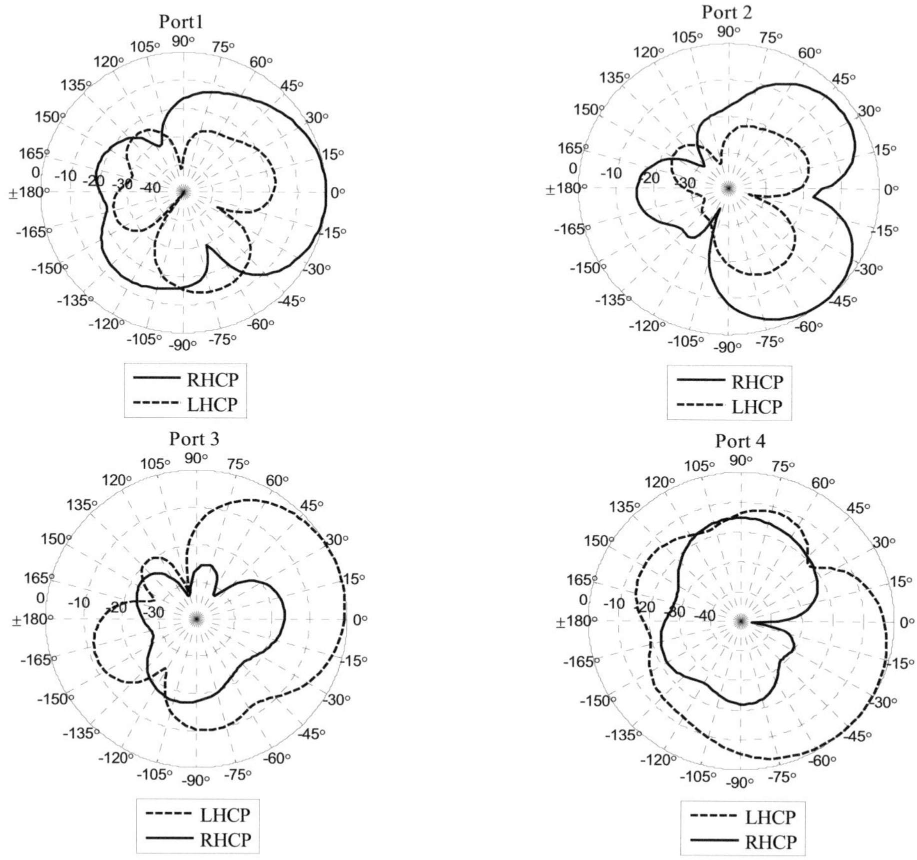

Fig. 29. Measured normalized radiation pattern of the planar microstrip antenna array with 4×4 Butler matrix at 2.437 GHz

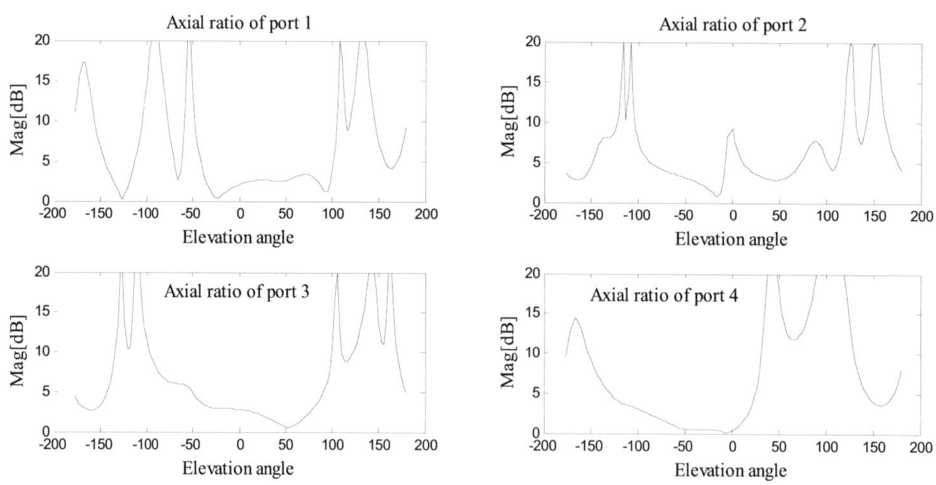

Fig. 30. Measured axial ratios of the planar microstrip antenna array with 4×4 Butler matrix at 2.437 GHz

Fig. 31. Measuring a planar microstrip antenna with Butler matrix inside near field Satimo chamber

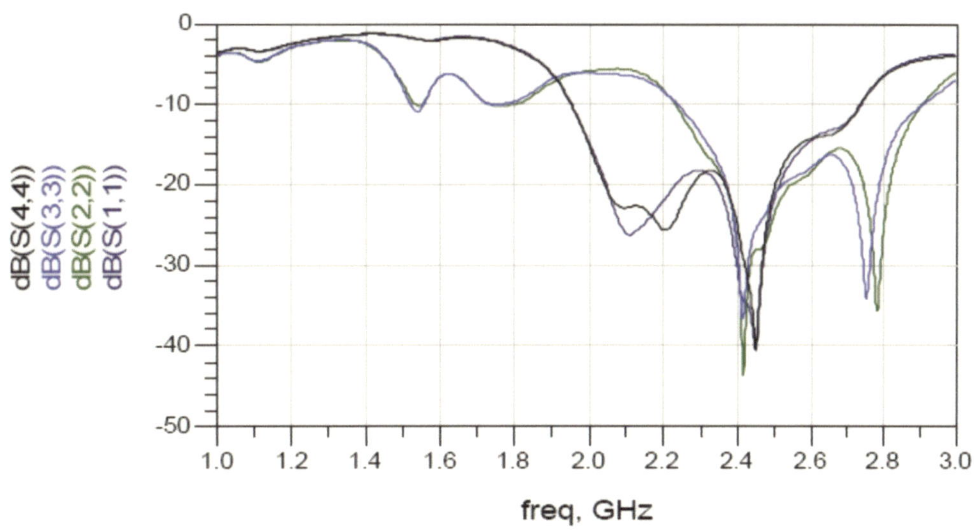

Fig. 32. Simulated reflection coefficients versus frequency for the planar microstrip antenna array with 4×4 Butler matrix using FR4 substrate

3. Analysis, design, and implementation of an aperture-coupled microstrip antenna

The circular polarization diversity with the square patch microstrip antenna is obtained by the implantation of the quadrature hybrid as a microstrip feed line. The small size antenna is achieved by using the aperture coupled structure. In the present study, the ACMSA is analyzed based on cavity model because the TLM is useful only for patches of rectangular shape [14]. The cavity model is used to design the ACMSA and determine the length of the open microstrip stub line in order to match microstrip antenna. The Matlab and ADS are used for designing and simulating the ACMSA.

3.1 Analysis of ACMSA using cavity model

3.1.1 Resonant Frequency

The cavity model is based on treating the microstrip antennas as cavities formed by microstrip lines as shown in Figure 33. The region between the patch and ground plane may be treated as a cavity bounded by electric walls above and below, and magnetic walls along the edges. The fields inside the antenna are assumed to be the fields inside this cavity. Due to the electrically thin substrate the fields in the interior region do not vary with z. Also inside the cavity the electric field has only a z component (E_z) and the magnetic field has H_x and H_y components [14]. Figure 33 shows the cavity model of the ACMSA.

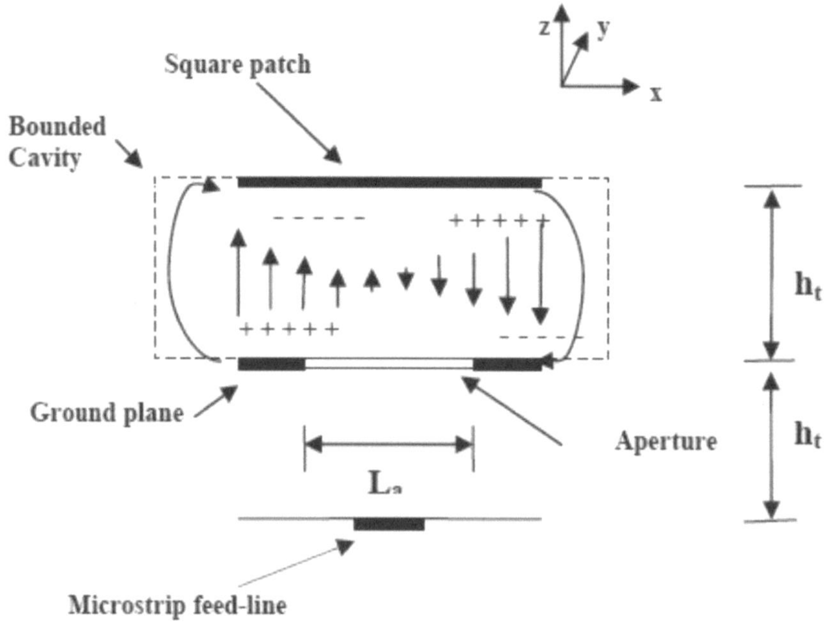

Fig. 33. Cavity model of the ACMSA

The magnetic field H inside the cavity volume can be developed by Maxwell's equations [16]:

$$\nabla \times H = J_e + j\omega\varepsilon_o\varepsilon_r E \quad (13)$$

$$\nabla \times H = \begin{vmatrix} \hat{x} & \hat{y} & \hat{z} \\ \frac{\partial}{\partial x} & \frac{\partial}{\partial y} & \frac{\partial}{\partial z} \\ H_x & H_y & 0 \end{vmatrix} \quad (14)$$

$$\nabla \times H = \left(\frac{\partial H_y}{\partial x} - \frac{\partial H_x}{\partial y} \right) \hat{z} \quad (15)$$

$$\nabla \times (\nabla \times H) = \begin{vmatrix} \hat{x} & \hat{y} & \hat{z} \\ \frac{\partial}{\partial x} & \frac{\partial}{\partial y} & \frac{\partial}{\partial z} \\ 0 & 0 & \left(\frac{\partial H_y}{\partial x} - \frac{\partial H_x}{\partial y} \right) \end{vmatrix} \quad (16)$$

$$\nabla \times (\nabla \times H) = \left(\frac{\partial^2 H_y}{\partial x \partial y} - \frac{\partial^2 H_x}{\partial y^2} \right) \hat{x} + \left(\frac{\partial^2 H_x}{\partial x \partial y} - \frac{\partial^2 H_y}{\partial x^2} \right) \hat{y} \quad (17)$$

$$\nabla \times \left(J_e + j\omega \varepsilon_o \varepsilon_r E \right) = \left(j\omega \varepsilon J_{mx} + \omega^2 \mu \varepsilon H_x \right) \hat{x} + \left(j\omega \varepsilon J_{my} + \omega^2 \mu \varepsilon H_y \right) \hat{y} \quad (18)$$

$$K = \omega \sqrt{\mu \varepsilon} \quad (19)$$

where K is the propagation constant inside the cavity, $\varepsilon = \varepsilon_o \varepsilon_r$. Equations (20) and (21) are obtained by substituting the right hand side of equation (17) into the left hand side of equation (18);

$$\frac{\partial^2 H_y}{\partial x \partial y} - \frac{\partial^2 H_x}{\partial y^2} - K^2 H_x = j\omega \varepsilon J_{mx} \quad (20)$$

$$\frac{\partial^2 H_x}{\partial x \partial y} - \frac{\partial^2 H_y}{\partial x^2} - K^2 H_y = j\omega \varepsilon J_{my} \quad (21)$$

In order to solve equation (20), the cavity model assumes that the tangential magnetic field is zero, so that J_{mx} equals to zero.

$$\frac{\partial^2 H_y}{\partial x \partial y} - \frac{\partial^2 H_x}{\partial y^2} - K^2 H_x = 0 \quad (22)$$

Letting the Eigen functions of the homogeneous wave equation (22) be Ψ_{mn} and K_{mn} be the eigenvalues of K, K_{mn} can be obtained from equation (23) as indicated in Appendix B.

$$K_{mn}^2 = K_m^2 + K_n^2 \tag{23}$$

The resonance frequency of (m, n) mode can be obtained by equation (24);

$$f_{mn} = \frac{cK_{mn}}{2\pi\sqrt{\varepsilon_r}} \tag{24}$$

where c is the speed of light, ε_r is the relative dielectric constant.

3.1.2 Magnetic field component in x-direction

The solutions of equation (20) can be expressed in the following eigenfunction expansion form [16];

$$H_x = \sum_{mn} \beta_{x,mn} \psi_{x,mn} \tag{25}$$

where $\beta_{x,mn}$ is the modes coefficients and given by (see Appendix C)

$$\beta_{x,mn} = \frac{j\omega\varepsilon}{(K_{mn}^2 - K^2)K_n^2} \int_0^{W_a}\int_0^{L_a} J_{mx}.\psi_{x,mn}^* dxdy \tag{26}$$

The propagation constant (K) in equation (26) is replaced by the effective propagation constant (K_{eff}) because the losses of the cavity are included [2,17].

$$K_{eff}^2 = \omega^2 \mu\varepsilon_o(\varepsilon_r(1 - \frac{j}{Q})) \tag{27}$$

where Q is the quality factor. It is assumed that the electric field distribution in the aperture paralleled to x axis (E_{ay}) is in the form of a single piece-wise sinusoidal mode [18]. The distance from the square patch edge to the center of the aperture is defined as X_1 and from the square patch edge to the center of the aperture as Y_1 as shown in Figure 34.

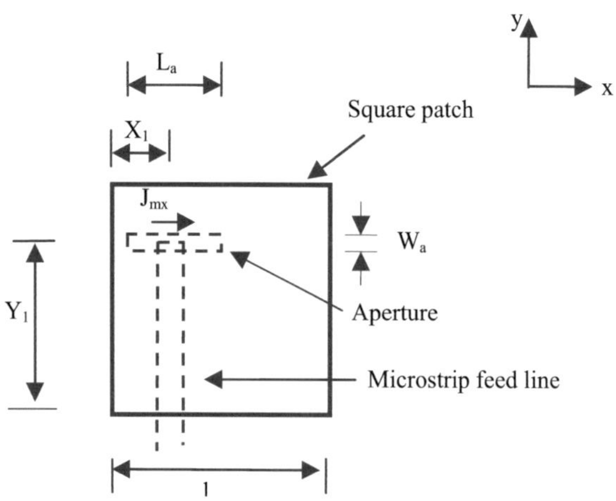

Fig. 34. Position of the aperture paralleled to x-axis

$$E_{ay} = \frac{V_{ox}}{W_a} \frac{\sin\left[K_a\left(\frac{L_a}{2} - (X_1 - x)\right)\right]}{\sin\left(K_a \frac{L_a}{2}\right)} \qquad X_1 - \frac{L_a}{2} \le x \le X_1 \tag{28}$$

$$E_{ay} = \frac{V_{ox}}{W_a} \frac{\sin\left[K_a\left(\frac{L_a}{2} - (x - X_1)\right)\right]}{\sin\left(K_a \frac{L_a}{2}\right)} \qquad X_1 \le x \le X_1 + \frac{L_a}{2} \tag{29}$$

V_{ox} is the voltage at the middle of the aperture paralleled to the x axis. K_a is the wave number of the aperture [19]. The magnetic current in the aperture parallel to x axis given by [16,17]:

$$M_{ax} = -2\, E_{ay} \tag{30}$$

The corresponding current density J_{mx} is given by:

$$J_{mx} = \frac{-2\, E_{ay}}{h} \tag{31}$$

By substituting for J_{mx} from equation (31) into equation (26) the modes coefficients $\beta_{x,mn}$ are determined. Then the magnetic field H_x in the cavity is given by

$$H_x = \sum_{mn} A_{mn} \beta_{x,mn} K_n \cos(K_m x) \sin(K_n y) \tag{32}$$

3.1.3 Magnetic Field Component in y-direction

The eigenfunction expansion for the magnetic field component in y direction H_y is used for solving equation (21) [16];

$$H_y = \sum_{mn} \beta_{y,mn} \psi_{y,mn} \tag{33}$$

Figure 35 shows the aperture paralleled to y axis; the electric field distribution (E_{ax}) can be determined using the following equations:

$$E_{ax} = \frac{V_{oy}}{W_a} \frac{\sin\left[K_a\left(\frac{L_a}{2} - (Y_2 - y)\right)\right]}{\sin\left(K_a \frac{L_a}{2}\right)} \qquad Y_2 - \frac{L_a}{2} \le y \le Y_2 \tag{34}$$

$$E_{ax} = \frac{V_{oy}}{W_a} \frac{\sin\left[K_a\left(\frac{L_a}{2} - (y - Y_2)\right)\right]}{\sin\left(K_a \frac{L_a}{2}\right)} \qquad Y_2 \le y \le Y_2 + \frac{L_a}{2} \tag{35}$$

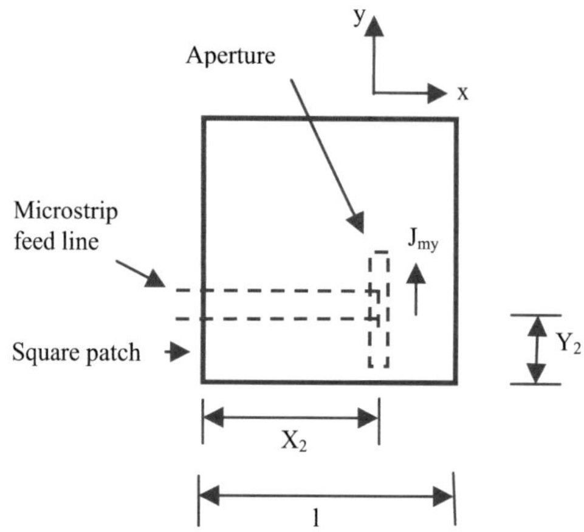

Fig. 35. Position of the aperture paralleled to y-axis

where V_{oy} is the voltage at the middle of the aperture paralleled to the y axis. The corresponding current density J_{my} is given by:

$$J_{my} = \frac{-2\, E_{ax}}{h} \tag{36}$$

The modes coefficients $\beta_{y,mn}$ are obtained by;

$$\beta_{y,mn} = \frac{j\omega\varepsilon}{(K_{mn}^2 - K^2)K_m^2} \int_0^{L_a}\int_0^{W_a} J_{my}.\psi_{y,mn}^* dxdy \tag{37}$$

The magnetic field H_y in the cavity is given by:

$$H_y = \sum_{mn} A_{mn}\beta_{y,mn}K_m \sin(K_m x)\cos(K_n y) \tag{38}$$

3.1.5 Input impedance

If the aperture is parallel to x direction; the admittance of the patch can be obtained by the following equation [16,20]:

$$Y_{x,ant} = \frac{\iiint_V H_x J_{mx}^* dV}{|E_z h|^2} \tag{39}$$

where V is the volume occupied by the source within which the magnetic currents exist. The electric field E_z inside the cavity is determined based on equation (13) and equation (15) as follows:

$$E_z = \frac{1}{j\omega\varepsilon}\left(\frac{\partial H_y}{\partial x} - \frac{\partial H_x}{\partial y}\right) \tag{40}$$

But the admittance of the patch in case of the aperture is parallel to y direction can be obtained from equation (41):

$$Y_{y,ant} = \frac{\iiint_V H_y J_{my}^* dV}{|E_z h|^2} \quad (41)$$

The admittance value of the aperture Y_{ap} can be obtained from the transmission line model of the aperture by considering the aperture as two short circuited slot lines [2].

$$Y_{ap} = \frac{-2j}{Z_{ca}} \cot(K_a \frac{L_a}{2}) \quad (42)$$

where Z_{ca}, L_a are, respectively, the characteristic impedance, the length of the aperture. The input impedance of the ACMSA is given by:

$$Z_{in} = \frac{n^2}{Y_{x,ant} + Y_{ap}} + \frac{n^2}{Y_{y,ant} + Y_{ap}} \quad (43)$$

The open microstrip stub line (L_{os}) which is used for forcing the imaginary part of the input impedance to be zero, is obtained by;

$$L_{os} = \frac{1}{K_F} \cot^{-1}\left(\frac{imaginary(Z_{in})}{2\,Z_F} \right) \quad (44)$$

The input impedance of the ACMSA with the open microstrip stub line is given by

$$Z_{in\_stub} = \frac{n1^2}{Y_{x,ant} + Y_{ap}} + \frac{n1^2}{Y_{y,ant} + Y_{ap}} - 2jZ_F \cot(K_F L_{os}) \quad (45)$$

Z_F and K_F, are, respectively, the characteristic impedance and the wave number of the microstrip feed line; n1 is the transformation ratio which describes the coupling between the microstrip feed line and the patch [19].

3.2 Circular polarization diversity with ACMSA

The electric field components in x-direction (E_x) and y direction (E_y) are generated by putting an aperture parallel to y axis and x axis respectively as shown in Figure 36. The circular polarization diversity requires the phase shift between E_x and E_y to be $\pm 90^\circ$. By using the quadrature hybrid as a feeder for the square patch microstrip antenna the circular polarization diversity can be achieved. When the right port of the hybrid is selected, E_x will lead E_y by 90° and the RHCP will be generated. If the left port of the hybrid is selected, E_x will lag E_y by 90° and the LHCP will be generated [20].

3.3 Geometry of ACMSA

The geometry of the proposed ACMSA with a circularly polarized diversity is shown in Figure 37. The square patch is printed on Rogers's substrate (patch substrate) with a thickness h_t=0.85 mm, a loss tangent=0.0021 and a dielectric constant $\varepsilon_r = 3.55$. Based on the cavity model theory and ADS optimization the length of the square patch is 31.599 mm at a resonant frequency of 2.437 GHz. The quadrature hybrid is printed on the Roger's substrate

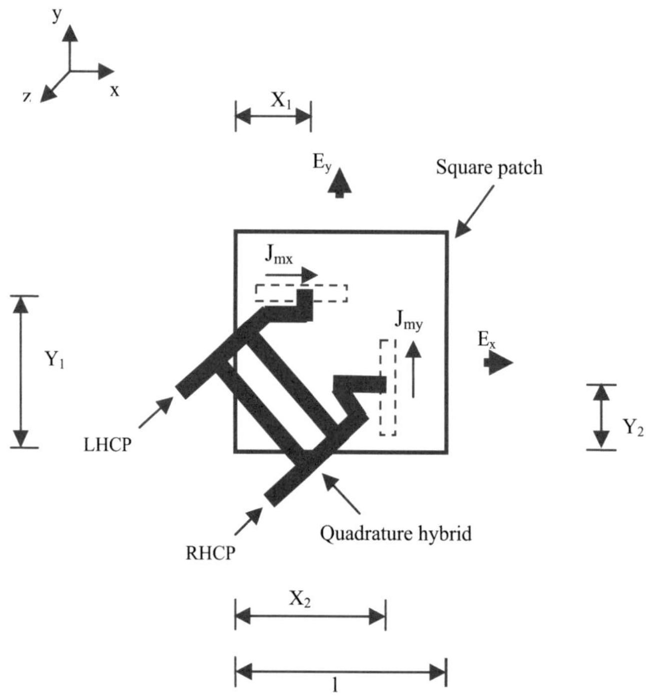

Fig. 36. Circular polarization diversity with ACMSA

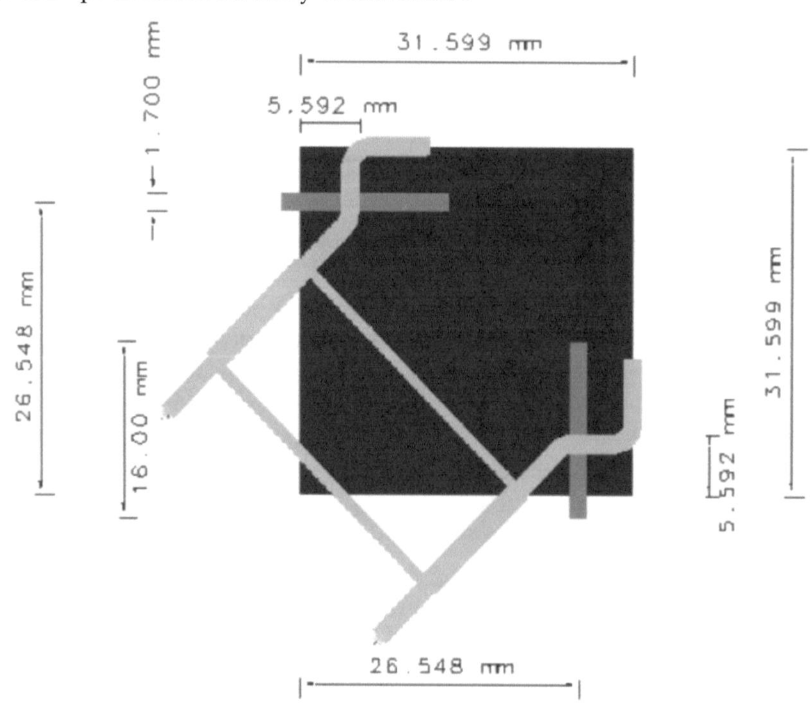

Fig. 37. Geometry of the proposed ACMSA

(feed substrate) which has the same parameters as those of the patch substrate. The ground plane exists in between the patch substrate and the feed substrate. There are two identical apertures printed on the ground plane, one parallel to x-axis and the other paralleled to y-axis. The length and the width of the aperture are 16 mm and 1.7 mm respectively [2, 14, 16-19, 21].

The size of the geometry with bended open microstrip stub line (L_{os}) as shown in Figure 37 is about 22% smaller than that of the other geometry shown in Figure 38 which has straight open microstrip stub line (L_{os}).

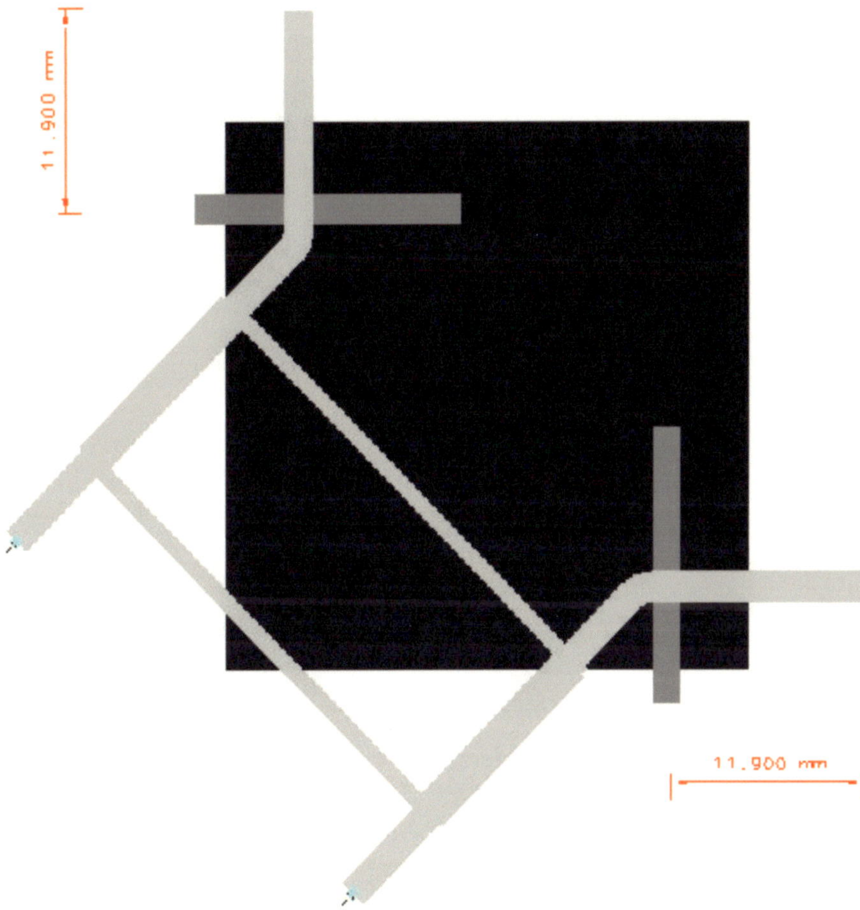

Fig. 38. Geometry of ACMSA with straight open microstrip stub line

The ADS is used to convert the straight L_{os} into its equivalent bended L_{os} by modeling the straight L_{os} using MLOC with length 11.9 mm in ADS schematic as shown in Figure 39. The ADS schematic as shown in Figure 40 is implemented to model the bended L_{os} using two MLIN of length 3 mm, MCURVE and MLOC with variable length.

The ADS schematic as shown in Figure 41 is used to find the optimum length of the bended L_{os}.

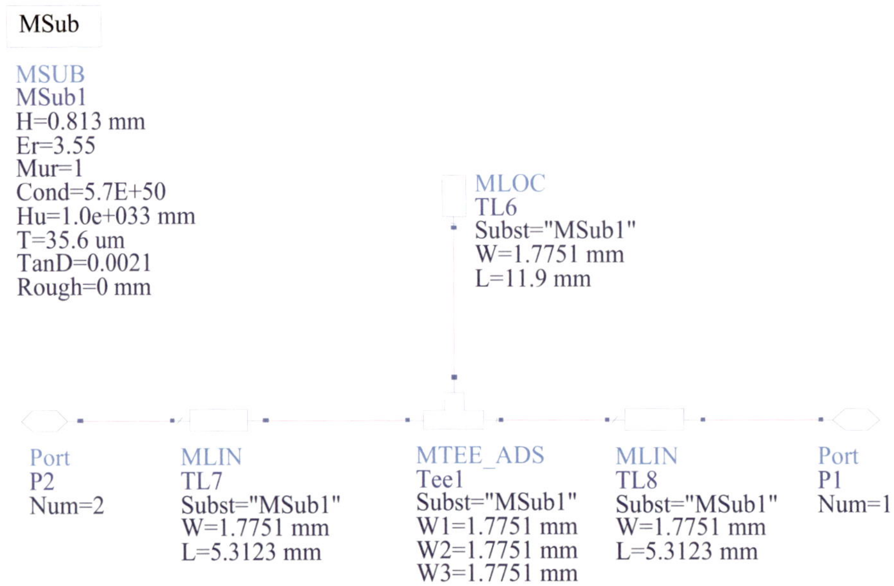

Fig. 39. ADS schematic model for straight open microstrip stub line

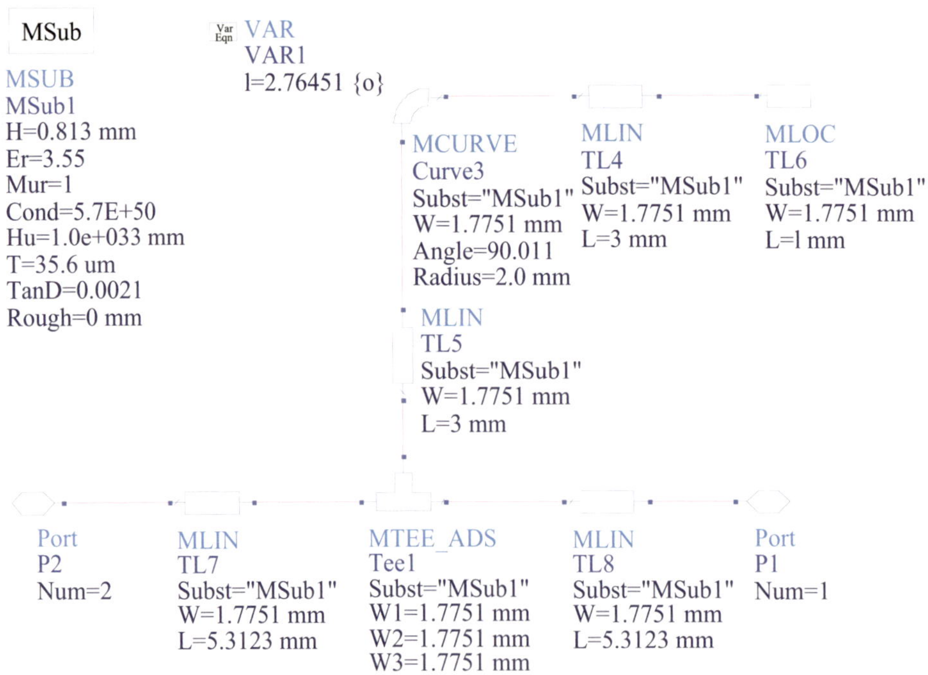

Fig. 40. ADS schematic model for bended open microstrip stub line

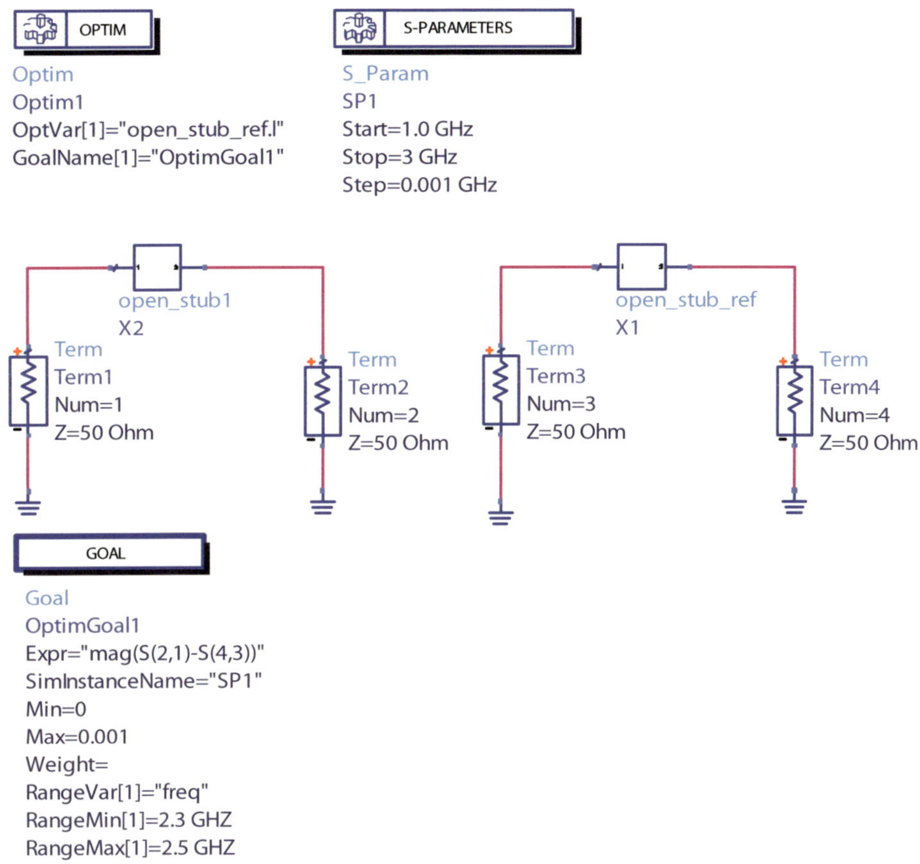

Fig. 41. ADS schematic to find the optimum length of the bended open microstrip stub line

Figure 42 shows that the bended L_{os} is equivalent to straight L_{os} when the variable length is 2.76451 mm.

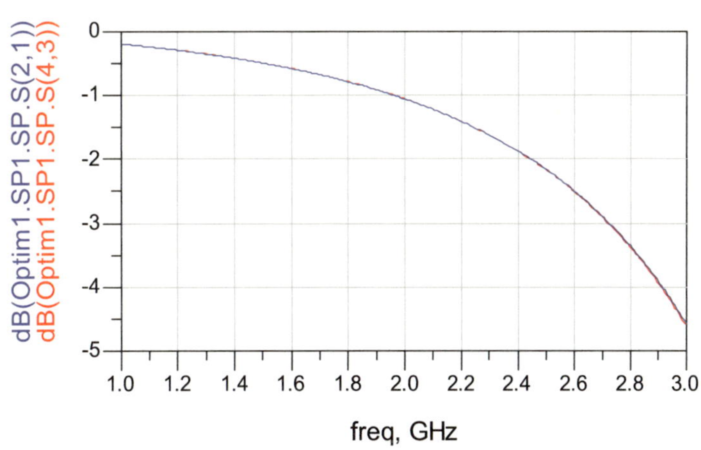

Fig. 42. Simulated responses for straight and bended open microstrip stub line

3.4 Simulation and measured results

The value of Y_1 and X_2 is the same which is selected as 26.548 mm to place the quadrature hybrid at a further distance into the square patch in order to minimize the size of the ACMSA. The value of X_1 and Y_2 is the same and equals to the distance at which the real part of the input impedance is approached to 50 Ω at the center of each aperture. Figure 43 shows the input impedance of the ACMSA without the implementation of the open microstrip stub line (L_{os}). In this case the imaginary part of the input impedance is 42.86 Ω when X_1 is 5.688 mm. The length of the open microstrip stub line is determined in order to force the imaginary part of input impedance to be zero.

Figure 44 shows the input impedance of the ACMSA with L_{os} equals to 11.9 mm. The results for X_1, X_2, Y_1 and Y_2 are obtained using the Matlab simulation and these results are optimized using ADS as shown in Figure 37.

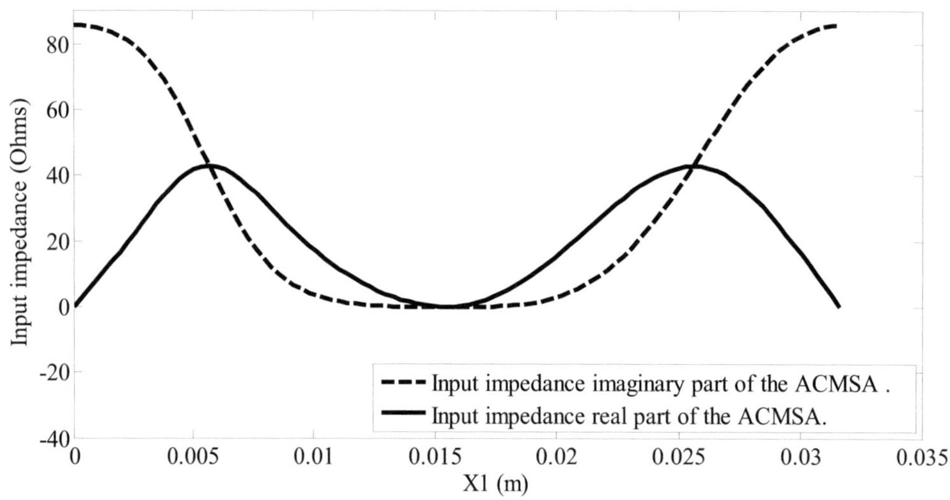

Fig. 43. Simulated input impedance of the ACMSA without L_{os}

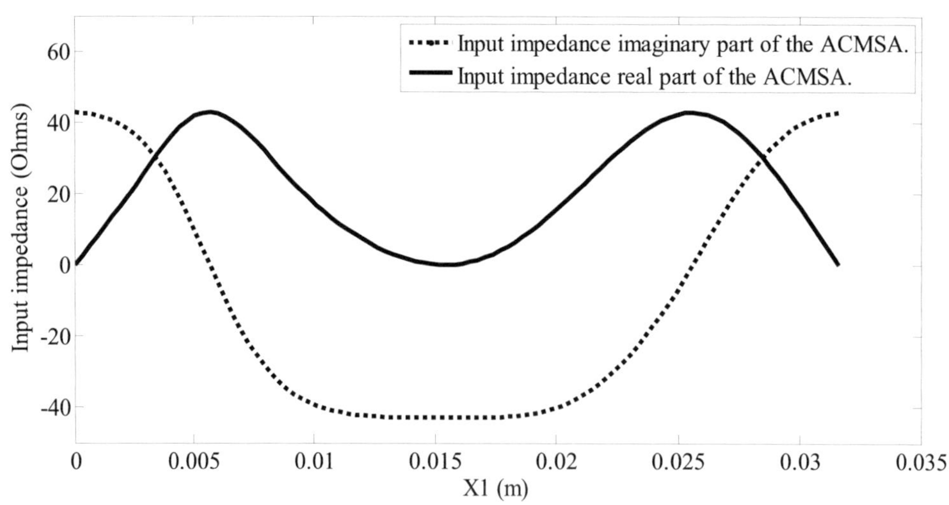

Fig. 44. Simulated input impedance of the ACMSA with L_{os}

If the right port of the quadrature hybrid is selected, the reflection coefficient (S_{11}) versus the frequency and the normalized measured radiation pattern of the ACMSA are shown in Figures 45 and 46 respectively. Figure 45 shows that, the proposed ACMSA provides an impedance bandwidth of 20.1% at port 1 (for S_{11} < -10dB).

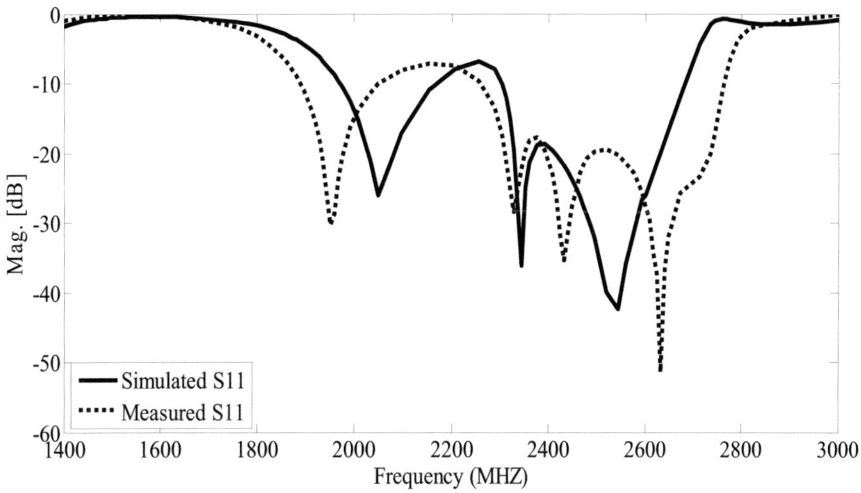

Fig. 45. Reflection coefficient versus frequency when the right port of the quadrature hybrid is selected

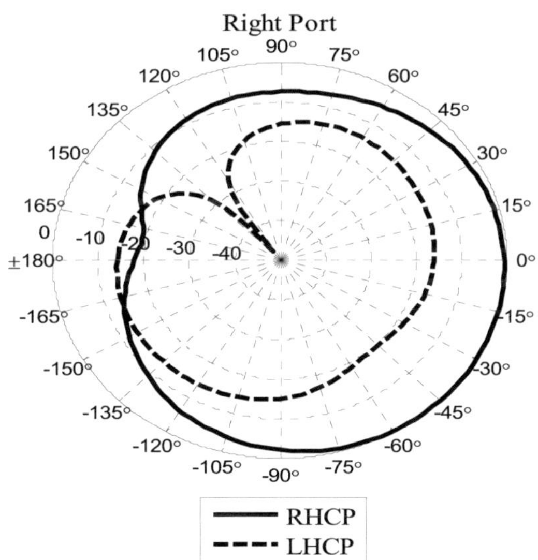

Fig. 46. Normalized measured radiation pattern when the right port of the quadrature hybrid is selected

Figures 47 and 48 respectively show the reflection coefficient (S_{22}) versus the frequency and the normalized measured radiation pattern of the ACMSA, when the left port of the quadrature hybrid is selected. The impedance bandwidth of 20 % is obtained at port 2 (for S_{22} < -10dB) as evident from Figure 47.

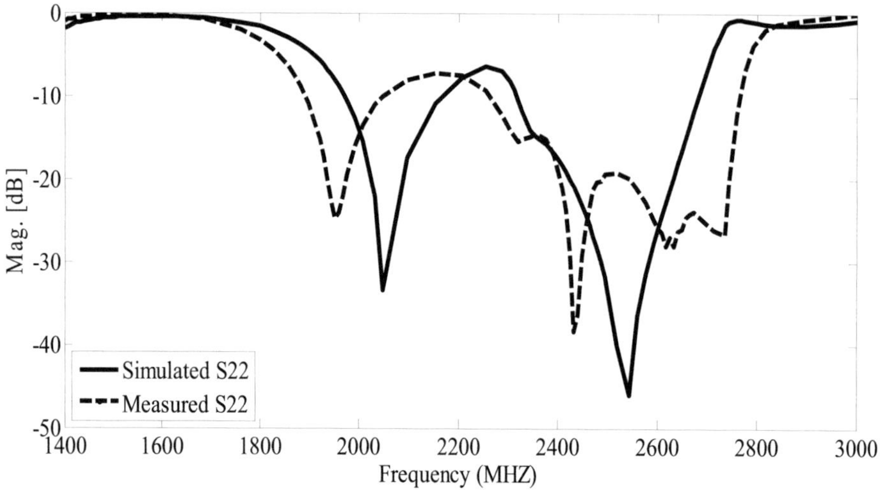

Fig. 47. Reflection coefficient versus frequency when the left port of the quadrature hybrid is selected

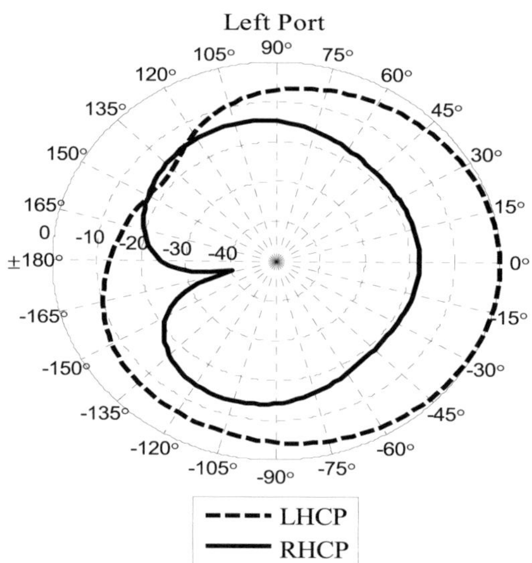

Fig. 48. Normalized measured radiation pattern when the left port of the quadrature hybrid is selected

The fabrication of one ACMSA requires two substrate slices (patch substrate and feed substrate). The parameters setup in ADS Momentum for substrate layers is shown in Figure 49.

The top side and the bottom side of each substrate are covered by metal layer. Parameters setup in ADS Momentum for metallization layers is shown in Figure 50. For patch substrate, the square patch is printed on the top side while the metal layer on the bottom side is removed. For the feed substrate the quadrature hybrid is printed on the bottom side while the two apertures are etched on the top side which is considered as a ground plane.

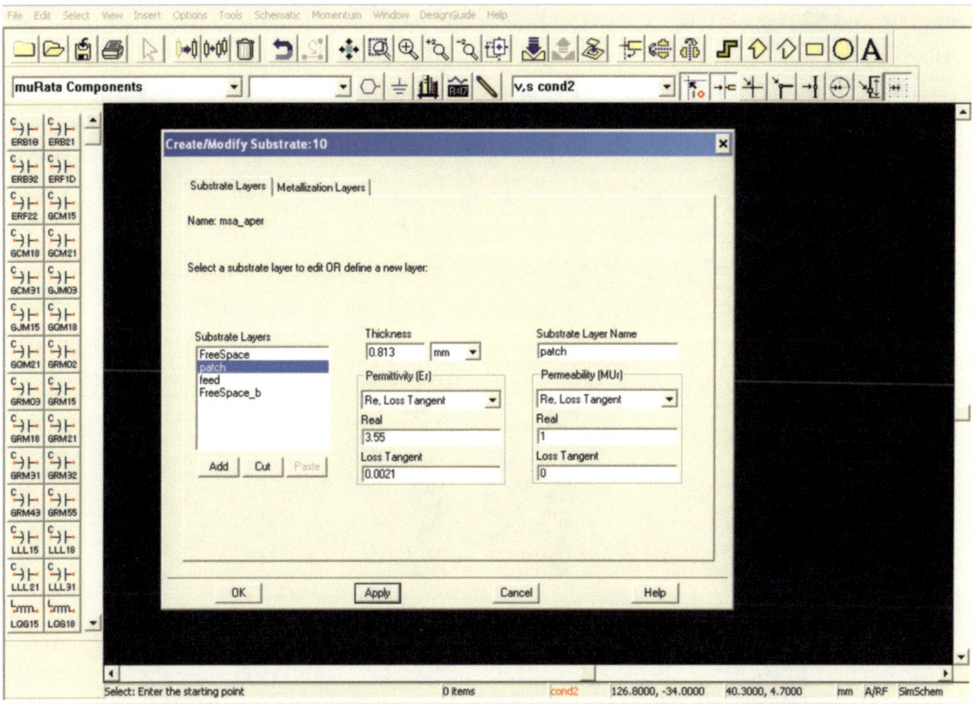

Fig. 49. Parameters setup in ADS Momentum for substrate layers of the ACMSA

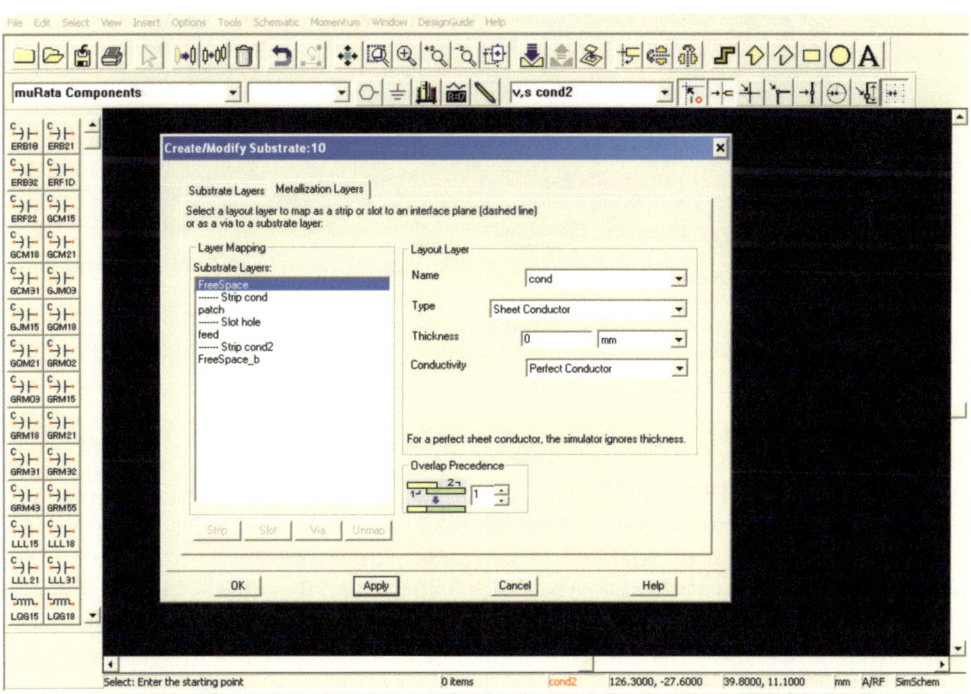

Fig. 50. Parameters setup in ADS Momentum for metallization layers of ACMSA

Figures 51 and 52 respectively show the fabricated patch substrate and feed substrate of the ACMSA.

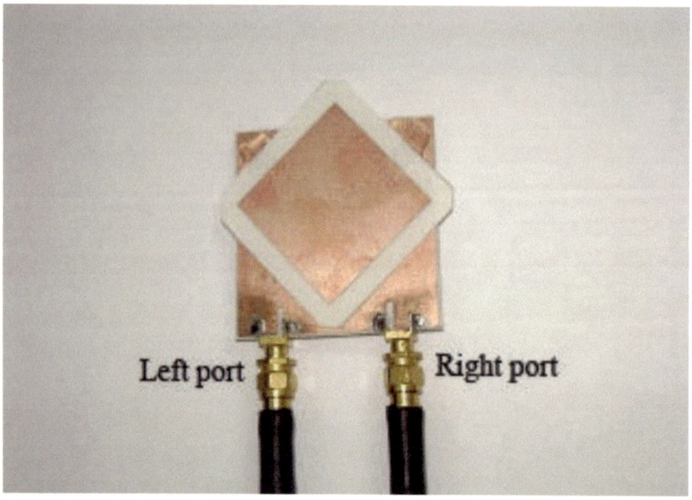

Fig. 51. Fabricated patch substrate

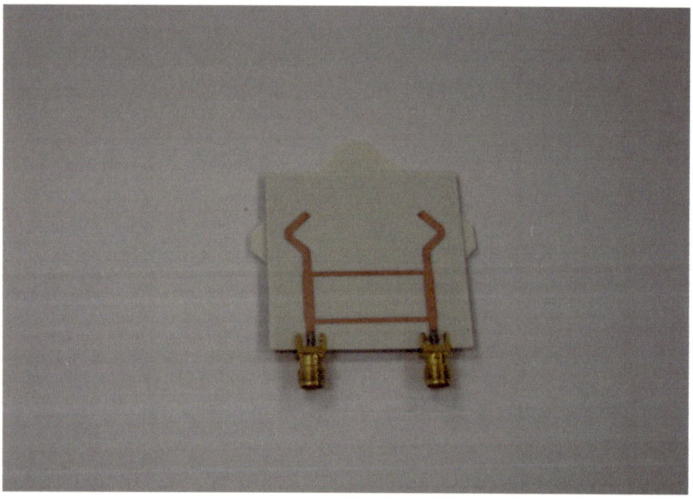

Fig. 52. Fabricated feed substrate

The measured gain at 2.437 GHz for right port and left port are 5dBi and 4.6 dBi respectively. The gain for the left port is measured using left hand circularly polarized standard antenna (2.4 GHz, 8 dBi, LHCP, Flat Patch Antenna), while the gain for the right port is measured using right hand circularly polarized standard antenna (2.4 GHz, 8dBi, RHCP, Flat Patch Antenna). Figure 53 shows the measured axial ratio at both ports of the ACMSA.

The ACMSA produces two beams, one of which is RHCP and the other is LHCP as shown in Figures 46 and 48 respectively. The axial ratio and the radiation pattern are measured in Satimo chamber. The measurement system consists of probe antennas mounted with equal spacing on a circular arch as shown in Figure 54. The measured data is collected automatically and saved in MS Excel format.

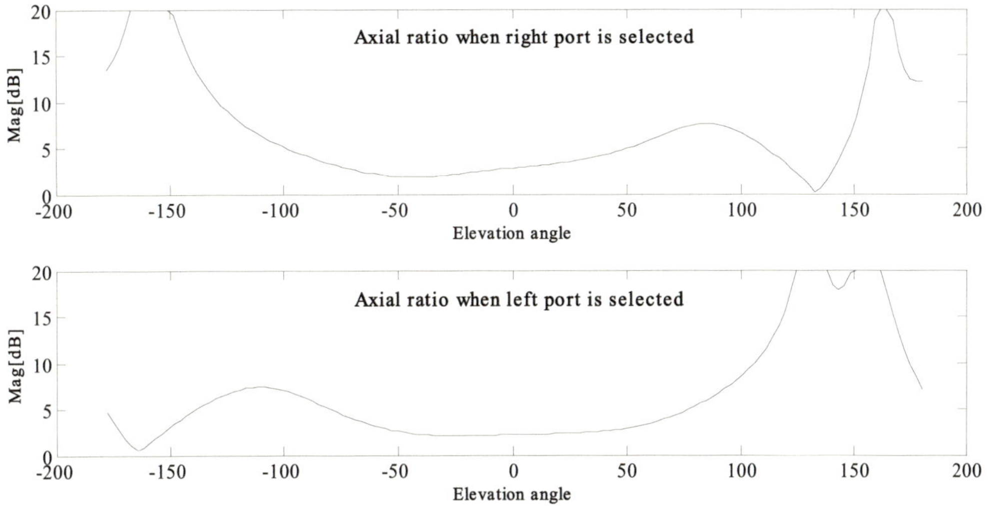

Fig. 53. Measured axial ratio at both ports of the ACMSA versus an elevation angle

Fig. 54. Measuring aperture-coupled microstrip antenna inside Satimo chamber

The size of the ACMSA is much smaller than that of the microstrip antenna which is fed by a quadrature hybrid without using an aperture coupled structure. The utilization of the aperture coupled structure could reduce the antenna size substantially compared with the one without aperture coupled structure. For instance, size reductions of 72% and 60 % are achieved compared to the cases of [22] and [23] respectively. The proposed ACMSA has a better bandwidth and good axial ratio compared to the designs presented by [22, 24-26]. Further, it can receive the RHCP and the LHCP simultaneously whereas the reconfigurable patch antennas could receive only one polarization type at a time [22, 24-26]. Above all, the proposed geometry can provide circular polarization diversity with good axial ratio over a broad angular range, wide bandwidth and small size.

4. Conclusions

A circularly polarized microstrip antenna array could be generated with linearly polarized patches. The separation distance between the patches in the x-direction has a strong effect on the E-plane of the planar microstrip antenna array pattern, whereas the separation distance between the patches in the y-direction has a weak affect on the H-plane of the planar microstrip antenna array pattern. Four narrow beams at four different directions are obtained through the excitation of a planar microstrip antenna array by a 4×4 Butler matrix. These four beams possess circular polarization diversity, good axial ratio and high gain.

Circular polarization diversity could be generated by coupling the quadrature hybrid to the square patch through two apertures. The aperture coupled structure minimized the size of the microstrip antenna which is fed by the quadrature hybrid. The proposed geometry provides good axial ratio over a broad angular range and wide bandwidth. The analysis of the ACMSA using the cavity model is comparable with the full-wave analysis and the experimental results.

APPENDIX A

ADS schematics to design path 3 and path 4 of the planar microstrip antenna array with 4×4 Butler matrix

Figure A.1 shows the ADS schematic for path 3 which connects between antenna 3 and 4×4 Butler matrix. The ADS schematic to modify the length of path 3 is shown in Figure A.2.

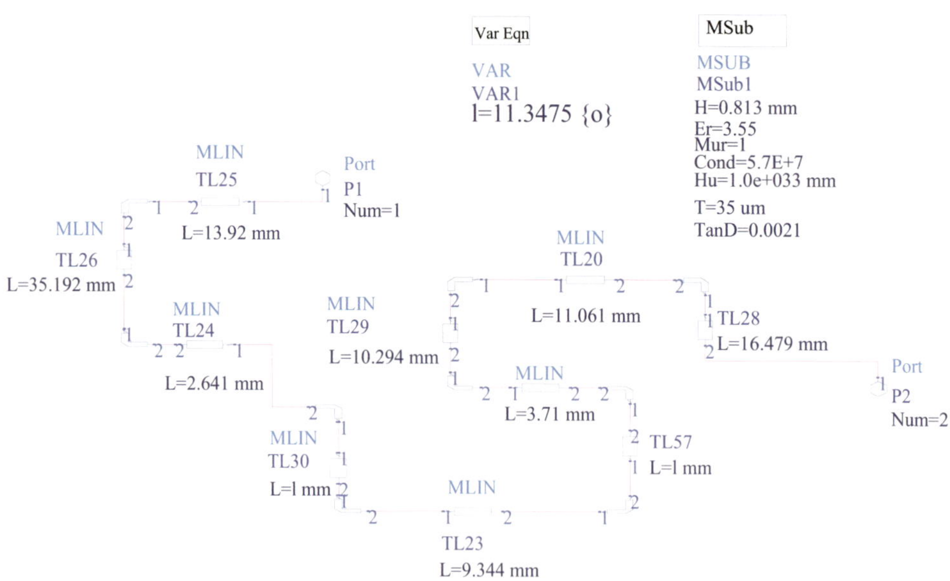

Fig. A.1 ADS schematic for path 3

Fig. A.2 ADS schematic to modify the length of path 3

The ADS simulated results for phases of path 3 and reference path are shown in Figure A.3.

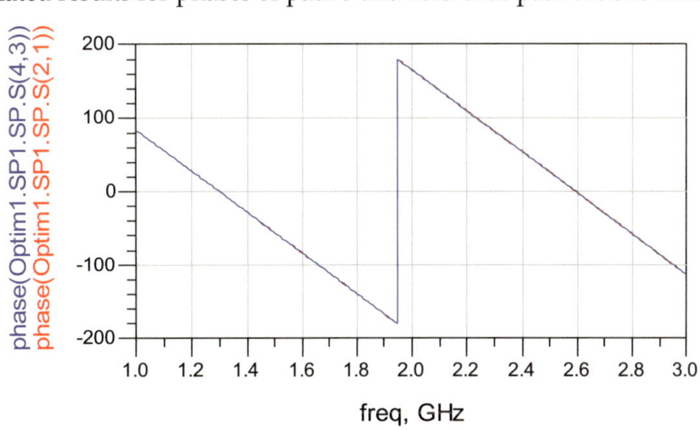

Fig. A.3 ADS simulated results for phases of path 3 and reference path

The ADS schematic for path 4 which connects between antenna 4 and 4×4 Butler matrix is shown in Figure A.4.

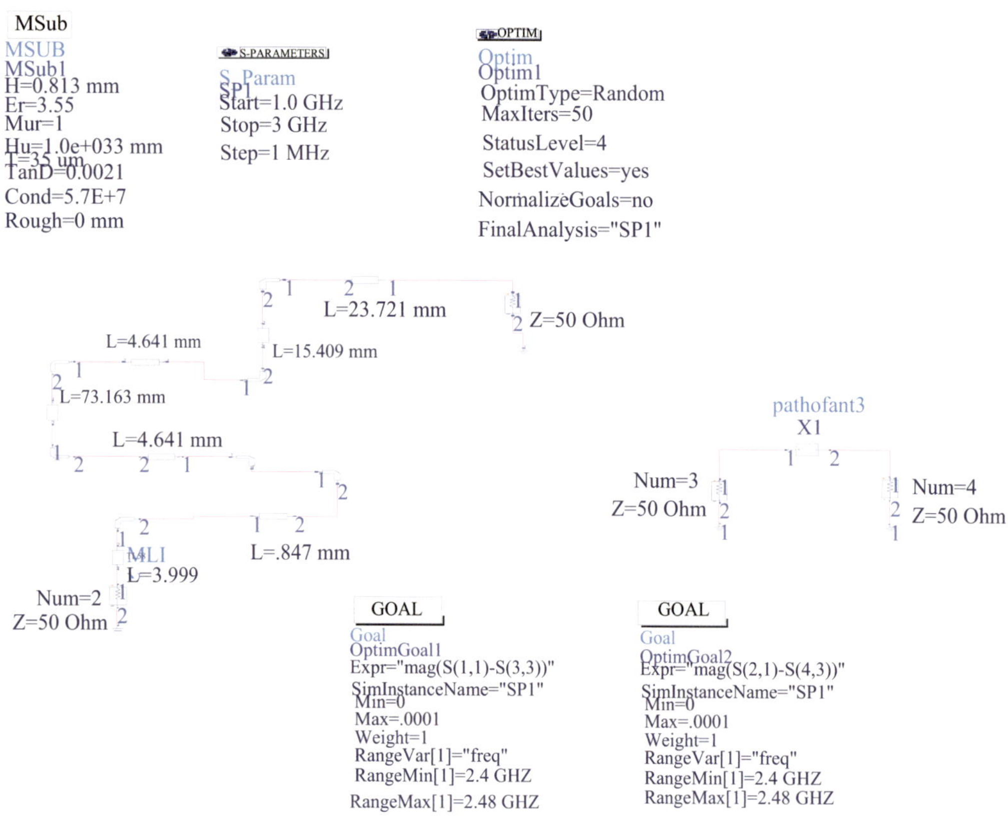

Fig. A.4 ADS schematic for modifying length of the path connecting antenna 4 and 4×4 Butler matrix

The length of the path 4 is modified by using the ADS schematic shown in Figure A.5. The ADS simulated results for phases of path 4 and reference path are shown in Figure A.6.

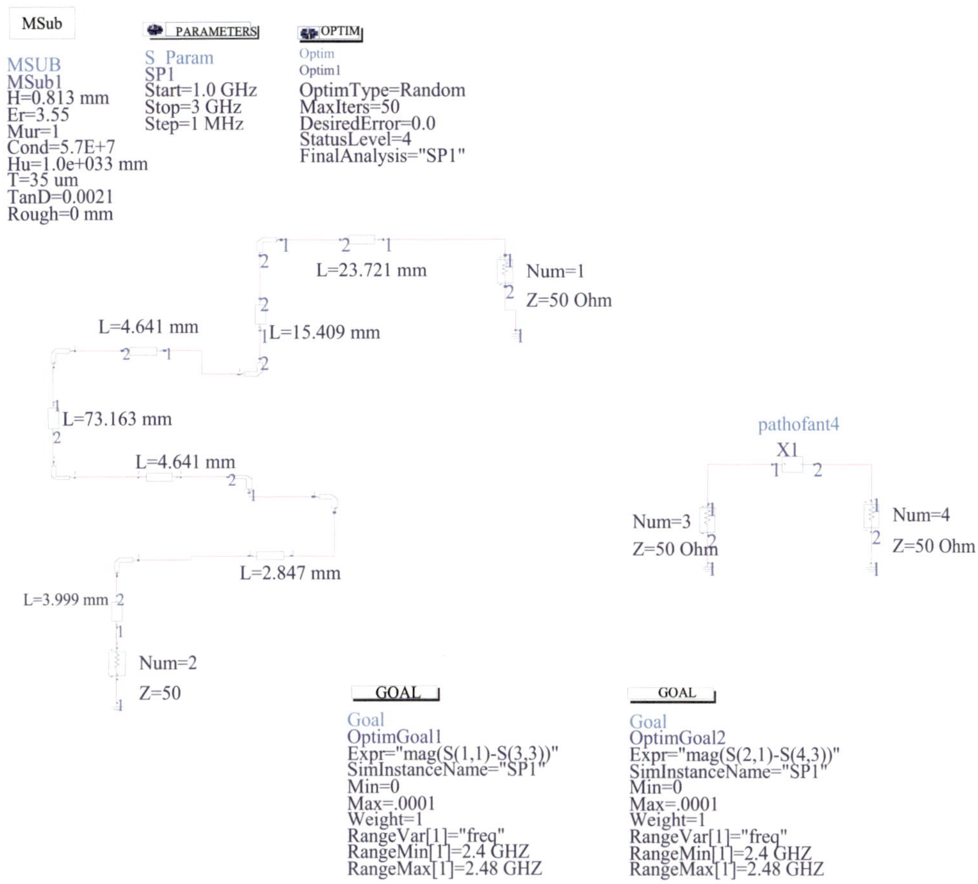

Fig. A.5 ADS schematic to modify the length of path 4

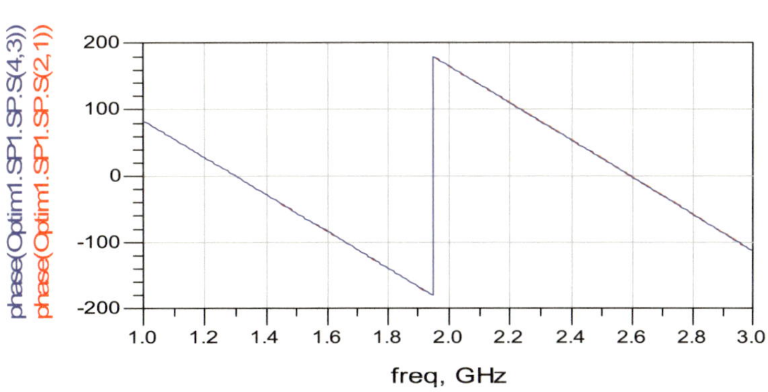

Fig. A.6 ADS simulated results for phases of path 4 and reference path

APPENDIX B

Propagation constant inside the cavity

$$\frac{\partial^2 \psi_{y,mn}}{\partial x \partial y} - \frac{\partial^2 \psi_{x,mn}}{\partial y^2} - K_{mn}^2 \psi_{x,mn} = 0 \tag{B.1}$$

$$\psi_{x,mn} = A_{mn} K_n \cos(K_m x) \sin(K_n y) \tag{B.2}$$

$$\psi_{y,mn} = -A_{mn} K_m \cos(K_n y) \sin(K_m x) \tag{B.3}$$

where $A_{mn} = \dfrac{\chi_{mn}}{\sqrt{\varepsilon . l^2 . h_t}}$

$$\chi_{mn} = \begin{cases} 1, & m = 0 \, and \, n = 0 \\ \sqrt{2}, & m = 0 \; or \; n = 0 \\ 2, & m \neq 0 \, and \, n \neq 0 \end{cases}$$

For a non radiating cavity

$$K_m = \frac{m\pi}{l}, K_n = \frac{n\pi}{l} \tag{B.4}$$

$$\frac{\partial \psi_{y,mn}}{\partial x} = -A_{mn} K_m^2 \cos(K_n y) \cos(K_m x) \tag{B.5}$$

$$\frac{\partial^2 \psi_{y,mn}}{\partial x \partial y} = A_{mn} K_m^2 K_n \sin(K_n y) \cos(K_m x) \tag{B.6}$$

$$\frac{\partial \psi_{x,mn}}{\partial y} = A_{mn} K_n^2 \cos(K_m x) \cos(K_n y) \tag{B.7}$$

$$\frac{\partial^2 \psi_{x,mn}}{\partial y^2} = -A_{mn} K_n^3 \cos(K_m x) \sin(K_n y) \tag{B.8}$$

$$A_{mn} K_m^2 K_n \sin(K_n y) \cos(K_m x) + A_{mn} K_n^3 \sin(K_n y) \cos(K_m x) = K_{mn}^2 A_{mn} K_n \sin(K_n y) \cos(K_m x) \tag{B.9}$$

$$K_{mn}^2 = K_m^2 + K_n^2 \tag{B.10}$$

APPENDIX C

Mode coefficients in *x*-dirextion inside the cavity

$$\frac{\partial^2 H_y}{\partial x \partial y} = \sum_{mn} \beta_{y,mn} \frac{\partial^2 \psi_{y,mn}}{\partial x \partial y} \tag{C.1}$$

$$\frac{\partial^2 H_x}{\partial y^2} = \sum_{mn} \beta_{x,mn} \frac{\partial^2 \psi_{x,mn}}{\partial y^2} \tag{C.2}$$

By substituting from equation (C.1) and (C.2) into equation (20) the following equation is obtained:

$$\sum_{mn} \beta_{y,mn} \frac{\partial^2 \psi_{y,mn}}{\partial x \partial y} - \sum_{mn} \beta_{x,mn} \frac{\partial^2 \psi_{x,mn}}{\partial y^2} - K^2 \sum_{mn} \beta_{x,mn} \psi_{x,mn} = j\omega\varepsilon J_{mx} \tag{C.3}$$

$$\frac{\partial^2 \psi_{y,mn}}{\partial x \partial y} - \frac{\partial^2 \psi_{x,mn}}{\partial y^2} = K_{mn}^2 \psi_{x,mn} \tag{C.4}$$

$$\sum_{mn} \beta_{y,mn} = \sum_{mn} \beta_{x,mn} \tag{C.5}$$

By substituting from equations (C.4) and (C.5) into equation (C.3) the equation (C.6) is obtained:

$$\sum_{mn} \beta_{x,mn} K_{mn}^2 \psi_{x,mn} - \sum_{mn} \beta_{x,mn} K^2 \psi_{x,mn} = j\omega\varepsilon J_{mx} \tag{C.6}$$

$$\sum_{mn} \beta_{x,mn} (K_{mn}^2 - K^2) \int_0^l \int_0^l \psi_{x,mn} . \psi_{x,mn}^* dxdy = j\omega\varepsilon \int_0^{W_a} \int_0^{L_a} J_{mx} . \psi_{x,mn}^* dxdy \tag{C.7}$$

Where $\psi_{x,mn}^*$ is the complex conjugate of $\psi_{x,mn}$

$$\int_0^l \int_0^l \psi_{x,mn} . \psi_{x,mn}^* dxdy = K_n^2 \tag{C.8}$$

5. References

[1] L. Basilio, M. Khayat, J. Williams, and S. Long, "The dependence of the input impedance on feed position of probe and microstrip line-fed patch antennas," *IEEE Transactions on Antennas and Propagation,* vol. 49, pp. 45-47, 2001.

[2] M. Mathian, E. Korolkewicz, P. Gale, and E. Lim, "Design of a circularly polarized 2x2 patch array operating in the 2.45 GHZ ISM band," *Microwave Journal,* vol. 45, pp. 280-286, 2002.

[3] T. Denidni and T. Libar, "Wide band four-port butler matrix for switched multibeam antenna arrays," in *14th IEEE Proceedings on Personal, Indoor and Mobile Radio Communications.* vol. 3, 2003, pp. 2461-2464.

[4] G. Srivastava and V. Gupta, "Microwave devices and circuit design," New Delhi: Prentice-Hall of India, 2006, p. 339.

[5] N. Jean-Sébastien and D. Gilles, "Microstrip EHF Butler matrix design and realization," *ETRI Journal,,* vol. 27, pp. 788-797, 2005.

[6] M. Halim, "Adaptive array measurements in communications," Boston: Artech House, 2001, pp. 170-177.

[7] J. Huang, "A technique for an array to generate circular polarization with linearly polarized elements," *IEEE Transactions on Antennas and Propagation,* vol. 34, pp. 1113-1124., 1986.
[8] J. Huang, "C.P. microstrip array with wide axial ratio bandwidth and single feed L.P. elements," in *International Symposium on Antennas and Propagation*. vol. 23, 1985, pp. 705-708.
[9] A. Rudge, "The handbook of antenna design. Vol.1," in *IEE electromagnetic waves series, 15* London: Peregrinus on behalf of the Institution of Electrical Engineers, 1982, pp. 1-29.
[10] K. Chang, "RF and microwave wireless systems," New York: Wiley, 2000, pp. 98-105.
[11] L. Johan, "Design and analysis of an electrically steerable microstrip antenna for ground to air use," in *Computer Science and Electrical Engineering*. vol. MSc: Luleå University of Technology, 2002.
[12] W. Ismail, "Active integrated antenna (AIA) with image rejection," in *Electrical and Computer Engineering*. vol. PhD: University of Birmingham, 2003.
[13] L. Ronglin, B. Pan, J. Laskar, and M. Tentzeris, "A novel low-profile broadband dual-frequency planar antenna for wireless handsets," *IEEE Transactions on Antennas and Propagation,* vol. 56, pp. 1155-1162, 2008.
[14] K. Carver and J. Mink, "Microstrip antenna technology," *IEEE Transactions on Antennas and Propagation,* vol. 29, pp. 2-24, 1981.
[15] H. John and J. Encinar, "Reflectarray antennas," Piscataway, N.J.; Hoboken, N.J.: IEEE Press ; Wiley-Interscience, 2008, p. 143.
[16] B. Al-Jibouri, H. Evans, E. Korolkiewicz, E. Lim, A. Sambell, and T. Viasits, "Cavity model of circularly polarised cross-aperture-coupled microstrip antenna," *IEE Proceedings -Microwaves, Antennas and Propagation,* vol. 148, pp. 147-152, 2001.
[17] S. Kanamaluru, L. Ming-Yi, and C. Kai, "Analysis and design of aperture-coupled microstrip patch antennas and arrays fed by dielectric image line," *IEEE Transactions on Antennas and Propagation,* vol. 44, pp. 964-974, 1996.
[18] M. Himdi, J. Daniel, and C. Terret, "Analysis of aperture-coupled microstrip antenna using cavity method," *Electronics Letters,* vol. 25, pp. 391-392, 1989.
[19] R. Garg, "Microstrip antenna design handbook," Boston, MA: Artech House, 2001, pp. 542,786-788.
[20] A. Petosa, "Dielectric resonator antenna handbook," in *Artech House antennas and propagation library* Boston: Artech House, 2007, pp. 49-50, 171-178.
[21] G. Parker, "A dual polarized microstrip ring antenna with very good isolation," in *Electrical Engineering*. vol. MSc: Kingston Ontario: Royal Military College, 1997.
[22] G. Wu, W. Mu, G. Zhao, and Y. Jiao, "A novel design of dual circularly polarized antenna eed by L-strip," *Progress In Electromagnetics Research,* vol. PIER pp. 39-46, 2008.
[23] Z. Harith, "Design of a circular polarization microstrip antenna at 2.4GHZ," in *Electronics and Telecommunication,*. vol. MSc: Universiti Teknologi Malaysia, 2005.
[24] K. Chung, Y. Nam, T. Yun, and J. Choi, "Reconfigurable microstrip patch antenna with switchable polarization," *ETRI,* vol. 28, pp. 379-382, 2006.
[25] Y. Sung, T. Jang, and Y. Kim, "A reconfigurable microstrip antenna for switchable polarization," *IEEE Microwave and Wireless Components Letters,* vol. 14, pp. 534-536, 2004.
[26] Y. Fan and Y. Rahmat-Samii, "A reconfigurable patch antenna using switchable slots for circular polarization diversity," *IEEE Microwave and Wireless Components Letters,* vol. 12, pp. 96-98, 2002.

17

DBDP SAR Microstrip Array Technology

Shun-Shi Zhong
School of Communication and Information Engineering
Shanghai University, Shanghai 200072
China

1. Introduction

1.1 Background

Since the American satellite SEASAT was launched in June 1978, an era for the space-borne synthetic aperture radar (SAR) systems has been started. Over 30 years from then on, more and more SAR systems have been proposed and manufactured, such as: ALMAZ SAR system ejected by the former U.S.S.R in 1991, ERS-1 of European Space Agency (ESA) in 1991, JERS-1 of Japan in 1992, RadarSat-1 and RadarSat-2 of Canada in 1995 and 2005, Terra SAR-X of German in 2007 and RiSat-SAR of India in 2007, etc. Another milestone is that the SIR- C/X SAR of American Space Shuttle Endeavour completed the high resolution three-dimensional (3D) imaging all over the globe in Feb. 2000.

The overall performances of a SAR system such as azimuth/elevation resolution, imaging ambiguity, width of mapping area and so on are directly determined by the performance of its antenna. Since the response of an object to the electromagnetic wave with different polarization is different, the dual-polarization operation of the antenna can provide additional information and thus the probability of detecting and identifying a target is enhanced. Because that the reflection characteristic of an object is various for the incident electromagnetic wave with different frequency, its multi-band operation will provide more information such as the back-scattering data and the penetration data of objects. The common bands of space-borne SARs are L (center at 1.275 GHz), S (3.0 GHz), C (5.3 GHz) and X (9.6 GHz) bands. For example, the SIR-C/X-SAR system uses L/C/X tri-bands with dual-polarization, whose advantages were proved and the function of 3D mapping was realized by the use of interference imaging [1]. However, as seen from Fig. 1, its antenna consists of three sub-arrays separately for L, C, X bands with 12 m × 4.1 m area, resulting in a bulky structure of about 3000 kg weight. The realization of a shared-aperture configuration will minimize the volume and weight of the antenna and share the sub-systems behind the array as well. Therefore, the dual-band dual-polarization (DBDP) shared-aperture antenna array has been proposed and studied. In China, following many researches on the dual-polarized planar antenna arrays [2−6], the development of DBDP arrays has also been carried out [7−9].

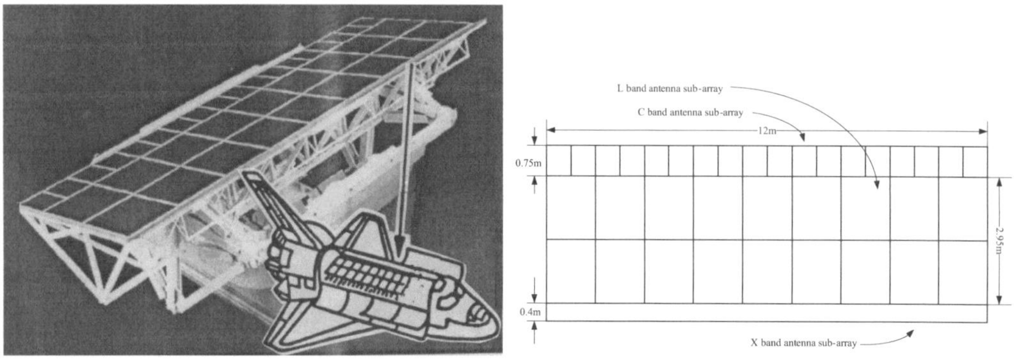

Fig. 1. Antenna of SIR-C/X SAR

1.2 Design principles

The dual-band and dual-polarized operation of an antenna has been applied in a lot of systems, such as wireless communications, radars, deep space communications and radio astronomical telescopes, etc. The requirements for antennas are varied in different missions, and thus the designs are different, especially for the latter two where the reflector antennas are preferred while a horn or a small patch array may be used as the feed [10, 11].

This paper emphasizes the DBDP antennas for SAR application. For the future remote sensing mission, two-plane phased scanning characteristics are needed, so generally the possibility of using reflector antennas is excluded. Meanwhile, although the waveguide slotted antenna array has better efficiency and sidelobe characteristics, it is still not preferred in future SAR application, because it is hard to realize two-plane phased scanning/ beam-forming DBDP operation and its mass is large [6], then the microstrip patch array becomes the best choice. Due to that the operation frequencies of different bands in the DBDP SAR application are widely separated, it is difficult to use dual-frequency radiator element to construct a DBDP antenna array. Thus it comes as an obvious alternative to construct a DBDP antenna by integrating the radiators for each individual band. And apparently, the multi-layer configuration is required to contain two individual band elements and their feed networks. In addition, the cross-polarization of the array should less than −30 dB or it will cause imaging ambiguity, and the isolation between dual polarization ports should also be considered carefully. Since the bandwidth impacts the elevation resolution directly, a wide bandwidth is more an important target.

According to the configuration, the typical DBDP arrays include: the perforated patch array, the interlaced elements array and some other configuration arrays derived from them. The review and comparison among them are presented in the following sections. Some advanced techniques to enhance DBDP antenna performances also are introduced.

2. DBDP perforated patch arrays

The DBDP antennas for SAR application are usually operating on two widely separated bands, causing great difference in dimensions of the radiators for dual bands. In order to keep the higher frequency (HF) radiator within a wavelength to suppress the grate lobes, the lower frequency (LF) radiators are perforated and the HF radiators are placed in the perforations. This type of array configuration is usually adopted when the frequency ratio of HF to LF is even. The example is an L/C-band perforated patch array by Shafai et al. (see

Fig. 2 and Fig.3) [12], where a perforated L-band patch and 16 C-bands patches form a unit cell with the frequency ratio of 4 : 1. The patches co-planarly configured in two bands are stacked to broaden the bandwidth, achieving a bandwidth for RL (return loss) ≤ −14 dB of 300 MHz at C-band (about 5.7%) and 90 MHz at L-band (about 6.4%). The patches of two bands are orthogonally fed to realize dual-polarization operation. The cross-polarization level in the bandwidth is lower than −30 dB. As seen from Fig.4, two slots of the C-band element are passive, but help in reducing the cross polarization to below -30dB.

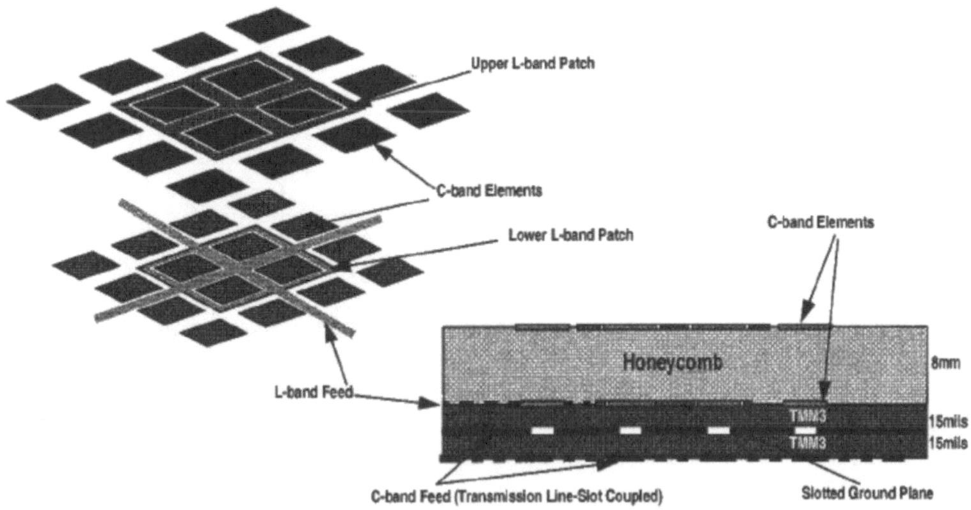

Fig. 2. L/C band perforated patch array[12]

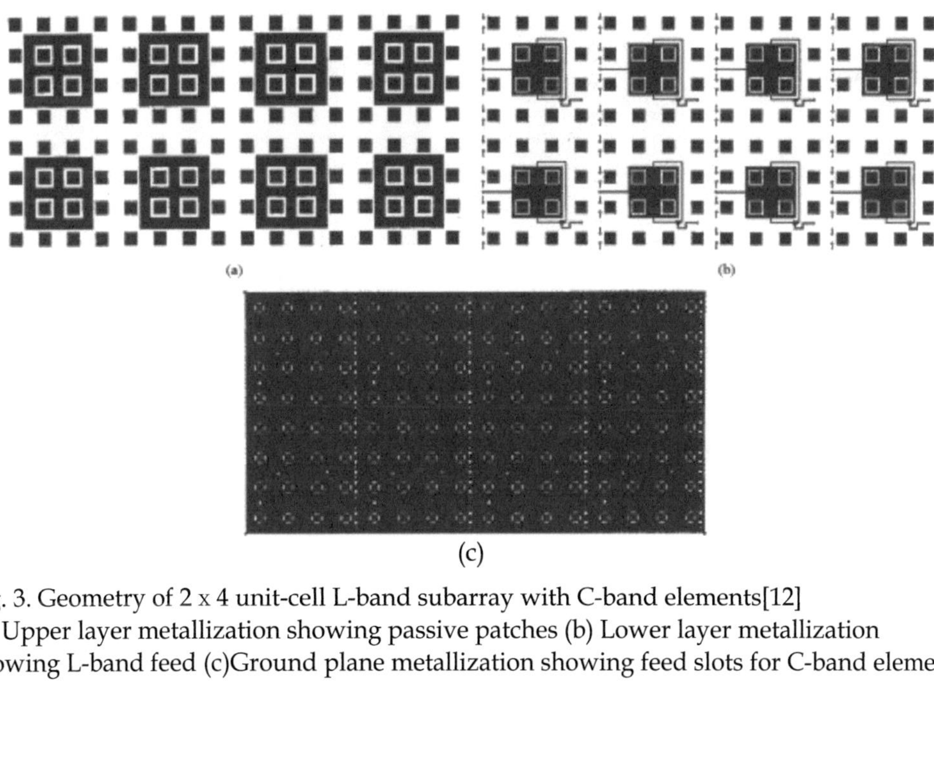

(c)

Fig. 3. Geometry of 2 x 4 unit-cell L-band subarray with C-band elements[12]
(a) Upper layer metallization showing passive patches (b) Lower layer metallization showing L-band feed (c)Ground plane metallization showing feed slots for C-band elements

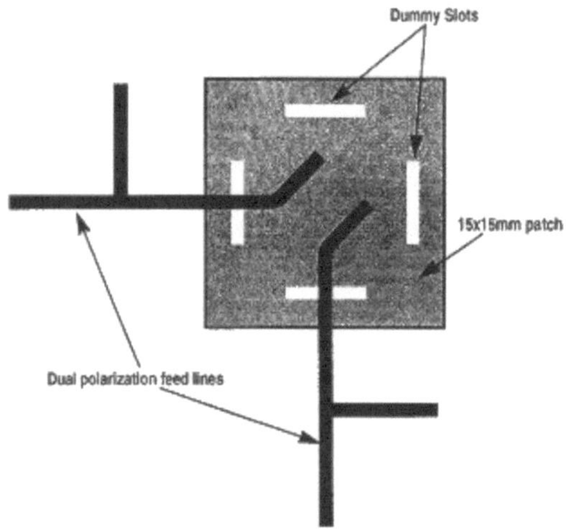

Fig. 4. Geometry of dual-polarized C-band element[12]

Another L/X DBDP array designed by Pozar and Targonski [13] uses the similar array configuration as shown in Fig. 5. The main advantage of this configuration is its wider bandwidth at LF compared with the other array designs, such as that of dipole/patch. But this will cause the difficulty in designing the antenna due to the dimension of perforation, which should be large enough to avoid hindering the radiation from HF elements. A larger

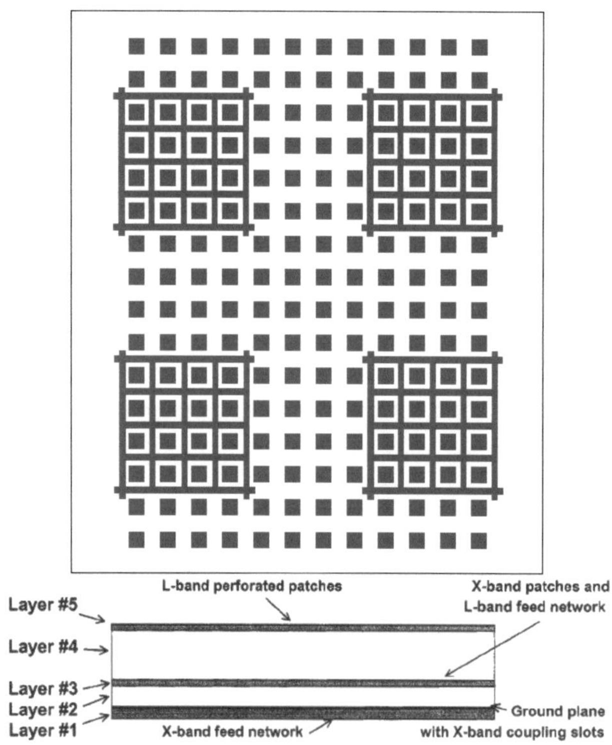

Fig. 5. Top view and side view of L/X-band DBDP antenna [13]

perforation size is also favored in terms of good isolation performance between the two bands and a more symmetric pattern at the cost of bandwidth at LF. Then a tradeoff must be made between the bandwidth and the isolation performance.

Recently a new element design of the DBDP perforated patch array has been reported, as shown in Fig.6 [14]. Its L-band antenna element is designed as a stacked perforated patches to ensure the required bandwidth(about 8%) and to allow for achieving integration with four C-band antenna elements. The perforated driven patch is excited by four probes, two of them generating vertical polarization and two generating horizontal polarization ,as shown in Fig 6b.Anti-phase excitations of the appropriate probes have been achieved with the use of 180°phase shifters realized as sections of transmission lines. Good return losses better than 15 dB and high isolation better than 30 dB have been achieved within the required broad frequency range. Two stacked square patches operating at C-band are fed through two orthogonally placed H-shaped slots as seen in Fig. 6c. Due to the limited space the feeding network of the vertical polarization is etched on the lower layer than that of the horizontal polarization. Good return losses better than 15 dB and high isolation better than 40 dB have been measured over the desired bandwidth (about 7 %) for a single C-band antenna element. The obtained cross-polarization level equals -20 dB.

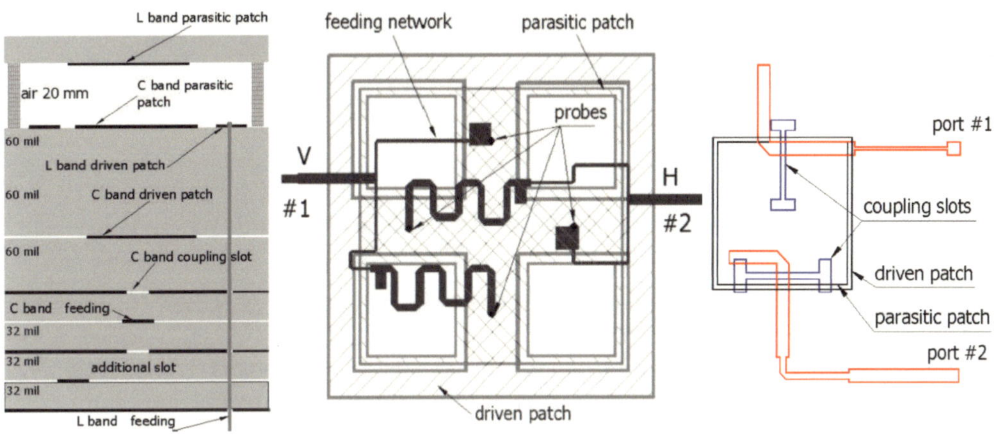

Fig. 6. L/C-band DBDP antenna element with the use of 180⁰ phase shifters [14](a) Cross-sectional view (b) Layout of L-band antenna element (c) Layout of C-band antenna element

3. DBDP interlaced elements arrays

3.1 Interlaced slot/dipole with patches

The DBDP array configuration of interlaced elements is attractive, where the element may be slot/dipole/ring or patch. Though an interlaced elements array employing either slots or printed dipoles for both bands and the architecture using microstrip patches for both bands may be considered [15], the shared-aperture DFDP array of interlaced slot/dipole with patches is often preferred. Because of its smaller size in one dimension, the slot/dipole of LF can be easily placed between two HF elements, thus meeting the interelement-distance requirement. An example is the DBDP L/C band array of the interlaced L-band slots with C-band patches designed by Pozar et al. (see Fig. 7) [16], where C-band square patches are located on the top layer and L-band slots on the ground. The slots are orthogonally placed to realize the dual-polarization in L-band. This design achieves a bandwidth more than 5% in both bands and the isolation is better than 24 dB in C-band and 23 dB in L-band. But to suppress the backward radiation of the slots, a reflector is placed 1/4 wavelength behind the slot and therefore the dimension of the array is increased.

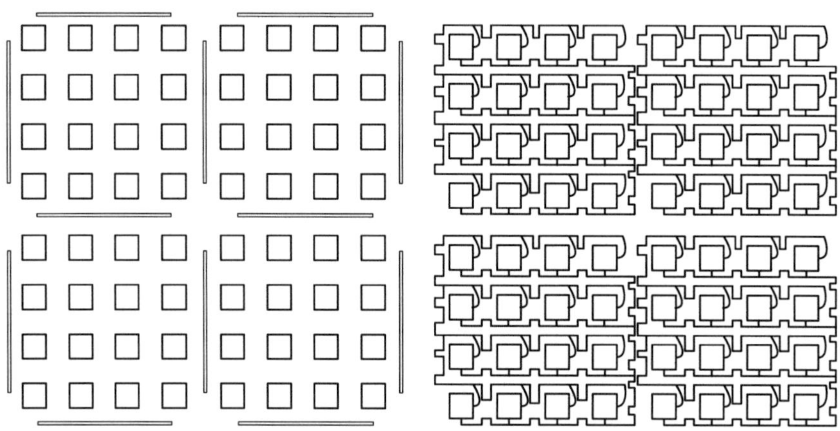

(a) (b)

Fig. 7. L/C band Interlaced slots with patches[16] (a) Top layer parasitic patch; (b) driven patch with feed

In another example, S/X band interlaced dipoles with patches are proposed by our group [7,8] as shown in Fig. 8, where both stacked patches and stacked dipoles are adopted, which achieves a measured bandwidth of 8.9% and 17% in S-band and X-band, respectively ,with its frequency ratio of about 1 : 3, as shown in Fig. 9. Since the S-band dipoles are cross-placed and located on another side of the substrate, the cross-polarization and the isolation between two ports at S-band are minimized, resulting in the cross-polarization levels of better than −26 dB and −31 dB at S-band and X-band, respectively. In addition, by using dipoles instead of slots, the serious backward radiation is avoided. However, the bandwidth of dipole is usually narrower than that of slot, so a stacked dipole is used and another substrate is needed compared to the interlaced slot/patches array aforementioned. From the radiation patterns of both bands depicted in Fig.10, it is obvious that the arrays of both bands have little impact on each other radiation patterns--from the shape and null points of patterns, so they look like working "separately".
Recently, a similar configuration [17] is reported to apply to the millimeter wave band as a feed of a parabolic cylindrical reflector antenna. The Ka-band microstrip patches working at 35.5GHz are in the upper layer and interlaced with the Ku-band cross slots working at 13.6 GHz. Dual-polarizations have realized for the Ku-band, but only single-polarization has realized for the-Ka band since space constraint. By sharing the same antenna aperture, the problem of placing two feeds at one focus is solved.

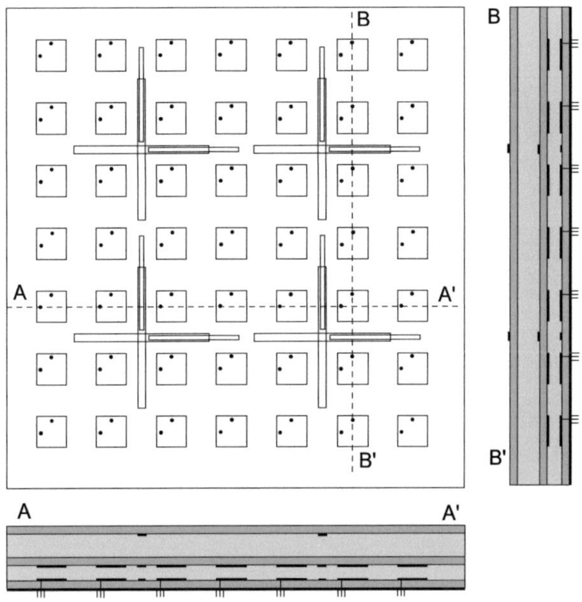

Fig. 8. S/X-band interlaced dipole with patches

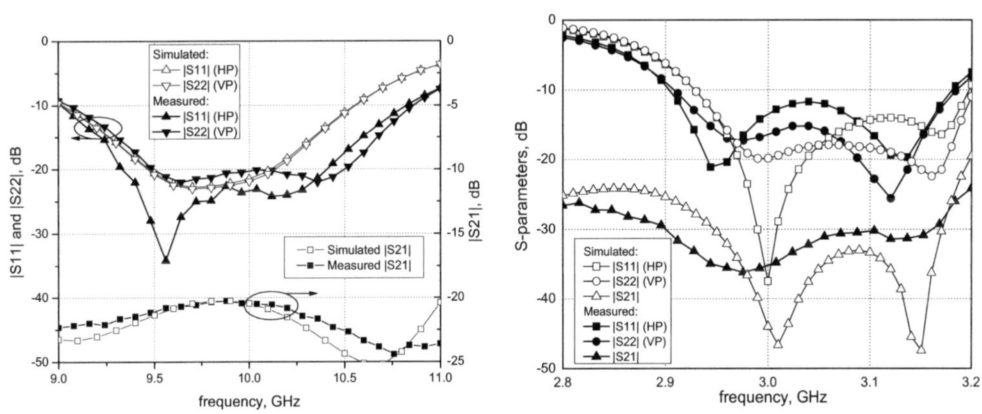

(a) (b)

Fig. 9. Measured *S* parameter of S/X-band array. (a) X-band element; (b) S-band element

The feed array consists of 8 Ku-band elements linear arrays and 16 Ka-band elements linear arrays, and top view of the half-model (4 Ku- and 8 Ka-band elements linear arrays) is shown in Fig.11a, where a Ku-band linear array includes 2 cross slots and a Ka-band one includes 4 microstrip patches. Fig.11b shows the cross-sectional view. Since the slot is bidirectional, 4 mm-thick reflector ground plane is used at a distance of about $\lambda/4$ behind the cross slot. This design has achieved satisfactory performances in both bands, verifying the validity for the millimeter wave band application.

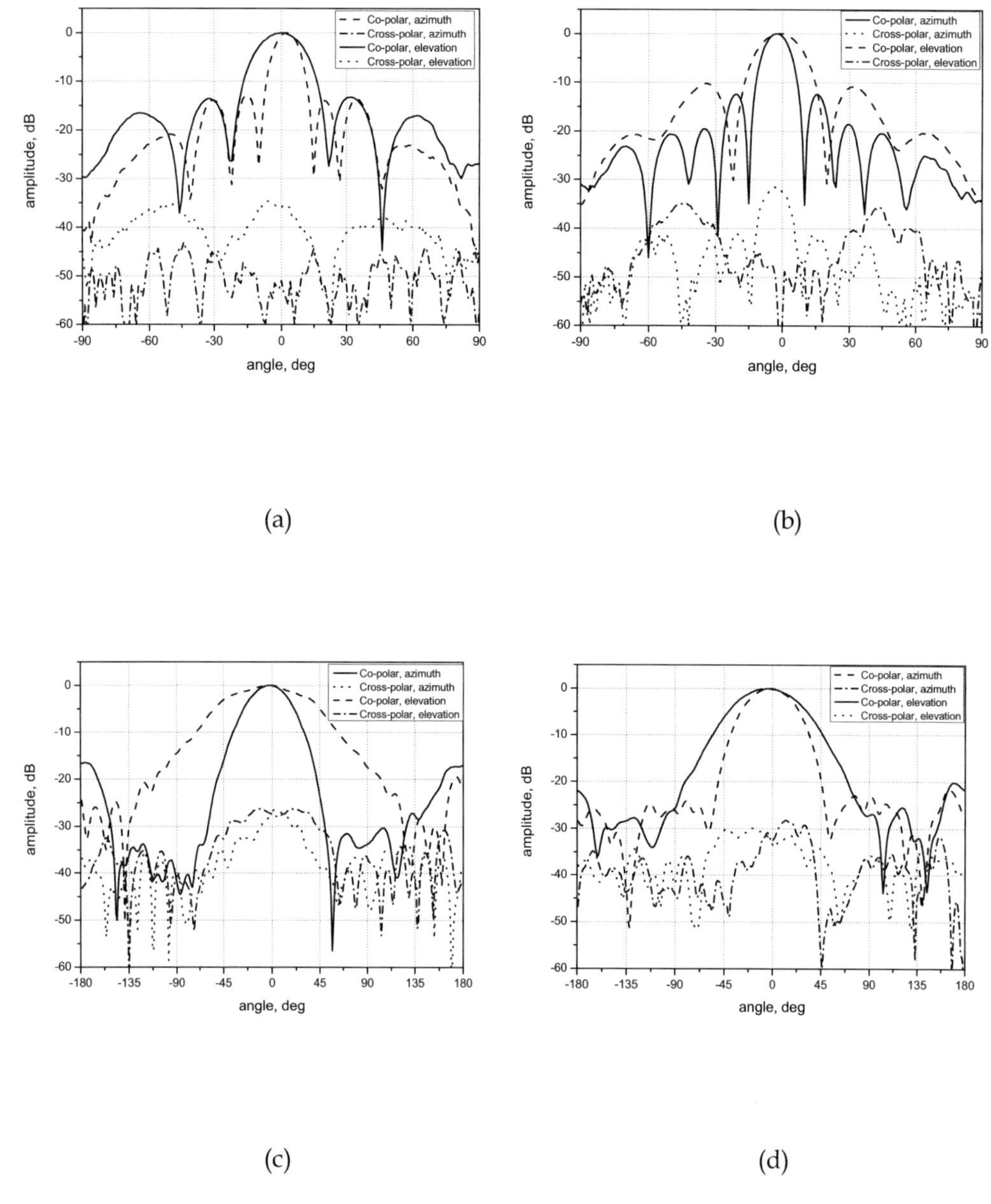

Fig. 10. Measured radiation pattern of S/X-band array. (a) H-polarization of X-band; (b) V-polarization of X-band; (c) H-polarization of S-band; (d) V-polarization of S-band

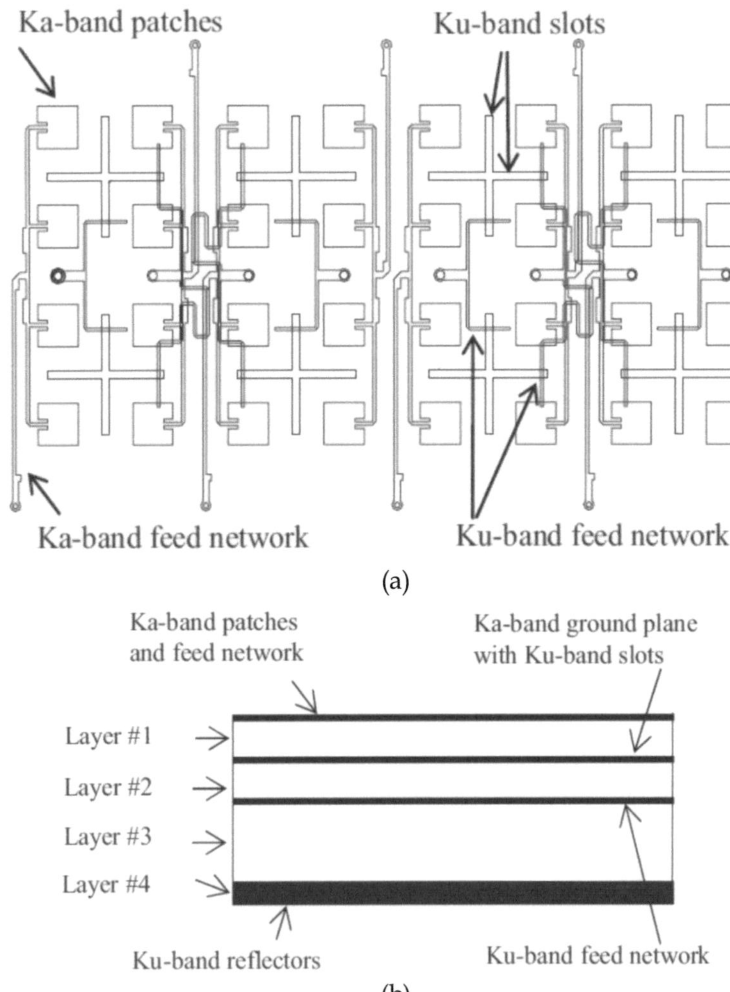

Fig. 11. Ka/Ku array of interlaced cross-slots with patches [17] (a) Top view (b) cross-sectional view

3.2 Interlaced ring with patches

A DBDP array consisting of interlaced ring with patches for airborne applications is presented in [18]. The S-band antenna elements sit on the top layer, and the X-band antennas are on the bottom layer. These two planar arrays with thin substrates are integrated to provide simultaneous operation at S-band (3 GHz) and X-band (10 GHz), as shown in Fig.12a. The X-band antenna elements are circular patches. They are combined to form a 4×8 array with a gain of 18.3 dBi, using the 4 element series-fed resonant type arrays to save the space of the feeding line network, as shown in Fig. 12b. The S-band element is a large rectangular ring-resonator antenna. The four sides of the square-ring element are laid out in such a way that they only cover part of the feeding lines on the bottom layer, but none of the radiating elements. The antenna has a mean circumference of about $2\lambda_g$ and has a 9.5 dBi gain that is about 3 dB higher than the gain of an ordinary ring antenna. The ring is loaded by two gaps of its parallel sides, and these make it possible to achieve a 50Ω input match at the edge of the third side.

Measured normalized radiation patterns for both bands are shown in Fig.13 and Fig.14.The measured and simulated specifications of the S/X-band array are summarized in Table 2.It is seen that its performances are quite good, but the bandwidths are narrow due to its thin structure, while the thin and lightweight structure is attractive for airborne applications.

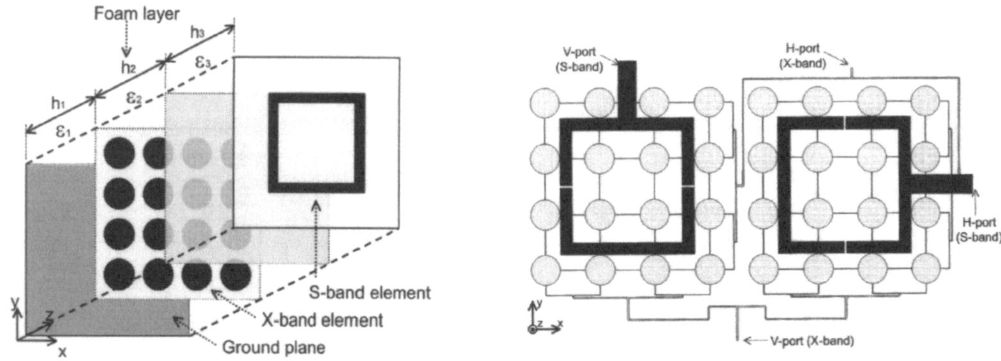

(a) (b)

Fig. 12. S/X-band array of interlaced rings with patches[18]
(a) multilayer structure (b) Top view

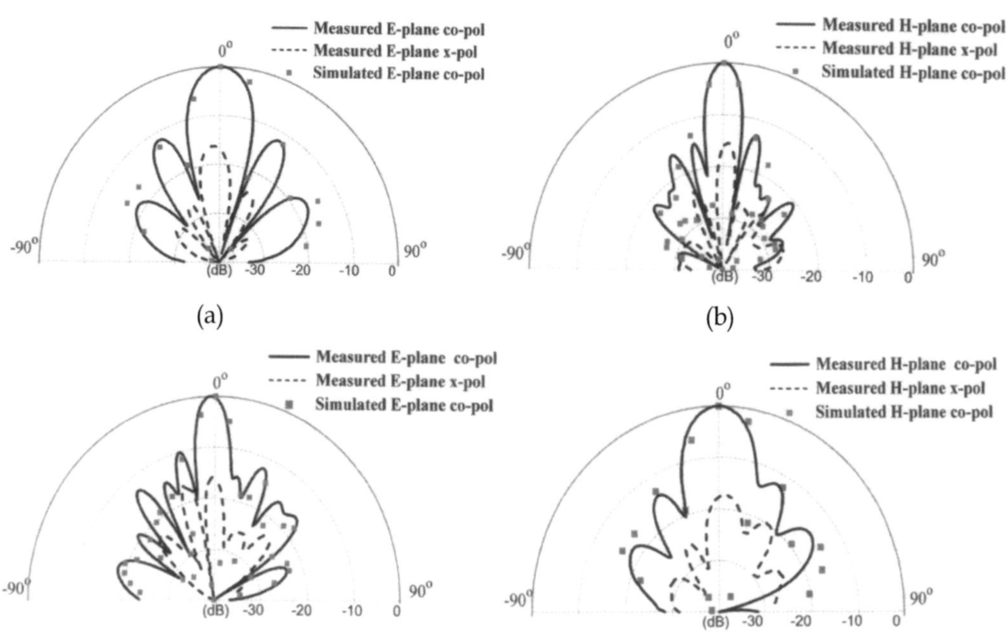

(a) (b) (c) (d)

Fig. 13. Radiation patterns of the X-band array[18]
(a) V-port feed, E-plane (b) V-port feed, H-plane
(c) H-port feed, E-plane (d) H-port feed, H-plane

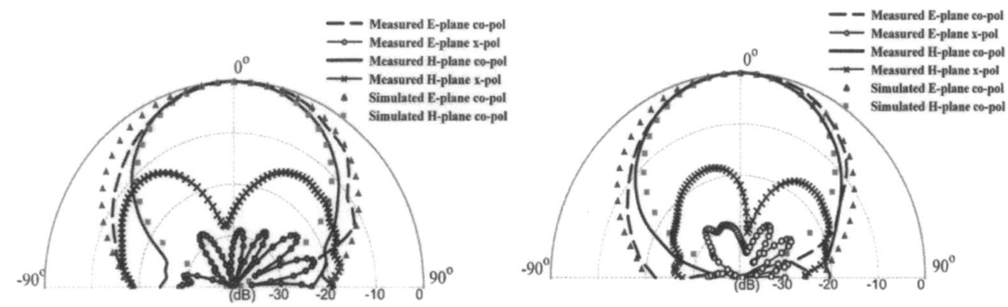

(a) (b)

Fig. 14. Radiation patterns of the S-band array[18] (a)V-port feed (b) H-port feed

		X-Band		S-Band	
	Polarization	V Port	H Port	V Port	H Port
Frequency (GHz)	Measurement	9.8-9.98	9.81-10.0	2.94-2.96	2.94-2.96
	Simulation	9.91-9.99	9.9-10.0	2.94-2.96	2.94-2.96
Bandwidth (%)	Measurement	1.8	2.3	1.03	1.03
	Simulation	1.0	1.0	1.03	1.03
Gain (dBi)	Measurement	18.3	17.3	9.5	7.9
	Simulation	17.6	17.6	8.72	8.73
Efficiency (%)	Measurement	32.4	28.8	96.8	66.8
	Simulation	27.5	27.5	80.9	80.9
HPBW (degrees)	E plane	17°	8.9°	58°	59°
	H plane	9.5°	17.4°	53°	52°
Peak SLL (dB)	Measurement	−12.3	−10.0	None	None
	Simulation	−13.0	−13.0		
FBR (dB)		−30	−30	−34.8	−25.5
Isolation (dB)		> 31.1 X to S band	> 25.3 X to S band	> 36.4 S to X band	> 33.8 S to X band

Table 2. A summary of the measured and simulated results for the S/X -band array[18]

3.3 Interlaced cross-patch with patches

An S/X-band cross-patch/patch array is proposed by Salvador *et al* in Ref. [19] (See Fig. 15). Its S-band cross-patch and X-band patches are co-planarly interlaced. It may be seen as the corner-removed perforated patch (L-band perforated patch in Fig. 2) or the co-plane cross-placed dipole (S-band dipole in Fig. 8). The bandwidth of the cross-patch is proportional to the width of its "leg", which is constrained by the inter-element distance. An obvious drawback is that if the gap between two bands is narrow (the leg of cross-patch is too wider), serious inter-band couple may be caused and the radiation pattern will be distorted.

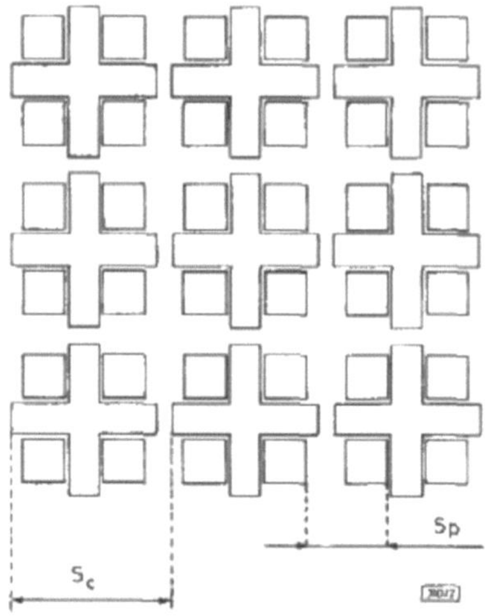

Fig. 15. S/X-band array of interlaced cross-patches with square patches [19]

4. DBDP other configuration arrays

Some other array configurations have also been proposed, most of which can be classified as the deriving forms of the two basic types mentioned in Sects. 2 and 3. One example is the L/C-band array of interlaced L-band perforated cross-patches with C-band patches, whose top view and cross section are shown in Fig. 16[10]. The LF perforated cross-patch on the top layer is interlaced and rounded by 9 HF patches located on another substrate behind it to form a unit cell. The unit cells are cascaded-fed to make up a traveling wave linear sub-array [11], and then a linear sub-array is used to construct a sub-array. Benefiting from its "H"-shape slot aperture coupled in both bands, the simulated cross-polarization at dual bands are claimed to be less than -40 dB, and the backward radiation is also small enough. However, the isolation performance may probably be a problem.

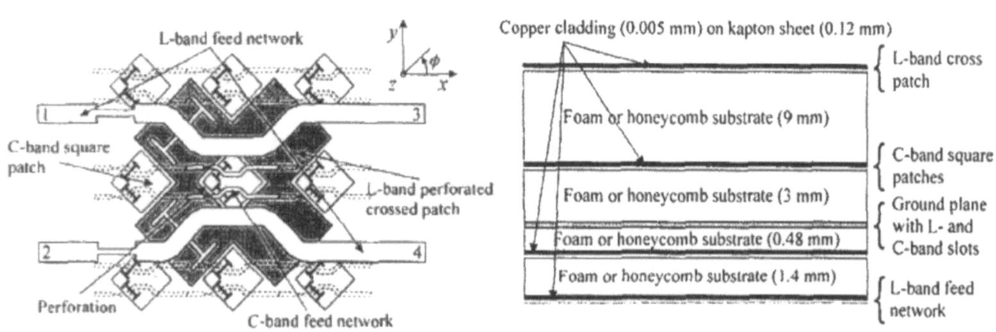

(a) (b)

Fig. 16. L/C-band perforated cross-patch/patches array[10] (a) top view (b) cross-sectional view

As an additional example, a DBDCP(dual-band dual-circular-polarization) element is shown in Fig.17[20] [21]. A square shorted-annular-ring element operates at a low frequency band,

and using notches at two opposing corners of the element's outer ring produces a circular polarization with a single-point feeding. Shorting the inner square ring of the shorted-annular-ring element creates an area that can be used for a printed square slot that operates at a higher frequency band. The slot element can be perturbed by notching two opposite corners to produce a circular polarization. The printed slot can be fed with a stripline that runs under the annular ring structure.

An element was designed with the goal to cover the 2.45 GHz and 5.8 GHz ISM (Industrial, Scientific and Medical) bands with dual-CP operation at each band [21]. The simulated S-parameters show that the isolation between the high and low band ports is better than 25dB. The isolation between the two high band ports has a maximum value greater than 40dB at the center of the band, while that between the low band ports is lower than the former.

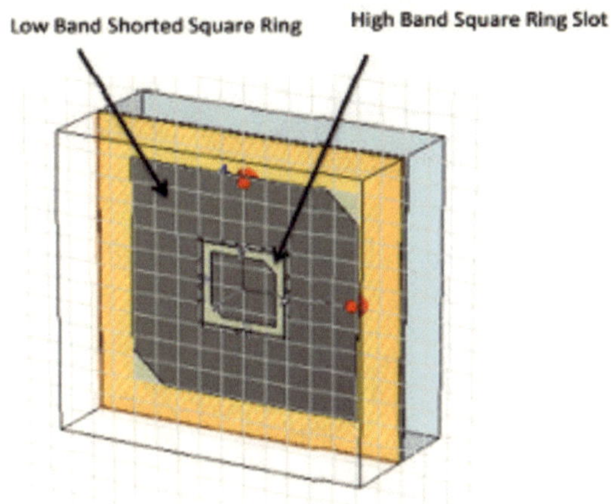

(a)

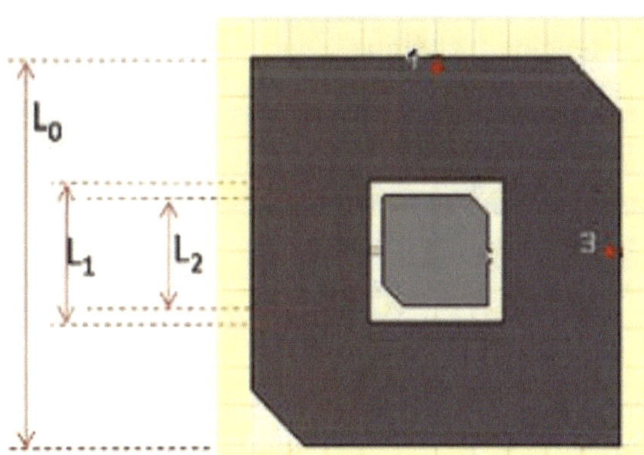

(b)

Fig. 17. Geometry of the dual-band, dual- CP element [21] (a) Isometric view (b) Top view

5. Comparison of DBDP arrays

A comparison of several DBDP designs is listed in Table 2, where some measured results of a new design with the "interlaced dipole and patch" configuration [22] is also included.

Generally evaluating from the listed three performances, it is seen that the "interlaced dipole and patch" configuration may be one of the best choices. In general, the arrays of interlaced slots/dipoles/rings with patches are preferred. Moreover, their flexibility in array configuration makes them more attractive.

array configuration	bandwidth	cross-polarization	port isolation	cross-band isolation
perforated patch [12] (in Fig. 2)	L: 6.4% C: 5.7%	L: about −32 dB (peak) C: ≤−30 dB	– –	– –
perforated patch [13] (in Fig. 5)	L: ≥6% X: ≥10%	L: about −22 dB (peak) X: ≤−20 dB	L: ≤−20 dB X: ≤−18 dB	≤−40 dB in both bands at both polarization
interlaced slot and patch [16] (in Fig. 7)	L: ≥5% C: ≥5%	L: ≤−23 dB C: ≤−24 dB	– –	L: ≤−15 dB C: ≤−40 dB
interlaced dipole and patch [7] (in Fig. 8)	S: ≥8.9% X: ≥17%	S: ≤−26 dB X: ≤−31 dB	S: ≤−20 dB X: ≤−20 dB	– –
interlaced dipole and patch[22]	S: ≥11.4% X: ≥13.9%	– –	S: ≤−35 dB X: ≤−25 dB	– –
interlaced slot and slot [15]	C:5% X: 2%-3% (VSWR≤1.5)	C: ≤−21 dB X: ≤−18 dB	– –	– –

Table 2. Comparison of various DBDP designs

6. Techniques of enhancing DBDP antenna performances

6.1 Pair-wise anti-phase feeding technique

The cross-polarization level of a DBDP antenna is influenced by the figure of its element, the feeding form and the array configuration. In general, more symmetric element shape and thinner substrate (for the patch element) will lead to a lower cross-polarization level. Besides these, the "pair-wise anti-phase feeding technique" is proposed (see Fig. 18) [23, 24]. The neighboring patches are mirror configured and anti-phase fed in H-port or in V-port, and thus all elements in subarray are of same effective excitation and the cross-polarization level is obviously improved at the boresight. As to the cost, in the area out of main beam, the cross-polarization level is raised.

6.2 Symmetrical feeding technology

As introduced in 6.1, the cross-polarization level can be improved by the "pair-wise anti-phase feeding technique" in array, while the port isolation of array is just about completely decided by element port isolation itself. Therefore, the main task in element designing is to achieve good port isolation and after that is the low cross-polarization characteristic. Since a circular patch generally works at TM_{11} mode, its current components orthogonal to the primary components will result in worse cross-polarization characteristic. Thus, the square patch is preferred. For its dual-polarized operation, a symmetrical feeding technology [25]

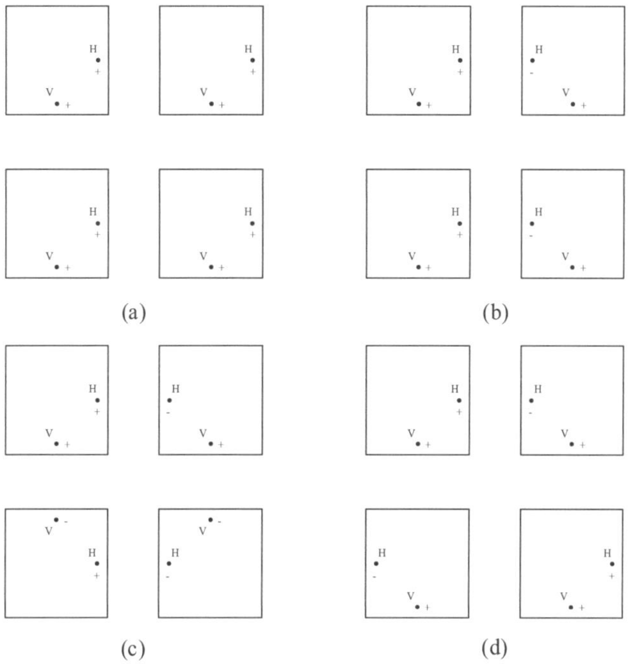

Fig. 18. Subarray configuration [23] (a) distant HH− (b) close HH− (c) HH−& VV− (d) distant & close HH−

is shown in Fig.19, where the stacked parasitic patch has been used to broaden the bandwidth. From the view of symmetry, an aperture-coupled feed located at the center of patch is chosen firstly. In order to increase the coupling between feed line and patch, the "H"-shaped slot is adopted. The simulated results in [25] show that the cross-polarization level of the patch with a pair of edge feeds is -36dB and that with a center feed is -42dB, about 6dB better than the former. Another merit of the center feed is that the slot at center can give a good coupling level and the front-back ratio of patch is improved. As for the another polarization, the balanced edge feeding is adopted to keep the symmetry. Two balanced pin-feeds are connected by the feed network which is carefully tuned to realize accurate "equal and anti-phase feed". Then a microstrip line is used to realize impedance matching.

Fig.20 is the simulated S parameters for two ports. It is seen that the balanced pin-feed port achieves an impedance bandwidth (Return loss≤ -10dB) of 840MHz (5.01GHz-5.85GHz) and the aperture couple port of 850MHz (5.09GHz -5.94GHz), while the port isolation keeps under -43dB .The simulated cross-polarization within the main lobe remains less than -37dB in whole bandwidth. The calculated gain of antenna is stable at the 9dB while the front-back ratio keeps better than 22dB in the whole bandwidth.

6.3 Slot-loaded patch for improving port isolation

It is proposed to etch a slot in the corner of a driven patch by our group [26] (see Fig. 21). The effect of using the slot-loaded method can be seen from Fig. 22, where the isolation level between two ports is improved for at least 5 dB. However, the field under the patch is

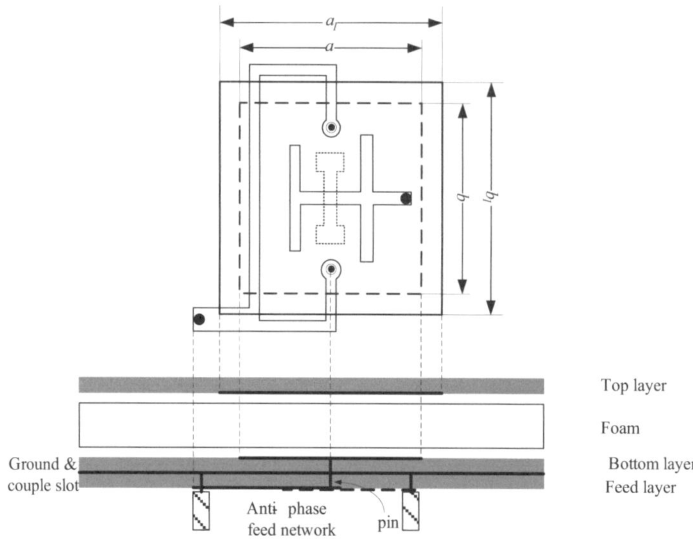

Fig. 19. Configuration of antenna element

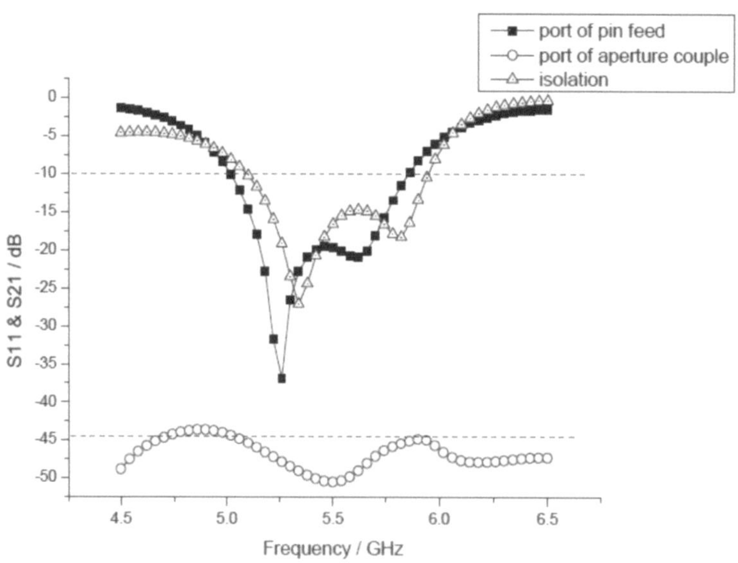

Fig. 20. S parameters of antenna element

disturbed by the slot, then the cross-polarized field is brought out and the cross-polarization level deteriorates. In Ref. [27], a similar method is adopted. The only difference is that "T"-shaped slot and some edge-slot are etched on the driven patch, which also achieves a good isolation.

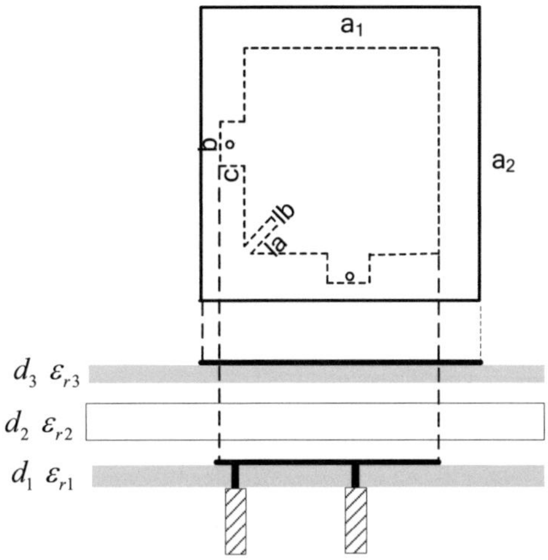

Fig. 21. Slot-loaded patch

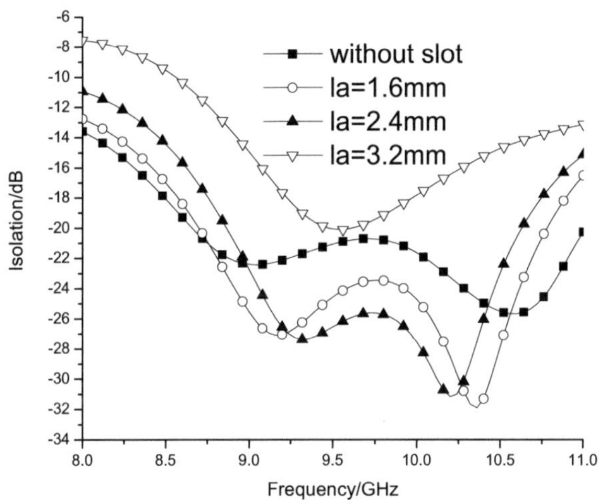

Fig. 22. Isolation S_{12} for various slot lengths

6.4 Bandwidth enhancement technique

The SAR systems use the impulse compression technology to realize the high-resolution at elevation direction and thus a wider bandwidth is required for the antenna to radiate narrower impulse (in time domain). An antenna system can not broaden bandwidth by means of array synthesis, and the antenna bandwidth generally lies on its element bandwidth. Therefore, a lot of bandwidth enhancement methods for antenna elements have been proposed, such as co-planar/stacked parasitic patches. In DBDP antenna array design, the distance between elements is limited by the scanning requirement and the using of stacked parasitic patches is probably the most effective broadening method for its room saving structure, which can achieve at least a bandwidth of 15% [26], about three times of that of a conventional patch.

7. Conclusion

The dual-band dual-polarization (DBDP) shared-aperture microstrip array technology for the synthetic aperture radar (SAR) application in the last decade has been reviewed. Several designs of DBDP SAR antenna arrays are introduced with their main performances, then their comparison is summarized. According to the configuration, the basic DBDP arrays include two types: the perforated patch array and the interlaced elements array. In the general, the arrays of interlaced slots/dipoles/rings with patches are preferred. Moreover, their flexibility in array configuration makes them more attractive. In addition, tri(or more)-band dual-polarization shared-aperture microstrip arrays for the SAR applications may be formed by means of different DBDP arrays[28]. Finally, some techniques enhancing DBDP antenna performances also have been presented.

8. Acknowledgements

This work was supported by the National Natural Science Foundation of China (Grant No. 60871030), and the National High-Technology Research and Development (863) Project of China (Grant No. 2007AA12Z125).

9. References

[1] Zhang Z Z. Guide of Airborne and Spaceborne SAR system. Publish House of Electronics Industry, Beijing, 2004 (in Chinese)

[2] Zhong S S, Cui J H. A new dual-polarized aperture-coupled printed array for SAR application. Journal of Shanghai University (English Edition), 2001, 5(4): 295-298

[3] Du X H, Li J X, Zheng X Y. Design of X-band dual-polarized active phased array. Modern Radar, 2002, 18(5): 67-70 (in Chinese)

[4] Zhong S S, Yang X X, Gao S C, Cui J H. Corner-fed microstrip antenna element and arrays for dual-polarization operation. IEEE Transactions on Antennas and Propagation, 2002, 50(10): 1473-1480

[5] Wang W, Zhong S S, Liang X L. A dual-polarized stacked microstrip antenna subarray for X-band SAR application. IEEE Antennas and Propagation Society International Symposium, Monterey, CA, 2004: 1603-1606

[6] Wang W, Li L, Zhang Z H. Dual-polarized space-borne SAR antenna array. Remote Sensing Technology and Application, 2007, 22(2): 166-172 (in Chinese)

[7] Qu X, Zhong S , Zhang Y, and Wang W. Design of an S/X dual-band dual-polarised microstrip antenna array for SAR applications. IET Microwave,Antennas,and Propagation, 2007, 1(2): 513-517

[8] Zhong S S, Qu X, Zhang Y M, and Liang X L. Shared-aperture S/X dual-band dual-polarized microstrip antenna array. Chinese Journal of Radio Science, 2008, 23(2): 305-309

[9] Zhai G H, Hu M C, Li J X. A novel dual-polarization microstrip patch antenna for space-borne SAR application. Modern Radar, 2007, 29(4): 72-75 (in Chinese)

[10] Vallecchi A, Gentili G B, Calamia M. Dual-band dual polarization microstrip antenna, IEEE Antennas and Propagation Society International Symposium, Columbus, OH, 2003: 134-137

[11] Granet C, Zhang H Z, Greene K J, James G L, Forsyth A R, Bird T S, Manchester R N, Sinclair M W, Sykes P. A dual-band feed system for the Parkes radio telescope, IEEE Antennas and Propagation Society International Symposium , Boston., MA, 2001: 296-299

[12] Shafai L L, Chamma W A, Barakat M, Strickland P C, Seguin, G. Dual-band dual-polarized perforated microstrip antennas for SAR applications. IEEE Transactions on Antennas and Propagation, 2000, 48(1): 58-66

[13] Pozar D M, Targonski S D. A shared-aperture dual-band dual-polarized microstrip array. IEEE Transactions on Antennas and Propagation, 2001, 49(2): 150-157

[14] Wincza K, Gruszczynski S, Grzegorz J ,Integrated dual-band dual-polarized antenna element for SAR applications, IEEE 10th Annual Wireless and Microwave Technology Conference (WAMICON'09), Clearwater, FL, 2009: 1 - 5

[15] Pokuls R, Uher J, Pozar D M. Dual-frequency and dual-polarization microstrip antennas for SAR applications. IEEE Transactions on Antennas and Propagation, 1998, 46(9): 1289-1296

[16] Pozar D M, Schaubert D H, Targonski S D, Zawadski M. A dual-band dual-polarized array for spaceborne SAR. IEEE Antennas and Propagation Society International Symposium, Atlanta. GA, 1998: 2112-2115

[17] Gao G.M., Y.M. Zhang Y.M., Li Ang, Zhao J.M., Cheng H. Shared-aperture Ku/Ka bands microstrip array feeds for parabolic cylindrical reflector, 2010 International Conference on Microwave and Millimeter Wave Technology (ICMMT2010), Chengdu, China, 2010 : 1028-1030

[18] Hsu S.H., Ren Y.J., and Chang K. A dual-polarized planar-array antenna for S-band and X-band airborne applications, IEEE Antennas and Propagation Magazine, 2009，51(4):70-77

[19] Salvador C, Borselli L, Falciani A, Maci S. Dual frequency planar antenna at S and X bands. Electronics Letters, 1995, 31(20): 1706-1707

[20] Zaghloul A. I. and Dorsey W. M. Evolutionary Development of a Dual-Band, Dual-Polarization, Low-Profile Printed Circuit Antenna, International Conference on Electromagnetic in Advanced Application, Torino, Italy, 2009: 994-997

[21] Dorsey W.M. and Zaghloul A.I. Dual-polarized dual-band antenna element for ISM bands, IEEE International Symposium on Antennas and Propagation, Charleston, South Carolina, 2009:1-4

[22] Zhong S-S, Sun Z. S/X dual-band dual-polarization microstrip dipole/stacked patch array antenna, Invention Patent of China (Applied), No.201010275934.8, Date:08-09-2010 (in Chinese)

[23] Woelders K, Granholm J. Cross-polarization and sidelobe suppression in dual linear polarization antenna arrays, IEEE Transactions on Antennas and Propagation, 1997, 45(12): 1727-1740

[24] Liang X-L, Zhong S-S , Wang W. Cross-polarization suppression of dual-polarization microstrip antenna arrays, Microwave and optical technology Letters, 2004, 42(6): 448-451

[25] Sun Z, Zhong S-S, Tang X-R, Liu J-J. C-Band Dual-Polarized Stacked-Patch Antenna with Low Cross-Polarization and High Isolation, 3rd European Conference on Antennas and Propagation(EuCAP2009), Berlin, Germany, 2009:2994-2997

[26] Chen K D, Zhong S-S, Yan X-R. Design of S/X dual-band dual-polarized shared-aperture microstrip antenna array. Journal of Microwaves, 2008, 24(6):65-67 (in Chinese)

[27] Zaman A U. Dual polarized microstrip patch antenna with high port isolation. Electronics Letters, 2007, 43(10): 551-552

[28] Zhong S-S, Sun Z and Kong L-B. L/S/X tri-band dual-polarization planar array antenna, Invention Patent of China (Applied), No.201010275940.3, Date:08-09-2010 (in Chinese)

18

Microwave Properties of Dielectric Materials

JS Mandeep[1] and Loke Ngai Kin[2]
[1]*Department of Electrical, Electronic & Systems Engineering*
[1]*Institute of Space Science,*
Faculty of Engineering & Built Environment
Universiti Kebangsaan Malaysia
43600 UKM Bangi
Selangor Darul Ehsan,
[2]*School of Electrical and Electronic Engineering,*
Universiti Sains Malaysia, Nibong Tebal,
Engineering Campus 14300, Pulau Pinang,
Malaysia

1. Introduction

In recent years, the study of newer types of dielectric materials and compositions has been of great interest. The quest for new, innovative and easily obtainable dielectric materials that yield predictable and controllable permittivity and permeability values with very low dielectric loss has always been fruitful. New ideas and designs to implement these materials in microwave devices and structures with the most efficiency and performance are also of equal importance.

The book chapter covers the synthesis and characterization of new dielectric material compositions and the design, implementation and testing of a prototype dielectric resonator antenna and filter utilizing the fabricated dielectric material. It also covers the use of different design and testing techniques in this research. By studying different methodologies and new material types that were previously published in cited technical journals such as Science Direct, different types of materials and synthesis methods were able to be identified. The appropriate materials and method of synthesis were then derived and utilized in the fabrication of the new type of dielectric ceramic substrate material.

After extensive literature reviews, research and analysis, various methods used in dielectric resonator designs were studied and analyzed. An overview of the synthesis process associated with the fabrication of the new dielectric composition will be discussed. Next, with the primary objective of investigating the microwave properties of a new composition of high permittivity ceramic dielectric substrate at microwave frequencies, research is then carried out on designing an electrical model in order to utilize the newly fabricated dielectric material in a microwave environment. The electrical model that is proposed will be of dielectric resonator devices which will incorporate the new dielectric material type. Tests and measurements will also be carried out at various microwave frequencies in order to study the behaviour of the dielectric material. Different shapes and thicknesses of the dielectric material will also be studied to observe the effects they have towards the characteristics of the electrical model at microwave frequencies. Lastly, analysis will be carried out for these variations in order to determine the characteristics and the traits of the dielectric designs at microwave frequencies.

2. Microwave properties of dielectric materials

2.1 Barium strontium titanate

Two batches of pre-synthesized Barium Strontium Titanate (BST) pellets were provided by the School of Material Engineering for dielectric characterization and testing at microwave frequencies. They will also provide a starting template as a dielectric model in CST in order to design the dielectric resonator antenna and filter that will be used with the Bismuth Lanthanum Titanate pellets. These BST samples will provide an overview on the dielectric characteristics and properties of similar ceramic substrates. Thus, it will be easier to transfer the BST dielectric initial designs to accommodate uncharacterized BLT materials that were not ready during the electrical model design process. The most important reason for the analysis of these pre-made samples is to study the effects of different variations in the synthesis process towards the performance and dielectric properties of the material.
For Barium Strontium Titanate, there are two batches of samples which are Type A and Type B. Type A being the batch of samples synthesized via mixed oxide heat treatment method and Type B is the batch that was prepared using wet chemical method. While it is observed that the value of permittivity varies in relation with the measurement frequency (especially in the high frequency range), all measurements were carried out at 1.4GHz as it best represents the dielectric properties in the range of operating frequency where it is used in the electrical model.

2.2 Type A (mixed oxide heat treatment) 1.4GHz

By using the HP 4291B Material Analyzer connected to the permittivity test head, the dielectric properties of the samples are measured at the sweep frequency of 1.4GHz. It is also noted that the maximum measurement frequency for the analyzer is limited to 1.8GHz, thus the dielectric properties of the test material at frequencies higher than 1.8GHz could not be measured. The samples were first cleaned and polished at the contact planes for maximum contact with the test probes with minimal air gaps. The measured dielectric properties are presented in the following table and figures.

Type A	Permittivity, ε_r	Loss Tangent	Thickness (mm)
1	110.74	5.58E-02	2.85
2	299.62	2.25E-01	2.74
3	-120.34	5.98E-01	2.30
4	64.364	1.3568	2.48

Table 1. Dielectric Properties of Type A Samples.

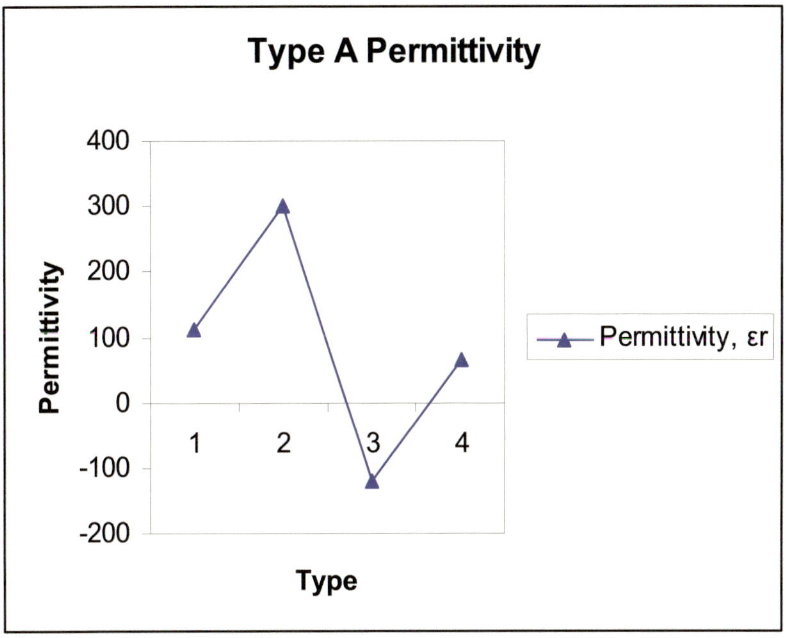

Fig. 1. Type A Permittivity Trend.

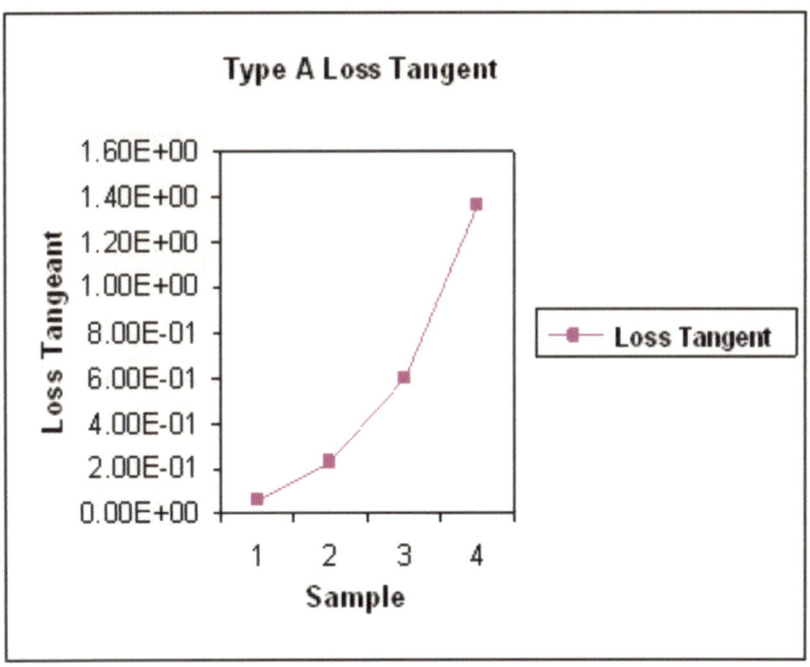

Fig. 2. Type A Loss Tangent Trend.

Observing the permittivity characteristics obtained from the Material Analyzer, the Type A samples has an ideal maximum permittivity condition at Sample 2. The Samples 1, 2, 3 and 4 are each synthesized at 1100°C, 1200°C, 1300°C and 1400°C respectively. Interestingly, Sample 3 is observed to be measured with a real negative vector value of permittivity. A dielectric material measured having these negative ε or μ values indicates that electric fields (ε) or magnetic fields (μ) entering the material has a negative refraction angle, meaning the field lines do not penetrate the material, but reflecting off it and is opaque to electromagnetic radiation. Rare types of materials with both negative values of permittivity and permeability are known as metamaterials where the dielectric properties are now affected by the structural properties of the material itself and not the composition of molecules. [19-21] As the permeability of the sample in test cannot be measured at the moment using the existing analyzer without the permeability test head, it is uncertain if Sample 3 is of metamaterial characteristics. However it is known that Sample 3 exhibits a negative real part of permittivity thus meaning that the refraction angles of penetrating electromagnetic field lines are negative as well rendering the material opaque to electric fields. The study of this material however, will not be covered in this project as further studies are still needed in this relatively new topic of research.

2.3 Type B (wet chemical method) 1.4GHz

Dielectric measurements were taken for the Type B batch of samples that were synthesized using wet chemical method on the same measurement setup. Similarly, the samples were cleaned and polished at the contact planes for maximum contact with the test probes with minimal air gaps. Variations of dopant ratios were introduced into the individual samples in order to observe the effect on the dielectric properties at microwave frequencies. It is noted that the thickness of each sample is kept constant at 1.15mm for this batch. However, it is understood that the thickness of the samples will not have any effect on the measurement results. The measured dielectric properties are presented in the following table and figures.

Type B	Strontium Dopant % Ratio	Permittivity, ε_r	Loss Tangent	Thickness (mm)
1	1	103.34	1.52E-01	1.15
2	0.9	43.95	7.91E-02	1.15
3	0.7	118.23	8.35E-02	1.15
4	0	60.658	2.46E-03	1.15

Table 2. Dielectric Properties of Type B Samples.

Dielectric properties for Type B samples which are synthesized via wet chemical method shows normal distribution of permittivity and loss relative to the composition ratio of Strontium dopants in the samples. It is noted that the maximum permittivity is obtained from the percentage ratio of 0.7 and the maximum loss, meanwhile is obtained from the ratio of 1. This indicates that the introduction of Strontium dopants in the dielectric material increases the loss tangent of the electric fields passing through the sample. The excess Strontium atoms doped in the crystalline lattice might have introduced dampening to the electromagnetic waves that were propagating through the material. By using wet chemical method in synthesizing the samples, no negative permittivity or permeability properties

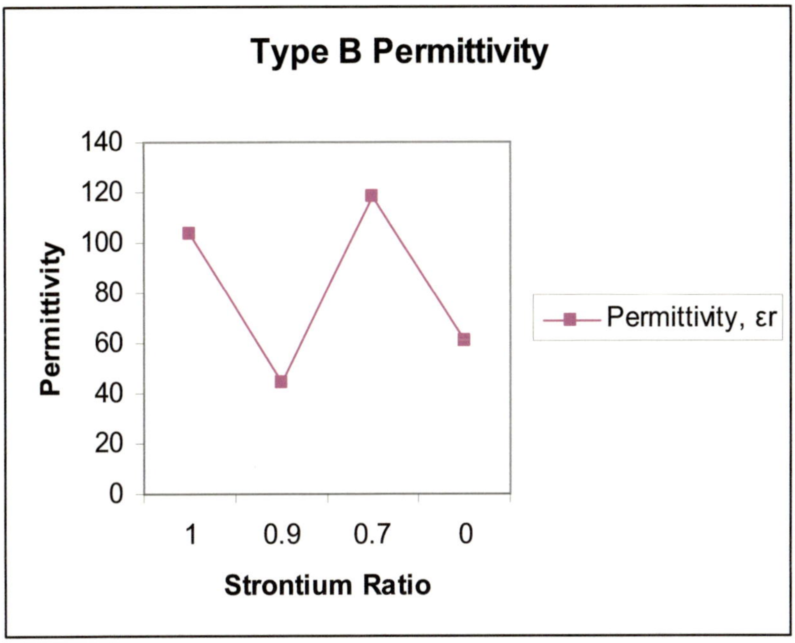

Fig. 3. Type B Permittivity Trend.

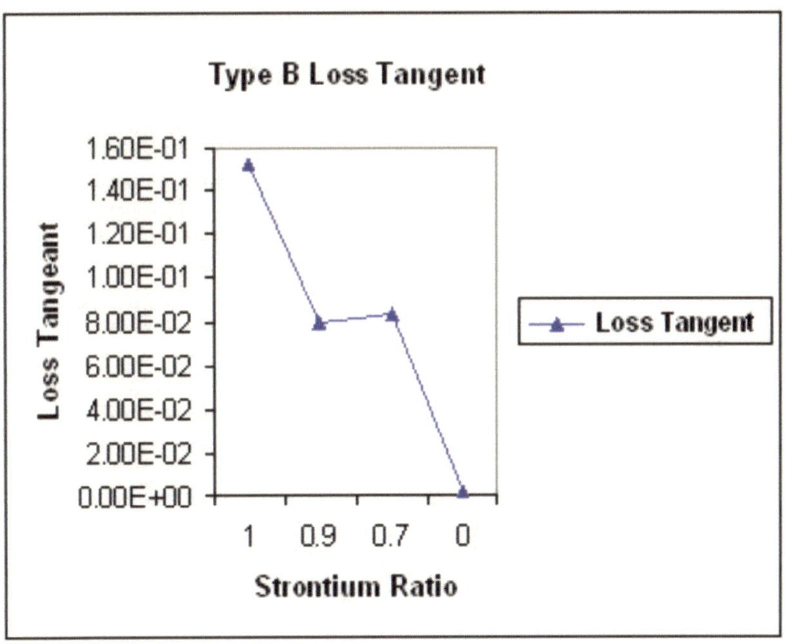

Fig. 4. Type B Loss Tangent Trend.

were observed among the samples. This also suggests that by using the wet chemical method, no significant structural atomic properties were formed during synthesis as opposed by the heat oxide treatment method.
Unlike the mixed oxide synthesis method which requires considerable amount of heat to form crystalline molecular bonds inside the material, no significant atomic alteration occurred for these samples thus retaining more typical properties. From the observation of the samples in comparison with the previous Type A samples, it is obvious that wet chemical method is more suitable as a synthesis process for high permittivity, doped, ceramic dielectric materials. It is also observed that keeping the dopant ratio below 0.7 yields the best dielectric characteristics for the material. From the results and analysis of the dielectric properties displayed by BST synthesized via different techniques, the best method and process was able to be roughly determined for the synthesis of Bismuth Lanthanum Titanate.

2.4 CST simulation and measurement results and comparison

2.4.1 DRA results

The design that was constructed in CST was exported into DXF format to be photo-etched onto a Rogers type double-sided copper-clad substrate. The substrate measured with permittivity of 3.5398, loss of 4.5364E-03 and thickness of 0.80mm. A 50Ω SMA connector is used at the feed of the strip line. Meanwhile for consistency, the sintered and polished pellets are then measured again using a HP Material Analyzer for their permittivity and dielectric loss values at 1.4GHz. The sample with 0.5 doping ratio was selected for the ideal test sample as it has the best solid form and also good dielectric properties. The sample is very dense and is not brittle or porous compared to the other samples which are less desirable. Its measured permittivity and loss are ε_r = 94.243 and tanδ = 6.654E-03 respectively. The dielectric samples will be shaped into the various shapes previously discussed for measurements and comparison in the following sections. Plate 4.3 shows the fabricated DRA on a Rogers double sided substrate board.

Fig. 5. Dielectric Resonator Structure.

2.4.2 Cylindrical pellet

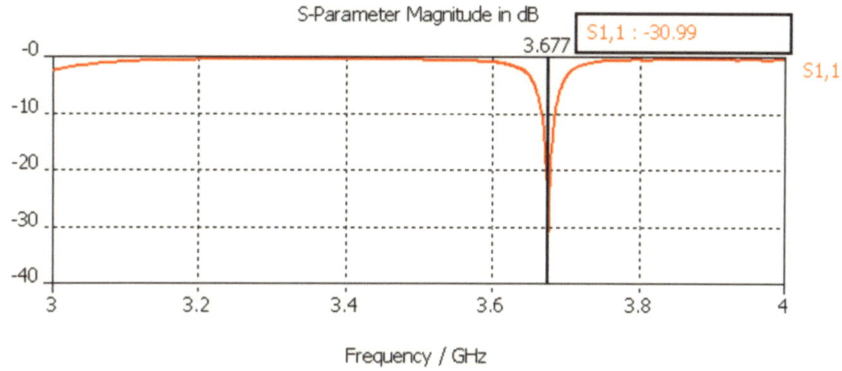

Fig. 4.5. S(1,1) Frequency Response of DRA Structure.

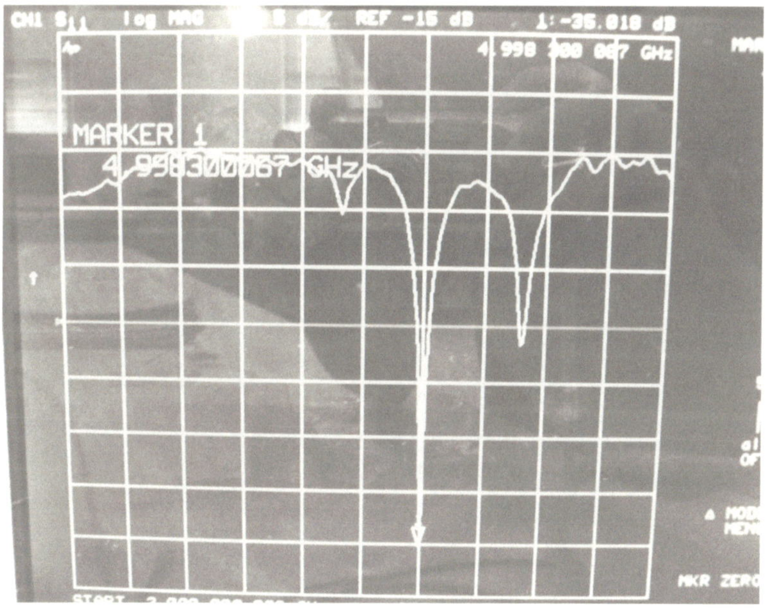

Fig. 6. Measured S(1,1) Frequency Response of DRA Structure.

Figure 6 shows the simulated and measured S(1,1) frequency response of the resonator structure at a sweep frequency range of 3GHz to 4GHz. The resonant frequency is iteratively tuned to 3.677 GHz. By varying the strip length and width, the resonant frequency can be shifted accordingly. Comparison with the simulated response showed a 35% frequency shift at 4.998GHz on the actual measured response at -35.010dB. This is expected as there is some slight known tolerance in the fabrication of the PCB and also of the non constant microwave characteristics of the dielectric material at different microwave frequencies. Input impedance at the input port is at 52.4 - 1.605j Ω at resonance frequency of 3.677 GHz which is close to the desired 50Ω. Simulated and measured impedance response for the frequency range is shown below in Figure 7 and Figure 8 respectively.

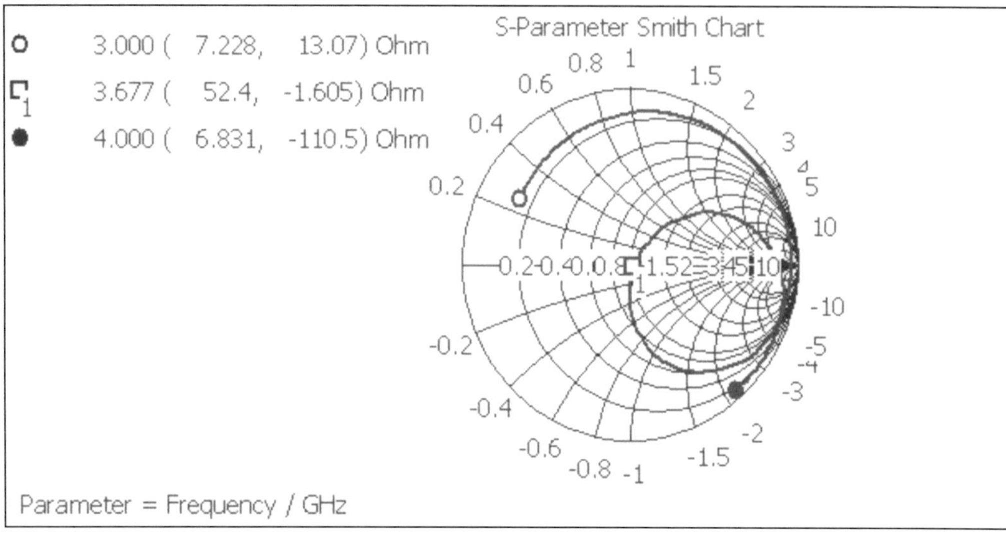

Fig. 7. Input Impedance Response of DRA Structure.

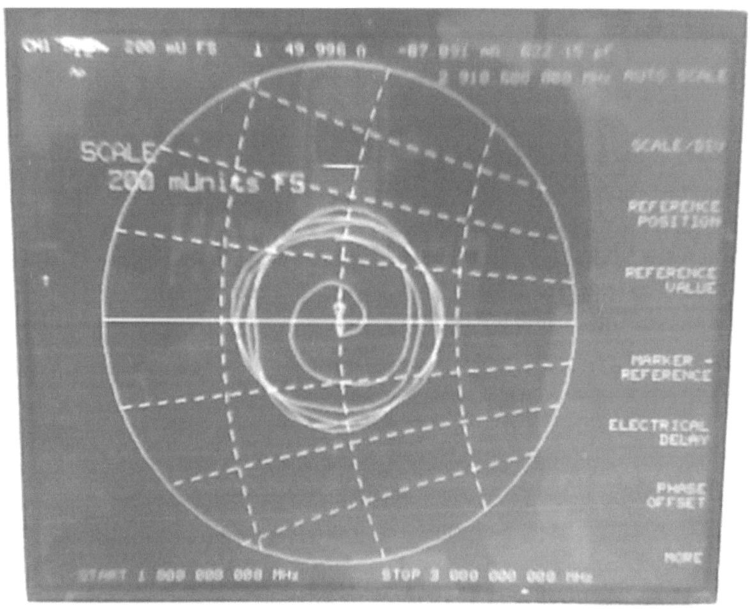

Fig. 8. Measured Input Impedance of DRA Structure.

Figure 9 shows the simulated radiation pattern and characteristics for the resonator operating at its simulated resonant frequency at 3.677GHz. The radiation pattern is a ϕ-Plot in which the radiation is observed at a parallel plane with the DRA. As observed, the attenuation characteristic of this DRA towards the radiation reduces radiation efficiency and transmission power. Thus, it is noted that this DRA design has a tendency to suppress or attenuate electromagnetic radiation. It is in essence, not an efficient electromagnetic radiator structure. However, it is greatly preferred in pure resonator or RF load applications where electromagnetic radiation is undesirable. The actual radiation pattern is not measured due to its poor radiation properties. However, from this simulation we know that this particular type of dielectric resonator design can be modified into an efficient microwave filter structure that does not require electromagnetic radiation. Further explanation on this is as follows.

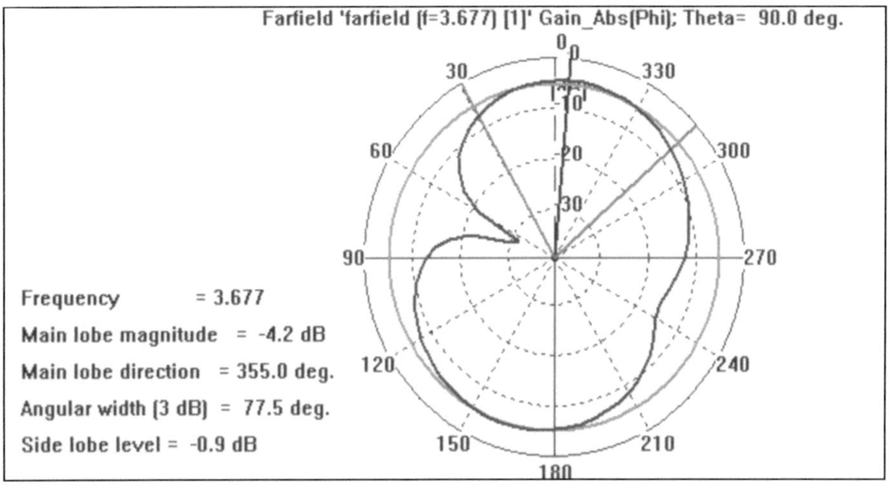

Fig. 9. Radiation Pattern and Properties of DRA Structure (ϕ-Plot).

Most dielectric resonator antennas or array antennas radiate from a single sided plane, in which fields fringing from the conductor edges towards the ground plane transmit radiation energy. The gain or directivity in dBi can be altered so that the radiation energy is focused into a specific direction or pattern, as shown in radiation pattern plots. The value of 0dBi refers to the absolute gain value of 1, which is the gain produced by an isotropic antenna in which radiation power is uniformly distributed at all angles. However, values of dBi which are negative in this case mean that the DRA is attenuating the radiation energy making it less efficient as a radiating device. In this case, it is an ideal configuration for a non-radiating resonator device. Dielectric resonator designs like this can sometimes be derived into filters, couplers and mixers that do not require energy radiation into free space. Thus, the need for extensive RF shielding can be minimized for this type of device. This means that costly metallic shielding enclosures are no longer needed to ground the electromagnetic fields that radiates from the resonator structure.

From further analysis, the high permittivity dielectric material placed in the center of the microstrip ring alters the natural input impedance of the microstrip patch and matches it to 50Ω at its resonant frequency. This enables the mismatched microstrip patch to resonate at its natural resonant frequency. The design of a dielectric resonator filter was also able to be derived from this basic dielectric resonator antenna structure as the excitation mode that is generated by the microstrip ring coupling creates confined $TM_{01\delta}$ radiation. This design will be tested and measured in the later DRF section. The use of various dielectric shapes and variations is observed in the following sections below.

The measurement tests carried out on the dielectric resonator filter device is similar to the DRA with the exception of the filter being a two port device. Thus, the S-parameters will be the measured for S(1,2) and S(2,1) in order to obtain the frequency response of the filter at the operating frequency range of 1-3GHz. The HP Network Analyzer shown in Figure 10 is used to measure the S-parameters and also the impedance of the filter device at the determined frequency range. Additionally, a power transmission measurement will be carried out on the filter device to further observe the actual application and response of the filter in a proper transmission circuit. The circuit diagram and also the actual test configuration for this measurement setup are illustrated and shown in Figure 11 and Figure 12 on the following page.

The design that was constructed in CST was exported into DXF format to be photo-etched onto a Rogers type double-sided copper-clad substrate. The substrate measured with permittivity of 3.5398, loss of 4.5364E-03 and thickness of 0.763mm. A 50Ω SMA connector is used at the feed of the microstrip transmission line at both ports. Meanwhile, the sintered and polished pellets are then measured again using a HP Material Analyzer for their permittivity and dielectric loss values at 1.4GHz for consistency. The sample pellet piece

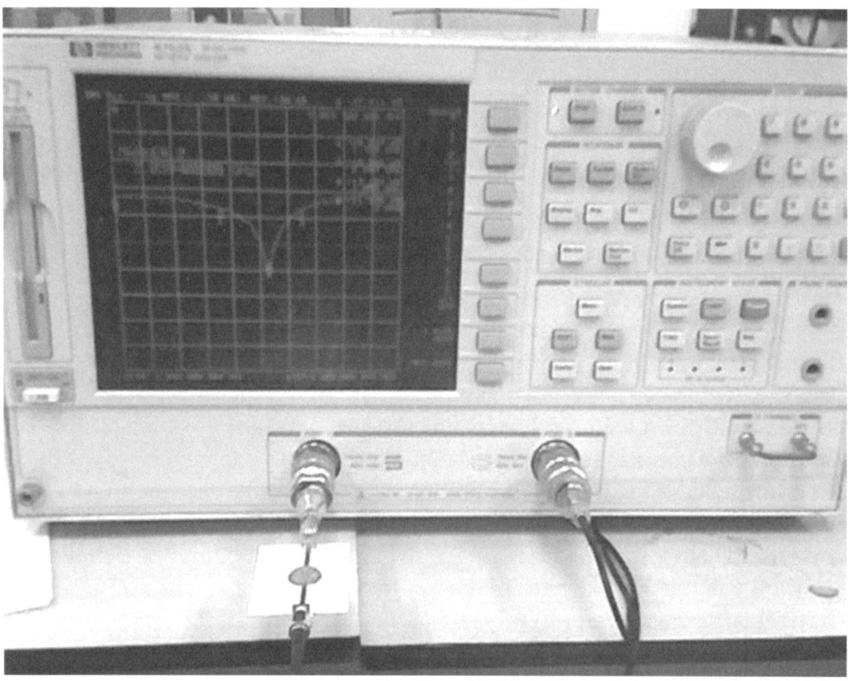

Fig. 10. DRF S(1,2) Measurement Setup.

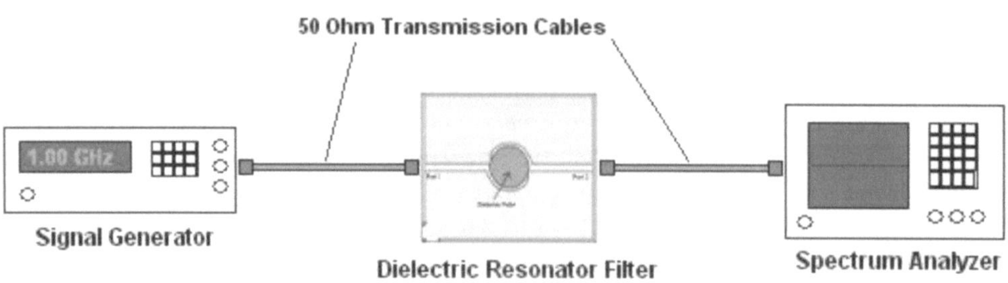

Fig. 11. DRF Power Transmission Measurement Setup Diagram.

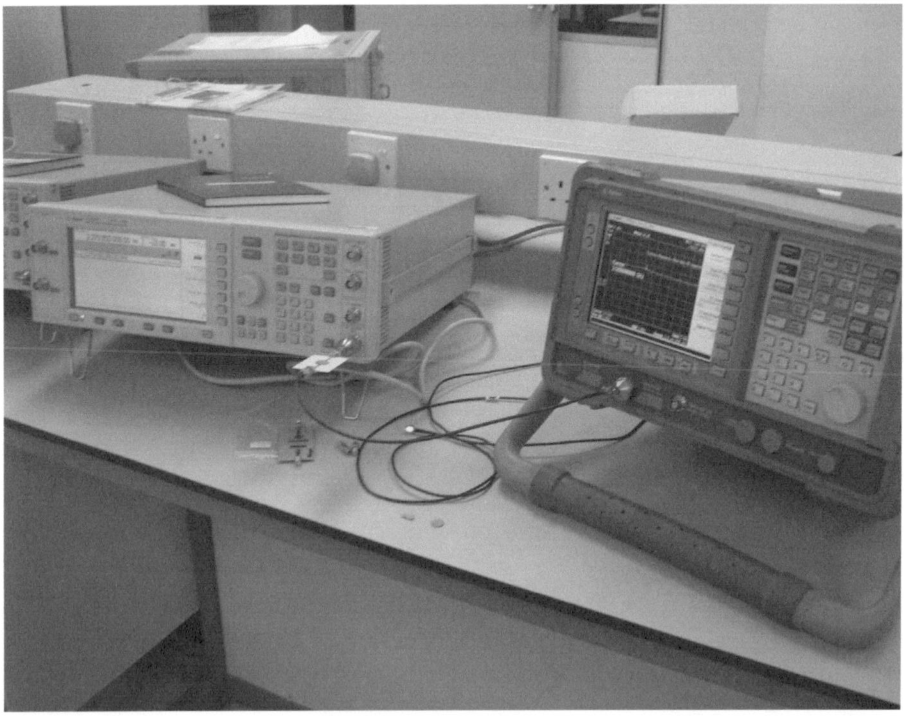

Fig. 12. DRF Power Transmission Measurement Setup.

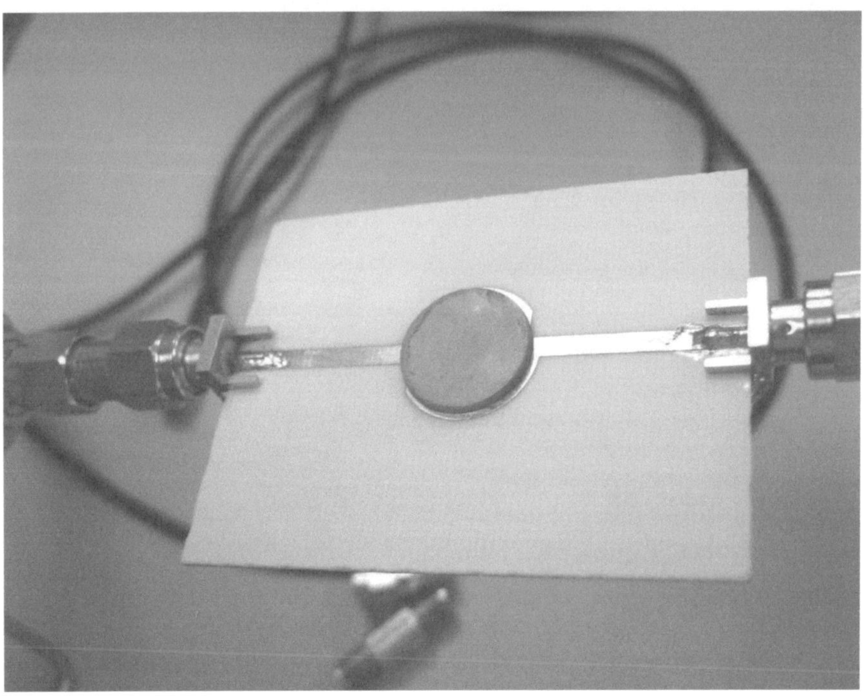

Fig. 13. Dielectric Resonator Filter Structure.

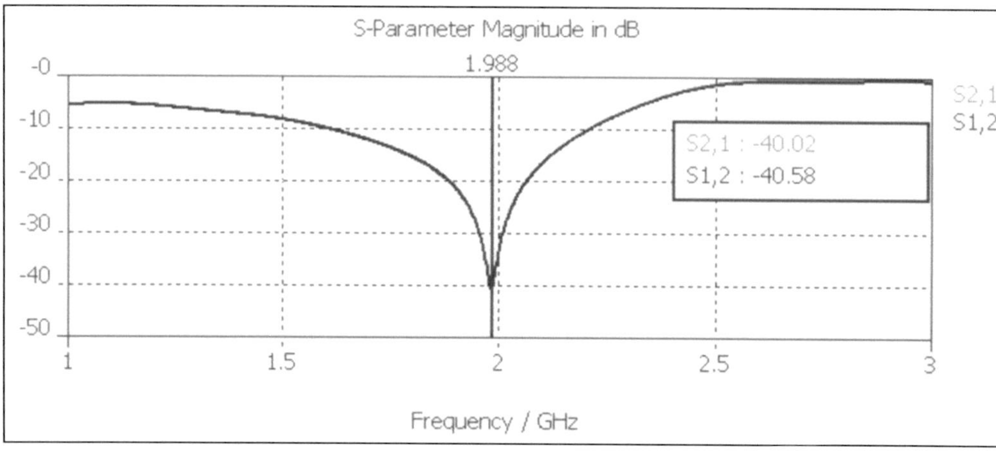

Fig. 14. S(1,2) and S(2,1) Frequency Response of DRF Structure.

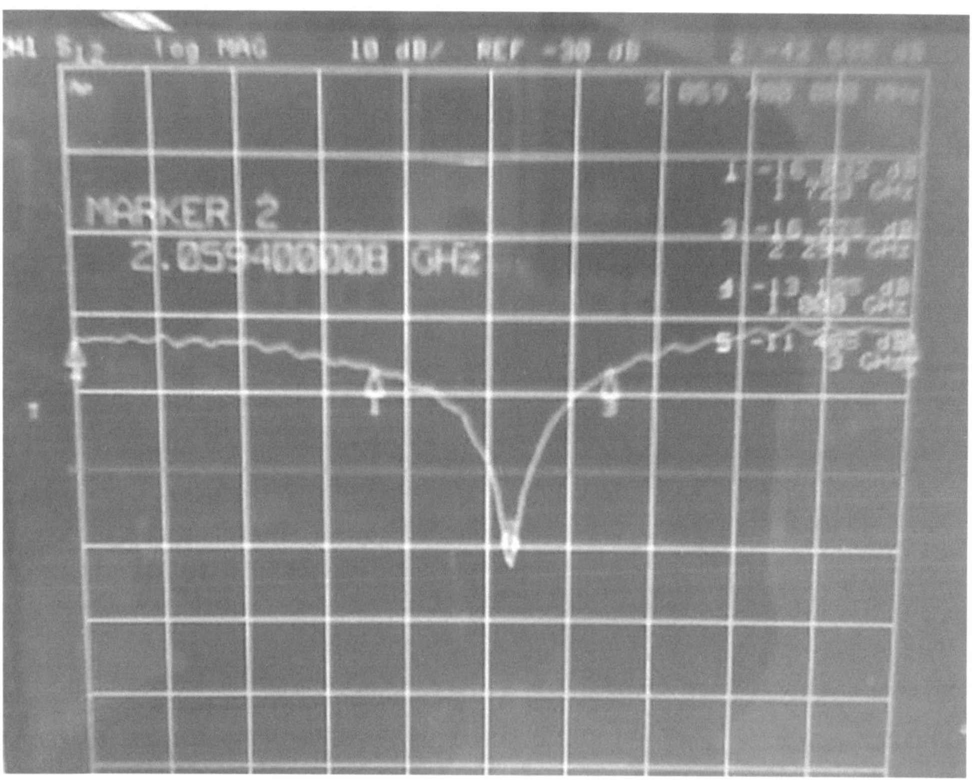

Fig. 15. S(1,2) Measured Frequency Response of DRF Structure.

with 0.5 doping ratio that was sintered at 1000°C was observed and measured to have a good solid form and also good and consistent dielectric properties. The sample is very dense and is not brittle or porous. Its measured permittivity and loss are εr = 49.416 and $\tan\delta$ = 20.296E-03 respectively. The thickness of the sample is measured at 1.15mm. It is selected for use with the filter structure.

Figure 14 above shows the simulated S(1,2) and S(2,1) frequency response of the resonator structure at a sweep frequency range of 1GHz to 3GHz. The DRF structure exhibits a bandstop type filter response. The bandstop center frequency was fine tuned to 1.988 GHz by varying the strip lengths and width during the iterative tuning process in CST. The filter is connected to the HP Network Analyzer to measure its S(1,2) and S(2,1) response at a test frequency range 1.5GHz to 2.5GHz. The test setup used was shown in Figure 12. As the dielectric resonator filter in test is symmetrical, thus only the S(1,2) response data is collected since it will be the same for S(2,1). The actual S-parameter measurement obtained from the network analyzer is shown in Figure 14. The actual measurements are verified to be the same as the simulated response and by comparison, the actual measurements show a better symmetry at both the low and high bandpass frequency spectrum of the filter. The notch filter center frequency is measured at 2.059GHz at -42.579dB. The key measurement points and -3dB levels at 1.723GHz and 2.294GHz for the S(1,2) response curve of the filter are listed in Table 3.

Frequency (GHz)	**S(1,2) (dB)**
1.000	-13.208
1.723 (-3dB)	-16.839
2.059	-42.579
2.294 (-3dB)	-16.786
3.000	-11.495

Table 3. S(1,2) Measurement Values

Based on the measurement values obtained from the network analyzer, the bandwidth and also the quality factor (Q-factor) for the DRF can be calculated. The bandwidth for all microwave filters is obtained from the frequency range of the -3dB gain cutoff levels derived from their respective S(1,2) or S(2,1) plots shown in Table 4.4. The bandwidth of a filter represents the range of frequencies in which the device passes or stops, depending on the type of filter being used. The Q-factor however defines the quality of frequency separation for microwave filter in question. For example, a microwave bandpass filter with a high Q-factor enables the cutoff frequencies to be more precise and distinct, which in graphical terms mean a steeper cutoff frequency slope. As an example, a Butterworth filter has the lowest Q-factor of all other known filter types because of its more gradient cutoff. For this notch filter DRF, the bandwidth is calculated at 571MHz while the Q-factor is at a very acceptable value of 3.6. The calculation for the bandwidth and Q-factor of the DRF are shown in the following expressions.

$$Bandwidth\ (MHz) = 2.294\ GHz - 1.723\ GHz = \underline{\underline{571\ MHz}} \tag{1-a}$$

$$Q\text{-}Factor = \frac{Center\ Frequency}{Bandwidth} = \frac{2.059\ GHz}{0.571\ GHz} \tag{1-b}$$

$$= \underline{\underline{3.6}}$$

In order to measure the power transmission for the filter, the filter is connected to a signal generator and spectrum analyzer as shown previously. Figure 16 shows the readout on the spectrum analyzer obtained from the power transmission measurements performed on the DRF. A high frequency signal generator is connected to the filter at Port 1 via a 50Ω rated transmission line and connector and the output from Port 2 is connected to the spectrum analyzer using the same method. The signal generator is swept from 1GHz to 3GHz at a power output of -10dBm. The null reference of the spectrum analyzer is at approximately -57dBm before input signal is fed into the filter. The key measurement values from the analyzer are as listed in Table 4.

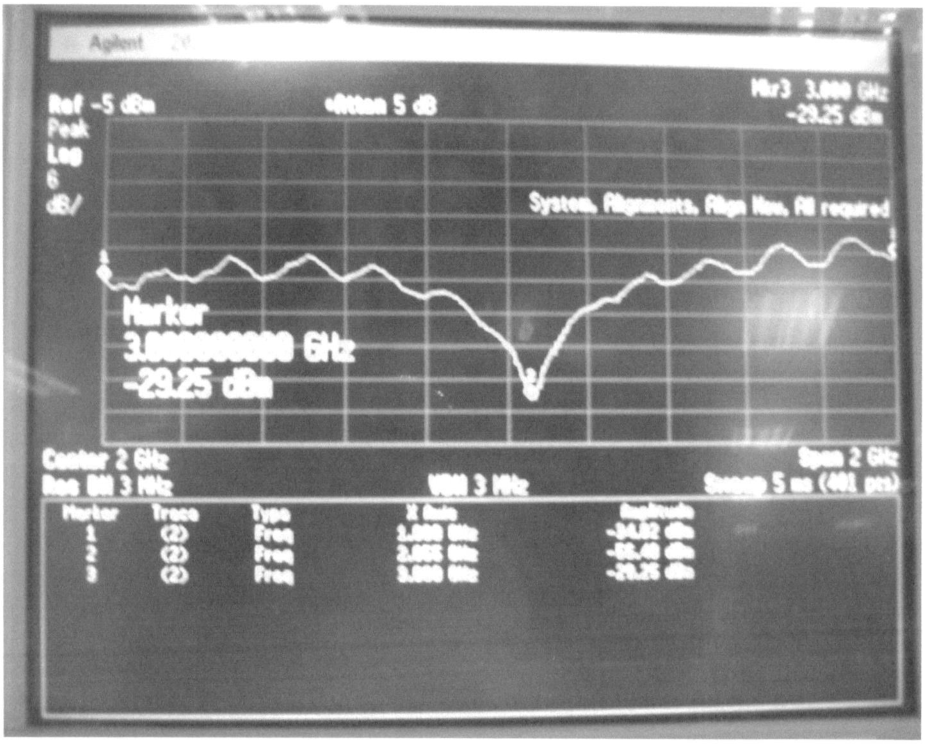

Fig. 16. Measured Power Transmission (1 - 3GHz).

Frequency (GHz)	Power (dBm)
1.000	-34.82
2.055	-56.48
3.000	-29.23

Table 4. Transmission Power Measurement Values.

Power delivery and attenuation is then observed at the spectrum analyzer. Comparison with the measured S(1,2), S(2,1) response showed a near accurate bandstop center frequency at 2.055GHz. Power attenuation due to lossy transmission line medium and also dielectric effects is at a combined approximate of -19.23dBm. As observed from the scattering response and measurements, there is slightly higher attenuation at the lower frequency range below 2GHz. This is the maximum insertion loss of the filter and can be calculated by using (*Eq. 2-a*). The calculations for the insertion loss of the filter at 1GHz are presented below.

$$Insertion\ Loss\ (dB) = 10\ \log_{10} \frac{P_R}{P_T} \qquad (2\text{-a})$$

$$Transmission\ Line\ Loss = \ -19.23\ dBm = \left[10^{\left(-\frac{19.23}{10}\right)}\right] \times 10^{-3} = 1.194 \times 10^{-5}\ Watts$$

$$Input\ Power = \ -10\ dBm = \left[10^{\left(-\frac{10}{10}\right)}\right] \times 10^{-3} = 10^{-4}\ Watts$$

$$Transmitted\ Power,\ \boldsymbol{P_T} = \ -34.84\ dBm = \left[10^{\left(-\frac{34.84}{10}\right)}\right] \times 10^{-3} = 3.296 \times 10^{-7}\ Watts$$

$$Received\ Power,\ \boldsymbol{P_R} = (Input\ Power) - (Transmission\ Line\ Loss)$$

$$= \ 10^{-4} - 1.194 \times 10^{-5} = 8.806 \times 10^{-5}\ Watts$$

$$\therefore \ Insertion\ Loss\ (dB) = 10\ \log_{10} \left(\frac{8.806 \times 10^{-5}}{3.296 \times 10^{-7}} \right) = \underline{\underline{24.27\ dB}}$$

When compared to the insertion loss that is simulated in CST which is below 10dB, the insertion loss calculated from real world power transmission measurements for the filter is higher than the simulated value. There are a few external factors that lead to the deviation of the measured results from the simulated data. The insertion loss for filters, as mentioned before is caused by lossy dielectric medium and also a non unity value of the magnitude of reflection coefficient between the two ports. The latter can also be analyzed from the Smith Chart plot for the design. The impedance curve seen in the plot in Figure 17 shows the impedance of the filter from 1GHz to 3GHz. If the filter is acting as a true short circuit for both the input and output ports at the pass band frequencies, the impedance line should be ideally running along the outer most circumference of the Smith Chart. The lost signal power is converted into heat, most of it in the dielectric medium and some from the microstrip substrate. The unwanted attenuation and insertion loss can be lowered by using high permittivity dielectric materials with lower dielectric loss on the DRF structure. However, the dielectric resonator filter works as expected with the design derived from the DRA structure. The presence of a high permittivity dielectric in the center of a radiating strip ring excites the transverse magnetic (TM) mode of resonance, thus creating a central magnetic field that induces surface current on the coupled adjacent microstrip ring.

From the Smith Chart plot obtained from the analyzer in Figure 18, the key measurement impedance values are identified respectively in Table 6 below. Comparisons from the simulated and measured impedance response of the DRF at 1-3GHz shows similar matched impedance of 50Ω at the center frequency of 2GHz. Although the overall plot pattern of the measured impedance response have less similarity towards the simulated data, both of them shows that the filter circuit is acting as a matched termination load at the notch center frequency of 2GHz. The impedance of the filter moved towards the outer radius of the Smith Chart at the other frequencies.

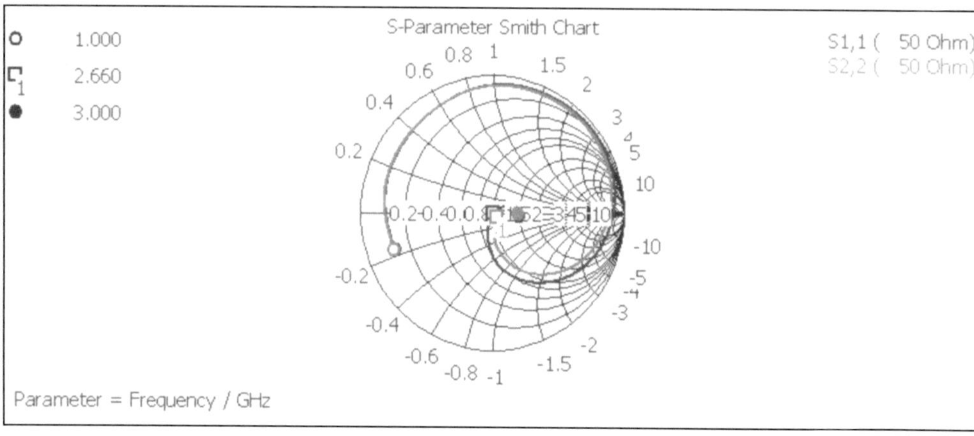

Fig. 17. Impedance Response of Structure.

Fig. 18. Measured Impedance Response of Structure.

Frequency (GHz)	Impedance (Ω)
1.000	57.166 + j 22.856
1.723	66.398 – j 2.9104
2.059	50.250 + j 0.7012
2.294	39.457 – j 7.5371
3.000	30.729 + j 9.9124

Table 6. Measured Impedance Response Values.

2.4.3 Half cylindrical pellet

In order to study the effect of the half cylindrical pellet towards the frequency response of the filter, the cylindrical pellet is replaced with a split pellet from the same batch of samples. The thickness for this pellet is sanded to the same measurements as the previous sample. Furthermore, the dielectric properties are also re-measured to make sure that it is consistent and similar to the previous pellet sample. Thus this will give a good representation of the variation of shape in a test to determine the effect of various shapes of dielectrics towards the microwave properties of the electrical model or structure. Dielectric material orientation will be similar to the previous DRA structure and the test setup for this is also shown in Figure 19.

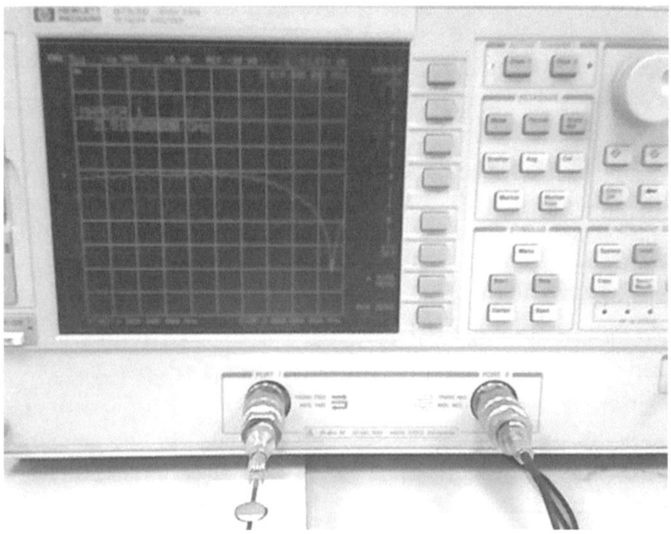

Fig. 19. DRF Measurement (Half Cylindrical Pellet).

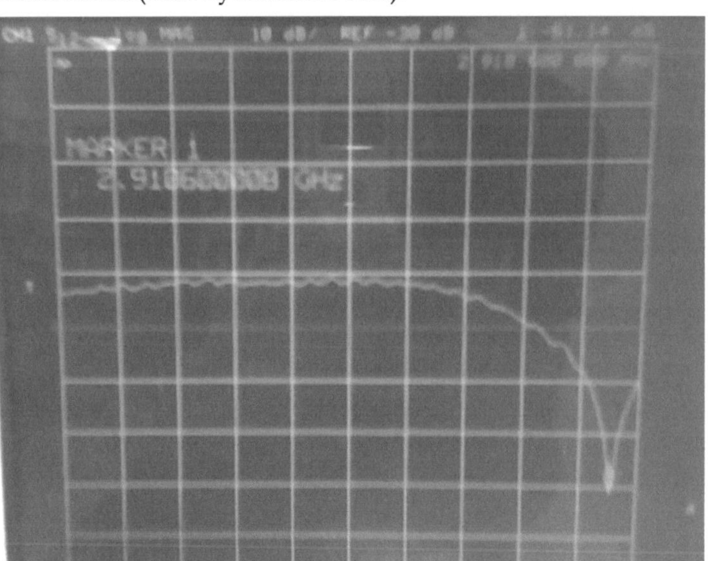

Fig. 20. Measured Frequency Response (Half Cylindrical Pellet).

The measured S(1,2) frequency response for the half cylindrical pellet is shown in Figure 20. As observed, the bandstop response curve has shifted to 2.91GHz. The return loss is measured at a lower -61.14dB compared to the previous DRF utilizing the full cylindrical dielectric pellet. From this test, it is observed that the half cylindrical dielectric material shape increases the stop frequency by almost 1GHz and lowered the stopband attenuation by 20dB. Although the network analyzer that was used for the measurements was limited to a maximum sweep frequency of 3GHz, it can be observed from the S(1,2) plot that the bandwidth of the filter remains consistent for both dielectric shapes. Thus, it can be deduced from this that splitting the dielectric material in half on the filter generally increases the notch filter center frequency by 50% and also lowers the attenuation at the center frequency also by 50%.

3. Conclusion

Results from the dielectric measurements suggest that by using 2 different synthesis techniques, different dielectric behavior of the dielectric material can be achieved. By using heat treatment, the structural properties of the material will be altered thus giving the material more radical dielectric properties. Wet chemical method on the other hand, produces normal dielectric properties on the samples with varied dopant compositions. Both methods also indicate that there is respectively, an ideal temperature and composition value for the materials to achieve their maximum dielectric properties.

The study of the trial batch of Barium Strontium Titanate samples suggests that dielectric properties of the pellet increased proportionally to the relative percentage of Strontium atoms inside the material. Measurements also indicated that the density of atoms inside the pellet affects the dielectric properties in a positive way. Another factor that greatly affects the dielectric properties of the material is the use of different degrees of heat treatment synthesis. As seen from initial tests, at 1300°C, the atomic structural properties of dielectric material is directly affecting the dielectric permittivity of the sample. In this case, the sample has a negative real value of permittivity, also meaning that penetrating electrical fields are refracted off the material.

By using the ideal synthesis methods obtained from the study of the Barium Strontium Titanate samples, the new composition of Bismuth Lanthanum Titanate was able to be produced. Wet chemical method that was used to synthesize the BLT power was favoured over other methods as it is an easier reaction to start and volatile elements of the composition can be safely retained. The drawback of this method is that the resultant product needs to go through a calcination process in order to purify the material. This is because of the presence of organic chemicals used as solvents and chelating agents in the raw material. However this preferred method is the least complicated and can be done at typical laboratory environments without needing specialized equipment. Lanthanum ions were selected as a dopant for Bismuth Titanate because of its rare earth mineral properties. Most mineral and metals of this class exhibits very good magnetic properties such as Neodymium alloys. In conclusion, a new type of dielectric material composition has successfully been fabricated using the least complex and feasible synthesis method available. The dielectric properties of the material are also desirable and effective in microwave applications.

4. Acknowledgment

The authors would like to thank Universiti Kebangsaan Malaysia and Universiti Sains Malaysia for their technical help and support in materials procurement, DRO design and fabrication.

5. References

[1] Petosa, A., *Dielectric Resonator Antenna Handbook*, Artech House, 2007.
[2] Peixeiro, C., "Microstrip rectangular ring bandpass filter elements for GSM," *APMC2000*, 1273–1276, 2000.
[3] Wong, K.-L., *Compact and Broadband Microstrip Antennas*, John Wiley and Sons, 2002.
[4] James, J. R. and P. S. Hall, *Handbook of Microstrip Antennas*, IEE, 1989.
[5] Martin, M. A., B. S. Sharif, and C. C. Tsimenidis, "Dual frequency microstrip antenna with V slot," *Wideband and Multiband Antennas and Arrays*, IEE (Ref. No. 2005/11059), 183–184, 2005.
[6] Ping, D. E. and R. B. Dybdal, "An efficient dual frequency antenna feed," 0-7803-8302-8/04/$20.00, IEEE, 2004.
[7] Mandeep, J., N.K. Loke, S. Hassan, M. Ain, S. Sreekantan and K.Y. Cheong "Investigation of microwave properties of high permittivity ceramic substrate," *J. of Electromn. Waves and Appl.*, Vol. 22, 1873–1882, 2008.
[8] Rajabi, M., M. Mohammadirad, and N. Komjani, "Simulation of ultra wideband microstrip antenna using EPML-TLM," *Progress In Electromagnetics Research B*, Vol. 2, 115–124, 2008.
[9] Mandeep, J., A. Lokesh, S. Hassan, M. N. Mahmud, and M. Ain, "Design of Cartesian feedback RF power amplifier for L-band frequency range," *Progress In Electromagnetics Research B*, Vol. 2, 207–222, 2008.
[10] Hassani, H. R. and M. Jahanbakht, "Method of moment analysis of finite phased array of aperture coupled circular microstrip patch antennas," *Progress In Electromagnetics Research B*, Vol. 4, 197– 210, 2008.
[11] Khalaj-Amirhosseini, M., "Wideband or multiband complex impedance matching using microstrip nonuniform transmission lines," *Progress In Electromagnetics Research*, PIER 66, 15–25, 2006.
[12] Wang, S., X. Guan, D.Wang, X. Ma, and Y. Su, "Electromagnetic scattering by mixed conducting/dielectric objects using higherorder MOM," *Progress In Electromagnetics Research*, PIER 66, 51–63, 2006.
[13] Liao, S. L. and R. J. Vernon, "On the image approximation for electromagnetic wave propagation and PEC scattering in cylindrical harmonics," *Progress In Electromagnetics Research*, PIER 66, 65–88, 2006.
[14] Moeckly, B. H. and Y. Zhang, "Strontium Titanate thin films for tunable YBCO microwave filters," IEEE Trans. Appl. Supercond., Vol. 11, 450-453, 2001.
[15] Boutayeb, H., A.-C. Tarot, and K. Mahdjoubi, "Focusing characteristics of a metallic cylindrical electromagnetic band gap structure with defects," *Progress In Electromagnetics Research*, PIER 66, 89–103, 2006.
[16] Khalaj-Amirhosseini, M., "Microwave filters using waveguides filled by multi-layer dielectric," *Progress In Electromagnetics Research*, PIER 66, 105–110, 2006.

[17] Rezaei, P., M. Hakkak, and K. Forooraghi, "Design of wideband dielectric resonator antenna with a two-segment structure," *Progress In Electromagnetics Research*, PIER 66, 111–124, 2006.
[18] Chen, H., L.-X. Ran, B.-I. Wu, J. A. Kong, and T. M. Grzegorczyk, "Crankled S-ring resonator with small electrical size," *Progress In Electromagnetics Research*, PIER 66, 179–190, 2006.
[19] Sihvola, A. H., "Peculiarities in the dielectric response of negative permittivity scatterers," *Progress In Electromagnetics Research*, PIER 66, 191–198, 2006.

19

Hybrid Microstrip Antennas

Alexandre Perron, Tayeb A. Denidni and Abdel R. Sebak
Institut national de la recherche scientifique
Concordia University
Canada

1. Introduction

Various techniques can be used to increase the gain of an antenna. For example, two or more uniformly distributed radiating elements can be used to create an antenna array. Array antennas have been successfully used for several years, but, unfortunately, they still suffer from a fundamental disadvantage: their feeding structure is inevitably more complex and may generate considerable losses (Horng & Alexopoulos, 1993). This is especially true at millimeter-wave frequencies, where losses of 40% to 50% are not uncommon (Uchimura et al., 2005; Huang & Wang, 2006; Weily & Guo, 2009). This type of antenna is also somewhat limited by the mutual coupling of its individual elements (Mohammadian et al., 1989; Malherbe, 1989). These factors, combined with the greater quantity of materials required to manufacture an array, can be responsible for a longer time-to-market and higher production costs.

Nevertheless, array antennas can offer useful functionalities such as electronic beam steering, introduction of nulls in specific directions (to counter a source of interference, for example) and suppression of secondary lobes by adjusting the phase and/or amplitude of the signals feeding the individual elements. However, if these characteristics are not necessary for a given system, an array antenna might not be the best choice, especially if complexity and cost are issues.

The single element hybrid microstrip antennas presented in this chapter rely on a large electrical size to increase the gain. This is achieved by exciting a higher order mode inside a dielectric ring resonator. Two hybrid antennas that were designed and fabricated using this approach will be presented in this chapter. The first antenna is linearly polarized and the second has two orthogonal linear polarizations.

2. Linearly polarized hybrid microstrip antenna

2.1 Antenna configuration

Figure 1 shows the general shape of the first designed antenna. In essence, it is a hybrid antenna combining an aperture-fed circular microstrip patch with a ring-shaped dielectric resonator (DR). As opposed to other reported hybrid antennas (George et al., 1998; Esselle & Bird, 2005; Oh et al., 2007), the proposed antenna has a directive radiation pattern. Moreover, its design process is relatively straightforward, making it an interesting candidate for various commercial applications.

The DR ($\varepsilon_{r1} \approx 10$) is fed by a circular microstrip patch etched on a thin grounded dielectric layer with a low permittivity ($\varepsilon_{r2} \approx 2$). The patch itself is fed via a resonant slot in the ground

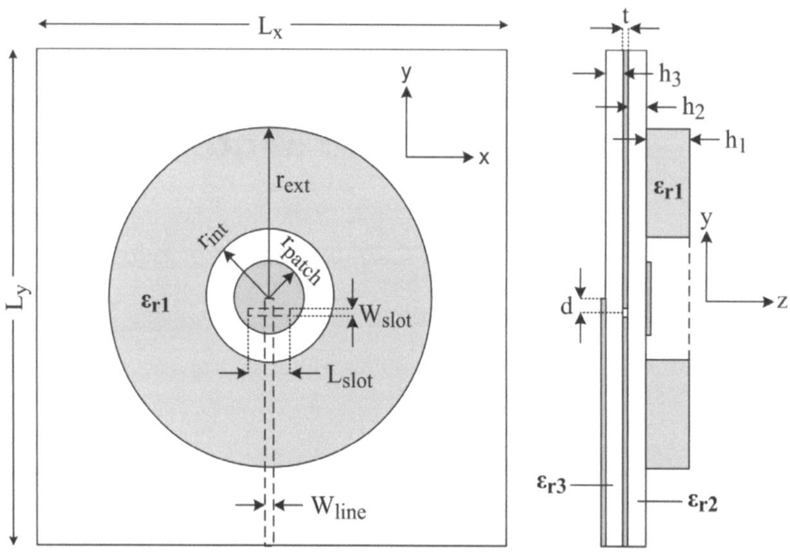

Fig. 1. Top and side views of the linearly polarized hybrid microstrip antenna

plane by a 50 Ω microstrip line sitting on the backside of a second thin dielectric layer with a higher permittivity ($\varepsilon_{r3} \approx 10$). The difference in permittivity ensures that the feed line is not as large as the patch ($W_{line} < 2\, r_{patch}$), a common problem encountered at millimeter-wave frequencies, where the physical dimensions of the conventional feeding structures are often comparable to those of the radiators.

2.2 Design procedure

The design process is relatively simple. The aperture coupled microstrip patch is first designed (without the DR) using conventional techniques - such as those described in this book - and its resonant frequency (TM_{110} mode) is adjusted so that the lower portion of the desired impedance bandwidth is covered. To reinforce the resonance, the length of the slot (L_{slot}) is approximately set to half a guided wavelength at a frequency just below the resonant frequency of the patch. During this first step, the width of the slot (W_{slot}) and other critical dimensions are chosen in function of the limitations of the PCB fabrication process. In our case, a minimum width of 0.1524 mm (6 mils) was tolerable.

For the DR, a $HEM_{1n\delta}$ (with n = 1, 3, 5 ...) hybrid mode must first be selected and its resonant frequency is set so that the upper portion of the desired band is covered. We have found that the field configurations of this family of hybrid modes are well suited for coupling with a microstrip patch. However, care must be taken when choosing the order n of the hybrid mode as it has a direct effect on the shape of the far-field radiation pattern of the antenna. Fabrication constraints must also be considered.

Cohn's model (Cohn, 1968) is used to develop an estimation of the resonant frequency of the $HEM_{mn\delta}$ hybrid modes of a ring-shaped DR placed on a thin layer of grounded substrate:

$$f_{mn\delta}(GHz) = \frac{c}{2\pi\sqrt{\varepsilon_{eff}}}\sqrt{\left(\frac{\chi_{mn}}{r_{out}}\right)^2 + \left(\frac{\delta\pi}{2h_{eff}}\right)^2} \qquad (1)$$

where c is the speed of light and χ_{mn} is the zero of derivative $J'_m(\chi_{mn}) = 0$ of the Bessel function $J_m(x)$ for odd values of n and the zero of $J_m(\chi_{mn}) = 0$ of the same Bessel function for even values of n. Table 1 lists the values of χ_{mn} for hybrid modes $HEM_{1n\delta}$ (n = 1 to 8).

Hybrid mode	χ_{mn}	Hybrid mode	χ_{mn}
$HEM_{11\delta}$	1.8412	$HEM_{15\delta}$	8.5363
$HEM_{12\delta}$	3.8318	$HEM_{16\delta}$	10.1735
$HEM_{13\delta}$	5.3315	$HEM_{17\delta}$	11.706
$HEM_{14\delta}$	7.0156	$HEM_{18\delta}$	13.3237

Table 1. Values of χ_{mn} for hybrid modes $HEM_{1n\delta}$ (n = 1 to 8)

The factor δ in Equation (1) is a quantity between 0 and 1 representing the number of half-wavelength variations of the field in the z direction. It can be calculated by solving the following set of equations:

$$\beta \cdot 2h_{eff} = 2 \cdot \tan^{-1}\left(\frac{\alpha}{\beta}\right) \quad (2)$$

$$\alpha = \sqrt{\left(\frac{\chi_{mn}}{r_{out}}\right)^2 - k_0^2} \quad (3)$$

$$\beta = \sqrt{k_0^2 \cdot \varepsilon_{eff} - \left(\frac{\chi_{mn}}{r_{out}}\right)^2} \quad (4)$$

$$\delta = \beta \cdot \frac{2h_{eff}}{\pi} \quad (5)$$

Equation (2) is transcendental in nature and can be easily solved numerically using a mathematical tool such as Matlab®. The effective height and the effective permittivity are defined as:

$$h_{eff} = h_1 + h_2 \quad (6)$$

$$\varepsilon_{eff} = \frac{V_{total}}{\frac{V_{hole}}{\varepsilon_{air}} + \frac{V_{ring}}{\varepsilon_{r1}} + \frac{V_{substrate}}{\varepsilon_{r2}}} = \frac{r_{out}^2 \cdot h_{eff}}{\frac{r_{in}^2 \cdot h_1}{\varepsilon_{air}} + \frac{h_1(r_{out}^2 - r_{in}^2)}{\varepsilon_{r1}} + \frac{r_{out}^2 \cdot h_2}{\varepsilon_{r2}}} \quad (7)$$

where ε_{air} is the permittivity of air (≈ 1), V_{hole} is the volume of the DR hole, V_{ring} is the volume of the dielectric ring itself (without the hole), $V_{substrate}$ is the volume of the dielectric layer directly underneath the DR, and V_{total} is the sum of these three volumes.

2.3 Simulation results

Using the commercial electromagnetic simulator CST Microwave Studio®, the proposed antenna is simulated and optimized for operation in the 57 GHz to 65 GHz band. The final optimized values for the parameters of Figure 1 are listed in Table 2.

Parameter	Value (mm)	Parameter	Value (mm)
ε_{r1}	10.2	r_{ext}	3.6
ε_{r2}	2.2	r_{patch}	0.78
ε_{r3}	10.2	L_{slot}	0.835
h_1	0.635	W_{slot}	0.1524
h_2	0.254	d	0.2
h_3	0.254	L_x	10.5
r_{int}	1.3	L_y	10.5

Table 2. Parameter values (Figure 1) used for the simulations

Simulated S_{11} parameters of the aperture coupled microstrip antenna (without the DR) and of the proposed hybrid antenna (with the DR) are shown in Figure 2. The negative peak near 59 GHz (for both the microstrip and the hybrid antenna) is the result of the combination of

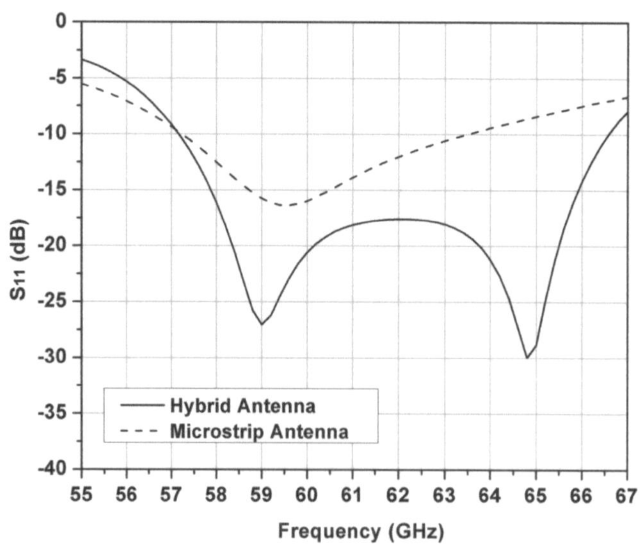

Fig. 2. Simulated S_{11} parameter of the linearly polarized hybrid antenna

two closely spaced resonances: one from the patch itself and another one from the resonant coupling slot. For the microstrip antenna (without the DR), the effect of this combination can be observed as the upper portion of the bandwidth of the patch is extended.

For the hybrid antenna, a second negative peak (near 65 GHz) is observed. It is the result of the excitation of the $HEM_{15\delta}$ hybrid mode inside the DR. Indeed, solving Equations (1) through (7), we find that h_{eff} = 0.889 mm, ε_{eff} = 3.522, δ = 0.6537, and $f_{15\delta}$ = 67.1 GHz (a difference of about 3%). For some designs, it may be possible to excite another mode within the $HEM_{1n\delta}$ (n = 1, 3, 5 ...) family, given that the r_{out}/h_{eff} ratio and inner radius (r_{in}) of the DR are properly adjusted to maximize the coupling with its feeding structure. Generally, a ratio between 1 and 4 is expected for n = 1, 3, and 5, whereas modes with higher indices require a larger ratio, making the antenna too large for typical applications.

Three normalized radiation patterns at the lower edge, the center and the upper edge of the impedance bandwidth are shown in Figure 3. The first observation is that the main lobe is always directed at broadside and that secondary lobes are generated and get more visible as the frequency increases. These side lobes are a direct consequence of the excitation of the

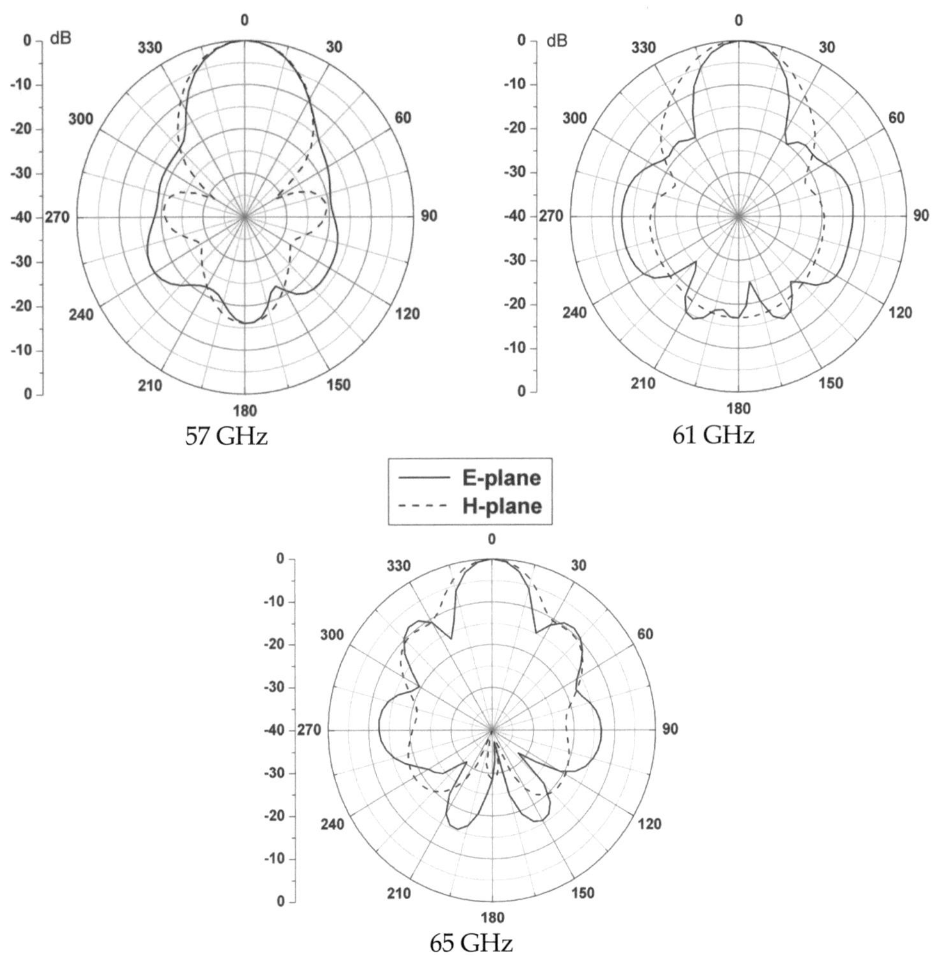

Fig. 3. Simulated radiation patterns of the linearly polarized hybrid antenna

$HEM_{15\delta}$ hybrid mode inside the DR. Nevertheless, they remain within acceptable levels, i.e. more than 9 dB lower than the main beam (at 65 GHz).
The half-power beamwidth is reduced as the frequency increases from 57 to 65 GHz (32.1° to 22.1° for the E-plane and 32° to 26° for the H-plane). However, this does not translate into a gain increase due to the increase in the side lobe levels. Another interesting quality of the proposed antenna is its very low cross polarization level, which is less than -27 dB (not shown).
The most appealing feature of this antenna resides in its high gain, as depicted in Figure 4. Indeed, it can be noted that the simulated gain is very high for such a simple structure i.e. more than 13 dB (with a simulated radiation efficiency between 86.8 and 89.7 % over the whole bandwidth). The gain is also stable over the bandwidth of interest, with a variation of about 0.35dB over the whole frequency band. The theoretical gain in Figure 4 is calculated with the following approximation formula (with a radiation efficiency η = 1 or 100%):

$$Gain(dB) = 10\log\left(\frac{\eta \cdot 4\pi \cdot A}{\lambda_0^2}\right) \tag{8}$$

where $A = \pi \cdot r_{out}^2$ is the physical area of the antenna and λ_0 is the free space wavelength. Since Equation (8) is only an approximation, it is not abnormal to see the simulated gain exceed the maximum theoretical value; this does not imply that the efficiency is over 100% for the lower portion of the bandwidth. Nonetheless, this theoretical approximation gives an upper limit for the gain we can expect from the proposed antenna.

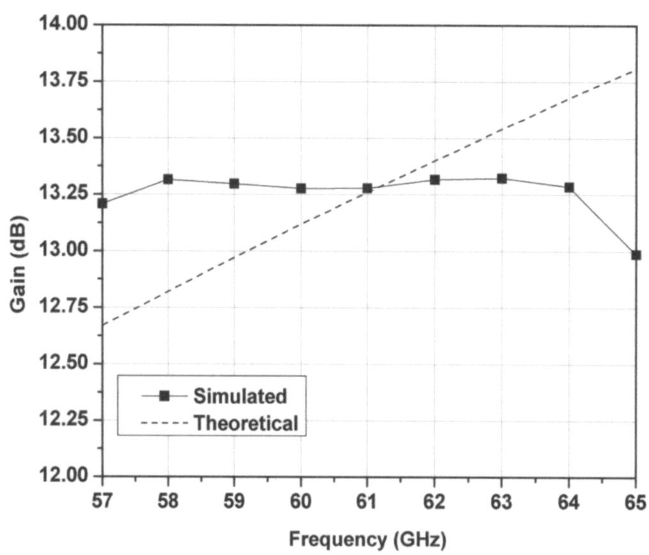

Fig. 4. Simulated gain of the linearly polarized hybrid antenna

2.4 Experimental results

A standard low cost PCB process is used to fabricate most of the antenna. Alignment holes are drilled outside the ground plane of the antenna to ensure that different components of the aperture coupled microstrip antenna are correctly aligned. The laminates are then

"glued" together using a thin film of bounding material and cut to the desired size, leaving out the alignment holes. The DR is cut using a computer controlled Nd-Yag Q-switched laser (a photograph of many fabricated samples is shown as an inset in Figure 5).
Figure 5 shows photographs of the fabricated antenna. The widely used RT/Duroid® 5880 is chosen as the substrate of the patch (ε_{r2} = 2.2, h_2 = 0.254 mm) whereas the feed line is etched on a thin layer of RT/Duroid® 6010LM (ε_{r2} = 10.2, h_3 = 0.254 mm).

Fig. 5. Photographs of the linearly polarized hybrid microstrip antenna

All the dimensions are as listed in Table 2, except for L_x (25 mm), L_y (30 mm) and r_{out} = 3.2 mm. A larger ground plane was necessary to install the required housing to hold a 1.85 mm connector, for measurement purposes. A slightly smaller DR was chosen to suppress the undesired $HEM_{14\delta}$ hybrid mode, for which the calculated cutoff frequency fell too close to the operational band (56.7 GHz).
The S_{11} parameter of the prototype (with and without the DR) was measured using a Anritsu® 37397C Vector Network Analyzer (VNA). The results are presented in Figure 6.

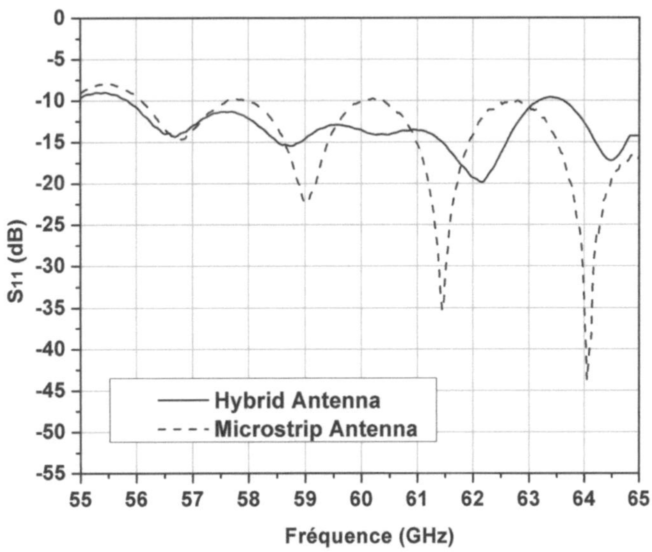

Fig. 6. Measured S_{11} of the microstrip (without the DR) and hybrid (with the DR) antennas

Considering the fact that the prototype is fabricated using a standard PCB process, that the two substrate layers are bounded together with a material of non-zero thickness, and that a connector is manually added, some discrepancies between the simulated and measured results are expected. However, after a thorough investigation, it was found that the coaxial-to-microstrip transition (the connector) was responsible for the oscillations in the S_{11} curves. A simple setup is used to measure the radiation patterns of the prototype. A standard gain V-band horn antenna (24 dB gain at mid-band) is connected to port 2 of the VNA and the antenna under test (AUT) is connected to port 1. Both antennas are vertically and horizontally aligned and are placed at a distance of 30 cm. The AUT is then manually rotated around its vertical axis by increments of 10-degrees and the S_{12} parameter is saved for post-processing. The measured H-plane normalized patterns (microstrip antenna and hybrid antenna) are plotted in Figure 7.

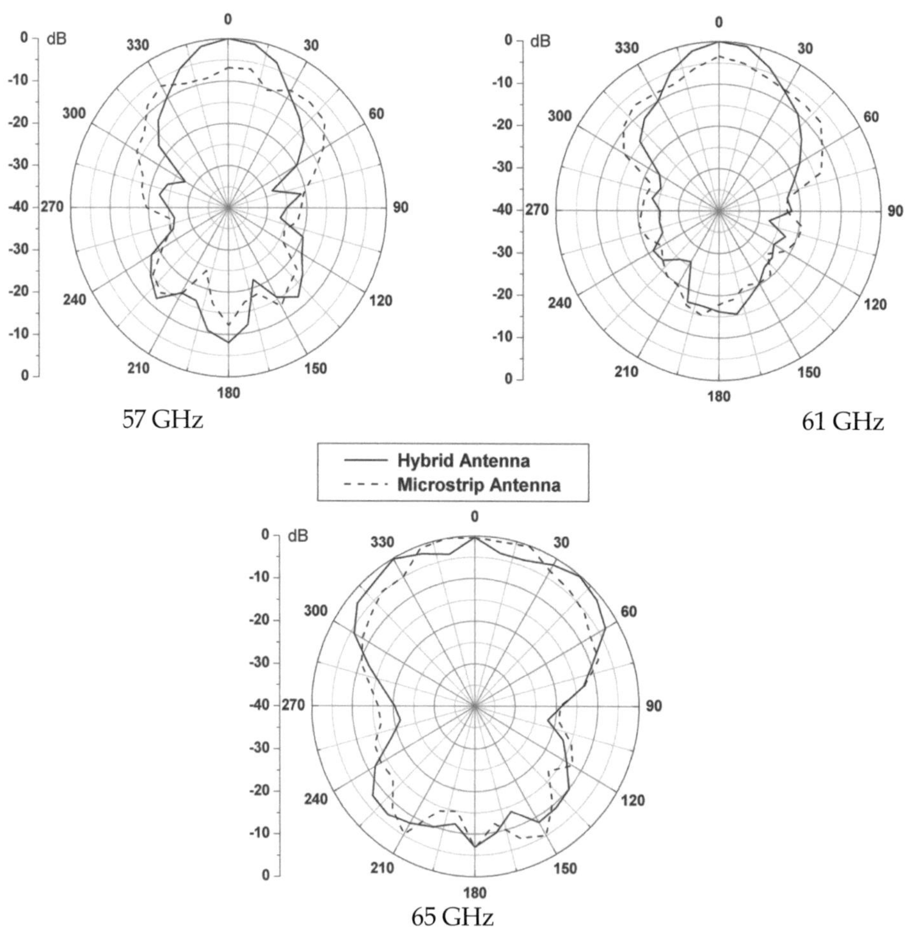

Fig. 7. Measured radiation patterns of the microstrip and hybrid microstrip antennas

The first observation we can make from these patterns is that, all things considered, the hybrid antenna is more directive than the aperture coupled microstrip antenna, except at the upper limit of the band (65 GHz), where the $HEM_{14\delta}$ mode manifestly takes over to generate a monopole-like pattern with a small peak at broadside. Secondly, the patterns of the hybrid antenna are stable over a large frequency range and are very similar to the simulated H-plane patterns presented as *dashed lines* in Figure 3. Indeed, the main lobe is dominant for most of the bandwidth and the front-to-back ratio is between 10-15 dB. The backward radiation is higher than expected due to the strong resonance of the slot, especially near 57 and 65 GHz. Small side lobes at 90° and 270° can also be observed at some frequencies.

Using two identical standard gain horns, the gain is measured by applying the gain-transfer technique (Balanis, 2005) and removing the reflection losses with the aid of Figure 6. The measured results are presented in Figure 8. New simulated results taking into account the larger ground plane are plotted. The theoretical results are also recalculated based on the new outer radius of the DR (r_{out} = 3.2 mm).

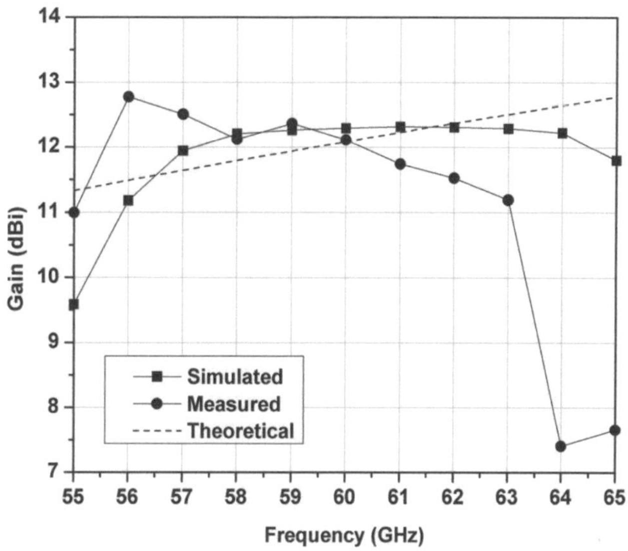

Fig. 8. Measured gain of the linearly polarized hybrid microstrip antenna

The measured gain is quite close to the simulated values. It is also stable over a large portion of the impedance bandwidth, as predicted. The theoretical gain is lower for r_{out} = 3.2 mm due to the reduced area occupied by the DR. Nonetheless, the measured gain is over 11 dB from 55 to 63 GHz, a bandwidth that would be large enough to cover the whole 57-65 GHz ISM band. The realized gain (not shown) is over 10.5 dB from 55 to 63 GHz.

2.5 Discussion

This section highlighted an innovative high-gain hybrid antenna for millimeter wave applications. A thorough theoretical study of the behavior of this antenna made it possible to define a simple design procedure. The proposed configuration achieves double digit gain by coupling a microstrip patch with a ring-shaped DR, thus increasing the electric size of the antenna. Indeed, simulated and measured results confirmed that the gain is clearly improved when the patch is correctly exciting an appropriate hybrid mode inside the dielectric structure. Furthermore, measurements from a prototype have also established that the major part of the antenna can be manufactured with a simple PCB process, that it is more directive than a conventional microstrip antenna, and that its impedance bandwidth would be large is enough to cover the whole ISM 57-65 GHz band.

3. Dual-polarized hybrid microstrip antenna

In this section another high gain and wide-band antenna designed from the same hybrid configuration will be presented. This second antenna could be used to transmit and/or receive signals by using two orthogonal linear polarizations (vertical and horizontal), via two independent feed ports.

3.1 Antenna configuration

The proposed antenna geometry is shown in Figure 9. A ring-shaped dielectric resonator (DR) (ε_r = 10.2; tan δ = 0.0023) is fed by a square patch etched on a dielectric substrate (thickness = 0.254 mm; ε_r = 2.2; tan δ = 0.001). The patch is fed via two orthogonally disposed slots of identical dimensions in the ground plane by two 50 Ω microstrip lines sitting on the backside of a second substrate layer (thickness = 0.254 mm; ε_r = 10.2; tan δ = 0.0023).

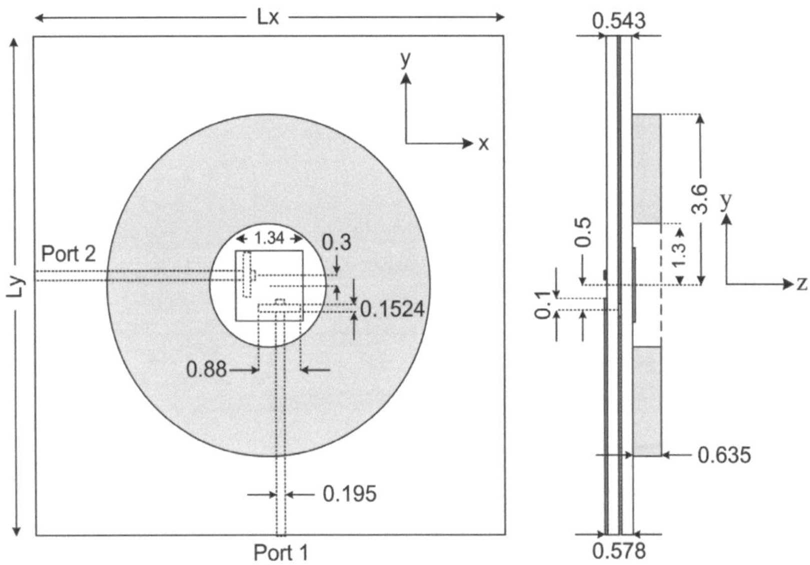

Fig. 9. Top and side views of a dual-polarized hybrid microstrip antenna

The L-shaped arrangement of the coupling slots ensures that both ports exhibit the same theoretical return loss, i.e. the same impedance bandwidth. A T-shaped disposition could be used to increase the isolation (Porter et al., 1999), but the impedance bandwidth of the two ports would then be dissimilar. The same observation has been made in (Perron et al., 2007) for a dual-polarized CPW-fed microstrip antenna operating at 2.45 GHz.

3.2 Results

A prototype of the proposed antenna was fabricated using Rogers Corporation® high frequency laminates (RT/Duroid® 5880 and 6010LM). The DR was easily cut using a computer controlled Nd-Yag Q-switched laser and bonded to the substrate with a few droplets of cyanoacrylate. Figure 10 shows photographs of the fabricated antenna. Again, note that a relatively large ground plane (Lx = Ly = 30 mm) is used to install a brass housing to hold two 1.85 mm connectors.

Fig. 10. Photographs of the dual-polarized hybrid microstrip antenna

The S-parameters of the antenna are measured using the VNA. The results are presented in Figure 11. For comparison, simulated results from CST Microwave Studio® are also plotted.

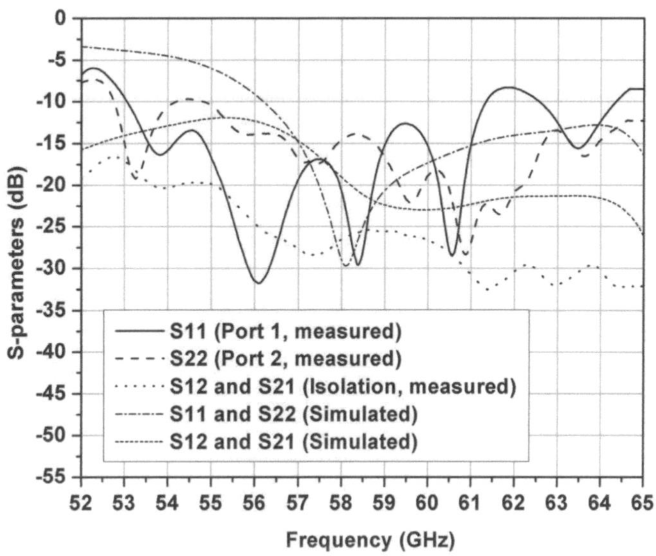

Fig. 11. Measured S-parameters of the dual-polarized hybrid microstrip antenna

Since the connectors and the brass support are not included in the simulation model (to decrease computational time), some discrepancies between measured and simulated data are expected. Nevertheless, the impedance bandwidth (VSWR < 2) of both ports of the antenna is fairly similar and is more than acceptable for many antenna applications. As the bandwidth is relatively large (more than 8 GHz), oscillations in the measured curves are also noticeable. This phenomenon is due to the inductive loading effect of the coaxial to microstrip transition. The measured isolation is better than expected, remaining over 20 dB over most of the bandwidth.

Using a three-antenna technique, the antenna gain for both polarizations is measured. Two result sets are presented in Figure 12: the gain of the antenna without the DR in place (the microstrip patch only) and the gain of the hybrid configuration (microstrip patch and DR). Alignment errors during the fabrication and variable transition losses introduced by the connectors are responsible for the differences between the two ports. Nevertheless, the hybrid configuration exhibits a higher gain, with a peak value of more than 10 dB at 56 GHz for both polarizations. The average simulated gain of the simplified model is 11.8 dB within the operational bandwidth of the antenna. This high-gain is due to the increase in the electrical size of the radiator when the DR is mounted on its surface.

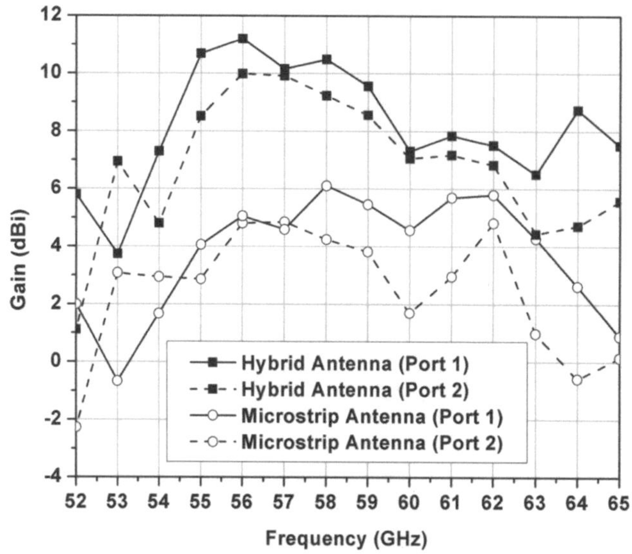

Fig. 12. Measured gain of the dual-polarized microstrip and hybrid configurations

The radiation pattern of the proposed antenna at 56 GHz (port 1) is presented in Figure 13. The pattern of port 2 (not shown) is almost identical. The patterns are measured manually using a simple yet precise mechanical setup covered with RF convoluted foam absorbers. A high-gain horn antenna is connected to the first port of the VNA while the second port is connected to the antenna under test. The S_{21} parameter at different angles is then stored for post-processing. Note that only half of the E-plane is measured since the connector would block the other half. For convenience, the other half is therefore estimated by mirroring the measured half. This is not the case for the H-plane which is completely measured. Again, the measured results incorporate both the microstrip patch and hybrid configurations.

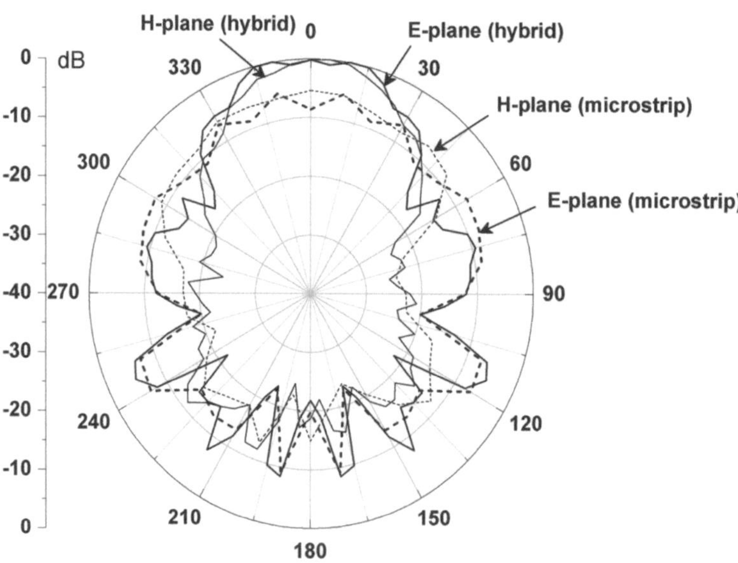

Fig. 13. Measured E- and H-plane of the microstrip and hybrid configurations (56 GHz)

The first observation is that the main beam of the antenna is directed at broadside and that the hybrid configuration is more directive than the microstrip patch antenna by itself. Also note the presence of small side lobes at 75° and 285° in the E-plane of the hybrid configuration. These side lobes are a result of the excitation of a higher order hybrid mode inside the DR (the $HEM_{13\delta}$ mode). In fact, by studying the electromagnetic distribution inside the ring-shaped DR, it can be seen that the latter acts as a dielectric waveguide that collects (or distributes) electromagnetic waves through a large aperture size. Unfortunately, as the frequency increases, the side lobes become more important and the gain decreases.

3.3 Discussion

A new hybrid microstrip antenna with two orthogonal linear polarizations has been presented in this section. Experimental results indicate that it could be used for applications requiring a high gain and a relatively broad bandwidth. Moreover, as this antenna is based on standard commercial components and that it is almost entirely manufactured with a conventional PCB process, it constitutes an interesting alternative to more complex configurations, such as array antennas. Finally, even if the prototype presented in this section exhibits a maximum gain at 56 GHz, it would undoubtedly be possible to optimize its dimensions so that the maximum gain is reached inside the 57-65 GHz band.

4. Possible improvements

The hybrid microstrip antennas presented in Sections 3 and 4 possess several desirable characteristics, making them excellent candidates for several millimeter-wave applications. In an effort to further improve their performance, a new modified geometry (with a higher gain and reduced side lobe levels) is presented in this section.

4.1 Antenna configuration

The modified configuration is presented in Figure 14. The dimensions of the radiating elements are the same as in Section 2. A brass support allows the antenna to be fitted with a 1.85 mm coaxial connector (note that the model in Figure 14 only shows the glass bead that is used to make the transition from the central conductor of the coaxial connector to the

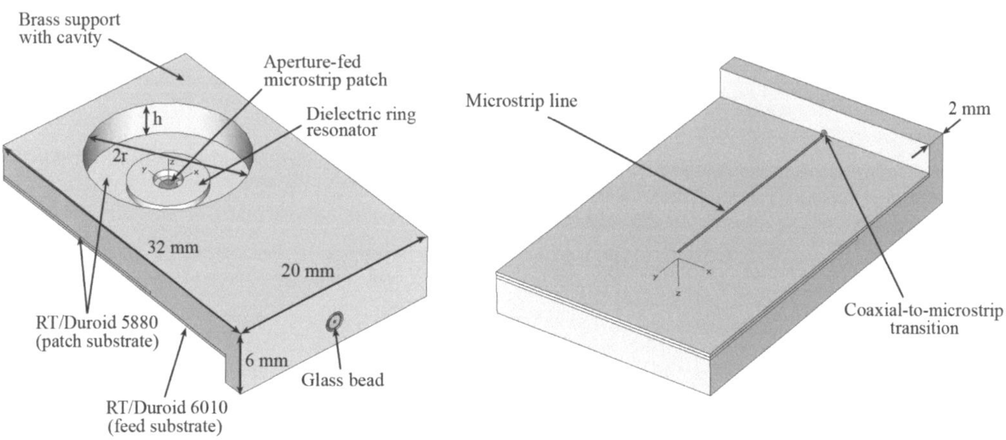

Fig. 14. Top and bottom views of the modified antenna geometry.

microstrip feed line). This housing adds a much needed structural support to the very thin PCB assembly. Indeed, both the feed substrate (RT/Duroid 6010, ε_r = 10.2) and the microstrip patch substrate (RT/Duroid 5880, ε_r = 2.2) are only 0.254 mm thick The brass support also features a circular opening (of height h and radius r) positioned directly above the DR and the patch. As it will be confirmed in the next section, it is possible to increase the gain and reduce the side lobe levels of the antenna by optimizing the dimensions of this metallic cavity.

4.2 Simulation results

In order to highlight the effects of adding a metallic cavity to the antenna, this section presents a number of antenna parameters simulated with and without the brass support (and cavity).
The simulated S_{11} of the proposed antenna (with and without the brass support) is shown in Figure 15. Note that the simulation port is located at the end of the microstrip feed line rather than at the end of the glass bead. This allows a better comparison of both antenna configurations and reduces computational time as the coaxial-to-microstrip transition is not unnecessarily simulated. The optimized dimensions of the cavity are h = 3 mm and r = 7.25 mm. In both cases, the impedance bandwidth (VSWR < 2) is wide, ranging from 56.5 to 66.5 GHz (16.3 %), approximately. The strongest resonances come from the patch and the DR, but the cavity itself affects the S_{11} parameter of the antenna by introducing small oscillations in the curve, especially at the upper end of the impedance bandwidth. By studying the input impedance of the antenna, we conclude that these oscillations are a result of a loading effect introduced by the brass support. Nonetheless, the reflection coefficients of both configurations are similar.

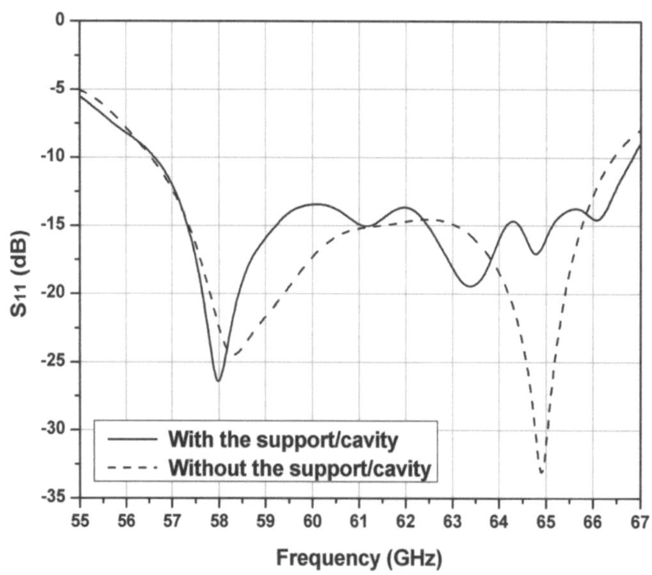

Fig. 15. Simulated S_{11} of the antenna (with and without the support and cavity).

Figure 16 and Figure 17 show the simulated gain and side lobe suppression levels of both configurations, respectively. From Figure 16, it is evident that the gain is improved when the support and cavity are present. This is most likely due to the constructive interference of waves reflected by the wall of the latter. Figure 17 confirms that the side lobe suppression is better for most of the impedance bandwidth.

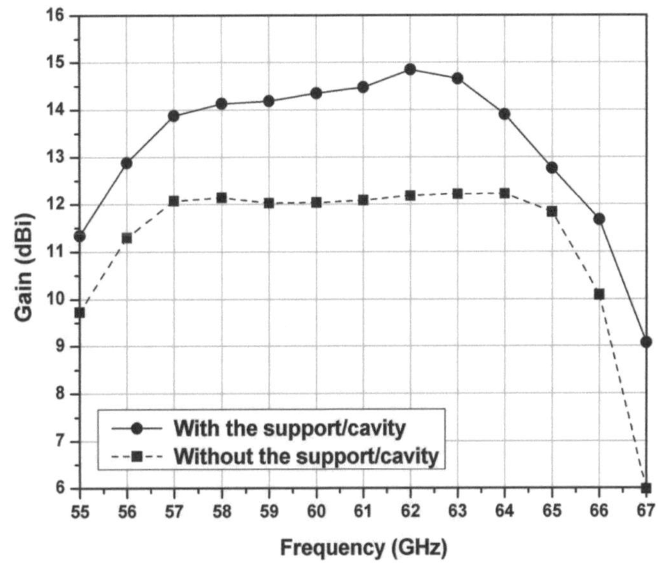

Fig. 16. Simulated gain of both configurations (with and without the brass support and cavity).

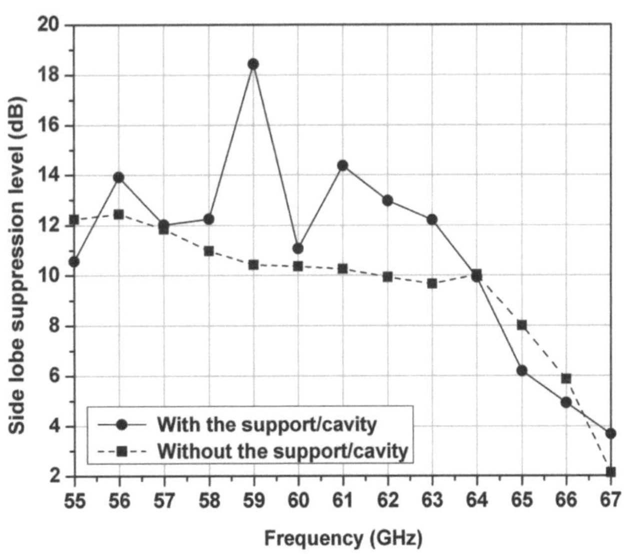

Fig. 17. Simulated side lobe suppression levels of both configurations (with and without the brass support and cavity).

The radiation patterns of both configurations near mid-band (61 GHz) are shown in Figure 18. The side lobe levels of the antenna fitted with the brass housing are clearly lower. Also note that the cross-polarized field levels are very low, especially the cross-polarized E-field, which is negligible (thus not visible in Figure 18).

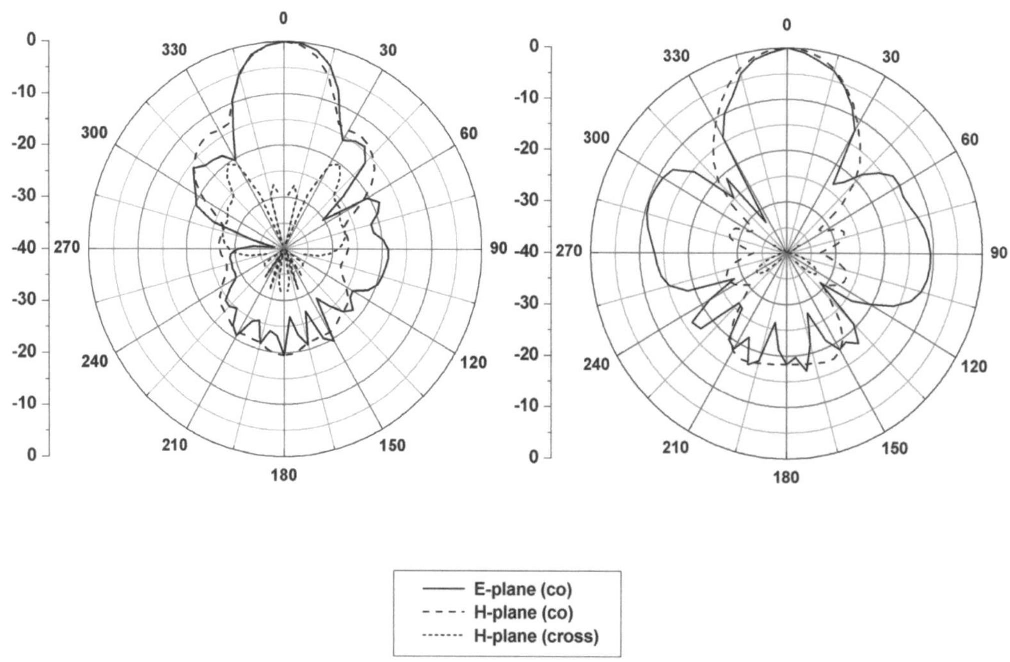

(a) (b)

Fig. 18. Simulated normalized radiation patterns at 61 GHz: (a) with the brass support and cavity and (b) without the brass support and cavity.

4.3 Parametric study

The results of a short parametric study are presented in Figure 19. In Figure 19a, the radius (r) of the cavity is varied from 6.75 to 7.75 mm. In Figure 19b, the height (h) of the cavity is varied from 2.5 to 3.5 mm. From these results, it is obvious that gain is directly affected by the dimensions of the metallic cavity, especially its radius. By conducting similar studies for the side lobe levels and impedance bandwidth, it is possible to optimize the cavity and reach the best side lobe suppression/gain/bandwidth tradeoff.

4.4 Discussion

In this section, a new modified geometry for a hybrid microstrip/dielectric resonator antenna has been presented. The simulated gain remains over 14 dB for a large portion of the 10 GHz impedance bandwidth. The side lobe levels are also significantly lowered with the addition of a metallic cavity.

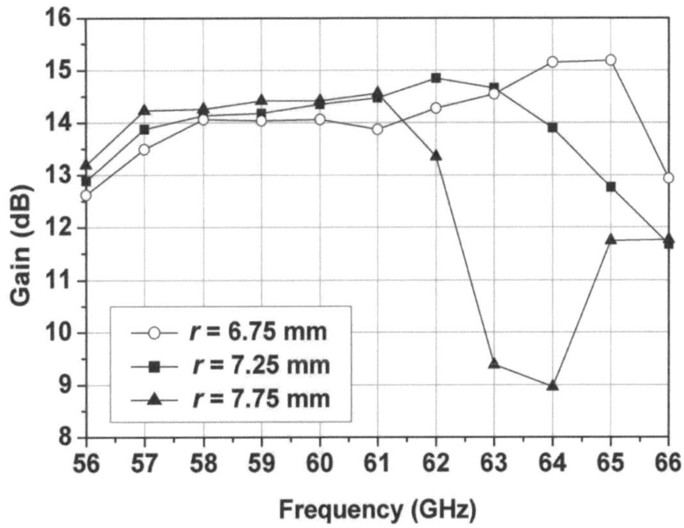

(a)

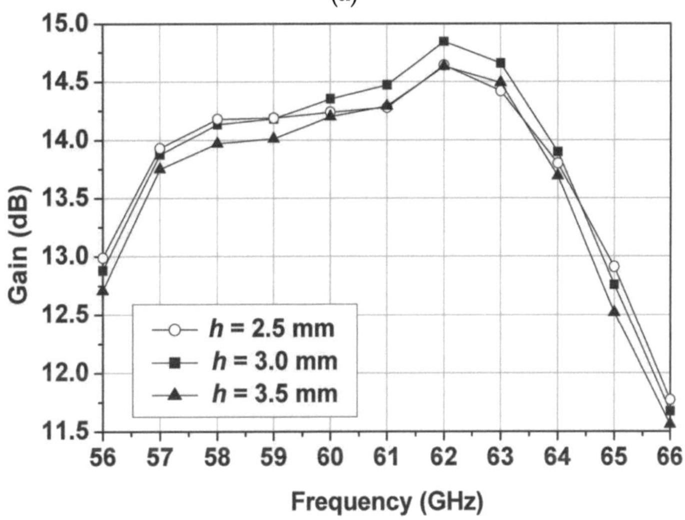

(b)

Fig. 19. Parametric study: (a) radius of the cavity and (b) height of the cavity.

5. Conclusion

In this chapter, two hybrid microstrip antennas were briefly introduced. In each case, it was possible to enlarge the impedance bandwidth and increase the electrical size (thus enhancing the gain) of the antenna by using two different radiating elements (a microstrip patch and a dielectric resonator). These examples show how hybrid antennas can be used to meet various requirements and still remain relatively easy to design and manufacture.

In addition to fabricating prototypes based on the improved antenna geometry of Section 4, other projects could be undertaken in the future. Initially, an optimization campaign would probably help to limit the losses generated by the 1.85 mm connector (by improving the coaxial-to-microstrip transition). Indeed, though this stage is not absolutely necessary, that would facilitate the performance evaluation of the antennas. Finally, within the framework of a project of greater scale, it would be possible to integrate various active components (a low noise amplifier, for example) on the feed side (with the microstrip line), to design a compact antenna front-end. The metallic support could then cover the back of the antenna and protect these components. The dimensions of this protective cover could even be optimized to reduce the undesired backward radiation of the antenna.

6. References

Balanis, C. A. (2005) *Antenna Theory: Analysis and Design*, 3rd ed. Hoboken, NJ: Wiley-Interscience.

Cohn, S. B. (1968) Microwave bandpass filters containing high-Q dielectric resonators. *IEEE Transactions on Microwave Theory and Techniques*, Vol. 16, No. 4, pp. 218-227.

Esselle, K. P. & Bird, T. S. (2005). A Hybrid-Resonator Antenna: Experimental Results. *IEEE Transactions on Antennas and Propagation*, Vol. 53, No. 2, pp. 870-871.

George, J.; Aanandan, C. K.; Mohanan, P.; Nair, K. G.; Sreemoolanathan, H. & Sebastian, M. T. (1998). Dielectric-resonator-loaded microstrip antenna for enhanced impedance bandwidth and efficiency. *Microwave and Optical Technology Letters*, Vol. 17, No. 3, pp. 205-207.

Horng, T.-S. & Alexopoulos, N. G. (1993). Corporate feed design for microstrip arrays. *IEEE Transactions on Antennas and Propagation*, Vol. 41, No. 12, pp. 1615-1624.

Huang, K. C. & Wang, Z. (2006). Millimeter-wave circular polarized beam-steering antenna array for gigabit wireless communications. *IEEE Transactions on Antennas and Propagation*, Vol. 54, No. 2, pp. 743-746.

Kajfez, D. & Guillon, P. (1986) *Dielectric Resonators*, Dedham, MA: Artech House, Inc.

Malherbe, J. A. G. (1989). Analysis of a linear antenna array including the effects of mutual coupling. *IEEE Transactions on Education*, Vol. 32, No. 1, pp. 29-34.

Mohammadian, A. H.; Martin, N. M. & Griffin, D. W. (1989). A theoretical and experimental study of mutual coupling in microstrip antenna arrays. *IEEE Transactions on Antennas and Propagation*, Vol. 37, No. 10, pp. 1217-1223.

Oh, J.; Baek, T.; Shin, D.; Rhee, J. & Nam, S. (2007). 60 GHz CPW-fed dielectric-resonator-above-patch (DRaP) antenna for broadband WLAN applications using micromachining technology. *Microwave and Optical Technology Letters*, Vol. 49, No. 8, pp. 1859-1861.

Perron, A., Talbi, L. & Denidni, T. A. (2007). Dual-polarised CPW-fed microstrip antenna for wireless applications at 2.45 GHz. *IEE Electronics Letters* Vol. 43, No. 14 (2007), 740-742.

Perron, A.; Denidni, T. A. & Sebak, A. R. (2009). High-gain hybrid dielectric resonator antenna for millimeter-wave applications: design and implementation. *IEEE Transactions on Antennas and Propagation*, Vol. 57, No. 10, pp. 2882-2892.

Porter, B. G., Rauth, L. L., Mura, J. R. & Gearhart, S. S. (1999). Dual-Polarized Slot-Coupled Patch Antennas on Duroid with Teflon Lenses for 76.5-GHz Automotive Radar Systems. *IEEE Transactions on Antennas and Propagation*, Vol. 47, No. 12, pp. 1836-1842.

Uchimura, H.; Shino, N. & Miyazato, K. (2005). Novel circular polarized antenna array substrates for 60GHz-band. *2005 IEEE MTT-S Digest*, p. 1875-1878, Long Beach, CA, June 2005.

Weily, A. R. & Guo, Y. G. (2009). Circularly polarized ellipse-loaded circular slot array for millimeter-wave for WPAN applications. *IEEE Transactions on Antennas and Propagation*, Vol. 57, No. 10, pp. 2862-2870.

20

Integration of 60-GHz Microstrip Antennas with CMOS Chip

Gordana Klaric Felic and Efstratios Skafidas
National ICT Australia, The University of Melbourne
Australia

1. Introduction

New emerging wireless systems that operate at millimeter wave frequencies, such as high data rate 60-GHz transceivers for wireless personal area networks (WPAN), use integrated antennas. The small wavelength at 60-GHz enables the antenna to be integrated either in the package or on the chip. However, on-chip integrated antennas for 60-GHz systems, built on a conductive substrate, such as complementary metal-oxide-semiconductor (CMOS) or silicon germanium (SiGe), results in poor radiation efficiency. Therefore, antennas for these systems are commonly implemented on in-package solutions. The integration of antenna-in-package can be achieved by using wire bonding or flip-chip bonding interconnections. Although bonding wires are typically employed to connect passive devices to chip modules since they are robust, inexpensive and tolerant of chip thermal expansion they increase reflections, impedance mismatches and power loss at millimeter wave frequencies. The flip-chip bonding of antenna to chip module can significantly reduce these effects as it uses metallic bumps for device connections which are kept small compared to the length of the bond wire that results in better impedance matching and reduced interconnection losses.
In this chapter two types of antenna integration methods are presented; a microstrip antenna for flip-chip integration and a coplanar waveguide (CPW) fed antenna for wire bonding. The flip-chip integrated antenna compatible with CMOS technology achieves good performance and is mounted on the CMOS chip using standard printed circuit board and flip-chip bonding technology. This approach is compact and inexpensive. The microstrip patch is printed on one substrate layer with a CPW feeding line through the ground patch that permits direct attachment of the patch to the die. The use of low dielectric constant and low dissipation loss substrate material improves antenna radiation efficiency and input impedance bandwidth. The flip-chip approach does have some limitations. The flip-chip interconnection parameters, such as bump height and diameter, vary with the fabrication and cannot be accurately predicted in the design stage. The supporting under fill layer between the antenna and chip influences antenna performance. In this chapter we also investigate the effects of flip-chip interconnection on microstrip antenna performance and explore the relationship of the flip-chip interconnection parameters on bandwidth, radiation efficiency and gain. Our design methodology shows that microstrip antenna bandwidths of 15% can be achieved with careful flip-chip interconnection design and fabrication.

The second integrated antenna described is a coplanar waveguide fed antenna with wire bond interconnections. The use of bonding wires for antenna integration with a semiconductor device suits coplanar antenna structures since the bonding wires can directly connect the coplanar output pads of the device. In our approach, the coplanar bonding wires are modeled as transmission lines. This permits us to use the wire bond interconnections in the CPW-fed antenna design at mm-wave frequencies. This chapter describes the techniques that utilise wire bond interconnection in the CPW-fed patch/slot antenna design as a part of the antenna circuit avoiding the need for tuning capacitors or reconfiguration of the packaging structure. In this method reactance of the bonding wires is accounted for in the design, hence lessens the reduction of antenna matching bandwidth. The described technique permits for bonding process variation to be accounted for in the design stage. The CPW-fed antennas have a 10 dB return loss bandwidth of 10% and a measured gain varies from 6-7.8 dBi across the impedance-matched bandwidth. These results indicate that the CPW-fed slot/patch antenna type and wire bonding technology can be adequately used for integrated 60 GHz transceivers.

The techniques described in this chapter have been used to implement integrated antennas. The use of standard bonding technologies to achieve low cost packaging systems is feasible if the antenna-chip interconnections are properly accounted for in the design process and the successful integration and acceptable performance at mm-wave frequencies can be achieved using both, flip-chip or wire bond antenna integration methods.

2. Microstrip patch antennas

2.1 Patch Antenna on CMOS chip

The previous research verified the use of flip-chip technology for the integration of antenna with a CMOS transceiver at 60-GHz frequency band (Felic & Skafidas, 2008). The antenna and flip-chip interconnection configurations are seen in Fig. 1. It shows a single patch antenna and the flip-chip interconnection between the CMOS die and antenna input pads.

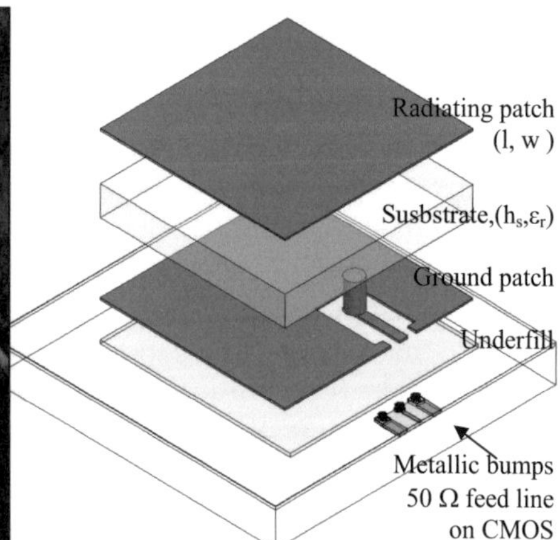

Fig. 2.1. Patch antenna structure on CMOS chip (photograph and geometry of fabricated antenna). (Parameters: l=1.65mm, w=1.67mm, ε_r=2.2, y'=0.435mm, h_s=0.254mm)

The shape of the patch is nearly square with the feed location selected along the diagonal starting at the lower right corner of the ground plane. This type of connection allows the antenna to be fabricated by using standard PCB and flip-chip bonding technology. The radiating patch is connected to the input signal by via and coplanar waveguide feeding network in the ground patch. The interconnection between the antenna and microstrip line on CMOS die consists of three gold stud bumps which connect ground and signal conductors. Gold stud bumps are placed on the CMOS die pads through a modification of the "ball bonding" process used in conventional wire bonding. In order to mechanically support the antenna structure the underfill layer is placed in the gap between the antenna and die. Figure 2.2 displays the geometry of the flip-chip antenna transition and the photograph of the fabricated prototype.

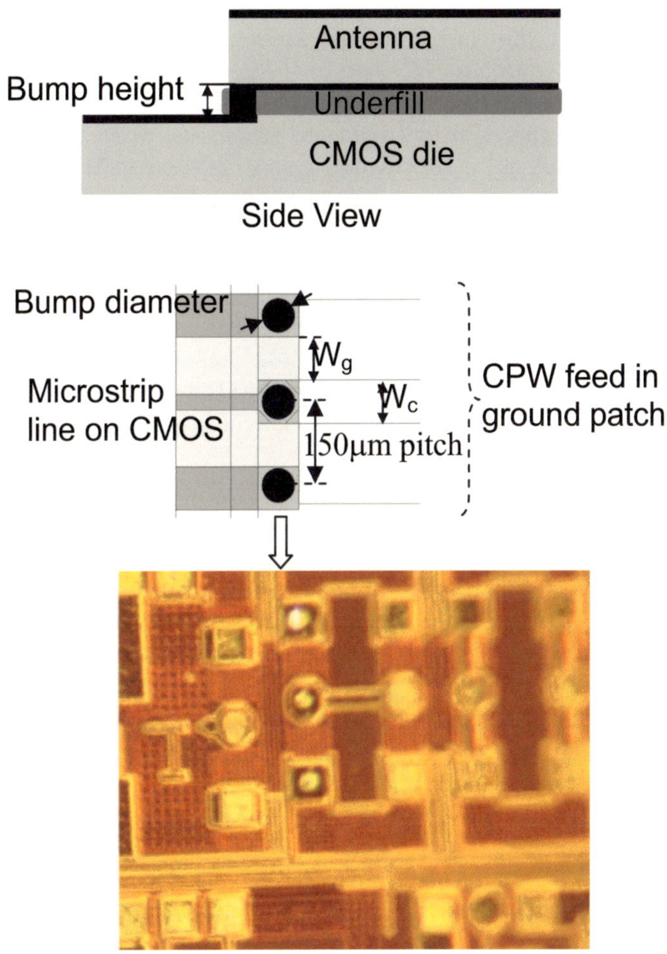

Fig. 2.2. Geometry of flip-chip transition (side and top view of the transition) and photograph of flip-chip gold bumps on CMOS chip. (Parameters: Wg=Wc=75 μm)

The optimized design of the antenna includes interconnections. The design value for bump diameter is 60 μm, the permittivity of underfill layer is 4.1 (Epotek, U300) and bump height is 20μm. However, 20μm bump height was not achievable in our manufacturing process (stud bump technology) and final fabricated bump height was 40μm (+/- 5μm tolerance). The main parameters for this study are the bump height, bump diameter and permittivity of underfill layer. Other parameters such as the bump pitch and parameters of microstrip line on CMOS are predetermined by the chosen technology and required 50-Ω output impedance of the chip design (Felic & Skafidas, 2008). Although the values of the bump pitch can be selected in the early stage of chip-package co-design, in this antenna-chip configuration the minimal bump pitch was limited by the capabilities of PCB manufacturing technology.

2.2 The effects of flip-chip parameters on antenna performance

In particular the effects of flip-chip bump parameter variations on the antenna input impedance bandwidth and gain are examined in (Felic, G. & Skafidas, S. 2009). The wide bandwidth can be achieved with very short bumps and thin underfill layer at the expense of decreased efficiency and gain. Bump diameter variations provided they are within acceptable tolerance don't affect the bandwidth, gain or efficiency. The low permittivity of underfill layer increases gain and efficiency but only slightly reduces the bandwidth. This investigation shows that standard and inexpensive flip-chip bonding process can be used to achieve antenna-die integration provided that careful consideration is given to flip-chip parameters (e.g. bump size and tolerance) during the antenna design stage. By selecting flip-chip bonding, antenna integration is compatible with standard CMOS technology. Moreover, flip chip mounting enables a low loss connection between the coplanar waveguide formed on the ground plane to a microstrip transmission line or other type of connection as may be formed upon a CMOS die upon which the antenna element is flip chip mounted.

2.3 Experimental validation

The input impedance matching was measured on a probe station using a GSG probe. Figure 2.3 shows the simulated and measured magnitude of the reflection coefficient (input impedance matching) of the fabricated antenna with flip-chip interconnection (Fig. 2.1).

In the frequency band from 60 GHz to 69 GHz, measured reflections lower than -10 dB were achieved (15% bandwidth). Simulation results show that when the bump height increases from 20 to 40 μm the -10dB bandwidth decreases by 1.5 GHz. This is due to the increased impedance (reactance) of extended bump length which increases input impedance mismatch. To fully cover the 57-66 GHz band this reactance can be compensated by redesign of the antenna with larger patch size. Figure 2.4 shows the measured gain, simulated gain and efficiency.

The gain was measured at angle of $\theta=0^{\circ}$ and $\phi=0^{\circ}$ (Fig. 2.1). The measurement set-up was calibrated using coaxial calibration kit and V-band horn antennas (Balanis, 1997). Then, one port is connected to GSG probe (calibrated at the tips) and test antenna and the other port to the horn antenna. Fabricated antenna achieves gain of 2-4 dBi inside the -10 dB impedance bandwidth at $\theta=0^{\circ}$. This level of gain is low, however, an array can be formed by a number of these antenna elements to enhance the radiation and compensate the path loss for 60-GHz radio.

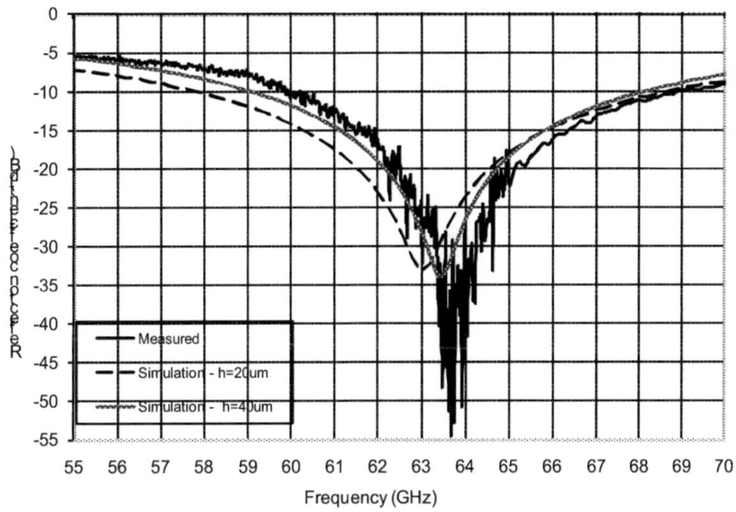

Fig. 2.3. Matching impedance performance of an integrated antenna with flip-chip interconnect. (Simulation parameters: ε_r=4.1, bump diameter=60 μm)

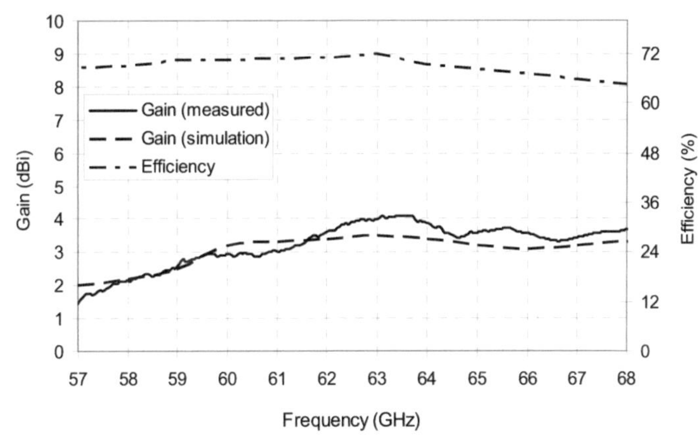

Fig. 2.4. Gain and radiation efficiency of an integrated microstrip antenna with flip-chip interconnect. (Simulation parameters: h=40 μm, bump diameter=60μm)

2.4 Patch Array on CMOS chip

As it can be determined from Fig. 2.4 the realised gain of on-chip mounted antenna is low and an array must be formed to enhance the radiation and compensate the path loss for 60-GHz radio. In order to meet the link budget of a 60GHz system over a range of 10 m using 16QAM modulation (Wicks, B. N., Ta, C. M., Zhang, F. et al., 2009) one requires an antenna gain exceeding 12dBi at both Tx and Rx. There are several ways of achieving such gains and the implications and engineering tradeoffs that come into play when designing the antenna array should be taken into account. Fig. 2.5 shows a possible solution with 2x4 antennas array on a substrate designed for on-chip mounting. The array is built on the three layers of

Taconic substrate with relative permittivity of 2.2. The patch array is designed to suit the flip-chip integration with CMOS shown in Section 2.1. The antenna elements are patch type and they can be connected to CMOS chip with via connections and metallic bumps as it is shown in Fig. 2.1. The simulated radiation pattern show that the on-chip mounted 2x4 array can achieve maximal gain of 14 dBi.

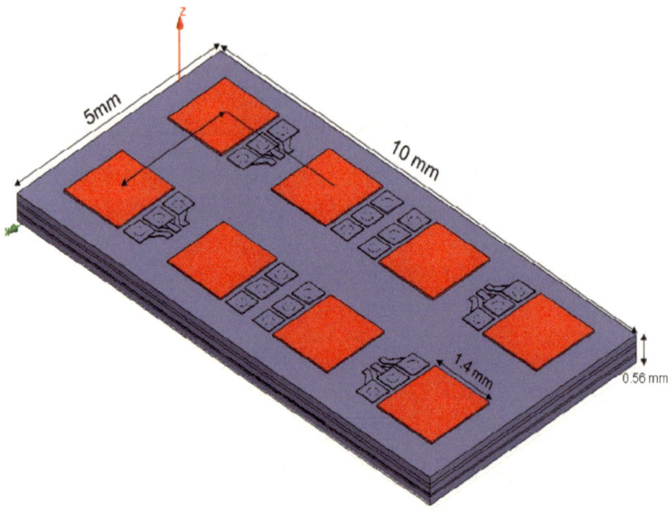

Fig. 2.5. Patch array test structure for CMOS chip mounting. (Overall dimensions: l=10 mm, w=5 mm, ε_r=2.2, h=0.56 mm)

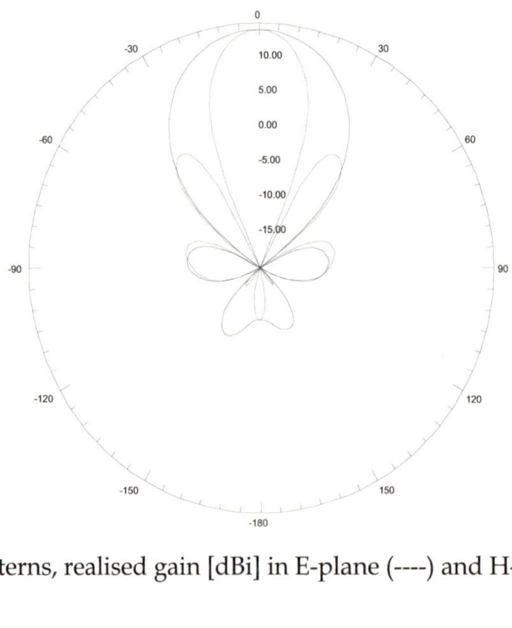

Fig. 2.6. Radiation Patterns, realised gain [dBi] in E-plane (----) and H-plane (__) at 60 GHz.

3. CPW patch antennas

3.1 The use of wire bonding for antenna integration

Bonding wire is typically employed to connect passive devices to chip modules since it is robust, inexpensive, and tolerant of chip thermal expansion. However, this integration approach has undesirable consequences of increasing impedance mismatches and power losses at mm-wave frequencies and the requirement for impedance compensation networks. To improve the performance of wire bond interconnects and enable the use of longer bonding wires the design techniques applied to mm-wave frequencies focus on two main methods; the inclusion of low pass filters on chip (Budka, 2001) and design of specific compensation networks (Karnfelt at al., 2006). The first method requires the co-design of the chip and package, and it is not suitable for connecting a radio chip to the antenna when a DC blocking capacitor is required between them. The second method includes the compensation network (impedance transformer and double stub) on the package that takes a large package area. Recently, a novel bond wire compensation structure was proposed in (Zhang et al., 2009) that uses a compensation capacitor. A series connected capacitor to the antenna tunes the inductance of the wire bond wire to a resonant condition, thus compensating the high inductance of the bond wire at the resonant frequency. This method is specifically suitable for CPW-fed coplanar antenna structures on package. The compensation structure (capacitor) is implemented on package between the wire bond pad and CPW transmission line (signal trace) at the antenna input. Although this does not enlarge the package surface area, it requires an additional dielectric substrate layer for implementation of the compensation capacitor.

The use of bonding wires for antenna integration with a semiconductor device suits coplanar antenna structures since the bonding wires can directly connect the coplanar output pads of the device to the coplanar antenna input. The other interesting aspect of coplanar bonding wires is that they can be thought of as transmission line (Goosen, 1999). This consideration opens the opportunity for exploitation of wire bond interconnection in the CPW-fed antenna design at mm-wave frequencies.

A CPW-fed loop slot antenna for integration with a mm-wave CMOS transceiver chip has been presented (Felic & Skafidas, 2009). A distinguishing feature of this design is that the chip-antenna interconnections are achieved using standard bonding wires. The designed antenna including wire bond interconnections is verified with measurements and as part of an operational 60 GHz wireless transceiver system fabricated on CMOS (Wicks et al., 2009). Although many papers have been published for mm-wave antennas, only (Zhang et al., 2009) and (Felic & Skafidas, 2009) reported the results accomplished using wire bond antenna-chip integration.

This section elaborates on the exploitation of wire bond interconnection in the CPW-fed antenna design to avoid the need for a compensation network, tuning capacitors or reconfiguration of the packaging structure. This technique is based on the transmission line model of the bonding wires and the compensation of its reactance by the antenna design. As the reactance of the bonding wires is compensated by the design, the reduction of antenna matching bandwidth is less significant.

3.2 CPW-fed patch/slot antenna

A number of papers have been published dealing with the parameters that affect the design and matching of these antennas. A brief summary follows.

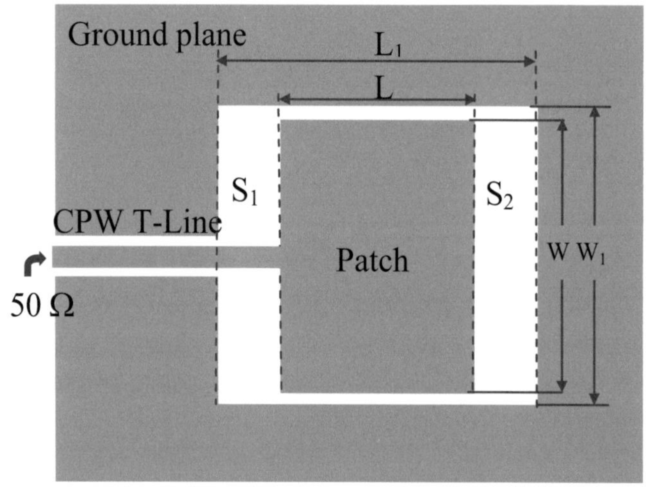

Fig. 3.1. A CPW-fed patch/slot antenna for wire bond integration (Greiser, 1976)

The original coplanar waveguide (stripline) patch/slot antenna structure was first introduced in (Greiser, 1976), Fig. 3.1. This design was based on the concept of an open ended coplanar waveguide resonator which operates in the fundamental mode, when its circumference is approximately one wavelength. However, in (Tong et al., 2004) it is reported that coplanar waveguide antenna structures behave more like microstrip patch antennas with the resonant frequency primarily determined by the patch length (which is approximately equal to half guided wavelength) and not by the circumference of the slot. A parametric study of a similar coplanar waveguide antenna structure is introduced by (Chen, 2003) to achieve wide impedance matching bandwidth. By tuning the spacing between the tuning stub and ground plane and having the length of the tuning stub equal to about half of the slot length, an impedance bandwidth of more than 50% has been achieved. This can be increased by selecting the width of the tuning stub to be about 0.59 to 0.9 times the slot length. A coplanar waveguide slot antenna with wideband feeding mechanism in the form of a radial tuning stub is also presented by (Rao, 2006). The design of this antenna structure is based on the optimal length of the radial stub protrusion which should be in the range of 0.45-0.6 times the slot's width. Further, the impedance matching is achieved by increasing the width of printed stub. In addition, the edge treatment of the stub has an impact on the bandwidth improvement. Transmission line (T-line) models for the design of microstrip patch antennas are studied and used in (Balanis, 1997) and (Watson et al., 1999). In the same way, a transmission line model for CPW-fed patch antennas is used in (Watson, 2001) to design the essential antenna elements: radiating patch, slots in ground plane and input feed line. The radiating patch is represented by a uniform length of transmission line in communicating the energy from the input edge to the distal edge of the patch. The feeding T-line element transfers the energy from the input port (50 Ω connector) to the radiating element (patch edge). It also serves as an impedance transformer since it alters the relatively low impedance of the transmission line at the input to the higher impedance at the edge of the patch. The slot region between the antenna input and the edge of the patch and the distal slot region are represented by lumped admittances (capacitance and conductance). Thus for the selected size of ground plane, good input impedance matching can be obtained by selecting the lengths of the patch and slot that enhance the coupling between the patch and ground plane. Further improvement of impedance matching can be achieved by adjusting the patch width.

3.3 Utilisation of wire bond Interconnection in antenna design

The antenna with wire bond interconnections can be designed without a wire bond compensation network if the wire bond interconnection is understood as a high impedance T-line extension of the antenna structure and is then included in the design. Therefore, the input impedance seen at 50 Ω input (source) is:

$$Z_{in} = Z_0 \frac{Z_A + jZ_0 \tan \beta l}{Z_0 + jZ_A \tan \beta l} \tag{1}$$

where Z_A is the antenna input impedance, Z_0 is the characteristic impedance of the T-line (wire bond interconnection), $\beta = 2\pi/\lambda$, and l is the length of the T-line. The assumption is that the T-line is lossless. As the overall electrical length of the antenna-feed line structure is now increased by β λ, the input impedance resonance moves to lower frequencies. The reflection coefficient at the wire bond antenna junction is:

$$\Gamma_A = \frac{Z_A - Z_0}{Z_A + Z_0} \tag{2}$$

while the overall reflection at 50 Ω source input is:

$$\Gamma = \frac{Z_{in} - 50\Omega}{Z_{in} + 50\Omega} \tag{3}$$

Under the view of conjugate matching the maximum power transfer from 50 Ω source is achieved when R_{in}=50 Ω and X_{in}=0. Therefore, for a fixed wire bond interconnection (Z_0) the reflections defined in (2) and (3) and impedance matching can be improved by adjusting the input impedance Z_{in} and antenna impedance Z_A. These impedances can can be adjusted by decreasing the length of the patch or by increasing the length of the slot. Moreover, by properly selecting the length of the patch and the length of the slot, a wide bandwidth operation can be achieved.

The antenna structure without wire bond interconnection shown in Fig. 3.1 is designed for 50 Ω input impedance matching of better than -10 dB from 57 to 66 GHz. The antenna uses a single layer of Rogers RT/Duroid 5880 substrate with relative permittivity equal to 2.24 and dissipation factor equal to 0.004 (at 60 GHz). The thickness of the dielectric layer is 508 μm. The ground plane reflector is placed at the back of the substrate to obtain a uni-directional radiation pattern. The top ground plane dimensions are 3 mm x 3 mm. The patch of length L=1 mm and width W=1.7 mm is connected to the CPW feed line. The rectangular radiating slot dimensions, W_1 x L_1 are 1.83 mm x 1.14 mm. The length of the slot between the input feed line and radiating edge, S1 is 50 μm. The CPW feed transmission line is designed with strip and gap width of 50 μm, corresponding to a characteristic impedance of Z_0= 72 Ω. The CPW strip and gap size of 50 μm matches with the coplanar wire bonds and pads on CMOS die (pitch of 100 μm). The input impedance matching results (HFSS) for the antenna (antenna without wire bond) are shown in Fig. 3.2.

The next step is to model the bonding wires for electromagnetic simulations (HFSS). The bonding wires are wedge/wedge type made of aluminium (Al), the diameter of the wires is 24.5 μm (1mil) and the length of the wire is 450 μm. The centre wire-wire spacing, s is 100 μm and substrate-wire gap, h is 30 μm. Figure 3.3 shows the approximated geometry of a wedge/wedge bond for electromagnetic simulation.

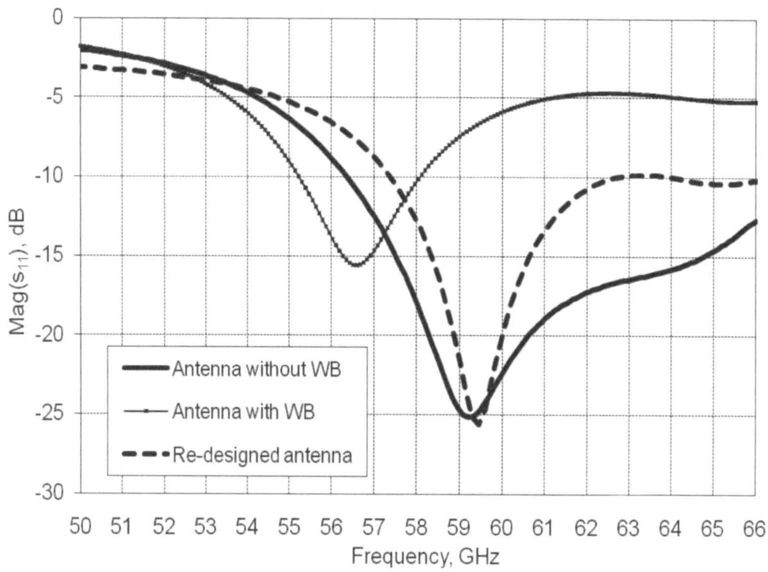

Fig. 3.2. Magnitude of reflection coefficient (simulation) for antenna without wire bond, antenna with wire bond and re-designed antenna.

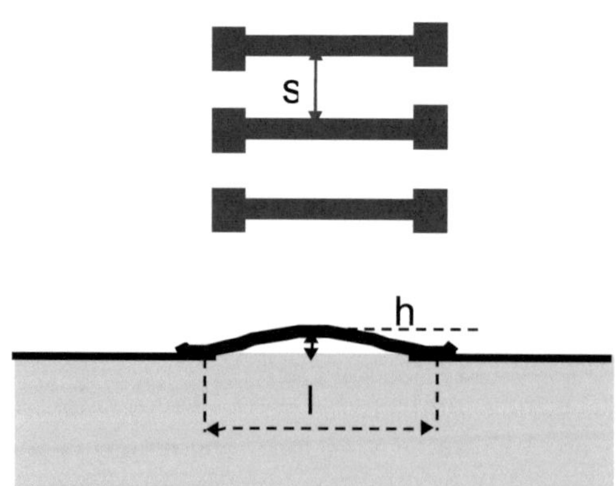

Fig. 3.3. Geometry of wire bond transition implemented for EM simulation (top and side views of the transition), s=100 μm, h=30 μm, l=450 μm.

Although this type of wire bonding, with nearly flat wires (low loop) has lower impedance than ball-wedge type wire bonds which have larger loop height, they can still be seen as an additional high impedance T-line (numerically estimated characteristic impedance is 137 Ω). The numerically estimated input impedances of the antenna with the wedge-wedge bonds and the antenna without wire bond are shown in Fig. 3.4.

After adding the wedge type wire bonds, the resonance frequency shifts from 60 GHz to around 56.5 GHz (Fig. 3.4) and the input impedance matching deteriorates.

The final step is to modify the antenna structure. The input impedance matching can be adjusted by decreasing the patch length, L. When the patch length is decreased from 1 mm to 0.9 mm the input impedance of the antenna (including wire bond) decreases and improves the 50 Ω impedance matching as is evident in Fig. 3.2 (re-designed antenna). The magnitude of S_{11} is now lower than -10 dB form 57.5-62 GHz, indicating improved matching to the 50 Ω source (CMOS chip).

It is also possible to achieve input impedance matching by increasing the length of the slot, L_1 while the patch size and all other parameters are kept constant. This decreases the capacitance of the distal slot, S2 and shifts the resonance to a higher frequency. Figure 3.5 shows the input impedance matching for the antenna without bond wires and re-designed antenna (L_1=1.2 mm).

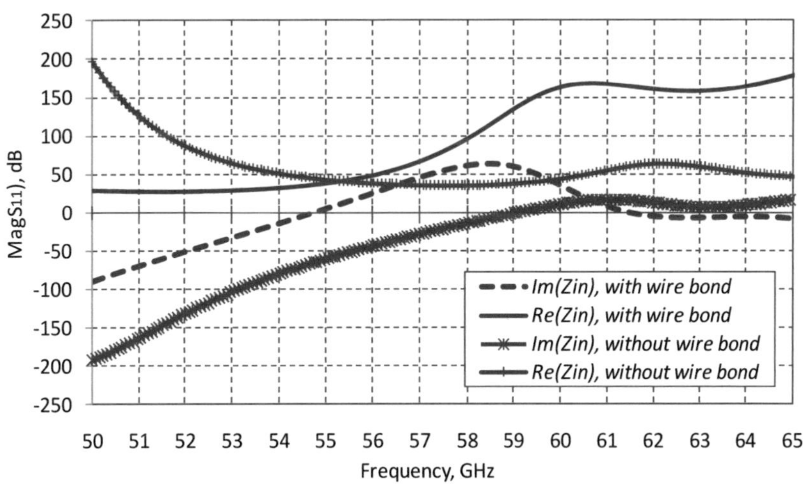

Fig. 3.4. Real and imaginary components of input impedance (antenna with and without wire bond).

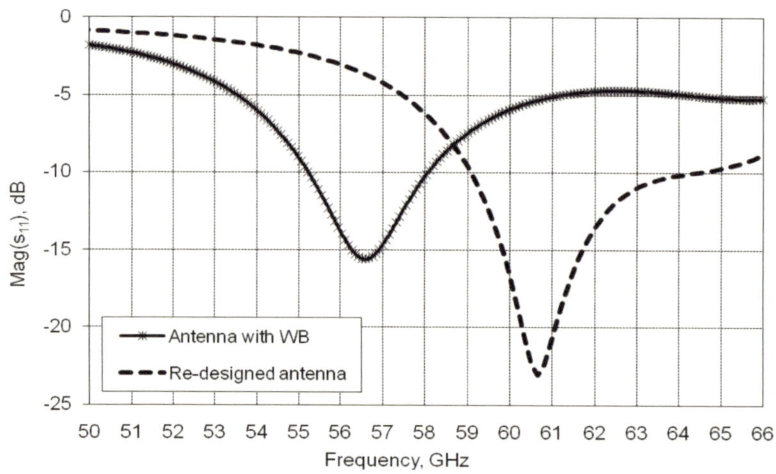

Fig. 3.5. Reflection coefficient for the antenna with wire bond and re-designed antenna.

3.5 Experimental validation

To validate the proposed antenna design technique by measurements the antenna including wire bond interconnections with an attached CPW line is fabricated (Fig. 3.6). The measurements were conducted on a probe station using coplanar microprobes (200 μm pitch). Figure 3.7 shows the simulated and measured input impedance matching. The simulations results are obtained for the prototype antenna and for the integrated antenna. In the frequency band from 57.3 GHz to 65 GHz, measured return losses lower than -10 dB were achieved (reflection coefficient S_{11} lower than -10dB). The simulation results are obtained for

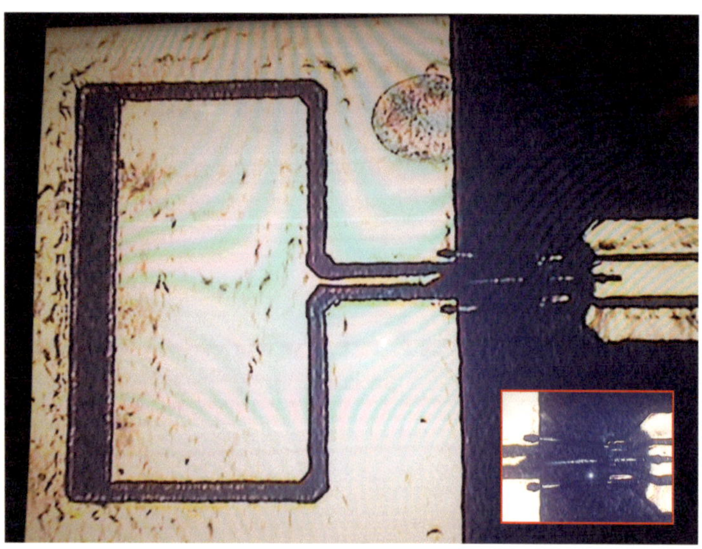

Fig. 3.6. Prototype of the antenna including wire bond interconnections.

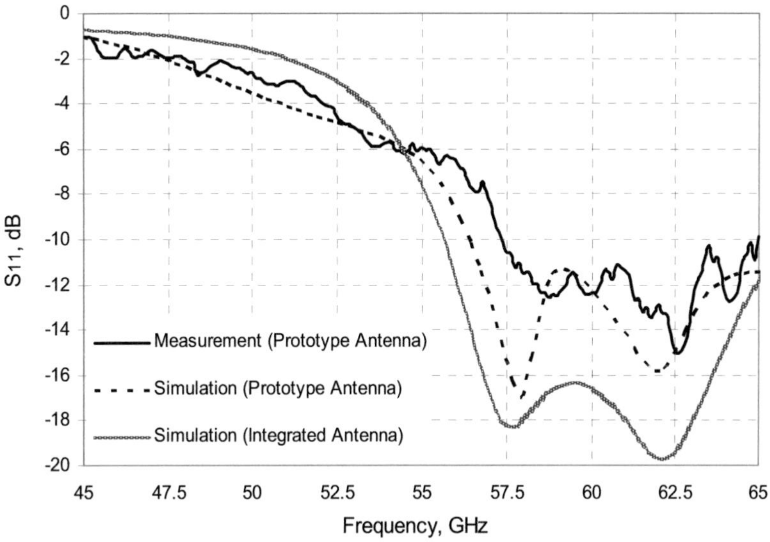

Fig. 3.7. The measured and simulated magnitude of reflection coefficient for the re-designed antenna (wire bond with s=90 μm, l=450 μm, h=30 μm).

the parameters of the fabricated sample (s=100 μm and l=450 μm). The discrepancy between the measurement and predicted results is most likely due to the variation of fabricated bond wire-wire spacing which is not uniform over entire length and less than 100 μm at the bonding point at CPW line.

The gain measurement set-up was calibrated using a V-band coaxial calibration kit and two identical WR-15 V-band horn antennas. The gain measurement followed the procedure described in (Balanis, 1997). The measured maximum antenna gain varies between 6-7.5 dBi within the -10 dB impedance matching bandwidth (Fig. 3.8). The estimated efficiency of the antenna at 60 GHz is 91%.

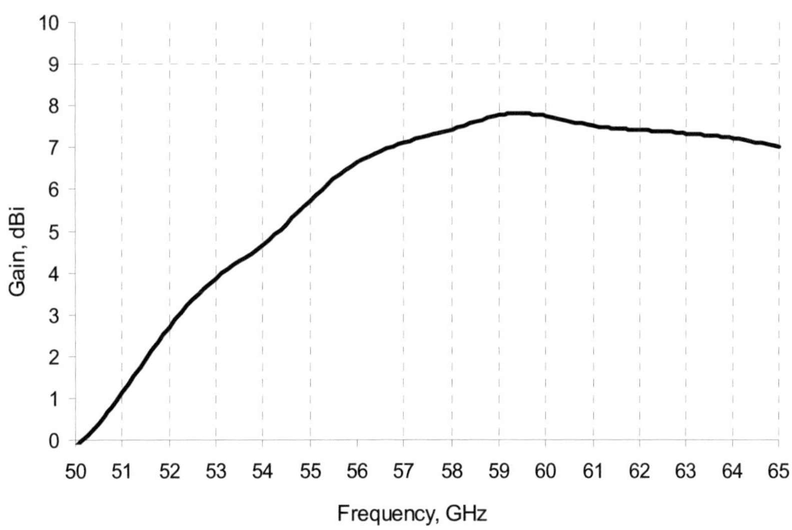

Fig. 3.8. Measured and simulated peak gain versus frequency.

3.5 Integrated antenna

Figure 3.9 shows the antenna integrated with 60-GHz direct-conversion transceiver on 130-nm CMOS. The 130-nm CMOS transceiver demonstrates very high level of integration incorporating transmitter, receiver and digital control interface on a single silicon die (Wicks, B. Et al., 2009). The successful implementation of this transceiver is an important step to move the mm-wave circuits onto the low cost CMOS chip.

As it is illustrated in Fig. 3.9, the antenna CPW input is connected to the coplanar pads on CMOS die with wedge-wedge type wire bonding. To further compensate for the inductive component of the wire bond interconnection, in the integrated configuration, an additional matching stub which is printed on the substrate under the wire bond structure is introduced and directly connected to the antenna patch. The printed stub serves to add distributed capacitance, with respect to the ground, to cancel the distributed inductance of the signal bond wires. The dimensions of the rectangular stub (230μm x 80μm) were determined with HFSS simulations. The bonding wires are made of gold (Au) and the diameter of the wires is 24.5μm (1mil). The pitch of the coplanar wires varies from 100 μm to 125μm.

The CPW-fed receiving and transmitting antennas are directly connected to the coplanar outputs of the CMOS transceiver device by using wedge-wedge type wire bonding. Additional set of wire bonds connect the ground pads on CMOS with top antenna ground plane.

The above results demonstrate the effectiveness of a design approach that understands the wire bond interconnections as extended antenna feed lines. The antenna for wire bond packaging can be designed without a wire bond compensation network. Furthermore, this work confirms the feasibility of low cost standard wire bonding technology at mm-wave frequencies, provided that the wire bonding interconnections are appropriately exploited in the design process. It also offers an easy and convenient integration for planar components on the top of the substrate.

Fig. 3.9. Micrograph of the antenna integrated with CMOS transceiver on the PCB test board.

4. Conclusion

The techniques described in this chapter have been used to integrate antennas with CMOS chip at mm-wave frequencies. The use of standard bonding technologies to achieve low cost packaging systems is feasible if the antenna-chip interconnections are properly accounted for in the design process and the successful integration and acceptable performance at mm-wave frequencies can be achieved using both, flip-chip or wire bond antenna integration methods. At mm-wave frequencies interconnections are seen as integral part of antenna design and the antenna for wire bond packaging can be designed without an additional compensation network. Although the wire bonds are robust, inexpensive, and tolerant of chip thermal expansion this integration approach requires the antenna feed (CPW-fed) loacted on the top of substarte and particular on-side connection to the CMOS chip. Furthermore, a low cost standard flip-chip technology for antenna integration with CMOS chip is more flexible since both, the coplanar and microstrip antenna feeds are available with flip-chip interconnections.

5. References

Balanis, C. (1997). *Antenna theory, analysis and design*, Wiley, Second Edition, 0-471-59268-4

Budka, T.P. (2001). Wide-bandwidth millimeter-wave bondwire interconnects. *IEEE Transactions on Microwave Theory and Techniques*, 49, 4, Apr. 2001, pp. 715-718, 0018-9480

Chen, H-D. (2003). Broadband CPW-Fed Square Slot Antennas With a Widened Tuning Stub. *IEEE Transactions on Antennas and Propagation*, Vol. 51, No. 8, August 2003, pp. 1982 - 1986, 0018-926X

Felic, G. & Skafidas, E. (2008). A microstrip antenna with flip-chip interconnect for millimeter wave transceiver on CMOS. *Proceedings of IEEE International AP-S Symposium*, San Diego, CA, July 2008.

Felic, G. & Skafidas, S. (2009). Flip-Chip Interconnection Effects on 60-GHz Microstrip Antenna Performance. *IEEE Antennas and Wireless Propagation Letters*, Vol., 8, 2009, pp. 283 - 286, 1536-1225

Felic, G. & Skafidas, E., (2009). A CPW-Fed Loop Slot Antenna for Integration with Millimeter Wave CMOS Transceiver. *European Microwave Week, EuMW2009*, Rome, Italy, September 2009

Goossen, K. W. (1999) On the Design of Coplanar Bond Wires as Transmission Lines. *IEEE Microwave and Guided Wave Letters*, Vol. 9, No. 12, 1999, pp. 511-513, 1051-8207

Greiser, J. W. (1976). Coplanar Stripline Antenna. *Microwave Journal*, Vol. 19, No. 10, pp. 47-49, October 1976.

Karnfelt, C., Hallbjorner, P., Zirath, H., & Alping, A. (2006). High Gain Active Microstrip Antenna for 60-GHz WLAN/WPAN Applications. *IEEE Transactions on Microwave Theory and Techniques*, Vol. 54, No. 6, Jun. 2006, pp. 2593-2603, 0018-9480

Rao, P. H. (2006). CPW-Fed Octave band Slot Antenna. *Microwave and optical Technology Letters*, VOL. 48, No. 12, December 2006.

Tong, K. F., Li, K., Matsui, T. & Izutsu, M. (2004). Broad-Band Double Layer Coplanar Patch Antennas with Adjustable CPW Feeding Structure. *IEEE Transactions on Antennas and Propagation*, Vol. 52, No. 11, pp. 3153-3156, Nov 2004.

Watson, P. M., Creech, G. L. & Gupta, K.C. (1999). Knowledge Based EM-ANN Models for the Design of Wide Bandwidth CPW Patch/Slot Antennas. *IEEE AP Symposium* 1999, Vol. 4, 11-16 July 1999, pp-2588-2591.

Watson, P. M. & Gupta, K.C. (2001). Broadband Patch/Slot Antenna. United States Patent, US 6,198,437 B1, Mar. 6, 2001.

Wicks, B. N., Ta, C. M., Zhang, F. et al. (2009). 60-GHz Direct-Conversion Transceiver on 130-nm CMOS with Integrated Digital Control Interface. *European Microwave Week, EuMW2009*, Rome, Italy, September, 2009

Zhang, Y.P., Sun, M., Chua, K.M., Wai, L.L. & Liu D. (2009) Antenna-in-Package Design for Wirebond Interconnection to Highly Integrated 60-GHz Radios. *IEEE Transactions on Antennas and Propagation*, Vol. 57, No. 10, 2009, pp. 2842-2852

21

A Practical Guide to 3D Electromagnetic Software Tools

Guy A. E. Vandenbosch and Alexander Vasylchenko[1]
Katholieke Universiteit Leuven
Belgium

1. Introduction

This chapter considers the numerical analysis of planar antennas. First the fundamental theoretical techniques widely used in the general area of computational electromagnetics are discussed. The focus is on the specific case of planar antennas. Second, several simulation tools implementing these techniques are overviewed. Both commercial and non-commercial tools are considered. Examples found in literature are given. Third, three planar antennas are benchmarked, using a variety of tools, in order to show the reader the quality and matureness of the existing tools, and to prove that the analysis of planar antennas is by no means always straightforward. Remaining challenges that still need to be faced and "missing links" are identified, and suggestions are given for the future.
In the strictest sense a planar antenna means an antenna flat compared to the wavelength. Although the height of the antennas considered is different from traditional antennas, in most cases no special modeling techniques or software tools are used for planar antennas. Almost all planar antennas reported in literature have been designed / analyzed with the well-known techniques and (commercial) software packages. This means that in this chapter the major numerical techniques and software tools used can be overviewed in a general sense, referring to standard literature. These techniques will not be derived or explained here in detail. In stead, this chapter focuses on those aspects that come into the picture when the antenna is planar.
After the section on techniques and tools, this chapter will focus on the performance of these techniques and tools for planar antennas. This is done through an overview of benchmarks available in literature.

2. Modeling techniques

In this section the full wave solvers are introduced and categorized on the basis of their solution method: Integral Equations (IE) solved by Method of Moments (MoM), Finite Elements (FE), Finite Differences in the Time Domain (FDTD), and Finite Integration Technique (FIT). Based on their theoretical specificities, the application of each method in

---

[1] Based on "Quality assessment of computational techniques and software tools for planar antenna analysis", by A. Vasylchenko, Y. Schols, W. De Raedt, and G. A. E. Vandenbosch which appeared in IEEE Antennas Propagat. Magazine, Vol, 51, No. 1, pp. 23-38, Feb. 2009.

the case of planar antennas is discussed. A history and a comprehensive overview of the different numerical techniques and of their application in computational electromagnetics (CEM) may be found in [1]. Recent work on the last developments in CEM [2] concentrates on the two main approaches of differential and integral methods. A Good review and perspectives concerning the relationship between differential and integral equations (IE) modeling is recommended in a review paper by Miller [3]. A thorough discussion on the different techniques with clarifying examples is also given in [4].

2.1 Integral equation techniques

Integral equation methods make use of Maxwell's equations in integral equation form to formulate the electromagnetic problem in terms of unknown currents flowing on the object to be described. The kernel of the integrals, or in other words the coupling between these currents and the fields, is formed by a Green's function tensor which includes the electromagnetic influence of the complete infinite so-called "background" environment, such that the problem formulation automatically covers the entire surrounding space without making any fundamental approximations. As a consequence, the corresponding solution is automatically valid in every point of the background medium. Far field radiation phenomena, surface waves in layered structures, etc., that are vital for efficient and accurate antenna analysis, are analytically included in the solution.

By straightforward application of the classic Method of Moments (MoM), as introduced in the basic book [5], boundary equations expressing the physical nature of the object to be described (conductivity on a conductor, permittivity in a dielectric part of the object), are enforced. This is either done at the boundaries of volumes utilizing Surface Integral Equations (SIE) [6] or inside the entire volumes themselves, applying Volume Integral Equations (VIE) [7] on the inside of the components. IE-MoM gives rise to a dense matrix equation, which can be solved using standard matrix algebra technology.

IE-MoM is normally applied in the frequency domain, i.e. for time-harmonic problems. This means that the solution has to be determined at each frequency of interest.

Intrinsically, IE formulations are very powerful, since they inherently minimize the number of unknowns for a certain problem. IE can be formulated also very general. Volumetric MoM for example in principle can describe any arbitrary structure, involving the most complex features. However, the resulting dense MoM matrix invokes a large memory demand. Also, there is a lack of matrix solvers comparable in efficiency to the solvers used in the differential equation techniques, which yield sparse matrices. This has precluded the use of volumetric MoM for modeling complex structures on widespread computer systems.

In general, it can be stated that state-of-the-art IE-MoM-based solvers are more efficient for open region problems, possibly involving multiple layers, involving impenetrable or homogeneous objects, since IE-MoM satisfies the radiation condition automatically through the Green's function.

It has to be noted that the integral equations can be written down in different forms (dyadic form, mixed-potential form, hybrid forms, etc. [8]), which give rise to specific implementations, as found in the different commercial solvers.

In the following sections, several aspects linked with the use of integral equation techniques for planar antennas are discussed.

2.1.1 Occurrence of layer structures

As explained above, IE technique solvers are formulated making use of Green's functions. These Green's functions can be formulated for multi-layered structures, such that the background medium of the antenna may consist of an arbitrary number of horizontal, infinitely stretched, dielectric and metallic layers, which are taken into account analytically. This environment is particularly interesting for the numerous families of planar antennas and circuitry designs, because the horizontally large components such as dielectric substrates or conducting ground planes can be contained in the environment. The only remaining components are local scattering objects with medium or small dimensions such as radiating patches, apertures, or volumetric inhomogeneities. These components are replaced with equivalent surface/volume currents, which appear as the primary unknowns in the resulting integral equations.

However, in many planar antennas extended layer structures do not occur, and the computational advantages linked to the use of IE for this type of structure in many cases simply are irrelevant. Only in cases where planar antennas are used for example on larger PCB structures, the PCB substrate may be modeled efficiently by IE-MoM [9]. Nevertheless, even then, in the benchmarking section, it will be shown that solvers based on IE techniques are in many cases still very fast.

2.1.2 Complex environment

In case the environment is really complex, with inhomogeneous lossy dielectrics (e.g. biological tissue, [10]), the IE-MoM technique is inferior to the differential equation techniques discussed in the next section. The reason is that, in case a pure IE-MoM technique is used, the inhomogeneity of the dielectrics of the environment has to be described by Volume Integral Equations, leading to a number of unknowns proportional to the size of the volume of the object + environment. Even if the number of unknowns in these cases is still below the number of unknowns for differential equation techniques, the dense coupling matrix of the IE-MoM technique requires much higher computational resources and in practice prohibits its use.

This drawback may be partially overcome by developing special techniques, taking into account the specificities of different regions [11], but their implementation and application in the case of planar antennas is still in its infancy.

2.2 Differential equation techniques

Differential equation methods are derived directly from Maxwell's curl equations or the Helmholtz wave equations with little analytical preprocessing. The most popular differential equation-based methods are the Finite Element Method (FEM), for example utilized in Ansoft's HFSS software package, and the Finite-Difference Time Domain method (FDTD), which is employed for example by CST's Time Domain transient solver (in the particular case of Cartesian grids).

Since the number of unknowns is proportional to the volume and the resolution considered, differential equation methods are particularly suitable for modeling small full three-dimensional volumes that have complex geometrical details, for example smaller closed-region problems involving inhomogeneous media [12]. Intrinsically, differential equations are less suited for open problems. The reason is that in principle they require a discretization of the entire space under consideration. This space is limited in case of closed problems, but corresponds to infinite space in case of open problems. In practice, this problem is solved by the introduction of techniques like Absorbing Boundary Conditions, and Perfectly Matched Layers (PML) [13]. The quality of these truncating techniques nowadays is so high that, in practice, this fundamental problem has been overcome, albeit it in an approximate numerical way.

Since many techniques to make planar antennas small involve complex topological 3D shaping, the use of non-homogeneous dielectrics, etc., differential equation techniques have a clear inherent advantage for these structures over integral equation techniques.

2.2.1 The Finite Element Method (FEM) [13], [14]

FEM is a method based on solving partial differential equations. It is most commonly formulated based on a variational expression. It subdivides space in elements, for example tetrahedra. Fields inside these elements are expressed in terms of a number of basic functions, for example polynomials. These expressions are inserted into the functional of the equations, and the variation of the functional is made zero. This yields a matrix eigenvalue equation whose solution yields the fields at the nodes.

Its first formulations were developed as matrix methods for structural mechanics. This lead to the idea to approximate solids and Courant (1942) introduced an assembly of triangular elements and the minimum of potential energy to torsion problems [15]. The first paper on the application of FEM to electrical problems appeared in 1968 [16]. An extensive review on the history of FEM in electromagnetics was published in an issue of the Antennas and Propagation Magazine [17]. FEM normally is formulated in the frequency domain, i.e. for time-harmonic problems. This means that, as for IE-MoM, the solution has to be calculated for every frequency of interest.

Numerous references can be given developing, explaining, and using FEM. A good book to start with is [14]. A software tool using FEM and very widely spread is Ansoft HFSS.

2.2.2 The Finite-Difference Time-Domain technique (FDTD) [18], [19], [20]

The nature of Maxwell's differential equations is that the time derivative of the H-field is dependent on the curl of the E-field, and the time derivative of the H-field is dependent on the curl of the E-field. These basic properties result in the core FDTD time-stepping relation that, at any point in space, an updated value of an E/H-field in time is dependent on the stored value of the E/H-field and the numerical curl of the local distribution of the H/E-field in space. The numerical translation into a time-stepping algorithm was introduced by Yee in 1966. Indeed, swapping between E-field and H-field updates allows to define a marching-on-in-time process wherein sampled fields of the continuous electromagnetic waves under consideration are used. These waves can be seen to propagate in the Yee lattice, a numerical three-dimensional space lattice comprised of a multiplicity of Yee cells. More specifically, Yee proposed a leapfrog scheme for marching-on in time wherein the E-field and H-field updates are staggered so that E-field updates are observed midway during each time-step between successive H-field updates, and vice versa. Positive is that this explicit time-stepping scheme avoids the need to solve simultaneous equations. It also yields dissipation-free numerical wave propagation. Negative is that this scheme results in an upper bound on the time-step to ensure numerical stability. This means that simulations may require many thousands of time-steps for completion. The use of the Yee lattice has proven to be very robust in numerical calculations.

FDTD is extremely versatile since the interaction of an electromagnetic wave with matter can be mapped into the space lattice by assigning appropriate values of permittivity to each electric field component, and permeability to each magnetic field component. This can be done without seriously compromising the speed of the method.

The fact that time is observed directly, and the versatility of the method, make it the most efficient technique for complex 3D transient problems. A very efficient implementation of the FDTD technique can be found in the commercial tool Empire [21].

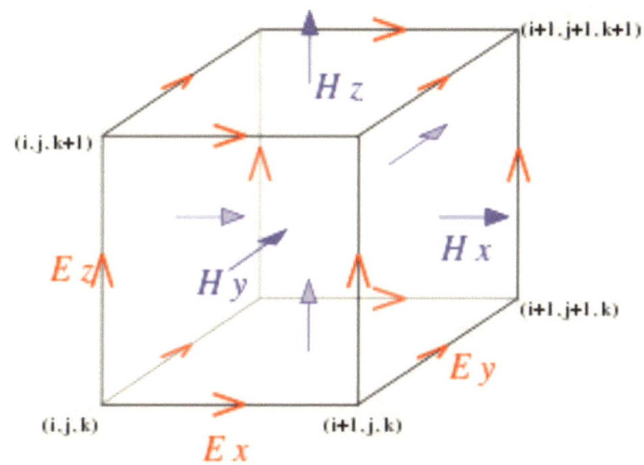

Fig. 1. Standard Cartesian Yee cell used in the FDTD technique.

2.2.2 The Transmission Line Matrix method (TLM) [22]

The TLM method is a special Finite-Difference Time-Domain technique. It is based on elementary parallelepiped cells obtained with an interconnected transmission lines network. The propagation and scattering of voltage impulses through this network is equivalent to the propagation of electromagnetic fields in arbitrary media. At each time step, a scattering matrix associated to each cell computes the reflected voltage impulses from the incident ones. This reflected voltage impulses become incident voltage impulses on the six neighboring nodes for the next time step. The six electromagnetic fields are directly related to the incident voltage impulses at the center of the cell. Capacitive, inductive and resistive stubs connected to the nodes allow to take into account variable cell sizes, permittivity, permeability, electric and magnetic conductivities. Free space is simulated by using a modified TLM node allowing implementation of PML.

2.2.2 The Finite Integration Technique (FIT) [23]

The Finite Integration Technique was introduced by Weiland in 1977. The word integration does not imply any relation with integral equations. FIT first describes Maxwell's equations on a grid space. The matrix equations for the electromagnetic integral quantities obtained by FIT possess some inherent properties of Maxwell's equations, for example with respect to charge and energy conservation. This makes them very attractive from a theoretical point of view. FIT can be formulated on different kinds of grids, e.g. Cartesian or general non-orthogonal ones, which is a clear advantage. In the time-domain, which is our field of interest, the resulting discrete grid equations of FIT are, at least in "some" cases, identical to the discrete equations derived with the classical Finite-Difference Time-Domain (FDTD) method. In contrast to FIT, which is applied to the integral form of the field equations, FDTD (as a subset of the finite integration method) is applied to the differential form of the governing Maxwell curl equations. Theoretical links between FDTD and the FIT approach realized in CST may be found in [24]. A comparative study on Time Domain methods was presented in [25]. In some sense, FIT can thus be considered as a powerful generalization of the FDTD technique. A software tool using FIT and very widely spread is CST Microwave Studio.

3. Software tools

In the following sections, several solvers are briefly described with more specific information on the implemented solution method. Commercial solvers and academic solvers are considered. For some of the commercial solvers an application section is given, where references to literature can be found. In these papers designs based on these solvers are reported. It has to be emphasized that by no means this overview is complete. The references given are just mere illustrations. A wealth of information concerning this issue can be found in the IEEE Antennas and Wireless propagation Letters, where over the years numerous designs of planar and integrated antennas can be found.

3.1 Commercial software tools

Since the cost of a commercial solver in many cases is high, the choice of the commercial solvers considered was based on their availability to the authors, directly or through cooperation with others. The fact that a specific solver is missing does not represent any statement about its quality. Nevertheless, the authors believe that the solvers selected do represent the global landscape in this area.
The IE solvers are roughly ordered according to increasing complexity of the geometries that can be handled and in addition to that, meshing abilities are clarified for every software package.

3.1.1 Momentum [26]: IE-MoM

HP-Momentum is the IE-MoM solver integrated within the ADS system of Agilent. The integral equations are formulated in mixed potential form and the matrix elements are evaluated completely in the spatial domain. Momentum was originally developed to analyze planar circuitry. The latest version of Momentum can model vertical currents per layer as well as horizontal side-currents on thick conductors. A combination of rectangles and triangles represent cells of Momentum's mesh grid. The mesh reduction option allows the shape of basis functions to incorporate some of the physics of the design and the mesh cell can be any polygon compounded by rectangles and triangles. Nevertheless, the modeling of finite dielectric volumes is not included, which limits the modeling capabilities for full 3D structures. These limitations can be resolved by running EMDS (the FE based solver included into the ADS package) without having to run a stand-alone tool. Mesh frequency and number of cells per wavelength are used for determining the mesh density of the entire circuit or any single object. The edge meshing option adds a relatively dense mesh along the edges of objects. Since the current density is higher along the edges of objects, the edge mesh can improve the accuracy and speed of a simulation.

3.1.2 IE3D [27]: IE-MoM

The integral equations are formulated with a full dyadic Green's function and the matrix elements are computed completely numerical in the spatial domain. IE3D can model truly arbitrary 3D metal structures. Since 2006 also finite 3D dielectric volumes can be modeled with a VIE approach (Volume Integral Equation). IE3D performs automatic generation of a

non-uniform mesh with both rectangular and triangular cells. The user can control the highest meshing frequency and the number of cells per wavelength. An automatic edge cell feature is available for accurate simulation of the currents concentrated near the edges of metallic surfaces.
IE3D has been successfully used in the design of small antennas for mobile phones. Specific topologies can be found in [28], [29], [30], [31], [32]. In [30] for example, an internal dual-band patch antenna is described. The antenna is simulated and measured and the agreement is excellent for this topology. An external multi-band antenna can be found in [33]. In general, impedance, radiation patterns, and radiation efficiency seem to be well predicted.

3.1.3 FEKO [34]: IE-MoM, in combination with other techniques

FEKO is based on the IE-MoM method, which can be combined with other techniques, like the geometric optics approach (GO), the unified theory of diffraction (UTD), and the multilevel fast multimode method (MLFMM). GO and UTD are at the moment the only practical approaches for solving a class of very large problems, the size of which exceeds the handling capabilities of MoM, FEM or FDTD [35]. The matrix elements are computed using a mixed-potential formulation and a spatial domain approach. The solver can model truly arbitrary 3D structures. Dielectric volumes can be modeled in three different ways: with a SIE approach, with a VIE approach, or with a hybrid approach with the FE method. The surface of the structure is discretized using a triangular mesh, while tetrahedra are used for volumetric discretization. In order to allow flexible control of the mesh, a user can specify different cell dimensions for any selected region, face or edge.

3.1.4 HFSS [36]: FEM

Since it was one of the first tools in the market, and also due to its generality and flexibility, HFFS is one of the tools heavily used in industrial design environments. The purpose of HFSS is to extract parasitic parameters (S, Y, Z), visualize 3D electromagnetic fields (near- and far-field), and generate SPICE models, all based on a 3D FEM solution of the electromagnetic topology under consideration. Very useful features of HFSS are its automatic adaptive mesh generation and refinement, which in most cases frees the designer of worrying about which mesh/grid to choose.
This software is extremely popular and is used for all kinds of purposes. Specific results for small planar antenna topologies can be found in [37], [38], [39], [40], [41], [42]. Input impedance and radiation patterns are generally predicted very well. Few results are found about the efficiency.

3.1.5 Empire [21]: FDTD

Empire XCcel is based on the FDTD technique. Due to adaptive on-the-fly code generation it comes with a highly accelerated kernel, providing very fast simulations. It features the Perfect Geometry Approximation (PGA) algorithm to yield more accurate results for curved structures, frequency dependent loss calculation and special algorithms for modeling thin conducting sheets. Several structure import and export formats are supported. EMPIRE XCcel's applicability ranges from analyzing planar, multi-layered and conformal circuits, components and antennas to multi-pin packages, waveguides, and SI/EMC problems including the device's operational environment. Time signals, scattering parameters, and field animations are generated for a broad frequency range within only one simulation run.

Monitoring and animation capabilities give physical insight into electromagnetic wave phenomena while accurate results are obtained with little effort.

3.1.6 CST [43]: FIT

CST Microwave Studio (CST MWS) is based on the finite integration technique (FIT). It allows to choose the time domain as well as the frequency domain approach. Despite the presence of transient, eigenmode, and frequency domain solvers within CST MWS, the transient solver was examined for benchmarking further in this chapter as the flag ship module of CST MWS. The Time Domain Solver calculates the broadband behavior of electromagnetic devices in one simulation run with an arbitrarily fine frequency resolution. The modeling of curved structures using the Perfect Boundary Approximation® technique and the modeling of thin perfectly electric conducting sheets with the Thin Sheet Technique® tries to cope with the typical difficulties inherent to classical FDTD methods. The transient analysis of the proposed antennas is done utilizing the hexahedral mesh type. The automatic mesh generator detects the important points inside the structure (fixpoints) and locates mesh nodes there. The user can manually add fixpoints on a structure, as well as fully control the number of mesh lines in each coordinate with regards to the specified wavelength. Energy based adaptation of the mesh allows to refine it in a predefined number of passes, providing a mesh refinement of sophisticated design features for the price of longer overall simulation time. The analyses in this chapter use automatic direct meshing without any local settings. Although the FIT in principle can handle material parameters changing over the dielectric volumes defined, this is not implemented yet.

CST, as a general purpose software package being a real competitor for HFSS, has gained popularity in the last few years. Also for the analysis and design of planar and small antennas, more and more results obtained with CST can be found in literature, for example in [70]. A problem sometimes observed with CST is a ripple in the frequency response in case the tool settings are not appropriate. This is due to the fact that the flagship of CST is inherently a time domain solver.

3.2 Non-commercial software tools

3.2.1 MAGMAS 3D [44], [45] [46], and [47]: IE-MoM

MAGMAS 3D is the IE-MoM code developed at the Katholieke Universiteit Leuven, Belgium. It was developed in cooperation with the European Space Agency for 2D, 2.5D and quasi-3D structures. Specific in comparison with other MoM codes is that the matrix elements are computed using a hybrid dyadic-mixed potential formulation and a combined spectral-space domain approach. This allows to perform a large part of the computation procedure for these matrix elements analytically in the spectral domain. This makes the code computationally more efficient. Surface and volume currents are decomposed in horizontal and vertical currents (= quasi 3D approximation), which are both expanded using generalized rooftop functions. A full mesh control of combined rectangular and triangular mesh cells is available in manual meshing mode. Exact coordinates and dimensions can be set for every single mesh cell. A Graphical User Interface is available.

4. Benchmarking

Nowadays, antenna engineers rely heavily on highly specialized full wave electromagnetic (EM) field solvers to develop and optimize their designs. Computer-aided analysis and optimization have replaced the design process of iterative experimental modifications of the initial design [48]. The operation of these EM solvers is based on the numerical solution of Maxwell's equations in differential or integral form. It is evident that the underlying solution method significantly influences the efficiency and accuracy by which certain structure types are analyzed. Nevertheless, the commercial focus increasingly switches from such key theoretical considerations to improvements in the area of layout tools and system-level design tools. Therefore, users may get the wrong impression that a given solver is automatically suited to solve any kind of problem with arbitrary precision. This is of course not true.

This section verifies the plausibility of such expectations by presenting an extensive benchmark study for planar antennas which focuses on the capabilities and limitations of the applied EM modeling theories that usually remain hidden for the antenna designer. Many challenging problems for EM solvers, which turn on the fundamental capabilities of modeling techniques, were highlighted in papers in open literature [49].

To ensure a well-founded result, the scope of the investigation is extensive: different commercial packages and a home-made academic solver are used to analyze very different but characteristic planar antenna structures, which are subsequently manufactured and measured. This work also leads to establishing guiding rules to assist users in choosing the appropriate simulation software for a specific antenna design. It is important to note that only theoretical considerations will be taken into account: the benchmark study does not include the comparison of Graphical User Interfaces (GUI), price/performance ratio, user friendliness, etc.

To our knowledge, such an extensive benchmark study over a large variety of solvers and for several structures was only reported in [50], which is the core paper providing most of the basic material for this section. In [51] and [52] a comparable analysis was conducted using four different solvers, but only for one structure. An accuracy and speed benchmark on six EM solvers was presented in [53] for four different antennas, but the physical link between simulation results for a specific antenna problem and the applied solution method has not yet been generally established. [54] and [55] present comparison of commercial EM solvers with measurements in predicting S-parameters for resonating structures.

Each method described above is represented in the benchmark by at least one widely known commercial solver. Due to the much higher level of analytical pre-processing for the IE-MoM solvers, they show a huge range of possibly implemented algorithms. That is why in this benchmark the differential solvers (FE and FDTD) are outnumbered by the IE-MoM solvers.

Since commercial solvers are often very expensive, although they are often available with substantial discounts to academic institutions, their availability is limited. Apart from the academic solver MAGMAS 3D, the other IE solvers included in this benchmark are IE3D of Zeland Software, FEKO of EM Software & Systems, and ADS Momentum of Agilent. Three of the leading differential EM tools: HFSS of Ansoft for the FE method, Empire and CST Microwave Studio for the FDTD method were also introduced. Good reviews of these CEM techniques and solvers may be found in [56]. The versions used for the respective software tools are given in Table 6.

4.1 Antenna structures

4.1.1 Line fed planar patch with local dielectric inhomogeneity

The antenna depicted in Fig. 2 is ideally suited to serve as a first reference benchmark structure. It is a classical, well-known structure. It is the basis for many small planar antennas on which to apply a single or a combination of miniaturizing techniques (PIFA, S and E antenna, H antenna, patches with slots, etc.). It contains elements that are naturally unsuited for both the differential and the integral equation solver families.

The structure consists of a line-fed patch antenna, situated on an infinitely stretched, grounded dielectric substrate of thickness d = 1.52 mm and permittivity $\varepsilon = (3.38 - j0.0074)\ \varepsilon_0$. A local dielectric inhomogeneity with permittivity ε_d is added to the infinitely stretched substrate in the form of a finite dielectric volume under the patch. Such modifications to the substrate structure can be used to improve the radiation pattern [57], to control the surface wave coupling with other elements [58], or to miniaturize the patch. The square patch (W = L = 15.2 mm, H = 5.07 mm) is symmetrically fed by a simple microstrip line of which the dimensions L_1 = 30.4 mm, L_2 = 5.7 mm and W_s = 3.8 mm are indicated in Fig. 2.
The antenna is analyzed on a completely homogeneous substrate, i.e. for $\varepsilon_d = \varepsilon = (3.38 - j0.0074)\ \varepsilon_0$, and also analyzed on an inhomogeneous substrate, with a dielectric volume of $\varepsilon_d = (3.00 - j0.0039)\varepsilon_0$ included under the patch.

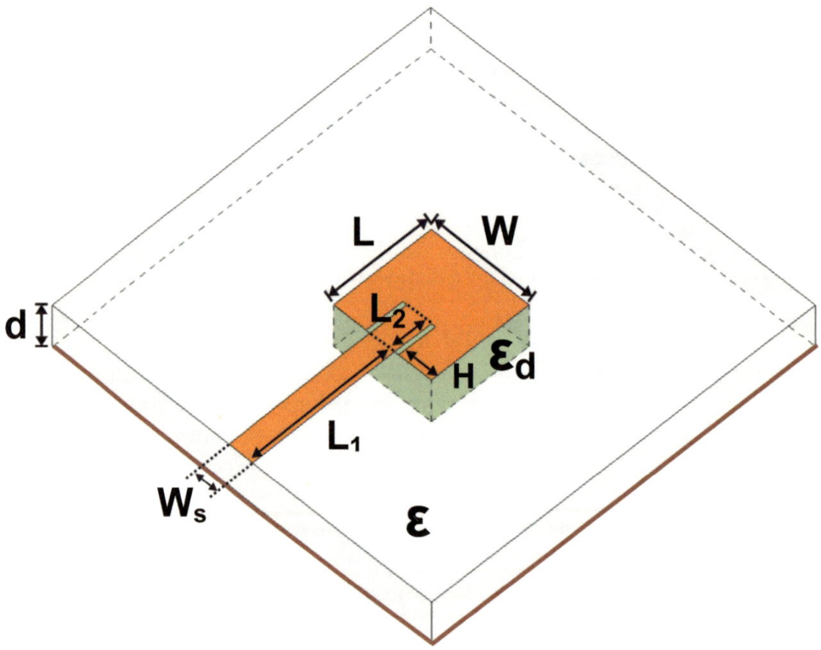

Fig. 2. Line-fed patch antenna on an inhomogeneous substrate.

4.1.2 UWB antenna

Small-size, low-cost planar UWB antennas have recently received considerable interest due to the new release of the frequency band from 3.1 to 10.6 GHz for short-range low-power communication. The Printed Tapered Monopole Antenna (PTMA), see Fig. 3, is a typical representative of contemporary UWB designs [59]. The structure with maximum dimensions of 25.0 x 28.5 x 1.27 mm^3 (w x l x h) consists of a tapered radiating element fed by a microstrip line. In order to obtain broad band behavior two slots are added in the radiating element and in the ground plane, respectively. All metal parts are simulated as perfect electric conductors. The antenna prototype was etched on a piece of PCB (RO3210) with ε_p = 10.2 (tanδ=0.0027). The dimensions are: W_{UWB} = 24 mm, L_{UWB} = 20 mm, S= 1 mm, l_a = 18 mm, and w_t = 4 mm. Fig. 4 shows the configuration of the microstrip feeding with W_f = 2 mm, W_m =3 mm, $L_{m1} = L_{m2}$ = 4 mm, W_c = 4 mm, L_c = 4 mm, and L_{gr} = 8 mm. The UWB antenna topology has a symmetrical design. Hence, a symmetry plane can be defined, in order to reduce the simulation time.

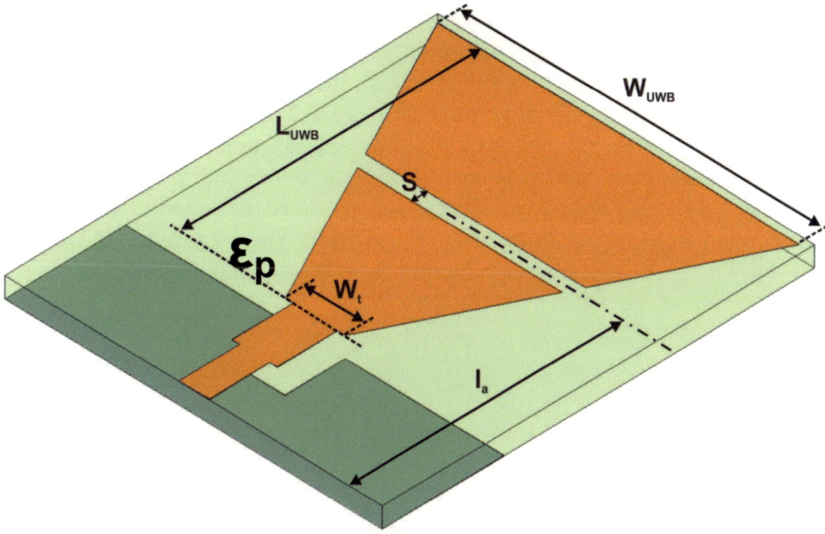

Fig. 3. UWB monopole antenna topology.

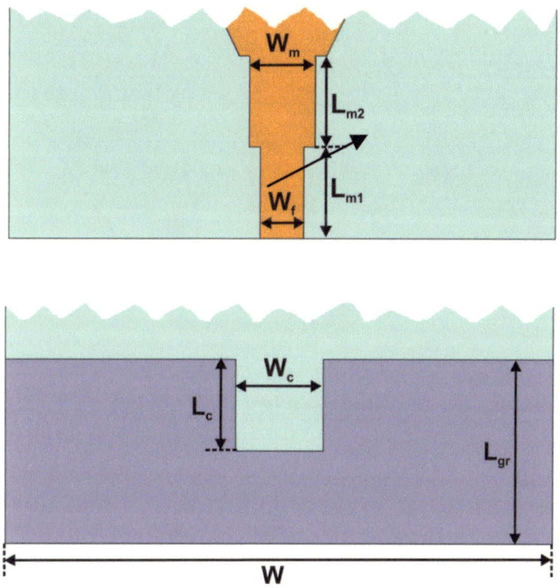

Fig. 4. UWB monopole antenna feeding system, top and bottom view.

4.1.3 Folded monopole antenna for GSM application

The RangeStar Ultima™ 'World GSM antenna' (P/N 100709), covering the three GSM frequency bands, is an example of a planar monopole antenna. Its specifications indicate a 2.5:1 VSWR in the frequency bands 880–960 MHz and 1710–1990 MHz. The substrate is assumed to be a typical PCB material such as FR4, with dielectric permittivity ε_g = 4.4 and tanδ=0.02. An integrated planar test version of this antenna using monopoles, ground plane and coplanar feeding line printed on the same interface was manufactured. The test PCB has outer dimensions 37.6 x 46.275 mm^2. The substrate thickness is d = 1.57 mm. A sketch of the full topology is given in Fig. 5. The antenna (represented in orange) with the outer dimensions 36.88 x 9.42 mm^2 (W_{GSM} x L_{GSM}) is fed in the center. The feeding CPW conductor, which is separated from the pair of ground planes (15.2 x 32.2 mm^2 (W_{GP1} x L_{GP}) for GP1, and 19.541 x 32.2 mm^2 (W_{GP2} x L_{GP}) for GP2) with S = 0.218 mm width for the slots, has W_s = 1.7 mm width. The left-hand part is a short monopole which mainly resonates in the upper frequency band, while the right-hand part is longer and resonates in the lower frequency band. On the lower right of the antenna there is a short circuit to match the antenna to a 50 Ω feed. The antenna is at a distance G = 2.2 mm from the ground planes.

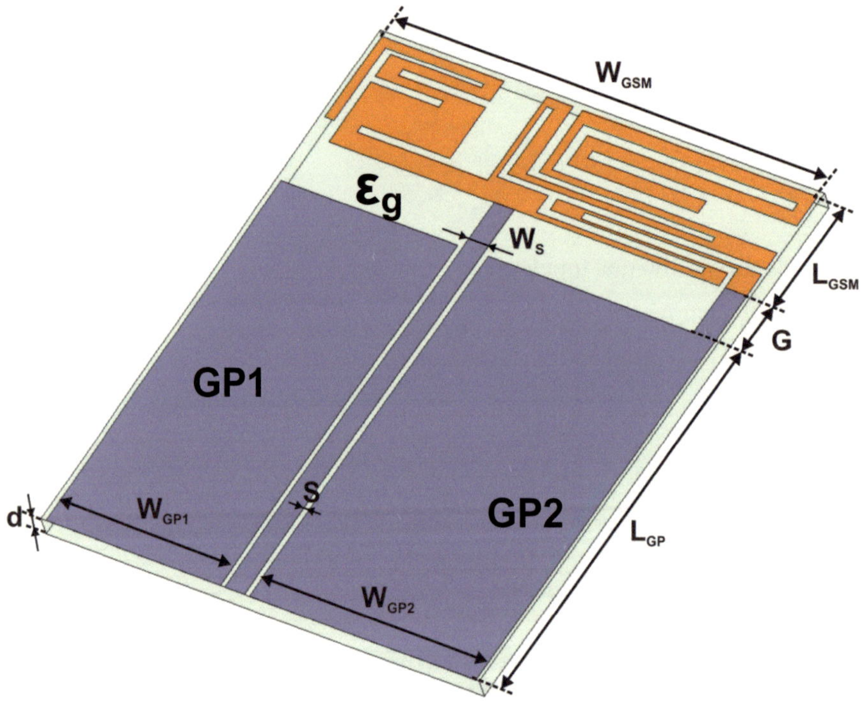

Fig. 5. CPW-fed GSM antenna integrated on a test PCB.

4.2 Measurement setup

The antennas were fabricated, see a photo in Fig. 6, and measured. A 50 Ω SMA was connected to the end of the feeding strip/probe and grounded to the ground plane. An RF cable from the Vector Network Analyzer (HP85107B) was connected to the SMA to excite the Antenna Under Test (AUT). It is well-known that the measurement of a small planar antenna is very sensitive to the presence of the RF cable located in the near field of the AUT [60]. The additional radiation from the outer surface of the signal cable can have a significant impact on the measured parameters such as operating frequency, radiation pattern, and gain. Ferrite chokes around the front end of the cable were used to reduce the additional radiation from the signal cable by forming a high impedance, in this way reducing the current on the outer cable surface [61].

Since our point of interest is the return loss only, during the measurements an anechoic box was used in order to suppress the influence of any metallic object in the surroundings. The box has a cubical form with an edge size of 80 cm and the inner faces are covered with the absorbing material HR-05 from Emerson & Coming, with absorber length 610 mm. These absorbers do not reveal any measurable effect on the radiation characteristics of the electrically small patch antenna around 1.8 GHz, as long as they are kept further than 100 mm away from the AUT [62]. Every AUT was put into the middle of the anechoic box and hung there by the RF cable only. Thus, the distortions in the measurement of the S11 parameter were minimized due to the absence of any unnecessary object in the near field of the AUT.

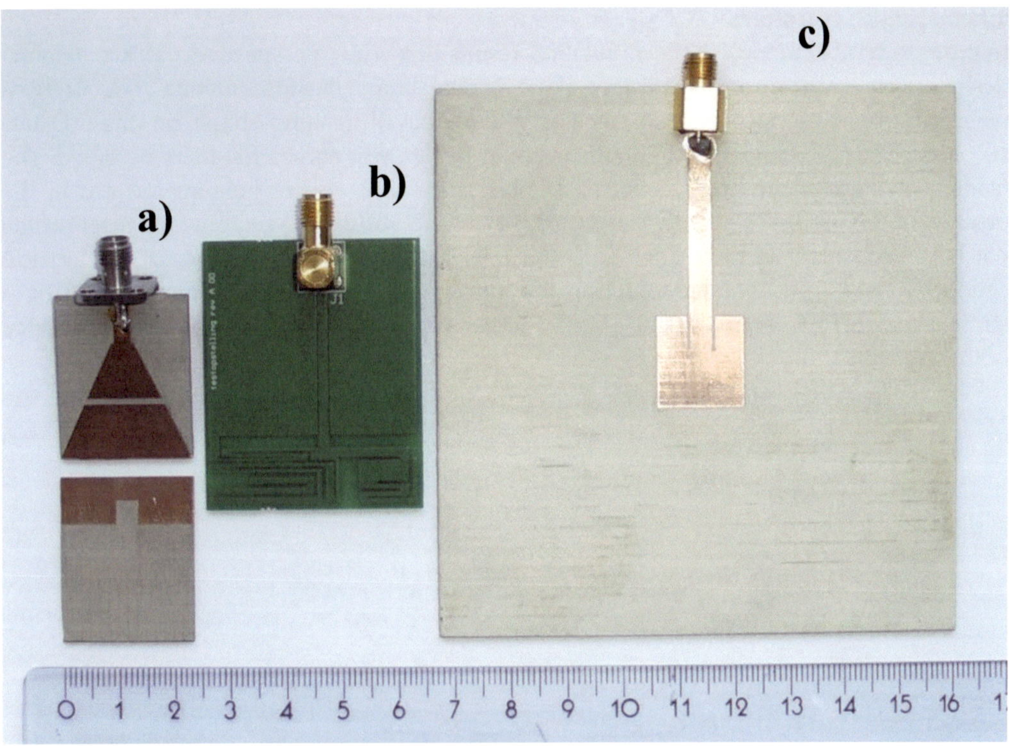

Fig. 6. Antenna prototypes: a) UWB monopole (top and bottom view); b) GSM RangeStar UltimaTM integrated on test-PCB with ground plane; c) line-fed patch on the homogeneous substrate.

4.3 Benchmarking results

4.3.1 Line fed patch antenna

The line-fed patch antenna is first simulated for a completely homogeneous substrate with $\varepsilon = (3.38 - j0.0074)\,\varepsilon_0$. For the solvers MAGMAS 3D, IE3D, FEKO and MOMENTUM, the homogeneous substrate is modeled as an infinitely stretched dielectric layer and included as such in the background. This implies that only the metallic parts give rise to unknowns. In MAGMAS 3D, IE3D and MOMENTUM, the stripline feed is modeled by a horizontal port model which assumes an incident voltage wave arriving from infinity. In a second analysis the dielectric inhomogeneity is included as a volume with permittivity $\varepsilon_d = (3.00 - j0.0039)\,\varepsilon_0$ and dimensions 15.2 × 15.2 × 1.52 mm³. This volume is described by unknown polarization currents. Only MAGMAS 3D and IE3D can calculate Green's functions for layered media with discretization of only a small dielectric volume with permittivity different from the host medium. A multilayer substrate and dielectric bodies cannot be combined in FEKO. Therefore, FEKO models the patch on the inhomogeneous substrate using free-space Green's functions with discretized finite volumes for both dielectrics. This volumetric discretization of both dielectrics dramatically increases the number of unknowns of the whole problem and significantly rises its computational time. Since finite dielectric volumes are not supported by MOMENTUM, it is impossible to simulate the patch antenna on the inhomogeneous substrate.

In order to be able to look at the simulation results in a wider perspective, the two antenna designs were realized and measured (Fig. 7). The homogeneous antenna (Fig. 6c) was realized simply by mechanically etching the required metallic shape on the RO4003 substrate. The fabrication of the inhomogeneous design was somewhat more involved. The feeding line was etched on the RO4003 substrate with a square hole spared out at the position of the patch. An equally sized piece of RO3003 substrate, with only the insets of the patch etched out, was then fitted into the hole to construct the antenna. The electrical conductivity was ensured by soldering the metal junctions. The strip lines are fed by a straight coaxial PCB connector of Huber & Suhner: type SMA-S50-0-45, mounted at the edge of the substrate.

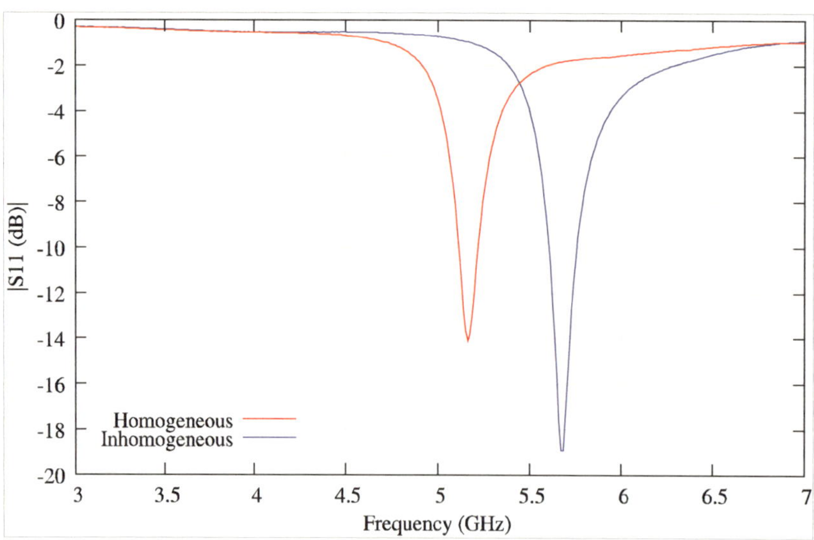

Fig. 7. Measurement for line-fed patch on homogeneous and inhomogeneous substrate.

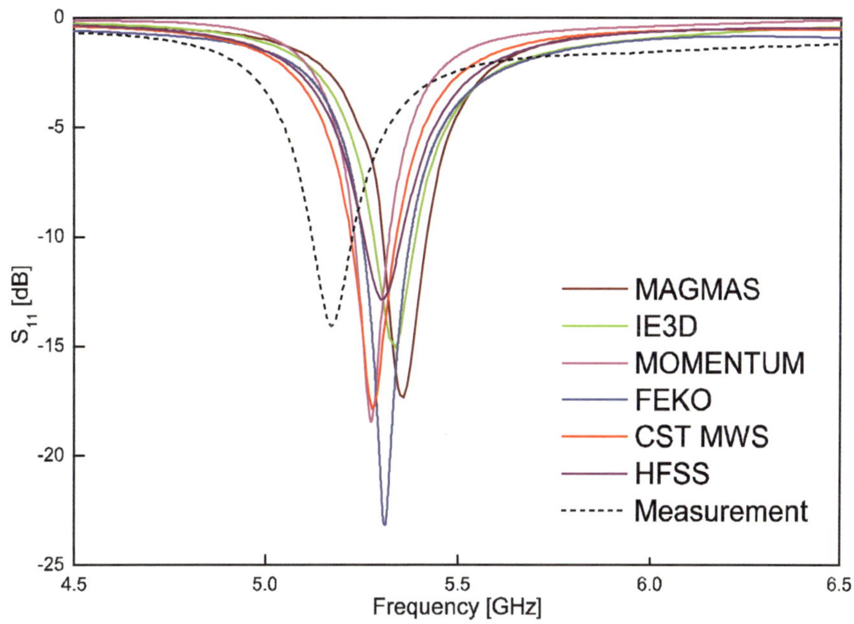

Fig. 8. Simulations and measurements for line-fed patch antenna on homogeneous substrate.

Software Package	Fmin (GHz)	Fmax (GHz)	Fc (GHz)	Fc-shift (%)	BW (GHz)	FcD (%)
MAGMAS	5.305	5.421	5.363	3.67	0.116	0.96
IE3D	5.28	5.39	5.335	3.13	0.11	0.44
FEKO	5.25	5.375	5.313	2.7	0.125	0.01
MOMENTUM	5.23	5.321	5.276	1.98	0.091	-0.68
HFSS	5.256	5.353	5.305	2.54	0.097	-0.14
CST MWS	5.22	5.34	5.28	2.07	0.12	-0.60
Measurement	5.12	5.226	5.173	0.0	0.106	-2.61

Table 1. Nominal frequency values for patch antenna on homogeneous substrate at return loss level of -10 dB.

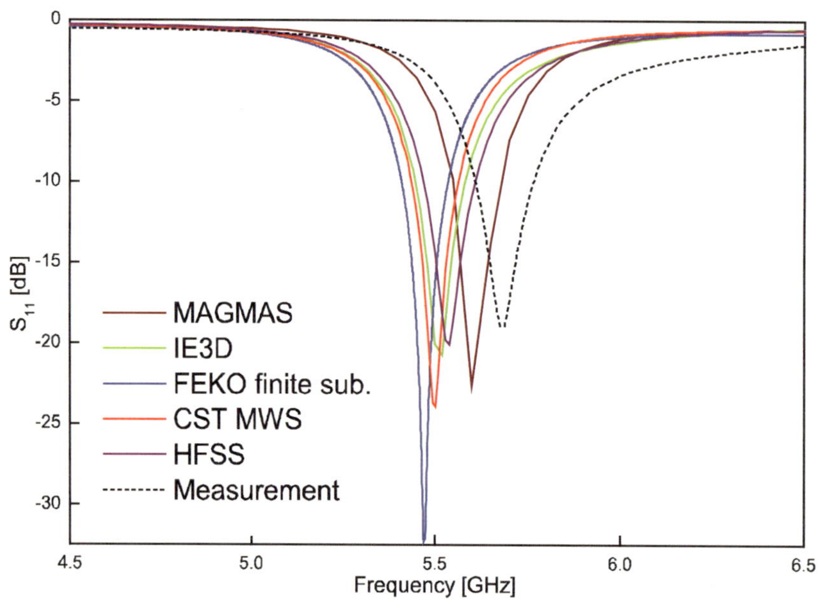

Fig. 9. Simulations and measurements for line-fed patch antenna on inhomogeneous substrate.

Software Package	Fmin (GHz)	Fmax (GHz)	Fc (GHz)	Fc-shift (%)	BW (GHz)	FcD (%)
MAGMAS	5.55	5.68	5.615	-1.25	0.13	1.58
IE3D	5.445	5.581	5.513	-3.04	0.136	-0.26
FEKO	5.408	5.535	5.472	-3.77	0.127	-1.01
MOMENTUM	NA	NA	NA	NA	NA	NA
HFSS	5.47	5.608	5.539	-2.59	0.138	0.21
CST MWS	5.435	5.563	5.499	-3.29	0.128	-0.52
Measurement	5.61	5.762	5.686	0.0	0.152	2.87

Table 2. Nominal frequency values for patch antenna on inhomogeneous substrate at return Loss level of -10 dB.

The simulation results in Fig. 8 and 9 show S_{11} from 4.5 to 6.5 GHz. Tables 1 and 2 give the data in absolute values. The minimal and maximal frequencies crossing the Return Loss level at -10 dB are Fmin and Fmax respectively. The difference between Fmin and Fmax defines the bandwidth (BW) at S11 = -10 dB. The central frequency (Fc) is determined by Fc = (Fmax + Fmin)/2. Hence, the Fc-shift is the deviation between Fc measured and Fc simulated. The central frequency deviation (FcD) shows the dispersion with respect to the mean of the simulated results only. It has the purpose to exclude fabrication and measurement inaccuracies. It is defined as FcD = Fc - FcAV, where

$$FcAV = \frac{1}{n}\sum_{i=1}^{n} Fc \tag{1}$$

is the arithmetic average of the set of simulation results. The simulation results predict a shift to a higher resonance frequency, when the RO4003 substrate is locally replaced by the RO3003 substrate. This agrees with the fact that material of lower dielectric density is added to the structure, such that the overall antenna becomes electrically smaller. This is also confirmed by the measurement results in Fig. 7. Closer investigation of the simulation results reveals that the mutual correspondence between the different solvers is not perfect. The accuracy for the simulated center frequencies is comparable for all solvers and is generally better than 3.8%. Possible explanations for the observed deviations are the differences in feed model used and imperfections in the realizations. Namely, air gaps between RO3003 and RO4003 generated during fabrication give a resonance shift towards the higher frequencies for the patch on the inhomogeneous substrate. Also, the coax-to-microstrip transition realized by soldering the 50 Ohm SMA connector causes a feeding line discontinuity. This discontinuity moves the measured resonance of the patch on the homogeneous substrate to lower frequencies by adding some capacitive and inductive parasitics. During simulations, ports are actually exited using a calibration process that removes any undesired reactive effects of the mode mismatch at the port boundary. In the case of measurements, a connector characterization with a following de-embedding from the measurements has to be executed in order to minimize the differences between the antenna model and its practical realization [63].

It is very important to point out that, considering the FcD values, the simulation results have a relatively small dispersion. The maximum deviation of 1.58 %, observed for the antenna on the inhomogeneous substrate, proves the good agreement for all software packages.

4.3.2 UWB antenna

The return loss simulated in between 2 and 9 GHz (50 Ohm port reference) is presented in Fig. 10. The IE-based solvers show a very good agreement. Solely, the FEKO result has a slight shift to higher frequencies. This difference is probably due to the different modeling of the feed. For both HFSS and CST MWS, the use of symmetry planes reduced the calculation time. The central frequencies (Fc) and bandwidths at S11 < - 10 dB are predicted with a tolerance up to 7% (Table 3). Having a non-resonant nature the UWB antenna does not have a clearly defined resonance. Therefore, the estimated bandwidth (at S11 = -10 dB) deviation about the mean may be more indicative. This BW deviation lays within 0.01% for all solvers, which is remarkably low. It shows that the main difference lies in the estimation of the border frequencies for the working band.

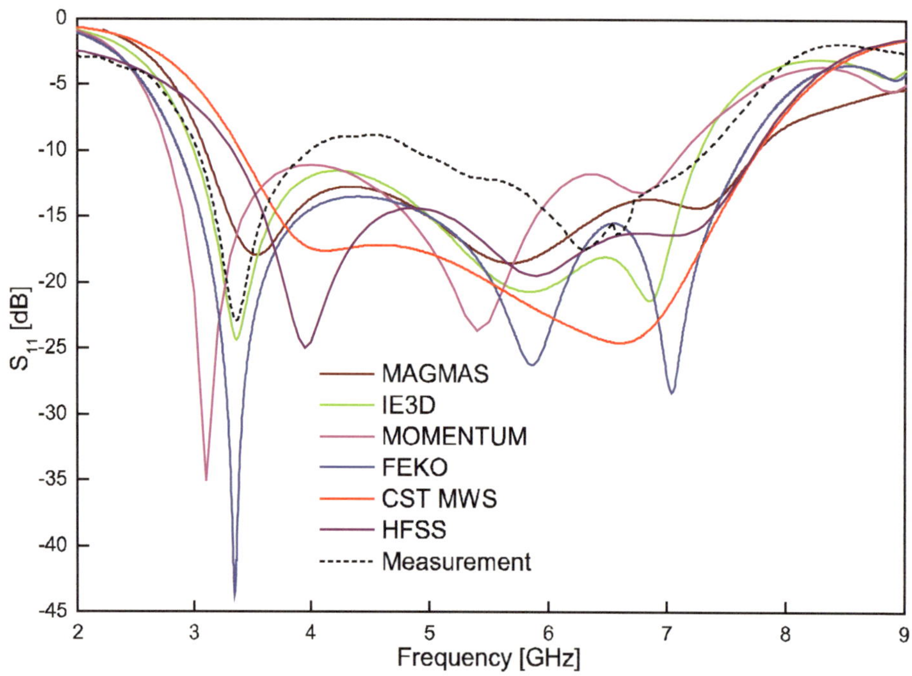

Fig. 10. Simulation and measurement results for UWB monopole antenna topology.

Software Package	Fmin (GHz)	Fmax (GHz)	Fc (GHz)	Fc-shift (%)	BW (GHz)	FcD (%)
MAGMAS	3.11	7.74	5.425	4.73	4.63	2.25
IE3D	3.0	7.27	5.135	-0.87	4.27	-3.22
FEKO	2.87	7.648	5.259	1.53	4.778	-0.88
MOMENTUM	2.77	7.13	4.95	-4.44	4.36	-6.71
HFSS	3.34	7.635	5.488	5.94	4.295	3.42
CST MWS	3.4	7.757	5.579	7.69	4.357	5.14
Measurement	3.03	7.33	5.18	0.0	4.3	-2.37

Table 3. Nominal frequency values for UWB antenna at -10 dB power level.

4.3.3 Folded monopole antenna for GSM applications

The results for the GSM antenna are given in Fig. 11. In the figure, nine results are given, eight calculations and a measurement. Six calculations have been performed at the Katholieke Universiteit Leuven, Belgium, with the several solvers available there. One calculation has been performed at IMST, Germany, with their Empire software tool, and one calculation has been performed at the University of Sienna, Italy, with CST Microwave Studio. The measurement was also performed at the Katholieke Universiteit Leuven.
These results reveal that the solvers really differ, for example concerning the location of the predicted resonant frequencies. It is seen that the curves of all MoM-based solvers agree quite well for the lower resonance. FEKO outlines the middle resonance dissimilar to other solvers, completely merging it with the higher resonance. Among the IE-based software packages, the -5 dB power level in the lower GSM band (900 MHz) was attained by IE3D only, see Table 4. In case the -5 dB criterium was not reached NA is used in Table 4. The volumetric mesh refinement available in CST MWS and Ansoft HFSS helped to uncover the low band resonance. HFSS like CST MWS shows a resonance shift to higher frequencies.

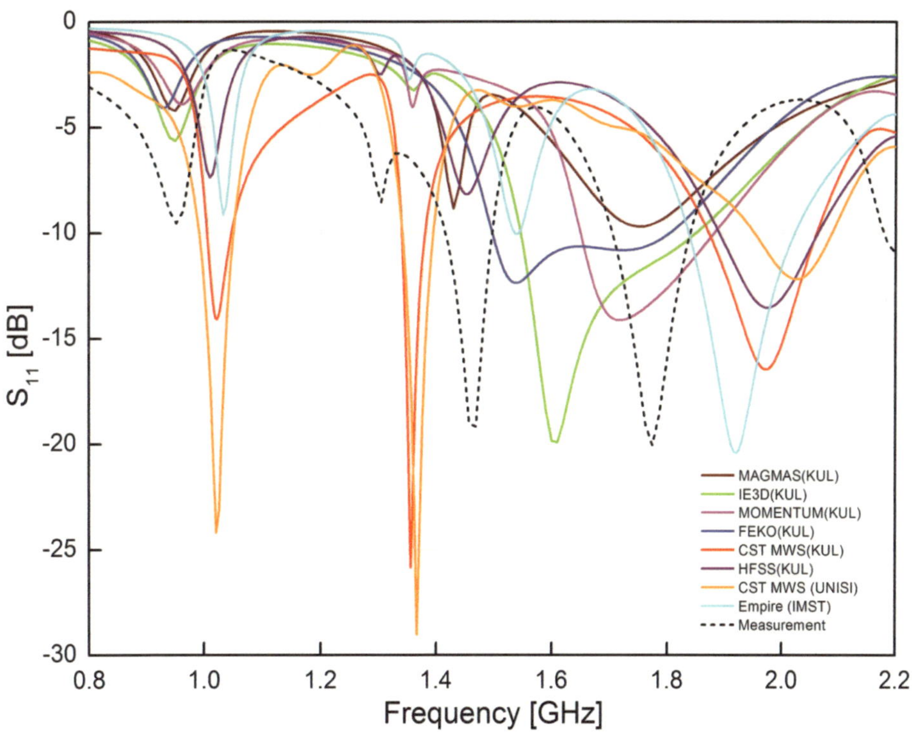

Fig. 11. Simulations and measurement for CPW-fed GSM antenna integrated on a test PCB.

Software Package	1st resonance					
	Fmin (GHz)	Fmax (GHz)	Fc (GHz)	Fc-shift (%)	BW (GHz)	FcD1 (%)
MAGMAS	NA	NA	NA	NA	NA	NA
IE3D	0.932	0.963	0.948	1.45	0.031	-5.80
FEKO	NA	NA	NA	NA	NA	NA
MOMENTUM	NA	NA	NA	NA	NA	NA
HFSS	0.99	1.029	1.01	8.08	0.039	0.36
CST MWS	0.986	1.135	1.061	13.54	0.149	5.43
Measurement	0.882	0.986	0.934	0.0	0.104	-7.14

Software Package	2nd resonance					
	Fmin (GHz)	Fmax (GHz)	Fc (GHz)	Fc-shift (%)	BW (GHz)	FcD2 (%)
MAGMAS	1.404	1.457	1.431	1.96	0.053	0.53
IE3D	NA	NA	NA	NA	NA	NA
FEKO	NA	NA	NA	NA	NA	NA
MOMENTUM	NA	NA	NA	NA	NA	NA
HFSS	1.41	1.505	1.458	3.88	0.095	2.42
CST MWS	1.325	1.437	1.381	-1.57	0.112	-2.95
Measurement	1.271	1.535	1.403	0.0	0.264	-1.41

Software Package	3rd resonance					
	Fmin (GHz)	Fmax (GHz)	Fc (GHz)	Fc-shift (%)	BW (GHz)	FcD3 (%)
MAGMAS	1.577	1.986	1.782	0.39	0.409	-2.75
IE3D	1.493	2.04	1.767	-0.45	0.547	-3.57
FEKO	1.44	1.982	1.711	-3.58	0.542	-6.6
MOMENTUM	1.596	2.033	1.815	2.25	0.437	-0.95
HFSS	1.77	2.19	1.98	11.58	0.42	8.09
CST MWS	1.737	2.138	1.938	9.19	0.401	5.77
Measurement	1.622	1.927	1.775	0.0	0.305	-3.13

Table 4. Nominal frequency values for GSM antenna at -5 dB power level.

4.4 Discussion of the results

The differences are small for the reference antenna. All solvers predict the correct resonant frequency and bandwidth within a small tolerance. The differences are acceptable for the UWB antenna. Also here, the bandwidth is very well predicted by all solvers. For the GSM antenna however, the differences are considerable and not really acceptable. There are several possible causes for this discrepancy. In the following sections, all of them are discussed.

4.4.1 Validity of the input

For the calculations performed at the Katholieke Universiteit Leuven, the input was submitted to and in most cases indeed checked by the technical team of the software vendor responsible for the tool. In some cases, small improvements were suggested and then taken into account. In other cases, an upgrade of the version of the software yielded a slightly different result. For the calculations performed at the other two locations, this check was not performed. Interesting to see is that the two CST calculations are in quite good agreement.

The main conclusion however is that, as far as we can tell, the "quality of the input" is not the cause of the discrepancies.

4.4.2 Meshing

It is well known that the mesh quality and resolution are key factors in the accuracy of any solver. In the case of the GSM antenna for example, the electromagnetic coupling between nearby segments may differ considerably due to the specific meshing used, especially in the meandering parts of the low band monopole and matching circuit. The question is whether the meshes used are adequate. This issue was investigated in the following way. First, simulations were done using standard mesh settings. Second, a convergence study was performed in terms of the resolution of the meshes (see Fig. 12). The high resolution meshes and the lower resolution meshes in practice showed nearly the same results, ensuring that convergence was reached. Third, for each software package, as already mentioned in the previous section, the meshing scheme was submitted to and checked by the software vendor. The general conclusion is that, as far as we can tell, the meshing as used to produce the results shown does not cause the discrepancies.

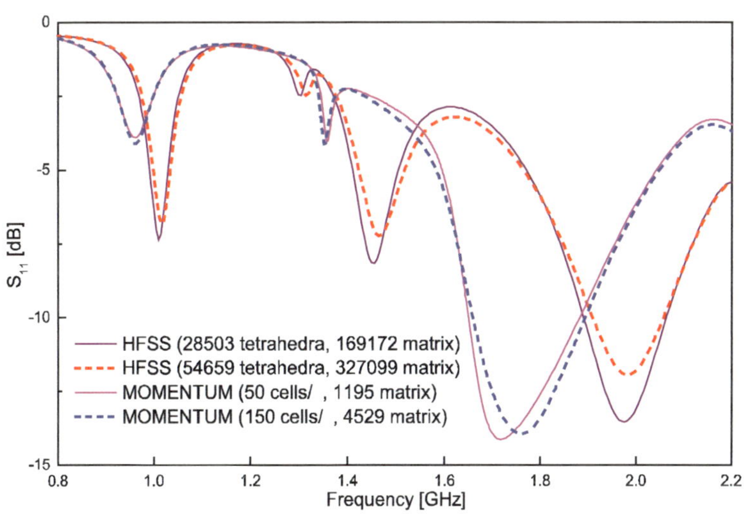

Fig. 12. Solvers convergence for GSM antenna simulation.

4.4.3 Modeling of losses

The importance of taking into account the substrate losses, especially for a lossy material like FR4 (tanδ = 0.02), is obvious. The effect for the GSM and UWB antennas is illustrated in Fig. 13. At the same time, it is known that solvers sometimes tend to show difficulties in accurately modeling losses. This may explain some of the discrepancy between the solvers. It is important to point out that the PCB manufacturing process, although it does not provide adequate uniformity of losses and dielectric constant over the substrates, is not the cause of the discrepancy between the solvers. It could explain some of the discrepancy between the solvers on the one hand, and the measurement on the other hand.

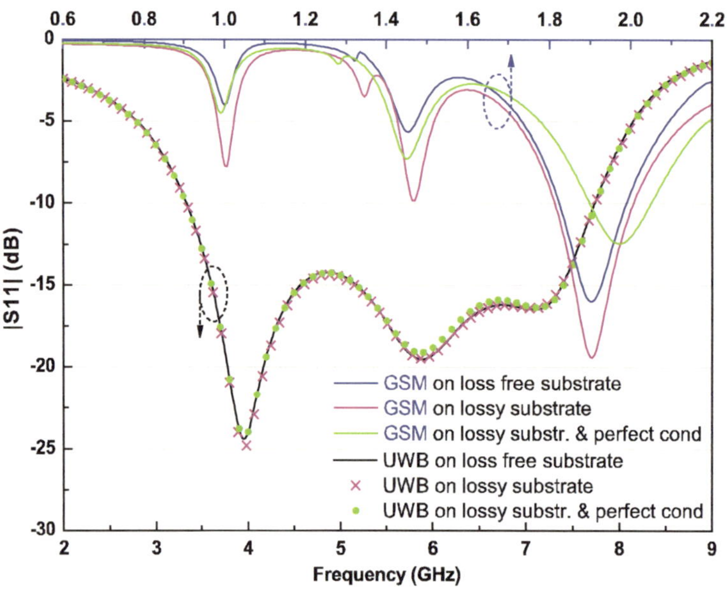

Fig. 13. Simulation results (HFSS) for UWB and GSM antennas considering dielectric and conductor losses.

4.4.4 Implementation of the frequency sweep

Except for MAGMAS all the solvers use fitting functions to model and interpolate the parameters of interest, such as input impedance or S-parameters, in a certain frequency range. Such an adaptive frequency sampling technique automatically selects the sample points. It enables fast and accurate solutions without any a priori knowledge of the parameter of interest. The response is sampled more densely where needed, for example at resonances, to ensure that important details are included. This usually results in run-time savings over discrete frequency sampling. The only observation we want to make is that compared to the standard settings, a quite dense frequency sweep had to be set for HFSS in order not to miss any resonance. Although CST does not need this precaution, its predictions for high-Q resonant antennas are less accurate and slower. The use of the built in auto regressive (AR) filter in CST MWS leads to a faster convergence and helps to avoid the typical ripples, sometimes observed in the more attenuated zones.

4.4.5 Finite or infinite layer structures

Although all MoM solvers used an infinite substrate in the simulations, FEKO and IE3D are able to model finite substrates. Both finite and infinite substrate models have their advantages and drawbacks. Finite substrate models are supposed to be able to include the effects at the edges. However, they enforce the boundary conditions on the top and bottom surfaces of the substrate in a numerical way, in this way losing accuracy on these surfaces. Infinite substrate models enforce the boundary condition on the top and bottom surfaces analytically, and thus rigorously. However, they neglect the edge effects of the finite substrates. A comparison of MoM-based simulation results for both the finite and infinite substrate model is depicted in Fig. 14. The differences are considerable, and the conclusion is clear. Accurate modeling of the finiteness of a layer structure is necessary. Since this was not done for the MoM solvers in Fig. 11, this may explain part of the discrepancy. The non-MoM solvers have another issue. They use a box of finite dimensions to contain the structure. However, from literature, it is well-known that the boundary conditions used on the surface of this box nowadays are of high quality, provided that the box is large enough.

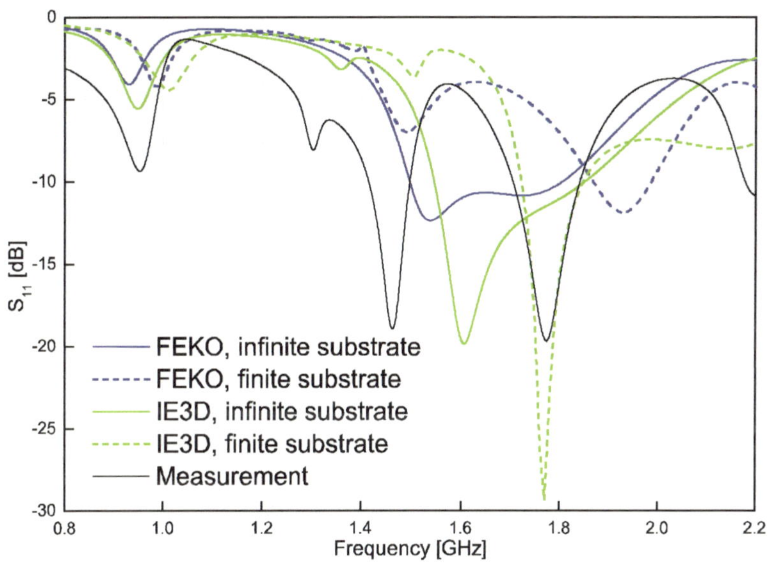

Fig. 14. Simulation results of GSM antenna for both finite and infinite substrate models.

4.4.6 Excitation of the slotline mode in the CPW

The topology is clearly not symmetric with respect to the CPW. In particular, the signal line of the GSM antenna has a meandered matching line which is short circuited to only one ground plane (GP2 in Fig. 5). This asymmetry may give rise to a radiating slotline mode. In practice, the slotline mode is suppressed by connecting the two ground planes of the feeding line with air bridges [64], [65]. The ground planes on the test PCB are connected only at the side of the SMA connector. Inevitably, the slotline mode will appear along with the coplanar mode at the antenna side. The vector E-field plot in Fig. 15a shows the asymmetric slotline mode launched by a lumped port in HFSS, whereas the use of the air bridge in Fig. 15b suppresses the parasitic slotline mode and gives rise to the symmetric CPW mode. Does this contribute to the observed discrepancy? In theory, it should not. The reason is that as long as a unique and identical topology is inserted in several solvers, the user has the theoretical right to expect identical output. However, it is a fact that the presented EM solvers use different theoretical techniques to derive the scattering parameters of CPW fed structures. It is clear that this may considerably contribute to the simulation results diversity. This is also clearly confirmed by simulations, which exhibit a significant difference when the transversal air bridge is added at the end of the CPW line (Fig. 16). The air bridge was modeled as a bond wire whose ends were soldered to both ground planes in order to equalize their potentials. The position and length of the bond wires have a big influence on the resonant frequency of the two lowest resonances. Normally, the power leaking to the slotline mode can be identified as the main contribution to the losses of the CPW feeding line [66].

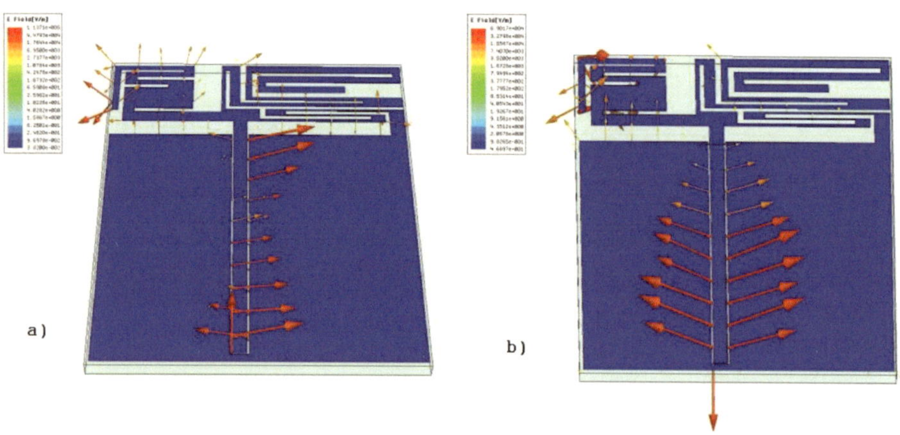

Fig. 15. E-field plot for GSM antenna simulated at 1.8 GHz in HFSS: a) without the air bridge, b) with the air bridge at the end of the CPW.

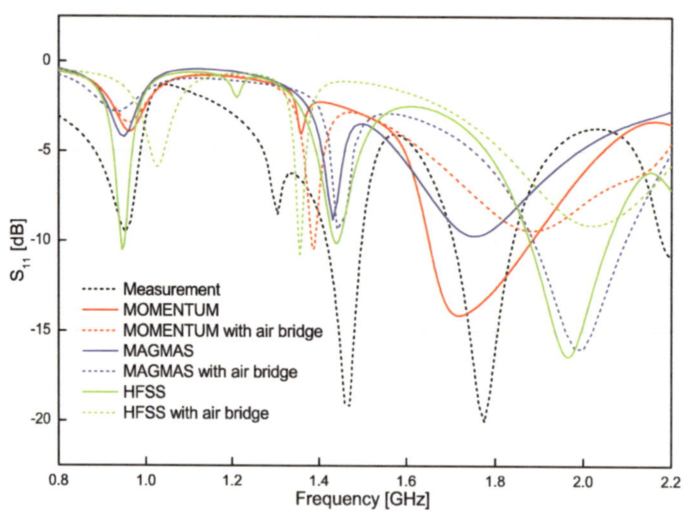

Fig. 16. Simulation results for GSM antenna with air bridge placed at the end of the CPW.

4.4.7 Modeling of the feed

In the case of small planar antennas, the issue of the modeling of the antenna feed is really essential. Since a different feed topology may result in completely different performance, a proper choice of the feed model is absolutely crucial. The basic rule is that the model to select has to correspond as closely as possible to the actual feed topology that is going to be used later in practice.

Most software packages offer several feed topologies. In general, a distinction can be made between the "local" feeds and the "transmission line" feeds. The local feed in essence consists of a localized current or voltage source, possibly with inner impedance. During the simulation, an imposed current or voltage is applied to the structure and its effect is analyzed. It is evident that this feeding topology is really a local one, not able to incorporate with high accuracy the effect of any transmission line that maybe in practice is feeding the structure. The transmission line feed assumes a transmission line (which can be very general, involving multiple conductors, waveguides, etc.) supporting incoming waves exciting the antenna structure. If necessary, the software calculates the eigenmodes of the two-dimensional feed face region. Any actual excitation may be written as a superposition of the eigenmodes. In many software packages there is also a relation between the "local" feeds and the "transmission line" feeds. A transmission line feed in many cases is derived from a local feed by a so-called de-embedding procedure [67].

Which feed is chosen best depends on the structure that has to be modeled. If it concerns an antenna embedded within a mobile device, not fed by a long coaxial cable, the discrete port may be the most appropriate. If it concerns a small antenna embedded within a PCB, fed by a microstrip line, the transmission line feed is probably the proper choice.

Also, in between software packages, there are significant differences in the possible range of models offered for the feed of the antenna. This will certainly contribute to any discrepancy. The question is how much? Let us concentrate on the case of the GSM antenna. The fabricated structure has a coax to coplanar waveguide transition and utilizes the right-angle SMA connector (R125680000 by Radial®) in the measurement setup. Such a feeding setup inevitably induces unwanted parasitics, which affect the characteristics of the antenna.

It has to be emphasized that only in the Empire simulation the connector feeding the fabricated structure was modeled in more detail. There a coaxial feed model was used similar to this connector. The concentrated port impedance of this coaxial feed was set to 50 Ω. The other simulations used the implemented feeding techniques embedded within the solvers, which is the standard practice in most designs reported in literature. These simulations with Empire showed that the connector feeding the antenna has a strong influence on the results.

This puts the antenna design community in a tough position. If the mere presence of the connector indeed alters the results beyond expectation for this type of complicated antenna, this means that it has to be taken into account always in the analysis. This is not possible up to now with the IE solvers and would lead to much larger calculation times for HFSS and CST. Even the slightest difference between the feed model defined by the software user and the actual setup can lead to radical differences in results [68]. Two different feed models were studied in MAGMAS (Fig. 17). Both of them apply the horizontal current source, which consist of a user-defined electric current imposed on a so called active patch. The first topology has the active patch placed between the antenna feeding point and its ground plane (Fig. 17a). The second topology fully models the CPW feed with the active patch set on the actual feeding point of the SMA connector in the measurements setup (Fig. 17b). Simulated S11 results highlight the substantial importance of the feed modeling (Fig. 18).

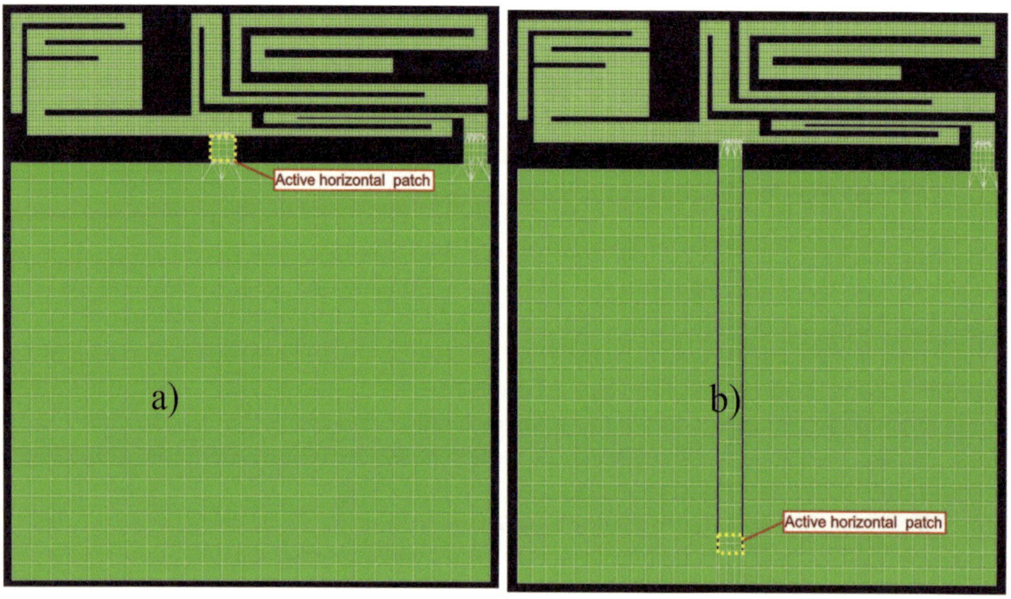

Fig. 17. Different feed topologies for GSM antenna analysed in MAGMAS and HFSS: a) simplified model with localized active patch; b) more advanced CPW feed model.

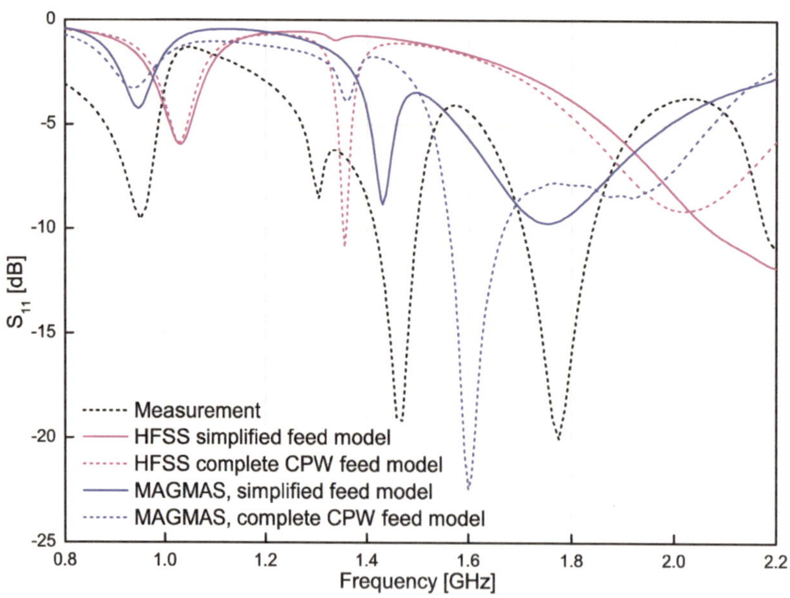

Fig. 18. Simulation results of GSM antenna for different feeds.

4.5 Calculation speed

The simulation time per frequency point (SPFP) and total simulation time (TST) in seconds are presented with the number of unknowns (NU) employed for every structure in Table 5. Several computers with different specifications and operation systems were used to run all the simulations. Table 6 offers an overview of the software tools and the computers used. For the UWB and both square patch antennas a symmetry plane was applied in HFSS and CST MWS. If the symmetry plane is specified, the calculation domain is cut by half. This size reduction of the problem helps to reduce the solution time. It is noticeable, that IE3D performs remarkably fast among the MoM-based solvers. It is only fair to mention that this was only attained after the installation of a "solver version" dedicated to the specific Win64 architecture of the processor used. This IE3D version was provided by the Zeland company during our discussion phase with the software vendors. Without installing the dedicated software version, the times were comparable to the times of the other MoM software packages. It also has to be pointed out that the number of frequency points used was not the same for all software packages, explaining why, comparing two packages, a SPFP time can be lower, while the TST is higher In order to compare the inherent speed of packages, the SPFP times are more appropriate. For the GSM antenna, since we have complete control over its source code, the MAGMAS results were calculated as reference results. They were consequently obtained with a very high number of unknowns. This explains the large calculation times for this topology.

Antenna type		Software package					
		MAGMAS	IE3D	FEKO	Momentum	CST	HFSS
Patch on an homogeneous substrate	NU	579	403	598	384	353912	60718
	SPFP	3.2	1	9.4	4.3	_	19.7
	TST	132	13	122	65	111	237
Patch on an inhomogeneou s substrate	NU	3627	999	13801	NA	353912	73394
	SPFP	55	84	929	NA	_	24.6
	TST	2255	1093	12075	NA	107	296
UWB tapered monopole	NU	1663	1595	1367	1748	673507	68076
	SPFP	19	10	63	31.2	_	32.6
	TST	696	169	1136	375	217	391
GSM folded monopoles	NU	6076	740	3879	1195	172422 5	169172
	SPFP	367	5	433	15.2	_	126
	TST	26077	76	6062	350	5664	2144

Table 5. Comparative simulation time for different software packages.

Software package	CPU type	RAM installed	Operation System
MAGMAS 3D	Intel Xeon ™ 2.66 GHz	2 Gb	Linux Fedora Core 6
IE3D ver. 12.22	Intel Xeon ™ 2.8 GHz	4 Gb	Windows XP Professional 2002 SP2
FEKO Suite 5.3	AMD Athlon ™ 64 3800+	3.5 Gb	Linux x86_64 2.6.11.9
ADS 2006A.400 Momentum	2 x AMD Opteron ™ 250 (2.4 GHz) 64 bit	4 Gb	Red Hat Linux 2.4.21-27.ELsmp
CST MWS 2006B.03	Intel Xeon ™ 2.8 GHz	4 Gb	Windows XP Professional 2002 SP2
Ansoft HFSS ver. 10.1	2 x Dual-Core AMD Opteron ™ 285 (2.6 GHz) 64 bit	8 Gb	Windows Server 2003, Standard x64 Edition, SP1
Empire XCcel	2 x Intel Xeon ™ 5472 3 GHz	4 Gb	Linux

Table 6. EM solver versions and computer configurations used for the simulations.

5. Conclusions

Planar antennas have to be designed taking into account many aspects. Not only the flatness and size, but also the weight, the ease of manufacturing and the way of mounting become very important. The reason is that in many cases, they have to be integrated in correspondingly small stand-alone systems. This is why the design of planar antennas has become a critical issue in modern telecom system design. In this chapter, an overview has been given on the use of existing computational techniques and software tools for the analysis and design of planar antennas. The available commercial software tools are based on different electromagnetic simulation techniques. They provide the end-user with an intuitive and clear interface. The benchmarking of a wide range of representative electromagnetic simulation tools was carried out. The results are also compared to measurements of the antennas produced. From the study, important conclusions can be drawn.

- As a reference case, the classical patch antenna can be predicted accurately by all simulation programs. Moreover, while having a more attractive market price, MoM based programs perform the simulations inherently faster. Unlike MoM, FEM and FIT are more "brute force" techniques, inherently able to analyse much more general structures. However, they generally require the inversion of much larger, but sparse matrices. This requires the implementation of dedicated inversion techniques, which makes in many cases FEM and FIT based programs memory intensive. Although the calculation times are not that different at present, dedicated inversion techniques for MoM solvers are nowadays fully in development. This fact maybe will change the picture ...

- A proper mesh generation and a correct feeding model are two crucial issues predetermining the successful simulation in the software packages reviewed. In general, a very neat, adaptive mesh refinement implemented in Ansoft HFSS and as an option in CST MWS allows better handling of a design with difficult electromagnetic coupling between its different parts. Such characteristics pertain to applications in mobile gadgets such as the examined GSM antenna. Having no mesh refinement option, MoM based programs require more careful consideration of the initial meshing. MoM solvers can provide an improvement in simulation results and time using so-called edge meshing features, while avoiding excessive meshing on the bulk of the metal structure.
- The more challenging designs such as the GSM and the small UWB antenna require finite substrate effects such as space and surface wave diffraction from substrate edges to be taken into account. MoM-based solvers show better convergence when a dielectric substrate is infinite because then, it is modeled exploiting Green's functions. The practical trend to miniaturize antennas diminishes the advantage of using MoM-based solvers. Having a small size, finite substrate effects such as diffraction and reflection at the edges become more pronounced. Thus, at present FEM and FIT-based programs inherently are better suited for modeling small sized antennas.

In this chapter characteristics important to the customer such as interface, price/performance ratio, user friendliness, and so forth are not considered. However, they are factors influencing the end user's personal choice. The results shown also suggests that the feeding models as implemented today in the widespread commercial solvers are probably unsatisfactory in the case of small structures with complicated electromagnetic coupling behaviour. The slight grouping of results according to the basic theoretical technique used may also be explained by feed modeling, since the feeds used in the three MoM solvers are more alike then the feeds used in HFSS and CST MWS. The study suggests further that the meshing schemes used in all solvers, certainly those including adaptive meshing refinement, are adequate if used properly.

This leads to the following final guideline. The use of two different solvers, based on different theoretical methods (integral and differential) may provide a means to characterize the quality of simulation results. If the two results are in good agreement, it is reasonable to expect that the results can be trusted. If the two results are in disagreement, a deeper investigation of the structure and its modeling is absolutely necessary.

6. Acknowledgements

The authors gratefully acknowledge the following persons:
Dr. Vladimir Volski, and Mr. Soheil Radiom, Katholieke Universiteit Leuven, Belgium, for providing suggestions and figures; Prof. Raphael Gillard from IETR, Rennes, France, for interesting discussions; Mr. Winfried Simon and Mr. Andreas Wien, of the Empire Support Team, IMST, Germany, for providing additional analyses; Dr. Dr. Jian-X. Zheng from Zeland Software, Inc., Mr. David Prestaux and Mr. Alain Michel from Ansoft France, Mr. Werner Soergel from EM Software & Systems GmbH, Dr. Tilmann Wittig from CST GmbH, and Dr. Davy Pissoort from Agilent Technologies for their help in excluding the human factor by sharing their expertise to obtain quality input into their software products.
We also would like to express our gratitude to ACE (http://www.antennasvce.org), the European Network of Excellence on Antennas, and to AMICOM, the European Network of

Excellence on RF-MEMS, whose activities created the proper environment to complete this work.

7. References

[1] D. B. Davidson,"A review of important recent developments in full-wave CEM for RF and microwave engineering," IEEE 3rd Int. Conf. Comp. Electromagnetics and Its Applications, pp. PS/1-PS/4, Nov. 2004.

[2] C. W. Townbridge and J. K. Sykulski,"Some Key Developments in Computational Electromagnetics and Their Attribution," IEEE Antennas and Propag. Magazine, vol. 42, no. 6, pp. 503 - 508, Apr. 2006

[3] E. K. Miller,"A Selective Survey of Computational Electromagnetics," IEEE Trans Antennas Propag, vol. 36, no. 9, Sept. 1988, pp. 1281 - 1305

[4] F. Peterson, S. L. Ray, and R. Mittra, "Computational methods for electromagnetics", IEEE Press - Oxford University Press, 1998.

[5] R. F. Harrington, "Field Computation by Moment Methods", New York: Macmillan, 1968.

[6] T. K. Sarkar, E. Arvan, and S. Ponnapalli, "Electromagnetic scattering from dielectric bodies", IEEE Trans. Antennas Propag., vol. 37, pp. 673-676, May 1989.

[7] Y. Schols and G. A. E. Vandenbosch, "Separation of horizontal and vertical dependencies in a surface/volume integral equation approach to model quasi 3-D structures in multilayered media ", IEEE Trans. Antennas Propagat., vol. 55, no. 4, pp. 1086-1094, April 2007.

[8] M. Vrancken and G.A.E. Vandenbosch, Hybrid dyadic-mixed-potential and combined spectral-space domain integral-equation analysis of quasi-3-D structures in stratified media, IEEE Trans Microwave Theory Tech, vol 51, pp. 216-225, 2003.

[9] G. A. E. Vandenbosch and S. Mestdagh, "Conceptual topology for the integration of planar and quasi 3D antennas in chip packages", IEEE Trans. Antennas Propagat., Oct. 2008.

[10] P. S. Hall and Y. Hao, "Antennas and propagation for body-centric wireless communications", Artech house, 2006.

[11] G. D. Braaten, R. M. Nelson, M. A. Mohammed, "Electric field integral equations for electromagnetic scattering problems with electrically small and electrically large regions", IEEE Trans Antennas Propag, Vol. 56, No. 1, pp. 142-150, Jan. 2008.

[12] Awadhiya.; P. Barba, ; L. Kempel,; "Finite-element method programming made easy???", IEEE Antennas and Propagation Magazine, vol 45, pp.73 - 79, Aug. 2003.

[13] J. M. Jin, The Finite Element Method in Electromagnetics, second edition, John Willey & Sons, Inc., New York, 2002.

[14] J. L. Volakis, A. Chatterjee, and L. C. Kempel, Finite element method for electromagnetics, IEEE Press, Oxford University Press, 1997.

[15] R. L. Courant, "Variational methods for the solution of problems of equilibrium and vibration," Bulletin of the American Mathematical Society, 5, pp. 1-23, 1943.

[16] S. Ahmed, "Finite-element method for waveguide problems," Electronics Letters, vol. 4, Issue 18, pp.387 - 389, Sept. 1968.

[17] R. Coccioli, T. Itoh, G. Pelosi, P. P. Silvester, "Finite-Element Methods in Microwaves: A Selected Bibliography," IEEE Antennas and Propag. Magazine, vol. 38, Issue 6, pp.34 - 48, Dec. 1996.

[18] K. S. Yee, "Numerical Solution of Initial Boundary Value Problems Involving Maxwell's Equations in Isotropic Media," IEEE Trans on Antennas Propag, vol 14, pp. 302-307, 1966.

[19] Taflove, Computational electrodynamics: the finite difference time domain method, Artech House, 1997.

[20] D. M. Sullivan, "Electromagnetic simulation using the FDTD method", Wiley - IEEE Press., 2000, ISBN 978-0-7803-4747-2.

[21]Empire, Version 4.2, IMST GmbH, http://www.empire.de/main/Empire/en/home.php.

[22] Maurice Weiner, "Electromagnetic analysis using transmission line variables", World Scientific Publishing Company, 2001, ISBN 13: 978-9810244385

[23] T. Weiland, "A discretization method for the solution of Maxwell's equations for six-component fields", Electronics and Communications AEÜ, vol 31, No. 3, 116–120, 1977.

[24] Bossavit and L. Kettunen,"Yee-like schemes on a tetrahedral mesh, with diagonal lumping," Int. J. Numer. Model., vol. 42, pp. 129 - 142, 1999.

[25] M. Celuch-Marcysiak and W. K. Gwarek,"Comparative study of the time-domain methods for the computer aided analysis of microwave circuits," Int. Conf. on Comp. in Electromagnetics., pp. 30-34, Nov. 1991.

[26] Agilent Technologies, EEsof EDA, Momentum, http://eesof.tm.agilent.com/products/momentum_main.html

[27] Zeland Software, Inc., IE3D v. 11.2 user manual, Jan 2006, http://www.zeland.com

[28] P. Ciais, R. Staraj, G. Kossiavas, C. Luxey, "Compact Internal Multiband Antenna for Mobile Phones and WLAN Standards", Electronics Letters, Vol. 40, N°15, pp. 920-921, July 2004.

[29] J. Ollikainen, O. Kivekäs, A. Toropainen and P. Vainikainen, "Internal dual-band patch antenna for mobile phones", Millenium Conference on Antennas & Propagation, Davos, April 2000, CD-ROM SP-444, session 3A9.

[30] J. Villanen, J. Ollikainen, O. Kivekäs and P. Vainikainen, "Coupling Element Based Mobile Terminal Antenna Structures", IEEE Transaction on Antennas and Propagations, Vol. 54, N°7, pp. 2142-2153, July 2006.

[31] M. K. Kärkkäinen, "Meandered Multiband PIFA with Coplanar Parasitic Patches", IEEE Microwave Wireless Components Letters, Vol. 15, N°10, pp. 630-150, April 2004.

[32] M. K. Kärkkäinen, "Meandered Multiband PIFA with Coplanar Parasitic Patches", IEEE Microwave Wireless Components Letters, Vol. 15, N°10, pp. 630-150, April 2004.

[33] Z. Ying, "Some Important Antenna Innovations in the Mobile Terminal Industry in the Last Decade", Antenn 06 Conference, 30th May-1st June, Linköping.

[34] EMSS - EM Software & Systems Ltd, FEKO Suite 5.3 user manual, Jan 2006, http://www.emss.co.za

[35] R. Mittra,"A Look at Some Challenging Problems in Computational Electromagnetics," IEEE Antennas and Propag. Magazine, vol. 46, no. 5, pp. 18 - 32, Oct. 2004.

[36] Ansoft Corporation, HFSS v. 10.1.1 user manual, Jul 2006, http://www.ansoft.com

[37] S-H. Yeh, K-L. Wong, T-W. Chiou and S-T. Fang, "Dual-Band Planar Inverted F Antenna for GSM/DCS Mobile Phones", IEEE Transaction on Antennas and Propagations, Vol. 51, N°5, pp. 1124-1126, May 2003.

[38] Y-S. Shin, B-N. Kim, W-I. Kwak and S-O. Park, "GSM/DCS/IMT-2000 Triple-Band Built-In Antenna for Wireless Terminals", IEEE Antennas and Wireless Propagation Letters, Vol. 3, pp. 104-107, 2004.
[39] K-L. Wong and Y-C. Lin, "Thin Internal Planar Antenna for GSM/DCS/PCS/UMTS Operation in a PDA Phone", Microwave and Optical Technology Letters, Vol. 47, N°5, pp. 423-426, December 2005.
[40] W-I. Kwak, S-O. Park and J-S. Kim, "A Folded Planar Inverted-F Antenna for GSM/DCS/Bluetooth Triple-Band Application", IEEE Antennas and Wireless Propagation Letters, Vol. 5, pp. 18-21, 2006.
[41] P-L. Teng, T-W. Chiou and K-L. Wong, "Planar Inverted-F Antenna with a Bent Meandered Radiating Arm for GSM/DCS Operation", Microwave and Optical Technology Letters, Vol. 38, N°1, pp. 73-75, July 2003.
[42] K-L. Wong, S-W. Su, C-L. Tang and S-H. Yeh, "Internal Shorted Patch Antenna for a UMTS Folder-Type Mobile Phone", IEEE Transaction on Antennas and Propagations, Vol. 53, N°10, pp. 3391-3394, October 2005.
[43] CST GmbH - Computer Simulation Technology, CST Microwave Studio 2006 user manual, 2006, http://www.cst.com
[44] MAGMAS 3D, http://www.esat.kuleuven.be/telemic/antennas/magmas/
[45] G. A. E. Vandenbosch and A. R. Van de Capelle, Mixed-potential integral expression formulation of the electric field in a stratified dielectric medium - application to the case of a probe current source, *IEEE Trans. Antennas Propagat.*, vol. 40, pp. 806-817, July 1992.
[46] F. J. Demuynck, G. A. E. Vandenbosch and A. R. Van de Capelle, The expansion wave concept, part I: efficient calculation of spatial Green's functions in a stratified dielectric medium, *IEEE Trans. Antennas Propagat.*, vol. 46, pp. 397-406, Mar. 1998.
[47] Y. Schols and G. A. E. Vandenbosch, Separation of horizontal and vertical dependencies in a surface/volume integral equation approach to model quasi 3-D structures in multilayered media, *IEEE Trans. Antennas Propagat.*, vol. 55, no. 4, pp. 1086-1094, April 2007.
[48] K. G. Gupta, T. Itoh and A. A. Oliner, "Microwave and RF Education - Past, Present, and Future," IEEE Trans. Microwave Theory Tech., vol. 50, pp. 1006–1014, Mar. 2002.
[49] R. Mittra,"A Look at Some Challenging Problems in Computational Electromagnetics," IEEE Antennas and Propag. Magazine, vol. 46, no. 5, pp. 18 - 32, Oct. 2004.
[50] Vasylchenko, Y. Schols, W. De Raedt, and G. A. E. Vandenbosch, "Quality assessment of computational techniques and software tools for planar antenna analysis", IEEE Antennas Propagat. Magazine, Feb. 2009.
[51] M. Vrancken, Y. Schols, W. Aerts and G.A.E. Vandenbosch, Benchmark of full Maxwell 3-dimensional electromagnetic field solvers on prototype cavity-backed aperture antenna, AEU - Int J of Electronics and Communications, 2006.
[52] M. Vrancken, W. Aerts and G.A.E. Vandenbosch, Benchmark of full Maxwell 3-dimensional electromagnetic field solvers on SOIC8 packaged and interconnected circuit, Int J RF Microwave Comput Aided Eng, vol 16, 143-154, 2006.
[53] D. M. Pozar, S. Duffy and A. Herscovici,"A comparison of commercial software packages for microstrip antenna analysis," IEEE Ant. and Propag. Soc. Int. Symp., vol. 1, pp. 152-155, Jul. 2000.

[54] V. Radonic, V. Cronojevic-Bengin and L. Zivanov,"Comparison of Commercially Available Full-Wave EM Simulation Tools for Microwave Passive Devices," EUROCON 2005 "Computer as a Tool" Int. Conf., vol. 2, pp. 1699-1702, Nov. 2005.
[55] V. Radonic, V. Cronojevic-Bengin, B. Reljic and B. Jokanovic,"Accuracy of EM Simulation Tools in Modeling of Resonant Left-Handed Microstrip Lines," EUROCON 2007 "Computer as a Tool" Int. Conf., pp. 2104-2109, Sept. 2007.
[56] D. B. Davidson, Computational electromagnetics for RF and Microwave Engineering, Cambridge, UK: Cambridge University Press, 2005
[57] M.J. Vaughan, K.Y. Hur and R.C. Compton, Improvement of microstrip patch antenna radiation patterns, IEEE Trans Antennas Propag, vol 42, pp. 882-885, 1994.
[58] J.-G. Yook and L.P.B. Katehi, Micromachined microstrip patch antenna with controlled mutual coupling and surface waves, IEEE Trans Antennas Propag, vol 49, pp. 1282-1289, 2001.
[59] J. R. Verbiest and G.A.E. Vandenbosch, "A Novell Small Size Printed Tapered Monopole Antenna for UWB WBAN", IEEE Antennas and Wireless Propagation Letters, vol 5, pp. 377-379, 2006.
[60] Z. N. Chen, N. Yang, Y. X. Guo, and M. Y. W. Chia, "An investigation into measurement of handset antennas," IEEE Trans. Instrum. Meas., vol. 54, no. 3, pp. 1100-1110, Jun. 2005.
[61] C. Icheln J. Ollikainen and P. Vainikainen,"Reducing the influence of feed cables on small antenna measurements," Electron. Lett., vol. 35, no. 15, pp. 1212-1214, Jul. 1999.
[62] C. Icheln J. Ollikainen and P. Vainikainen, "Effects of RF Absorbers on Measurements of Small Antennas in Small Anechoic Chambers," IEEE AESS Systems Magazine, pp. 17-20, Nov. 2001.
[63] M.A. Goodberlet and J.B. Mead, "Microwave Connector Characterization," IEEE Microwave Magazine, vol. 7, no. 5, pp. 78-83, Oct. 2006.
[64] M. Rittweger, M. Abdo, and I. Wolf, "Full-wave analysis of coplanar discontinuities considering three dimensional bond wires," IEEE MTT-S Dig., pp. 465–468, 1991.
[65] Omar and Y. L. Chow, "A solution of coplanar waveguide with airbridges using complex images," IEEE Trans. Microwave Theory Tech., vol. 40, pp. 2070–2077, Nov. 1992.
[66] E. A. Soliman, P. Pieters, E. Beyne and G. A. E. Vandenbosch, "Suppression of the Parasitic Modes in CPW Discontinuities Using MCM-D Technology - Application to a Novel 3-dB Power Splitter," IEEE Trans. Microwave Theory Tech., vol. 46, pp. 2426–2430, Dec. 1998.
[67] M. Farina and T. Rozzi, "A short-open deembedding technique for method-of-moments based electromagnetic analysis", IEEE Trans. Microwave Theory Tech., vol. 49, pp. 624–628, Apr. 2001.
[68] H. ElKamchouchi and G. Abouelseoud, "Automating the Electromagnetic Simulation Procedure and Its Possible "Fatal" Consequences," IEEE Antennas and Propagation Magazine, vol 49, pp. 133–142, Apr. 2007.
[69] P. Ciais, C. Luxey, A. Diallo, R. Staraj, G. Kossiavas, "Pentaband internal antenna for handset communication devices", Microwave and Optical Technology Letters, Vol. 48, N°8, pp.1509-1512, August 2006.

[70] S. Radiom, H. Aliakbarian, G. A. E. Vandenbosch, and G. Gielen, "An Optimized Small-Size Tapered Monopole Antenna for Pulsed UWB Applications Designed by a Genetic Algorithm", IEE Proc. Microwaves, Antennas Propagat., vol. , no. , pp. -, accepted Oct. 2008.